Pump User's Handbook

Life Extension

4th Edition

Pump User's Handbook
Life Extension

4th Edition

By
Heinz P. Bloch & Allan R. Budris

LONDON AND NEW YORK

Published 2020 by River Publishers
River Publishers
Alsbjergvej 10, 9260 Gistrup, Denmark
www.riverpublishers.com

Distributed exclusively by Routledge
4 Park Square, Milton Park, Abingdon, Oxon OX14 4RN
605 Third Avenue, New York, NY 10158

First published in paperback 2024

Library of Congress Cataloging-in-Publication Data

Bloch, Heinz P., 1933-
Pump user's handbook : life extension / by Heinz P. Bloch & Allan R. Budris. -- 4th edition.
pages cm.
Includes bibliographical references and index.
ISBN 0-88173-720-8 (alkaline paper) -- ISBN 978-8-7702-2311-9 (electronic) -- ISBN 978-1-4822-2864-9 (Taylor & Francis distribution: alkaline paper) 1. Pumping machinery--Handbooks, manuals, etc. I. Budris, Allan R. II. Title.

TJ900.B648 2014
621.6'9--dc23

2013032243

Pump user's handbook: life extension/by Heinz P. Bloch & Allan R. Budris. Fourth edition.
First published by Fairmont Press in 2014.

Routledge is an imprint of the Taylor & Francis Group, an informa business

ISBN: 978-0-88173-720-2 (The Fairmont Press, Inc.)
ISBN: 978-8-7702-2311-9 (online)

ISBN: 978-1-4822-2864-9 (hbk)
ISBN: 978-87-7004-505-6 (pbk)
ISBN: 978-1-003-15180-7 (ebk)

DOI: 10.1201/9781003151807

Table of Contents

Preface to the Fourth Edition

Again, not much has changed since the third edition was released in 2010-. As before, next to electric motors, centrifugal pumps still represent the most frequently utilized machine on earth. It has been estimated that over 10,000,000,000 of them are in use worldwide. (Ref. I-1). And pump life extension, maintenance cost reduction, and safety and efficiency enhancement opportunities seem to grow.

Pumps certainly are simple machines, for quite unlike an aircraft jet engine that consists of somewhere between 6,000 and 9,000 parts, a centrifugal pump is made up of a rotor, two or three bearings, a few casing parts, perhaps a mechanical seal and a bunch of fasteners. And yet there are, in the United States alone, many thousands of pumps that achieve mean-times-between-failures (MTBF's) of only a year or less, whereas in numerous other identical services MTBF values of over eight years are not uncommon.

This text will explain just how and why the best-of-class pump users are consistently achieving superior run lengths, low maintenance expenditures, and unexcelled safety and reliability. Written by practicing engineers whose working career was marked by involvement in pump specification, installation, reliability assessment, component upgrading, maintenance cost reduction, operation, troubleshooting and all conceivable facets of pumping technology, this text endeavors to describe in detail how you, too, can accomplish best-of-class performance and low life cycle cost. Or, how your facility will get away from being a 1.1, or 2.7, or 3.9 year MTBF plant and will join the plants that today enjoy a demonstrated pump MTBF of 8.6 and, in at least one case further described in Chapter 16, over 10 years.

WHAT PUMPS DO

Pumps are used to feed liquids from one place to another. There is no liquid that cannot be moved by pumps. If pumps cannot move a product, the product is probably not a liquid. Pumps are used in every industry conceived by man and are installed in every country in the world.

But pumps are machines and machines need to be properly designed. The parts for the pumps need to be correctly manufactured and assembled into a casing. The assembled pump may have to be inspected and tested; it certainly has to be properly installed. It also needs to be serviced or maintained with appropriate care and knowledge. And, it needs to be operated within the intended design envelope.

In other words, pumps can, and usually will fail, if one or more of seven important criteria are not met. It has been proven (Ref. I-2) that:

- The *design* has to be correct
- The various components have to be *fabricated* just right
- Pumps must be *operated* within an intended service condition envelope
- Pumps must be *maintained* as required
- Pumps have to be *assembled and installed* correctly
- Pumps will not tolerate certain types of *operator errors*
- Component *materials* must be without defect

WHAT PUMP FAILURES REALLY COST

When pumps fail, the cost can be staggering. In 1996, a South American refinery repaired 702 of their 1,012 installed pumps. In that year, these pumps consumed 1,420 rolling element bearings. In 1984, a U.S. oil refinery performed close to 1,200 repairs on 2,754 installed pumps. Of these, 40% were done in the shop and 60% in the field. Based on work order tracking, the direct charges per repair amounted to $5,380, or $6,456,000 per year. The true costs, however, amounted to $10,287 per repair. Here, yearly pump maintenance expenditures were calculated to exceed $12,000,000 when incremental burden consisting of employee benefits and refinery administrative, plus overhead and materials procurement costs were taken into account.

Statistics from a plant with 3,300 centrifugal pumps installed indicate that 30% of the plant's yearly repair events are traceable to maintenance deficiencies. Neglect and faulty procedures fall into this category. Another refinery supplied Figures I-1 and I-2, pointing to bearing failures (40%) and lubrication issues as being the main culprits. Quite obviously, these illustrations show that considerable resources are expended on bearing and lube-related maintenance. Yet, speaking of lubrication, a report from a Mid-Eastern refinery with 1,400 pumps on dry sump oil mist specifically pointed out that there were no bearing failures in the year 2004 (Ref. I-3). Zero failures were also claimed by an oil refinery in Western Australia.

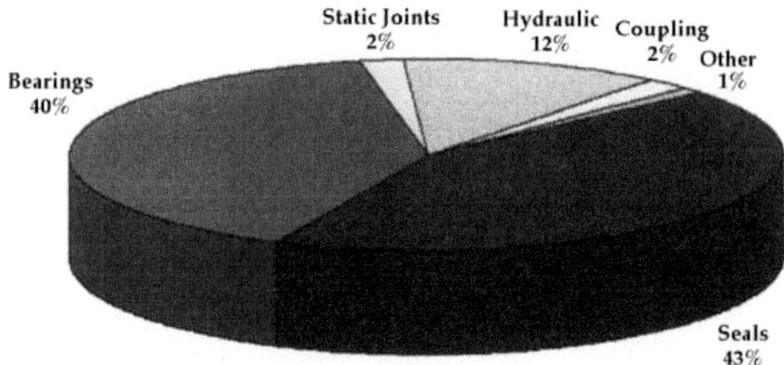

Figure I-1: Causes of equipment outage at a major refinery by component description, years 2003 and 2004

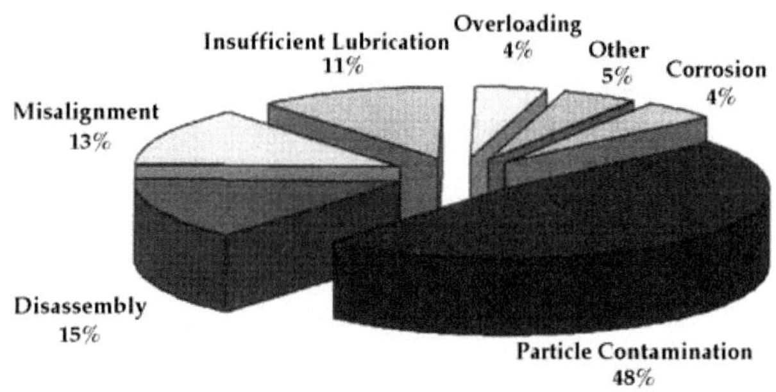

Figure I-2: Causes of bearing failures at a major refinery by cause category

Assembly and installation defects are responsible for 25% of pump failures at that facility, and 15% of the pump problems encountered were attributed to operation at off-design or unintended service conditions.

Improperly operated pumps constitute 12% of pump failures here. Fabrication and processing errors cause 8% of this plant's pump failures; faulty design was found responsible for 6% of pump failures, and 4% of the failure population suffered from material defects.

These statistics convey some very important facts:

- Most pumps fail because of maintenance and installation-related defects
- Since pumps generally represent a mature product, fundamental design defects are relatively rare
- Pump failure reductions are largely achieved by appropriate action of plant reliability staff and plant or contractor maintenance work forces. The pump manufacturer rarely deserves to be blamed, although it is unfortunate that the pump manufacturer is not always knowledgeable in pump failure avoidance.

In the mid-1980's, a chemical plant in Tennessee had over 30,000 pumps installed and a large facility near Frankfurt, Germany, reported over 20,000 pumps. However, the largest industrial pump user appeared to be a city-sized plant situated on the banks of the Rhine river south of Frankfurt. There were approximately 55,000 pumps installed at that one location alone.

United States oil refineries typically operate from 600 pumps in small, to 3,600 pumps in large facilities. Among the old refineries are some that have average pump operating times of over 8 years. However, there are also some that achieve an average of only about a year. Some of the really good refineries are new, but some of the good ones are also old. Certain bad performers belong to multi-plant owner "X" and some good performers also belong to the same owner "X". It can therefore be said that equipment age does not preclude obtaining satisfactory equipment reliability.

But it can also be said that facilities with low pump MTBF are almost always repair-focused, whereas plants with high pump MTBF are essentially reliability-focused. Repair-focused mechanics or maintenance workers see a defective part and simply replace it in kind. Reliability-focused plants ask why the part failed, determine whether upgrading is feasible, and then calculate the cost justification or economic payback from the implementation of suitable upgrade measures. Needless to say, reliability-focused plants will implement every cost-justified improvement as soon as possible.

Again, why pumps fail and how to avoid failures will be discussed in this text. Why the same pump model does well at one plant and does not do well at another will be described and analyzed. Pump life extension and energy cost reduction will be the overriding concerns and the collective theme of this book. This updated and expanded fourth edition contains experience-based details, data, guidance, direction, explanations, and firm recommendations. The material will assist all interested facilities to move from the unprofitable repair-focus of the mid-20th century, to the absolutely imperative reliability focus of the 21st century.

Finally, if you are a manager: Consider rewarding diligent readers and implementers of pump failure prevention. Value them, develop them, nurture them. Consider promoting competent leaders instead of favoring those who can glibly explain away repeat failures. Understand that house-of-sand next quarter profits are crumbling at random, whereas house-of-concrete organizations have few surprises and will keep standing. This book will help solidify things.

Acknowledgements

While compiling the material for this text, we looked at well over one thousand illustrations of pumps and related components. Picking the ones we did should in no way be inferred as disqualifying similar products from other manufacturers or suppliers. We carefully chose from readily available commercial information that was judged useful for conveying the technical points we wanted to make.

With this in mind, we gratefully acknowledge the cooperation of the many competent U.S. and overseas companies and individuals that granted permission to use portions of their marketing literature and illustrations. These include:

- Afton Pumps, Inc., Houston, Texas (process pumps, 13-4)
- A-Line Manufacturing Company, Liberty Hill, Texas (dial-type alignment fixtures, 3-47, 3-48)
- Allis-Chalmers, Milwaukee, WI (pumps, 13-25)
- AESSEAL, plc., Rotherham, UK, and Rockford, Tennessee (mechanical seals, bearing isolators, 6-11, 6-17 to 6-21, 7-59, 7-60, 8-13, 8-15, 8-62 to 8-74, 8-77 to 8-79, 8-89, 9-11, 9-36, 10-17 to 10-19.
- AEGIS® Electro Static Technology, Mechanic Falls, ME (Bearing Protection Rings, Chapter 4)
- API (American Petroleum Institute), Alexandria, Virginia (Equipment Standards, 2-1, 2-2)
- A.W. Chesterton Company, Stoneham, Massachusetts (mechanical seals, 8-9, 8-16, 8-17, 8-25, 8-26, 8-35 to 8-45)
- BaseTek® Polymer Technology, Chardon, Ohio (cast polymer baseplate technology, 3-30, 3-31)
- Borg-Warner and parent company John Crane, Temecula, California (mechanical seals, 8-5)
- Burgmann Seals America, Houston, Texas, and its parent company Dichtungswerke Feodor Burgmann, Wolfratshausen, Germany (mechanical seals, 2-7, 8-2, 8-3, 8-6, 8-7, 8-34, 8-60, 8-61, 9-35)
- Byron Jackson Division of Flowserve Pumps, Kalamazoo, Michigan (process pumps, 15-17 to 15-19)
- Carver Pump Company, Muscatine, Iowa (process pumps, 6-22)
- Coupling Corporation of America, York, Pennsylvania (shaft couplings, 11-11)
- CPC Pumps, Mississauga, Ontario, Canada (process pumps, 6-23)
- DuPont Engineering Polymers, Newark, Delaware (Vespel® wear materials, 6-28, 6-29)
- Emile Egger & Co., Ltd., Cressier, NE, Switzerland (ISO-compliant centrifugal pumps, 2-11, 13-37, 13-38)
- Enviroseal Corporation, Waverley, Nova Scotia, Canada (seal protectors, 8-75, 8-76)
- ExxonMobil, Lube Marketing Division, Houston, Texas (premium lubricants, 9-4, 9-7 to 9-9)
- FAG Corporation, Stamford, Connecticut (rolling element bearings, 9-2, 9-38, 9-41, 9-48)
- Falk Corporation, Milwaukee, Wisconsin (shaft couplings, 11-2, 11-6 to 11-8, 11-13)
- Flexelement Coupling Corporation, Houston, Texas (couplings, 11-9)
- Flowserve Corporation, Kalamazoo, Michigan (pumps, seals, 8-11, 8-12, 8-18, 8-19, 8-21 to 8-24, 8-39)
- Garlock Sealing Technologies, Palmyra, New York (sealing products, 9-33)
- General Electric Company, Schenectady, New York (electric motors, 11-4)
- Hermetic Pump, Inc., Houston, TX, and Gundelfingen, Germany (process pumps, 13-10, 13-11, 13-16, 13-17)
- HydroAire, Inc., Chicago, Illinois (pump repair and upgrading, 15-21 to 15-23)
- Inductive Pump Corporation, Frankfort, New York (inductive pumps, 9-29, 9-30)
- Ingersoll-Dresser, Allentown, Pennsylvania (process pumps, 9-23)
- INPRO/Seal Company, Rock Island, Illinois (bearing isolators, 6-15, 9-36)
- Isomag Corporation, Baton Rouge, Louisiana (magnetic bearing housing seals, 6-9, 9-34)

- ITT/Goulds, Seneca Falls, New York (ANSI and API-compliant centrifugal pumps, 3-2 to 3-4, 3-10, 3-29, 6-2 to 6-7, 6-11, 6-24 to 6-27, 11-1, 11-3, 13-13 to 13-15, 13-18, 13-19, 13-21 to 13-24, 13-26, 13-28 to 13-36, 13-39 to 13-44, 14-1, 14-6, 14-7, 15-1 to 15-14)
- John Crane Company, Morton Grove, Illinois and Slough, UK (Mechanical Seals, 8-27 to 8-33, 11-9, 16-1 to 16-11)
- Kingsbury Corporation, Philadelphia, Pennsylvania (hydrodynamic bearings, 7-44 to 7-58, 9-39)
- Koch Engineering Company, Inc., Wichita, Kansas (flow straighteners, 3-6)
- KSB A.G., Frankenthal/Kaiserslautern/Pegnitz, Germany (process pumps, 5-6, 9-39, 13-5)
- Lucas Aerospace, Utica, New York (couplings, 11-10)
- Lubrication Systems Company, Houston, Texas (oil mist lubrication and equipment preservation technology, 10-1, 10-5, 10-6, 10-8 to 10-15, 10-20 to 10-26)
- Ludeca, Inc., Miami, Florida (Laser-Optic Alignment, 3-39, 12-7, 12-17)
- Magseal, Inc., West Barrington, Rhode Island (magnetic seals, 6-9)
- Mechanical Solutions, Inc., Parsippany, New Jersey (rotor dynamics studies, 12-12)
- MRC Division of SKF USA Corporation, Jamestown New York (rolling element bearings, 9-15, 9-28)
- Noria Corporation, Tulsa, Oklahoma (lube analysis and application-related training, trade journals, 10-2 to 10-4, 10-7)
- Oil-Safe® Systems Pty., Ltd., Bibra Lake, Western Australia (lube transfer containers, 10-27 to 10-30)
- Pacific-Wietz and parent Flowserve, Dortmund, Germany (mechanical seals, 8-4, 8-10, 8-20, 15-15)
- Pompe Vergani, S.p.A., Merate/Como, Italy (process pumps, 13-3)
- Piping Technology and Products, Inc., Houston, Texas (pipe support design and manufacturing, 3-8, 3-9)
- Prueftechnik A.G., Ismaning, Germany (alignment systems, vibration monitoring instruments, induction current-type bearing mounting and dismounting tools, 3-37 to 3-40, 3-46, 3-49, 3-50, 12-7, 12-17, 15-20)
- Rexnord Corporation, Milwaukee, Wisconsin (shaft couplings, 11-12)
- Roto-Jet® Pump, Salt Lake City, Utah (pumps, 6-12, 13-20)
- Royal Purple, Ltd., Porter, Texas (premium synthetic lubricants, 9-5, 9-6)
- Ruhrpumpen, Germany (process pumps, 5-7, 9-14, 13-1, 13-5, 13-7, 13-9, 13-12)
- Safematic Oy, Muurame, Finland (central lubrication systems, mechanical seals, 8-1, 8-4, 9-49, 15-20)
- Siemens A.G., Erlangen, Germany (electric motors, 10-16)
- SKF USA, Kulpsville, Pennsylvania (rolling element bearings, 2-13, 7-1 to 7-42, 9-1, 9-10, 9-12, 9-21, 9-31, 9-40, 9-50, 9-52)
- SKF Condition Monitoring, San Diego, California (equipment condition monitoring technology, 12-3, 12-4)
- SPM Instruments, Inc., Marlborough, Connecticut (vibration and shock pulse monitoring devices, 12-5, 12-6)
- Stay-Tru® Corporation, Houston, Texas (pre-filled baseplate and grouting technology, 3-11 to 3-28, 3-32 to 3-34)
- Sulzer Company, Winterthur, Switzerland (process pumps, 9-16, 13-6, 13-8, 13-27)
- Sundyne Fluid Handling, Arvada, Colorado (process pumps, 13-16)
- Trico Manufacturing Company, Pewaukee, Wisconsin (lubricant application equipment, 6-8, 9-13, 9-18, 9-19, 9-25, 9-27, 9-51)
- Vibracon® Machine Support B.V., Ridderkerk, The Netherlands (self-level, adjustable chocks, 3-35, 3-36)
- Waukesha Bearings Corporation, Waukesha, Wisconsin (hydrodynamic bearings, 7-43)

Very special thanks go to Uri Sela, who reviewed both our draft manuscript and page proofs. With his unsurpassed machinery know-how, he tracked down and commented on the occasional misunderstandings that find their way into this kind of text.

Westminster, Colorado, and Washington, New Jersey
Heinz P. Bloch and Allan R. Budris

Chapter 1

Pump System Life Cycle Cost Reduction

The primary objective of life cycle costing is to evaluate and/or optimize product life cost while satisfying specified performance, safety, reliability, accessibility maintainability, and other requirements.

Pumping systems account for an estimated 25%-50% of the energy usage in many industrial plants, and perhaps 20% of the world's electric energy demand (Ref. 1-1). Centrifugal pumps rank first in failure incidents and maintenance costs. That is why centrifugal pumps in critical applications are installed in identical pairs, one serving as the operating, the other one serving as the standby or spare pump.

Despite these statistics, many pump purchase decisions are still made solely on the basis of lowest initial purchase and installation cost. The notion exists that, if a cheap pump doesn't perform well, it can always be upgraded. While this may be true in those pumps that suffer from installation errors or component defects, it is not true for pumps that suffer from fundamental design compromises. Moreover, these decisions seem to disregard that initial purchase price is generally only a small part of pump life cycle cost in high usage applications. Market conditions, short-term financial considerations, and organizational barriers are to blame for this short-sighted approach.

Conventional Wisdom: *You can always upgrade an inferior pump*

Fact: *Certain bad choices defy cost-effective upgrading. A plant may have to buy a better pump or suffer through and endure its bad decision for years.*

Progressive, reliability-focused pump users who seek to improve the profitability of their operations will have to consider using Life Cycle Costing, or LCC. The conscientious application of LCC concepts will help reliability-focused plants minimize waste. LCC will also dramatically reduce energy, operating and maintenance costs.

Life cycle pump cost is the total lifetime cost to purchase, install, operate, maintain (including associated downtime), plus the cost due to contamination from pumped liquid, and the cost of ultimately disposing of a piece of equipment.

A simplified mathematical expression could be

$$LCC = C_{ic} + C_{in} + C_e + C_o + C_m + C_{dt} + C_{env} + C_d$$

where:

- LCC = Life Cycle Cost
- C_{ic} = Initial Cost, purchase price (pump, system, pipe, auxiliary services)
- C_{in} = Installation and commissioning cost
- C_e = Energy costs (pump, driver & auxiliary services)
- C_o = Operation costs
- C_m = Maintenance and repair costs
- C_{dt} = Down time costs
- C_{env} = Environmental costs
- C_d = Decommissioning and/or disposal costs

Energy, maintenance and downtime costs depend on the selection and design of the pump, the system design and integration with the pump, the design of the installation, and the way the system is operated. Carefully matching the pump with the system can ensure the lowest energy and maintenance costs, and yield maximum equipment life.

When used as a comparison tool between possible design or overhaul alternatives, the Life Cycle Cost process will show the most cost effective solution, within the limits of the available data. Figure 1-1 shows a typical breakdown of pump life cycle costs. In this case, the initial pump purchase cost represents only nine percent of the total life cycle cost. Ref. 1-2 offers more details on Life Cycle Cost analysis.

Conventional Wisdom: *There's not enough data to calculate life-cycle cost*

Fact: *There's always enough data for a reasonably close estimate of pump life-cycle cost.*

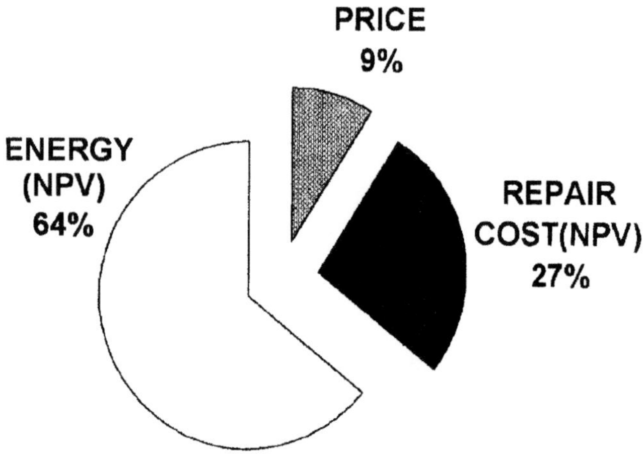

Figure 1-1: Typical life cycle cost breakdown

INITIAL COSTS

The initial investment costs include the original pump and pumping system costs. Initial costs also include engineering, bid process ("bid conditioning"), purchase order administration, testing, inspection, spare parts inventory, training and auxiliary equipment. The purchase price of the pumping equipment is typically less than 15% of the total ownership cost. Initial cost is also influenced by such critical factors as the size and design of the pump piping, pump speed, the quality and/or duty rating of the equipment being selected, materials of construction and control system. All of these choices can substantially affect the life cycle cost and working life of the pump.

INSTALLATION COSTS

Pump installation and commissioning costs include the foundations, grouting, connecting of process piping, connecting electrical wiring, connecting auxiliary systems, equipment alignment, flushing of piping and performance evaluation at startup. The care and effectiveness in executing these installation activities will have a great impact on subsequent reliability, maintenance and down time costs, during the life cycle of the pump. Unless shown to be outdated, the equipment manufacturer's installation, start-up and operation instructions should be adhered to. A checklist should be used to ensure that equipment and the system are operating within specified parameters.

ENERGY/OPERATING COSTS

Pump/system energy consumption is often one of the larger cost elements and may dominate the total life cycle costs, especially if pumps run more than 2,000 hours per year. Energy and maintenance costs during the life of a pump system are usually more than 10 times its purchase price (Ref. 1-1). Energy costs are dependent not only on the best efficiency of the pump(s), but also on the energy consumed by the pump system (pipe size, etc.), and by how much time and how far the pump spends operating away from the best efficiency flow rate. Additional influencing factors include minimum flow bypasses, control valve pressure breakdown, auxiliary service energy consumption, and driver selection and application.

Operating costs are labor costs related to the operation of a pumping system. These vary widely depending on the complexity and duty of the system. Regular observation of how a pumping system is functioning can alert operators to potential losses in system performance. Performance indicators include changes in vibration, shock pulse signature, temperature, noise, power consumption, flow rates and pressure.

***Conventional Wisdom*:** *Pump initial cost is the most important selection factor.*
Fact: *With rare exceptions, pump initial cost should be the least important selection factor.*

MAINTENANCE AND REPAIR COSTS

Obtaining optimum working life from a pump requires special care in the design of the system (pump piping, etc.), design and selection of the pump, installation, and regular and efficient servicing. The cost depends on the time and frequency of service and cost of materials. Pump design can greatly influence these costs through the hydraulic selection, materials of construction, components chosen, and the ease of access to the parts to be serviced.

Downtime can be minimized by optimized preventive maintenance programs, and programming major maintenance during annual shutdown or process changeover. Although unexpected failures cannot be predicted precisely, they can be estimated statistically by calculating mean time between failures, or possibly avoided through continuous monitoring fault tolerant smart pump control systems.

LOSS OF PRODUCTION

The cost of unexpected downtime and lost production is a very significant item in total life cycle cost, and can rival the energy costs and replacement parts costs. All of the above factors affecting the working life of a pump can impact downtime and loss of production. Despite the design or target life of a pump and its components, there will be occasions when an unexpected failure occurs. In those cases where the cost of lost production is unacceptable, a spare pump may be installed in parallel to reduce risk. If a spare pump is used, the initial cost will be greater, but the cost of lost production will be avoided, or at least minimized.

PUMP RELIABILITY FACTORS

Figure 1-2 summarizes the many factors that influence pump reliability. These, obviously, include selection (type chosen), installation, usage (application), operation, and maintenance-related parameters. While each of these will be covered in much more detail throughout this text, it should also be noted that correct operating instructions are often lacking. It is certainly obvious that pump start-up, operating surveillance, shut-down and related procedures can and will influence pump reliability. Detailed guidance on the correct procedures to be employed can be found in Appendix 1.

While it is beyond the scope of even this comprehensive text to deal with the many conceivable ways in which centrifugal pumps can be abused or incorrectly operated, the issue of operating with blocked-in flow is of major interest. As will be seen later in this book, pump shafts can (and will) deflect when operating outside a safe design and/or operational envelope. Radial and thrust bearings can be overloaded at non-optimized flow conditions and the contributing causes—rightly or wrongly—attributed to lubrication failure (Ref. 1-3).

Operator-induced failures may be as straightforward as the risky and questionable decision to run pumps, even briefly, against a closed discharge block valve or, worse yet, with both suction and discharge block valves closed. Over the years, one experienced consulting engineer analyzed several centrifugal pump failures that were the result of a blocked-in discharge line. The cause was usually a procedural or initial design error and was corrected after the damage was done. Cracked mechanical seals and vibration occurred in some instances and a total seizing of the pump in others. In all cases the potential for overheating the pumped fluid is one of many justified concerns (Ref. 4).

When a centrifugal pump operates in the no-flow or dead-ended condition, it is still delivering energy to the trapped liquid. In rough terms the no-flow power demand of many pumps is 40% to 60% of that needed at full flow. Of course, this energy has to go somewhere and much of it goes into heating up the confined liquid in the pump casing.

An approximation of the temperature rise of the liquid trapped in the casing of a blocked-in centrifugal pump can be obtained from the following equation:

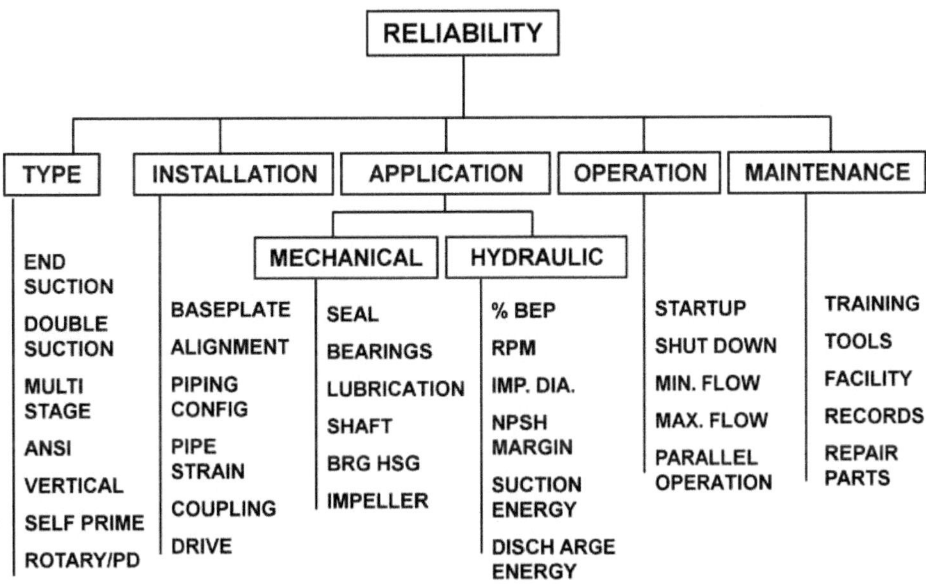

Figure 1-2: Pump reliability factors

$$\Delta T = 81 * HP_{NF} * t_{min} / [D^3 * \rho * C], \text{°F, where}$$

ΔT = Temperature rise of liquid in pump casing, °F.

HP_{NF} = No-flow or blocked-in horsepower

t_{min} = Time (duration) of pump operating at blocked-in condition, minutes

D = Impeller diameter, ft

ρ = Liquid density, lb/ft^3

C = Specific heat of liquid, Btu/lb-°F

The reader may note that this is just a rearrangement of the specific heat equation (Ref. 1-5). Instead of trying to determine the weight of liquid being heated, the pump casing is considered to be spherical. Realizing that the piping also contains liquid and treating the pump casing as a sphere, one only needs to consider the impeller size. In all then, this approximation seems adequate for educational purposes.

Consider the example of a 100 HP (75 kW) centrifugal water pump with a one-foot diameter impeller, absorbing a blocked-in horsepower that averages 50% of the normal power draw. Assume that it is blocked-in for two minutes so as to switch remote pumps or to commission standby pumps.

$$\Delta T = 81 * 50 * 2 / [1^3 * 62.4 * 1] = 130 \text{ °F.}$$

The water trapped in the casing would rise 130°F (72°C), which may or may not cause problems, depending on the pump. What is important is that there are five variables in this simple equation. Just saying a pump can be run for some time with no flow would be irresponsible. As can be seen by the equation, a pump with low no-flow horsepower and a large casing probably would not seize, but a high energy pump with a smaller volume might. Hydrocarbons would have a larger temperature rise than water since hydrocarbons have smaller ρ and C values. Vaporization in the casing and run-dry conditions would be of concern.

Eventually the heat is dissipated through the casing, piping and shafting to the atmosphere and the temperature reaches steady state conditions. After reaching steady state it will no longer increase, a situation which is not considered here. Our model only considers a blocked-in discharge for a short period of time and is based on the casing being filled with a liquid. Remember also that this example applies to radial flow centrifugal pumps only. Mixed-flow, axial flow and other types of pumps can have completely different blocked-in characteristics.

Starting up with a slightly open (slightly or "cracked open" typically means a 10% opening) discharge valve is standard practice on many low and medium specific speed pumps. However, if a pump must be operated blocked-in for any length of time, the safest approach is to discuss this with the pump manufacturer and to obtain concurrence in writing. Manufacturers are presumed to have experience in determining how long their product can be operated in this condition or if a valved bypass piping arrangement is required (Ref. 1-6).

Chapter 2

How to Buy and Ship a Better Pump

As observed earlier in this text, of the many machinery categories represented in the petrochemical industry, centrifugal pump sets, i.e. pumps and their respective drivers viewed as an assembled set or equipment train, rank first in number. This fact is sometimes cited to defend the high rate of failure incidents and sizeable maintenance costs associated with centrifugal pumps.

Centrifugal pumps are generally installed in identical pairs; one serving as the operating, the other one serving as the standby or spare pump. While certainly routine in the United States, the practice of installing full spares is not as universally accepted as we might be led to believe. For instance, many reputable German chemical companies install only one pump casing and keep spare internal assemblies on hand for emergencies. They find this practice feasible and profitable for many services in a typical plant.

We suspect that the successful lengthening of mean-time-between-failures (MTBF) at many European facilities could be attributed to careful operation and careful equipment selection. The merits of conscientious operation are intuitively evident and require no further elaboration. They are summarized in Appendix 1. There are also a hundred or more hydraulic and / or mechanical issues of interest. These, of course, are the subject of this book. Many of these points are recapped in Appendix 2.

But there are certain pump selection guidelines which we should consider as well. Their implementation can be expected to result in fewer failures and reduced maintenance expenditures for centrifugal pumps. The first of these selection guidelines is aimed towards the development of a good specification.

SPECIFICATION OVERVIEW

The determination of pump flow, pressure, and pumpage-related physical properties is usually left to the process engineer. This person fills in the applicable contractor or vendor-supplied data sheet, although using the current edition API-610 data sheets is highly recommended and always represent your best choice. They must be filled in as diligently and completely as possible. The layout of the API data sheets can be gleaned from two sample pages, Figures 2-1 and 2-2, taken from a previous edition of API-610 so as not to infringe on the API's copyright. For API pump specifications, be sure to always use the current edition of API-610.

However, it is the job of the machinery engineer to review these standard data sheets for accuracy and completeness. The various API specification clauses stipulate minimum requirements and even at that the American Petroleum Institute (API) informs industry that these are not mandatory compliance items. The machinery engineer's review is greatly facilitated if he or she has a thorough understanding of previous experience with equipment in similar service under similar operating conditions. This knowledge also makes it easier to review special component requirements, seal types, metallurgy, construction features, etc. The type of pump best suited for the particular service should now emerge in the reviewer's mind. Next, he or she can direct attention to the minimum flow and control aspects of the pump and pumping system.

ANSI VS. API—HOW DO THEY DIFFER?

Having mentioned the two principal pump types, we might explain their main areas of difference and their main areas of application.

Compared to an API pump, the typical ANSI pump exhibits

- A thinner casing, i.e. less corrosion allowance.
- Reduced permissible nozzle loads. It is even more sensitive to pipe-induced stresses than the API pump.
- Smaller stuffing box size. Unless a large bore option is chosen, an ANSI pump may not be able to accommodate the optimum mechanical seal for a given service.
- Impellers designed and manufactured without wear rings. Many ANSI pump impellers are open

PAGE 1 OF 5
JOB NO. ______ ITEM NO. ______
PURCH. ORDER NO. ______ DATE ______
INQUIRY NO. ______ BY ______
REVISION ______ DATE ______

CENTRIFUGAL PUMP DATA SHEET
CUSTOMARY UNITS

APPLICABLE TO: ○ PROPOSAL ○ PURCHASE ○ AS BUILT

FOR ______ UNIT ______

SITE ______ NO. REQUIRED ______

SERVICE ______ PUMP SIZE, TYPE & NO. STAGES ______

MANUFACTURER ______ MODEL ______ SERIAL NO. ______

NOTE: ○ INDICATES INFORMATION COMPLETED BY PURCHASER □ BY MANUFACTURER ⊡ BY MANUFACTURER OR PURCHASER

GENERAL

PUMPS TO OPERATE IN (PARALLEL) (SERIES) WITH ______ NO. MOTOR DRIVEN ______ NO. TURBINE DRIVEN ______

PUMP ITEM NO. ______ PUMP ITEM NO. ______

GEAR ITEM NO. 1 ______ MOTOR ITEM NO. ______ TURBINE ITEM NO. ______

GEAR PROVIDED BY ______ MOTOR PROVIDED BY ______ TURBINE PROVIDED BY ______

GEAR MOUNTED BY ______ MOTOR MOUNTED BY ______ TURBINE MOUNTED BY ______

GEAR DATA SHEET NO'S ______ DRIVER DATA SHEET NO.'S ______ TURBINE DATA SHEET NO'S ______

○ OPERATING CONDITIONS

○ CAPACITY, NORMAL ______ (GPM) RATED ______ (GPM)

OTHER ______

○ SUCTION PRESSURE MAX/RATED ______ / ______ PSIG

○ DISCHARGE PRESSURE ______ (PSIG)

○ DIFFERENTIAL PRESSURE ______ (PSI)

○ DIFFERENTIAL HEAD ______ (FT) NPSH AVAILABLE ______ (FT)

○ HYDRAULIC POWER ______ (HP)

SERVICE: ○ CONTINUOUS ○ INTERMITTANT (STARTS/DAY ______)

○ SITE AND UTILITY DATA

LOCATION:

○ INDOOR ○ HEATED ○ UNDER ROOF

○ OUTDOOR ○ UNHEATED ○ PARTIAL SIDES

○ GRADE ○ MEZZANINE ○ ______

○ ELECTRIC AREA CLASSIFICATION CL ______ GR ______ DIV ______

○ WINTERIZATION REQD. ○ TROPICALIZATION REQD.

SITE DATA:

○ ELEVATION ______ FT BAROMETER ______ (PSIA)

○ RANGE OF AMBIENT TEMPS: MIN/MAX ______ / ______ °F

○ RELATIVE HUMIDITY: % MAX/MIN ______ / ______

UNUSUAL CONDITIONS: ○ DUST ○ FUMES

○ OTHER ______

○ UTILITY CONDITIONS:

STEAM: DRIVERS HEATING

MIN ______ PSIG ______ °F ______ PSIG ______ °F

MAX ______ PSIG ______ °F ______ PSIG ______ °F

ELECTRICITY: DRIVERS HEATING CONTROL SHUTDOWN

VOLTAGE ______ ______ ______ ______

HERTZ ______ ______ ______ ______

PHASE ______ ______ ______ ______

COOLING WATER:

TEMP. INLET ______ °F MAX RETURN ______ °F

PRESS NORM ______ (PSIG) DESIGN ______ (PSIG)

○ SITE AND UTILITY DATA (CONT'D)

COOLING WATER:

MIN RETURN ______ PSIG MAX ALLOW Δ P ______ (PSI)

WATER SOURCE ______

INSTRUMENT AIR: MAX/MIN PRESS ______ / ______ (PSIG)

○ LIQUID

○ TYPE OR NAME OF LIQUID ______

○ PUMPING TEMPERATURE

NORMAL ______ °F MAX ______ °F MIN ______ °F

○ SPECIFIC GRAVITY ______ @ MAX TEMP

○ SPECIFIC HEAT ______ Cp (BTU/LB °F)

○ VISCOSITY ______ (cP) @ ______ °F

○ MAX. VISCOSITY @ MIN. TEMP. ______ (cP)

○ CORROSIVE/EROSIVE AGENT ______

○ CHLORIDE CONCENTRATION (PPM) ______

○ H_2S CONCENTRATION (PPM) ______

LIQUIDS: (3.5.2.11) ○ TOXIC ○ FLAMMABLE ○ OTHER

□ PERFORMANCE

□ RPM ______

PROPOSAL CURVE NO. ______

□ IMPELLER DIA RATED ______ MAX ______ MIN ______ (IN)

□ RATED POWER ______ (BHP) EFFICIENCY ______ %

□ MINIMUM CONTINUOUS FLOW:

THERMAL ______ (GPM) STABLE ______ (GPM)

□ MAX HEAD RATED IMPELLER ______ (FT)

□ MAX POWER RATED IMPELLER ______ (BHP)

□ NPSH REQUIRED AT RATED CAP. ______ (FT H_2O)

□ SUCTION SPECIFIC SPEED ______

□ MAX SOUND PRESSURE LEVEL ______ dBA

REMARKS: ______

3L/B5390?00835
Printed in U.S.A.

DS-610 – 1 11/88

Figure 2-1: API-610 Pump data sheet (customary units)

CENTRIFUGAL PUMP DATA SHEET
SI UNITS

PAGE 2 OF 5
JOB NO. ____ ITEM NO. ____
REVISION ____ DATE ____
BY ____

☐ **CONSTRUCTION**

☐ MAIN CONNECTIONS:

	SIZE	ANSI RATING	FACING	POSITION
SUCTION				
DISCHARGE				
BAL. DRUM				

☐ OTHER CONNECTIONS

SERVICE	NO.	SIZE	TYPE
DRAIN			
VENT			
PRESSURE GAUGE			
TEMP GAUGE			
WARM-UP			

CASING MOUNTING:
☐ CENTERLINE ☐ NEAR CENTERLINE
☐ FOOT ☐ SEPARATE MOUNTING PLATE
☐ VERTICAL ☐ SUMP
☐ IN-LINE

CASING SPLIT:
☐ AXIAL ☐ RADIAL

CASING TYPE:
☐ SINGLE VOLUTE ☐ DOUBLE VOLUTE
☐ BARREL ☐ DIFFUSER
☐ STAGGERED VOLUTES ☐ VERTICAL DOUBLE CASING

IMPELLER MOUNTED:
☐ BETWEEN BEARINGS ☐ OVERHUNG
○ IMPELLERS INDIVIDUALLY SECURED (2.5.4)

CASE PRESSURE RATING:
○ SUCTION PRESS. REGIONS OF MULTISTAGE OR DOUBLE CASING PUMP DESIGNED FOR MAXIMUM ALLOWABLE WORK PRESSURE.

ROTATION: (VIEWED FROM COUPLING END)
☐ CW ☐ CCW
REMARKS: ____

SHAFT:
SHAFT DIAMETER AT SLEEVE ____ mm
SHAFT DIAMETER AT COUPLING ____ mm
SHAFT DIAMETER BETWEEN BRGS. ____ mm
SPAN BETWEEN BEARINGS ℄ ____ mm
SPAN BETWEEN BEARING & IMPELLER ____ mm
REMARKS: ____

COUPLINGS: DRIVER-PUMP
○ MAKE ____
☐ MODEL ____
☐ CPLG. RATING (kW / 100 RPM) ____

COUPLINGS: (CONTINUED)
○ LUBRICATION ____
◙ LIMITED END FLOAT REQUIRED ____
◙ SPACER LENGTH ____ mm
◙ SERVICE FACTOR ____
○ DYNAMIC BALANCED AGMA BALANCE CLASS ____
DRIVER HALF COUPLING MOUNTED BY
○ PUMP MFR. ○ DRIVER MFR. ○ PURCHASER
○ COUPLING PER API 671
REMARKS: ____

MATERIAL

○ TABLE H-1 CLASS ____
☐ BARREL/CASE ____ IMPELLER ____
☐ CASE/IMPELLER WEAR RINGS ____
☐ SHAFT ____ SLEEVE ____
☐ DIFFUSERS ____
☐ COUPLING HUBS ____
☐ COUPLING SPACER ____
☐ COUPLING DIAPHRAGMS ____
☐ API BASEPLATE NUMBER / MATERIAL ____ / ____
○ VERTICAL LEVELING SCREWS (3.3.1.15) ____
○ HORIZONTAL POSITIONING SCREWS (3.3.1.14) ____
REMARKS ____

BEARINGS AND LUBRICATION

BEARING: (TYPE / NUMBER)
☐ RADIAL ____ / ____
☐ THRUST ____ / ____
○ REVIEW AND APPROVE THRUST BEARING SIZE

LUBRICATION:
☐ GREASE ☐ FLOOD ○ RING OIL
☐ FLINGER ☐ PURGE OIL MIST ○ PURE OIL MIST
○ CONSTANT LEVEL OILER
○ PRESSURE ○ API-610 ○ API-614
○ OIL VISC. ISO GRADE ____
◙ OIL HEATER REQ'D ○ ELECTRIC ○ STEAM
○ OIL PRESSURE TO BE GREATER THAN COOLANT PRESSURE
REMARKS ____

3L/B5390/1559
Printed in U.S.A.
DS-610 – 2 SI 11/88

Figure 2-2: API-610 Pump data sheet (SI units)

or semi-open, whereas API pumps feature closed impellers with replaceable wear rings.

- Foot mounting, whereas the API pump will be centerline mounted, as is best shown in Figure 2-26 near the end of this chapter. In foot-mounted pumps, depicted later in Figure 2-11, casing heat tends to be conducted into the mounting surfaces and thermal growth will be noticeable. It is generally easier to maintain alignment of API pumps since their supports are surrounded by the typically moderate-temperature ambient environment.

API pumps are simply pumps that comply with the safety and reliability-focused stipulations promulgated by American Petroleum Industry Standard 610. The decision on API vs. ANSI construction is experience-based; it is not governed by governmental or regulatory agencies. However, rules-of-thumb are shared by many reliability professionals.

Conventional Wisdom: *A governmental agency prescribes when an API-style pump must be used*

Fact: *The user decides if he wants to purchase an API-style pump. Once that decision is made, it is customary to adhere to, or even exceed, the minimum requirements stipulated in API-610.*

For toxic, flammable, or explosion-prone services, especially at on-site locations in close proximity to furnaces and boilers, a large number of reliability-focused machinery engineers would use the latest Edition API-610 pumps if one or more of the following were to exist:

- Head exceeds 350 ft (106.6m)
- Temperature of pumpage exceeds 300°F (149°C) on pumps with discharge flange sizes larger than 4 in., or 350°F (177°C) on pumps with 4 inch discharge flange size or less.
- Driver horsepower exceeds 100 hp (74 kW)
- Suction pressure in excess of 75 psig (516 kPa)
- Rated flow exceeds flow at best efficiency point (BEP)
- Pump speed in excess of 3,600 rpm.

We have seen exceptions made where deviations from the rule-of-thumb were minor, or in situations where the pump manufacture was able to demonstrate considerable experience with ANSI pumps under the same, or even more adverse conditions.

Conventional Wisdom: *API-compliant pumps are always a better choice than ANSI or ISO pumps.*

Fact: *Unless flammable, toxic or explosion-prone liquids are involved, many carefully selected, properly installed, operated and maintained ANSI or ISO pumps may represent an uncompromising and satisfactory choice.*

PROCESS INDUSTRY PRACTICES (PIP)

Process industry voluntary, or recommended practices are found to cover elements of detailed design, procurement of many manufactured products. Among other things, they harmonize non-proprietary individual company engineering practices, or engineering standards. Occasionally, this work product leads to the standardization of specifications, data sheet format, installation methods, design practices and, at times, even the final equipment configuration. These voluntarily shared practices can cover machinery, baseplates, piping layout, and a host of other issues.

To be technically relevant and to reflect the indispensable needs of mature, reliability-minded organizations, a modern pump specification must go beyond the minimum requirements stipulated in API or related industry standards. Supplementary specifications are preferably based on an existing Industry Standard and must be

1. Periodically updated to reflect state-of-art components and executions.
2. Structured to reflect owner's needs in areas of spare parts commonality, etc.

An "industry-composite," or "Corporate Engineering Practice"—essentially an example of a supplementary specification for Heavy Duty Pumps, can be found in Appendix 3. It was derived from various non-proprietary user company and contractor sources.

SUPPLEMENTARY SPECIFICATIONS

In addition to both the API-610 Standard and the "Corporate Engineering Practice" described in Appendix 3, competent users will often devise and invoke a third document. This site-specific, or project-specific supplementary specification document may reflect unusual site preferences or requirements that are unique to just the particular service considered. It may also cover items relevant to site component standardization, use of existing common spare parts, etc. An example of such a supplement is highlighted in Figures 2-3 through 2-5. Similarly, a mechanical seal specification sheet such as shown in Figures 2-6 and 2-7 may be provided by an experienced engineer. Together with relevant data sheets and checklists found in the latest edition of API-610, these user-developed addenda would complete the entire specification package for pumps intended to achieve extended mean-times-between-failures.

SPECIFICATION CHECKLISTS

Specification checklists are an excellent memory jogger. They can be used to verify that the specification is complete and may be as brief or as elaborate as deemed appropriate by the specifying engineer. He or

Specification for Deethanizer Overhead Pumps

1.0 SCOPE

.01 These specifications establish the minimum requirements for the design, fabrication, inspection, and testing of two pumps with electric motor drivers for deethanizer overhead service.

2.0 GENERAL

.01 The pumps shall be furnished in accordance with this document and the attached centrifugal pumps data sheets, dated February 12, 1986.

3.0 APPLICABLE STANDARDS

The following Engineering Standard (ES) and their associated Specific Requirements Sheets (SRS) are attached and are part of this specification.

ES No.	Title	Rev.	Date	SRS
	Heavy Duty Centrifugal Pumps	2	12/83	2/12/86
	A-C Motors	0	11/75	2/12/86

4.0 DESIGN REQUIREMENTS

.01 Flexible couplings shall be Elastometric type.

.02 Tandem mechanical seals shall be provided. The second flush shall be API Plan 52 with a fluid system provided by the pump vendor per Drawing CE.2.86. The tandem seals shall be per the attached Mechanical Seal Specification Sheet Type H.

.03 The angular contact bearings should be FAG Series 7 _ _ _ B.TVP.UO with a K5 Shaft fit or TRW Pumppac PP8000 series. Radial ball bearings should be specified with loose internal clearance (C3) and mounted using a K5 shaft fit.

.04 High efficiency, mill and chemical type motor drivers with through-the-bearings, dry sump, oil mist lubrication and a 1.15 service factor shall be provided. Acceptable vendors are

Figure 2-3: Supplementary pump specification, part 1 of 3

4.0 DESIGN REQUIREMENTS (cont'd)

.05 Pump and motor noise shall not exceed 85 dBA at any point three feet from the machine when operating at design conditions. Noise testing is not required.

.06 Pump and motor vibration shall not exceed 0.15 inch/sec unfiltered measured on the bearing housings when running at design and minimum flow conditions.

.07 Motors are to be sized so as not to exceed nameplate horsepower of the motor at the end of pump curve. No credit for continuous overload capability or service factor shall be taken.

.08 All welds on seal pot must be per ASME code Section IX. All welders must be ASME code tested and certified. Pot pressure rating must be equal to the pump maximum allowable pressure with 500 psig minimum. Hydrostatic test shall be 1.5 times maximum allowable pressure. All threaded fittings shall be at least 3,000 lbs. rating.

.09 Pump throat bushings, balance holes, and flush arrangement shall be designed to maintain the proper seal cavity pressure and provide adequate cooling to the seal as recommended by the seal vendor.

5.0 PROPOSAL INFORMATION

.01 The Vendor's proposal shall include the following information:

a. Copies of Purchaser's data sheets with complete Vendor's information entered

b. Typical cross sectional drawings and literature to fully describe details of the offerings.

c. A specific statement that the system and all components are in the strict accordance with the Purchaser's specifications. If they are not in strict accordance with the specifications, the Vendor shall include a specific list detailing and explaining every deviation. Deviations may include alternative designs or systems equivalent to and guaranteed for the specified duties.

d. An explicit statement of any deviations from the specified guarantee and warranty.

e. A statement detailing the number of weeks to effect shipment after receipt of the order.

f. A list of the Vendor' recommended spare parts, priced individually along with standard delivery time. List separately spare parts needed for start-up and long term operation.

Figure 2-4: Supplementary pump specification, part 2 of 3

she can let background and experience be the relevant guides.

It is important that the specification review encompasses the applicability of ANSI pumps versus API pumps. The reviewer will generally know the principal differences between ANSI pumps and API pumps: corrosion allowance, permissible nozzle loads, stuffing box dimensions, wear rings, foot mount vs. centerline mount, bearing housing seals, and bearing style, type and rating. However, unless his company has in-house specifications to determine whether ANSI or API pumps should be selected, he may wish to use selection criteria similar to those given earlier in this chapter (page 8).

Once the specification package has been prepared and reviewed, we come to the task of defining potential vendors. Let's see how we do this.

SELECTING A KNOWLEDGEABLE COOPERATIVE VENDOR

Selecting from among the many pump manufacturers may be tricky. That said, picking the right bidders may well be an important prerequisite for choosing the best pump. Note that three principal characteristics identify a capable, experienced vendor (Ref. 2-1):

5.0 PROPOSAL INFORMATION (cont'd)

g. Cross sectional drawing, make, model, materials, balance ratio, and face dimension of seals. Pumping ring capacity with associated gland through 15 ft. of 3/4-inch tubing with a 0.8 S.G. 10W lube oil.

h. Shaft diameter, bearing spacing, and impeller overhang dimensions. Impeller eye area.

i. Thrust bearing load at design flow, BEP, and minimum flow. Size and number of balance holes.

j. Seal assembly price and API Plan-52 price must be broken out and listed separately.

SAMPLE

Figure 2-5: Supplementary pump specification, part 3 of 3

- He is in a position to provide extensive experience listings for equipment offered and will submit this information without much hesitation.
- His centrifugal pumps enjoy a reputation for sound design and infrequent maintenance requirements.
- His marketing personnel are thoroughly supported by engineering departments. Also, both groups are willing to provide technical data beyond that which is customarily submitted with routine proposals.

Vendor competence and willingness to cooperate are shown in a number of ways, but data submittal is the true test. When offering pumps which are required to comply with the standards of the American Petroleum Institute, i.e. the latest Edition of API-610, a capable vendor will make a diligent effort to fill in all of the data requirements of the API specification sheet shown earlier, in Figure 2-1. However, the real depth of his technical know-how will show in the way he explains exceptions taken to API-610 or supplementary user's specifications. Most users are willing to waive some specification requirements if the vendor is able to offer sound engineering reasons, but only the best qualified centrifugal pump vendors can state their reasons convincingly.

Pump assembly drawings are another indispensable documentation requirement. Potential design weaknesses can be discovered in the course of examining dimensionally accurate cross section view drawings. There are two compelling reasons to conduct this drawing review during the bid evaluation phase of a project: First, some pump manufacturers may not be able to respond to user requests for accurate drawings after the order is placed, and second, the design weakness could be significant enough to require extensive redesign. In the latter case, the purchaser may be better off selecting a different pump model (Ref. 2-2).

The merits of a drawing review prior to final pump selection are best explained by looking at several examples of weaknesses in pumps.

It is intuitively evident that purchasing the least expensive pump will rarely be the wisest choice for users wishing to achieve long run times and low or moderate maintenance outlays. Although a new company may, occasionally, be able to design and manufacture a better pump, it is not likely that such newcomers will suddenly produce a superior product. It would thus be more reasonable to choose from among the most respected existing manufacturers, i.e. manufacturers with a proven track record.

The first step should therefore involve selecting and inviting only those bidders that meet a number of predefined criteria. Here's the process:

- Determine acceptable vendors
  - — Acceptable vendors must have experience with size, pressure, temperature, flow, speed, and service conditions specified
  - — Vendors must have proven capability in manufacturing with the chosen metallurgy and fabri-

MECHANICAL SEAL SPEC SHEET
TYPE H

Seal Type:	Tandem spring pusher seal.
Face Materials:	Carbon-graphite versus fine grained, reaction bonded silicon carbide, Refel or equivalent.
Secondary Sealing Elements:	Viton preferred.
Face Width:	0.15 inch (maximum)
Balance Ratio:	0.8 - 0.95
Flush Plan:	API Plan 2, 11, 13 as recommended by seal vendor subject to purchaser's approval.
Quench Plan:	None
Special Features:	Tangential outlet port for barrier fluid piping. High capacity pumping screw.
Barrier Fluid System:	Per Tandem Seal Barrier Fluid Circulation System drawing DS.14-281.
Barrier Fluid:	Ethylene glycol antifreeze, 70%/30% ethylene glycol/water by volume.
General Notes:	Primary seal flush must enter stuffing box within 3/4" of sealing faces. If possible seal gland shall be designed for full face contact with mating surface on pump to prevent gland distortion when gland bolts are tightened. If dimensional constraints permit, a close-fitting bronze throttle bushing as per API 610 Seventh Edition paragraph 2.7.1.13 shall be provided. Diametral clearance between sleeve O.D. and bushing bore to be no greater than 0.025". Inner and outer seals must be equivalent in temperature and pressure rating and preferably of the same size. A complete single assembly cartridge seal execution is preferred. Semi-cartridge type executions are acceptable. Approval drawing shall be submitted to purchaser for review prior to manufacture of sleeves and glands.

Figure 2-6: Mechanical seal specification sheets lead to more reliable pump components

cation method, e.g. sand casting, weld overlay, etc.

— Vendor's "shop loading" must be able to accommodate your order within the required timeframe (time to delivery of product)

— Vendors must have implemented satisfactory quality control (ISO 9001, etc.) and must be able to demonstrate a satisfactory on-time delivery history over the past two years

— If unionized, Vendor must show that there is virtually no risk of labor strife (strikes) while manufacturing of your pumps is in progress

The second step would be for the owner/purchaser to

- Specify for low maintenance. As a reliability-focused purchaser, you should realize that selective upgrading of certain components will result in rapid payback. Components that are upgrade candidates are identified in this book. Be sure you specify those. To quote an example: Review failure statistics for principal failure causes. If bearings are prone to fail, realize that the failure cause may be incorrect lube application, or lube contamination. Address these failure causes in your specification.

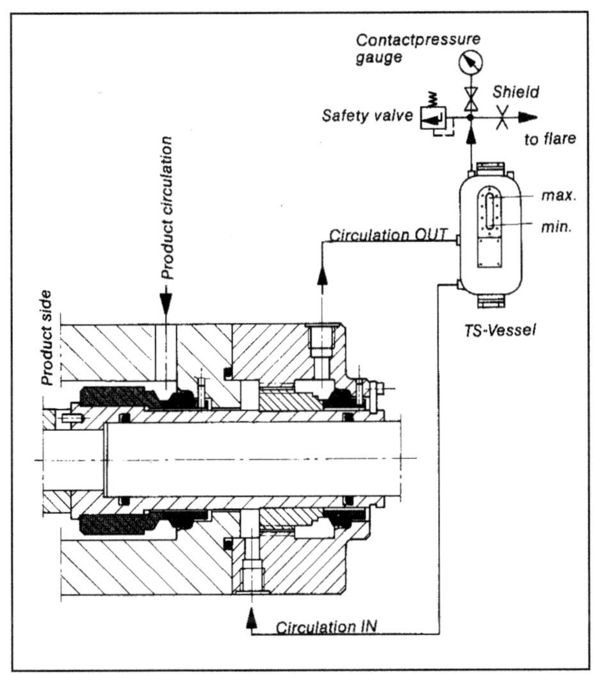

Figure 2-7: An illustrative sketch will assist in describing desired seal auxiliaries (Source: Burgmann Seals America, Inc., Houston, TX 77041)

- Evaluate vendor response. Allow exceptions to the specification if they are both well-explained and valid.

- Future failure analysis and troubleshooting efforts will be greatly impeded unless the equipment design is clearly documented. Pump cross-section views and other documents will be required in the future. Do not allow the vendor to claim that these documents are proprietary and that you, the purchaser, are not entitled to them.

  Therefore, place the vendor under contractual obligation to supply all agreed-upon documents in a predetermined time frame and make it clear that you will withhold 10 or 15 percent of the total purchase price until all contractual data transmittal requirements have been met. (See "Documentation is Part of Initial Purchase," later in this chapter).

- On critical orders, *contractually* arrange for access to factory contact or even designation of "Management Sponsor." A Management Sponsor is a Vice President or Director of Manufacturing, or a person holding a similar job function at the manufacturer's facility or head offices. You will communicate with this person for redress on issues that could cause impaired quality or delayed delivery.

The merits of a drawing review prior to final pump selection are best explained by looking at several examples of weaknesses in pumps.

DESIGN WEAKNESSES FOUND IN API CASING CONSTRUCTION

Figures 2-8 and 2-9 show two-stage overhung pump designs by two major manufacturers (Ref. 2-3). When purchased in the 1970's, each of these pumps complied with API Standard 610 and supplementary specifications then issued by the engineering offices of many contractors or major petrochemical companies. However, the pump illustrated in Figure 2-8 gave poor service and required frequent replacement of thrust bearings. Also, the pump experienced rapid performance deterioration which prompted the user plant to replace the motor driver (which had seemed adequate at plant start-up) with a higher horsepower motor.

After several such repair events, the pump design depicted in Figure 2-8 was analyzed in more detail and basic deficiencies found in the cover-to-casing fit-up at points (1) and (2). This design utilized a slip-fit engagement which allowed pumpage at elevated pressure to flow towards regions of lower pressure. In this specific case, high velocity flow through an average annular gap of 0.032 inch (0.8 mm) caused progressive erosion and performance drop off. In addition, rotor thrust could not be kept within original design limits and tended to shorten bearing life.

In contrast, the two-stage overhung pump of Figure 2-9 employed a gasketed step fit at internal joints (1) and (2). Flow bypassing was effectively prevented by this design feature and pump mean-time-between-failures significantly extended.

Conventional Wisdom: *You cannot go wrong buying from a reputable manufacturer.*

Fact: *Even reputable manufacturers are known to have produced non-optimized pumps and components.*

CASING DESIGN COMPROMISES IN ANSI PUMPS

The design conservatism found in pumps made to comply with ANSI Standards was not necessarily intended to match that of API pumps. It is not unusual,

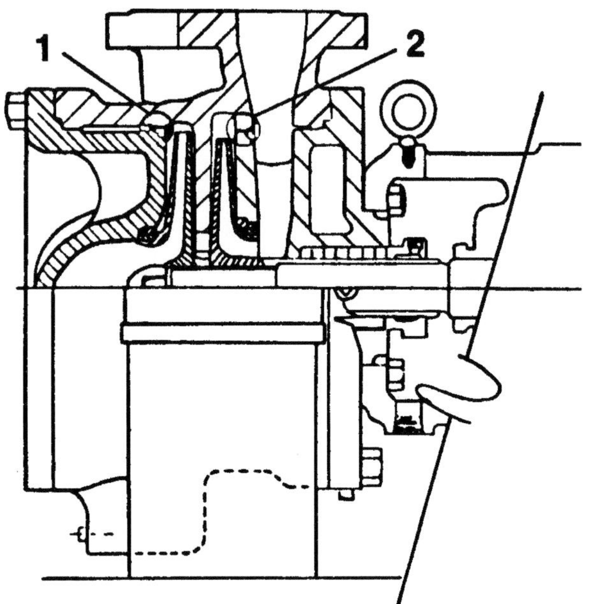

Figure 2-8: Vulnerable Two-Stage Overhung Pump Design. (Source: Ref. 2-3)

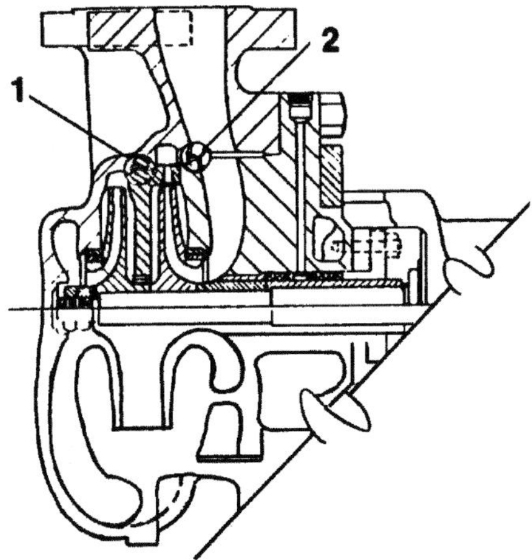

Figure 2-9: Preferred Two-Stage Overhung Pump Design. (Source: Ref. 2-3)

therefore, that the reviewer may find interesting differences among the various ANSI pumps offered on a given project. And, while these differences may not be very significant in terms of initial purchase cost, they could have a measurable impact on future maintenance costs and even the safety performance of the equipment.

As is the case with so many potential component design problems, the casing design weakness inherent in the pump shown in Figure 2-10 will become critically important only in the event of operating or maintenance errors. In one documented failure event, the pump was shut down and left full of liquid. It was then inadvertently started against a closed discharge block valve which, following the laws of thermodynamics, caused temperature rise and vaporization to take place. The casing was over-pressured and blew apart.

Failure analysis quickly showed that this particular pump model depended upon the cast iron bearing housing adapter (1) to hold cover (2) against the pump casing (3). The casing could blow apart abruptly because the bearing housing adapter was never designed to serve as a pressure containment member. The same design also invites uneven torque application to containment bolts. This could cause breakage of the unsupported bolt hold flange incorporated in the cast iron distance piece. Leaving a gap "F" between parts (1) and (3) as was done in this pump design is simply not good practice.

Figure 2-11 depicts a similar pump which should be given preference in procurement situations concerned with superior reliability. The superior design incorporates design features which avoid the above problems. Here, the pump cover (2) is fastened to the casing (3); the adapter piece (1) is fastened to the cover.

BEARING HOUSING DESIGNS MERIT CLOSER ATTENTION

Pump manufacturers have occasionally offered basic designs that worked well in theory, or under ideal operating conditions. Unfortunately, the user's site conditions or installation and maintenance practices will often differ from ideal and the user will now be faced with a marginal design. Since the manufacturer can always point to a few successful installations, and with the legal implications of admitting fault having potentially far-reaching effects, seeking redress from the manufacturer will rarely bring tangible or satisfactory results.

In those instances then, the user may attempt to obtain partial cooperation from the manufacturer and then proceed with implementing the changes himself. One of the authors has had success with an up-front acknowledgment that the manufacturer's design was deemed adequate, while at the same time requesting feedback on the feasibility of modifying the design "for site-specific reasons."

Safe bearing housing designs are a case in point. Figures 2-12 through 2-15 relate to the bearing housing portion of a generally well-designed single stage overhung pump. In the mid 1970's this pump complied with the then applicable revision of API-610 and was widely used by the petrochemical industry in the U.S.

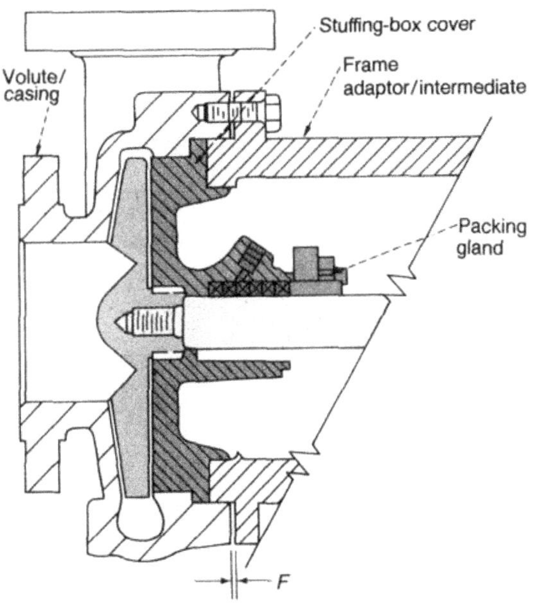

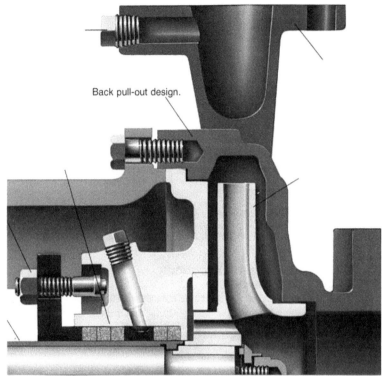

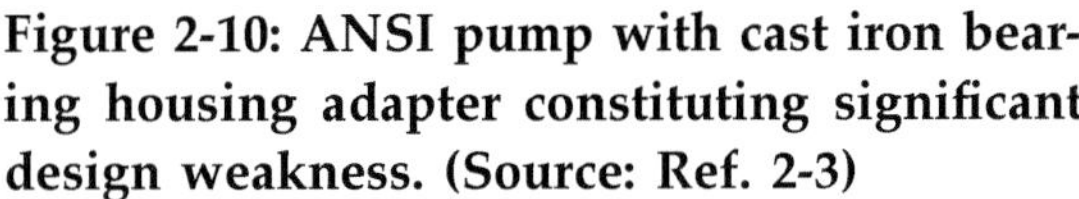
Figure 2-10: ANSI pump with cast iron bearing housing adapter constituting significant design weakness. (Source: Ref. 2-3)

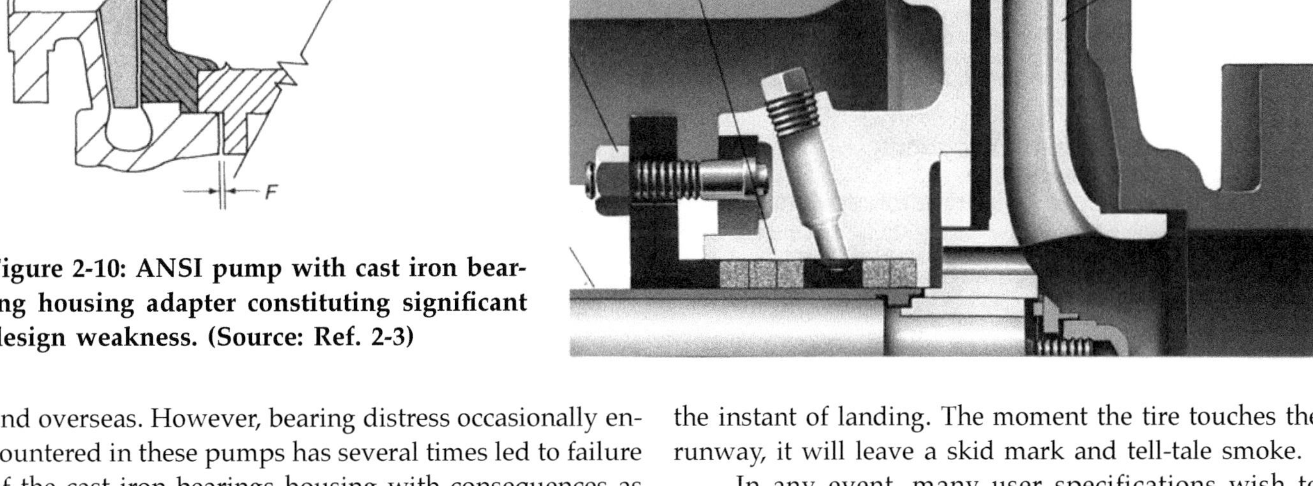

and overseas. However, bearing distress occasionally encountered in these pumps has several times led to failure of the cast iron bearings housing with consequences as far-reaching as major fires. For this pump, Figure 2-12 illustrates the standard execution utilizing a double row thrust bearing (1) inserted into a cast iron bearing housing (2). A snap ring (3) determines the installed location of the thrust bearing.

Double row thrust bearings in centrifugal pumps have generally not performed as well as two single row thrust bearings mounted back-to-back (Ref. 2-4). The reason for this is probably not a fundamental weakness in the design of double row bearings. It may be suspected that the need for using proper radial as well as axial internal fit double row bearings, as pointed out in some of the better pump repair manuals is sometimes disregarded or overlooked by the user's repair facility. On the one hand, loose radial internal fits allow for such contingencies as unexpected thermal expansion of the bearing inner ring and higher-than-intended interference fit between shaft and inner ring bore. However, with axial (thrust) loading, the loaded side of the bearing will undergo component deflection, whereas the unloaded side is now prone to skid. This is graphically represented in Figure 2-13 for a specific double row bearing in the size range found in centrifugal pumps and is further highlighted in the chapter dealing with bearings. A skidding bearing can be likened to an airplane tire at the instant of landing. The moment the tire touches the runway, it will leave a skid mark and tell-tale smoke.

In any event, many user specifications wish to avoid skidding and heat-related lubricant degradation. Therefore, they often disallow this bearing construction and mandate, instead, that two single row angular contact bearings be installed back to back, as shown in Figure 2-14. However, to comply with this requirement in the case illustrated, the pump manufacturer elected to accommodate the two single row thrust bearings by fabricating a retainer bushing (4) whose outside diameter would engage the bearing housing bore over a very narrow distance "D" only. While this practice would be of little consequence in numerous pumps operating within the limits of process and mechanical design intents, it proved costly for at least one user whose pump shaft was mechanically overloaded. Excessive forces and moments caused the retainer bushing to deflect and the cast iron bearing housing to break.

Whether or not a retainer bushing designed for the wider engagement shown in Figure 2-15 would have prevented fracturing of the bearing housing is difficult to determine. On the other hand, the uncompromising conventional design of Figure 2-16 utilizes a cast steel bearing housing and duplex bearings without retainer bushing. It could hardly have cost significantly more to execute and has since proven to give superior service in many installations.

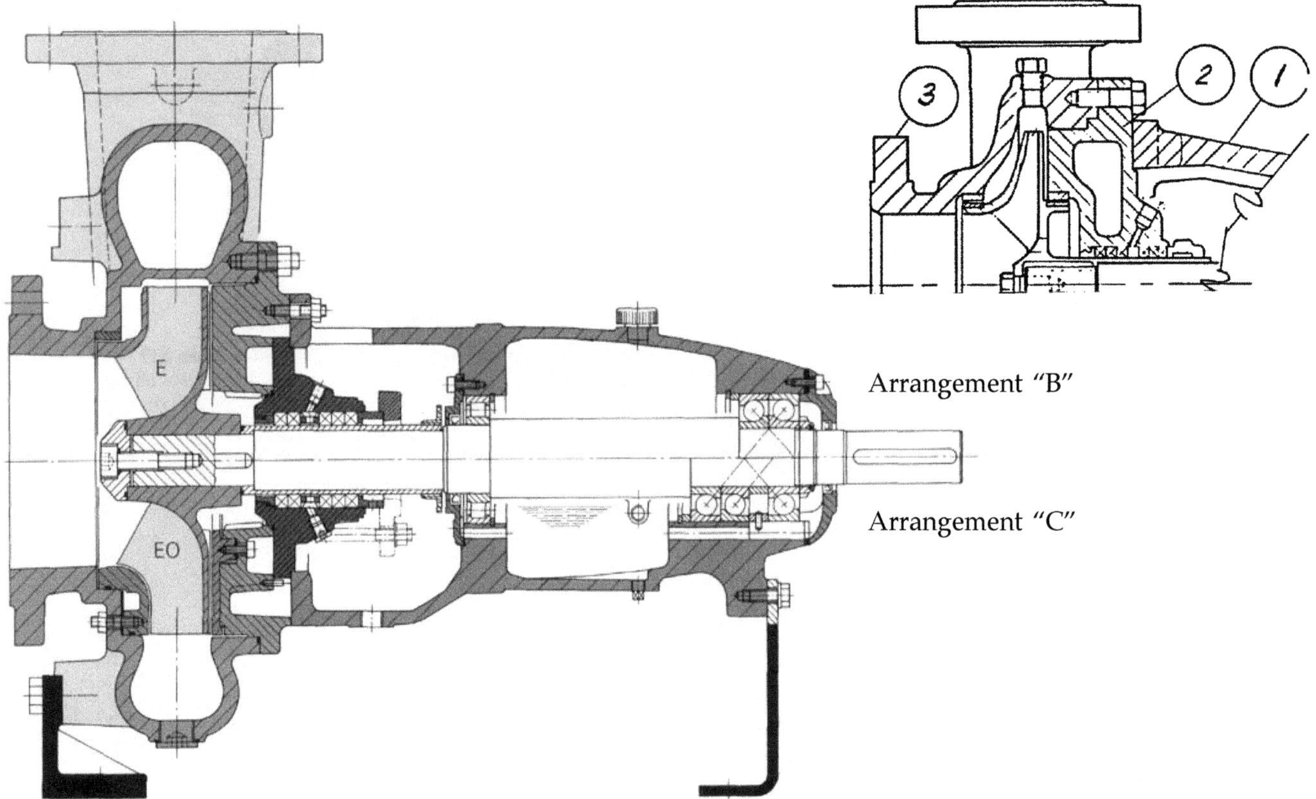

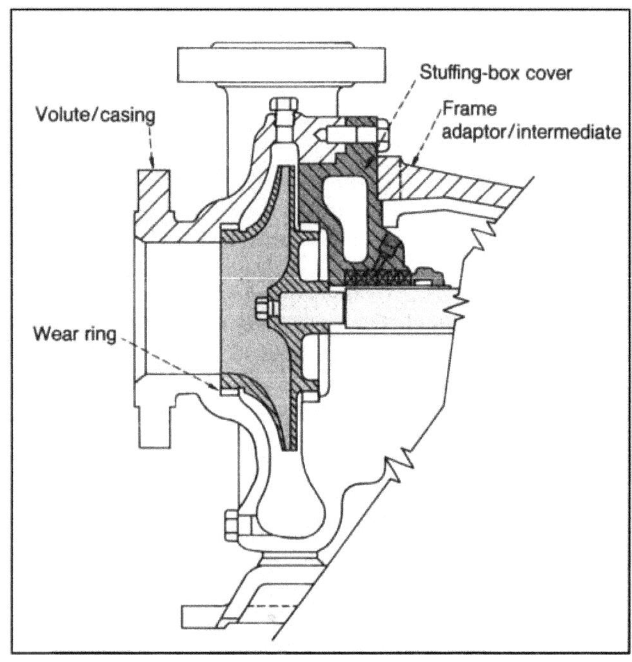

Figure 2-11: ANSI/ISO pump designs with cover directly bolted to casing overcome design weakness (Source: Emile Egger & Co., Ltd., Cressier, NE, Switzerland)

In the 1980's, observation of these and similar mechanical vulnerabilities prompted a major petrochemical company to develop a specification for what can be loosely called an Upgraded Medium Duty ("UMD") pump (Ref. 2-5). From failure statistics, field observations and reviews of available literature, it was found that seven improvements in the design of medium-duty process pumps in certain plant environments could lengthen mean-time-between-failures from the average of 13 (actual operating) months for existing pumps to a probable 25 on-line months for the UMD pump. Hydraulic performance—hence, energy efficiency—could be enhanced by allowing pump external dimensions to deviate from the constraints imposed by the present ANSI standards. The design changes needed were identified by the application of reliability engineering concepts.

Satisfactory performance and long periods between failures or overhauls can generally be achieved as long as centrifugal pumps are operating at near-design conditions. It should be noted, however, that "design" refers not only to pressures, flows, temperatures and other process parameters, but includes flange forces and moments, coupling forces and moments, bearing lubrication and similar mechanical considerations. Conservatively designed pumps will tolerate a certain amount

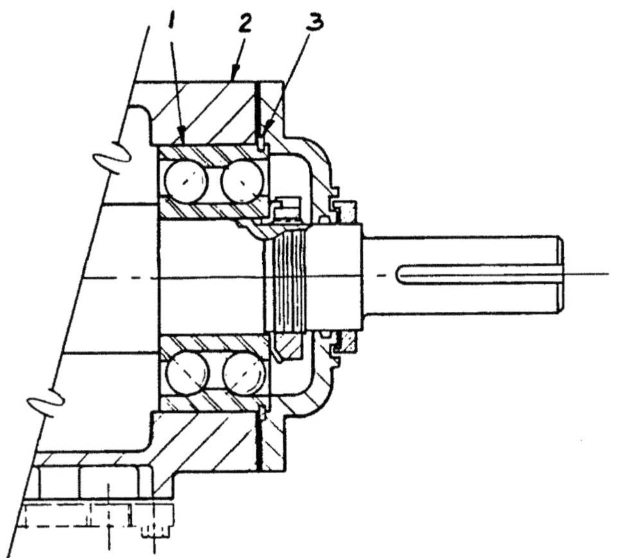

Figure 2-12: Bearing housing with double row thrust bearing equipped with snap ring groove. (Source: Ref. 2-3)

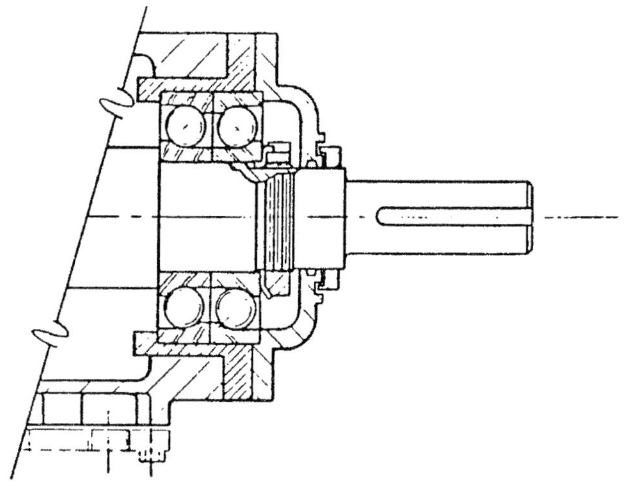

Figure 2-15: Improved adapter bushing inserted in bearing housing. (Source: Ref. 2-3)

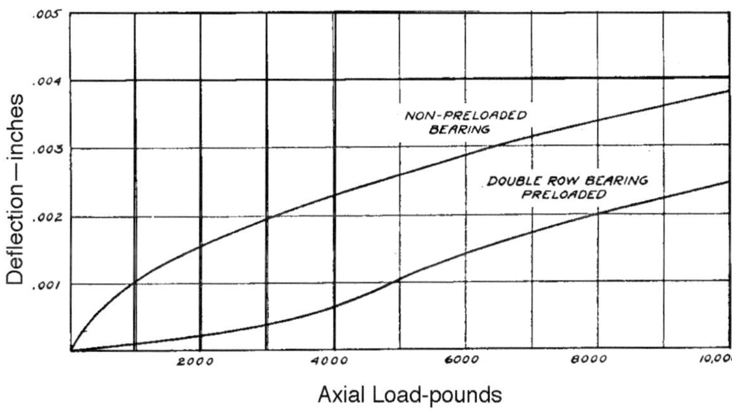

Figure 2-13: Load deflection curves for typical double-row bearings. (Source: SKF USA, Kulpsville, PA)

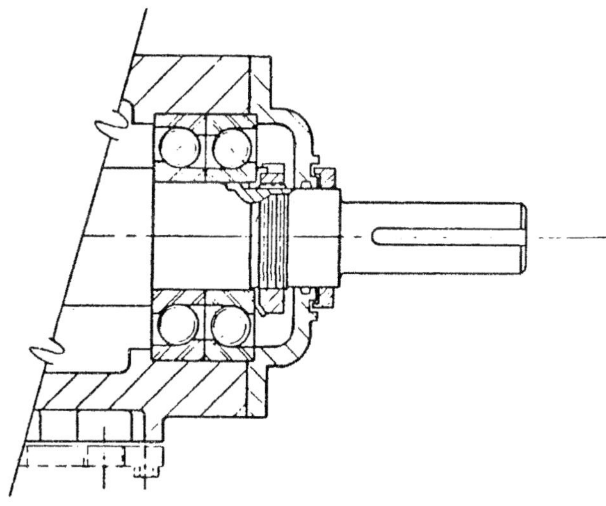

Figure 2-16: Least vulnerable execution of a pump bearing housing. (Source: Ref. 2-3)

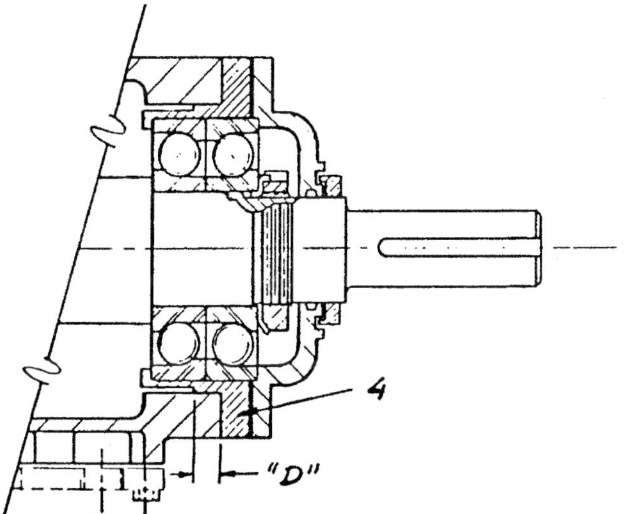

Figure 2-14: Vulnerable bearing housing modification

of off-design operation, either process or mechanically induced, before reaching the distress level.

Putting it another way, it is reasonable to expect that deviations from anticipated process operation, installation oversights and improper maintenance will adversely affect the ultimate life expectancy or mean-time-between-failure (MTBF) of centrifugal pumps. Conservatively designed centrifugal pumps should not, however, experience near-instantaneous catastrophic failure simply because one or more of the design conditions are exceeded or violated by small margins.

UPGRADING OF CENTRIFUGAL PUMPS

The vast majority of pumps in chemical process plants and oil refineries operate below 300 psig and 350°F. Even at these conditions, an inordinately high percentage of the maintenance expense for rotating equipment goes into pump repairs. In the late 1980's, a typical work order for pump repair in a United States Gulf Coast process plant exceeded $5,000. The reduction of such repair costs through the improvement of pump reliability should, obviously, rank high on any list of plant priorities.

Computerized record-keeping has made it possible to identify when and why a particular piece of equipment failed, and to compare its failure frequency and repair costs with similar equipment operating under similar process conditions. Such record-keeping has revealed the following :

- Although there were no doubt many exceptions, in the late 1990's standard ANSI (American National Standards Institute) pumps had a mean-time-between-failures (MTBF) of only 26 months in what were considered well-maintained facilities in North America. The actual total industry average was probably closer to 12 months (see Note 1, below).

- Attempts to correct the causes of pump failures have traditionally been repair-focused. Parts break, and parts are being replaced in kind. This old-style organizational approach is no longer sufficient. Plants must purchase better pumps and plant personnel must take the lead in implementing the reliability-focused approach. The reliability-focused approach implements systematic upgrading wherever feasible and cost-justified.

- Allowing pumps to leave the dimensional constraints of the ANSI specifications is sometimes justified. In other words, some of the ANSI pumps now installed could be replaced with types having an average efficiency that could add 10 or even more percentage points to the present hydraulic efficiency.

- API pumps in medium-duty service are not always cost-effective. Therefore, before choosing an API pump for a mild service, consider using an "in-between" pump. We might say that ISO and modified ANSI pumps fit this description.

- Typical shortcomings of certain ANSI pumps needed to be identified. The study described in Ref. 2-5 revealed the following weaknesses:

  (1) shaft deflection is often excessive

  (2) the dimensional limits imposed by ANSI B73.1 for a standard stuffing box do not generally allow sufficient space for the application of the superior mechanical seals now available and often curtail the attainment of optimum efficiency. (ANSI B73-1 large bore seal chambers do provide significantly more space.)

  (3) bearing life is shorter than it could be because of weaknesses in present bearing designs and lubrication systems

  (4) frangible pressure-containment sealing devices can create an unnecessary safety hazard, and

  (5) the average ANSI-design baseplate does not always provide adequate structural integrity and load-bearing capability.

Note 1: Unless otherwise noted, MTFB calculations were made by dividing the number of all installed pumps by the number of repairs per year. Also, every incident of parts replacement is counted as a full-fledged pump repair.

Conventional Wisdom: *An API-style pump is always a better life-cycle cost choice than an ANSI or ISO-style pump.*

Fact: *For light or medium duty services, non-API pumps may give equally satisfactory service at lower initial cost.*

EXCESSIVE SHAFT DEFLECTION

Pumps of overhung impeller construction are prone to excessive shaft deflection, which often leads to internal contacting of wear rings, bushings and sleeves, and frequently reduces the life expectancy of mechanical seals. In extreme instances, frequent stress reversals cause shaft fatigue. Because operating and maintenance costs tend to rise with increasing shaft deflection, reduced deflection should be a key feature of any new pump design.

The amount of shaft deflection can be calculated from the following equation (Ref. 2-6):

$$Y = F/3E(N^3/I_N + (M^3 - N^3)/I_M + (L^3 - M^3)/I_L + L^2X/I_X)$$

Where,

- Y = shaft deflection at impeller centerline, in.;
- F = hydraulic radial unbalance, lbs;
- M and N = distances from impeller centerline to the steps on the shaft, in.;
- L = distance from impeller centerline to centerline of inboard bearing, in.;
- X = span between bearing centerlines, in.;
- I_L, I_M, I_N and I_X = moments of inertia of the various diameters, in^4 and
- E = modulus of elasticity of shaft material, psi (see Figure 2-17).

For a reasonably accurate approximation, let

$$y = FL^3/3EI_M$$

Note that, because $I_M = D_M{}^4/64$, shaft deflection is a function of L^3/D^4. At least one major oil company makes use of this fact in its engineering specifications (Ref. 2-7). The pump data sheets of this company include spaces for the L^3/D^4 ratio to be provided by the vendor. This ratio is called the shaft flexibility factor, or SFF. In competitive bidding, the SFF values given by bidders are compared against the lowest SFF value, and the higher values are assigned a "maintenance assessment" of a certain percentage of their bid price (see Figure 2-18).

Shaft deflection changes as a function of the fluid flowrate through the pump. As the throughput capacity of a pump increases or decreases and thus moves away from the best efficiency point, the pressures around the impeller become unequal, tending to deflect it. In an overhung impeller pump having a standard single volute casing, this deflection can reach serious magnitudes (Figure 2-19).

Special casings—such as diffusers and double-volute and concentric casings—can greatly reduce the radial load and, hence, the deflection. However, not even the best-designed casing can completely eliminate pressure-induced shaft deflections.

That a single volute pump will not give satisfactory long-term service if operated too far from its best efficiency point is indicated by Figure 2-19. Regardless of the volute design, the mechanical strength of a pump tends to be improved by low actual shaft deflections. While low shaft flexibility factors tend to indicate low shaft deflections, it should be pointed out that full-fledged deflection calculations should be made.

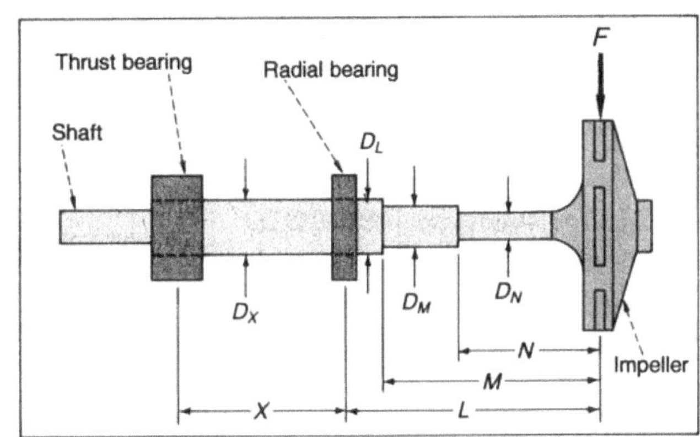

Figure 2-17: Shaft deflections vary with shaft diameter and overhung impeller distance

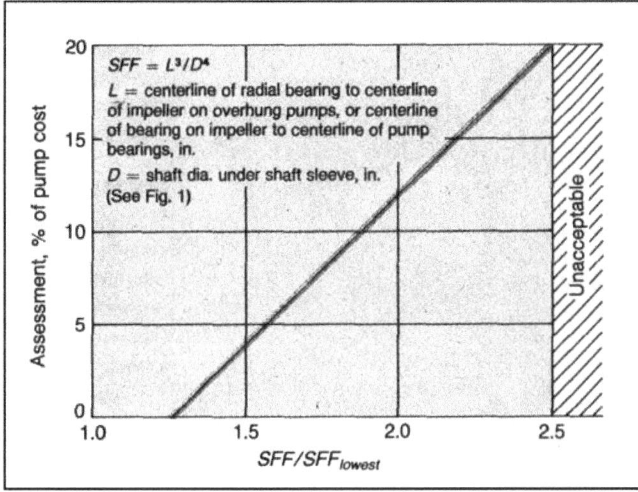

$$SFF = \frac{L^3}{D^4}$$

Where:

- SFF = Shaft flexibility factor
- L = C* of radial bearing to C* of impeller on overhung pumps; or C* to C* of bearings on impeller - between - bearings pump, inches
- D = Shaft diameter under shaft sleeve, inches

*C = Center

Figure 2-18: "Maintenance Assessment" allows credits and/or debits for L^3/D^4 ratios of pump shafts

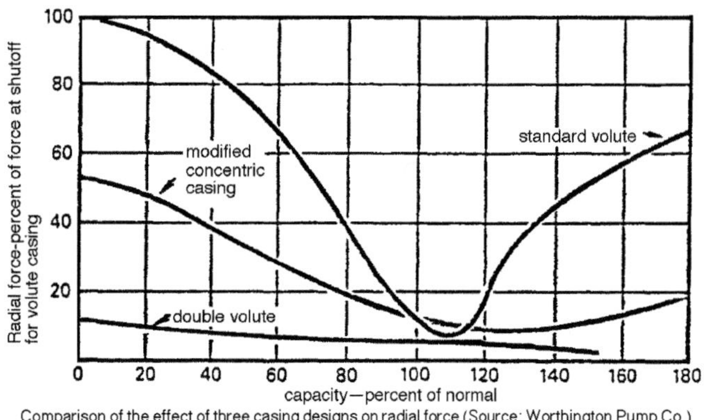

Comparison of the effect of three casing designs on radial force (Source: Worthington Pump Co.)

Figure 2-19: Shaft deflections are potentially high for single volute pumps

Suppose the detailed calculation for one manufacturer's pump model "A," indicates an actual shaft deflection of only 0.0001 inches. In that case, even a five-fold deflection of 0.0005 inches in the competitor's pump "B" will be so low that it should be of no concern. The upshot of this observation deals with marketing strategies. The manufacturer of pump "A" will proudly point out that his pump shaft exhibits one fifth of the shaft deflection of "B." While this may be fact, both "A" and "B" should certainly be considered acceptable selections.

FACTORS AFFECTING BEARING LIFE

Less than 10% of all ball bearings run long enough to succumb to normal fatigue failure. According to the Barden Corporation (Ref. 2-8), most bearings fail at an early age because of static overload, wear, corrosion, lubricant failure, contamination, or overheating. Skidding (caused by a bearing operating without load) is a frequent cause of failure in angular contact ball bearings operating as pairs. As will be shown later in Chapter 7, this problem can be solved by using a matched pair of bearings and proper shaft fits that result in a slightly preloaded condition after installation.

Actual operations have shown that better bearing specification practices will avert the majority of static overload problems. Other problems caused by wear, corrosion, lubricant failure, contamination and overheating can be prevented by the proper selection, application and preservation of lubricants. Oil viscosity and moisture contamination are primary concerns, and higher-viscosity lubricants are generally preferred (Ref. 2-9). The detrimental effects of moisture contamination are indicated in Table 2-1 (Ref. 2-10).

Unlike API pump bearings, which petrochemical companies often specify for an L_{10} life of 40,000 hours, ANSI pump bearings are selected on the basis of an expected 24,000-hour life. Nominally, this means that 90% of the ANSI pump bearings should still be serviceable after approximately three years of continuous operation. However, the failure statistics quoted in Ref. 2-5 indicate that conventionally lubricated ANSI pump bearings do not even approach this longevity. Lack of lubrication, wrong lubricants, water and dirt in the oil, and oil-ring debris in the oil sump all cause bearing life expectancies to be substantially less. It must be assumed that similar findings by other major users of ANSI pumps prompted the search for "life-time lubricated" rolling element bearings (Ref. 2-11).

Problem incidents caused by dirt and water have been substantially reduced by the pressure differential at the bearing housing seal provided by oil mist lubrication. However, serious failure risk can also be introduced by certain specification practices—including some contained in API-610. Without going into the many possible reasons for bearing life reductions, we can list several observations which should be considered when preparing a modern pump specification. These observations supplement, and summarize, the more detailed application criteria and data given later in Chapter 7 of this text:

- Deep-groove Conrad-type ball bearings with loose internal clearance (C3) are more tolerant of off-design pump operation than bearings with normal internal clearance. With few exceptions, this clearance should be specified for pump bearings.

- In centrifugal pumps that occasionally see temperature excursions, destruction of the cage may be experienced. Phenolic cages are generally limited to operation below 100°C (212°F). They tend to become brittle at both high and excessively low temperatures. New cage materials, such as polyamide 66, provide a marginally higher temperature limit and excellent lubricity at somewhat higher operating temperatures.

- All plastic cage material must be avoided in equipment where vibration data acquisition and analysis are used as the predominant predictive maintenance approach. Cage deformation and wear-related distress will not show up in the overwhelming majority of vibration data acquisi-

Table 2-1: The effect of water on the fatigue life of rolling contact bearings

Base Oil Description	*Water Content of Wet Oil*	*Fatigue Life Reduction, %*	*Test Equip. % Hertzian Stress*	*Author(s) and and References*
Base mineral Oil dried over Sodium	0.002% 0.014% 3.0% 6.0%	48 54 78 83	Rolling Four-Ball 8.6 GPa 1.25 × 10E6 psi	Grunberg & Scott 1960
Squalene and 0.001% water	0.01%	32-43	Rolling Four-Ball 6.34-8.95 Gpa 0.92-1.3 × 10E6	Schatzberg & Felsen
Dried mineral oil	not stated, moist air environm.	80	Rolling Four-Ball	Ciruna & Szieleit, 1972
Two formulated tetraester oils-0.005% water	0.05%	29; 73	#204 Brgs, 2.48 Gpa 0.36 × 10E6 psi	Feinstein 1969
Emulsifying hyd. fluid purged w/ argon	1% sea water	45	Rolling Four-Ball, 6.89 Gpa 1.00 × 10E6 psi	Schatzberg 1971
Mineral oil-based emulsifying hyd. oil, 0.02% water	0.1%	45	Angular Contact Bearing 2.27 Gpa 0.33 × 10E6 psi	Felsen, et al 1972

tion and detection instruments used by maintenance and reliability engineering work forces.

- The relative temperature sensitivity of both thermosetting (phenolic) and thermoplastic (polyamide) cages makes it impractical to entrust shaft mounting via heat dilation to anyone other than a select few, conscientious and highly experienced repair shops.

- Metallic cages are least temperature-sensitive. Deterioration of metallic cages will generally show up in vibration spectra displaying vibration amplitude vs. vibration frequency. Therefore, rolling element bearings with metallic cages (brass, steel) are the generally preferred repair, replacement, and/or retrofit component.

- The API-610 requirement to utilize duplex 40° angular contact angle thrust bearings was prompted by the desire to obtain maximum load capacity from a given size bearing. Similarly, the requirement of installing these bearings back-to-back with a light axial preload was aimed at reducing axial shuttling of rotors to prevent brinelling of contact surfaces (raceways) and to prevent ball skidding. The FAG Bearing Corporation has shown that, for the same load, 40° angular contact bearings generate less heat than thrust bearings with lower contact angles. However, preloading adds to the heat load, and using an interference fit between shaft and inner ring generally compounds the problem. Unless lubricant application methods take into account all of the above, bearing life and reliability may be severely impaired.

- Double or single-row "filling notch" bearings are considerably more vulnerable in pump thrust applications than other bearing types and should not be used.

- Ball bearings are sensitive to misalignment and must be mounted properly to eliminate this mode of failure. Misalignment must be no greater than .001" per inch of shaft diameter. Bearings operating in a misaligned condition are subject to failure regardless of cage type.

Conventional Wisdom: *Assuming the same external load is acting on a high load capacity bearing and an externally equivalent-sized lower capacity bearing, the high capacity bearing will give longer life.*

Fact: *If the high load capacity is achieved by accommodating an additional bearing ball via filling notch, such bearings will give inferior life and should not be used in process pumps.*

As mentioned at the outset, improvements can come from better specifications, more careful engineering and better dissemination or presentation of pertinent data. It is in these areas where close cooperation between user, contractor, pump manufacturer and component suppliers (bearings, seals, couplings) could prove highly advantageous to all parties.

The upgraded medium-duty (UMD) bearing housing shown in Figure 2-20 incorporates the various bearing-related features that have been discussed: (1) a deep-groove Conrad-type, loose-fitting radial bearing; (2) a set of duplex angular contact (40°), lightly preloaded back-to-back-mounted thrust bearings; (3) a vent port that remains plugged for dry-sump oil-mist lubricated bearings and that can be fitted with an expansion chamber, if conventionally lubricated; (4) a set of magnetic bearing housing seals; and (5) bearing housing end covers made to serve as directed oil-mist fittings. (It should be noted that there is no official industry standard covering UMD pumps. The term was coined by one of the authors while employed by a major multinational petrochemical company, as described in Refs. 2-5 and 2-6.)

Tables 2-2 through 2-4 illustrate different studies aimed at determining the approximate cost justification for upgrading a certain pump type. Table 2-2 uses typical statistics to determine the payback of initially spending an extra $1,600 for pumps with superior bearing housing protection ("hermetic seals") and finds this payback to occur in roughly 13 months. After that, the plant saves over one million dollars each year due to avoided repair incidents.

In Table 2-3, an attempt was made to determine the cost and benefit of retroactively upgrading a plant's pump population. Here, the payback period was six years. (It should be noted that the MTBF calculation was made on the average number of operating pumps rather than the total number of pumps installed.)

Finally, and perhaps most realistically, the plant described in Table 2-3 is again represented in Table 2-4. However, a projection is made of the value of systematically upgrading only those pumps that are sent to the shop. At an incremental cost of $1,120 per repair, the MTBF of upgraded pumps is thought to exceed that of the standard pumps by a factor of 1.9. Monthly repair averages are reduced from previously 17 to perhaps only 9 and in less than two years these routine upgrades or conversions will have paid for themselves.

Conventional Wisdom: *Pumps are shipped with the required documentation.*

Fact: *Pumps often leave the factory without proper installation and maintenance documentation. It is important to link final payment to fulfillment of all obligations, including documentation.*

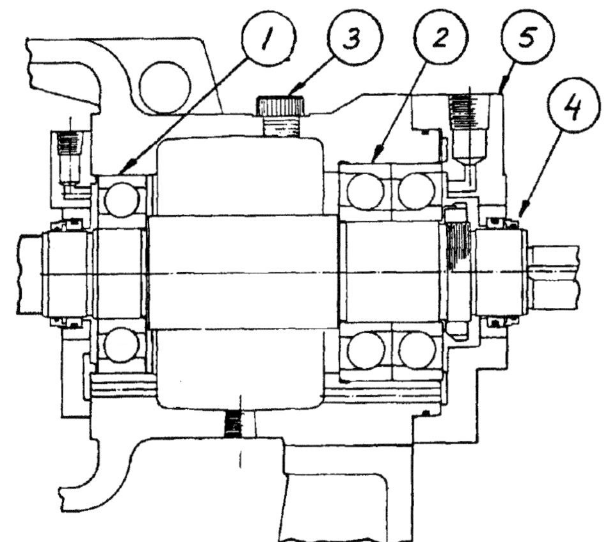

Figure 2-20: Upgraded medium duty (UMD) pump incorporating some maintenance reduction features

Table 2-2: "Method I" for estimating the economic justification of "ANSI-Plus" retrofits

Method I		
ESTIMATING THE ECONOMIC JUSTIFICATION FOR ANSI-PLUS RETROFITS*		
	Example Plant	*Your Plant*
1. Total number of ANSI pumps in plant/ mill/ unit	872	
2. Pump repairs due to bearing failures, per year	521	
3. Failures which could be avoided by selecting hermetically sealed bearing housing, per year	387	
4. Cost of each bearing failure (including cost of removal, re-installation, alignment, planner's and supervisor's time, parts and direct labor)	$2,679	
5. Process/production loss debits attributable to failure of unprotected bearings, each year (average)	$28,372	
6. Yearly combined costs due to failure of unprotected bearings (Item 3 × 4) + Item 5	$1,065,145	
7. Pump repairs due to mechanical seal failures, per year	217	
8. Failures which could be avoided by selecting Big Bore or taper-bore seal chamber	52	
9. Cost of each seal failure (including cost of pump removal, reinstallation, alignment check, planners and supervisor's time, parts and direct labor)	$3,011	
10. Process / production loss debits attributable to seal failures induced by non-optimized seal environment, each year (average)	$52,738	
11. Yearly combined costs due to failures initiated by non-optimized seal environment, each year (average)	$209,310	
12. Average cost of purchasing ANSI-PLUS Retrofit Kit	$1,600	
13. Estimated cost of retrofitting all ANSI pumps with ANSI-PLUS retrofit (Item 6 + Item 12)	$1,395,200	
14. Anticipated savings attributable to ANSI-PLUS retrofits (Item 6 + Item 11)	$1,274,455	
15. Payback Period—Item 13 divided by Item 14	~13 Months	
16. Anticipated savings - Years 2, 3, etc.	>$1,000,000 per year	

*An upgraded ANSI pump

DOCUMENTATION IS PART OF INITIAL PURCHASE

We had started out by highlighting the importance of documentation. Efficient and accurate repair and troubleshooting of machinery requires good documentation. Also, it is important that plant maintenance and technical service personnel be given ready access to this documentation. Experience shows that major contractors and owner's project engineers orient their initial data collection efforts primarily towards construction-related documentation. As the project progresses, more emphasis is placed on obtaining design-related machinery documentation as listed in the appendices of API specifications for major machinery. These requirements are generally understood by contractors and machinery manufacturers. Before the project is closed out, this collection of data is assembled in many electronic volumes of mechanical catalogs.

Unfortunately, data collection using API guidelines alone does not result in a complete data package or

Table 2-3: "Method II" for estimating the economic justification of "ANSI-Plus" retrofits

Method II

ESTIMATING THE ECONOMIC JUSTIFICATION FOR ANSI-PLUS RETROFITS

	*Example Plant**	*Your Plant*
1. Total number of ANSI pumps installed	417	
2. Number of ANSI pumps operating on any given day	238	
3. Number of ANSI pumps repaired each year, all causes	212	
4. Mean-Time-Between-Failures, all causes (Item 2 divided by Item 3) months	13.5	
5. Number of ANSI pumps requiring bearing and/or seal repairs each year	172	
6. Mean-Time-Between-Repairs, bearing and seal-related failures only (item 2 divided by Item 5)	16.6	
7. Average cost of bearing and/or seal related repairs (including cost of removal, re-installation, alignment, planner's and supervisor's time, parts and direct labor)	$3,122	
8. Anticipated Mean-Time-Between-Repairs, bearing and seal-related failures only, after retrofitting with ANSI-PLUS kits (1990 calculations project 1.9 - fold increase in MTBR) (Item 6 × 1.9)	31.5 months	
9. Number of ANSI-PLUS retrofitted pumps requiring bearing and/or seal repairs each year (Item 2 × 12 divided by Item 8)	91	
10. Number of pump repairs avoided, per year (item 5 - Item 9)	81	
11. Avoided repair costs, per year (Item 7 × Item 8)	$252,882	
12. Assumed average cost of ANSI-PLUS retrofit kit (materials only)	$1,600	
13. Assumed average cost of ANSI-PLUS retrofit conversion (parts, labor, overhead)	$3,670	
14. Total cost of retrofitting entire plant (item 1 × Item 13)	$1,530,390	
15. Payback period (item 14 divided by Item 11)	6 years	

*Using savings, Mean-Time-Between-Failures (MTBF), and cost projections published by a major chemical company in 1990.

adequate data format. Moreover, data collection in itself does not automatically provide means of letting plant maintenance and technical personnel have ready access to all pertinent machinery data. This is where machinery data file folders, or their computer-based equivalents fit in (Ref. 2-12).

Reliability-focused plants will insist on the development and compilation of machinery data file folders, whether on paper, as shown in Figure 2-21, or in a computer memory using the data listed in Figure 2-22. Some of the leading owner-purchasers have made this a contractual requirement. They have put pump manufacturers on notice that an amount of perhaps 10% or even 15% of the agreed-upon purchase price of the pump will be held in escrow until the following nine key items have been provided in an electronic format that is predefined and acceptable to all:

1. Installation, operating surveillance and maintenance instructions. These could be instruction manuals routinely available from equipment vendors, or instruction sheets and illustrations specifically prepared for a given machine. All of this information is generally intended to become part of the plant's equipment reference data library. However, some specific instructions may also be required for field posting, as illustrated in the steam turbine pump drive latching diagram example of Figure 2-23.

Table 2-4: "Method III" for estimating the economic justification for "ANSI- Plus" retrofits

Method III		
ESTIMATING THE ECONOMIC JUSTIFICATION FOR ANSI-PLUS RETROFITS		
	Example Plant	*Your Plant*
1. Total number of ANSI pumps installed	417	
2. Number of ANSI pumps previously repaired each year, all causes	212	
3. Average cost of each repair (direct labor, materials, associated costs, overhead)	$3,122	
4. Number of pumps failing, Month #1, and being retrofitted with ANSI-PLUS parts	18	
5. Failure projections of remainder of 2-year conversion period: 17 repairs Month #2 16 repairs Month #3 16 repairs Month #4 16 repairs Month #5 15 repairs Month #6 15 repairs Month #7 14 repairs Month #8 14 repairs Month #9 13 repairs Month #10 13 repairs Month #11 12 repairs Month #12 12 repairs Month #13 12 repairs Month #14 12 repairs Month #15 11 repairs Month #16 11 repairs Month #17 11 repairs Month #18 11 repairs Month #19 10 repairs Month #20 10 repairs Month #21 10 repairs Month #22 10 repairs Month #23 9 repairs Month #24 and every month thereafter.		
Total pump repairs in a 2-year period	308	
6. Cost difference, average, for repairing with ANSI-PLUS retrofit instead of conventional repair parts	$1,120	
7. Additional maintenance cost outlay for installing ANSI-PLUS retrofit parts instead of conventional repairs: (Item 6 × Item 6)	$344,960	
8. Avoided repairs in 24-month time period [(2 × Item 2) - Item 5]	116	
9. Value of avoided repairs In 24-month time period. (Item 3 × Item 8)	$362,152	

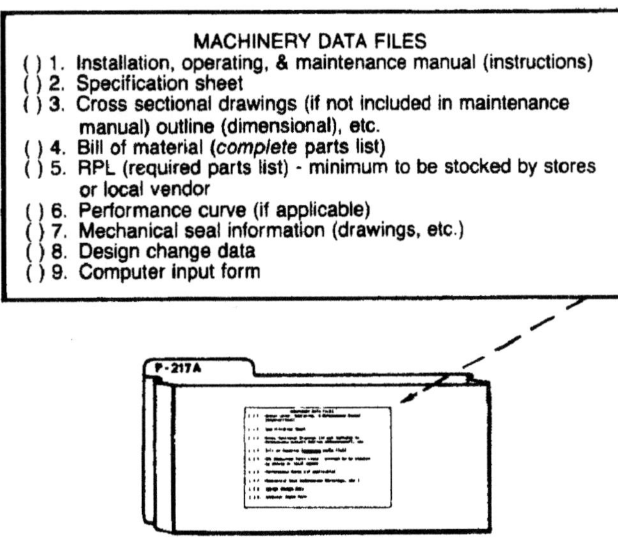

MACHINERY DATA FILES
( ) 1. Installation, operating, & maintenance manual (instructions)
( ) 2. Specification sheet
( ) 3. Cross sectional drawings (if not included in maintenance manual) outline (dimensional), etc.
( ) 4. Bill of material (*complete* parts list)
( ) 5. RPL (required parts list) - minimum to be stocked by stores or local vendor
( ) 6. Performance curve (if applicable)
( ) 7. Mechanical seal information (drawings, etc.)
( ) 8. Design change data
( ) 9. Computer input form
P-217A

Figure 2-21: Typical "old" machinery data file folder

2. Equipment specification sheets, such as API data sheets and supplementary data used at time of purchase.

3. Cross-section drawings showing the equipment assembly. These drawings must be dimensionally accurate. Dimensional outline drawings should also be included.

4. Bill of materials or complete parts list identifying components and materials of construction.

5. Tabulation of minimum number of parts to be kept on hand by storehouse or local vendor.

6. Performance curve.

7. Mechanical seal and seal gland drawings, if applicable.

8. Design change data. A typical design change form is shown in Figure 2-24. This particular form was used to document one of the many minor modifications that will inevitably be made after the machinery reaches the plant. Our example shows a change which had to be implemented on oil mist lubricated motor bearings. After modifying the upper bearing retainer to provide oil mist flow through the bearing instead of past the bearing, the design change form was placed in the file folder and a notation made on the front of the cover to let the users know how many design change forms they should find inside.

9. Computer input forms. These forms should be given to the equipment vendor as part of the specification package for pumps, motors and small steam turbines. Providing basic equipment data should be part of the vendor's contractual requirements. Some elements of a typical computer input form for pumps were highlighted in Figures 2-21 and 2-22. In addition to information given in the API data sheets, the equipment vendor must provide such important maintenance information as impeller and bushing clearances, manufacturing tolerances, and as-built internal dimensions. The computer input forms represent data base and nucleus of a computerized failure report system for a given plant. As a minimum, properly filled-in forms comprise a data bank of valuable maintenance and interchangeability information. Large petrochemical companies may derive additional benefits from using these computerized data sheets for intra-affiliate analyses of commonality of spare parts, etc.

BID TABULATIONS LEAD TO FINAL SELECTION

As mentioned earlier, it is essential that completed API data sheets be submitted with each proposed pump. The proposal package must also include performance curves and typical pump cross-sectional drawings. In addition to the above, the vendor must state minimum allowable flow and NPSH required (NPSHr) for the entire capacity range. Since minimum allowable flow listings could be governed by either thermal or recirculation-induced mechanical considerations, the vendors should be asked to specify their basis.

With the above data and any notable exceptions given by the various bidders, the comprehensive bid tabulation shown in Figure 2-25 can now be constructed. Careful review will narrow this bid tabulation to two or three principal manufacturers. These would be manufacturers whose equipment more consistently offers improved performance by demonstrating such features as low risk suction energy and operation near the best efficiency point (BEP).

We have alluded to cost considerations other than pure bid prices. When energy costs are high, pump efficiency becomes a major factor. However, the reviewer might have justifiable concern that quoted efficiencies could be falsely inflated. On large pumps, mere consideration of quoted efficiencies should be replaced by certified test-stand efficiencies, field feedback, and perhaps contractual bonus/penalty clauses.

INFO. ON EQUIPMENT LOCATION ONLY—SEE NOTE 1									PROCESS EQUIPMENT NAME OR SERVICE DESCRIPTION
UNIT LOCATION	COST CENTER	UNIT NO.	MAINT.		PROCESS		COMP		
			Sect.	Area	Dept.	Team			
16–24	25–29	30–32	33–34	35–38	39–42	43–44	45		46–72

UNIT ITEM NO.		SHUT DOWN CODE	CURR. USE	PREV. USE	WEIGHT k lbs		WARRANTY EXP. DATE			PURCHASE COST, $	MANUFACTURER	MANUFACTURER SERIAL NO.
							Mo.	Day	Yr.			
16–21	22–23	24–25	26–27	28–29	30–34	35–36	37		42	43–50	51–58	59–72

MANUFACTURER'S TYPE	START-UP DATE				APPROPRIATION NO.	SEE NOTE 1		DRAWING NO. (REF.)
	Mo.	Day	Yr.			PUMP TYPE		
16–27	28		33	34–36	37–51	52–53		60–72

MAIN DRIVER		ALTERNATE DRIVER		FLUID PUMPED				SPECIAL FLUID DATA			DESIGN FLOW RATE GPM	MIN. ALLOW. FLOW RATE GPM	SUCT. PRESS. PSIG	DIFF. PRESS. PSI
Type	Yard No.	Type	Yard No.	Code	Temp. °F	Viscosity @ Pt, CP	Sp. Gr. @ Pt.	Corrosive Code	Erosive Code	Fouling Code				
16–17	18–23	24–25	26–31	32–34	35–38	39–43	44–47	48–50	51–53	54–56	57–61	62–65	66–69	70–72

MAX. SUCT. PRESS. PSIG	DISCH. PRESS. PSIG.	DIFF. HEAD		NPSH (H20)		DESIGN SPEED RPM	ROTATION VIEWED FROM COUPLING	DRIVER HP	PUMP CASE					
		Design Ft.	Shut-off Ft.	Req'd. Ft.	Avail. Ft.				Mounting	Split	Type	Design Thk. In.	Retire Thk. In.	MAWP** 650°F (Psig)
16–19	20–23	24–27	28–31	32–34	35–37	38–43	44–46	47–50	51–54	55–57	58–60	61–64	65–68	69–72

H20 COOLING				INLET NOZZLE			OUTLET NOZZLE			THRUST BEARING ARRANGEMENT	VERTICAL PUMPS				LUBE SYS. TYPE	LUBE OIL TYPE
SB	BRG	PED	GLND	Nominal Size (in.)	Flange Rating	Position	Nominal Size (in.)	Flange Rating	Position		Length Ft.	Lift Mils	Line Shaft Bearing Lube.	Lower Bearing Lube		
16	17	18	19	20–23	24–27	28–31	32–35	36–39	40–43	44–49	50–52	53–55	56–63	64–66	67–68	69–72

STUFFING BOX			PACKING			MECHANICAL SEALS				SEAL FLUSH API PIPING PLAN	NO STAGES	IMPELLER DIAMETER - INCHES			
												FIRST STAGE		ALL OTHER STAGES	
SPARE LEAK	Cart Seal.	Press. Psig	Type	Size	No. Rings	TYPE	Seal Fluid	TA FL	Seal Alarm (PSIG)			Installed	Max.	Installed	Max.
16	17–18	19–21	22–29	30–32	33–34	35	36–38	39	40–43	44–50	51–52	53–57	58–62	63–67	68–72

DIAMETRAL CLEARANCE DATA (MILS)														
WEAR RING			IMPL. BACK VANES		OPEN FACE IMPL.		THROAT BUSHING		AXIAL		SLEEVE BEARING			Seal Face Comp. Mils
Dia.	Tolerance		To Casing	Tolerance	To Casing	Tolerance	Dia.	Tolerance	Movement	Tolerance	Dia.	Tolerance		
16–18	19 +	− 24	25 +	− 33	34 +	− 42	43 +	− 51	52 +	− 60	61 +	− 69		70–72

REMARKS
16–72

Figure 2-22: Computerized pump data input form

1. Open all 6 drains (casing, inlet, exhaust drains).
2. Verify operation of lube oil system and all steam traps.
3. ESTABLISH CONDENSER VACUUM, MOTIVE STEAM TO EJECTOR, COOLING WATER.
4. Turbine exhaust verified open.
5. Governor speed control knob turned "CCW" to minimum speed.
6. Handwheel turned clockwise to completely close trip valve.
7. Push S/U logic reset button on Local Panel.
8. Be sure ball "E" is pushed in to relatch overspeed trip.
9. If hydraulic pressure is established in line "A," rod "B" should have stroked in the direction of the arrow. This should allow latch "C" to engage lever "D."
10. Carefully open the handwheel and at the same time pull the reset lever "F" at the turbine governor valve to reset the governor hydraulic amplifier. Turbine should start. Slow roll for 15 min. at 500 RPM.
11. If rod "B" does not stroke to the right, check if electric solenoid is operating properly. If solenoid is open, hydraulic pressure cannot be established in line "A."
12. Verify that governor is functioning properly by observing movement of turbine inlet steam valves.
13. After governor has taken over, open handwheel fully. Verify turbine is operating at proper speed.

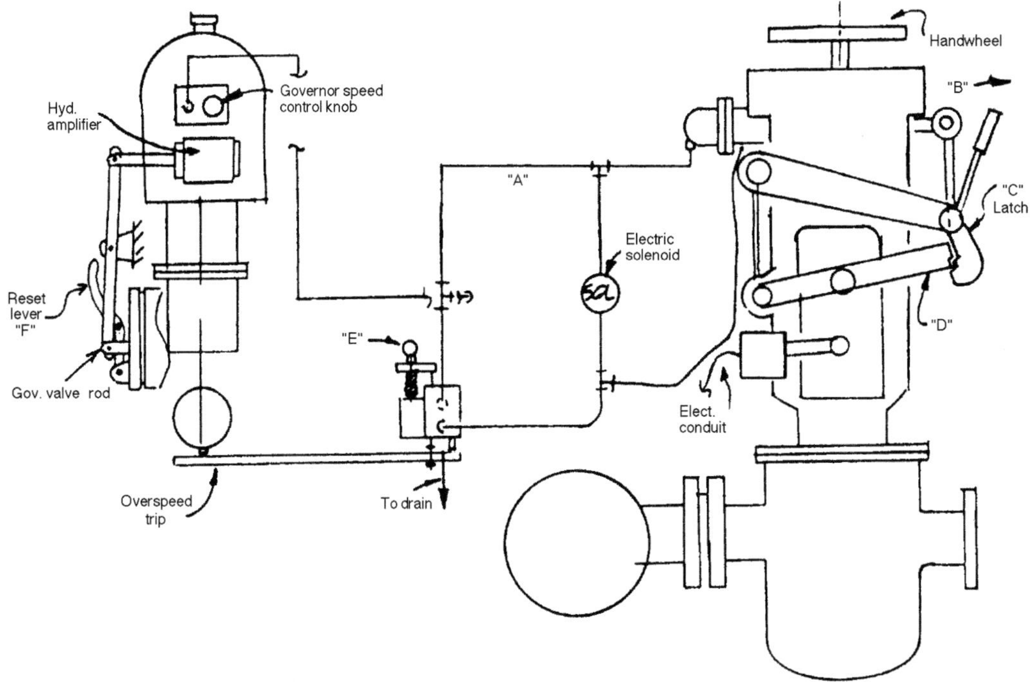

Figure 2-23: Steam turbine pump drive latching diagram intended to be laminated in plastic and mounted at the pump set

ROTATING EQUIPMENT CORRECTION/CHANGE

Letouch
CHEMICAL

DATE 8/17/79	REVISION 0

MATERIAL PROCUREMENT BY	WHEN TO IMPLEMENT	CHARGE NUMBER
☑ SHOP ☐ VENDOR/CONTRACTOR	☑ NOW ☐ (Specify when) ______	178A-1531

FOR IMMEDIATE DISTRIBUTION ☑ MECHANICAL SUPERVISOR ☑ SHOP FOREMAN

PLANT	EQUIPMENT NO.	EQUIPMENT CLASS
SEASIDE VIEW, TX	ZPM-18, 22, 23, 35 & 39	VERTICAL MOTORS

SIZE, TYPE, MODEL	SERIAL NO.	MANUFACTURER
FRAMES 254LP, 286LP, 364LP	SHOP TO PROVIDE	NORTH AMERICAN ELECTRIC

DESCRIPTION (Attach sketch, if helpful)

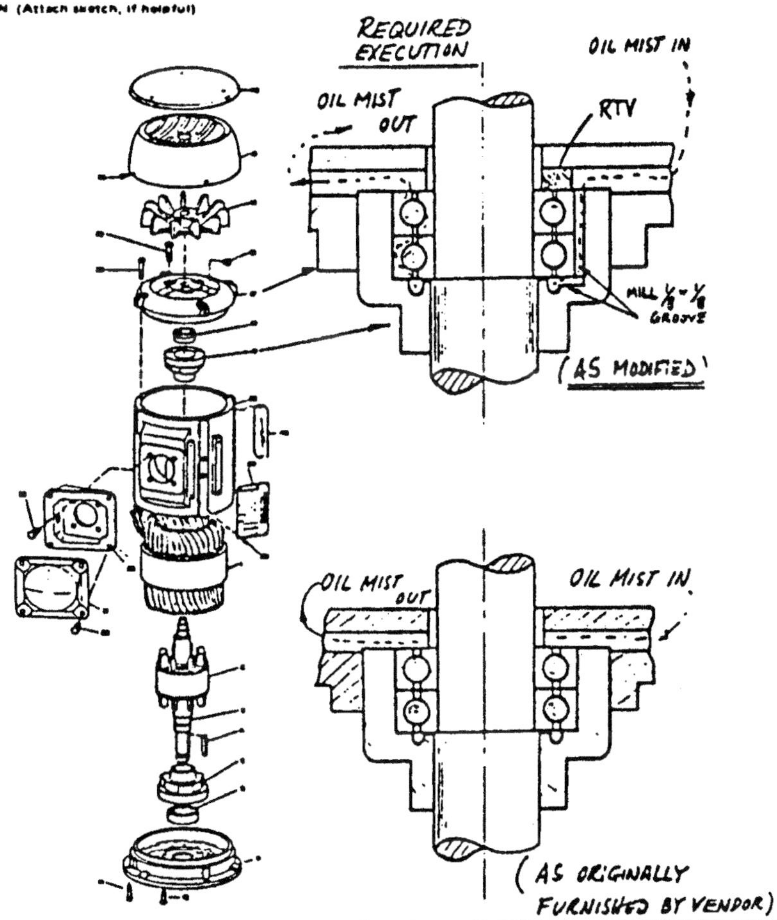

DESIGNER	APPROVED BY
JOHN A. BUDWEISER	E.A. Hungermaker Jr.

CC • Equipment Files • Eng. Spec. Sec. Supr. • Maintenance Section Supr. • Spare Parts Technician • Process Coordinator • Planner

Figure 2-24: Pump design change form

Credits and debits for efficiency deviations must compare the present value of money and the anticipated operating life of pumps.

Conventional Wisdom: *The average pump repair costs about $3,000 (in 2003).*

Fact: *Using appropriate calculation methods that include burden and overheard, many US process plants spend well in excess of $10,000 per average pump repair.*

Depending on the rate of return acceptable for energy conservation on the project, the value of one horsepower saved may be worth an incremental investment of several times the yearly cost of energy saved. Also, some users debit bids on the basis of excessively high suction specific speeds ("Nsss"). Although there seems to be no valid calculation basis for this debiting practice, it has been found that operation away from the best efficiency point (BEP) typically causes greater failure risk in high suction specific speed pumps than it does in lower suction specific speed pumps.

One user accepts pumps with suction specific speeds up to 9,000 without raising questions. Assuming now that he receives tree bids and calculates their respective Nsss values as (offer "A") 12,840, (offer "B") 8,790 and (offer "C") 14,886—the number listed in the example of Figure 2-25. This user would now use the bid cost for pump "B" as his comparison basis. He would use a cost multiplier of (12,840/9,000), or 1.43, in calculating the "imputed cost" of pump "A" and (14,886/9,000), or 1.65, in calculating the "imputed cost" of pump "C." Assuming, then, original as-bid costs of $43,000, $53,000, and $41,000 for pumps A, B, and C, respectively, this user's bid comparison would show ($43,000 x 1.43) = $61,490 for "A," ($53,000 x 1.00) = $ 53,000 for "B"' and ($41,000 x 1.65) = $67,650 for offer "C." He would purchase pump "B" for $ 53,000, reasonably assuming that repair cost avoidance will quickly make up for the ($53,000-$41,000) = $12,000 premium he is paying over the least cost bid.

Credits can be assigned also to recognize lower maintenance costs for superior lube application or sturdier couplings, to name just two of several parameters. If it is possible to make such an assessment based on comparisons of repair costs, the credit or debit number can be used outright, i.e. without attempting to calculate a multiplier.

If only repair frequencies are available for comparison, it should be remembered that for API pumps in an average size around 27 kW, the average repair cost was $5,000 per event in 1981. This figure included the cost of shop labor, materials, field labor, technical service coverage, warehousing, spare parts procurement and burden (overhead). In 1984, and for API pumps with an average size of 58 kW, the cost was slightly over $10,000 in the U.S. Gulf Coast area. It was $20k in 2011.

There is value also in demonstrably better, heavier, more easily groutable baseplates for large horizontal centrifugal pumps. A superior baseplate design may be worth $1,000 or more. An example is shown in Figure 2-26 where the circled portion discloses how reinforcing plates could serve to stiffen, and thereby upgrade, the pump support pedestals (Ref. 2-13).

Interestingly enough, the superior baseplate installation techniques described later in this text will lower the combined cost of baseplate and field work, while demonstrably increasing the strength and life expectancy of the entire installation.

Ponder, therefore, the probable cost of maintenance, or the value of self-venting and self-draining (Figure 2-27). Involve your maintenance and operating technician work forces in the final buy-in, which should be the very last step in the bid assessment and review process.

Unlike a bid tabulation sheet that lists only API data and bid cost, the so-called bid conditioning effort should serve to bring all offers on the same common denominator by assessing all relevant parameters. The best centrifugal pump for your application is often neither the most expensive nor the least expensive one on the bid slate. The best pump manufacturer is the one who realizes that addressing a user's concerns allows him to outflank the competition. And the most capable pump purchaser is the one who is always mindful of the advice given by a 19th century observer of England's industrial scene, John Ruskin: "There is hardly anything in the world that some man cannot make a little worse, and sell a little cheaper, and the man who buys on price alone is this man's lawful prey."

HOW TO SHIP PUMP SETS

In general, process pump manufacturers should be asked to provide pumps as a "set" or assembled package comprising pump, driver and baseplate. This is shown in Figure 2-26 and in several of the chapters that follow.

After ascertaining correct shaft separation to accommodate the selected coupling (see Chapter 11), and pre-aligning the two shaft centerlines within perhaps

Design Conditions	Normal	Max.	Fluid	C2/C3 HC
GPM	450			
Temp. (deg-F)	115	150	Sp. Gr.	0.48
Disch. Press. (psig)	570	600	Vap. Pr.	445 (psia)
Suct. Press. (psig)	460	540	Vis. (cp)	0.1
Diff. Press. (psi)	110	140	NPSHA	12.5
Differ. Head (ft)	529	674	Hydraulic H.P.	28.9

Project: | Item No.: P-623
Plant: | Revision: 1
Unit: | Date: 14-Apr-88

Service: Deethanizer Overhead

		Specified Requirements or Notes				
1	**Manufacturer**					
2	**Pump Size & Model**					
3	**Case:** Mount / Split / SV or DV		CL/RAD/DV			
4	Suct. Size / Rating / Face / Loc.		6/300/RF/END or TDP			
5	Disch. Size / Rating / Face / Loc.		4/300/RF/TDP			
6	MAWP / @ Max. Allowable Temp.		666/880			
7	**Hydro. Test Pressure**					
8	**Impeller:** Dia. Rated / Max.- Min.		11.5/12.5 - 10			
9	Type / Suction / Mount		Closed/Doub/OH			
10	**Bearings:** Radial-Type / I.D. No.		6314			
11	Thrust- Type / I.D. No.		7315			
12	**Shaft Design:** Brg. C.L. Distance		6.87"			
13	Shaft Dia. at Sleeve		2.125"			
14	Impeller C.L. to Bearing		11"			
15	Shaft Flex. (L^3/D^4)		65.3			
16	Relative Flexibility (% Min.)		1.0			
17	**Assessment for Mechanical Integrity**		$0			
18	**Materials: API Class Code**		5-6			
19	Case (bowls, diffusers, diaphrams		Steel			
20	Impeller(s)		Chrome Steel			
21	Wear Rings		Chrome Steel			
22	Shaft / Sleeve		Stainless Steel			
23	**Coupling: Mfr. & Model**		T.S. - B7			
24	Size / Spacer Length		12"			
25	Rtd. H.P. @ Design Speed					
26	**Seal:** Mfr. & Model					
27	API code / Mfr. Code					
28	Face width / Balance					
29	Face Material					
30	**Assessment for Seal Integrity:**					
31	**Performance:** No. Stages / RPM		1 Stg / 3550			
32	H.P. @ Design Point (%)	Dollars Per	46.6			
33	Efficiency @ Design Point (%)	H.P. Debit	62.0%			
34	Efficiency Debit	=$250	$0			
35	NPSH Margin @ Design (ft)		1.5			
36	Calculated Min. Flow (gpm)					
37	Quoted Min. Cont. Flow (gpm)		200.00			
38	Nss - Suction Specific Speed		14,886			
39	**Assessment for smooth & Flex Performance:**		$0			
40	Drive: Mfr. & Model		Reliance-XE			
41	Rated H.P. / S.F.		75/1.0			
42	**Weight:** Pump / Base / Driver		700 / 727 / 830			
43	**Size:** LxWxH Overall (inches)					
44	**Delivery:** Prints / Pump (wks)		12			
45	**Price:** Pump (each)					
46	Drive (each)					
47	Seal (each)					
48	Auxiliary Seal Plan each)					
49	Cost of Equipment (each)					
50	Testing & Inspection (total)					
51	Miscellaneous (total)					
52	Total Capital Cost:					
53	Penalties Listed Above:					
54	Other Penalties					
55	Total Evaluated Costs:					

At Design Point
Efficiency (%) 62%
NPSHR (ft) 11
NPSH Margin (ft) 1.5
Flow %BEP 55.2%
H.P. 46.6

At Shut Off:
Head 600
Head (% of Design) 113%

At BEP (Largest Impeller):
Head (ft) 590
Flow (gpm) 815
NPSHR (ft) 13
Speed 3570
Nss (Suction) 14,886
Ns (Pump head) 851

At end of curve
Head (ft) 450
Head (% of Design) 85.0%
Flow (gpm) 610
Flow (% of Design) 136%
H.P. 65

At min. continuous flow:
(per I.R. cal. method)
Class (see help)
Flow (gpm)
Flow (% design) 0.0%

Figure 2-25: Comprehensive bid comparison sheet

Figure 2-26: Pump support pedestal showing stiffening re-enforcement

0.020 inches (0. 5 mm), mounting holes are spotted from pump and driver to the baseplate's mounting pads. The pump and baseplate provider then proceeds to thread-tap bolt holes that have a diameter of about 0.060" (1. 5 mm) less than the mounting holes ("through-holes") in pump and driver. Mounting bolts are inserted at this stage of the pump assembly and the complete pump set is now considered ready for shipment as a pre-aligned "mounted" package to the designated recipient. Shipping crates are sometimes selected, as shown in the chapter on oil mist. External oil mist connections can be provided on these shipping crates to facilitate connection in outdoor storage yards; the details can be found in Chapter 10.

Seeing the conveniently mounted-for-shipping pump set (pump package) has, over the past few decades, led to misunderstandings. Contractors and plant personnel often (and erroneously) assume that the entire package can simply be hoisted up and placed on a suitable foundation. However, doing so is not best practice; in fact, how pumps are shipped has relatively little to do with how they should best be installed in the field. Decades ago, experienced OEM (Original Equipment Manufacturer) field service personnel knew much about this issue, but wise old field service folks are no longer

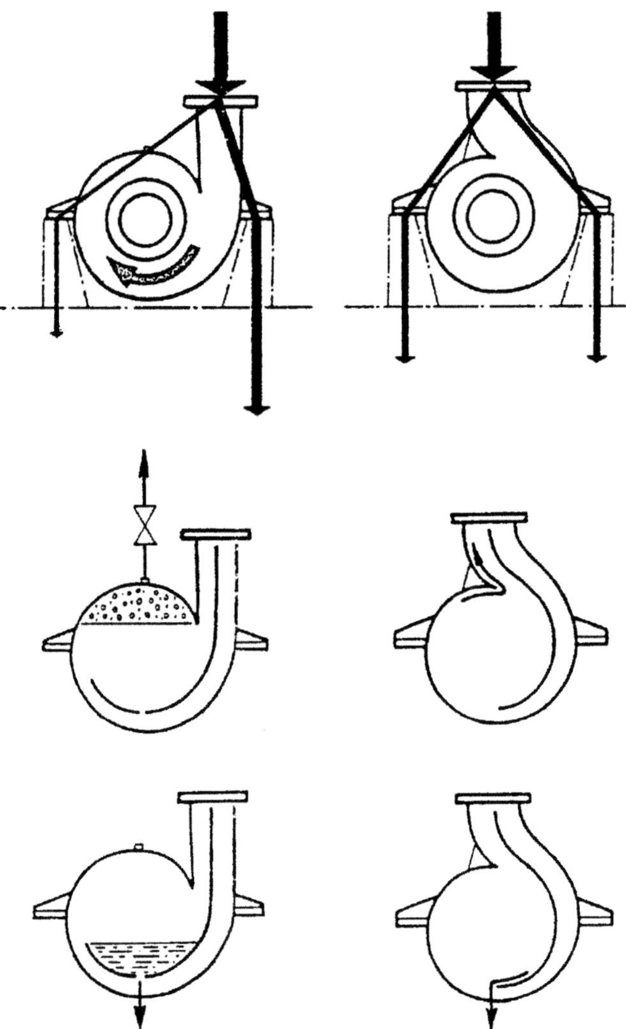

Figure 2-27: Principal features of tangential and radial discharge pumps

employed there. That is why we have to reinvestigate, read, and hope that certain erroneous ways we've become used to are rediscovered and corrected.

Contrary to the understanding of many of today's pump manufacturers' installation or service personnel, best practices companies (BPCs) will not install the equipment as a fully mounted "set." To ensure level mounting throughout, the baseplate (by itself, with pump and driver removed) is placed on the foundation into which hold-down bolts or anchor bolts (see Chapter 3) were encased when the re-enforced concrete foundations were poured. Leveling screws are then used in conjunction with optical laser tools or a machinist's precision level to bring the baseplate mounting pads into flat and parallel condition side-to-side, end-to-end and diagonally, within an accuracy of 0.002 inches/ft (~0.15 mm/m) or better. The nuts engaging the anchor bolts are now being secured and the baseplate perimeter, but not necessarily all the space between baseplate and foundation, filled with epoxy grout. If, as an example, a pump/driver/baseplate combination weighs 16,000 lbs and the pressure-on-grout should not exceed 100 psi, perimeter (inches) multiplied by the width of a "grout ribbon" should be 160 in^2 (or more).

After the epoxy grout has cured, pump and driver are aligned to criteria that harmonize with best practices—essentially the workmanship guidelines and reliability-focused practices of modern plants. More detailed information can be found in Chapter 3 and Appendix 4.

While being aligned (Chapter 3), dial indicators monitor soft-foot and pipe stress; sensitivity to piping being flanged up is closely monitored. Any dial indicator movement in excess of 0.002 inches will require making corrections to the piping. Virtually no process pump is designed to become an anchoring or support point for heavy pipe.

Finally, it should be noted that Best-in-Class users (or Best Practices Companies) specify and generally insist on epoxy pre-filled steel baseplates in certain size ranges. Decisions in that regard must be made during the specification and selection phases. Proper choices can eliminate labor-intensive conventional grouting procedures during initial installation. Even more important are the future maintenance savings that can result from maximizing the use of epoxy pre-filled baseplates. These baseplates represent a monolithic block that will never twist and never get out-of-alignment, as will also be discussed in Chapter 3.

Chapter 3

Piping, Baseplate, Installation, and Foundation Issues

Experience shows that many pump distress events have their root cause in the often neglected areas of piping, baseplates, installation and foundation soundness. These factors and issues are often intertwined and the pursuit of pump troubleshooting by focusing primarily on symptoms rather than root causes makes them elusive as well.

The pump engineer or reliability technician is understandably concerned with the physical pump. However, to achieve pump failure reductions, reliability-focused contributors must become familiar with these perceived "non-pump topics."

PUMP PIPING

The purpose of pump piping is to provide a conduit for the flow of liquid to and from a pump, without adversely affecting the performance or reliability of the pump. However, many pump performance and reliability problems are caused, or aggravated, by inadequate system piping.

Suction Piping

The function of suction piping is to supply an evenly distributed flow of liquid to the pump suction, with sufficient pressure to the pump to avoid cavitation and related damage in the pump impeller. An uneven flow distribution is characterized by strong local currents, swirls and/or an excessive amount of entrained air. The ideal approach is a straight pipe, coming directly to the pump, with no turns or flow disturbing fittings close to the pump. Furthermore, the suction piping should be at least as large as the pump suction nozzle and be sized to ensure that the maximum liquid velocity at any point in the inlet piping does not exceed 8 ft/sec (~2.5 m/s).

If the suction piping fails to deliver the liquid to the pump in this condition, a number of pump problems can result. More often than not, these include one or more of the following:

- Noisy operation.
- Random axial load oscillations.
- Premature bearing and/or seal failure.
- Cavitation damage to the impeller and inlet portions of the casing.
- Occasional damage from liquid separation on the discharge side.

Discharge Piping

System piping size is normally dictated by friction losses, which could have life cycle cost impact. System piping is also influenced by process considerations, with the maximum recommended velocity at any point in the pump discharge piping being 15 ft/sec (~4.5 m/s). Discharge piping flow characteristics normally will not affect the performance and reliability of a centrifugal or axial ("dynamic") pump, but a few exceptions exist nevertheless in a number of situations. Here are some examples:

- High-energy pumps, i.e. pumps with high values of "head times specific speed," as will be explained later, might be sensitive to flow-disturbing fittings mounted close to the pump discharge flange.

- Sudden valve closures might cause excessively high water-hammer-generated pressure spikes to be reflected back to the pump, possibly causing damage to the pump.

- Discharge piping might affect pump starting, stopping and priming.

- The discharge piping configuration also might alter any discharge flow recirculation that might extend into the discharge piping at very low flow rates. This could have a small effect on the head developed by the pump.

Piping Mechanical Considerations

Two of the more common detrimental effects from pump piping are the excessive nozzle loads the piping can place on a pump and the excessive nozzle loads that unsupported equipment such as valves or vertical in-line pumps can place on the piping. Excessive nozzle loads can be caused by thermal expansion of the pipe, unsupported piping and equipment weight, and misaligned piping (Ref. 3-1).

Proper piping and pump layout design and analysis prior to installation of a system are absolutely essential to the life and reliability of a pump. These steps can help ensure nozzle loads remain below acceptable limits for installed pumps. A properly supported piping system minimizes field adjustments during installation, saving both time and money. There has never been a process plant with high pump MTBF that allowed excessive pipe strain. Conversely, plants with low pump MTBF will not experience significant reductions in repair frequencies unless pipe stress has been eliminated or, at least, substantially reduced.

Final installation of restraints and anchors should take place after the pipe has been completely installed and all terminal and equipment connections are made. Provisions for a final piping field weld close to the pump are being used by the best performing reliability-focused user companies to ensure the piping system will be near a zero load condition, and to meet the offset and parallel recommendations listed below.

Conventional Wisdom: *Let the pipe fitter decide how to route the pipe and where to install pipe supports*

Fact: *Only a computer-model can accurately determine where and what types of supports must be installed for stress-free installation.*

More specifically, the process piping is led and installed to within 10-15 feet, or approximately 3-5 meters, of the pump nozzles. The remaining pipe runs then start with pipe flanges and reducers being placed at the pump nozzles, and "piping away" from the flanges toward the main pipe runs. Either welded or flanged connections are then provided at the points of juncture.

Flanges of connecting piping should never be allowed to be "sprung" into position. Pipe flange bolt holes must be lined up with nozzle bolt holes within the limits of the flange bolt hole manufacturing clearance, to permit insertion of bolts without applying any external force to the piping.

The use of "come-alongs," chainfalls, or other means of pulling piping into place to align the piping with the pump flanges is simply not acceptable. The pump and piping flange faces should be parallel within less than 0.001 inch per inch of pipe flange outer diameter, or 0.010", whichever is greater. Flange face separation should be within the gasket spacing, plus or minus 1/16 inch (1.5mm). This should ensure that the piping system will be within allowable pump nozzle load requirements, prior to hydro-test and start up.

Nozzle loads affect pump operation in various ways. At low levels, the effects might be insignificant. At high levels, nozzle loads can contribute to:

- Coupling or coupled shaft misalignment, which will inevitably load up the bearings. Rolling element bearings will thus be edge-loaded (Figure 3-1), and the reduced area in contact will almost always cause excessive pressure to exist between rolling elements and races. The oil film will rupture, metal-to-metal contact will be produced and high temperatures will result. This heat buildup will reduce oil film thickness, cause the lubricant to oxidize, and vastly decrease remaining bearing life. In many instances, fractures within the pump casing, decreased coupling life, increased noise and vibration levels, and breakage of the pump shaft have been experienced.

It is both intuitively evident, and field experience has proven, that high nozzle loads will cause:

- Shaft movement, which can lead to reduced mechanical seal life.
- Fatigue or failure of the shaft.
- Catastrophic structural failure of pump hold-down bolts/supports and pump nozzles.
- Pump casing gasket leaks.
- Pipe-to-pump flange leaks.
- Decreased mean time between repair, or failure.

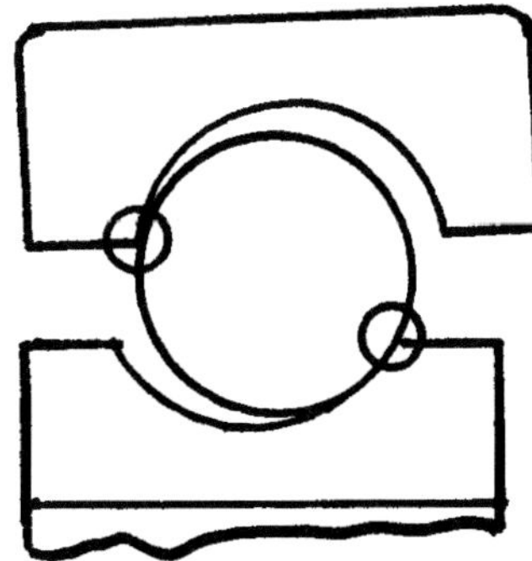

Figure 3-1: Pipe stress is likely to cause bearing inner and outer rings to be non- concentric. Edge loading will result in rapid failure

Hydraulic Considerations

Suction Piping

In general, pumps should have an uninterrupted and unthrottled flow into the inlet (suction) nozzle. Flow-disturbing fittings should not be present for some minimum length. Flow disturbances (Figure 3-2) on the inlet side of the pump can lead to:

- Deterioration in performance.
- Damage to the impeller and shortened impeller life (especially with high-suction-energy pumps).
- Shortened mechanical seal life.
- Shortened bearing life.

Isolation valves, strainers and other devices used on the inlet (suction) side of the pump should be sized and located to minimize disturbance of the flow into the pump. Eccentric reducers should generally be installed with the top horizontal and the bottom sloped, as noted in Figure 3-3 and the more detailed views of Figure 3-4. An exception is discussed in Chapter 5.

The most disturbing flow patterns to a pump are those that result from swirling liquid that has traversed several changes of direction in various planes. Liquid in the inlet pipe should approach the pump in a state of straight, steady flow. When fittings such as "T" fittings and elbows (especially two elbows at right angles) are located too close to the pump inlet, a spinning action or "swirl" is induced. This swirl could adversely affect pump performance by reducing efficiency, head and net positive suction head (NPSH) available. It also could generate noise, vibration and damage in high-suction-energy pumps.

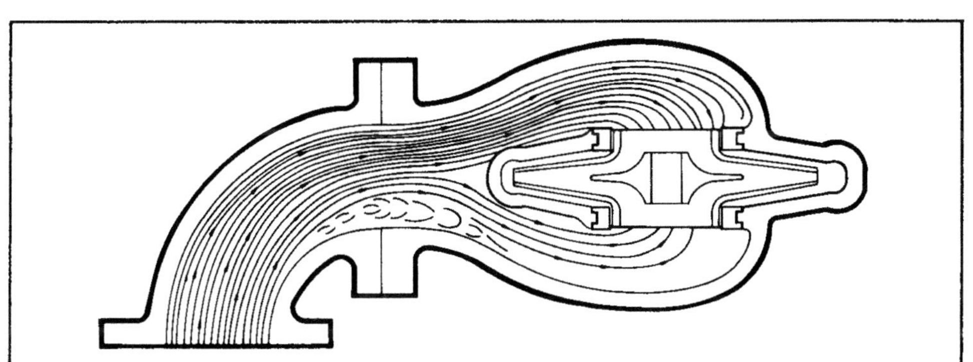

Figure 3-2: Flow disturbances at pump inlets will cause hydraulic load and force unbalance

Elbows must be vertical when next to a double suction pump (Figure 3-5). It is always recommended that a straight, uninterrupted section of pipe be installed between the pump and the nearest fitting. This should follow the minimum straight pipe length recommendations of the Hydraulic Institute Standards (Refs. 3-2 and 3-3), as summarized in Tables 3-1 and 3-2, and Figure 3-6. Essentials can be seen in Chapter

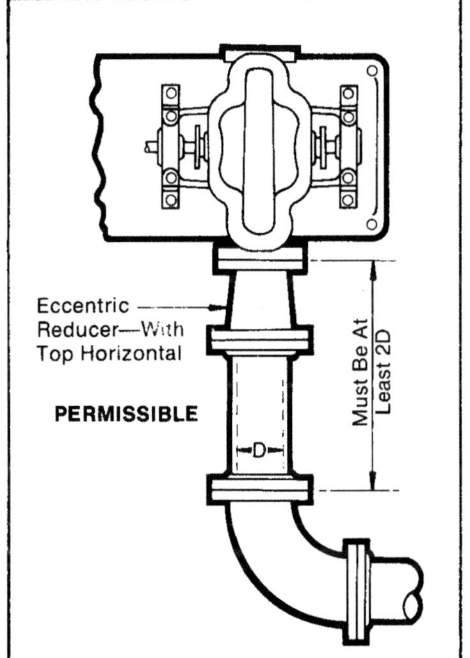

Figure 3-3: Eccentric reducer—with top horizontal

Table 3-1: Minimum recommended straight pipe length (L_1) before Pump Suction for Low Suction Energy/Low Specific Speed Pumps

Fitting	End Suction Pump	Double Suction Split Case Pumps	
	Fitting in Either Orientation	Fitting in Shaft Plane	Fitting Perpendicular to Shaft
Long Radius Elbow	1D	3D	1D
Short Radius Elbow	2D	5D	2D
45° Tee	1D	5D	1D
90° Tee	3D	8D	3D
Open Valves	2D	2D	
Check Valves	5D	5D	
Filters / Strainers	3D	3D	

Table 3-2: Minimum recommended straight pipe length (L_1) upstream of Nozzle Suction for High Suction Energy/High Specific Speed Pumps

Fitting	End Suction Pump	Double Suction Split Case Pumps	
	Fitting in Either Orientation	Fitting in Shaft Plane	Fitting Perpendicular to Shaft
Long Radius Elbow	5D	5D	3D
Short Radius Elbow	8D	8D	3D
45° Tee	8D	8D	3D
90° Tee	15D	15D	6D
Open Valves	3-5D	3-5D	
Check Valves	10D	10D	
Filters / Strainers	6D	6D	

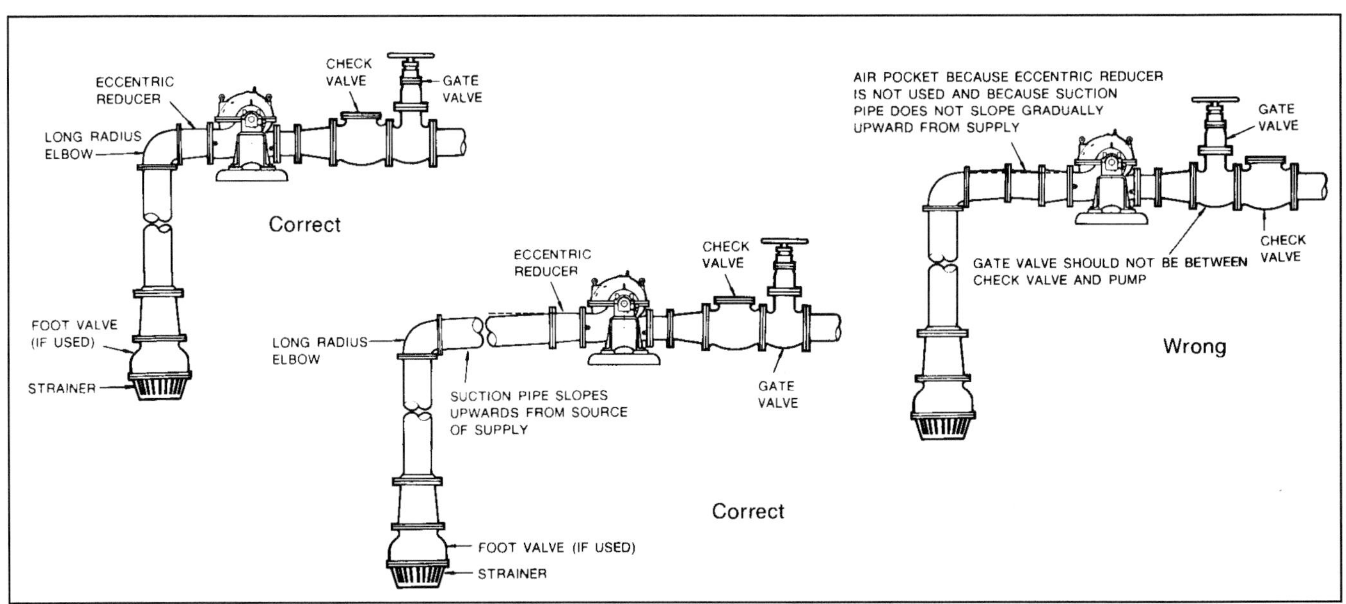

Figure 3-4: Eccentric reducer installations, correct vs. incorrect orientations

5; they range from:

- One to eight pipe diameters (for low suction energy/low specific speeds).
- Three to 16 pipe diameters (for high suction energy/high specific speeds).

The specific straight pipe length recommendation depends on the type of fitting(s), the pump type, the suction energy level and pump specific speed. Generally, high-suction-energy pumps have suction nozzle sizes of 10 or more inches at 1,800 revolutions per minute (rpm), and 6 in. or more at 3,600 rpm (see Chapter 5). High-specific-speed (Nss) starts above a value of 3,500. If the minimum recommended pipe lengths can't be provided, flow-straightening devices, Figures 3-6 and 3-7, should be considered. The contoured insert of Figure 3-6 is preferred over the somewhat more elementary welded-in-place insert shown in Figure 3-7. Table 3-3 represents a summary of the causes, effects, and anticipated results of installing Koch Engineering's "CRV® Flow Conditioning Unit" depicted in Figure 3-6 in certain pump suction lines.

In addition to the minimum required lengths of straight suction pipe, it is equally important that the NPSH margin ratio, i.e. net positive suction head available/net positive suction head required (NPSHa/NPSHr) guidelines listed in Chapter 5 be followed (Ref. 3-4).

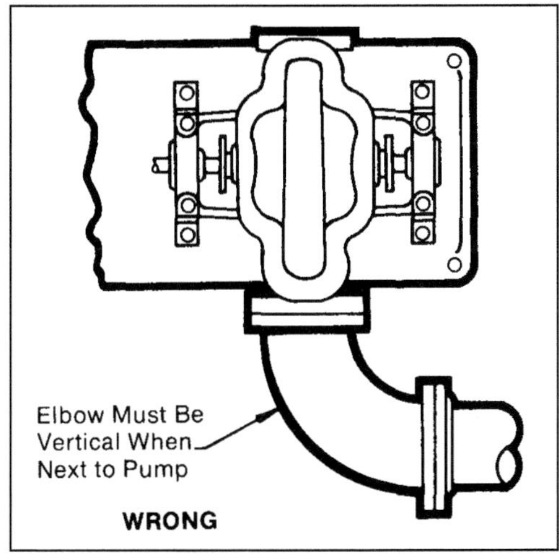

Figure 3-5: Elbows must be vertical when directly adjacent to pump

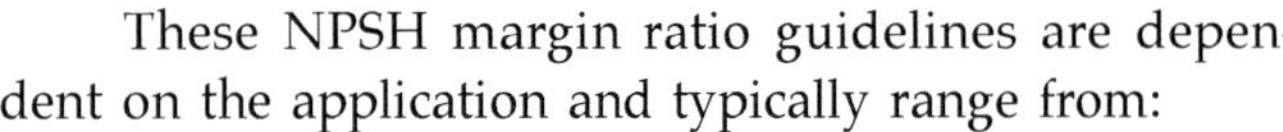

These NPSH margin ratio guidelines are dependent on the application and typically range from:

- 1.1 to 1.3 for low-suction-energy pumps.
- 1.3 to 2.5 for high-suction-energy pumps.

It is especially critical for plants to follow these NPSH margin recommendations for high-suction-energy pumps. By doing so, they avoid excessive pump noise, vibration and/or damage, particularly in the region of suction recirculation.

Discharge Piping

Pipe fittings mounted close to the outlet (discharge) flange normally will have a minimal effect on the performance or reliability of low-energy rotodynamic pumps. On the other hand, high-energy pumps can be sensitive to flow-disturbing fittings mounted close to the pump outlet flange. These fittings could result in increased noise, vibration and hydraulic loads within the pump.

According to the Hydraulic Institute, pumps with specific speed values below 1,300 that generate more than 900 feet of head per stage are considered to be high-energy pumps. The differential head value for high energy drops from 900 feet at a specific speed of 1,300 to 200 feet at a specific speed of 2,700.

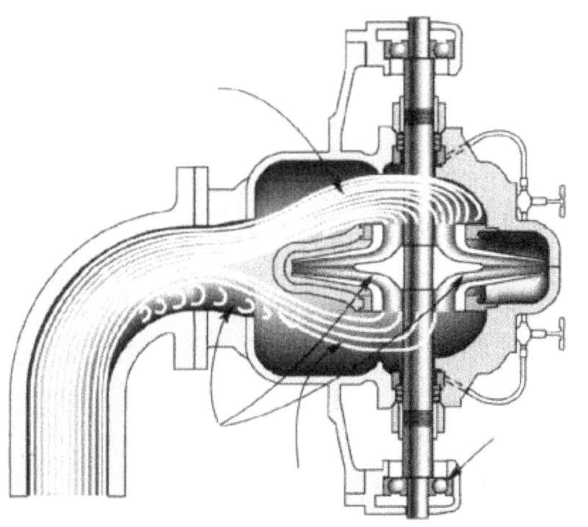

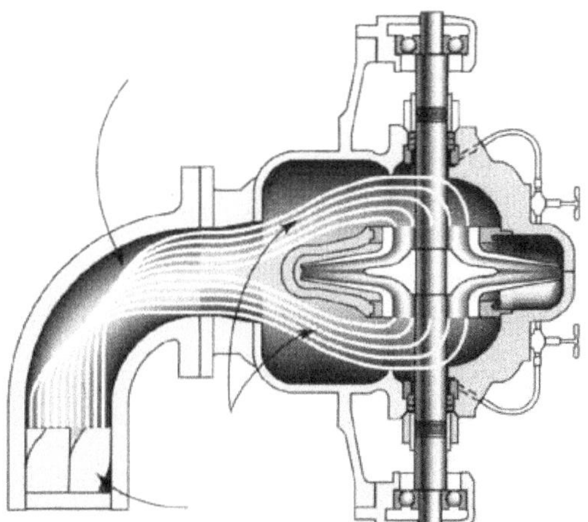

Figure 3-6: Contoured flow straightening insert for pump suction lines (Source: Koch Engineering Company Inc., Wichita, Kansas 67208)

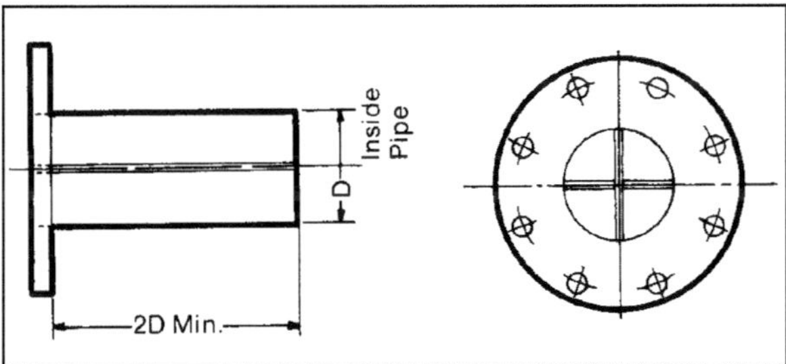

Figure 3-7: Elementary flow straightening insert welded into pump suction line

It is recommended that high-energy pumps meet experience-based minimum straight discharge pipe length requirements of, preferably, four to eight pipe diameters. The specific straight pipe length recommendation will depend on the type of fitting(s) downstream of the pump and pump discharge energy level.

For tracking and verifying proper field implementation, the following piping-related overview material and checklist will prove helpful. It should be noted that

Table 3-3: Summary of causes, effects, and anticipated results of installing engineered flow straightening devices in pump suction lines (Source: Koch Engineering Company, Inc., Wichita, KS 67208)

IF YOU HAVE...	WHICH CAUSES...	THE CRV® FROM KOCH IS THE ANSWER...
PUMP OR COMPRESSOR NON-UNIFORM SUCTION INLET FLOW	• Reduced Flow Rate • Cavitation • Noise • Bearing Wear • Vibration • Reduced Discharge Pressure	• Typically 5-10% Efficiency Improvement in Pressure and Flow • Extended Maintenance Period • Reduced Piping Cost (Space & Cost) • Reduction in Noise/Vibration
METER RUNS WITH CLOSE COUPLED UP/DOWN STREAM ELBOW	• Inaccurate Flow Measurement	• Shorter Straight Pipe Runs (Space & Cost) • Accurate Flow Measurement
ELBOW PRESSURE LOSS	• Reduction in Pump Horsepower	• Reduced Operating Cost • Capacity/Head Gain in Retrofit Applications
SLURRY, CATALYST, TWO PHASE OR PARTICULATE FLOW	• Elbow Erosion	• Extends Life of Elbow
WATER HAMMER	• Hydraulic Pressure Waves and Transients	• Minimize Damage to Piping, Instrumentation and Process Equipment Due to Rapid Valve or Pump Shutoff
NOISE & VIBRATION	• Elbow Generated Noise	• Reduction in Noise Level

all of these data and procedures reflect experience values that are routinely achieved by reliability-focused "Best Practices Companies"—high performers—in the United States and elsewhere.

FIELD INSTALLATION PROCEDURE FOR PIPE SPRING HANGERS AND PIPE SUPPORTS

Many different pipe hangers and pipe supports, Figure 3-8, are used in industry. To accommodate the wide-ranging needs of reliability-focused pump piping installations, Houston/Texas-based Piping Technology and Products, Inc., designs and manufactures variable springs, constant force supports, hydraulic shock and sway suppressors, clevis hangers, beam attachments, saddle supports, shoes, anchors, slide plates with and without coating, to name but a few. Hazardous situations are often created by improper piping analysis and some computer programs cannot handle hot condition calculations correctly.

Overlooking the need for pipe stress calculations and not using appropriately designed supports are known to defy attempts to reach high pump reliability.

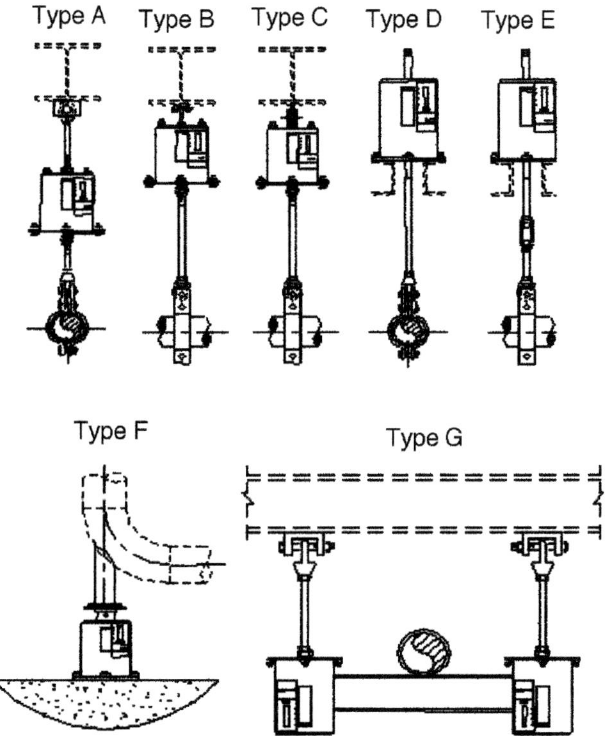

Figure 3-8: Different styles of variable spring pipe hangers (Source: www.pipingtech.com)

Hanger and spring support-related installation oversights are responsible for many excessive pipe stress incidents. Piping-induced loads will very often cause pump casings to deflect; internal parts, especially bearings and mechanical seals, become misaligned and fail prematurely. Pump reliability improvement efforts must include verification of acceptable pipe stress values.

Conventional Wisdom: *Piping installation is straightforward and is traditionally entrusted to a competent pipe fitter.*

Fact: *Pipe stress problems can be elusive. Important pumps and pumps that have experienced repeat failures should have their piping configuration and support systems analyzed using a pipe stress computer program.*

It is always wise to consult the plant's record system or asset virtualization program for listings of line numbers and system spring identification numbers, types and sizes of springs, calculated cold setting values and the calculated travel. This travel must never come close to the upper and lower limit stops shown in Figure 3-9. In new construction projects, such lists are generated for field follow-up and future checking of springs in their respective hot positions. Observe the following:

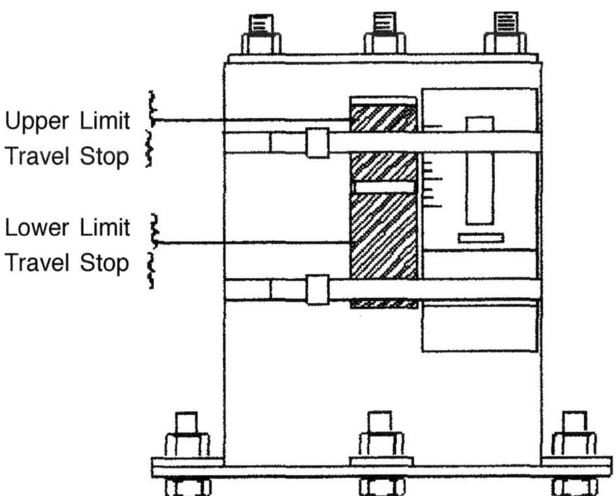

Figure 3-9: Travel stops for variable spring hangers

- Spring support/hanger stop pins or travel constraints affixed by construction contractors and/or plant mechanical work forces shall not be removed until

  (a) line hydrotesting, flushing and cleaning and, if the line requires insulation, with the pipe insulation either fully installed or replicated by a dummy load such as sand bags

  (b) flange alignment has been verified to be correct and free of excessive stress

- Special precautions are required for lines carrying liquids. Standard practice calls for spring hangers/supports to be calibrated such that when the piping is at its hot position, the supporting force of the hanger/support is equal to the calculated operational pipe load. Note that the maximum variation in supporting force will therefore occur with the piping in its cold position, when stresses added to piping and/or equipment nozzles are less critical.

- Weight calculations and spring calibration must, therefore, include pipe, insulation, and liquid totals. Verify that this was done.

- After the locking pins or constraints have been removed, the construction contractor or plant maintenance technician must adjust all springs to their required cold load marking values.

- Substantial forces could develop when machinery flanges have been aligned/bolted up on liquid-filled lines, the pins removed, and the spring travel indicator adjusted to the cold marking. Most of these forces would act directly on the pump nozzles, possibly exceeding allowable load limits. It is therefore recommended to provide sandbags or similar counterweights on 8" or larger liquid-carrying lines. This will account for the weight of the liquid *after* the lines have been connected to the pumps. (Alternatively, springs in those particular systems could be reviewed by responsible engineering groups and different load values calculated and provided for the simulation of startup conditions.)

- If the above step shows a particular spring load or travel indicator settling outside of the middle third of the total working range of the spring, the contractor's engineering staff and/or the plant mechanical work forces must determine appropriate remedial action.

- The proper cold setting of hangers/supports will now have been achieved. Equipment is typically turned over to plant operations personnel and piping is free of liquid. When flange bolts are *now* removed, a certain amount of pump misalignment may be introduced.

- Before a flange is opened up for any reason, all preloads on springs resulting from previous cold setting adjustments have to be eliminated first. Preferably, the spring settings should first be restored to the position they were in before the pins were pulled.

A Final Word

Following the above steps will go a long way towards ensuring that piping forces and moments will not be excessive. It must be re-emphasized that excessive pipe stress will definitely shorten machinery MTBF, the all-important mean time between failures. Neither centrifugal pumps nor pump support elements such as baseplates and pedestals are designed for zero deflection or warpage when subjected to excessive piping loads.

Centrifugal Pump Piping Checklist

1. The preferred method of piping a pump is to bring the piping toward the pump and then "piping away" the final 10-15 feet (3-5 meters). After the pump is grouted and aligned, suction and discharge flanges with gaskets are 4-bolted to pump. Piping is then field-fitted and welded to the flanges

2. Check pump discharge flange for level in two directions 90° apart. Maximum out of level .001"/foot (~0.1mm/m)

3. Check parallelism of suction and discharge pipe flanges to pump flanges. Maximum out of parallelism not to exceed .010" (0.25mm) at gasket surface

4. Check concentricity of suction and discharge piping to pump flanges. Flange bolts must slip into holes by hand

5. Check to see if proper pipe supports are installed.

6. Check for recommended length of straight pipe (Ref. 3-2) before suction and after discharge flanges of pump

7. Tighten piping flange bolts in crisscross pattern to proper torque. Coat bolt threads with anti-seize compound before tightening

8. Check pump shaft deflection using face and rim method during flange bolt tightening. Maximum allowed defection is .002" (0.05 mm) TIR (total indicator reading)

9. Provide a drop-out spool piece at the suction flange for a cone-type temporary mechanical strainer

10. Verify that a temporary mechanical strainer is installed. (Be sure not to leave temporary strainers in place permanently.) See also Figure 3-53.

Cleaning Mechanical Seal Pots and Seal Piping

Pump piping must be clean. Cleanliness is of greatest importance for auxiliary or otherwise pump-related piping. You or your piping contractor may wish to use any approved cleaning procedure. A typical approved procedure is given below. However, while useful for many pump applications, it is intended only as a typical example and must be reviewed by your designated representative—a person with corrosion know-how—before implementation. It is listed here in helpful checklist format.

Scope of Typical Cleaning Procedure

1. This specification covers the inspection, cleaning, and preservation of mechanical seal pots and their associated piping on new pumps prior to start-up and commissioning, and on existing pumps that have their mechanical seal piping modified.

2. This work is to be done concurrently with cleaning and setting of mechanical seals on new pumps.

Materials

The following materials are required for this work:

> Fresh water	> Sodium Hydroxide
> Sodium Metasilicate	> Trisodium Phosphate
> Non-Ionic Detergent	> Hydrochloric Acid
> Ammonium Bifluoride	> Armohib 28 or Equivalent
> Citric Acid	> Soda Ash
> Sodium Nitrite	> 10 Micron Filter Screens

Note the following:

- Special procedures are to be used to clean stainless steel piping or seal pots.
- Quantities will vary based on the size and amount of piping to be cleaned.
- Observe that it will be your responsibility to engage the services of a knowledgeable chemical cleaning company.

- Insist that this company submits a rigorous procedure for your review and gives you written assurances that their procedure will not harm either the piping or the connected appurtenances.

Disassembly

1. All instrumentation in the seal piping shall be removed prior to disassembly of the piping and seal pots.
2. The instrumentation shall be tagged and stored in a safe manner.
3. The seal piping shall be separated at the two closest flanged or union connections to the pump or gland and taken to the shop in one piece.
4. If the piping can be separated into shorter segments for easier cleaning it may be done at the shop.
5. Openings on the pump and/or gland shall be plugged or capped to prevent entry of dirt into the pump while the piping is in the shop.

Cleaning

1. All piping and seal pots shall be flushed with 77°-93°C (170°-200°F) alkaline solution containing three percent (3%) NaOH, one percent (1%) sodium metasilicate, one percent (1%) trisodium phosphate and one tenth percent (0.1%) non-ionic detergent in fresh water. This solution is to be circulated for at least six (6) hours. The piping and pots shall then be drained and flushed with fresh water until the pH level is less than 8.0. Water for stainless steel piping and seal pots shall contain less than 50 ppm chloride.

2. Stainless steel piping and seal pots shall be further flushed with 71°-93°C (160°-200°F) fresh water containing less than 50 ppm chloride at a high enough velocity (5-10 fps, or 1.5-3 m/s) to remove all dirt. The piping runs shall be positioned and supported so as to have no dead spots with the water entering at the lowest point in the piping and exiting at the highest point. The circulated water shall be filtered through ten (10) micron filter screens and returned to an open storage tank before being recirculated. When the filter screens remain clean for five (5) minutes, circulation may be terminated and the piping and seal pot air-dried.

3. Carbon steel piping and pots shall be flushed in the same manner as the stainless steel piping and seal pots except that six (6) to ten (10) percent inhibited hydrochloric acid by weight with twenty-five hundredths percent (0.0025:1 ratio) ammonium bifluoride by weight, inhibited with two (2) gallons per 1000 gallons Armohib 28, Rodine 213, or equivalent in fresh water, shall be used.

4. The acid solution shall be heated to 71° C (160° F) and circulated at a slow rate for six (6) hours or until the reaction is complete.

5. Samples of the acid solution shall be taken at hourly intervals and tested for acid strength and total iron concentration. If the acid concentration drops below three percent (3%), acid shall be added to raise the concentration to five percent (5%). Cleaning shall continue until iron and acid concentration reach equilibrium.

6. After acid flushing has been completed, the piping and seal pots shall be drained and flushed with fresh water containing one tenth percent (1000:1 ratio) by weight citric acid.

7. The piping and seal pots shall then be neutralized with one (1) percent by weight soda as $NaNO_2$ solution in fresh water circulated for two (2) hours heated to 79° to 88°C (~175°F to 190°F).

8. After neutralization the piping and seal pot shall be rinsed with fresh water and air-dried.

Preserving

1. Immediately after drying, the piping and seal pots shall be flushed or sprayed with the same oil as will be used in the seal system.

2. After the oil treatment the piping shall be reassembled if it was disassembled in the shop. All openings shall be plugged or capped to prevent entry of any contaminating matter.

3. As soon as all work on the mechanical seals has been completed, the piping and seal pots shall be re-installed and the lines filled with proper seal oil.

4. The instrumentation shall be re-installed at this time.

5. After filling the lines with oil and re-installing the instrumentation the pump shall be tagged to indicate that the seal system is ready for service.

Using Your Machinists on New Projects

Suppose you entrusted a contractor with the installation or pre-commissioning of pumps on new projects. Suppose further that during that time period you had a surplus of mechanical personnel since the plant is either not operating, or is operating at less than full capacity. This time period could be used to actively and productively engage your mechanical work force members in a wide range of tasks that will later result in improved pump reliability. Here, then, are some of these tasks:

- Verification of spare parts existence, condition, location and correctness in storehouse
- Witness and follow-up on equipment storage procedures
- Pre-grouting inspection checks:
  - Preliminary alignment of rotating equipment
  - Cleanliness and surface preparation of foundation
  - Witness grout pours on critical equipment
  - Ensure compliance with job grouting specifications
  - Witness leveling of equipment prior to grouting
- Check for pipe strain on rotating equipment
- Improve coordination and communication between contractor and equipment owner
- Coordination of protection and transportation of unit spare equipment to storehouse
  - Rotating equipment, gearboxes, pumps, motors, etc.
- Inspection of critical equipment piping for cleanliness
  - Suction and discharge piping
  - Lube oil system
  - Sealing fluid lines
- Check and accept final alignment of rotating equipment.
- Assist in pre-startup and run-in
- Assist in startup by training maintenance and process personnel (on-the-job training) on new equipment.

PUMP BASEPLATES AND GROUTING

Pump baseplates are available in cast iron, fabricated steel, "enhanced feature fabricated steel," molded polymer concrete, polymer shell repair elements, so-called "PIP-Compliant Fabricated Steel," and free-standing as well as stilt-mounted configurations. Torsional stiffness, rigidity and flatness are important for all of these and grouting and lifting issues must be addressed for most of these.

It is intuitively evident that every one of these configurations has its advantages and disadvantages (Ref. 3-5).

- Cast iron tends to warp, unless it has been allowed to age for a substantial time period. It must then be finish-machined, properly leveled, and grouted in place.
- Fabricated steel must be sufficiently heavy, properly welded, stress relieved and machined.
- Flimsy fabricated steel bases led to the development of "enhanced" versions.
- Molded polymer concrete has superior stiffness and warp resistance, polymer shell repair elements are hollow and fit over the old foundation. The voids are then being filled with concrete or epoxy grout.
- PIP-compliant fabricated steel baseplates are meant to represent a further enhancement of enhanced baseplates, or so it seems.
- Free-standing, or stilt-mounted baseplates (Figure 3-10) are available from pump manufacturers that wish to respond to user demands for low-cost installations. Stilt-mounted bases can involve a number of options, as shown in Figure 3-10, and the entire assembly is expected to move on the floor. The movement "tends" to equalize the forces exerted by cheaply-installed piping, and is expected to provide equal loads on both suction and discharge flanges. This assumes that all pump piping loads act in the same direction, which is seldom the case. While stilt-mounted bases may not have to contend with grout problems, they are rarely, if ever, the best choice for reliability-focused users.

Overcoming Insufficient Baseplate Stiffness and Rigidity

Needless to say, insufficient baseplate stiffness has plagued hundreds of thousands of pump installations. A weak, or improperly supported baseplate will make it impossible for centrifugal pumps to operate successfully over long periods of time. Installing better bearings, or upgrading one's mechanical seal selection, will often remain nothing but costly exercises in futility. The cycle of failures can be broken only if the basic installation is brought into compliance with applicable standards of

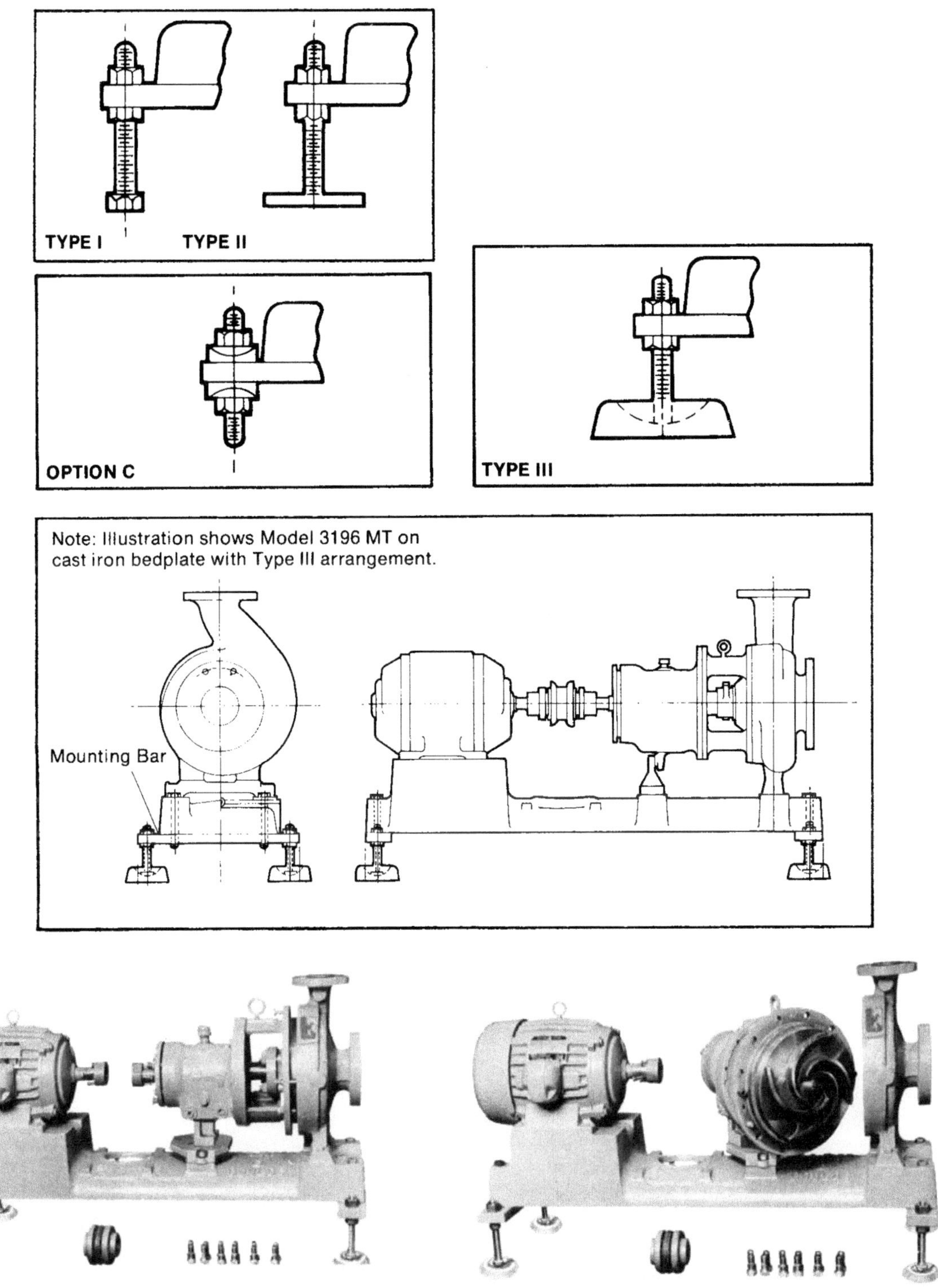

Figure 3-10: Free-standing, "stilt-mounted" baseplates (Source: ITT/Goulds, Seneca Falls, NY 13148

reliability engineering.

With the possible exception of large pumps, virtually all process pumps are typically mounted on baseplates. Steel is still the predominant material, although the non-metallic configurations mentioned above are making inroads and can be economically justified in well thought-out applications. A clean steel surface and the application of a suitable primer are needed before proceeding with epoxy grouting.

Specialty companies (Ref. 3-6) explain the importance of selecting an epoxy primer that will bond to the steel underside with sufficient tensile bond strength.

They reviewed a large number of primers with the manufacturer's laboratory to determine the tensile bond number. All products are based on proper surface preparation and blast profile, i.e. white metal blast with 0.003" (0.08mm) profile.

Ref. 3-7 determined the products listed in Table 3-4 to be among several well-bonding primers found acceptable at the time of this writing, but the same reference hastened to add that it should be the user's responsibility to obtain warranty information from the selected manufacturer.

Table 3-4: Epoxy primers that were found to properly bond to steel (Source: Ref. 3-7)

PRODUCT	*TENSILE STRENGTH (psi)*	*EVALUATION*
ITW Corrosion Inhibitive Primer	1,800	Acceptable
Rust Ban Penetrating Primer	1,500	Acceptable
Carboline 890	1,930	Acceptable
Tnemec Series 66 Epoxy Primer	800-1,200	Acceptable
Master Builders Ceilcote 680	2,000	Acceptable
Devoe 201 Primer	900	Acceptable

Optimized Method of Stiffening Baseplates

In the late 1990's Houston/Texas-based Stay-Tru® Services developed a highly effective method to minimize field installation costs and improve the reliability of pump baseplates. Their "Stay-Tru® Pre-Grouted Baseplate Procedure" accomplishes these goals and is able to provide a superior finished product. When coupled with their so-called Stay-Tru® Plus Machinery Quality Assurance Program, the pump train is 60% mechanically complete when it arrives at the plant site.

The Stay-Tru® procedure is a comprehensive pre-grout plan that includes a detailed inspection of the primer system used on the underside of the baseplate, proper preparation of this primer for grouting, grouting of the baseplate, post-curing of the grout material, and a detailed post-grout inspection of the baseplate mounting surfaces. If the mounting surface tolerances fall outside of the API 686 specifications for flatness, co-planar, and co-linear dimensions, precision grinding or machining is performed to restore or achieve the necessary tolerances.

The essential results of following this company's baseplate installation procedure include

- Void free grout installation.
- Proper surface preparation of baseplate underside.
- All machine surfaces are flat, co-planar, and co-linear, in accordance with API-686.
- Eliminating the need for costly and less effective field machining.
- 30% cost savings over traditional field installation procedure.

A brief description of the Stay-Tru® process is as follows:

Preparation and Grouting of Baseplate

The coating on the underside of the baseplate will be checked for bond, sanded, and solvent wiped prior to grouting. Any tapped bolt holes that penetrate through the top of the baseplate, such as the coupling guard hold down bolts, will be filled with the appropriate sized bolts and suitably coated to create the necessary space for bolt installation after the grouting of the baseplate. If anchor bolt or jack bolt holes are located inside the grouted space of the baseplate, provisions will be made for bolt penetration through the baseplate after grouting. Baseplate grout holes and vent holes will be completely sealed.

For applications using epoxy grout as the fill material, the epoxy grout will be poured directly into the baseplate cavity. For applications using a non-shrink cementitious grout, an epoxy adhesive will be applied to the surface prior to placement of the grout.

Post Curing of Grout Material

After the initial set of the grout material, a post curing procedure is used for both epoxy and cementitious grout, to assure that the grout has completely cured before any additional work is performed on the baseplate. For epoxy grout materials, the temperature of the baseplate will be elevated 20°F (12°C) above ambient temperature, for one day, to achieve full cure strength.

For cementitious grout materials, the surface of the grout will be kept wet for two to three days to facilitate hydration and reduce the potential for cracking of the grout. Plastic sheet will then be applied and taped to the baseplate to further assist the curing process. Curing agents will not be used as part of the post curing process. Curing agents prevent bonding to the grout material during the final grout pour in the field.

Post Grout Inspection of Baseplate

After the post curing process, the baseplate mounting surfaces will be inspected to assure conformance to API-686. Baseplate flatness, coplanar, and co-linear, or relative level, checks will be performed to the tolerances outlined below. Any deviation to the specifications for flatness, coplanar, and relative level, after the grouting process, will be corrected by having the baseplate surface ground. A detailed report of the inspection is provided as a record.

The Stay-Tru® process is utilized when only the bare baseplate has been received by this company. When the equipment is mounted on the baseplate, they employ their own, well-proven procedure. Called "Stay-Tru® Plus," this procedure has the added benefit of providing the opportunity for resolving all equipment deficiencies, such as motor soft-foot, bolt-bound alignment conditions, and coupling hub run-out, before the equipment ever reaches the field. This eliminates costly idle time for the field crews, and avoids the additional field engineering time required too resolve these problems.

A brief explanation of the Stay-Tru® Plus procedure is of value. Essentially, it consists of the following:

- *"As-Received" Inspection.* The "as-received" inspection is documented in a report. It includes a general condition review of the pumping system, and checks for bolt-bound condition, coupling hub run-out, baseplate mounting surface flatness, coplanar, and relative level checks, and a check for motor soft foot.

- *Disassembly of Pump/Motor Train.* Once the specifications for the "as-received" inspection have been satisfied, the pump/motor train will be disassembled. All components of the system are being tagged and stored on wooden pallets. The equipment number is then metal- stamped on the baseplate for tracking through the grouting process.

- *Re-Assembly and Final Alignment of Pump/Motor Train.* Once the Stay-Tru® process has been completed, the pumping system is reassembled on the baseplate and prepared for final alignment. The driver is centered in its bolt pattern, and the pump is moved to achieve final alignment. This will provide optimum flexibility for future motor alignments in the field. The tolerance for the final coupling alignment between the pump and motor will be 0.002" (0.05 mm) for parallel offset, and no more than 0.001" per inch (0.001mm/mm) of coupling spacing for angularity. Finally, a written record of the final alignment and soft foot readings will be provided with each pumping system.

It has been demonstrated that utilizing the combined hardware and procedural approach spearheaded by this company will save 30% of the cost associated with traditional machinery grouting methods, and provide a superior finished product. It can be inferred that superimposing the somewhat more elusive value of the resulting reliability improvement will substantially lower the life-cycle cost of pump installations that make use of this methodology.

Considerations for the Specification and Installation of Pre-grouted Pump Baseplates*

According to statistical reliability analysis, as much as 65% of the life cycle costs are determined during the design, procurement, and installation phases of new machinery applications (Ref. 3-5). While design and procurement are important aspects for any application, the installation of the equipment plays a very significant role. A superb design, poorly installed, will give poor results. A moderate design, properly installed, will give good results.

A proper installation involves many facets; good foundation design, no pipe strain, proper alignment, just to name a few. All of these issues revolve around the idea of reducing dynamic vibration in the machinery system. Great design effort and cost is expended in the construction of a machinery foundation, as can be seen in Figure 3-11. The long-term success of a proper installation and reduced vibration activity, is determined by how well the machinery system is joined to the foundation system. The baseplate, or skid, of the machinery system must become a monolithic member of the foundation system. Machinery vibration should ideally be transmitted through the baseplate to the foundation and down through the subsoil. "Mother Earth" can provide very effective damping (vibration amplitude attenuation) to reduce equipment shaft vibration. Failure to do so will result in the machinery resonating on the baseplate, as shown in Figure 3-12, very often causing consequential damage of the type indicated in Figure 3-13. Proper machinery installation will result in significantly increased mean time between failures (MTBF), longer life for mechanical seal and bearings, and a reduction in Life Cycle Cost.

---

*Excerpted, by permission, from commercial literature provided by Stay-Tru® Corporation, Houston, Texas

Figure 3-11: Good pump foundations are thoroughly engineered and well constructed

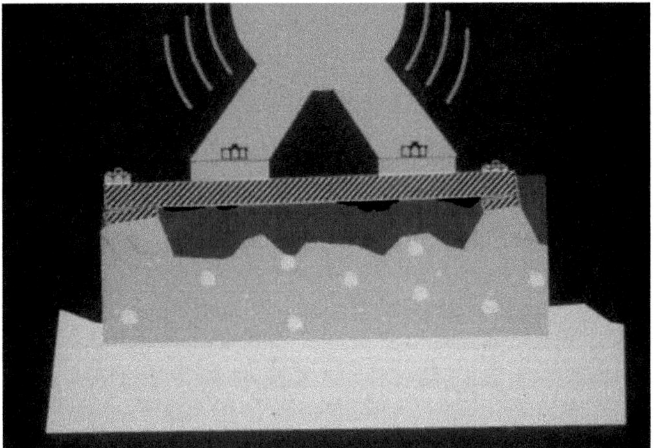

Figure 3-12: Void areas under pump baseplates often give rise to destructive resonant vibrations

The issue is to determine the most cost effective method for joining the equipment baseplate to the foundation. Various grouting materials and methods have been developed over the years, but the issue always boils down to cost: Life Cycle Costs versus First Cost. It's the classic battle between machinery engineers charged with achieving high equipment reliability, and project personnel whose career is advanced by cost and schedule considerations. The machinery engineer wants to use an expensive baseplate design with epoxy grout. The project engineer favors using a less expensive baseplate design with cementitious grout support.

Conventional Grouting Methods

The traditional approach to joining the baseplate to the foundation has been to build a liquid tight wooden

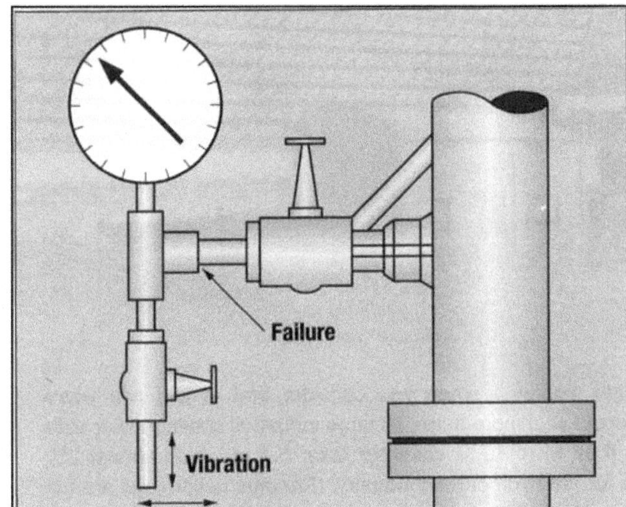

Figure 3-13: Lack of baseplate stiffness and the resulting vibration can lead to piping failures. Small piping is especially vulnerable

form around the perimeter of the foundation, and fill the void between the baseplate and the foundation with either a cementitious or epoxy grout. There are two methods used with this approach, the two-pour method, shown in Figure 3-14, or the one-pour method, shown in Figure 3-15.

The two-pour method is the most widely used, and can utilize either a cementitious or epoxy grout. The wooden grout forms for the two-pour method are easier to build because of the open top. The void between the foundation and the bottom flange of the baseplate is filled with grout on the first pour, and allowed to set. A second grout pour is performed to fill the cavity of the baseplate, by using grout holes and vent holes provided in the top of the baseplate.

The one-pour grouting method requires a more

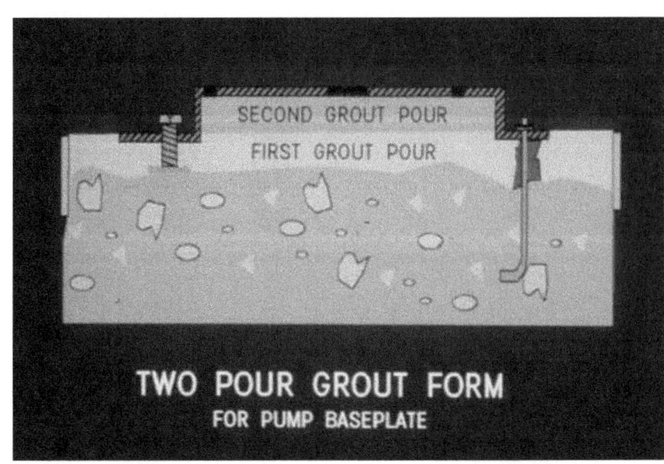

Figure 3-14: Two-pour grouting method

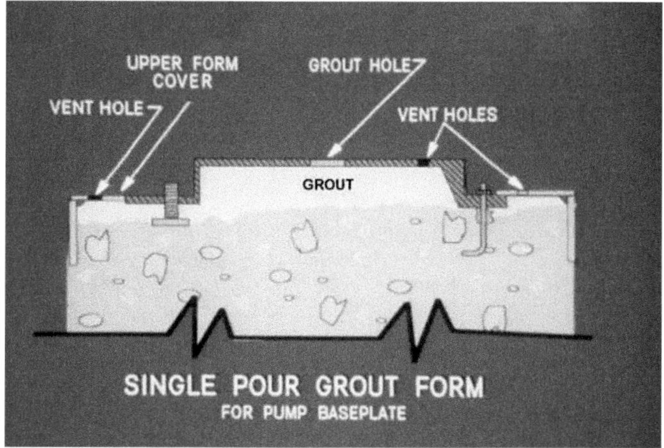

Figure 3-15: One-pour grouting method

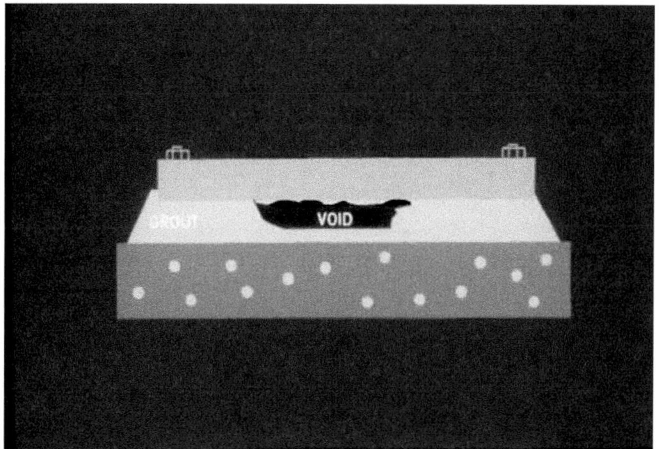

Figure 3-16: Void areas deprive the baseplate of needed support and often promote unacceptable vibration

elaborate form building technique, but does reduce labor cost. The wooden grout form now requires a top plate that forms a liquid tight seal against the bottom flange of the baseplate. The form must be vented along the top seal plate, and be sturdy enough to withstand the hydraulic head produced by the grout. All of the grout material is poured through the grout holes in the top of the baseplate. This pour technique requires good flow characteristics from the grout material, and is typically only used for epoxy grout applications.

Field Installation Problems

Grouting a baseplate or skid to a foundation requires careful attention to many details. A successful grout job will provide a mounting surface for the equipment that is flat, level, very rigid, and completely bonded to the foundation system. Many times these attributes are not obtained during the first attempt at grouting, and expensive field correction techniques have to be employed. The most prominent installation problems involve voids and distortion of the mounting surfaces.

Voids and Bonding Issues

As shown in Figure 3-16, the presence of voids at the interface between the grout material and the bottom of the baseplate negates the very purpose of grouting. Whether the void is one inch or one-thousandth of an inch deep, the desired monolithic support system has not been achieved. Voids inhibit the foundation system from damping resonance and shaft-generated vibration.

Voids can be caused by a variety of factors. These include:

- Insufficient vent holes in baseplate
- Insufficient static head during grout pour
- Grout material properties
- Improper surface preparation of baseplate underside
- Improper surface primer

Insufficient vent holes or static head are execution defects that, unless addressed through proper installation techniques, tend to leave large voids. The most overlooked causes of voids are related to bonding issues. These types of voids are difficult to repair because of the small crevices to be filled.

The first issue of bonding has to do with the material properties of the grout. Cementitious grout systems have little or no bonding capability. Epoxy grout systems have very good bonding properties, typically an average of 2,000 psi tensile to steel. However, as was mentioned earlier in this chapter, surface preparation and primer selection greatly affect the bond strength. The underside of the baseplate must be cleaned, and the surface must be free of oils, grease, moisture, and other contaminants. All of these contaminants greatly reduce the tensile bond strength of the epoxy grout system.

The type of primer used on the underside of the baseplate also affects the bond between epoxy grout and baseplate. Ideally, the best bonding surface would be a sandblasted surface with no primer. Since this is feasible only if grouting immediately follows baseplate sandblasting, a primer must be used, and the selection of the prime must be based on the tensile bond strength to steel. The epoxy grout system will bond to the primer, but the primer must bond to the steel baseplate to eliminate the formation of voids. The best primers will be epoxy based, and have minimum tensile bond strength of 1,000 psi. Other types of primers, such as inorganic zinc, have been used, but the results vary greatly with how well the inorganic zinc has been applied.

Figure 3-17 shows the underside of a baseplate sprayed with inorganic zinc primer. This primer has little or no strength, and can be easily removed with the tip of a trowel. Moreover, the inorganic zinc was applied too thick and the top layer of the primer is little more than a powdery matrix. The ideal dry film thickness for inorganic zinc is three mils (about 0.08 mm), and is very difficult to achieve in practice. The dry film thickness in this example is 9 to 13 mils (about 0.2-0.3 mm) as is being verified in Figure 3-18.

The consequences of applying epoxy grout to such a primer are shown in Figure 3-19. This is a core sample taken from a baseplate that was free of voids for the first few days. As time progressed, a void appeared, and over the course of a week the epoxy grout became completely separated ("disbonded") from the baseplate. The core sample shows that the inorganic zinc primer bonded to the steel baseplate, and the epoxy grout bonded to the inorganic zinc primer, but the primer delaminated. It sheared apart because it was applied too thick, and created a void across the entire top of the baseplate.

Figure 3-17: Underside of a baseplate sprayed with inorganic zinc primer

Figure 3-18: Verification of coating thickness is important

Figure 3-19: Core sample showing "disbonding" due to incorrect prime coating

Conventional Wisdom: *Epoxy will bond to anything. No priming is needed.*

Fact: *A suitable primer is required if grout jobs are to be long-term effective.*

Distortion of Mounting Surfaces

Another field installation problem with costly implications is distortion of the baseplate machine surfaces. Distortion can either be induced prior to grouting due to poor field leveling techniques, or the distortion can be generated by the grout itself.

Baseplate designs have become less rigid over time. Attention has been focused on the pump end of the baseplate to provide enough structural support to contend with nozzle load requirements. The motor end of the baseplate is generally not as rigid, as shown in Figure 3-20. The process of shipping, lifting, storing, and setting the baseplate can have a negative impact on the motor mounting surfaces. While these surfaces may have initially been flat, experience shows that there is work to be done by the time the baseplate reaches the field.

Using the system of jack bolts and anchor bolts, Figure 3-21, the mounting surfaces can be re-shaped during the leveling process, but the differing concepts of flatness and level have become confused. Flatness

Figure 3-20: The motor end of this pump baseplate lacks the strength of the pump end

Figure 3-21: Manipulation of jack bolts and anchor bolts can re-shape equipment mounting surfaces

cannot be measured with a precision level, and unfortunately this has become customary practice. A precision level measures slope in inches per foot. However, flatness is not a slope, it is a displacement. In the field, flatness should be measured with a straightedge, or with a ground bar and feeler gauge, as shown in Figure 3-22; flatness is thus not measured with a level. Once the mounting surfaces are determined to be flat, then the baseplate can be properly leveled. This confusion has caused many baseplates to be installed with the mounting surfaces out-of-tolerance for both flatness and level.

The other issue of mounting surface distortion comes from the grout itself. All epoxy grout systems have a slight shrinkage factor. While this shrinkage is very small, typically 0.0002"/in (0.2 mm/m), the tolerances for flatness and level of the mounting surfaces are also very small. A chemical reaction occurs when the epoxy resin is first mixed with hardener (to which the aggregate is added a little later so as to form what is commonly termed the epoxy grout). This reaction results in a volume change that is referred to as shrinkage. Chemical cross-linking and volume change occur as the material cools after the exothermic reaction. Epoxy grout systems cure from the inside out, as shown in Figure 3-23. The areas closest to the baseplate-to-grout interface experience the highest volume change.

Baseplates with sturdy cross braces are not affected by the slight volume change of the grout. For less rigid designs, the bond strength of the epoxy grout can be stronger than the baseplate itself. Referring back to Figure 3-23, after the grout has cured, the motor mounting surfaces often become distorted and are no longer coplanar. Acceptable tolerances for alignment and motor soft foot become very difficult to achieve in this scenario. This "pull down" phenomenon has been proven by FEA modeling and empirical lab tests performed by Stay-Tru® personnel working cooperatively with both a major grout manufacturer and a major petrochemical company.

Hidden Budget Impact

Correcting the problems of voids and mounting surface distortion in the field is a very costly venture. Repairing voids takes a lot of time, patience, and skill to avoid further damage to the baseplate system. Field machining the mounting surfaces of a baseplate also involves resources that are in short supply: time and money.

The real concern with correcting baseplate installation problems on site, is that repair-related issues are not reflected in the construction budget. Every field correction is a step backwards, in both time and money. On fixed-cost projects, the contractor must absorb the cost, whereas on cost-plus projects, the user-client must pay the bill. Either way, there will be extra cost, controversy and contentiousness.

Figure 3-22: Flatness being measured with ground bar stock and feeler gauges

Installing a Pre-grouted Pump Baseplate

Using the conventional method for installing a pre-grouted baseplate is no different from the first pour of a two-pour grout procedure. After the concrete foundation has been chipped and cleaned, and the baseplate has been leveled, grout forms must be constructed to hold the grout, see Figure 3-24. To prevent trapping air under the pre-filled baseplate, all the grout material must be poured from one side. As the grout moves under the baseplate, it pushes the air out. Because of this, the grout material must have good flow characteristics. To assist the flow, a head box should be constructed and kept full during the grouting process. (See Appendix 4 for additional information).

Hydraulic Lift of a Pre-grouted Baseplate

When using a head box, the pre-grouted baseplate must be well secured in place. The jack bolt and anchor bolt system shown later in this chapter must be tight, and the specific size anchor bolt nut in our example should be locked down to the equivalent of 30 to 45 ft-lbs (40-61 n-m).

The bottom of a pre-grouted baseplate provides lots of flat surface area. The specific gravity of most epoxy grout systems is in the range of 1.9 to 2.1. Large surface areas and very dense fluids create an ideal environment for buoyancy. Table 3-5 shows the inches of grout head necessary to begin lifting a pre-grouted ANSI baseplate. During the course of a conventional grouting procedure, it is very common to exceed the inches of head necessary to lift a pre-filled baseplate. For this reason, it is very important to ascertain that the baseplate is locked down. As a point of interest, the whole range of API baseplates listed in Appendix M of API-610 can be lifted with nine inches of grout head.

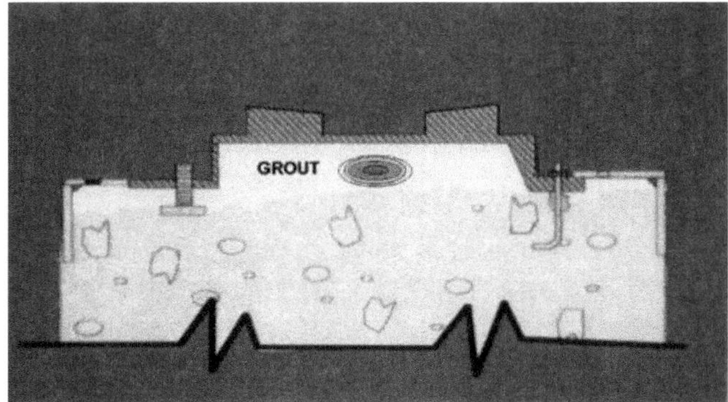

Figure 3-23: Epoxy grout systems cure from the inside towards the outside

New Field Grouting Method for Pre-grouted Baseplates

Conventional grouting methods for non-filled baseplates are, by their very nature, labor and time intensive. Utilizing a pre-grouted baseplate with conventional grouting methods helps to minimize some of the cost, but the last pour still requires a full grout crew, skilled carpentry work, and good logistics. To further minimize the costs associated with baseplate installations, a new field grouting method has been developed for pre-grouted baseplates. This new method utilizes a low viscosity high strength epoxy grout system that greatly reduces foundation preparation, grout form construction, crew size, and the amount of epoxy grout used for the final pour.

Figure 3-24: Grout forms are constructed to contain the grout. Note protective plastic on pump set

Table 3-5: Grout head pressure required to lift a pre-grouted baseplate

Grout Head Pressure Required to Lift a Pre-Grouted Baseplate										
ANSI Type Baseplates										
Base Size	Length (in)	Width (in)	Height (in)	Area (in^2)	Volume (in^3)	Exact Weight (lbs)	Base Weight (lbs)	Epoxy Grout. Weight (lbs)	Equalizing Pressure (psi)	Grout Head (in)
139	39	15	4.00	585	2340		93	169	0.45	6.22
148	48	18	4.00	864	3456	124	138	250	0.45	6.22
153	53	21	4.00	1113	4452		178	322	0.45	6.22
245	45	15	4.00	675	2700		108	195	0.45	6.22
252	52	18	4.00	936	3744		150	271	0.45	6.22
258	58	21	4.00	1218	4872		195	352	0.45	6.22
264	64	22	4.00	1408	5632	225	225	407	0.45	6.22
268	68	26	4.25	1768	7514		283	544	0.47	6.47
280	80	26	4.25	2080	8840		332	639	0.47	6.47
368	68	26	4.25	1768	7514		283	544	0.47	6.47
380	80	26	4.25	2080	8840		332	639	0.47	6.47
398	98	26	4.25	2548	10829		407	783	0.47	6.47

Density of Grout 125 lbs/ft^3
Specific Gravity 2.00

While there may be other low viscosity high strength epoxy grout systems available on the market, the discussion and techniques that follow are based on the flow and pour characteristics of Escoweld 7560. This type of low viscosity grout system can be poured to depths from 1/2″ to 2″ (13-50 mm), has the viscosity of thin pancake batter, and is packaged and mixed in a liquid container. As shown in Figure 3-25, this material can be mixed and poured with a two-man crew.

Concrete Foundation Preparation

Irrespective of baseplate style, i.e. pre-grouted or traditional unfilled, correct preparation of the top of the concrete foundation will have long-term reliability implications and is important. The laitance on the surface of the concrete must be removed for proper bonding, regardless of grouting method and material selected. Traditional grouting methods require plenty of room to properly place the grout, and this requires chipping all the way to the shoulder of the foundation. However, utilizing a low viscosity epoxy grout system will greatly reduce the amount of concrete chipping required to achieve a long-term satisfactory installation.

As illustrated in Figure 3-26, the new installation method allows for the chipped area to be limited to the footprint of the baseplate. Either a pneumatically or an electrically operated hammer can be used to remove the concrete laitance, and the required depth of the final grout pour is reduced to 3/4″ to 1″ (18-25 mm).

New Grout Forming Technique

With the smooth concrete shoulder of the foundation still intact, a very simple "2×4" grout form can be used, see Figures 3-27 and 3-28. One side of the simple grout form is waxed, and the entire grout form is sealed and held in place with caulk. While the caulk is setting up, a simple head box can be constructed out of readily available material. Due to the flow characteristics of the low viscosity epoxy grout, this head box does not need to be very large or very tall.

The low viscosity epoxy grout is mixed with a hand drill, and all the grout is poured through the head box to prevent trapping air under the baseplate.

This new installation method has been used for both ANSI and API-style baseplates with excellent results. With this technique, field experience has shown that a pre-grouted baseplate can be routinely leveled, formed, and poured with a two-man crew in three to four hours. Here, then, is the proof:

Field Installation Cost Comparison

The benefits of using a pre-grouted baseplate with the new installation method can be clearly seen when field installation costs are compared. This comparison

Figure 3-25: Epoxy grout being mixed

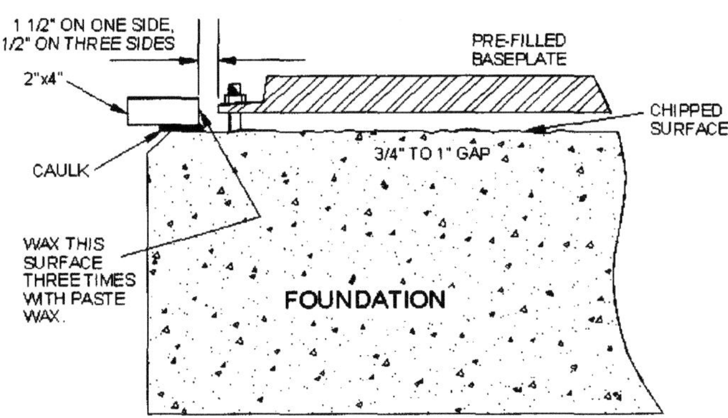

Figure 3-27: Simple grout forms are sufficient for new grout forming technique

applies realistic labor costs; it does not take credit for the elimination of repair costs associated with field installation problems, such as void repair and field machining.

Industry experience shows that eight men are typically involved in the average size conventional grouting job. An actual labor cost of $45 per man-hour must be used in U.S. installations when employee benefits and overhead charges are included as of 2004.

A cost comparison can be developed, based on the installation of a typical API baseplate using epoxy grout, for the conventional two-pour procedure and a pre-grouted baseplate using the new installation method. The following conditions apply:

Baseplate Dimensions:	72" × 36" × 6" (1.8 × 0.9 × 0.15 m)
Foundation Dimensions:	76" × 40" × 2" (1.93 × 1.0 (grout depth) × 0.05 m)

Figure 3-26: Foundation chipping in progress

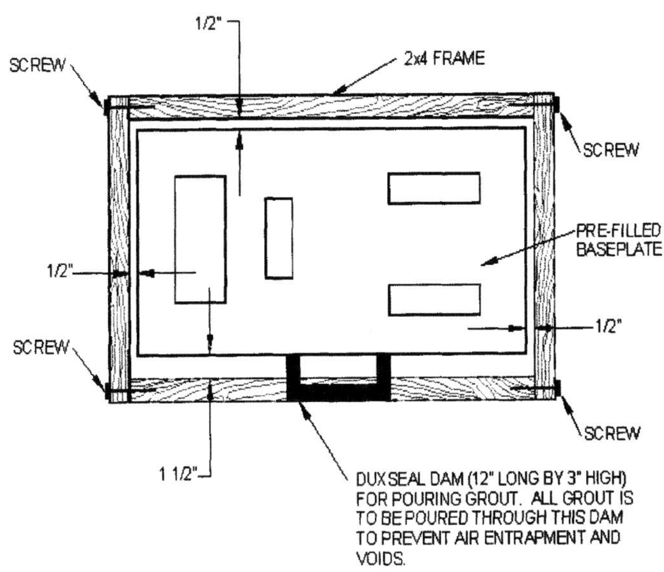

Figure 3-28: Typical grout form details

Labor Cost:	$45/hr
Epoxy Grout Cost:	$111/cu. ft. ($3,920/m^3)

In 2001, a baseplate with the listed dimensions could be pre-grouted for $2,969. This would include surface preparation, epoxy grout, surface grinding, and a guaranteed inspection.

Table 3-6 shows a realistic accounting of time and labor for the installation of a typical API baseplate. The total installed cost for a conventional two-pour installation is $6,259. The total installed cost for a pre-grouted

Table 3-6: Cost comparison two-pour conventional grouting vs. Stay-Tru® Pre-filled baseplate grouting system (Ref. 3-1). Use multiplier of millwright rate for current-year cost conversion.

Installation Labor Cost for Two Pour Procedure		*Installation Labor Cost for Stay-Tru System*	
Leveling of Baseplate		**Leveling of Baseplate**	
Millwright: 2 men × 4 hrs × $65/hr	520	Millwright: 2 men × 1 hr × $65/hr	130
Forming of Baseplate		**Forming of Baseplate**	
4 MEN × 4 hrs × $45/hr	720	2 MEN × 2 hours × $45/hr	180
First Pour			
Grout Set-Up Time		Grout Set-Up Time	
8 MEN × 1.0 hrs × $45/hr	360	2 MEN × 1.0 hrs × $45/hr	90
Grout Placement		Grout Placement	
8 MEN × 2.0 hrs × $45/hr	720	2 MEN × 2.0 hrs × $45/hr	180
Grout Clean-up		Grout Clean-up	
8 MEN × 1.0 hrs × $45/hr	360	2 MEN × 1.0 hrs × $45/hr	90
Additional Cost:		**Additional Cost:**	
Fork Lift & driver; 1.0 hrs × $45=	45	Wood Forming Materials =	50
Supervisor: 4.0 hrs × $45/hr	180		
Mortar Mixer =	100		
Wood Forming Materials=	100		
Second Pour			
Grout Set-Up Time			
8 MEN × 1.0 hrs × $45/hr	360		
Grout Placement			
8 MEN × 2.0 hrs × $45/hr	720		
Grout Clean-up			
8 MEN × 1.0 hrs × $45/hr	360		
Additional Cost:			
Fork Lift & driver; 1 hr × $45=	45		
Supervisor: 4.0 hrs × $45/hr	180		
Mortar Mixer =	100		
LABOR COST	4570	LABOR COST	670
ADDITIONAL COST	300	ADDITIONAL COST	50
GROUT COST	1389.56	STAY-TRU COST	2969
		GROUT COST (7560)	505.3
TOTAL PER BASE	$6,259.56	TOTAL PER BASE	$4,194.30

baseplate, installed with the new installation method, is $4,194. Aside from the very obvious cost savings, the reliability impact of this void-free and fully co-planar installation is of great importance to reliability-focused pump users.

While it must be pointed out that today's costs differ from the above, it is of interest to note that the reliability advantage of pre-grouted baseplates has been well proven and several pump OEM's are now routinely employing the method.

Polymer Composite Baseplates

Polymer composite baseplates (Figure 3-29) are a less expensive option that may be suitable for certain applications and services. ITT/Goulds offer their smaller pump sets on stilts and a "PermaBase®."

Another baseplate manufacturer, BASETEK® offers products claimed to outperform both steel and vinyl ester resin baseplates. The various BASETEK® products are primarily serving the ANSI pump sizes and can accommodate both close-coupled and spacer-type

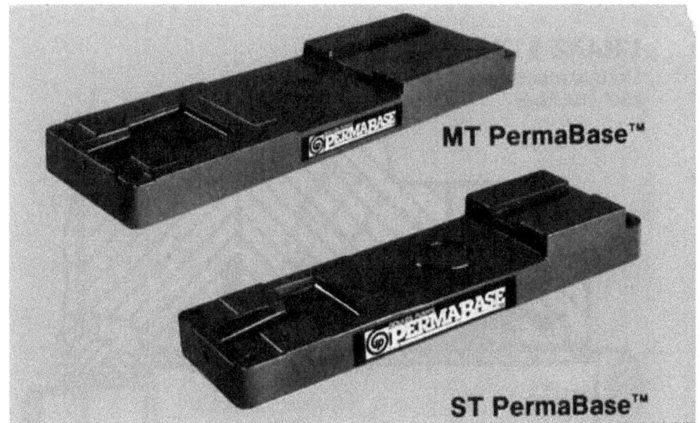

Figure 3-29: Composite baseplate (Source: ITT/Goulds, Seneca Falls, NY 13148

coupling installations (Figures 3-30 and 3-31). Threaded stainless steel inserts are cast into the thermally relatively stable Zanite®, a blend of silicon dioxide ceramic quartz aggregate, epoxy resin, and proprietary additives.

Requirements of Principal Foundation

Having dealt with baseplate issues and grout supports, we are now ready to work our way down to an overview of foundation-related topics.

Although generally falling within the jurisdiction of design contractors and civil engineers, pump users looking for years of trouble-free pump operation and lowest possible life cycle cost must understand the requirements of a principal pump foundation. All but the smallest pumps must be placed on foundations and the few process plants that decades ago opted for stilt-mounted baseplates, i.e. installations without pump foundations as shown earlier in Figures 3-10 and 3-29, have probably never been able to reach the MTBF of comparable installations with baseplates placed on solid foundations.

This overview is intended to provide a written specification or procedure for the installation of foundation chock plates for critical rotating equipment, including, of course, pumps. Unless otherwise noted, the overview and its guideline data pertain to foundations upon which either pre-grouted or unfilled baseplates will be installed. Following these instructions will lead to a foundation that will support the machinery at proper elevation, minimize vibration, and maintain alignment.

Foundation Prerequisites

The following assumptions are made concerning the principal foundation, Figure 3-32:

1. The foundation has been adequately designed to support the machinery

2. The foundation rests on bedrock or solid earth, completely independent of other foundations, pads, walls, or operating platforms

3. The foundation has been designed to avoid resonant vibration conditions originating from normal excitation forces at machinery operating speed or multiples of the operating speed

4. The driver, gearbox, and driven equipment rest on a common foundation

Figure 3-30: BASETEK® composite baseplate for close-coupled pumps

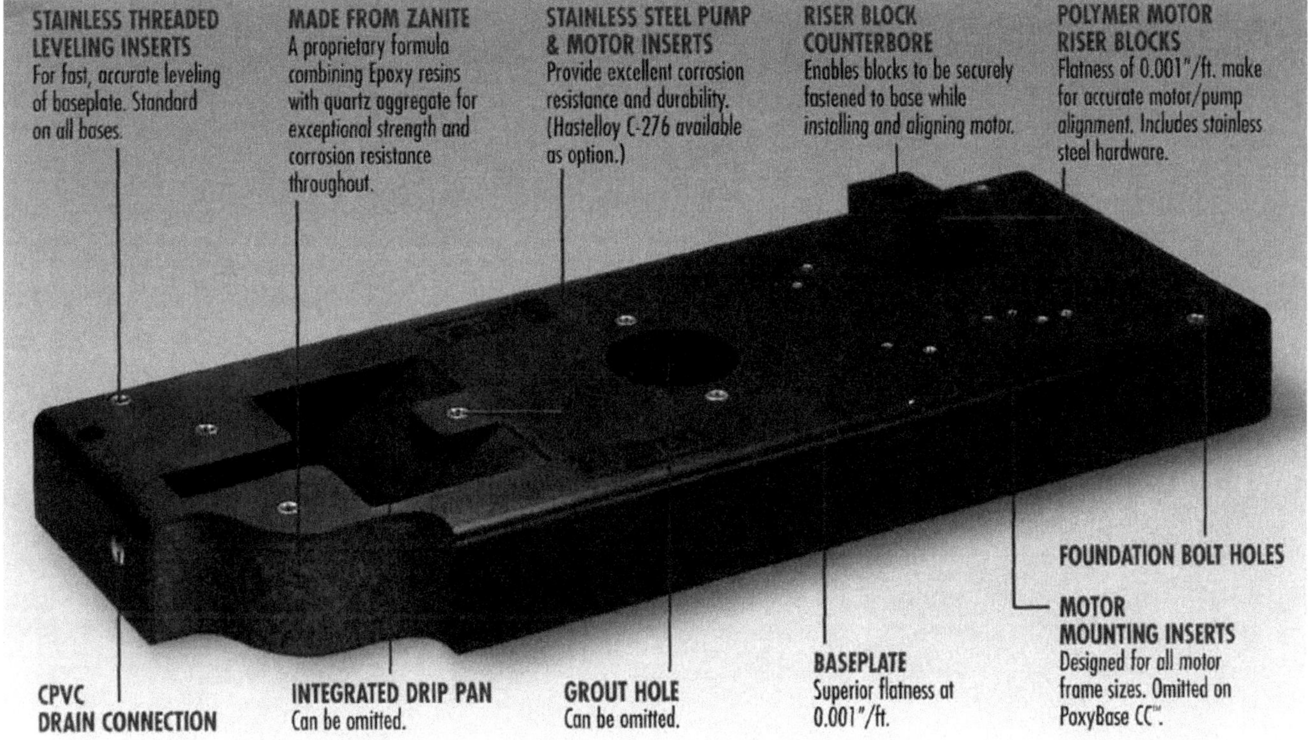
STAINLESS THREADED LEVELING INSERTS
For fast, accurate leveling of baseplate. Standard on all bases.
MADE FROM ZANITE
A proprietary formula combining Epoxy resins with quartz aggregate for exceptional strength and corrosion resistance throughout.
STAINLESS STEEL PUMP & MOTOR INSERTS
Provide excellent corrosion resistance and durability. (Hastelloy C-276 available as option.)
RISER BLOCK COUNTERBORE
Enables blocks to be securely fastened to base while installing and aligning motor.
POLYMER MOTOR RISER BLOCKS
Flatness of 0.001"/ft. make for accurate motor/pump alignment. Includes stainless steel hardware.
FOUNDATION BOLT HOLES
MOTOR MOUNTING INSERTS
Designed for all motor frame sizes. Omitted on PoxyBase CC™.
BASEPLATE
Superior flatness at 0.001"/ft.
GROUT HOLE
Can be omitted.
INTEGRATED DRIP PAN
Can be omitted.
CPVC DRAIN CONNECTION

Figure 3-31: BASETEK® composite baseplate with spacer coupling between pump and driver

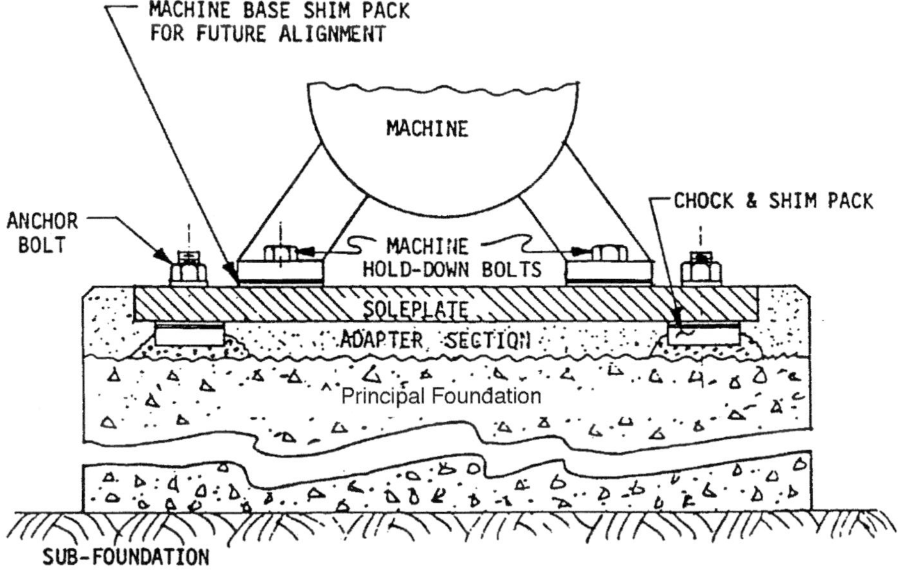
MACHINE BASE SHIM PACK FOR FUTURE ALIGNMENT
MACHINE
CHOCK & SHIM PACK
ANCHOR BOLT
MACHINE HOLD-DOWN BOLTS
SOLEPLATE
ADAPTER SECTION
Principal Foundation
SUB-FOUNDATION

Figure 3-32: The "Total Foundation"

5. The foundation is designed for uniform temperatures to reduce distortion and misalignment

A few definitions are in order.

- "The Total Foundation"—The complete unit assembly which supports a machine and is made up of a machine base, adapter section, chock plate, principal foundation, and sub-foundation
- "Machine Base"—The metal part to which a machine is bolted, sometimes called a sole plate, bed plate, or bed rail
  - — Baseplates are single-unit structural supports to which both pump and driver are bolted
  - — Sole plates are steel plates to which only the driver (e.g. steam turbine), or only the driven machine (e.g. large pump, compressor) is bolted. Separately sole plate-mounted driver and

driven machine bodies comprise a machinery train
— Bed plates or bed rails are generally used with positive displacement machines (e.g. reciprocating compressors, etc.)

- "Adapter Section"—The area filled with grout to join the machine base to the principal foundation

- "Chock Plate"—A metal plate machined parallel on two sides used to level and support the machine base prior to grouting

- "Principal Foundation"—A steel and concrete structure that supports the machine base and transmits the load to the sub-foundation

- "Sub-foundation"—The bedrock or solid earth and pilings (If required) on which the entire load bears.

Chock Plate Installation

There are two methods used to support the machine base (baseplate, sole plate, bed plate): the "grout only" and the chock "plate/grout" methods. The chock plate/grout method is preferred for precision leveling of baseplates and sole plates. Accordingly, it will be the only method considered here.

Quantity, Size, and Position of Chocks

1. Consult the machine manufacturer's drawings for required number of chocks and their locations.

2. If the vendor does not specify the location and number of chocks (Figure 3-33), use the following and/or calculate, as needed:
   — Determine the total operating load on the baseplate.
   — Calculate the minimum required area of chock plates based on a uniform loading of 300 psi (~2,070 kPa) on the concrete foundation, i.e. operating load (in pounds) divided by 300 psi equals chock plate area (square inches).
   — Chock plates to have a minimum thickness of 3/4" and be 4" square with the load bearing surfaces machined parallel and flat to within .002".
   — The minimum number of chocks is determined by dividing the minimum required total chock plate area (in square inches) by 16 square inches (the area of a 4" x 4" chock).

3. Location of Chock
   — A chock plate is to be located on both sides of each anchor bolt to reduce sole plate distortion due to tightening the anchor bolts.
   — Long, unsupported sole plate spans are to be avoided. Chocks will be provided on 18" centers regardless of loading calculations.
   — Have the survey crew locate the anchor bolts and elevation of the sole plate. Chip away defective concrete and laitance, leaving the entire surface of the foundation reasonably rough but level. Allow for a minimum of about 1.5" (37-40 mm) of grout.
   — Locate one chock plate at the correct elevation, using epoxy grout, or "dry pack" grout such as "Five Star" or a similar non-shrinking, ready-mix formulation. This chock is to be leveled to .0005" per foot in two directions. Allow the Portland cement grout to cure for 24 hours before installing any more chocks. Chock elevation tolerances are on the plus side and range from zero to 1/32" (0.8 mm).
   — Install the remaining chocks at an elevation within .002" of the reference chock and level to .0005" per foot. Either an optical level (K&E 71-3015) or transit (K&E 71-1010) is preferred for locating the chocks, however, a master level

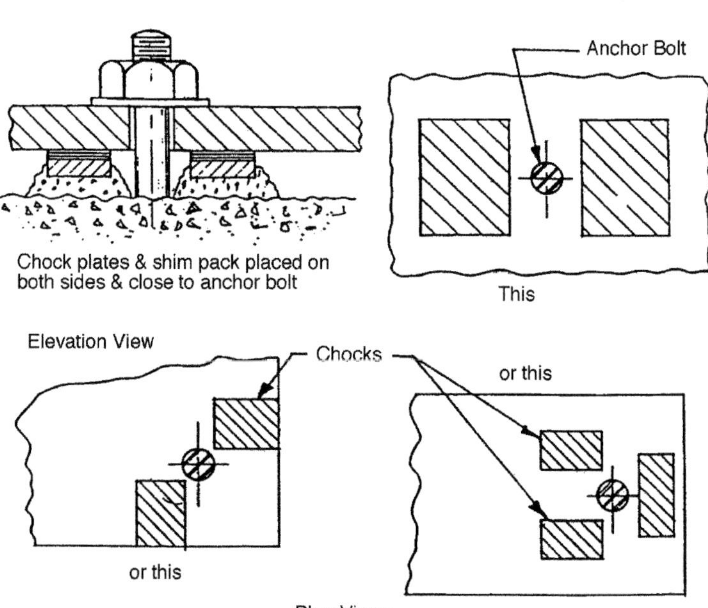

Figure 3-33: Typical supports at anchor bolts

(.0005"/div) and a straight edge can be used as well.

Elevation Adjusting Devices

1. Shim Pack. Figure 3-34 shows four methods that are sometimes used to provide vertical adjustment of the sole plate off the chocks: a single wedge (Figure 3-34a), parallel wedges (Figure 3-34b), jack screws (Figure 3-34c) and a shim pack (Figure 3-34d). The single wedge method *does not qualify* as a satisfactory vertical adjustment device. Any of the three remaining can be employed to level the sole plate.
   - All shims are to be made of 304 or 316 SS. In chlorine or hydrochloric acid units use non-metallic Mylar® shim packs because stainless steel will be chemically attacked.
   - Limit a "shim pack" to a reasonable maximum of four, but never more than six shims. Use the heaviest gauge shims possible in thick shim packs because a large number of thin shims will be "spongy."
   - Individual shims should not be less than .005" (0.12 mm) thick, with thin shims being sandwiched between heavier shims, i.e. shims .040" to .060", (~1-1.5 mm).
   - Start with a shim pack made up of one each 0.062", 0.031", 0.025", 0.015", 0.010", and .005" (~1.5, 0.7, 0.5, 0.3, 0.2 and 0.1 mm) shims at each chock.

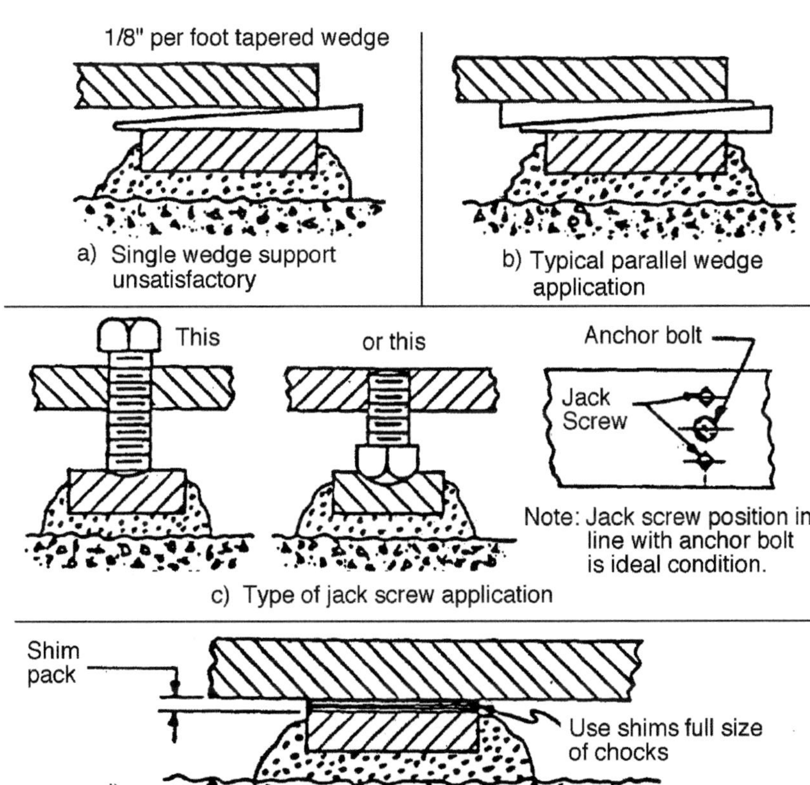

Figure 3-34: Shimming and leveling methods

Universal Adjustable Chocks

Although primarily developed for larger machinery, Vibracon® SM self-level, adjustable chocks have been used in pump installations, although the user must verify the absence of resonant vibration. These chocks (Figure 3-35) represent a mechanically stiff mount that serves in place of grouts, machined chocks or cut shims. Self-level adjustable mounting systems inherently eliminate component soft-foot while saving installation time and possibly even the need for coplanar milling of component supports. Pinning or dowelling (Figure 3-36) is still possible after installing these components.

Owners cite reusability as the best characteristic of the Vibracon® SM mount. Once the adjustable chock is installed the realignment of equipment is often facilitated (Ref. 3-8).

Baseplate or Sole Plate Leveling (See Table 3-7)

1. Clean the rough concrete surface of the foundation to remove any oil, grease, dirt, and loose particles.

2. Clean the surfaces of the baseplate that will be in contact with the finished grout to remove any oil, grease, dirt, and rust. (Sandblast and prime if necessary). De-grease anchor bolts and clear out bolt holes. The bottom of the baseplate could be painted with epoxy grout to provide a secure bond of the baseplate to the finish grout.

3. Set the baseplate on the chocks and shim packs and tighten all anchor bolts using a torque wrench. It is important to tighten all the bolts to the same value after each shim change. The torque values are based on standard published torque rating of A7 bolts for a given bolt diameter.

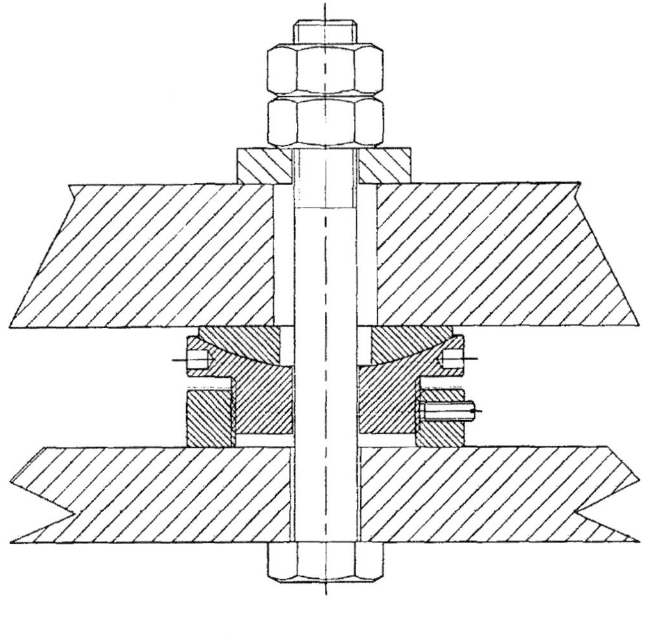

Figure 3-35: Vibracon® SM self-level, adjustable chocks

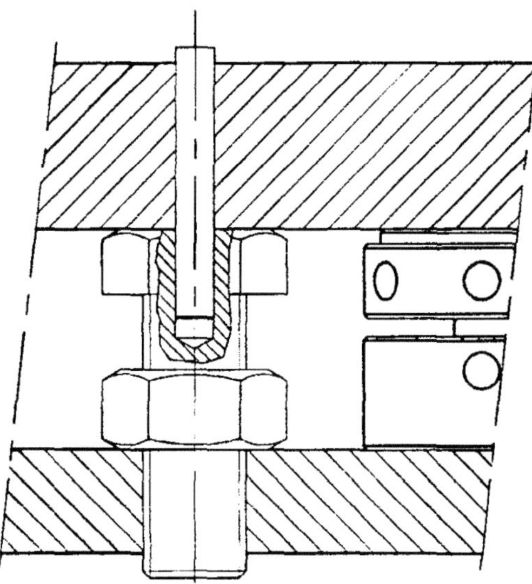

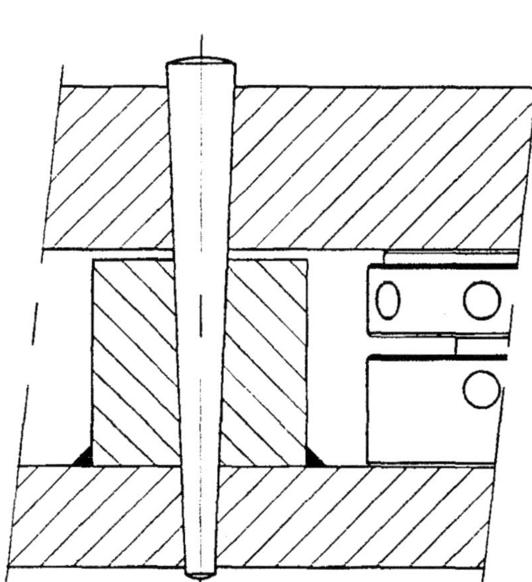

Figure 3-36: Dowel pinning options for equipment supported by adjustable chocks

4. Check the baseplate level at the machined equipment mounting surfaces using a "Master Level" (0.0005"/div). Make the necessary adjustments by adding or removing shims as required to obtain an overall level of 0.0005" per foot or less in two directions. If the overall elevation variation is 0.010" (0.25 mm) or more, consult the responsible machinery engineer.

5. After final anchor bolt tightening, recheck level. Maximum overall elevation variation of baseplate should not exceed 0.005" (0.12 mm).

6. Establish preliminary coupling alignment.

7. Grout the baseplate in accordance with separate specification "Precision Grouting of Rotating Equipment."

8. Mount the rotating equipment to the baseplate in accordance with vendor certified outline drawings.

9. Align the rotating equipment in accordance with separate specification "Precision Alignment of Rotating Equipment" (Ref. 3-9).

Portland Cement Grouting

There can be no question that epoxy grouts are superior to cement grouts. Cement grouts will degrade

Table 3-7: Horizontal pump baseplate checklist (Ref. 3-10)

	DATE/BY
1. Concrete foundation roughed up to provide bond for grout.	_______
2. Concrete foundation clean and free of oil, dust and moisture. Blown with oil free compressed air.	_______
3. Foundation bolt threads undamaged.	_______
4. Foundation bolt threads waxed or covered with duct seal.	_______
5. Baseplate underside clean and free of oil, scale, and dirt.	_______
6. Eight positioning screws, two per driver pad.	_______
7. Baseplate welds continuous and free of cracks.	_______
8. Mounting pads extend 1 inch beyond equipment feet each direction.	_______
9. Mounting pads machined parallel within 0.002 inch.	_______
10. Shim packs or wedges on two sides of each foundation bolt. shims are stainless steel.	_______
11. Baseplate raised to proper height per drawing.	_______
12. Pad heights permit 1/8″ minimum shim under driver feet.	_______
13. All leveling devices make solid contact with concrete and baseplate.	_______
14. All machined surfaces on base level with 0.0005 in/foot in two directions 90′ opposed using "master level" (0.0005 in./division) with anchor bolt nuts snugged down.	_______
15. Foundation and baseplate protected from dirt and moisture contamination, when all of above accepted, baseplate leveling is accepted and baseplate can be grouted.	_______

when allowed to come into contact with lubricating oils and many pumped products. While life cycle cost considerations will certainly favor epoxy grouts, the initial, as-installed cost of epoxy grouts can exceed that of cementitious grouts.

There will always be owner/purchasers that consider only initial cost, which is why the traditional cementitious grouts will probably still be with us for some time. Therefore, we would rather give an overview of an acceptable cementitious grout procedure than leave its placement to chance.

This specification/procedure covers Portland cement grouting of mechanical equipment on concrete foundations.

Materials

1. Portland cement shall conform to ASTM C-150.
2. Filler sand aggregate shall conform to ASTM C-33 fine aggregate.
3. Only clean, fresh water shall be used in preparing the grout.
4. All materials shall be stored indoors and kept dry, free of moisture, in its own original shipping containers.
5. Storage temperature shall be maintained between 40°F (5°C) and 90°F (32°C). Grouting shall not be done at ambient temperatures outside these limits.

Preparation of Foundation

Note: Preparation of foundation will be identical irrespective of grouting material. This procedure is therefore valid for either epoxy or cementitious grouts.

1. Foundation bolt threads shall be examined for stripped or damaged threads. These threads shall be re-chased to clean up, or the foundation bolts replaced if necessary. The foundation bolts and their threads shall be protected during the leveling and grouting operations.
2. The concrete foundation shall be dry and free of oil. If oil is present it shall be removed with solvent prior to chipping.
3. The concrete shall be chipped to expose its ag-

gregate, approximately 1/2 to 1-1/2 inch (13 to 37 mm) deep so as to remove all laitance and provide a rough surface for bonding. Light hand tools shall be used for chipping.

4. After chipping, the exposed surface shall be blown free of dust using oil-free compressed air from an approved source.

5. After the foundation has been chipped and cleaned, it shall be covered so as to prevent it from becoming contaminated with oil or dirt.

6. Foundation bolts shall be waxed or covered with duct seal over their entire exposed length. If the bolts are sleeved—and sleeved bolts are highly recommended over bolts that are mostly encased in concrete—the sleeves shall be packed with non-bonding material to prevent the annular space around the bolt from being filled with grout.

Preparation of Baseplate

The underside and side surfaces of the baseplate which will come in contact with the grout shall be completely clean and free of all oil, rust, scale, paint, and dirt. This cleaning shall be done shortly before the leveling of the baseplate and placement of the grout.

Forming

Note: This guideline apples to all grout types

1. All forming material coming into contact with the grout shall be coated with colored paste wax, the coloring to contrast with the forming material colors.

2. Forms shall be made liquid tight to prevent leaking of grout material. Cracks and openings shall be sealed off with rags, cotton batting, foam rubber or caulking compound.

3. Care shall be taken to prevent any wax from contacting the concrete foundation or the baseplate.

4. Leveling screws, if used, shall be coated with wax.

Wetting the Foundation

Note: Foundation wetting applies to cementitious grout only!

1. At least eight (8) hours prior to placement of the grout the concrete shall be wetted with fresh water and kept wet up to the actual placement of the grout.

2. Excess water is to be removed from the concrete just prior to the placement of the grout.

Mixing the Cementitious Grout

1. The mixing equipment shall be clean and free of all foreign material, scale, moisture, and oil.

2. The cement and sand filler shall be hand blended dry to form a homogeneous mixture.

3. Fresh water shall be added to the dry materials and hand blended to wet all the dry material. Water quantity shall be the lesser of the amount specified by the cement manufacturer or 50% of the cement by weight.

4. The rate of blending shall prevent air entrainment in the grout mixture.

Placement

1. A suitable head box shall be prepared to hydraulically force the grout into the baseplate cavities.

2. Grouting shall be continuous until the placement of grout is complete under all sections or compartments of the baseplate. If more grout is needed than can be prepared in one mixing box, a second box shall be employed. Subsequent batches of grout should be prepared so as to be ready when the preceding batch has been placed.

3. No push rods or mechanical vibrators shall be used to place the grout under the baseplate.

4. One 4" × 4" × 4" (100 × 100 × 100 mm) test cube shall be made from each unit of grout placed. The sample(s) shall be tagged with the equipment number on which the batch was used and where in the foundation the batch was placed.

Finishing

1. When the entire form has been filled with grout, excess material shall be removed and the exposed surfaces smoothed with a trowel. The trowel shall be wetted with fresh water. Care shall be taken to prevent excessive blending of water into the grout surface.

2. Forms shall be left in place until the grout has set and then removed.

3. If wedges were used to level the baseplate, they shall be removed when the grout has set but not cured. After the grout has cured the wedge voids shall be filled with Escoweld 7501 putty.

4. The remaining cement grout surfaces shall be made smooth with a trowel wetted with fresh water. Sharp grout edges shall be struck to give a 1/4" (6mm) chamfer.

5. The grout shall be kept wetted with fresh water for three (3) days.

6. The grout shall cure for seven (7) days with the ambient temperature not less that 50°F (10°C) prior to mounting any equipment on the baseplate.

7. The top of the baseplate shall be sounded for voids. After all the voids have been defined and the grout has cured, the baseplate shall have two holes drilled through to each void in opposite corners of the void. One of the pair of holes shall be tapped and fitted with a pressure grease fitting. Using a hand grease gun, the other hole shall be filled with pure (unfilled) Escoweld 7507 Epoxy with hand grease gun. Care must be taken to prevent lifting the baseplate should a blockage occur between the grease fitting and the vent. When the void has been filled, the grease fitting shall be removed and both holes dressed smooth.

8. All exposed grout and concrete surfaces shall be sealed with one coat of unfilled Escoweld 7507 Epoxy, said epoxy being hand-applied with a paintbrush. The epoxy sealer shall be tinted (contain a color) to contrast with the concrete and the grout.

Clean-up

1. Immediately after grouting is completed all tools and mixing equipment shall be cleaned using fresh water.

2. All unused mixed epoxy materials shall be disposed of in accordance with instructions on the epoxy containers.

Epoxy Grouting

The next specification and procedure covers epoxy grouting of mechanical equipment on concrete foundations. While our example deals here with a proprietary epoxy material, other materials may work as well and the user/purchaser may wish to explore their applicability with the supplier.

1. Epoxy shall be Escoweld 7505 two part liquid (epoxy resin and converter).

2. Filler shall be Escoweld 7530 sand aggregate.

3. All materials shall be stored indoors and kept dry. To prevent moisture intrusion, keep all materials in their original shipping containers.

4. Storage temperatures shall be maintained between 40°F (5°C) and 90°F (32°C). Grouting material shall be kept at 65°F (18°C) minimum for 48 hours prior to mixing and placement.

Preparation of Foundation

1. Foundation bolt threads shall be examined for stripped or damaged threads. These threads shall be re-chased to clean-up or the foundation bolt replaced if necessary. The foundation bolts and their threads shall be protected during the leveling and grouting operations.

2. The concrete foundation shall be dry and free of oil. If oil is present it shall be removed with solvent prior to chipping.

3. The concrete shall be chipped to expose its aggregate approximately 1.5" (37 mm) deep so as to remove all laitance and provide a rough surface for bonding. Light hand tools shall be used in chipping.

4. After chipping, the exposed surfaces shall be blown free of dust using oil free compressed air from an approved source.

5. After the foundation has been chipped and cleaned, it shall be covered so as to prevent it from becoming wet.

6. Foundation bolts shall be waxed or covered with duct seal over their entire exposed length. If the bolts are sleeved (much preferred over bolts that are largely encased in concrete), the sleeves shall be packed with non-bonding material to prevent the annular space around the bolt from being filled with grout.

Preparation of Baseplates Other than Stay-Tru® Epoxy Pre-filled Types

1. Surfaces of the baseplate which will come in contact with the epoxy grout shall be clean and free of all oil, rust, scale, paint and dirt. This cleaning shall be done immediately prior to the leveling of the baseplate and placement of the epoxy grout.

2. If the grouting is to be delayed by more than eight (8) hours after the baseplate is cleaned, the cleaned baseplate shall be painted with one coat of unfilled Escoweld 7505 to give a dry fill thickness of three (3) mils (0.07mm). This coat shall be fully dried prior to placement of grout.

3. If the epoxy coated baseplate is not grouted within thirty (30) days after being painted, the epoxy surface shall be roughed up with a wire brush to remove the bloom or shine. All dust produced by brushing shall be wiped off using a water dampened rag. These surfaces shall be air dried prior to placement of grout.

Forming

1. All forming material coming into contact with the grout shall be given a thick coating of colored paste wax. The coloring must contrast with the colors of the forming materials.

2. Forms shall be made liquid tight to prevent leaking of grout material. Cracks and openings shall be sealed off with rags, cotton batting, foam rubber or caulking compound.

3. Care should be taken to prevent any wax from contacting the concrete foundation or the baseplate.

4. Any leveling wedges that are to be removed after the grout has cured shall be coated with wax. Wedges and/or shims which are to remain embedded in the grout do not require coating.

5. Leveling screws, if used, shall be coated with wax.

Mixing

1. As epoxy compounds have limited pot life after mixing, both the elapsed time from beginning of mixing to the completion of the pour and the ambient temperatures at the beginning of mixing and at the completion of pour shall be recorded. The responsible machinery engineer shall place the data in the permanent equipment records.

2. Ambient temperature shall be at least 65°F (18°C) during mixing, placement and curing of the epoxy grout.

3. Mixing equipment shall be clean and free of all foreign material, scale, moisture and oil.

4. All personnel handling or working with the grouting materials shall follow safety instructions printed on the can labels.

5. Only *full units* of epoxy resin, converter and aggregate shall be used in preparing the grout.

6. The epoxy resin and the converter/hardener shall be hand-blended for at least three, but no more than five minutes. *Only then* shall the aggregate filler be added to this mix. (NOTE THAT FAILURE TO OBSERVE THIS REQUIREMENT WILL PREVENT THE GROUT FROM CURING.)

7. Immediately after the liquid blending has been completed, the aggregate shall be slowly added and gently hand blended so as to fully wet the aggregate. The rate of blending shall be slow and must prevent entrainment of air in the grout mixture.

Placement, Finishing and Clean-up

Placement, finishing and clean-up of grout can now proceed as mentioned earlier in this chapter.

PUMP ALIGNMENT

Pump shafts exist in three-dimensional space and misalignments can exist in any direction. Therefore, it is most convenient for the purpose of description to break-up this three-dimensional space into two planes, the vertical and the horizontal, and to describe the specific amount of offset and angularity that exists in each of these planes simultaneously, *at the location of the coupling*. Thus, we end up with four specific conditions of misalignment, traditionally called *Vertical Offset* (VO), *Vertical Angularity* (VA), *Horizontal Offset* (HO), and *Horizontal Angularity* (HA).

These conditions are described at the location of the coupling, because it is here that harmful machinery vibration is created whenever misalignment exists.

The magnitude of an alignment tolerance (in other words, the description of desired alignment quality), must therefore be expressed in terms of these offsets and angularities, or the sliding velocities resulting from them. Attempts to describe misalignment in terms of foot corrections alone do not take into account the size, geometry, or operating temperature of a given machine. It can be shown that accepting the simple "foot corrections approach" can seriously compromise equipment life and has no place in a reliability-focused facility. An illustration of the fallacy of the "foot correction approach" will be given later.

How much vibration and efficiency loss will result from the misalignment of shaft centers depends on shaft speed and coupling type. Acceptable alignment tolerances are thus functions of shaft speed and coupling geometry. It should be noted that high-quality flexible couplings are designed to tolerate more misalignment than is good for the machines involved. Bearing load increases with misalignment, and bearing life decreases as the cube of the load increase, i.e. doubling the load will shorten bearing life by a factor of eight.

Why, then, would high-quality flexible couplings (Chapter 11) be generally able to accommodate greater misalignment than is acceptable for the connected pumps? Well, a large percentage of pumps must be deliberately misaligned—sometimes significantly so—in the "cold" and stopped condition. As the pumps reach operating speeds and temperatures, thermal growth is anticipated to bring the two shafts into alignment. The following case history illustrates the point.

A refinery has a small steam turbine, foot-mounted, and enveloped in insulating blankets. The operating temperature of the steel casing is 455°F, and the distance from centerline to the bottom of the feet is 18 inches. The turbine drives an ANSI pump with a casing temperature of 85°F; its centerline-to-bottom-of-feet distance is also 18 inches. Both initially started up at the same ambient temperature. The differential in their growth is (0.0000065 in/inch-°F) × 18 inches × (455°F – 85°F) = 0.043 inches (1.1 mm).

If these two machines had their shafts aligned center-to-center, this amount of offset would be certain to cast the equipment train into the frequent failure (often called "bad actor") category. Using the "80/20" rule, it would be safe to assume that 20% of a facility's machinery population consumes 80% of the maintenance money. *This* pump train would almost certainly be a "bad actor" in the 20% group.

Pump alignment center-to-center without paying attention to thermal growth is surely one of the factors that keeps its practitioners in the repair-focused category. Using alignment tools and procedures that take into account all of the above is a mandatory requirement for reliability-focused companies (Ref. 3-9).

Misalignment Effects Quantified

The results of controlled tests found in maintenance publications (ICI study by J.C. Lambley) indicate that misaligned pumps can require up to 15% more energy input than well-aligned pumps. A statistical analysis involving many pumps at a plant in Rocksavage, UK, took into account local alignment conditions and determined an approximate power consumption increase of 8.3% attributable to misalignment. This indicated an average alignment error (vectorial average of parallel and angular offsets) of 0.038 inch (~1mm)—a rather drastic and, for reliability-focused plants, unacceptable value.

Even small pumps can generate big losses when shaft misalignment imposes reaction forces on shafts, even if the "flexible" coupling suffers no immediate damage. The inevitable result is premature failure of shaft seals and bearings. Performing precise alignment, therefore, pays back through preventing the costly consequences of poor alignment. It should also be pointed out that pump alignment procedures may differ with pump geometry.

The liquid-filled casing of a foot-mounted pump (Figure 2-11) may thermally grow at a rate that differs from that of the typically air-enveloped outboard support leg. Consequently, cold alignment should be done with the support leg temporarily disconnected from the bearing housing. After starting the pump and allowing it to operate for a while, the support leg should be reconnected. This type of care is essential in defining acceptable alignment procedures.

Indeed, using precise alignment methods is one of the principal attributes of a reliability-focused organization. Good alignment has been demonstrated to lead to:

- lower energy losses due to friction and vibration
- increased productivity through time savings and repair avoidance
- reduced parts expense and lower inventory requirements

More specifically, since the mid 1950's, hundreds of technical articles and presentations have made the point that serious problems are caused by incorrect alignment between pump and driver. To cite but a few:

- Coupling overheating and resulting component degradation
- Extreme wear in gear couplings and component fatigue in dry element couplings
- Pump and/or driver shaft fatigue failure
- Pump and driver bearing overload, leading to failure or shorter bearing life
- Destructive vibration events

Misinformed claims to the contrary, alignment accuracy is critical to pump and driver longevity. Figure 3-37 shows the best one can expect from the three most prevalent alignment methods practiced in the industrial world today. Moving from the straight-edge method shown on the left, to the laser-optic method on the right shows alignment accuracy improving by a factor of 100:1. It is noteworthy that the dial indicator method shown in the center of Figure 3-37 often suffers from one or more of the seven shortcomings shown in Figure 3-38. Since each sketch is self-explanatory and there's no sense belaboring the point, we might just agree that the devil is in the details. Details tend to be overlooked and are the most frequent source of pump unreliability!

Laser-optic shaft alignment methods (Figure 3-39) clearly offer unmatched speed and precision. Reliability-focused plants will easily justify and exclusively use this mature and highly cost-effective approach. It is not surprising that laser-optic alignment is a key ingredient of any successful pump reliability improvement program. Moreover, a good laser-optic alignment device will be useful to verify the line-bore concentricity, surface out-of parallelism and perpendicularity achieved for machinery during the various assembly and/or repair phases of the equipment (Ref. 3-15). It can be a tool that will be of lasting value for years and its true versatility often amazes even the experienced professional.

Thermal Growth Compensation

Thermal growth estimates are known to give inconsistent results. Over the years, and for foot-mounted ANSI and ISO pumps, these estimates have often been expressed as rules of thumb, such as:

- For pumping temperatures below 200°F (93°C)—set motor shaft at same height as pump shaft.
- Above 200°F (93°C)—set pump shaft lower, 0.001 inch per inch (1 mm per m) of centerline height above feet, for every 100°F (56°C) above 200°F (93°C).
- Vertical centerline offset = H × T × C, which is to say: (height from bottom of feet to pump centerline) × (coefficient of thermal expansion of pump casing material) × (difference between pump operating minus ambient temperature).

For centerline-mounted API pumps, the thermal rise is generally estimated as (1/3) × (H × T × C). However, estimates are sometimes not good enough, as can be gleaned from Figure 3-40. Small errors in either the calculation or the alignment correction moves can result in significant shaft misalignment.

For critical pumps the user should consider measuring the alignment changes that occur during operation. Instrumentation to measure on-stream the "dynamic" relative thermal growth has been available for many years and includes laser based devices such

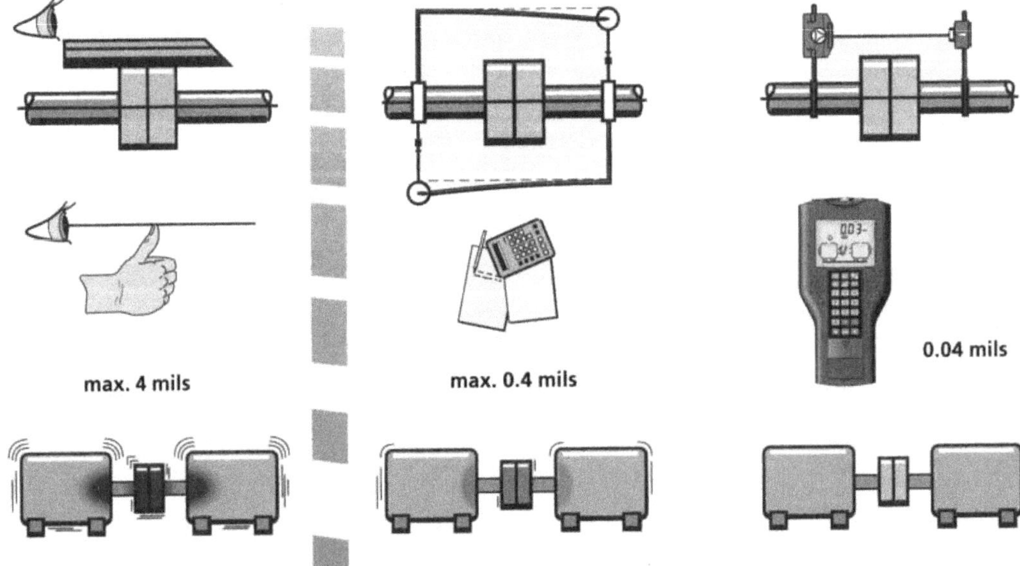

Figure 3-37: Alignment method affects alignment quality (Source: Prueftechnik Alignment Systems, D-85737 Ismaning, Germany)

Sagging indicator brackets

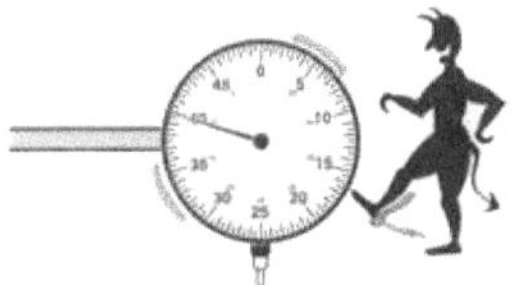

Sticking/jumping dial hands

Low resolution=rounding losses

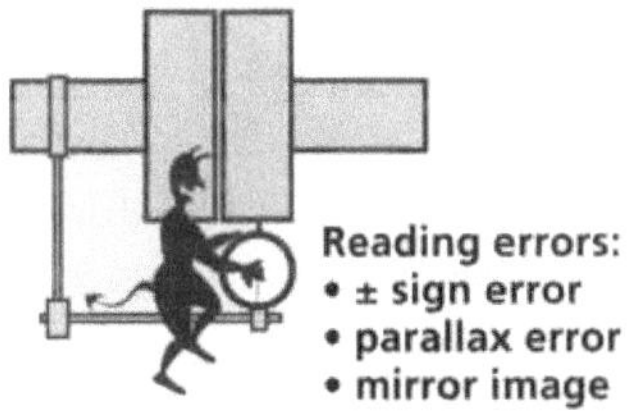

Reading errors:
• ± sign error
• parallax error
• mirror image

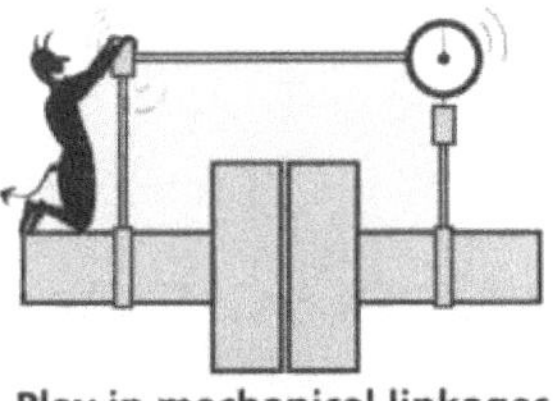

Play in mechanical linkages

Tilted dial indicator = offset error

Figure 3-38: Seven shortcomings of traditional alignment methods (Source: Prueftechnik)

Figure 3-39: Laser-optic shaft alignment in progress (Source: Ludeca, Inc., Miami, Florida 33172)

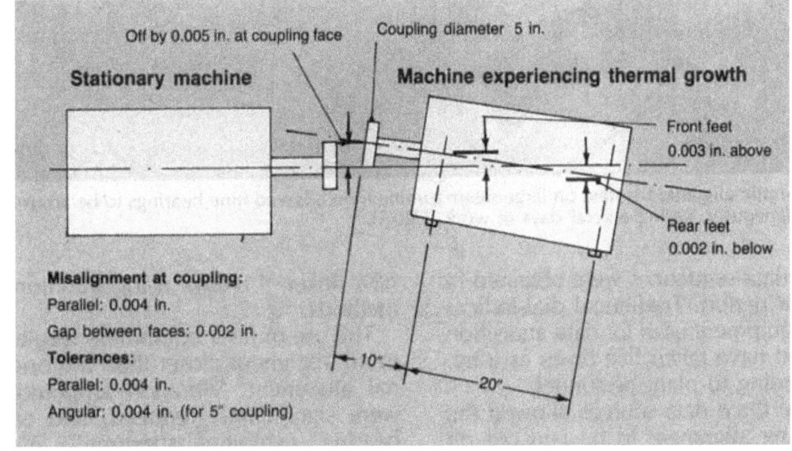

Figure 3-40: Small alignment correction error causing significant shaft misalignment

as "Permalign"—as will be discussed later—and older devices based upon proximity probes ("Dynalign bars," also called "Dodd bars").

Because rule-of-thumb expressions are somewhat imprecise estimates, plants have occasionally employed water-cooled pump support pedestals. Unfortunately, over the years, these water-cooled pedestals have proven to be more liability than advantage. Corrosion of the water-side has caused buckling, collapse, and catastrophic pump failures. In other instances, differential movement resulted in twisting of the top mounting plates leading, in turn, to pump-internal misalignment and component distress.

Reliability-focused pump users have disallowed water-cooled pump pedestals in since the mid-1950's when these facts first surfaced.

Conventional Wisdom: *Water-cooled pedestals keep pumps aligned*

Fact: *Water-cooled pedestals represent a serious safety and reliability risk. Alternative methods of ensuring pump alignment exist and should be implemented.*

Alignment Tolerances

As is often the case, there is little consensus among machinery manufacturers and users as to acceptable misalignment for pumps, or any other machinery for that matter. While it may be possible in some cases to have confidence in manufacturers' recommendations, we have often found these to reflect only the mechanical deflection capability of the coupling. Many manufacturers pay little, if any, attention to the resulting decrease in coupling life, or additional load-related decrease in bearing life, or vibration-related risk of inducing mechanical seal distress.

Suffice it to state that alignment tolerances are normally dependent on the three parameters coupling type, shaft speed, and distance between driver and pump shaft ends. That said, we believe a few "rules of thumb" and three alignment graphs will provide sufficient guidance for owner-operators who, for one reason or another, are unable to use advanced laser-optics. Truly reliability-focused pump users employ state-of-art laser optic alignment determination methods; they combine these with alignment verification devices. As will be described later, where both determination and verification are handled by well-designed laser optic devices, there are no longer any burning alignment issues.

Rule-of-thumb Thermal Offset Calculations

For pump applications, allow operation with 0.0005 inch maximum centerline offset per inch of shaft separation, or distance between shaft ends (DBSE). For a shaft separation of 24 inches (610mm), the allowable offset would thus be 0.012 in., or 0.3 mm.

Sliding Velocity Tolerances

Another approach for specifying tolerances is to describe the permissible limit of the velocity that the moving elements in a flexible coupling may attain during operation. This can be easily related to the maximum permissible offset and angularity through the expression in Eq. 3-1:

$$\text{Maximum allowable component sliding velocity} = 2 \times d \times r \times a \times \pi \qquad \text{(Eq. 3-1)}$$

where

d = coupling diameter
r = revolutions/time, and
a = angle in radians.

When offset and angularity exist, the flexible or moving coupling element must travel by double the amount of the offset and angularity, every half-rotation. Since the speed of rotation is defined, so must be the velocity that is achieved by the moving element in accommodating this misalignment as the shaft turns. In essence, when we limit the permissible sliding velocity, we have—by definition—also limited the offset and angularity (in any combination) that can exist between the coupled shafts as these turn. For 1,800 RPM, typically this limit is about 1.13 inches per second for excellent alignment, and 1.89 inches per second for acceptable alignment. A good laser alignment system will let reliability-focused users apply this approach as well.

The above guideline value (Eq. 3-1) is quite similar to that for gear couplings (Eq. 3-2) which is based on investigations of lubricant retention in the tooth mesh area (Ref. 3-9 and 3-11). Equation 3-2 recommends limiting the sliding velocity of engaged gear teeth to 120 inches (~3 m) per minute, or 2 inches (50 mm) per second. By way of additional explanation, this velocity could be approximated by the expression

$$\text{Vmax} = 120 \text{ in/min}, \qquad \text{(Eq. 3-2)}$$

where:

V = 2DN "tan alpha"
D = Gear pitch diameter, in.
N = Revolutions per minute
"2 tan alpha" = total indicator reading (TIR) obtained at hub outside diameter, divided by the

distance between indicator planes on driver and driven equipment couplings.

Thus, for equipment operating at 10,000 RPM with an 8 in. (~200 mm) pitch diameter coupling, and a 24 in. (610 mm) separation between indicator planes, the allowable offset would be based on the expression

$$2 \tan \text{alpha} = 120/[(8)(10{,}000)] = \text{TIR}/24$$

Hence, the permissible total indicator reading (TIR) = 0.036 in., and allowable offset would be 0.036/2, or 0.018 in (0.46 mm).

Another generalized guideline is graphically represented in Figure 3-41. It covers experience-based recommendations for pumps at speeds not exceeding 4,000 RPM. Figure 3-41 suggests, on the upper diagonal, barely operable allowable centerline offsets. Recommended target offsets are shown on the lower diagonal. Both diagonals are displayed as a function of the distance between flexing planes of spacer couplings.

For pumps operating at speeds higher than 4,000 RPM it would be wiser to recognize a speed dependency. This is reflected in Figures 3-42 and 3-43, and also in Figures 3-44 and 3-45.

Figure 3-42, a guideline used by many petrochemical companies, suggests that running misalignment be kept at or below the total reverse indicator readings (TIR) obtained from the expression shown on the illustration. After inserting L, P and shaft speed, (L = length between tooth centers—we could also say flexing planes—on coupling hub or spacer, and P = coupling tooth pitch diameter or distance between diagonally opposite bolts on dry couplings), one half of the resulting TIR would represent the allowable shaft centerline offset.

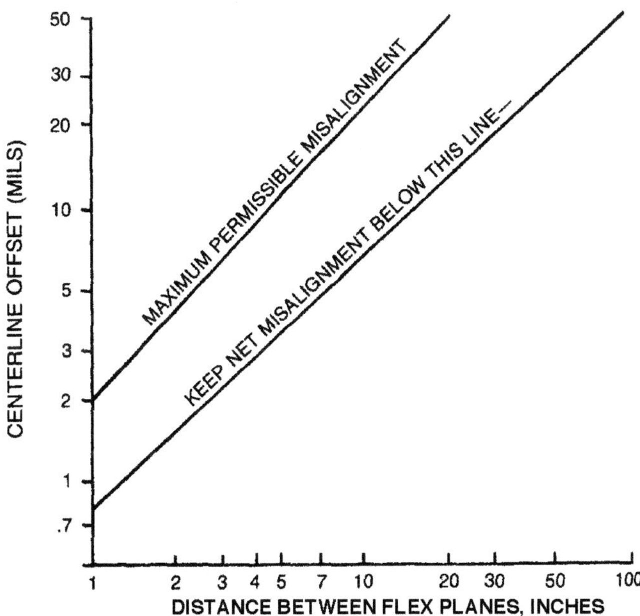

Figure 3-41: Misalignment tolerances for pumps operating at speeds below 4,000 rpm (Source: Bloch/ Geitner; "Machinery Component Maintenance and Repair," 2nd Edition, Gulf Publishing Company, Houston, Texas, 1990)

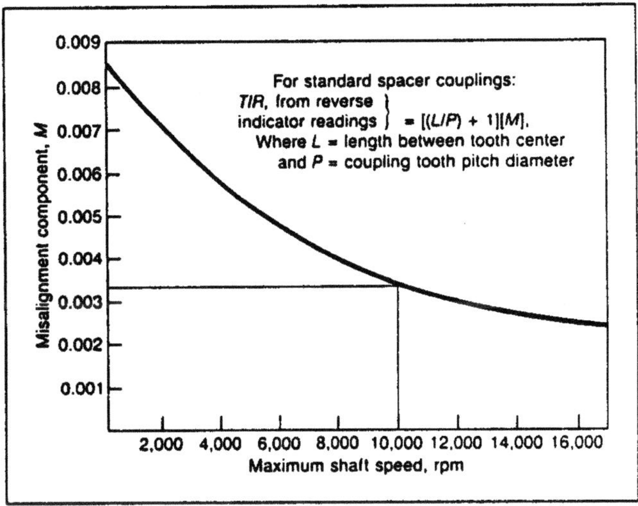

Figure 3-42: Alignment guidelines used by many petrochemical companies in the United States (Source: Ref. 3-7)

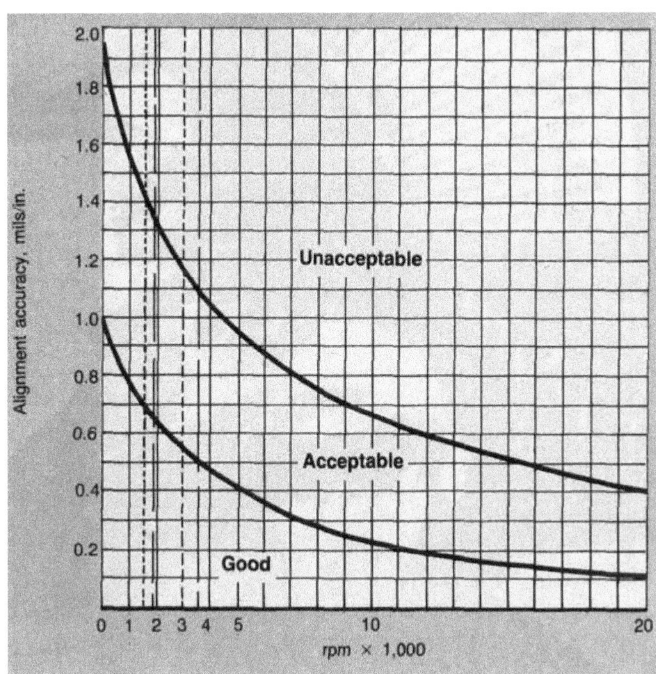

Figure 3-43: Alignment acceptability guidelines developed by John Piotrowski (Ref. 3-11)

The values developed by John Piotrowski (Figure 3-43) allow us to see experience-based as-new ("good"), acceptable (ok to keep in service) and unacceptable (in need of correction) misalignment values linked to speed (Ref. 3-11). Similar information, "excellent" vs. "acceptable" is conveyed in Figures 3-44 and 3-45 for short couplings and spacer couplings, respectively.

Laser-optic Devices

A state-of-the-art laser-optic alignment system emits a pulsating non-hazardous laser beam that automatically determines relative shaft positions and conveys this information to its microprocessor. The graphic display shows the numerical values for necessary alignment corrections (Ref. 3-12).

The advantages of modern laser-optic alignment devices (Figure 3-46) far outweigh the possible initial cost advantages of other, more conventional methods.

Well-proven Laser-optic Instruments

- Do away with dial indicator readings independent of mechanical bridging devices and are thus free of gravitational hardware sag problems
- Work with coupling in-place or uncoupled
- Use adjustable quick-fit brackets for fast and easy mounting on most every shaft diameter found on centrifugal pumps. Magnetic or chain brackets are available for vertical machinery or large diameters.
- Detect and measures extent of "soft foot" at machinery foundation, allowing user to correct condi-

Shaft alignment tolerances (short couplings)				
	Excellent		*Acceptable*	
RPM	Offset mils<>mm	Angularity (mils/in<>mm/m)	Offset mils<>mm	Angularity mils/in<>mm/m)
600	5.0<>0.12	1.0	9.0<>0.22	1.5
900	3.0<>0.07	0.7	6.0<>0.15	1.0
1200	2.5<>0.06	0.5	4.0<>0.10	0.8
1800	2.0<>0.05	0.3	3.0<>0.07	0.5
3600	1.0<>0.02	0.2	1.5<>0.03	0.3
7200	0.5<>0.01	0.1	1.0<>0.02	0.2

Figure 3-44: Shaft alignment tolerances for short couplings (Source: Prueftechnik Alignment Systems, D-85737 Ismaning, Germany)

Shaft alignment tolerances (spacer couplings)		
Angularity (angles α and β), or projected offset (offset A, offset B) expressed in mils per inch, also mm/m		
RPM	*Excellent*	*Acceptable*
600	1.8	3.0
900	1.2	2.0
1200	0.9	1.5
1800	0.6	1.0
3600	0.3	0.5
7200	0.15	0.25

Figure 3-45: Shaft alignment tolerances for spacer couplings (Source: Prueftechnik, D-85737 Ismaning, Germany)

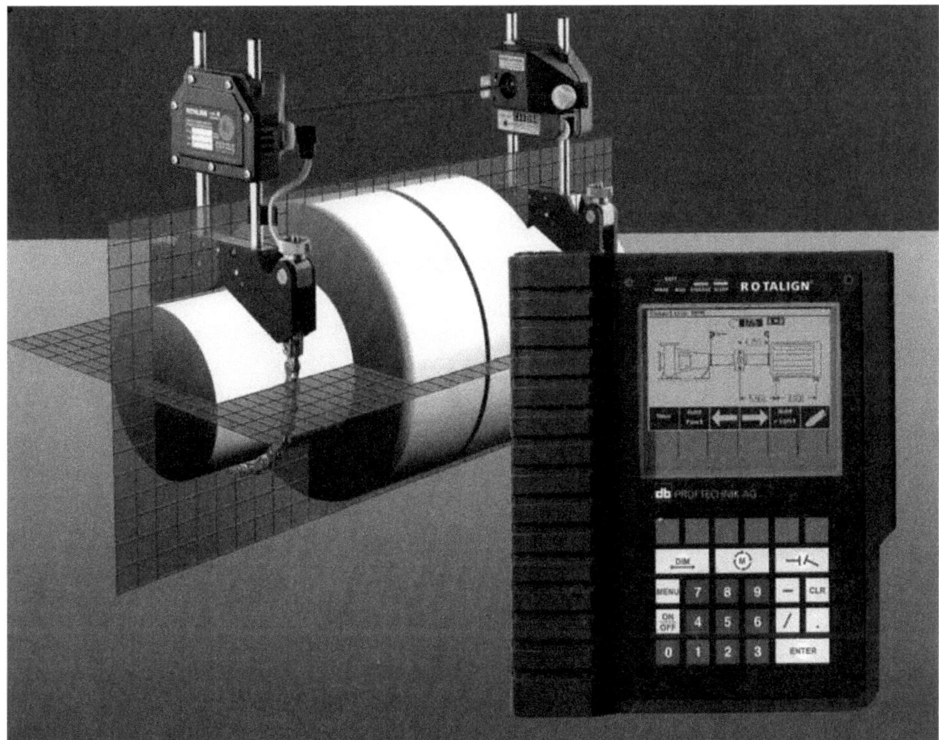

Figure 3-46: Modern laser-optic alignment system for rotating machinery (Source: Prueftechnik Alignment Sytems, D-85737 Ismaning, Germany)

tion in a fraction of time normally required.

- Permit entry of target specifications for thermal growth
- Feed shaft misalignment data directly into a microprocessor; display instructions for vertical and horizontal alignment corrections at front and back feet
- Display misalignment condition at rim and face of coupling or any other desired point along the shaft
- Require no more than 1/4 turn of the shaft to determine extent and location of alignment corrections needed
- Graphically guide user through procedure for corrective alignment moves without need for dial indicators
- Allow alignment if only one or neither shaft can be rotated, through use of special brackets
- Permit entry of bolt circle diameter and bolt spacing (angle) for alignment of vertical pump-motor sets and other vertically mounted equipment
- Are portable and independent of electrical power source; use normal "C" cell batteries.
- Determine dimensional concentricity and perpendicularity of equipment bore and surface dimensions at assembly, and during rebuilding or repair work.

Where Does This Leave Dial Indicator Methods?

Traditional reverse dial indicator alignment is recommended for facilities and at plant locations where laser-optic equipment is simply not available. In that instance, it makes eminent sense to purchase well-designed, sturdy, deflection-resistant brackets, as shown in Figure 3-47. A good carrying case, Figure 3-48, reduces the risk of damage to light-weight components and dial indicators. The vendor, A-Line Manufacturing Company, Liberty Hill, TX 78642, can provide training videos and associated reading material. Detailed and straightforward calculation or easy-to-learn graphic procedures can be used to determine alignment moves. These are also thoroughly explained in Ref. 3-10.

Not Acceptable: Tolerances Expressed as Corrections at the Machine Feet

While applying laser-optics with suitable software leaves little room for error, some plants cling to claims that alignment is sufficiently accurate as long as in the "cold," stand-still condition, the shaft centerlines are within 0.002 inches (0.05 mm) of each other. However, this approach is wrong. It is impossible to define the quality of the alignment between rotating shaft cen-

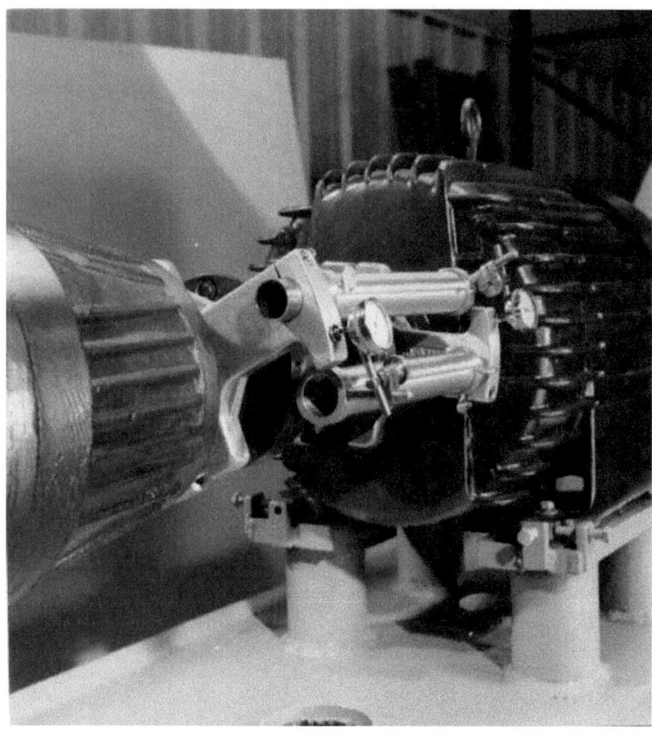

Figure 3-47: Mechanical dial indicator setup (Source: A-Line Mfg. Co., Liberty Hill, Texas 78642)

terlines in terms of correction values at the feet alone, unless one also specifies the exact dimensions related to these specific correction values, and does so each time (Ref. 3-9). This approach is too cumbersome and error-prone, since two machines will rarely share the same dimensions between the coupling and the feet, and between the feet themselves.

A tolerance that only describes a maximum permissible correction value at the feet without references to the operative dimensions involved makes no sense. This is because the same correction values can yield vastly different alignment conditions between the machine shafts with different dimensions. Such a tolerance simply ignores the effects of RISE over RUN, which is essentially what shaft alignment is all about. Furthermore, such a tolerance does not take into account the type of coupling or the rotating speed of the machines.

It should be noted that alignment "tolerances" specified *generically* in terms of foot corrections can have two equally bad consequences: the values may be met at the feet, yet allow poor alignment to exist between the shafts, or, these values may be greatly exceeded, while representing excellent alignment between the shafts! This means that, in the first scenario, the aligner may stop correcting his alignment before the machines are properly aligned, and in the second case may be misled into continuing to move machines long after they *have already arrived* in tolerance.

Let's take a look at an example that illustrates the fallacy of the generic foot correction approach. We will assume that the specified alignment tolerance for an 1,800 RPM machine is defined as a maximum correction value for the machine feet, ±2 mils (±0.05 mm). A machine is found to require a correction of –2 mils (–0.05 mm) at the front feet and +2 mils (+0.05 mm) at the back feet, therefore it is deemed by this method to be in tolerance. If the distance between the feet is 8 inches (200 mm), this would imply an existing angular misalignment of 0.5 mils per inch (0.5 mm/m).

If, say, the distance from the front bearing to the coupling center is 10 inches (254 mm), the resulting offset between the machine shafts at the coupling would be +7.0 mils (~0.18 mm)! This offset is considerably in excess of the ±3 mils (0.08 mm) offset considered maximum acceptable for an 1,800 RPM machine at the coupling. Yet, with the improperly specified foot correction tolerances, this alignment would—erroneously—be considered to be in tolerance!

This is a classic example where small correction values at the feet do not necessarily reflect good alignment at the coupling. Our earlier Figure 3-40 explains this quite well.

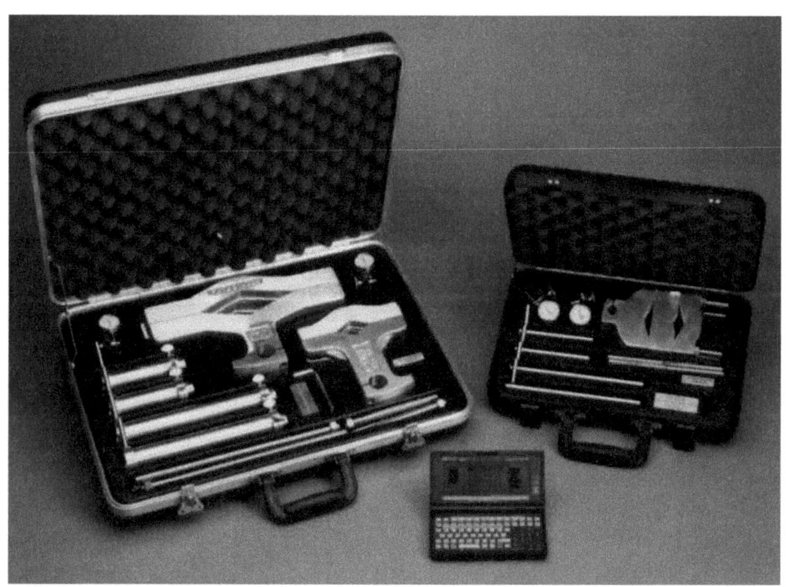

Figure 3-48: Reverse indicator devices and components in carrying case (Source: A-Line Mfg. Co., Liberty Hill, Texas 78642)

Laser-optics for On-stream Alignment Verification

As mentioned more than once in this text, one of the many significant factors influencing machinery reliability in process plants is alignment accuracy. When driven process machines operate in misaligned condition with respect to their drivers, equipment bearings are exposed to additional loads. Vibration severity may increase, bearings will be more highly loaded and equipment life expectancy will diminish. This is why there is hardly a major process plant that does not use some method to verify the initial, or standstill alignment between driver and driven shafts on critically important machinery.

Conventional Wisdom: *Flexible couplings compensate for misalignment*

Fact: *All flexible couplings allow misalignment forces to be super-imposed on normally acting bearing loads. Doubling the force reduces ball bearing life eight-fold; a three-fold load increase reduces ball bearing life to 1/27th of its normally anticipated operating life!*

Reliability-focused plants include pumps in their alignment program. Where they may have previously used careful measurements using dial indicators, they will now employ laser-optic devices for greater speed and precision of the initial alignment of pump and driver shafts in non-running condition. However, even the most knowledgeable pump users have frequently experienced machine movement after startup. Such movements can be caused by thermal, piping or baseplate related, and other mechanical or non-mechanical, static or dynamic load factors. Moreover, these movements are often of a transient nature and are thus extremely difficult and sometimes impossible to predict with purely analytical methods.

Consequently, alignment accuracy under operating conditions is often far from acceptable. Faced with inaccurate movement predictions many plants are simply not able to achieve the degree of running alignment accuracy and hence, equipment reliability that would otherwise be possible with sophisticated "cold," or standstill alignment techniques. These installations will benefit from continuous on-stream alignment monitoring (Ref. 3-13).

Operating Principles of Continuous On-stream Monitors

Since the 1990's monitoring of alignment changes has become routine for many reliability-focused pump users (Ref. 3-14). Laser optic on-stream alignment verification systems are used for this purpose. They are derived from their predecessor systems that use the laser optic technique for rapid and accurate shaft alignment in the standstill position. These on-stream or running alignment monitors thus represent an optical measurement system for the detection and display of relative positional changes between two points in terms of direction and magnitude. The operating principles of these novel measurement systems may be easily visualized using the following analogy.

When directed perpendicular to an ordinary mirror, a flashlight beam will be reflected straight back toward the flashlight. If the mirror then is tipped, the light beam is reflected in a different direction. The on-stream type laser optic device would detect and measure this change in the position of the mirror and calculate the exact magnitude and direction of the movement. This information is useful for monitoring any object to which the mirror is attached.

One such system, "Permalign," (Figure 3-49) can measure the absolute move of a machine or the relative movement between two coupled machines. It may be used for permanent monitoring or just the neces-

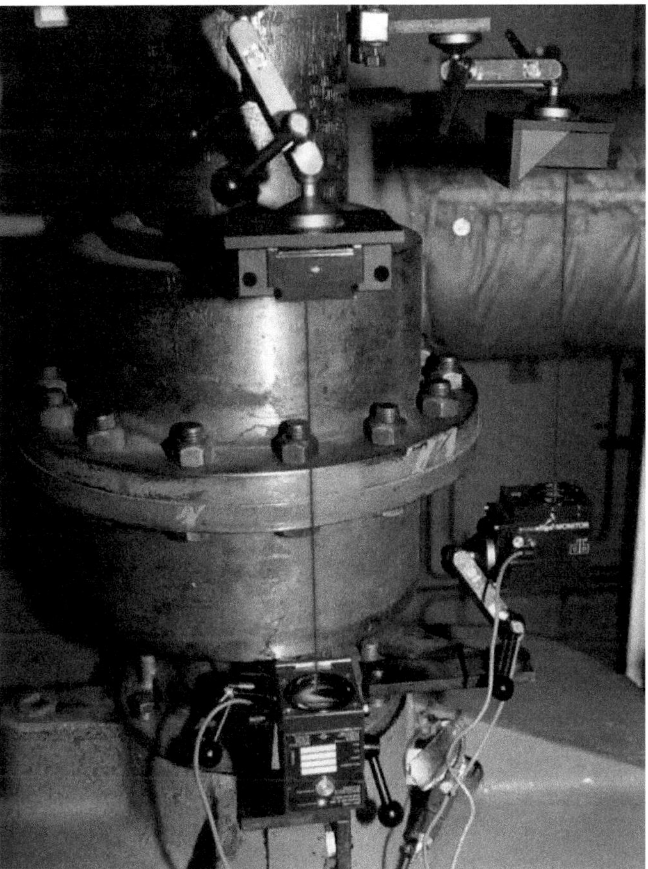

Figure 3-49: Laser-optic on-stream alignment verification system "Permalign" (Source: Prueftechnik Alignment Systems, D85737 Ismaning, Germany)

sary time to determine positional change from cold to hot and vice versa. The user may obtain a printout of numerical data, graphics and plot curves of positional changes (Figure 3-50). In contrast with other monitoring methods, "Permalign" allows the user to determine what movements have thermal origin and what movements are caused by dynamic influences, pipe strain or load variations, etc. It is thus possible to not only measure but also analyze the causes of measured movement.

Suffice it to say, the as-operating misalignment of otherwise well designed major process equipment is sometimes surprisingly severe and can result in considerable reductions of the mean-time-between failures (MTBF) of these machines. Laser optic running alignment monitoring, therefore, must be considered one of the more important pump reliability improvement tools (Ref. 3-15).

Some Final Comments on Piping and Flange to Nozzle Connections

As this chapter closes, the piping issue deserves to be revisited. Pump-to-driver alignment is difficult to maintain if there is pipe stress. Since piping is much stiffer than the typical pump casing to which it is being connected, casing-internal component misalignment is sure to result wherever pipe stress is excessive. Bearing inner rings will not remain concentric to bearing outer rings and money spent for better bearings will be wasted. The primary rotating components of mechanical seals will no longer be perpendicular to their respective mating stationary components. Money spent for superior mechanical seals would simply be wasted. All the care taken with pump foundations and baseplates will have been for naught where pipe stress is excessive.

Detecting Excessive Pipe Stress

Excessive pipe stress can be detected by a series of dial indicators with support stands temporarily placed on the baseplate. Each of these dial indicators is to be set at zero while touching the coupling hub and pump suction and discharge nozzles in the horizontal and vertical directions. During final flanging-up with the appropriate gaskets inserted and again during subsequent loosening of the flange bolts, none of the dial indicators should show movement in excess of 0.002 in (0.05 mm). A good pre-test would be for a maintenance worker to try pushing the disconnected pipe flange towards its mating pump nozzle with both hands while another worker is able to freely insert the flange bolts. If this pre-test is not successful, chances are the indicator deflection test will fail also.

Nozzle Connections

Proper pipe flange to pump nozzle bolt-up is also important. Unless the pump manufacturer gives precise procedural guidelines, the more typical installation and inspection procedure would call for the following:

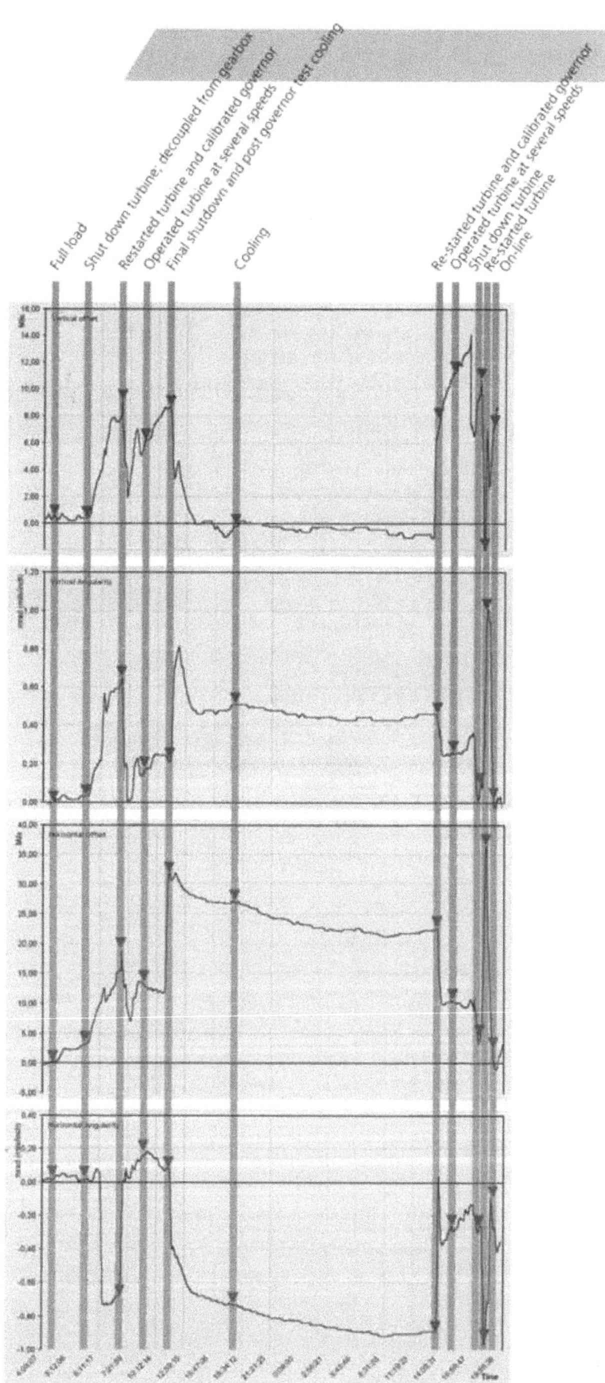

Figure 3-50: Tracking alignment changes of turbine-driven equipment showing severe operating misalignment (Source: Prueftechnik Alignment Systems, D-85737 Ismaning, Germany)

- Prior to gasket insertion, check condition of flange faces for scratches, dirt, scale, and protrusions (Ref. 3-10). Wire-brush clean as necessary. Deep scratches or dents (see Table 3-8 for acceptance criteria) will require re-facing with a flange facing machine.

- Check that flange facing gasket dimension, gasket material and type, and bolting are per specification. Reject non-specification situations. Improper gasket size is a common error.

- Check gasket condition. Only new gaskets should be used. Damaged gaskets (including loose spiral windings) should be rejected. The ID windings on spiral-wound gaskets should have at least three evenly spaced spot welds or approximately one spot weld every six inches of circumference (consult API-601 for exact specifications).

- Use a straightedge and check face flatness. Reject warped pipe flanges and pump nozzles.

- Check alignment of mating flanges. Avoid use of force to achieve alignment. Verify that the flange and nozzle faces are parallel to each other within 10 mils (0.25 mm) at the extremity of the raised face. Ascertain also that flange and nozzle centerlines coincide within 120 mils (3 mm).

Pipe flange to pump nozzle joints not meeting these criteria should be rejected.

Controlled Torque Bolt-up of Pipe Flange to Pump Nozzle Connections.

Experience shows that controlled torque bolt-up is warranted for certain flanged connections. These would typically include:

- All flanges (ratings/sizes) with a design temperature > 900°F (482° C).

- All flanges (all ratings) 12 in. diameter and larger with a design temperature > 650°F (343°C).

- All 6 in. diameter and larger 1500 pound class flanges with a design temperature > 650°F (343°C).

- All 8 in. diameter and larger 900 pound class flanges with a design temperature > 650°F (343°C).

- All flanges not accessible from a maintenance platform and > 50 ft (~15 m) above grade.

Table 3-8: Flange face damage acceptance criteria (Source: Ref. 3-10)

Gasket Type Used	*Damage*	*Critical Defect*	*Permissible Limits*
1) Ring joint	Scratch-like	Across seating surface	1-2 mils deep, one seating surface only
	Smooth depression		3 mils deep, one seating surface only
2) Spiral wound in tongue & groove joint	Scratch-like	>1/2 of tongue/groove width	1 mil maximum
3) Spiral wound in raised face joint	Scratches, smooth depressions & general metal loss due to rusting	>1/2 of seated width (min. of 1/4 inch intact surface left)	Up to 1/2 of serrated finish depth
4) Approved asbestos substitute	Same as above	> 1/2 of seated width	Same as above

For gasket types (1) and (2), resurfacing shall be required if more than 3-5 permissible defects are found. Seating surface is taken as center 50% of groove face.

In addition, it is reasonable to apply the above criteria to flanged connections on fluid machinery auxiliary equipment and other components such as:

- Valve bonnets, where the valve is positioned to include the above referenced design temperature/size/flange rating category.
- Flanged equipment closures where they qualify for inclusion in the above categories.
- All flanged connections which will eventually be covered with low temperature insulation within the above reference criteria.

Adherence to the following procedure is recommended for controlled torqueing of line flanges, bonnet joints, etc., when specified.

Preparation

Thoroughly clean the flange faces and check for scars. Defects exceeding the permissible limits given in Table 3-8should be repaired.

- *Check studs and nuts for proper size, conformance with piping and pump casing material specifications, cleanliness, and absence of burrs.*
- *Gaskets should be checked for size and conformance to specifications. Metal gaskets should have grease, rust, and burrs completely removed.*
- *Check pipe flange alignment. Out-of-alignment or lack of parallelism should be limited to the tolerance given in Figures 3-51 and 3-52.*
- *Number the studs and nuts to aid in identification and to facilitate applying criss-cross bolt-up procedure.*
- *Coat stud and nut thread, and nut and flange bearing surfaces with a liberal amount of bolt thread compound.*

Finally, on pump nozzles and piping that meet the elevated temperature criteria listed above, identify critical junctures and maintain records. One of many suitable record forms is shown in Figure 3-52. A suggested identification procedure is to use the line identification number and proceed in the flow direction with joints #1, #2, etc.

PUMP SUCTION STRAINERS

Suction strainers are often inserted in the piping just upstream of the pump suction nozzle. Not to be confused with product filters, suction strainers are used to protect pumps from unintended ingestion of tower packing, nuts, bolts, and other debris. While it would be smart to investigate why this stuff shakes loose, we will confine our comment to common principles and occasional misunderstandings about strainers.

Whenever strainers are used because the upstream equipment is flawed, be sure to understand the important requirements imposed on strainers by reliability-focused engineers. These engineers recognize, first and foremost, that a distinction is to be made between temporary and permanent strainers.

Temporary strainers are generally installed with the tip pointed in the upstream direction, which places at least some of the material in compression instead of tension. These temporary strainers must be removed about one week after commissioning the piping loop. They are often fabricated on-site, using the general configuration shown in Figure 3-35. However, variants do exist and are often constructed with support rods instead of the guard screen described below.

As a general rule, permanent strainer are designed to be left in place and must be cleanable without shut-

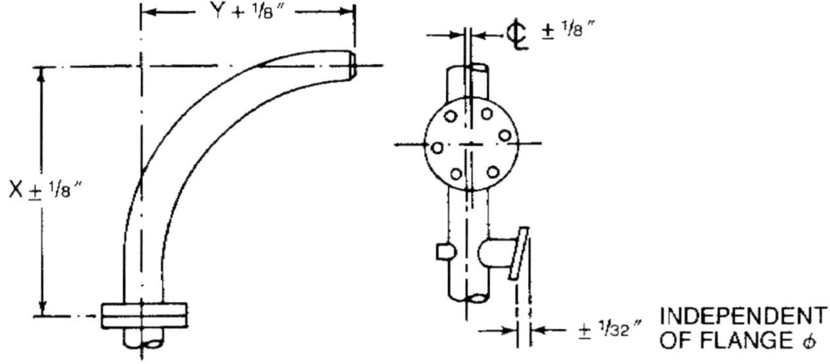

Figure 3-51: Dimensional variations permitted for piping and pipe flanges

DATE:____________________ CREW: ________________

RECORDER: ______________ LINE NO.: ________________

Joint No. (in Flow Direction Sequence)	Flange One		Flange Two		Joint Alignment (3)					Gasket Condition		Bolts			Remarks
	Face (1) Condition	Warpage Check (2)	Face (1) Condition	Warpage Check (2)	a	b	a-d	c	d	New Gasket?	Proper Type?	Any Defects	Lubri-cated?	Length Checked?	
1															
2															
3															
4															
5															
6															

Notes:

1) Note any scratches, dents, weld spatter or scale. Note if facing was wire brushed. Note if facing was refaced or any other repairs made.
2) Use a straight edge and check for any deviation of facing from straight edge.
3) See sketches.

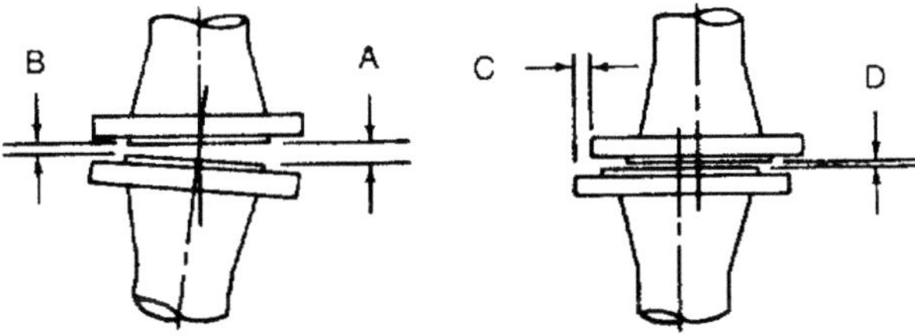

Figure 3-52: Typical pipe flange and flange-to-nozzle joint record form

ting down the pump. They are typically available from a variety of commercial sources, must be made of high-grade corrosion-resistant materials and, in the larger sizes, can be quite expensive.

Using the strainer guidelines found in Ref. 3-16,

1. Strainers (both temporary and permanent designs) may be cone or basket-shaped and shall be installed between the suction flange and the suction block valve. The preferred orientation is shown in Fig 3-53.

2. The mesh size of strainers shall be selected to stop all objects too large to pass through the pump main flow passage.

3. Temporary strainers shall be used during pre-startup flushing and initial (one week) operating periods, *unless permanent strainers* are specified.

4. Piping layout shall permit removal of strainers without disturbing pump alignment; spool pieces are typically used.

5. If *permanent* strainers are selected, the design and location of these strainers shall permit cleaning without removing the strainer body.

6. Arrangement of strainers shall permit cleaning without interrupting the pumping service.

7. For installations with permanent strainers and equipped with a spare pump, a permanent strainer shall be installed in the suction line of each pump.

8. Twin strainers or self-cleaning strainers may be used for pumps without spares.

9. Y-type strainers shall be restricted to 2-inch maximum size.

10. Suction lines for proportioning pumps shall be chemically or mechanically cleaned to permit operation without strainers.

There are some very important points we wish to re-emphasize.

First, best practices companies (BOCs) distinguish between *startup* strainers and *permanent* strainers. They insist on the removal of *startup* strainers long before they will have become a serious disintegration risk. Also, BPCs have established that strainers are *not* needed upstream of most conventional process pumps after the initial startup period.

Second, it follows that, once BOCs have determined strainers should be left in place for some reason, they allocate the resources needed to upgrade entire systems in order to reduce failure risk and maximize equipment uptime.

Third, because BOCs are serious about maximizing pump uptime, they insist on best practices being implemented at all times. At those facilities deviations from best practice have to be justified in writing and a manager is asked to accept responsibility in those instances.

Sizing of Strainers

Referring back to Figure 3-53, even an experienced reliability professional may find some customary terminology on *wire gauge* and *wire mesh* confusing.

Wire gauge (alternatively spelled "wire gage") is indicative of electrical conductor wire diameter, with primary implications for allowable current flow (amperes). According to working tables published for Standard Annealed Solid Copper Wire (Ref. 3-17), 20-gauge wire has a diameter of 0.032 inches and, at a temperature of about 40 degrees F, has a resistance of approximately 11 ohms per 1000 ft of length. While suction strainers have nothing to do with electric current, we assume that the same wire diameter (i.e., roughly 1/32 of an inch) also applies to the stainless steel wire used in wire mesh. For the pump suction strainer in Figure 3-53, an as yet undefined size of "mesh" is to be placed inside the three-mesh guard screen.

As regards terminology, wire mesh and wire cloth are terms used interchangeably in North America. Either term denotes a metal wire weave, or wire cloth, with its various implications regarding available flow-through area as a percentage of the total square-inches of wire cloth area. The notes on the illustration (Figure 3-53) recommend placing a *wire mesh made from 20-gauge wire inside a three-mesh guard screen*. So, to re-emphasize: While wire mesh made from 1/32 inch wire (20-gauge) is to be placed *inside a three-mesh* guard screen, nothing is said about 20-mesh wire cloth.

Next, an explanation of "mesh," with particular reference to the three-mesh guard screen, likely recommended to strengthen, or to back-up, an as yet unspecified finer mesh.

- "3-mesh" means wire cloth (or guard screen) on the sketch, Figure 3-53. That would be 3 wires per inch in the x-direction (horizontal), and 3 wires-per-inch in the y-direction (vertical). Similarly, 8-mesh would be 8 wires in the x-direction and 8 wires in the y-direction; or 200-mesh would be 200 wires in the x-direction and 200 wires in the y-direction, etc.
- In a 3-mesh screen there are, therefore, 3 x 3 = 9 openings. Moreover, in a 3-mesh screen, the distance from the center of one wire to the center of an adjacent wire is 1/3rd of an inch, or 0.333 inches.
- One of many manufacturers of wire cloth offers 3-mesh product made with wires ranging in diameter from 0.031" to 0.162." Irrespective of wire size, there would be 9 openings and the distance from the center of a wire to the center of an adjacent wire would always be 0.333 inches.
- If one were to pick the 0.162" diameter wire, one would have greater strength than if one chose the 0.031" wire.
- Likewise, if one picked the 0.162" wire diameter, the resulting open area would only be 26%, whereas if we had chosen the 0.031" wire diameter, we would have 82% open area (Ref. 3-18).

A reasonable choice would be a 3-mesh guard screen with a wire diameter of, say, 0.135." To re-iterate, the wire-to-wire center distance would be 0.333" and we would have 9 openings. For this wire diameter, the tables in Ref. 3-18 give an opening width of 0.198" (remember that opening width plus wire diameter equals 0.333"). Also, the tables tell us that we have ~35% of the area open for liquid flow and ~65% of the area would be taken up by the 0.135" diameter guard screen wires.

Next, one would determine the wire mesh of the cone-within-a cone, which was to utilize the 20-gauge wires. That particular wire cloth, to be placed inside the 3-mesh guard screen, should use 20-gauge—0.032"—wire. Ref. 2 shows this wire diameter about mid-way in its 6-mesh table. In 6-mesh wire cloth with 20-gauge wires, the openings are 0.1347." Adding the two numbers (0.032"+0.1347") = 0.1667"—six squares per inch.

Using the above explanations and Figure 3-53 will equip us to ask a draftsperson to design a pointed cone—the 3-mesh guard screen—with (based on experience) an open surface area in both guard screen and inserted finer mesh that's approximately 3 times the cross-sectional area of the pipe or spool piece. The 6-mesh metal cloth would go on the inside of the cone, at which time the device finally qualifies being called a strainer.

While it would be preferable to install the strainer such that the pointed tip of the cone encounters the flowing liquid first, it would still work with flow the other way around. But one would endeavor to ascertain the welder does the job properly and the materials of construction are picked in accordance with any corrosion concerns.

Regarding suction strainers it's worth recalling our basic premises: First, strainers are primarily installed to catch hard hats, welding rods, and beer bottles left in the piping. Second, strainers are intended to catch tower packing that dislodges from upstream equipment. Note that these different contingencies exist regardless of whether one is dealing with clean or dirty services. Experience shows the entire issue being more relevant during equipment commissioning. If the premise is different from what we have described here, there's probably a need for filtration equipment. In other words, strainers are not filters, and filters are not strainers.

Thus, permanent strainers would have the same mesh as temporary strainers. However, permanent strainers would be (a) made from corrosion-resistant materials; (b) if small enough, would be inserted in a "pipe-Y" with valved blow-down provision; or, (c) would be part of a redundant twin (parallel) installation.

On any permanent strainer installation, it would be wise to monitor the delta-p (the pressure drop) between the upstream and downstream sides of the mesh. Clogged strainers can cause serious machine malfunction and costly damage.

ECCENTRIC REDUCERS AND STRAIGHT RUNS OF PIPE AT PUMP SUCTION

Questions relating to the proper application of reducers in centrifugal pump suction lines date back many decades. Until his death (at age 84, in 1995), world-renowned pump expert Igor Karassik frequently commented on such issues (Ref. 3-19). In a sequence of letters to Karassik, pump users referred to Figure 3-54 and noted that this was quite typical of illustrations found in many textbooks. In essence, Figure 3-54 indicates that, with a suction line entering the pump in the horizontal plane, the eccentric reducer is placed with the flat at the top. Available texts often give no indication as to whether the pumpage comes from above or below the pump.

If the source of supply was from above the pump, the eccentric reducer should be installed with the flat at the bottom. Entrained vapor bubbles could then migrate back into the source instead of staying near the pump suction. If the pump suction piping entered after a long horizontal run or from below the pump, the flat of the eccentric reducer should be at the top.

Many older texts assume that the source of the pumpage originates at a level below the pump suction nozzle.

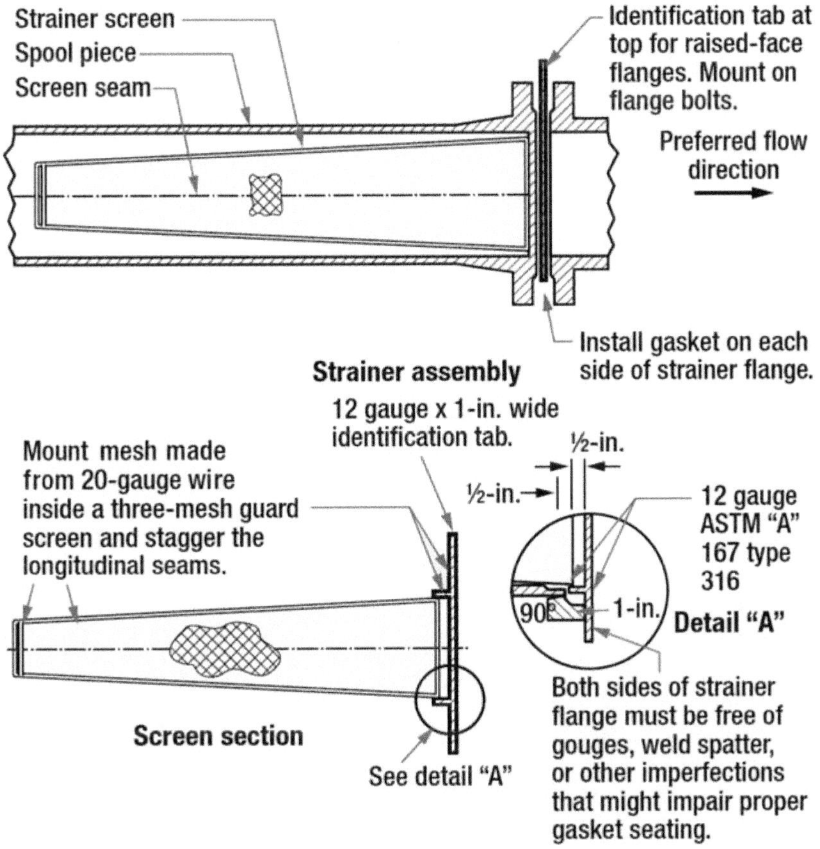

Figure 3-53: Typical temporary suction strainer for pumps

Also, older Hydraulic Institute Standards commented on the slope of the suction pipe:

> "...Any high point in the suction pipe will become filled with air and thus prevent proper operation of the pump. A straight taper reducer should not be used in a horizontal suction line as an air pocket is formed in the top of the reducer and the pipe. An eccentric reducer should be used instead."

This instruction applies regardless of where the pumpage originates. Depending on the particulars of an installation, trapped vapors can reduce the effective cross-sectional area of a suction line. Should that be the case, flow velocities would tend to be higher than anticipated. Higher friction losses would occur and pump performance would be adversely affected.

In the case of a liquid source above the pump suction and particularly where the suction line consists of an eccentric reducer followed by an elbow turned vertically upward and a vertical length of pipe—all assembled in that sequence from the pump suction flange upstream—it will be mandatory for the flat side of the eccentric reducer to be at the bottom. That said, Figure 3-55 should clarify what reliability-focused users need to implement.

Also, whenever vapors must be vented against the direction of flow, the size of the line upstream of any low point must be governed by an important criterion. The line must be of a diameter that will limit the velocity of pumpage to values below those where bubbles will rise through the liquid.

In general, it could be stated that wherever a low point exists in a suction line, the horizontal run of piping at that point should be kept as short as possible. In a proper installation, the reducer flange will thus be located at the pump suction nozzle and there is usually no straight piping between reducer *outlet* and pump nozzle. Straight lengths of pipe are, however, connected to the *inlet* flange of the eccentric reducer. On most pumps one usually gets away with 5 diameters of straight length next to the reducer. In the case of certain unspecified velocities and other interacting variables (e.g. viscosity, NPSH margin, style of pump, etc.), it might be wise to install as many as 10 diameters of straight length next to the reducer inlet flange. The two different rules-of-thumb explain seeming inconsistencies in the literature, where both the 5D and 10 D rules can be found.

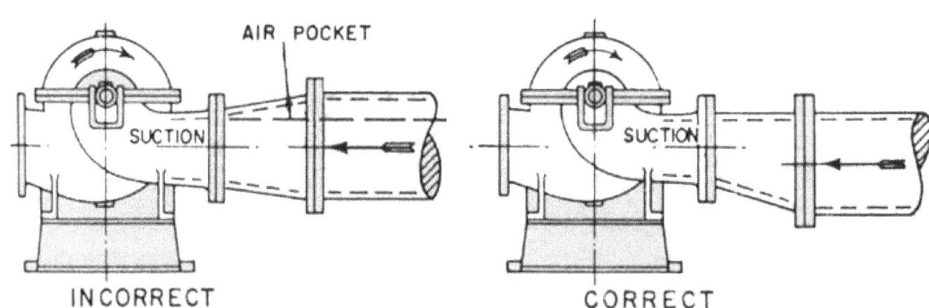

Figure 3-54. Illustration of eccentric reducer mounting from Hydraulic Institute Standards.

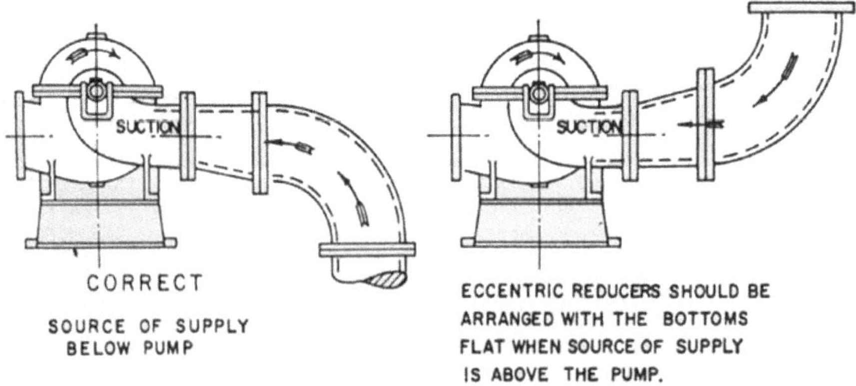

Figure 3-55. Suggested modifications for eccentric reducer mountings.

Chapter 4

Operating Efficiency Improvement Considerations

TOTAL SYSTEM EFFICIENCY

Proper pumping system design is the most important single element in minimizing Life Cycle Costs ("LCC"). Guaranteed or rated pump efficiency is only one component among many, and not necessarily the most important. However, when it comes to improving the operating efficiency of a pumping system, users often look for the pumps with the higher efficiencies, and favor the one which might exceed others by as little as 1/2 or 1 percent. All things being equal, there is some logic to this approach. But all things are seldom equal! To begin with, small differences in guaranteed efficiency may have been obtained at the expense of reliability, either by providing smaller running clearances or using lighter shafts, a point to note when comparing specifications and proposals. Moreover, the savings in power consumption obtained from a small difference in efficiency are rarely significant.

There are other, often more meaningful, savings to be realized by looking at the entire pump system, including the piping, fittings and valves before and after a pump, as well as the motor and motor driver. There can be multiple pumps, motors and drivers, and they can be arranged to operate in parallel or in series. Pump systems can have static head (pressure), or be circulating systems (friction only systems). Parallel pumps can be operated in a variety of combinations and flow rates. Pump operation can have a sizable impact on overall energy costs. Pump operation includes pump selection in parallel operation (selecting the most energy effective pumps), and operating the pumps at the most energy effective flow rates. Energy effectiveness refers to the input power requirement for a given system flow demand (gpm/kW or m^3/kWh). The greater the flow rate for a given power input the more energy effective a pump is for an application. This means that, for a specific flow rate, the lower the pump head, and/or higher the pump and/or drive efficiencies, the more energy effective the application. The maximum energy effectiveness for a constant speed pump always occurs at the highest flow rate that it can achieve in a system. This means that a lower efficient pump with a lower head could be more energy effective in a system than a more efficient pump with a higher head. The system may force the pump to operate far off the pump best efficiency, especially when oversized pumps are selected. System control may be achieved through large pressure breakdowns across a control valve (as opposed to variable speed drives).

ENERGY EFFECTIVENESS

Pump selection for parallel pump installations is key to minimizing energy costs. Many parallel pump systems aren't operated in the most energy efficient manner or combination. A new and more suitable pump selection and operating plan can, therefore, provide substantial cost savings opportunities. The way to achieve the optimum parallel pump selection is to calculate the Energy Effectiveness (gpm/kW) of each pump and then select the single pump or pump combination that yields the highest Energy Effectiveness. The Energy Effectiveness of a pump over its flow range can be determined from the following equation:

$$\text{gpm/kW} = 5{,}310 \times \text{Eff}_p \times \text{Eff}_d / H$$

where:

gpm/kW = Energy Effectiveness
Eff_p = Pump Efficiency
Eff_d = Drive Efficiency (motor + any VSD)
H = Head developed by pump (feet)

Figure 4-1 shows the results of using the above formula in a real-world installation handling potable water. Five different high specific speed pumps are investigated; in each case they would operate in parallel with additional pumps of the same type or style. As can be seen, the energy effectiveness of each pump increases with flow rate (for constant speed pumps), and is lower for the larger pumps, at a given flow rate. So as to make the optimum Energy Effective selection for a parallel pump installation, the following guidelines should be used.

1. Calculate the gpm/kW vs. Flow Rate curve for each pump and combination.

2. Operate the minimum number of pumps for any required system flow condition.

3. Select the highest Energy Effectiveness (normally lowest head/capacity) pumps for any flow requirement.

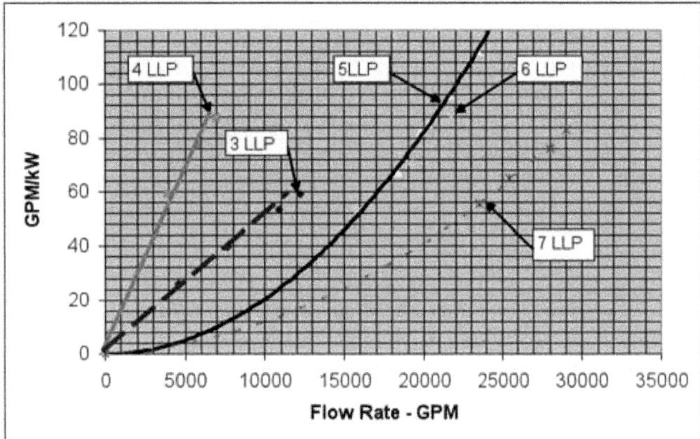

Figure 4-1: Energy effectiveness of five dissimilar pump styles. (Each style is operated in parallel with pumps of the same style)

Next, look at the chain of elements that are involved in the overall energy flow in a pumping system. Each element, and/or interface, adds some inefficiency. The goal should be to maximize the *overall* efficiency, which is the product of the efficiency of each of the system energy transfer or conversion devices. Note that the pump itself is only one of these elements.

- Electric utility feeder (higher voltages and power factors reduce these losses)
- Transformer
- Motor breaker/starter
- Adjustable speed drive (VFDs or variable speed drives can improve motor power factor)
- Motor (available in various efficiency ratings)
- Coupling
- Pump
- Fluid system (includes control valves and piping, which determine friction loss)
- Ultimate goal (flow rate at a given static head increase)

Computerized tools such as the U.S. Department of Energy, Office of Industrial Technologies, "Pumping System Assessment Tool" (PSAT) can be used to estimate pump system efficiency, based on a limited number of on-site measurements. PSAT assesses the overall efficiency of a pump system relative to its optimal performance and determines if further engineering analysis is justified to improve the pump and its system components and controls.

ENERGY OPERATING COST SAVINGS POTENTIAL

These days, most maintenance organizations are forced to operate with minimal manning levels. An effective way to maximize energy cost savings with these limited resources is to use the well-known 80/20 rule. Applied to pumps, this means that 80% of the savings can be achieved from 20% of the pump systems. A good way to identify the top 20% of the pump/systems which offer the greatest potential is to tabulate all pumps in a facility, 25 horsepower and above, along with the percentage of time each pump operates. Then, using Figure 4-2, determine the approximate annual energy cost for each of the pumps. As an example, the annual energy cost of a pump, which draws 250 horsepower and operates 60% of the time, is approximately $60,000. Experience with one large, older industrial plant has show that isolated energy savings can be as high as 70% of current costs, with average savings of 40% of current costs typically feasible. Energy costs should be adjusted to reflect the actual energy rates, since Figure 4-2 is based on an energy rate of $.05/kW-hr. Finally, the pumps should be sorted in descending order of annual energy cost. The top 20% of the pumps on this newly reordered list should represent approximately 80% of the total energy cost reduction potential for the facility. These are the pumps which should be further analyzed for possible energy cost savings.

Normally, field tests are required to determine the true current pump energy costs and actual savings potential for various options. Before conducting such testing, however, answers to the following questions will help further refine the pump selection list, and the energy savings potential as a percentage of current costs:

- Does the pump currently have a variable speed drive? Savings potential is normally small for a

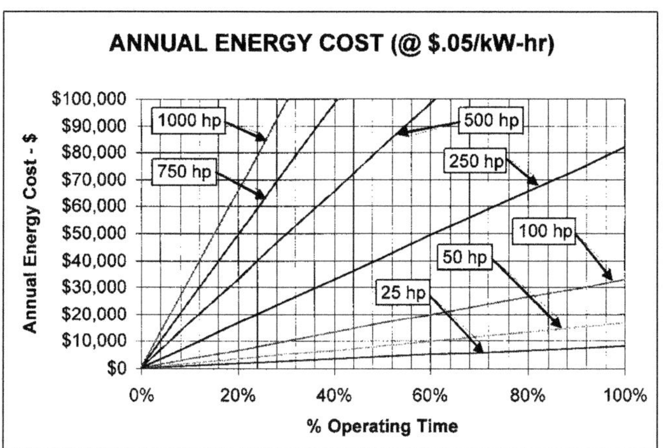

Figure 4-2: Annual Energy Cost/Maximum Savings Potential

pump that already has a VSD, unless the system head is comprised primarily of static head.

- Is the system performance curve comprised primarily of static head, with little friction head loss? It is hard to justify the cost of a variable speed drive on such a system, since speed reduction is limited. However, combining by-pass control with a VSD can lead to significant savings on high static head systems, especially for high specific speed pumps.

- Has the system demand been reduced since the pump was initially installed? If the answer to this question is yes, it probably means that there is excess throttling of the control valve and wasted energy.

- Is the pump control valve normally set at less then 80% open? This means that energy may be saved by trimming the pump impeller(s).

- Are there multiple pumps operated in parallel? Parallel pumps are often not operated in the most optimum manner and combinations, allowing opportunity for energy cost savings from pump selection.

- Are high specific speed pumps regulated with control valves? Pump input power demand and pump head increase quickly, as the flow rate is reduced for high specific speed pumps. This further increases the power loss when these pumps are throttled to lower pump capacities. By-pass control is often better for these pumps.

- Has the pump performance fallen off, or has it been a long time since the last pump overhaul? When pump degradation is in the form of opening up of internal hydraulic clearances (which can increase pump input power for a given net flow rate) energy savings will result from renewing these clearances. It's different when pump degradation is caused by roughening of the internal hydraulic passages. Reconditioning the pump through sand/shot blasting, and/or coating the casings and impellers can result in significant increases in pump flow and efficiency. However, in cases where the current flow rate is sufficient, unless reconditioning of the pump is coupled with an impeller trim, or the pump flow is controlled by a variable speed drive, little energy savings will result. Any gain in flow rate would then be lost across the control valve. Pump reconditioning may, however, still be required to meet system goals regardless of energy savings.

- Does the pump currently have an "Energy-efficient" or "Premium-efficiency" motor? Normally it is difficult to justify the cost of one of these high efficiency motors as a replacement, based solely on the improvement in motor efficiency. The reason is that high efficiency motors operate at higher speeds than standard efficiency motors. This higher speed increases the pump head, along with increased pump flow, which means that the control valve must be throttled back, if the system objective is to maintain the same flow rate. This would destroy some of the efficiency savings from the high efficiency motor. Energy-efficiency and premium-efficiency motors are easier to justify when the current motor is being replaced for some other reason, such as failure of the current motor, or replacement due to significant over-sizing or under-loading.

- Are low power factor motors (especially below 60% to 70%), and/or over-sized motors in service? Power factor correction (motor running) capacitors can increase the motor power factor to 95%, which reduce utility power factor penalty charges (due to the reduction of reactive current flow from the electric utility, through cables and motor starters). This reduces heat dissipation from these kW losses.

- Is the control valve under-sized, or does it have a high pressure drop (low CV) design, such as a globe or low CV butterfly valve? Properly sized ball control valves can result in sizable energy savings. Field testing may be required to determine the actual valve pressure drop/CV.

SYSTEM EFFICIENCY

It is essential to understand the goal of the fluid system to optimize it. Understand why the system exists, have clearly defined criteria for what the system must accomplish, and understand what is negotiable and what is not. There is a lot to gain by looking at the pump system efficiency, which looks at system static head as a system goal, and piping friction as an efficiency loss. The equation for the system efficiency is:

System Efficiency =
(Static Head)/(Static Head + Friction Head) Eq. 4-1

As an example, a system that has a static (elevation/pressure) head of 200 feet, and piping friction head loss of 200 feet would have a system efficiency of only 50%. Although there may be few piping changes which can be financially justified in an existing system, except possible valve changes or replacement with a variable speed drive, system efficiency should be given major attention during the initial system design.

System piping

Operational costs are strongly influenced by the selection of system piping components and their size. Such selection is important to determine the friction losses in a piping system, for sizing the pumps, reducing wasteful energy consumption, and to ensure that sufficient NPSHa is provided to the pump for an adequate NPSH margin.

Flow rate

When the flow rate increases, the flow velocity increases and so does the friction or resistance to flow caused by the liquid viscosity. The head loss is approximately related to the square of the velocity, so it will increase very quickly.

Pipe Inside Diameter

When the inside diameter is made larger, the flow area increases and the liquid velocity for a given capacity is reduced. When the liquid velocity is reduced there is lower head loss in the pipe. Conversely, if the pipe diameter is reduced, the flow area decreases, the liquid velocity increases and the head loss increases. The piping diameter should be selected based on the following factors:

- Economy of the whole installation (pumps and system)
- Required minimum flow velocity for the application (to avoid sedimentation)
- Required minimum internal diameter for the application (solids)
- Maximum flow velocity that will still provide the necessary available net positive suction head to the pump, and/or acceptable erosion in the piping and fittings
- Maximum flow velocity limits established by industry standards and specifications. The Hydraulic Institute specifies 8 ft/sec for the maximum suction pipe velocity and 15 ft/sec as the maximum discharge pipe velocity.
- Plant standard pipe diameters

Increasing the pipeline diameter has the following effects:

- Piping and component procurement and installation costs will generally increase, although fewer pipe supports may be needed for the larger diameter (hence, stiffer) pipe. Note that stiffer piping increases pump flange loading caused by piping weight and flange alignment tolerances.
- Pump installation procurement costs will decrease as a result of decreased flow losses with consequent requirements for lower head pumps and smaller motors. Costs for electrical supply systems will therefore decrease.
- Operating costs will decrease as a result of lower energy usage due to reduced friction losses.

Some costs increase with increasing pipeline size and some decrease. Because of this, an optimum pipeline size may be found, based on minimizing life cycle costs over the life of the system.

Length of Pipe

Head loss occurs all along a pipe. It is constant for each unit length of pipe and must be multiplied by the total length of pipe used. Figure 4-3 gives an example of how piping friction loss (pressure drop) changes with varying flow rate and pipe size, per 100 feet (~30.5 m) of pipe. At 5,000 gpm, pipe friction loss can be reduced from 4.8 ft to 1.4 ft/100 feet (a 70% drop) by increasing the pipe size from 12 inch to 16 inch (Ref. 4-1).

This reduction in pipe friction loss many not seem significant until we look at the annual cost of this energy loss, Figure 4-4. This annual cost is per 100 feet of pipe length and assumes an 80% combined pump and motor efficiency and $.10/kWh energy cost. As an aside, the cost of energy in parts of the West Coast of the United States has exceeded $ 0.14 per kWh in 2005.

Fittings

Elbows, tees, valves, strainers and other fittings will impede the liquid flow. Whenever the free flow of liquid is impeded or disrupted, it will suffer a head loss. Head loss values for most fittings have been tested and are published as "K" factors in such publications as the Hydraulic Institute Standards, based on the velocity head or equivalent straight pipe lengths.

Considerable system pressure loss is caused by valves, in particular control valves in throttle-regulated installations. Figure 4-5 shows the pressure drop from three different types of 12-inch control valves, at a flow rate of 2,500 gpm (577 m^3/hr) as a function of valve opening. It should be noted that valve type can have a significant impact on system head loss (energy cost), with the linear globe valve having the highest loss, followed by the butterfly, and the ball valve giving the lowest value. Control valves can have a marked negative impact on total pump system efficiency, which could justify the initial purchase cost of a variable speed drive.

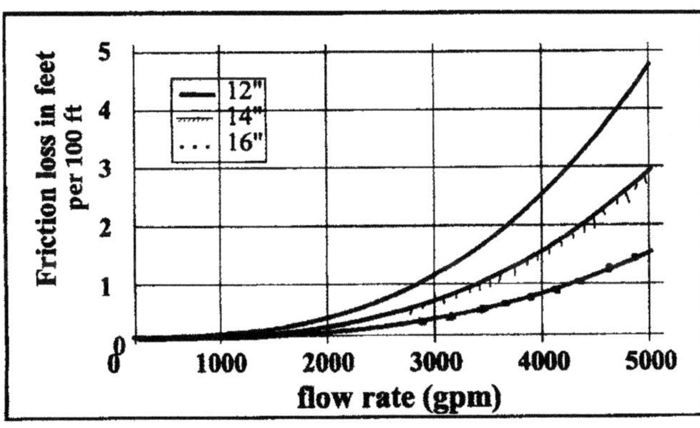

Figure 4-3: Effect of pipe size on friction loss

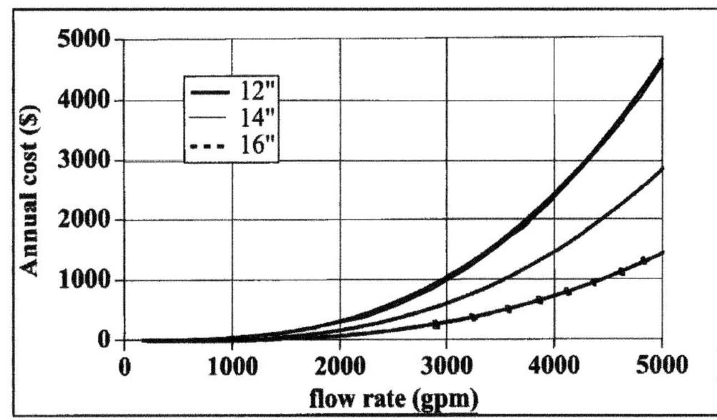

Figure 4-4: Annual friction loss cost per 100 feet of pipe

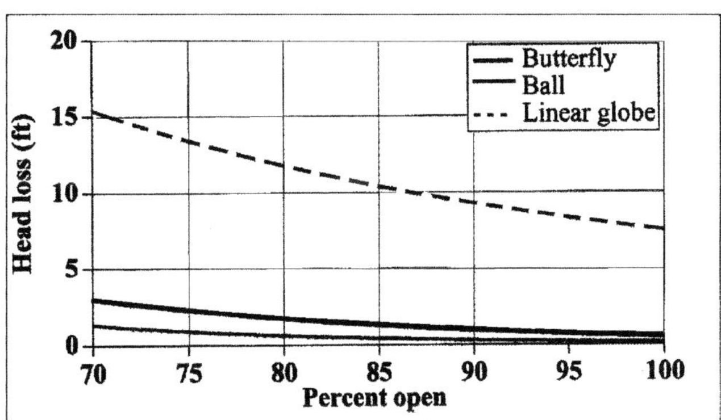

Figure 4-5: Pressure drop for 3 types of control valves

Figure 4-6 shows the annual cost of a 12-inch check valve and full open butterfly valve as compared to Schedule 40 new pipe for flow a rates to 2,500 gpm (577 m^3/hr), again assuming an 80% combined pump and motor efficiency and $0.10/kWhr electrical cost.

System Head, Power and Cost Formulas

Since the major friction loss in many pump systems is caused by the control valve, it is useful to apply the

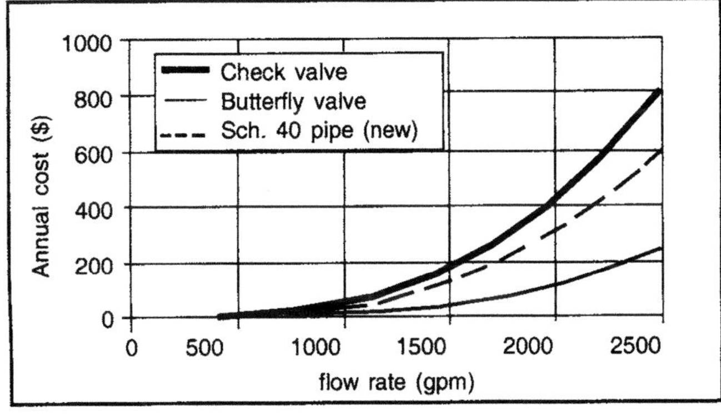

Figure 4-6: Annual cost of two 12-inch fittings vs. 100 ft of Schedule 40 pipe

valve flow coefficient CV (loss constant) to the entire system. The formula for the valve and/or system CV is:

$$CV = Q/(dP)^{0.5} \qquad \text{Eq. 4-2}$$

Where:

Q = Flow Rate (GPM)
dP = Valve or System Pressure Drop (psi)

Another way to look at the flow coefficient (CV) is that it is the flow rate that will cause a one psid pressure drop across the valve or system. According to equation 4-2, the lower the pressure drop the higher the CV valve. Therefore, to improved system efficiency, control valves and piping should be selected with the highest CV values when 100 percent open. Figure 4-7 shows the shape of typical "CV vs. % Open" curves for various valve types. Gate valves (Figure 4-8) are generally used for quick open applications, although they can and are used as control valves due to their relatively high 100% Open CV values. Ball valves have the highest 100% open CV values (Figure 4-9), but do come with higher price tags. Butterfly valves tend to have lower values of full open CV (Figure 4-10), but are often used for control valves due to their low cost. Globe valves are seldom used as control valves because of their higher cost and much lower CV valves (Figure 4-11). Control valves are often offered with different trims which can change the !00% open CV values and the shape of the CV characteristic (vs. percent valve opening). Actual valve CV values can be obtained from the valve manufacturer, or from field tests.

Taking the CV concept to the next step, and applying it to the entire pumping system, including the piping, control valve and other system hydraulic frictional components, yields a similar characteristic to the valve CV curves. Typical system "CV" flow coefficients are plotted in Figure 4-12. System characteristic curves can be developed from field testing, or by calculating the total system pressure drops by adding the individual component pressure drops, including the control valve value. The control valve may or may not dominate the system CV flow characteristic depending on the valve type and system design.

The system "dP" is basically the system friction head, except that the unit of measure is psi instead of feet. The formula for the total system head curve, in terms of "CV" can, therefore, be obtained as follows:

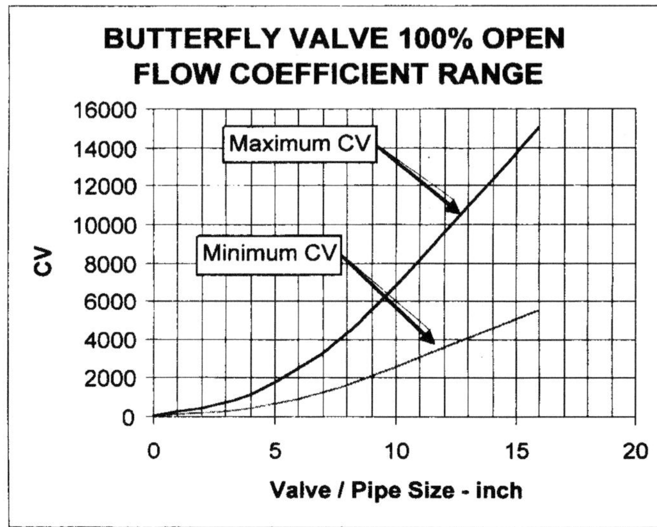

Figure 4-7: Typical control valve "CV" flow characteristics

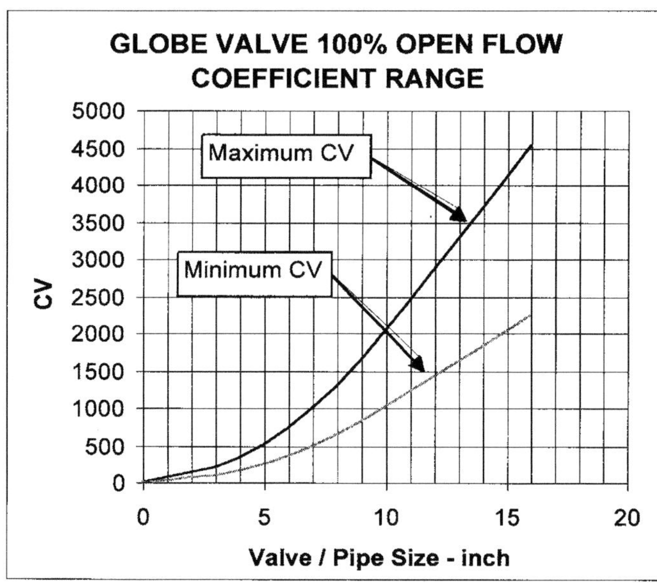

Figure 4-8: Gate valve 100%-open flow coefficient range, by pipe size

$$H_{Friction} = 2.31 * dP/S.G. \qquad \text{Eq. 4-3}$$

$$dP = (Q/CV)^2 \qquad \text{Eq. 4-4}$$

$$H_{friction} = 2.31/S.G. * (Q/CV)^2 \qquad \text{Eq. 4-5}$$

$$H_{System} = 2.31/S.G. * (Q/CV)^2 + H_{Static} \qquad \text{Eq. 4-6}$$

Where:

H_{System} = Total system head (feet)
$H_{friction}$ = System friction head (feet)

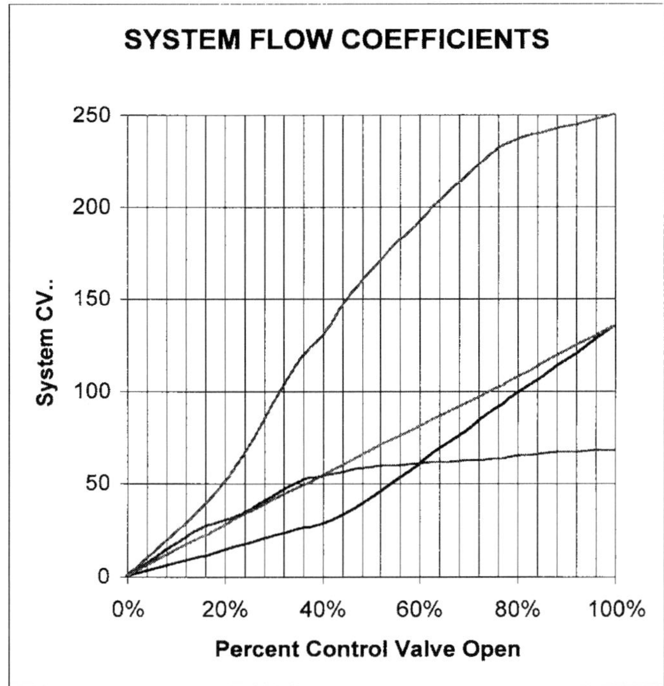

Figure 4-9: Ball valve 100%-open flow coefficient range, by pipe size

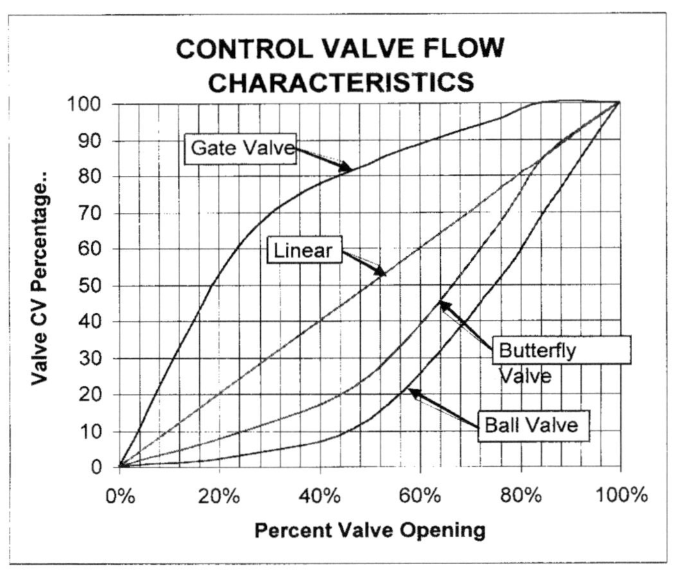

Figure 4-10: Butterfly valve 100%-open flow coefficient range, by pipe size

H_{Static} = System static head (feet)
S.G. = Specific Gravity
CV = System CV constant
Q = Flow rate (gpm)

If now, we use the above total system head formula (equation 4-6) to the linear system flow characteristic

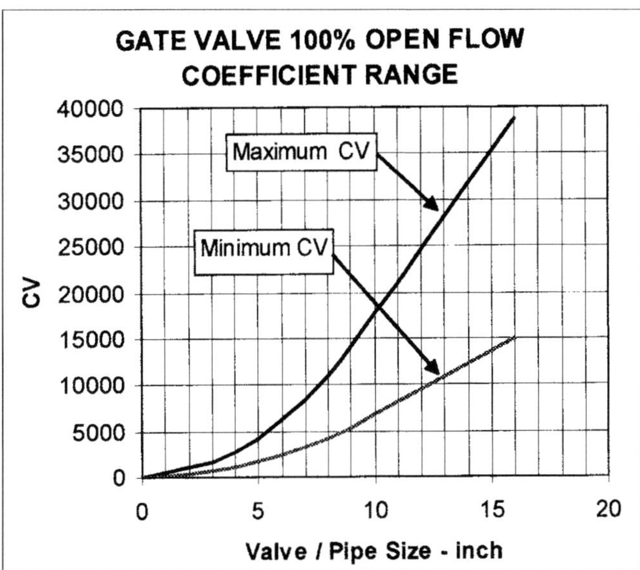

Figure 4-11: Globe valve 100%-open flow coefficient range, by pipe size

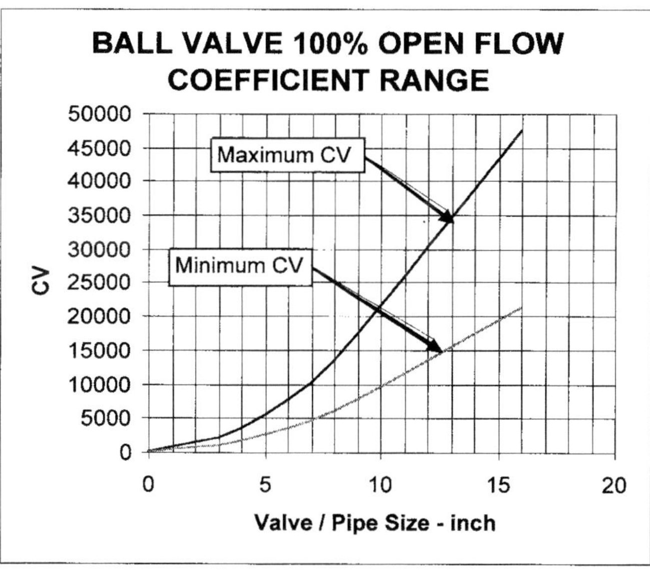

Figure 4-12: Typical System "CV" Flow Coefficients

shown in Figure 4-7, for 50%, 80% and 100% valve opening, we get the following set of head curves, shown in Figure 4-13. This curve shows the affect of valve throttling on the system curve, and when superimposed on the pump head-capacity curve, will show the effect of valve throttling on the resulting flow rate, Figure 4-14.

Equation 4-4 can also be used to approximate the cost and savings potential from throttling control valves for various system/valve CV ratings. If we combine equation 4-4 with the formula for pump input horsepower (equation 4-7), the formula for driver input power (equation 4-8), and energy cost formula (equation 4-9),

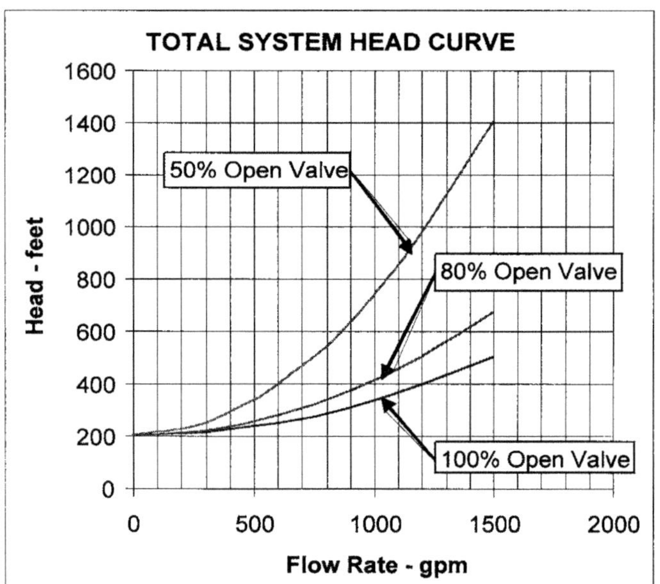

Figure 4-13: Total linear system head curve example (CV100% = 135)

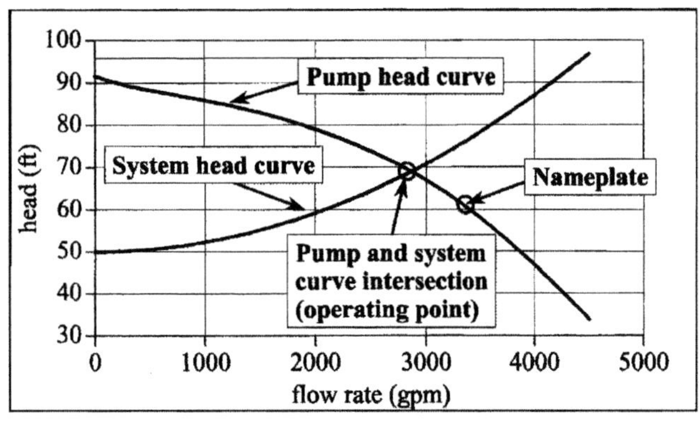

Figure 4-14: System impact on pump performance

we get the formula for the system energy cost in terms of system/valve CV (equation 4-10).

$$HP_{in} = Q * H_{friction} * S.G./3960/Eff_p \qquad \text{Eq. 4-7}$$

$$kW_{in} = .7457 * HP_{in}/Eff_d \qquad \text{Eq. 4-8}$$

$$Cost_{energy} = Rate * kW_{in} * Hrs \qquad \text{Eq. 4-9}$$

$$Cost_{energy} = .7457*Rate*Hrs*Q^3/(1714/Eff_p*Eff_d*CV^2) \qquad \text{Eq. 4-10}$$

Where:

HP_{in} = Horsepower in (hp)
Eff_p = Pump Efficiency – %
Eff_d = Driver Efficiency – %
kW_{in} = Driver input power – kW
Hrs = Operating hours (hr)
Rate = Electrical Energy Cost rate – $/kW-hr

In order to use equation 4-10 to determine system potential savings we need to make certain assumptions, such as the pump efficiency, driver efficiency, energy cost rate, operating hours, and the amount of control valve throttling. For estimating purposes, we have made the following assumptions for this approximation. If actual values are available, costs should be adjusted accordingly:

- Rate = $.05/kW-hr
- Hrs = 8,760 hours (100% operation)
- Pump Efficiency (Eff_p): Hydraulic Institute optimum obtainable efficiency for 1,500 specific speed pumps. End suction ANSI pumps were used for the "Low Flow" efficiency values (Figure 4-15), and double suction pumps were used for the "High Flow" efficiency values (Figure 4-16). It should be noted that field experience has shown that actual pump efficiencies are almost always less than these assumed optimum values, sometimes as low as 50% of these values. This could double the savings calculated in the following examples.
- Drive Efficiency (Eff_d): Standard efficiency electric motors are assumed, with "as-new" efficiencies.
- System CV Constants: The savings shown in the following figures assume that the average system flow demand results in a system CV that is 50% of the full open (100%) CV values, and that the optimum situation would be a 100% open valve. Obviously, if less valve throttling is required, on average, energy savings will be less. A 50% CV value results is four times the energy cost of a 100% open control valve system.

Figures 4-17 shows the potential energy savings for low flow applications, using end suction pumps, and with (100% open) system CV values of 100, 200 and 300. Figures 4-18 shows the potential energy savings for high flow applications, using double suction pumps, and with (100% open) system CV values of 1,000, 2,000 and 3,000. As can be noted, the savings can be substantial, especially with the lower CV values at the higher flow rates. In order to achieve all are part of these savings, the system CV should be increase by:

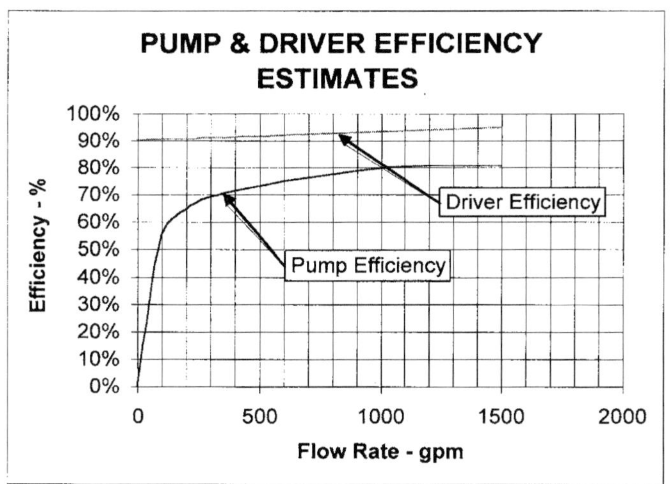

Figure 4-15: Low-flow pump and driver efficiency estimates (Ref. Hydraulic Institute Standards—ANSI end suction pumps)

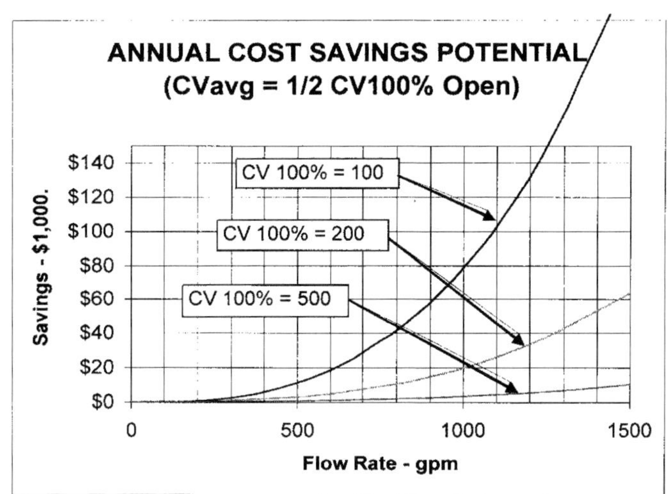

Figure 4-17: Low-flow annual cost savings potential (assuming average system CV equals half of the system CV with control valve 100% open)

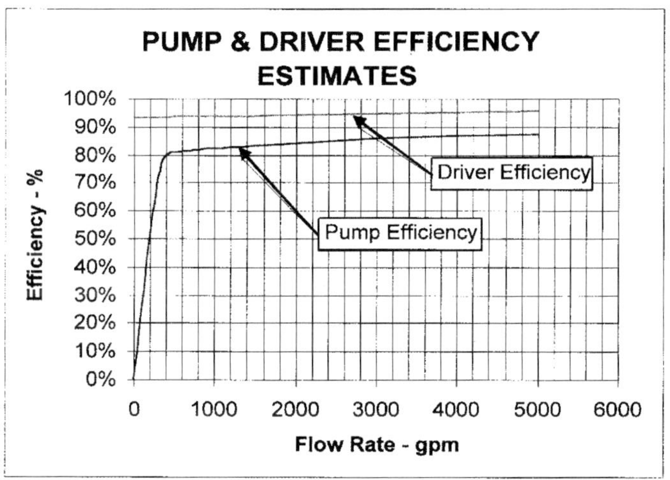

Figure 4-16: High-flow pump and driver efficiency estimates (Ref. Hydraulic Institute Standards—double suction pumps)

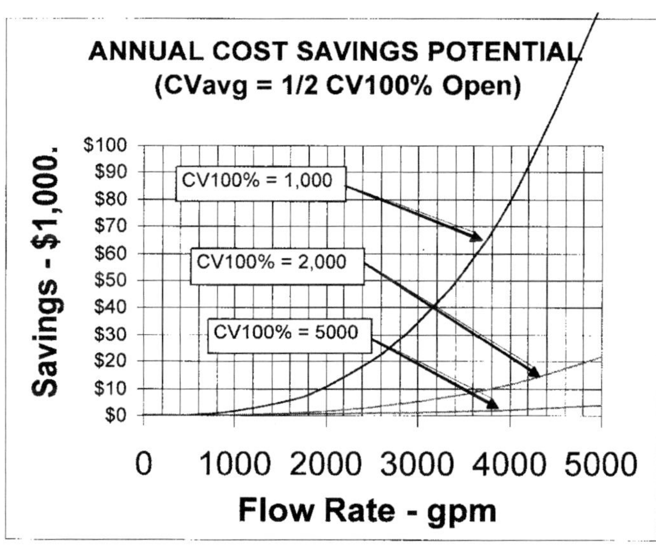

Figure 4-18: High-flow annual cost savings potential (assuming average system CV equals half of the system CV with control valve 100% open)

- Changing the control valve (if feasible).
- Trimming the pump impeller, if the maximum system requirements do not required the current impeller diameter for maximum demand (a trimmed pump impeller would require less valve throttling).
- Replacing the control valve with a variable speed drive.

Changes in Flow Rate Demand

It is important to determine the flow demand, in order to accurately calculate the potential savings and pump performance requirements. Flow rate demand may fluctuate on hourly, daily, weekly, or monthly bases (see Figures 4-19 and 4-20. To properly evaluate a pump system for potential energy savings, the historical or estimated future, flow demand range should be ascertained. The average, minimum and maximum flow demand rates can than be determined. The average flow demand can be used to determine current energy costs and potential savings, as discussed earlier. The minimum and maximum demand requirements establish the pump size, impeller trim diameter, and the potential for a variable speed drive.

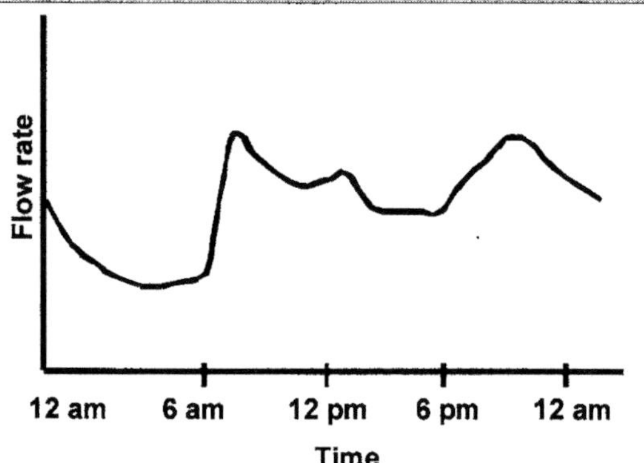

Figure 4-19: Daily flow fluctuation example

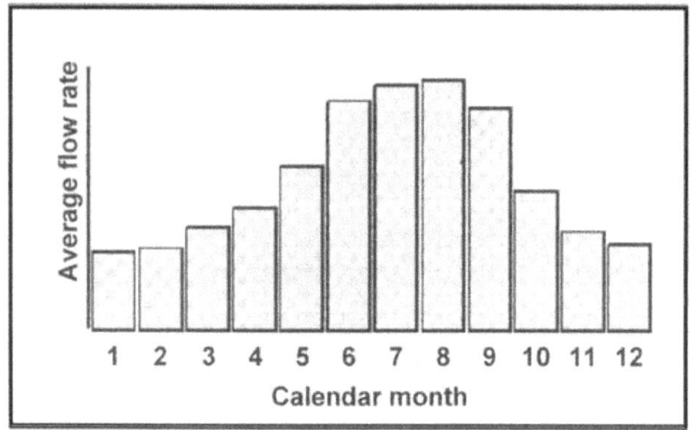

Figure 4-20: Annual flow fluctuation example

System Impact on Pump Performance

The point at which a pump operates on its performance curve is dependent on the intersection between the pump head-capacity curve and the system curve, Figure 4-14. Therefore, when planning for energy conservation, the system should be designed—or pumps selected—to force the pumps to operate close to their best efficiency flow rate. Avoid wasting power by oversizing pumps. Figure 4-21shows an over-sized pump application that results in an operating efficiency of 58%, compared to the pump BEP (best efficiency point) efficiency of 77%.

One of the most serious wastes of power involves the traditional practice of over-sizing a pump by selecting design conditions with excessively "conservative" margins in both capacity and total head. This practice can lead to the strange situation in which a great deal of attention is paid to a small 1/2 or 1 percentage point gain in pump efficiency, while ignoring a potential power savings of as much as 15 percent, through an overly conservative attitude in setting the required conditions of service. Figure 4-22 shows the loss of efficiency for different specific speeds. As can be seen, the lower the specific speed N_s the greater the efficiency loss at part load (capacity) performance.

It is true that some margin should always be included, mainly to compensate for the wear of internal clearances, which will, in time, reduce the effective pump capacity. The point here is that, traditionally, system designers have piled margin on top of margin, "just to be safe." Some of this margin can definitely be eliminated.

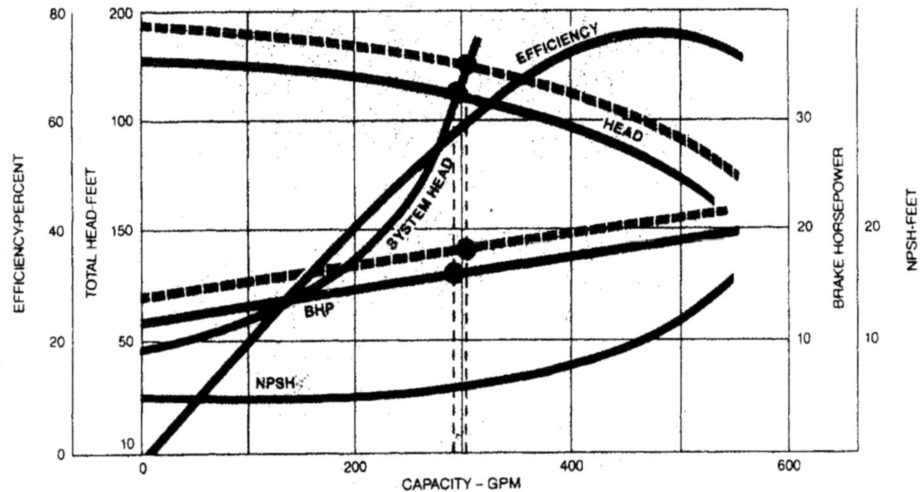

Figure 4-21: Impact of impeller trim on pump performance

A centrifugal pump operating in a given system will deliver a capacity corresponding to the intersection of its head-capacity curve with the system-head curve, providing the available NPSH is equal to or exceeds required NPSH (Figure 4-21). To change this operating point requires changing either the pump head-capacity curve, the system-head curve, or both. The first can be accomplished by varying the speed of the pump (Figure 4-23) while the second requires altering the friction losses by throttling a valve in the pump discharge (Figure 4-24).

A pump application might need to cover several duty points, of which the largest flow and/or head will determine the rated duty for the pump. The pump user must carefully consider the duration of operation at the individual duty points to properly select the number of pumps in the installation and to select the output control.

CREATING PUMP/SYSTEM HEAD-CAPACITY CURVES

Any serious pump energy cost reduction effort should start with the development (confirmation) of the true current pumping system "Head-Capacity" curve. The true system head-capacity curve not only permits the accurate determination of the current pump operating conditions, but it is also required to establish realistic potential energy cost savings for optional improvement actions.

As part of the design of most pumping systems, the initial head-capacity curve is normally calculated, based on either "as-new," or maybe 10 year old, pipe and fitting losses. However, as piping systems age they may not follow these theoretical friction loss estimates, which can push the pump to higher or lower flow rates then planned. A field test of the pump head at one or more measured flow rates can help determine the actual, current, pump and system H-Q curves.

A pump field test should include both pump suction and discharge pressure measurements (with a minimum of two pipe diameters of straight pipe in front of the gauges), taking into consideration the sizes (velocity heads) of the suction and discharge pipes (at the pressure gauge connections). Further, the heights of the suction and discharge gauges, with respect to the center line of the pumps, should be determined, since all pressure measurements must be adjusted to the pump center line. What should ultimately be measured is the total head developed by the pump, between the pump suction and the pump discharge. Total head is comprised of three components: Static Pressure (head), Elevation Head, and Velocity Head. The formula for the

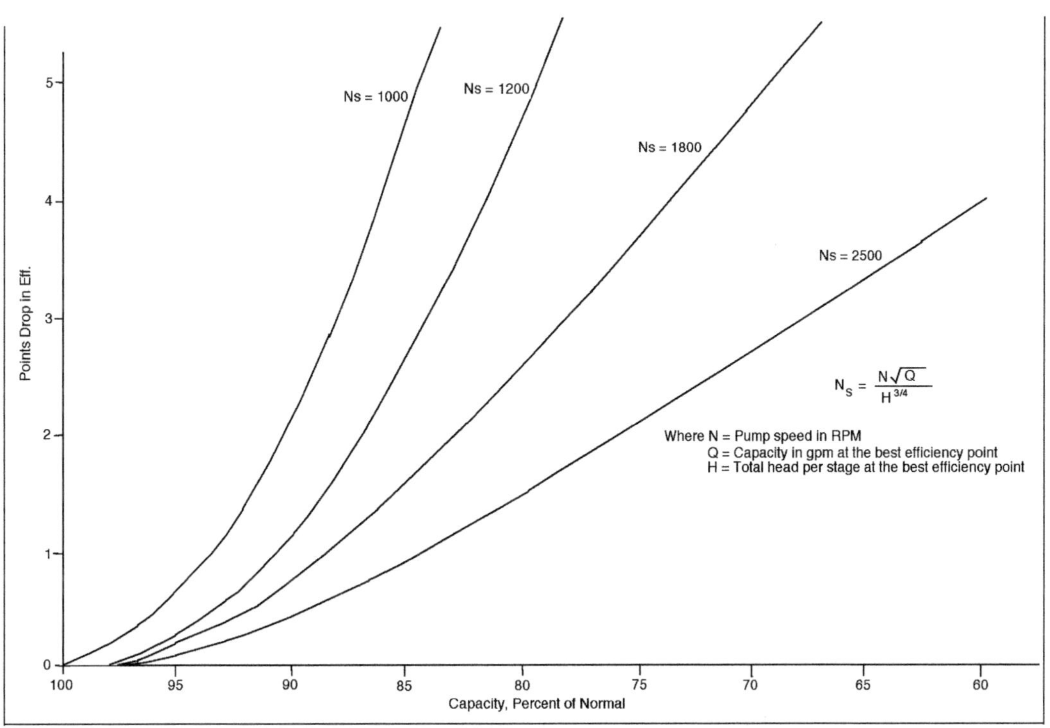

Figure 4-22: Performance at part load

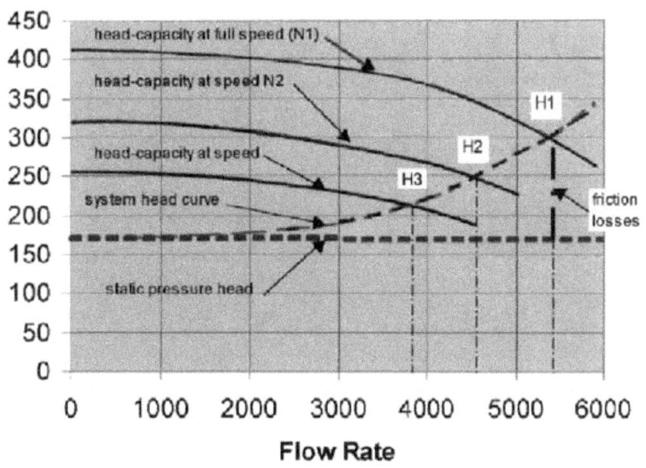

Figure 4-23: Effect of varying speed on pump performance

g = Acceleration due to gravity (32.2 feet/second²)

Z_d = Elevation discharge gauge distance above pump center line (feet)

Z_s = Elevation suction gauge distance above pump center line (feet)

S.G. = Liquid specific gravity (cold water is 1.0)

Ideally, the pump field (total head) data should be taken at both the average and full open throttle valve settings. The accuracy of these measurements is only as good as the accuracy of the gauges, so (if possible) calibrated pressure and flow gauges should be used. Field bourdon type pressure gauges are notoriously inaccurate. See the Hydraulic Institute Standards for more details on pump testing.

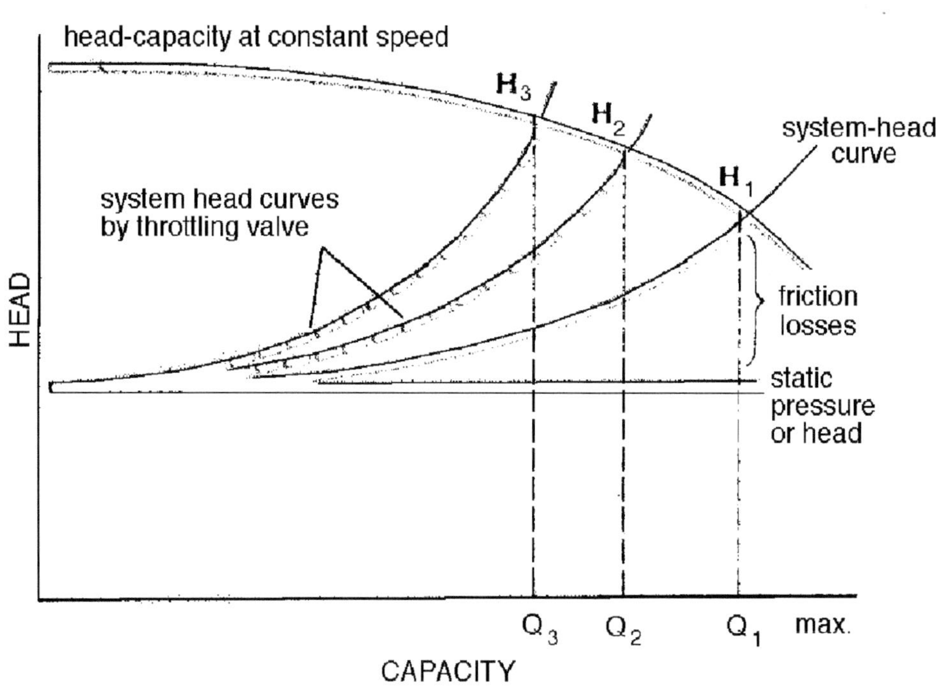

Figure 4-24: Effect of throttling valve on pump performance

Total head developed by a pump is:

$$H_{Pump\ Total} = (pd - ps)*2.31/S.G. + (hvd - hvs) + (Zd - Zs)$$

Where:

p_d = Discharge gauge pressure (psig)

p_s = Suction gauge pressure (psig)

h_{vd} = Discharge velocity head, $V_d^2/2g$ (feet)

h_{vs} = Suction velocity head, $V_s^2/2g$ (feet)

V_d = Discharge pipe velocity (feet/second)

V_s = Suction pipe velocity (feet/second)

System Control Methods

Several methods may be adopted to control a pump system, and each has its impact on life cycle cost.

- Control valve operation—Simple method for varying flow or pressure to a system. High losses, when throttling control valve (see Figure 4-13), and valve wear from cavitation and erosion.

- Variable speed operation (VFD)—Ideal for continuous flow or head variation in systems with relatively low static head. Good efficiency at part flow, versatile control. There are losses at full speed and little benefit in systems with high static head.

- Parallel operation on multiple pumps—Widely used for systems with varying flow demand. Low cost, simple technology, can readily provide stand-by pump facility. Only provides stepwise flow variation. Pump selection is key to energy cost reduction. Typically the minimum number and most energy effective pumps should be operated at every specific flow demand, for maximum savings.
- Stop-start—Used for sump emptying and small booster pumps. Low cost and simple technology. On-off flow and liquid storage usually required.
- Multiple speed motor (usually two-speed)—An efficient system when there are two very different pump duty requirements. Uses non-standard motor. Provides uneven flow if the duty is randomly varying.
- Bypass valve operation—Used for maintaining a pressure flow in a system when system load changes. It is a simple technology. Energy losses are high and it cannot handle solids in the liquid. The return of liquid to the pump suction can create temperature rise problems.
- "Hybrid Control" is a new (*combination*) concept, which involves adding "by-pass" flow control to a VSD control system to keep the pump operating close to its bep at all speeds. This new hybrid control concept should be considered for systems with a (relatively) high static head content, and a medium to high specific speed pump. It has the capability of allowing pumps to operate closer to their best efficiency point (bep) flow rates at lower speeds. This can also result in a marked improvement in pump reliability (under low system flow/ speed requirements), along with a substantial energy/cost savings. By having both the pump and driver operating closer to their best efficiency points, it can more than compensate for the energy loss from the bypass recirculation. An example of this new hybrid control technology, with a 3,350 specific speed pump, on a system where the static head represented 40% to 70% of the total system head, resulted in an energy savings potential of more then 20%, compared to a VSD only controlled system.

In the majority of pump installations, the driver is a constant speed motor, and throttling is the method used to change the pump capacity. Thus, if we have provided too much margin in the selection of the pump head-capacity curve, the pump will have to operate with considerable throttling at all times. In effect, we are first expending power to develop a much higher pressure than needed, and then wasting a part of it in frictional losses in order to reduce pump delivery to the desired value, an obvious waste of power, causing additional equipment wear and tear as well.

PARALLEL PUMP OPERATION

As mentioned in the prior section under "System Control Methods," parallel pump installations, where two or more pumps take suction from a common manifold and discharge into a common header, are widely used in order to handle varying system flow requirements. Pumps are operated in parallel as a means of flow control and for emergency backup (installed spare). However, if the pumps are not properly selected for parallel operation, or operated in the most optimum combinations, pump reliability and overall system energy efficiency can be compromised. Operating the wrong pumps in parallel can even cause one of the pumps to operate at or near shut off, resulting in overheating and possible failure.

When pumps run in parallel they operate against the same discharge head. The combined pump head-capacity (H-Q) curve is, therefore, determined by adding the respective flow rates of each operating pump, at a series of specific head values. This is demonstrated in Figure 4-25, where, at a head value of 240 feet, the flow generated by pump "A" at this head (320 gpm) is added to the capacity of pump "B" at this head (640 gpm) to create the combination H-Q pump curve point (960 gpm @ 240 ft).

The head (and therefore flow rate) that each individual parallel pump will operate at, in a given system, is then determined by the intersection between the system H-Q curve with the parallel pump combined H-Q curve. In Figure 4-25, the system and parallel pump combined (pump A + B) H-Q curves intersect at 575 gpm, at a head of 268 feet. The contributions of each of the pumps at 268 feet, are: pump A operating at 195 gpm (49% of BEP) and pump B operating at 380 gpm (48% of BEP). This compares with 455 gpm (114% of BEP) when only pump A operates on this system, and 542 gpm (68% of BEP) when only pump B operates on the system. As can be seen, very little is gained (+6%) in this example, by operating both of these pumps in parallel, compared to only operating pump B on this system. In addition, it can be noted that, parallel operation forces each of the pumps to operate at lower flow rates, as compared to individual pump operation on the same system. These lower flow rates can result in higher energy costs (because the individual pumps are also operated at lower efficiencies, higher heads, and lower energy effectiveness), when operated in parallel.

Further, as can also be seen in this example, the best efficiency point (BEP) flow rate for pump B, of 800 gpm, is twice the BEP flow rate (400 gpm) for pump A. This means that the pump B H-Q curve is the same as the combined H-Q curve of two pump A's operating in parallel. So it can be inferred that, by operating two (identical) pumps A in parallel on this system, it would increase the single pump A's flow rate of 455 gpm to 542 gpm (for two pumps A), a 19% increase. It should also be noted that, a system with a flatter system H-Q curve (higher static component) would yield an even greater flow increase with parallel pump operation.

In summary, it can be noted that when pumps are operated in parallel, their flow rates are pushed back to lower values, and these lower flow rates (often well below the pump bep), can result in reduced pump life/reliability. This reliability impact at low flow rates can be seen in Figure 5-2, from Chapter 5. This fact must be taken into account when selecting, and/or buying, pumps for parallel operation.

ADDITIONAL RELIABILITY ISSUES WITH PARALLEL OPERATION

Dissimilar head pumps can also be operated individually or in parallel, to provide increased flexibility in meeting changing system requirements (as shown in Figure 4-26). However, this creates a further reliability concern for parallel operation (when using dissimilar pumps}. That is because pumps operate at the same head when they operated in parallel, so a lower head pump could be pushed to shut off at higher system heads. It, therefore, becomes extremely important to use extra care when operating these different size pumps in parallel, in order to achieve maximum efficiency, without sacrificing pump reliability. As can be seen in Figure 4-26, if the system were to push the combined (total) pump flow rate below 1500 m^3/hr, the head would be increased above 45 meters. This would

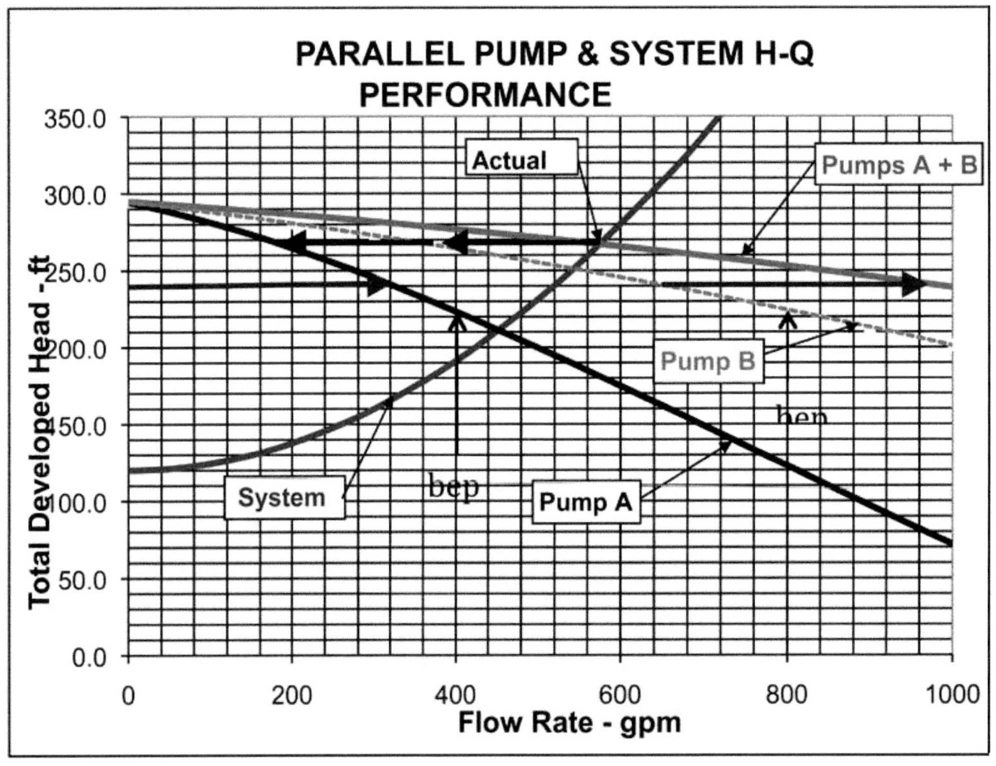

Figure 4-25: Parallel Pump Operation

push pump A to shut-off, which could cause the pump to overheat and fail. Below a combined (total) flow rate of 500 m^3/hr, pump B would also be pushed to shut-off.

When dissimilar pumps are operated in parallel, extreme care must be taken to avoid operation that pushes any pump below its minimum allowable flow rate. Further, the minimum number of pumps, with the highest "Energy Effectiveness" (see "Energy Effectiveness" section at the beginning of Chapter 4), should typically be operated at each specific system H-Q condition point, for maximum efficiency (and reliability).

Another issue to be considered when selecting pumps to operate in parallel is the shape of the pump head–capacity curve. If the pump H-Q curve droops (the head drops) as the flow is reduced towards shut-off (zero flow), see Figure 4-27, a second identical pump may not be able to come on line, and could, therefore, run at shut-off, overheat, and possibly fail. In the example shown in Figure 4-27, the system would cause one pump to operate at 242 feet and 300 gpm. However, this head is also the shut-off head for the other identical pump, which is where the second pump would operate. Even if the pump were to come on line, it would only increase the system flow rate by 5 gpm.

Further, if the Figure 4-27 system curve were mainly composed of static head, it could intersect the pump head-capacity curves at more than one flow rate, causing unstable operation (the pumps would not know which of the two flow rates to operate at).

Pumps with head-capacity curves that droop towards shut-off (or have a dip at a higher flow rate that yields more than one flow condition for a given head), should not be operated in parallel.

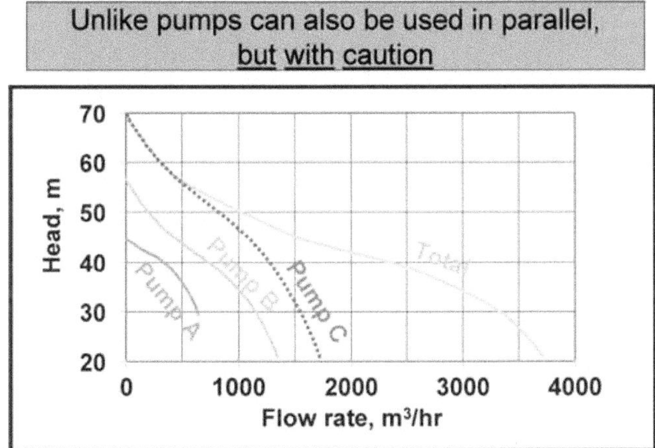

Figure 4-26: Dissimilar Pumps in a Parallel Installation

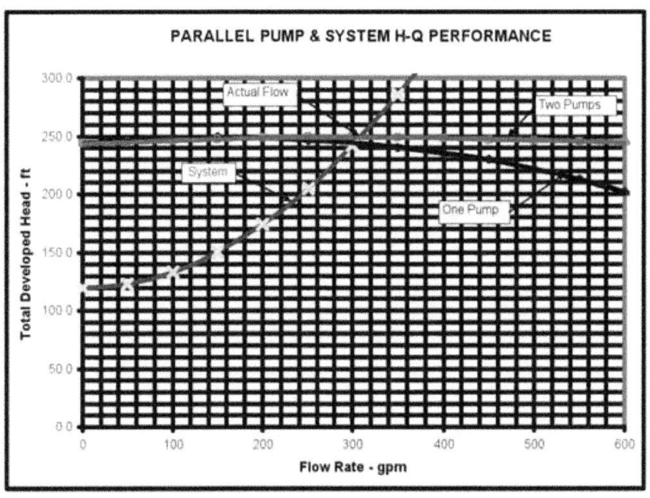

Figure 4-27: Parallel Pump and System Head-Capacity Performance with drooping pump H-Q curves

SPECIFIC PARALLEL PUMP GUIDELINES/PROCESS PUMPS

It is important to match the performance curves of pumps operating in parallel, in most cases. Unless there is a very close match, the loads will not be shared equally among operating pumps. The various editions of API-610 point out that "flat" performance curves will generally disqualify pumps from operating in parallel because one pump could drive the other(s) into a prohibited low-flow operating region, as discussed above. Reliability-focused users insist on a head rise of at least 10% (from operating point to shut-off) on pumps intended for parallel operation. If, say, the head rise from the operating point to shut-off is 4 ft and the operating point is at 200 ft, head rise would be only 2%. In that case, a risk-averse user would not allow the pump to operate in a parallel arrangement.

It should be noted that, for reasons of gaining small incremental power/efficiency benefits, the performance curves of many pumps were originally designed to be relatively flat at flows approaching shutoff. The resulting pressure rise from operating point to shutoff is then either insufficient or non-existent (see Figure 4-27). As one or more pumps are operating too close to shutoff (in addition to the hydraulic issues), shafts deflect and bearings are overloaded. The oil film that must separate bearing rolling elements from stationary elements becomes both too thin and too hot—a vicious cycle. High load, a thin oil film and high metal temperature combine, and bearings fail prematurely. Depending on the bearing cage type, these failures range from gradual and detectable to sudden, difficult-to-detect-in-advance, and plain catastrophic.

IMPACT OF VARIABLE SPEED CONTROL ON PARALLEL PUMP OPERATION

Besides the reduction in pump percent bep flow rate with speed reduction (which occurs with systems that include a "static" head component), parallel operation of two or more identical pumps will reduce the percent bep flow rate even further, as discussed above and shown in Figure 4-28 (Ref. 4-4). In this actual field, high suction energy (see Chapter 5), sewage lift pump application, reducing the pump speed from 60 Hz (1185 RPM) to 40 Hz (790 RPM), reduces the BEP flow rate from 97% to 80%. This, by itself, would not be a reliability issue, at least with sufficient NPSH margin, and/or a cavitation resistant impeller (stainless steel or hard coated). However, when the pump is then operated in parallel with another identical pump, the BEP flow rates are pushed back even further, to 54% and 46%, depending on the speed. This could push one or both pumps into "Suction Recirculation" (see Chapter 5), and if either pump were a "High Suction Energy" pump, it could result in excess vibration and/or cavitation damage.

For this application it would be best if these pumps were not operated in parallel, (unless the system flow requirement exceeds the one pump maximum speed/maximum flow rate of about 8,000 gpm), and then only for short periods of time (from both an efficiency and reliability standpoint).

MULTI-PUMP PARALLEL OPERATION—CONTROL

Control of parallel pump systems is one of the most challenging tasks pump users face, and unfortunately many operators choose to run all of their pumps, all of the time, rather than face the potential of missing process demands. As a result, many multi-pump systems have pumps that run outside their recommended minimum or maximum flows, ultimately leading to reduced pump reliability and efficiency.

Modern intelligent pump drives/controls can adjust the pump speeds accordingly, using pre-engineered flow balancing logic. At peak times, multiple pumps will be required to keep pace with the flow. Under these conditions the intelligent drives can automatically determine how many pumps should be in operation, and automatically sequence pump operation based on the hours of operation for each pump. This will equalize operating times for each pump to equalize wear. These intelligent drives can also balance the flows in order to

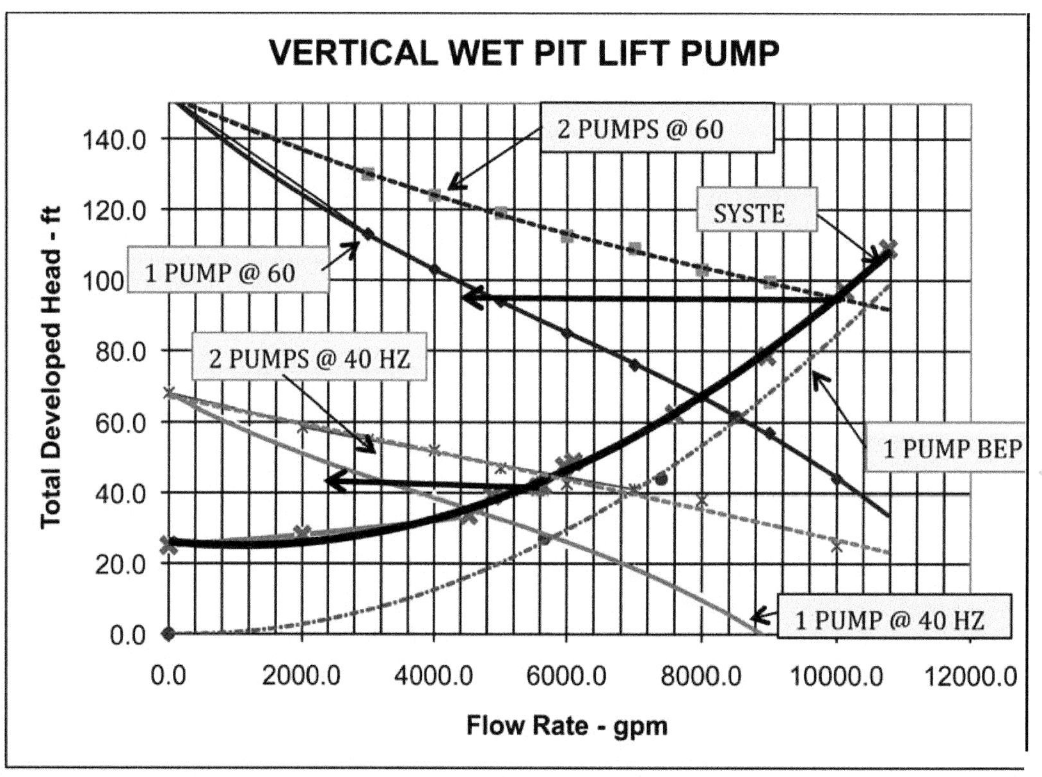

Figure 4-28: Parallel and Single Pump vs. System H – Q Curves

balance pump loads to each pump, rather than simply operating the pumps at the same speed, thus accounting for differences in pump wear. This results in improved pump reliability, efficiency and process control.

PARALLEL PUMP OPERATION CONCLUSIONS

The above demonstrates why it is essential that special care be taken when operating pumps, especially "High Suction Energy" pumps, in parallel, from both a reliability and efficiency standpoint. The bottom line recommendation is that, *"it is normally best to restrict parallel operation to identical pumps and to operate the minimum number of pumps necessary for the various system requirements, and finally that all parallel pumps operate within their allowable flow operating region."*

Variable Speed Drives

On the other hand some installations do take advantage of the possible savings in power consumption provided by variable speed operation. The typical variable speed drive system today is a squirrel-cage induction motor fed from a variable frequency drive, or VFD. The most common VFDs use "Insulated Gate Bi-polar power transistors" (IGBT) to create the voltage source "pulse width modulation" (PWM) to generate the variable voltage frequency for the motor. PWM/IGBT drives have the best overall performance, with high power factors throughout the speed range, and are the most common type used for small and medium horsepower motors. They generate low speed torque, quiet motor operation and improved low speed stability. PWM/IGBT drives do stress the motor and drive cable, due to the very fast output voltage rises; this may cause voltage doubling in the feeder cable from voltage reflections. An impedance load reactor should generally be used on the load side of the VFD when the motor lead length exceeds approximately 35 to100 ft, with the maximum length being dependent on the VFD design, size and cable type. Always consult the VFD manufacturer for more specific guidance on this issue. VFDs are by far the most efficient type of variable speed drive as shown in Figure 4-29. The most recent generations of VFDs do perform very well and have few complications, when properly applied and matched to the motor and electrical system.

The energy consumed by a pump varies as the third power of the speed, so an 80% reduction in speed will reduce the power consumed by 50%. It is then possible to match pump operating speed to the exact conditions of service without throttling.

For instance, consider a system-head curve for a new installation that corresponds to curve "D" in Figure 4-30. If we wish the pump to handle exactly 2,700 gpm and 152 ft. head, we could use a 14.75 in. impeller and run it at 87.5 percent of design speed. We could also use a 14-in. impeller at 92.2 percent of design speed. In either case, the pump would take 118 bhp. Compare this with the 165 bhp taken by a constant speed pump with a 14 $^{3}/_{4}$-inch impeller at the same 2,700 gpm.

Not all of this horsepower difference is savings, however, because a variable-speed device has its own losses. The efficiency of a variable frequency drive (inverter) is affected by operating speed, ranging from about 97% at 100% of rated speed to around 91% at 50% of rated speed (input frequency), for the latest generation of drives. In addition, a VFD causes harmonic losses in the motor, due to the imperfect sinusoidal waves from the VFD that supplies the motor. These losses cause the motor to heat up, which is the reason that the motor may have to be de-rated when it runs with a VFD. Inverter rated motor windings are required for 440 V and above. For PWM inverter operation, motor designs NEMA A is best, but B normally gives satisfactory service. Further, there are additional losses in the motor due to the movement of the motor duty point. The total VFD/motor efficiency drop can range from 65% to 83% at 40% of rated speed, to 97% at 100% of rated speed, depending on the type of voltage control.

In addition the large potential energy savings from VFDs, they offer many other benefits.

- The resulting lower operating speed will increase the life of the pump (See Chapter 5, Figure 5-1).

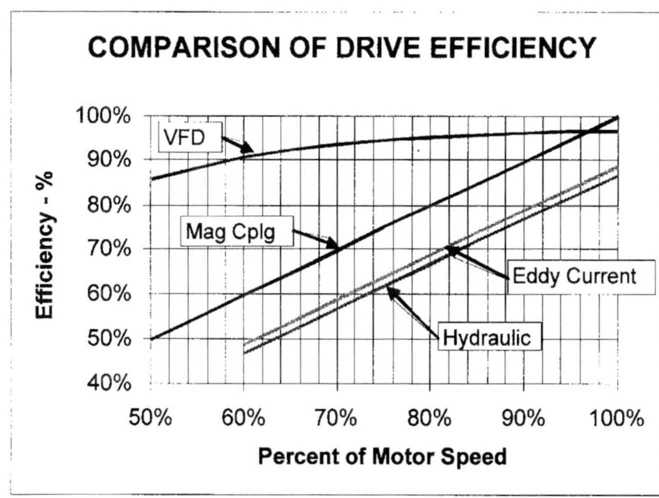

Figure 4-29: Efficiency comparison of various types of variable speed drives

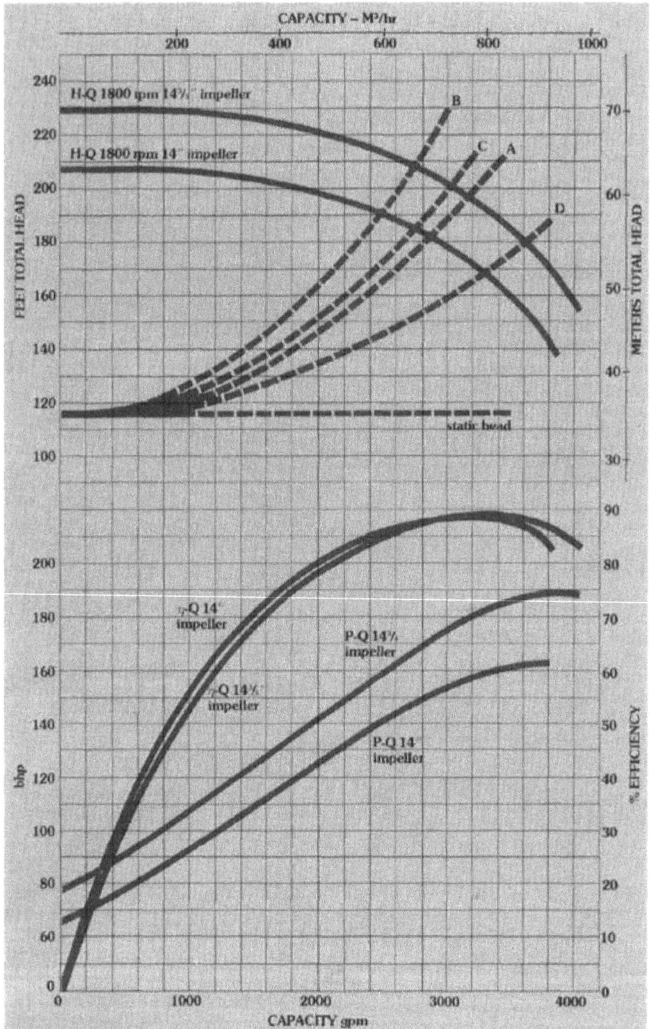

Figure 4-30: Impact of impeller trim on specific pump performance

- They eliminate the control valve (and associated losses), starter and any by-pass lines. They are easy to retrofit.
- They generate higher supply side power factors at all loads, unlike a constant speed motor.
- They eliminate the inrush current on startup. Soft starting reduces the stress on the motor, pump, coupling and the supply network.
- They can provide system control logic, fault protection and diagnostics.
- They can restart a spinning load.
- They can enhance the quality of the product pumped.

The non-sinusoidal VFD output voltage does lead to a number of undesirable consequences. Increased motor losses, noise and vibrations, a detrimental impact on the motor insulation systems, and bearing failures are examples of VFD related problems. VFD drives can generate stray motor currents. This may require grounding of the motor rotor and/or the use of a bearing with an insulating coating on the outer ring. Small electric discharges between the rolling elements and the bearing raceway can eventually damage (pit) the bearings, and/or cause them to run hotter, especially with larger motors (over 150 hp). The non-sinusoidal waves generated by a VFD can also pollute the supply grid with harmonics, although with today's VFDs most of the harmful harmonics are eliminated.

- An input filter is sometimes prescribed by the power company. This filter will decrease the available voltage by typically 5-10%. The motor will consequently run at 90-95% of nominal voltage. The consequence is additional heating. Further motor de-rating might be necessary. VFDs should also be located in safe/clean locations, which may require long cable runs that increase the peak voltage.

- Increased motor losses dictate a de-rating of the motor output power to prevent overheating. Temperature rises may be 40% higher with a VFD, compared with a conventional sinusoidal power grid. Ongoing VFD improvements have solved many of the problems. However, reducing the motor and VFD losses tends to increase the detrimental impact on the motor insulation. A de-rating of 10-15% is recommended for large motors. Motor manufacturers are, of course, aware of these VFD complications. New motor designs (inverter-resistant motors) are beginning to appear on the market. Better stator winding insulation and other structural improvements promise motors that can be better adapted for VFD applications. To ensure function and ample motor life, it is absolutely necessary that a winding be adapted for use with a VFD. Motors for voltages above 500 volts must have some form of reinforced insulation. It may be necessary to insert a filter between the VFD and the motor. Such filters are available from most well know VFD suppliers. Careful planning must be done to insure trouble free VFD operation.

Sometimes there is temptation to run the pump at frequencies above the commercial power grid frequency, in order to reach a duty point that would otherwise be impossible. Doing so calls for extra awareness. The shaft power for a pump will increase with the cube of the speed. An over-speed of 10% will require 33% more output power. Roughly speaking, we can expect that the motor temperature rise will increase by about 75%.

To decide whether such a VFD is advisable, plot the actual system-head curve. This will give you the speed required at various capacities over the operating range, and the motor horsepower input to the variable-speed device over this range. The difference between this horsepower ("BHP") and the pump BHP at constant speed represents potential power savings. Next, assign a predicted number of hours of operation at various capacities and calculate the potential yearly savings in hp-hrs or kW-hrs. Finally, compare these savings to the added cost of a VFD (minus the cost of a control valve) to determine whether the cost of variable-speed operation is justified. If it is, carefully proceed with the modification. If not, the options of smaller impellers or of narrower impellers are still available.

GROUNDING RING TECHNOLOGY FOR VARIABLE FREQUENCY DRIVES (VFDS)

Variable frequency drives (VFDs) provide substantial energy savings and control for commercial and industrial process applications. Unfortunately, previously unanticipated bearing failures are occurring because modern VFDs induce a harmful voltage on the motor shaft. As the resulting current travels through motor bearings it can cause catastrophic failure and costly downtime.

While there are compelling reasons to specify insulated (actually, aluminum oxide- coated) or ceramic ("hybrid") rolling element bearings for VFDs, there may be instances where bearing protector rings are well justified and might further reduce the risk of shaft current-induced bearing distress. When specifying such shaft grounding rings, steer clear of knock-off products that use carbon fibers and mounting methods in a manner that compromises long-term reliable service.

Reliability-focused VFD users would involve both VFD manufacturer and bearing suppliers in issues dealing with bearing failure avoidance strategies. Also, reliability-focused users would endeavor to become familiar sound with grounding ring technology.

A company that initially focused on mitigating static charges in the printing and imaging markets, Electro Static Technology Company (EST) has---since about 2005---been producing conductive micro fiber grounding rings for rotating equipment. EST's Bearing Protection Ring products are marketed as AEGIS™ products, representing proprietary technology said to provide a reliable and essentially maintenance free shaft grounding ring. Such rings may be needed to mitigate the issues of electrical erosion in motor bearings when electric motors are controlled by pulse width modulated (PWM) variable frequency drives (VFD).

Fundamentals of shaft grounding (SGR) rings

At the simplest level, an AEGIS™ SGR provides the "path of least resistance to ground" for VFD-induced shaft voltages. If these voltages are not diverted away from the bearings to ground, they may discharge through the bearings and cause damage known as electrical discharge machining (EDM), pitting, and fluting failure in bearings. These shaft grounding rings can be adapted as an integral part of the motor design. Good products meets both spirit and intent of the NEMA MG1 Part 31 specification, aimed at preventing bearing fluting failure in electric motors as well as their attached equipment. NEMA MG1 identifies induced shaft voltage in VFDs as a potential cause of motor failure and recommends shaft grounding as a solution to protect both motor bearings and attached equipment.

Properly designed shaft grounding rings must provide a large number of small diameter fibers to induce ionization; they must discharge voltages away from motor bearings and to ground. Selecting carbon fibers of specific mechanical strength and electrical characteristics is critically important to providing break-free and non-wearing service. The fibers must be allowed to flex within their elastic limit and while contacting the shaft with the proper overlap. For long-term reliability they must be placed in an engineered holder that protects against breaking and mechanical stress. Chances are that, without placement in protective channels, the reliability of shaft grounding rings is severely compromised.

Circumferential rows of fibers are the best available designs. Optimized fiber density maintains the required fiber flexibility. If too many fibers are bundled together (as may be the case in less-than-optimal designs) the fibers will break. A soundly engineered SGR has two full rows of fibers. The continuous circumferential "ring" design and fiber flexibility allows them to direct small amounts of oil film, grease, and dust particles away from the shaft surface. It was noted that EST's patents prevent others from copying technology which arranges one or more rows of fibers in a continuous fiber ring inside a protective channel completely surrounding the motor shaft. This design ensures that there are literally hundreds of thousands of fibers available to handle discharge currents from VFD-induced voltages at the various prevailing high frequencies. The fibers can then flex inside the channel while maintaining optimal contact with the motor shaft.

Pump Selection

Selecting the proper pump for an application is extremely important for the best overall operating efficiency. For general industrial service, the centrifugal is the largest single category of pumps. Reciprocating pumps are usually limited to low-capacity, high-pressure applications, and rotary pumps such as gear and vane types are more or less reserved for higher viscosity fluids. When it comes to clear, low-viscosity liquids such as water, the obvious choice is a centrifugal pump.

The major influences on centrifugal pump efficiency are specific speed (N_S), pump size, NPSHa, NPSHr, and the type of pump selected to meet the service conditions.

Specific Speed is a dimensionless factor that dictates the impeller geometry and best attainable efficiency for the head, flow rate and speed, as shown by the chart in Figure 4-31. Not to be confused with *suction* specific speed, *specific speed* is defined as:

$$\text{Specific Speed } (N_S) = \text{rpm x } (\text{capacity in gpm})^{1/2}/(\text{total head in feet})^{3/4}$$

The centrifugal pump is a hydrodynamic machine, with an impeller designed for one set of conditions of flow and total head at any given speed. Impeller geometry or shape runs the gamut from very narrow, large-diameter impellers for low flows, through much wider impellers for higher flow, to the specialized propeller for highest-flow, low-head conditions (Ref. 4-2). Unfortunately, not all designs can have equally good efficiency. In general, medium flow pumps are most efficient: extremes of either low or high flow will drop off in efficiency. The optimum efficiency occurs when the specific speed (N_S) is in the vicinity of 2,500 (English system).

While efficiency tends to drop off at high specific speed, the greater difficulty is at specific speeds below 1000 (English system). In Figure 4-31, the slope of the efficiency curve below 1000 becomes quite steep and efficiency falls off rapidly. And so, other factors being equal, for good efficiency it is best to avoid pumps designed for specific speeds below 1000.

So what can one do to avoid low specific speed pumps. Well, according to the equation for specific speed, for a given flow rate, either the speed must be increased or the total head reduced. Also, except for special pump designs, the maximum speed is limited by either the two-pole motor speed of 3,600, or the net positive suction head required (NPSHr) for the application. However, using a multistage pump with two or more impellers can reduce the pump total head.

The efficiency values shown in Figure 4-31 cannot be achieved for all centrifugal pump type and design due to other factors that also influence pump efficiency:

1. Desired curve shape (rise to shutoff, etc.)
2. Surface roughness (affects small, low specific speed pumps most)
3. Internal clearances (affects low specific speed pumps most)
4. Design compromise for manufacturability
5. Low NPSHr designs (special large impeller eye designs can reduce efficiency)
6. Low NPSHr designs also might increase suction specific speed to undesirable level leading to inlet flow recirculation problems
7. Mechanical losses (i.e. double mechanical seals on small (less than 15 hp, or ~11 kW) pumps)
8. Solids handling capability (large passage ways required in impeller and casing)
9. Impeller diameter trim (high specific speed designs have greater reductions in efficiency due to trim)
10. Viscous mixtures, entrained air and slurries.

Pump Viscosity Correction

The performance of a centrifugal pump on a viscous fluid will differ from the performance on water, which is the basis for most published curves. Head, flow rate and *efficiency* will normally decrease as viscosity increases. Power will increase, as will net positive suction head required in most circumstances. Starting torque may also be affected (Refs. 4-2 and 4-3).

The Hydraulic Institute has developed a generalized method for predicting performance of centrifugal pumps on viscous Newtonian fluids, by applying correction factors for head, flow rate and efficiency to the performance in water. It uses an empirical method

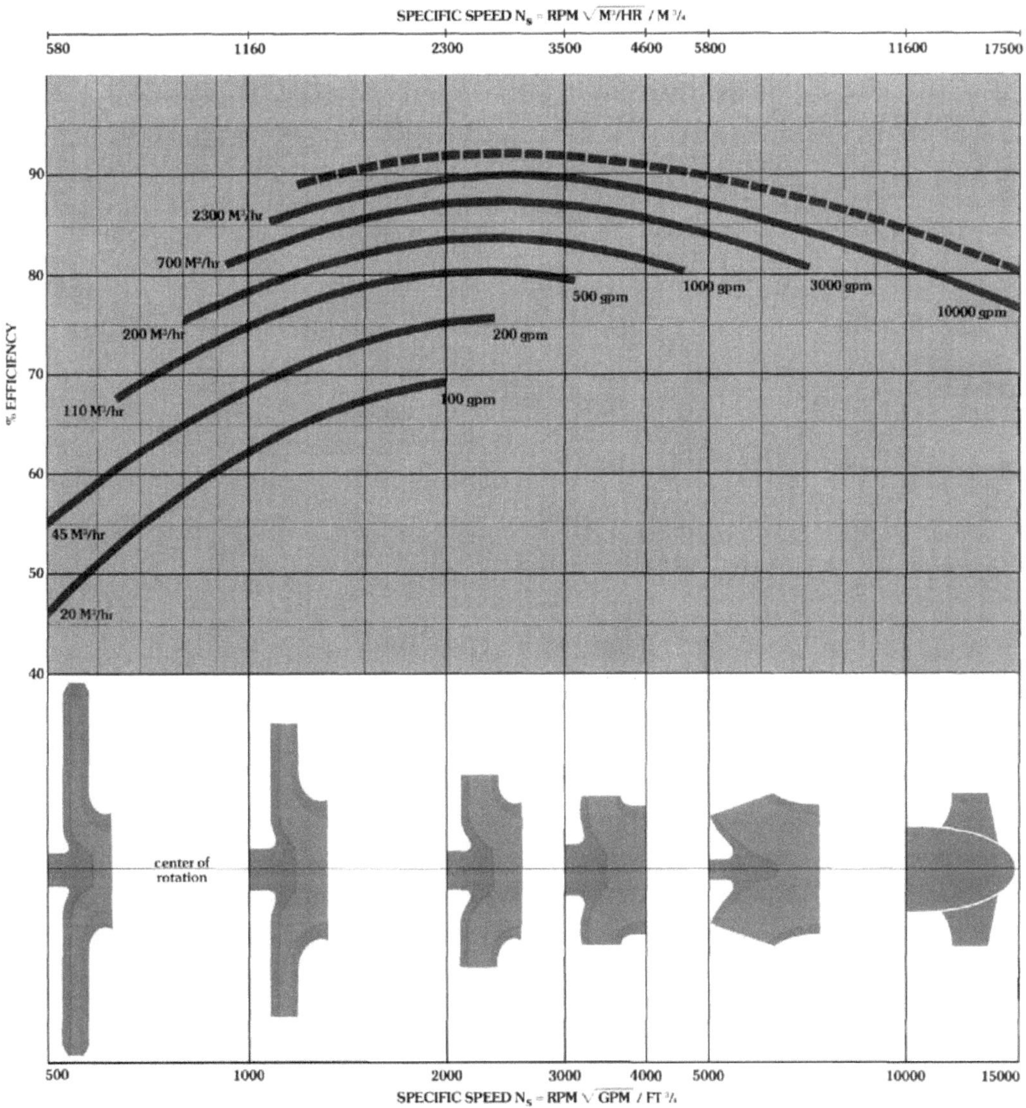

Figure 4-31: Pump efficiency chart

based on test data from sources throughout the world. Performance estimates using the HI method are only approximate. There are many factors for a particular pump geometry and flow conditions that the method does not take into account. It does not apply to high specific speed (axial flow pump types) or for pumps of special hydraulic designs. It applies only to the pumping element. Pumps which incorporate external piping, a suction barrel for vertical can type pumps, a discharge column or other appurtenances for fluid conveyance to or from the pumping element, require additional consideration for viscous losses.

Internal pump components, such as the pump shaft and associated drive mechanisms should be checked to assure they are adequate for the additional torque that the pump will experience. Externally, the pump driver, such as an electric motor, and the coupling between the pump and driver, need to be sized for the higher starting and operating torque, and starting cycles demanded by the service.

Mechanical seals or sealing devices must be capable of sealing the pump for the range of anticipated viscous conditions, including transient or upset conditions. Mechanical seal components may not perform as anticipated and may experience higher loads than with water. Orifices and filters in seal flush piping may plug or cease to function correctly when handling viscous liquids. The piping is normally external to the pump case and may require heat tracing or other considerations to assure proper seal flushing.

Sealless pumps require additional considerations when handling viscous liquids. The additional viscous drag due to the immersion of internal drive and bearing components will lead to higher losses, resulting in lower efficiency, increased power consumption and increased starting torque requirements. Furthermore, cooling/ lubrication flow to the motor or magnetic coupling and bearings will be decreased. The temperature rise caused by the increased losses and decreased cooling flow must also be considered.

Existing Installations— How to Change Pump Performance

But what of existing installations in which the pump or pumps are over-sized and have excessive margins? Is it too late to achieve these savings? Far from it! As a matter of fact, it is possible to establish the true system-head curve even more accurately by running a performance test once the pump has been installed and operated. Once a reasonable margin has been selected, four choices become available:

1. The existing impeller can be cut down to meet the real service conditions required for the installation

2. A replacement impeller with the necessary reduced diameter may be ordered from the pump manufacturer. The original impeller is then stored for future use if friction losses are ultimately increased with time or if greater capacities are ever required.

3. If two separate impeller designs are available for the same pump, one with a narrower width, as is sometimes the case, a replacement may be ordered from the pump manufacturer. Such a narrow impeller will have its best efficiency at a lower capacity than the normal width impeller, depending on the degree to which excessive margin had originally been provided. Again, the original impeller is put away for possible future use.

4. Replace the control valve with a VFD.

In addition to over-sized pumps which result from excessive margin, different industries with many different processes will have requirements for the same pump to operate at different capacities, different heads, and to have a different shape of the head-capacity curves. To ideally satisfy these requirements, one should have a variable speed pump with adjustable vanes in the impellers. But because most of the drivers in the process industries operate at constant speed, and the adjustable vanes cannot be produced economically, variable pump performance must be achieved by mechanical means without sacrificing efficiency.

In order to provide this flexibility at minimum cost, studies were made to change pump performance within a given pump casing. This can be accomplished by varying the impeller design, by impeller cuts, by changing the running speed, by modifying the impeller vane tips, by filling the volute cut-water lip, or by installing an orifice in the pump discharge.

Pump users would prefer to use the same casing for a wide variation of pump performance. The pump casing is usually the most costly part of the pump. To replace a pump casing means extensive and costly work on base plate and piping.

The prediction of pump performance by modifying parts other than the casing is largely based on experimentation. Many tests have been conducted by the various pump companies in such areas as:

1. Trimming the pump impellers
2. Removing metal from the tips of impeller vanes at the impeller periphery
3. Removing metal from the volute tip in the pump casing
4. Providing impellers of the same angularity, but different width
5. Providing impellers with different number of vanes and different discharge angles
6. Installing an orifice in the pump discharge casing near the nozzle.

Each of these means is discussed next.

Impeller Cuts

Assuming that the impeller represents a standard design, say specific speed below 2,500, and that the impeller profile is typically of average layout and not specifically designed for high NPSH, the pump performance with trimmed impellers will follow the affinity laws as long as some vane overlap is maintained.

Affinity laws (Ref. 4-3)

The affinity laws express the mathematical relationship between the several variables Involved In pump performance. They apply to all types of centrifugal and axial flow pumps and are reasonably accurate as long as there is some vane tip overlap. Although in cases with little or no vane tip overlap the affinity laws must be used with caution, they are worthy of consideration and are shown in the expressions and relationships that follow:

1. With impeller diameter, D, held constant:

$$Q_2 = (Q_1)(N_2/N_1)$$
$$H_2 = (H_1)(N_2/N_1)$$
$$BHP_2 = (BHP_1)(N_2/N_1)^3$$

Where

Q = Capacity. GPM
H = Total Head, Feet H1
BHP = Brake Horsepower
N = Pump Speed, rpm

2. With speed, N, held constant:

$$Q_2 = (Q_1)(D_2/D_1)$$
$$H_2 = (H_1)(D_2/D_1)^2$$
$$BHP_2 = (BHP_1)(D_2/D_1)^3$$

When the performance (Q_1, H_1 & BHP_1) is known at some particular speed (N_1) or diameter (D_1), the formulas can be used to estimate the performance (Q_2, H_2, & BHP_2) at some other speed (N_2) or diameter (D_2). The efficiency remains nearly constant for speed changes and for small changes (up to 5%) in impeller diameter.

To compensate for casting and mechanical imperfections, correction factors are normally applied to the impeller cuts, see Figure 4-32. The efficiency of the cut impellers (within a 25% cut) will usually drop about two points at the maximum cut. On high N_s pumps, the performance of the cut impellers should be determined by shop tests or vendor curves.

EXAMPLE

To illustrate the use of these laws, assume that you have performance data for a particular pump at 1,750 rpm with various impeller diameters. The performance data have been determined in actual tests by the manufacturer. Now assume that you have a 13" maximum diameter impeller, but you want to belt drive the pump at 2,000 rpm.

The affinity laws listed above will be used to determine the new performance, with N_1 = 1,750 rpm and N_2 = 2,000 rpm. The first step is to read the capacity, head, and horsepower at several points on the 13" diameter curve. For example, one point may be near the best efficiency point where the capacity is 300 gpm, the head is 160 ft, and the horsepower is approximately 20 hp. According to the affinity laws at N = 2,000 rpm, the new capacity is 343 gpm, the new head is 209 ft., and the new horsepower is 30 hp.

$$Q_2 = (300)(2{,}000/1{,}750)$$
$$Q_2 = 343 \text{ gpm}$$
$$H_2 = (160)(2{,}000/1{,}750)^2$$
$$H_2 = 209 \text{ ft.}$$
$$BHP_2 = (20)(2{,}000/1{,}750)^3$$
$$BHP_2 = 30 \text{ hp}$$

Removing Metal from Vane Tips

The pump performance can be changed by removing metal from the vane tips at the impeller periphery. Removing metal from the underside of the vane is known as *underfiling*. Removing metal from the working side of the vane is known as *overfiling*. The effect

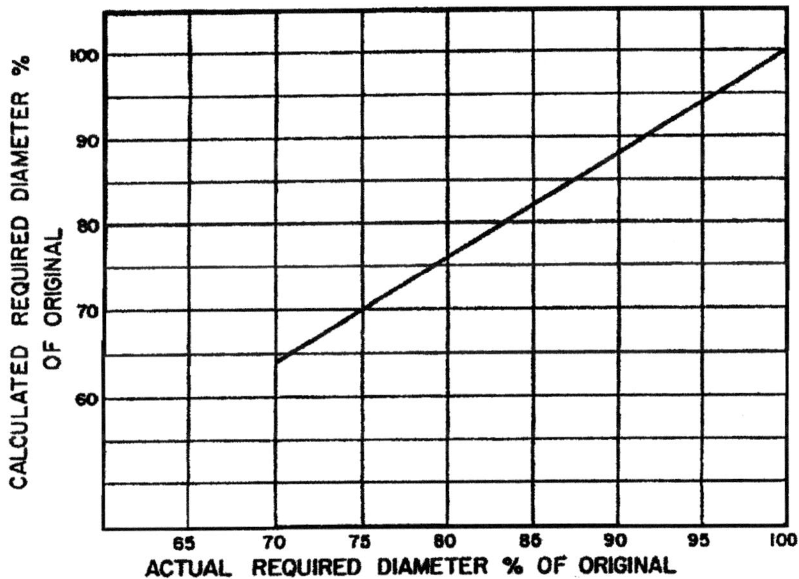

Figure 4-32: Impeller trim correction

of overfiling on pump performance is very difficult to predict and very difficult to duplicate. This is because filing vanes by hand on the working side changes the discharge angle of the impeller, and non-uniformity exists between each vane. Occasionally pump efficiency can be increased slightly by overfiling, with little change in total head.

"Underfiling" however, is more consistent, more predictable and easier to apply. Underfiling is most effective at peak efficiency and to the right of peak efficiency. Also, underfiling will be more effective where vanes are thick and specific speeds are high. Underfiling increases the area at the impeller discharge, thereby the capacity at peak is increased. This increase is directly proportional to the increase in area due to filing, or can be said to equal dimension "A" over "B" in Figure 4-33. The head rating will move to the right of peak efficiency in a straight line to the new capacity. With underfiling, the shutoff head does not change, therefore the change of performance by impeller underfiling is less effective to the left of peak efficiency (Ref. 4-3).

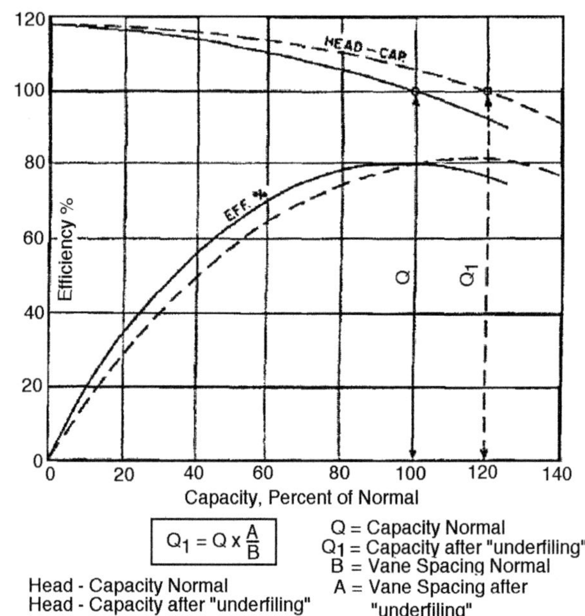

Figure 4-33b: Changing pump performance by underfiling

Removing Metal from the Volute Tongue

Capacity increase in a given pump can also be achieved by trimming of the tip of the volute tongue in the pump casing. This is illustrated in Figures 4-34 and 4-35. Removing metal at this point increases the total volute area. The peak efficiency and peak capacity will move to the right as the square root ratio of the new area divided by the original area. Pump efficiency at peak will normally drop one or two points.

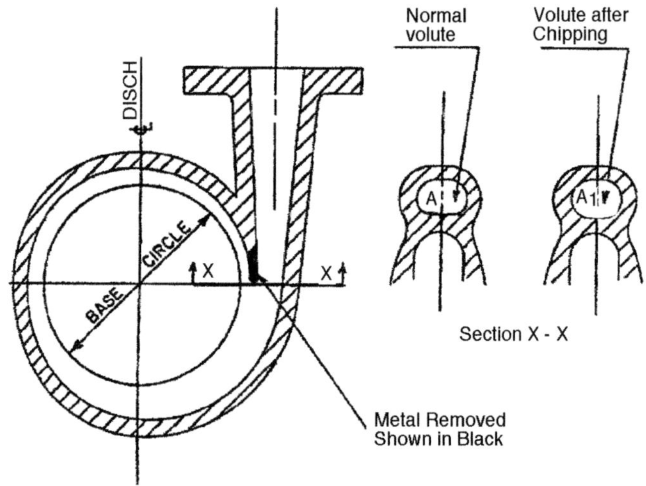

Figure 4-34: Volute tongue metal removal

Low and High Capacity Impellers

In the majority of pump casings we can install impellers of different width for low or high capacity performance. Because of the variations in the design of the impeller vanes (angularity and number of vanes), it is very difficult to predict their performance. However, if we take a given impeller with a given angularity and number of vanes, we can reasonably predict the performance of the narrow, medium, and wide impellers. Figure 4-36 shows actual test data of a two-stage 14" pipeline pump with a specific speed of 1,600. In this pump, the peripheral width of the normal impeller was 2," whereas the high and low capacity impellers were 2.75" and 1," respectively. Capacities ranged from 5,000 gpm to 9,000 gpm, and efficiencies bracketed 82% to 88%. The performance of the different impellers in the same casing is to some degree related to specific speed, with the efficiency increasing with higher capacity impellers.

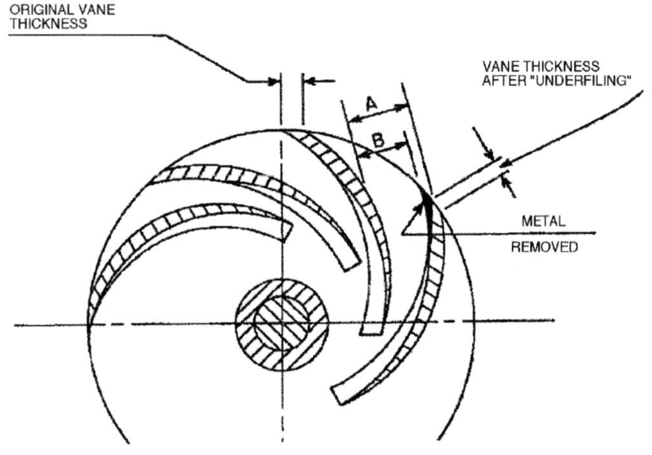

Figure 4-33a: Impeller vane underfiling

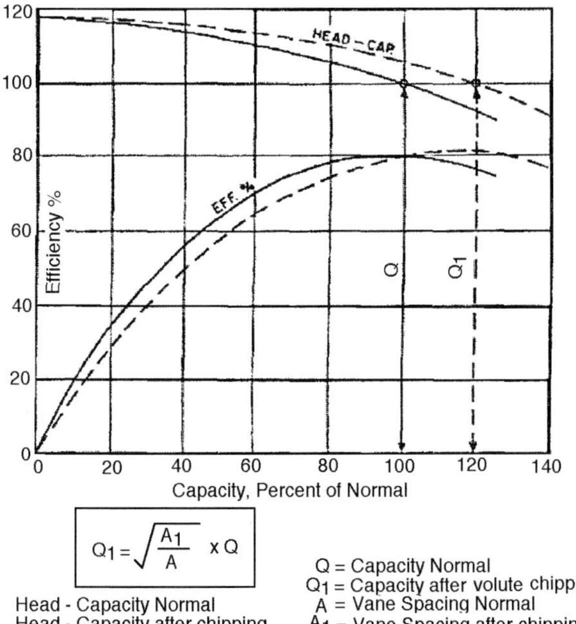

Figure 4-35: Changing pump performance by volute chipping

Impellers of Different Number of Vanes

Certain pump applications require the pump performance curves to have differently shaped head-capacity curves. For instance, to overcome friction only, as in pipeline service, the highest head per stage, or a very flat curve is desirable. To overcome static head or to have pumps run in parallel as is customary in process or boiler feed services, a continuously rising head-capacity curve is usually needed for highest possible efficiency.

There are different ways to vary the shape of a head-capacity curve (Ref. 4-3):

1. If the existing impeller has 6 or more vanes, removal of vanes and equally spacing the remaining vanes will produce a steeper head-capacity curve. The fewer the number of vanes, the steeper the curve. When vanes are removed, the effective fluid discharge angle decreases due to increased slip. This moves the peak efficiency flow point to the left, as shown in Figure 4-37. The efficiency will also drop, lowest being at the least number of vanes. In the case of a 7-vane impeller reduced to 4 vanes, the efficiency will drop about four points.

2. If a different head-capacity curve shape is required in a given casing and the same peak capacity must be maintained, a new impeller must be designed for each head-capacity shape. The steeper the head-capacity curve, the fewer will be the number of vanes, with a wider impeller used to maintain the best efficiency capacity. For example, a 7-vane/27-degree exit angle will have a flat curve and a narrow impeller, whereas a 3-vane/15-degree exit angle will have a steep curve and the widest impeller (refer to Figure 4-38). In other words, to peak at the same capacity the impeller discharge area must be the same, regardless of head-capacity relationships. Also, for a given impeller diameter, the head coefficient will be the highest for the flattest curve. The efficiency of the above impellers can be maintained within one point.

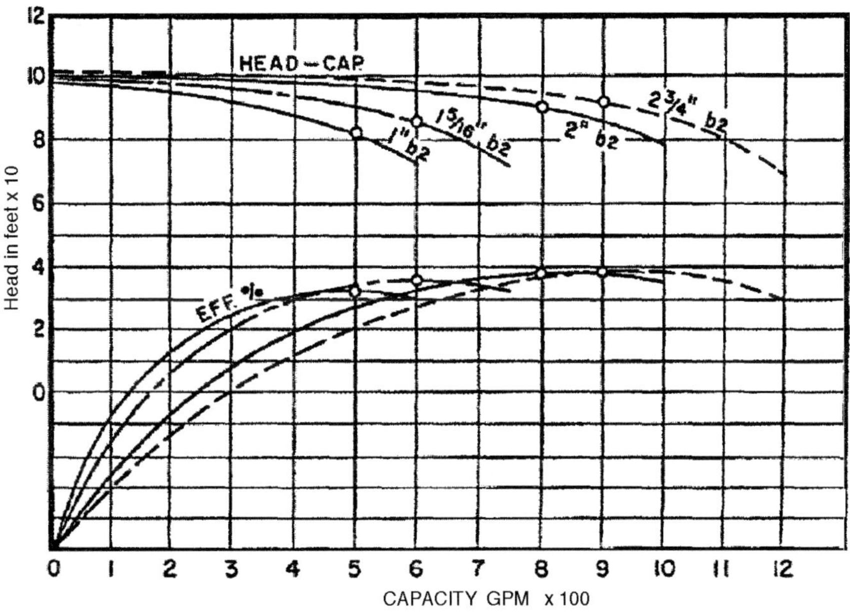

Figure 4-36: Effect of impeller width on pump performance

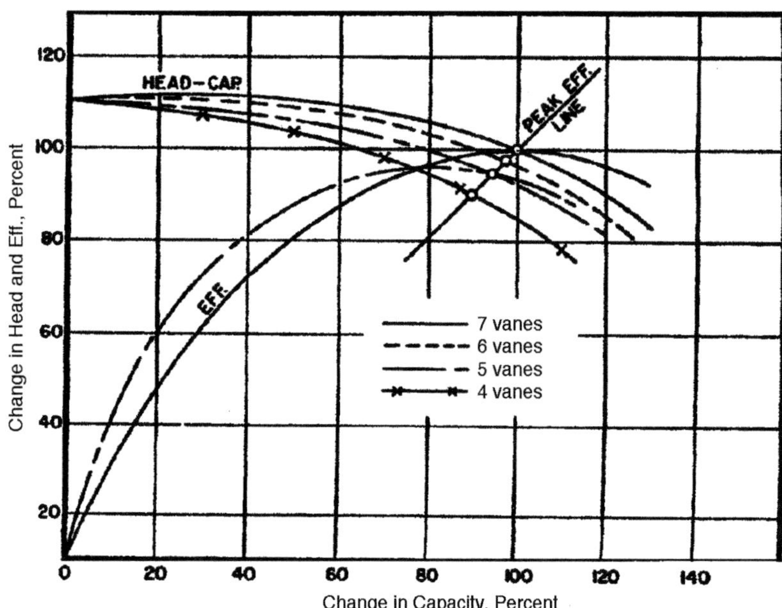

Figure 4-37: Impact of vane number on pump performance

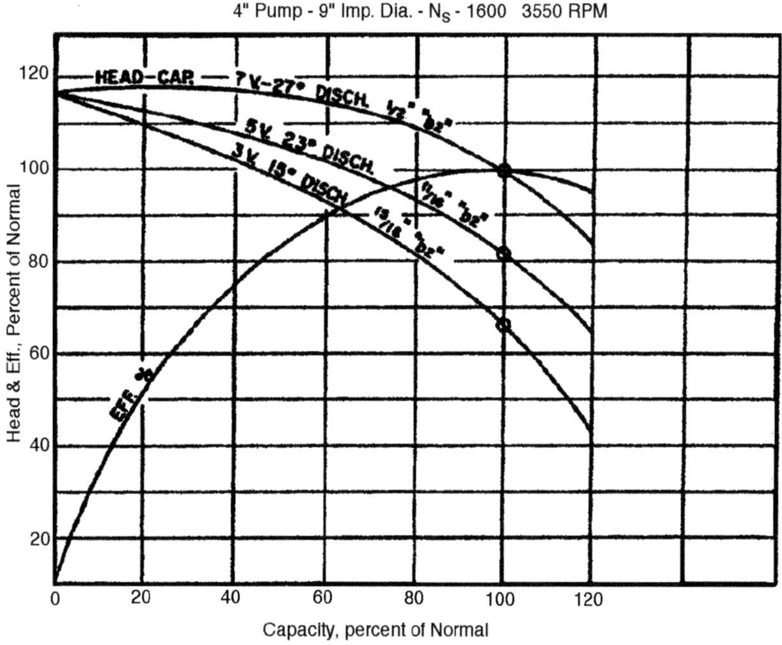

Figure 4-38: Comparison of various impeller designs

3. The slope of a head-capacity curve can also be increased by trimming the impeller outer diameter at an angle, with the front shroud diameter being larger than the back (hub) shroud, Figure 4-39.

4. Extending the impeller vanes further into the impeller eye can increase the slope of the head-capacity curve. An extreme version of this case is the addition of an inducer (Figure 4-40) in front of the impeller. The naturally steep head-capacity performance of the axial flow inducer is then added to the flatter performance of the lower specific speed impeller. Figure 4-41 compares the NPSHr trend of an inducer-less impeller with an impeller fitted with a "standard" inducer and an impeller with specially engineered inducer. It should be noted that off-the-shelf "standard" inducers may lower the NPSHr only in the vicinity of BEP operation.

Orificing Pump Performance

In low specific speed pumps, where impellers are already very narrow and low capacity or narrower impellers cannot be cast, capacity reductions can be obtained by using restriction orifices in the pump discharge nozzle. Figures 4-42 and 4-43 illustrate these points. Figure 4-42 shows the performance of a 2" pump where the discharge was throttled with different size orifices. Figure 4-43 shows the predicted performance of a throttled pump, and illustrates orifice size selections.

However, before embarking on orificing, impeller changes, or any other corrective

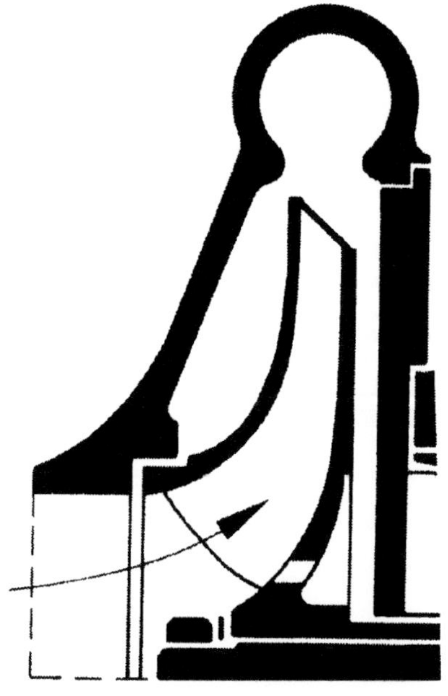

Figure 4-39: Impellers with angle-trim at the periphery exhibit steeper head-capacity curves

Figure 4-40: Inducers can be used to lower the required NPSH

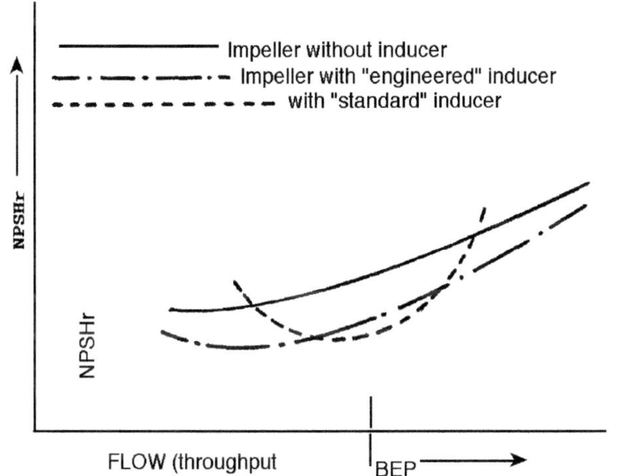

Figure 4-41: Inducers provide NPSH-lowering primarily near the best efficiency point (BEP)

measures, it would be wise to question the pump system design from the ground up. All too often, the actual friction differs from the originally calculated pipe friction. Some piping may have accumulated solids deposits and be smaller in diameter than originally intended. Elbows, filters and strainers may have been added somewhere in the history of the pumping circuit. The intersection of the head vs. flow curve with the systems resistance curve may be quite far from where it was when the pump was commissioned years ago. Again, and after thoroughly investigating the situation, it might be prudent to make rather fundamental changes in the overall design. In any event, thoroughness is one of the chief attributes of a good pump engineer, failure analyst or troubleshooter.

Saving Energy by Operating Pumps in Reverse as Hydraulic Turbines

Most centrifugal pumps can be run in reverse as hydraulic turbines to recover otherwise lost energy. Hydropower can be considered an excellent source of renewable energy. A good centrifugal pump operating with high efficiency may be expected to display good performance and reliability when the direction of flow is reversed and the pump is used as a driver. As a result, pumping equipment needed to satisfy the requirements of small hydropower users is readily available, unlike conventional turbines. Besides being readily available, pumps are also less complex than conventional turbines. They are more flexible in the sense that they can be mounted vertically or horizontally, wet pit, dry pit, and even in submersible mode. They can attain similar efficiencies to conventional turbines and, of prime importance to the small-site owner, they are normally less expensive and easier to cost-justify. Spare parts availability and shorter production lead times are also advantages for pumps.

The basic hydraulic behavior of centrifugal pumps operating as hydraulic power recovery turbines (HPRT or alternatively called "Pump as Turbine"—PAT) is not much different from that of a pump. Both follow the same basic affinity laws over narrow ranges. In most instances, no design changes or modifications are needed for a pump to operate as a turbine. The following simple relationships can be used to approximate the turbine best efficiency head, flow rate and range of efficiencies based on the known performance of the same unit operating as a pump. These relationships are reasonable approximations since, to date, no simple theoretical

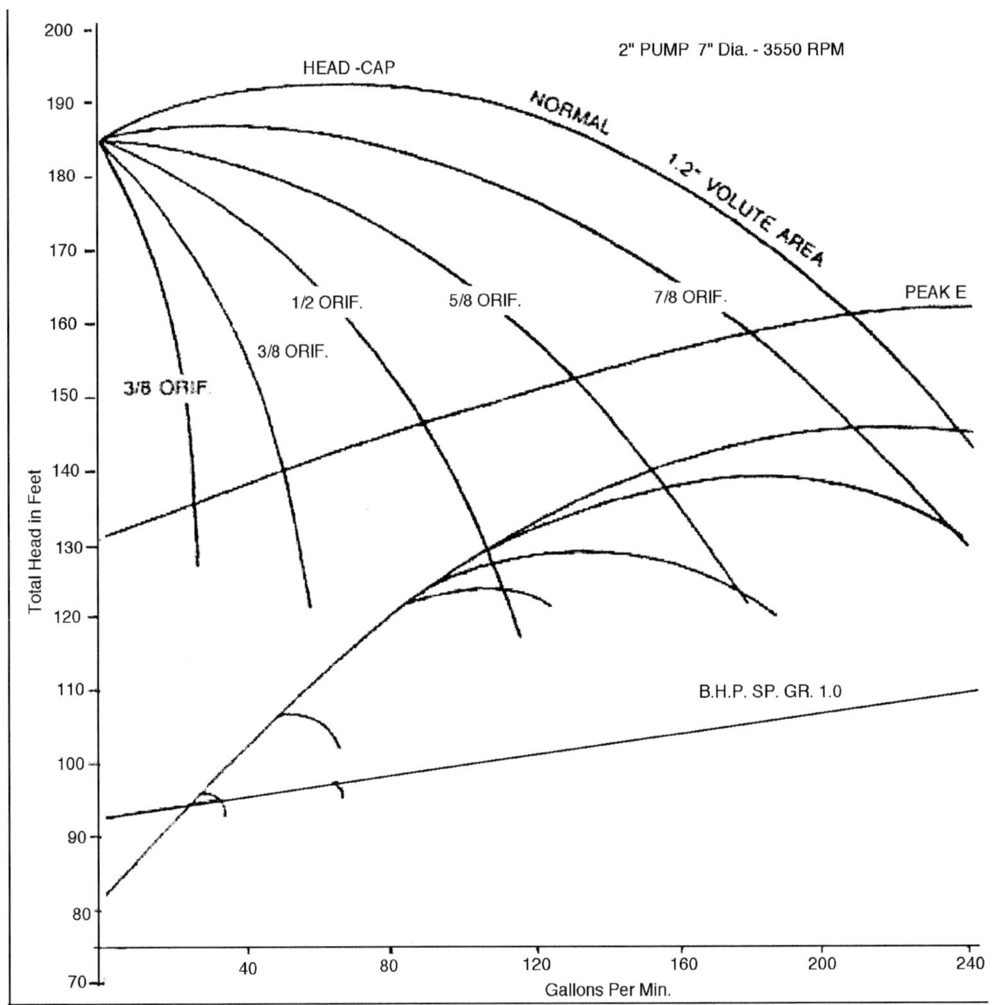

Figure 4-42: Performance with discharge orificing

method has been established for accurately predicting turbine performance. (Of course, there is always the option of actual physical testing.)

$$Q_{Turbine\ (bep)} = Q_{Pump\ (bep)} / Efficiency_{Pump\ (bep)}$$
$$H_{Turbine\ (bep)} = H_{Pump\ (bep)} / Efficiency_{Pump\ (bep)}$$
$$Efficiency_{Turbine\ (bep)} = Efficiency_{Pump\ (bep)}$$

where:

$Q_{Turbine\ (bep)}$ =
Best Efficiency Flow Rate as a Turbine in gpm

$Q_{Pump\ (bep)}$ =
Best Efficiency Flow Rate as a Pump in gpm

$H_{Turbine\ (bep)}$ =
Best Efficiency Head as a Turbine in feet

$H_{Pump\ (bep)}$ =
Best Efficiency Head as a Pump in feet

$Efficiency_{Turbine\ (bep)}$ =Best Efficiency as a Turbine

$Efficiency_{Pump\ (bep)}$ = Best Efficiency as a Pump

The pump manufacture should be contacted to obtain the actual performance of a particular pump operating as a HPRT. When a selection is made, a design review is also required, because when operating as a turbine the rotation is reversed and the operating heads, flow rates and power output are higher, as noted above. Consequently, a design review should include items such as:

- Check that threaded shaft components cannot loosen.
- Evaluate the adequacy of the bearing design.
- Shaft stress analysis.
- Checking the effect of increased pressure forces.

Pump-as-turbine performance can be presented in the form of either a constant head graph, as shown in Figure 4-44, or a constant speed curve, as shown in Figure 4-45. Of these two formats, constant speed curves are more common and readily accepted, since many

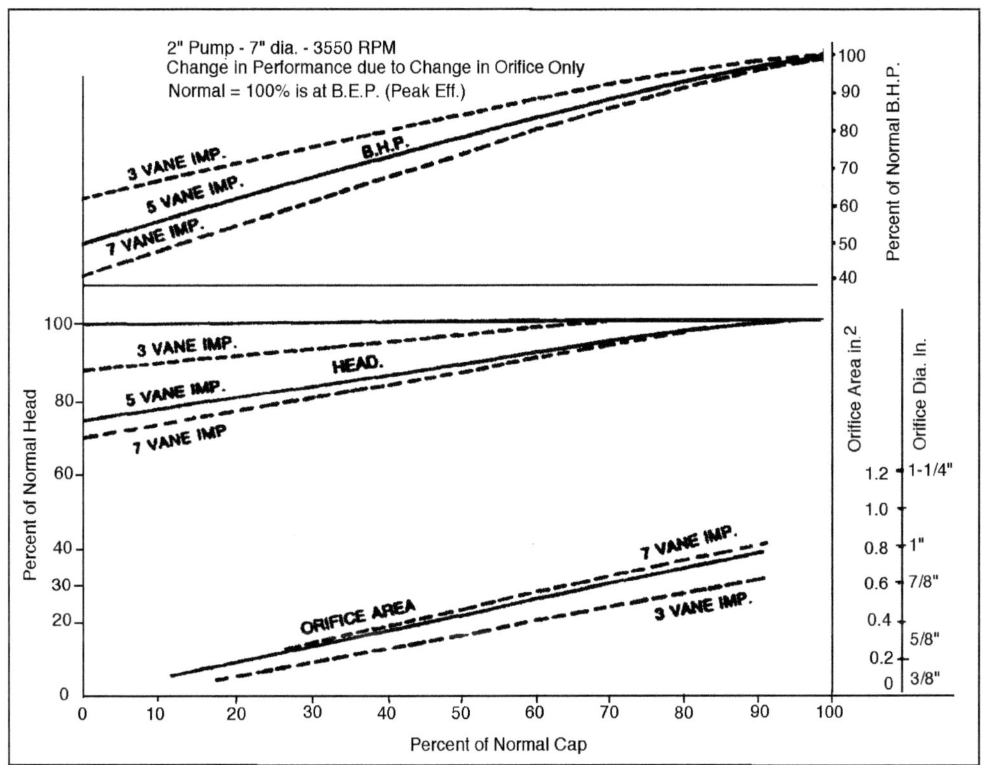

Figure 4-43: Change in performance due to change in orifice only

recovery turbines are operated in tandem with a steam turbine or electric motor, which holds the recovery turbine speed constant.

Figure 4-44 illustrates the variation of head-capacity, efficiency, brake horsepower and torque with speed for a typical PAT operating in a constant head system. Note that the torque available from the turbine reduces

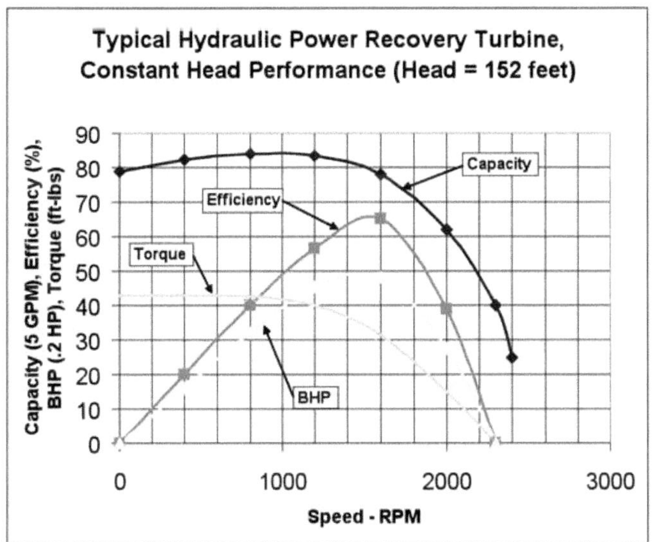

Figure 4-44: Typical PAT constant head performance, (2-1/2" axially split case pump, with 10" impeller)

to zero at approximately 146 percent of the design speed. This ratio will vary from 125 to 170 percent, depending on the pump. Thus, for mechanical considerations, a pump-as-turbine is not analogous to a steam turbine, or variable inlet vane conventional hydraulic turbine, in that elaborate over-speed trip devices may not be necessary.

From the PAT constant speed graph (Figure 4-45), it can be noted that the head-capacity curve is an exact opposite to what would be expected in a centrifugal pump, with the turbine shut-off head being approximately half the shut-off value obtained for the same unit operating as a pump. Also, it should be noted that the output power (bhp), and therefore the efficiency, are zero at approximately 40-42 percent of the turbine best efficiency flow rate. Energy must be added to the hydraulic turbine in order for it to rotate (at constant speed) at flow rates below this capacity. It is for this reason that hydraulic power recovery turbines tend to be selected to operate near or above the turbine best efficiency point (BEP), and that knowledge of the actual PAT best efficiency flow rate is critical. Conventional variable inlet vane hydraulic turbines have a definite advantage over PAT units below the BEP, since the power developed by conventional turbines does not fall off nearly as fast toward shut-off. Changes in casing nozzle throat size are the most effective way to change the performance of a PAT. On the other hand, changes in pump speed or impeller diameter are much less effective ways to change PAT head or output horsepower.

To develop the approximate constant speed performance of a pump operating as a turbine, Figure 4-45 can be coupled with the formulas above, which estimate the turbine best efficiency head, flow rate and efficiency. This can help with the initial pump selection.

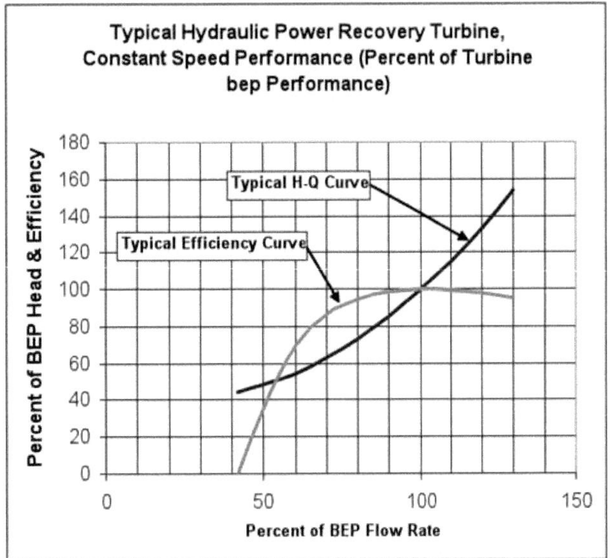

Figure 4-45: Typical Pump-as-Turbine Constant Speed Performance (Percent of Turbine BEP Performance)

In addition to using pumps operated in reverse to recover power from small waterfalls, or in place of throttle valves, PATs can be used in bypass flow control lines. An attractive applications for a PAT would be in the bypass line of the "hybrid control" system outlined earlier in this chapter. This would make the hybrid control system even more efficient when the system flow demand is low.

There are situations where step-less, infinitely variable flow, with constant speed and head, is required from a PAT installation. A common application for a pump as a turbine of this type would be as a pressure-reduction machine in a municipal water supply, or petrochemical process. A pump-as-turbine (PAT) should be installed in parallel with an existing pressure-reducing control valve, as shown in Figure 4-46, to recover the lost energy. The system head curve and PAT head curve, along with their efficiency curves, for this type of application are shown in Figure 4-47.

The system just described is quite sufficient for many applications. However, note that because some head is still wasted in valving, the system efficiency is not optimum. Further, as noted above, the output power (bhp), and therefore the efficiency of a constant speed, constant head PAT goes to zero at approximately 40-42 percent of the turbine best efficiency flow rate. Energy must be added to the hydraulic turbine in order for it to rotate (at constant speed) at flow rates below this capacity. To further improve the efficiency, multiple pumps-as-turbines could be used.

By installing multiple pumps, optimum efficiencies can be obtained from each on-line unit. Depending upon water flows, different pump units can be turned on or off to match inflows. This means that there are much lower losses from inefficiencies in throttling the pressure reducing or by-pass valves. Also, in case of repairs to one of the units, the only downtime will be the actual unit that is being serviced, allowing the others to keep working. With one main large PAT, if it breaks down, there is 100% loss in energy savings.

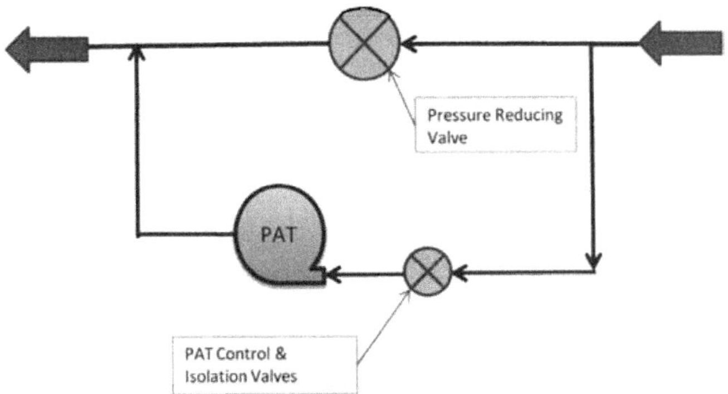

Figure 4-46: Schematic diagram for pump-as-turbine installed in parallel with a pipeline pressure reducing valve

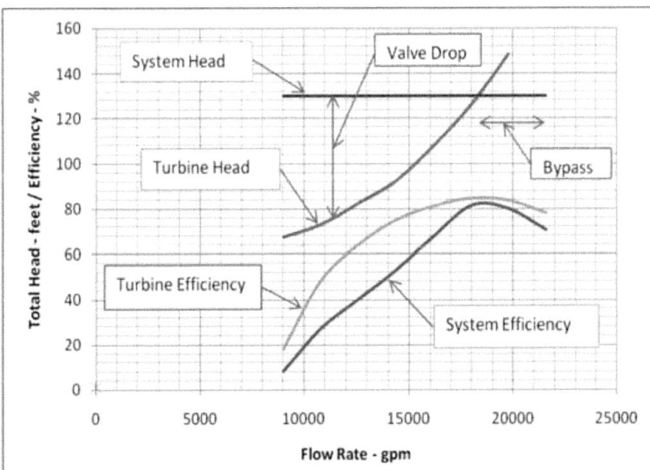

Figure 4-47: Variable-flow system with constant head, utilizing a single constant-speed pump-as-turbine

Figures 4-48 and 4-49 demonstrate how a system of this type could operate. Note the improvement in the system efficiency with this multiple pump-as-turbine configuration.

Constant speed applications are common, since the generators driven by a PAT are often interconnected to a utility, which means that the generators must operated at synchronous (60 Hz) speeds. This also allows for the use of an induction motor for the generator, which is very cost effective. It works quite well when interconnected to a utility, as an induction system requires no governor controls. The induction motor, instead of consuming energy, is driven at 50 RPM over its rated speed and the motor becomes a generator. Induction generators are much less expensive than other types of generators, but require excitation to operate. This is why they are ideally suited for interconnected utility applications.

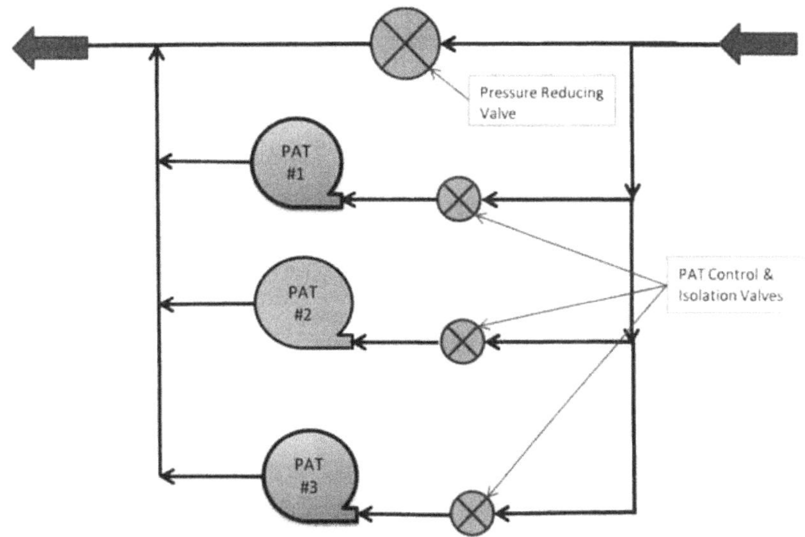

Figure 4-48: Schematic diagram for a multiple pump-as-turbine, variable flow system

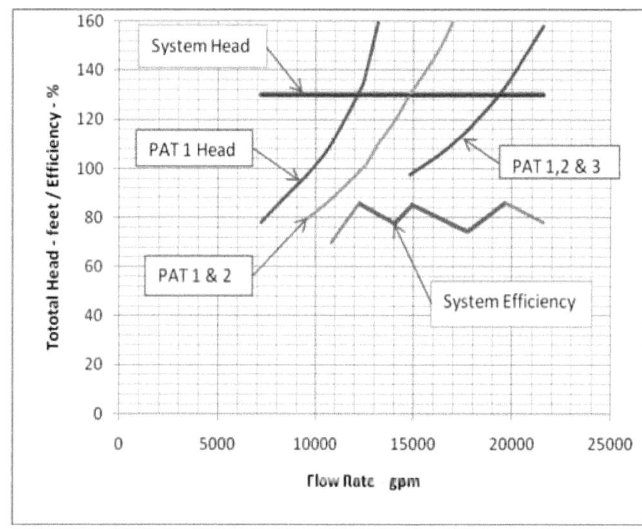

Figure 4-49: Variable-flow system with constant head, utilizing multiple constant-speed pumps-as-turbines

Another type of constant speed PAT application is where a PAT is installed in tandem, on the back end of a double extended motor shaft, and the motor is driving some other piece of equipment, such as a pump. The power generated from the PAT would, therefore, unload the electric driven motor. No separate speed control system would be required for such an installation, since the motor would control the speed. An overrunning clutch may want to be added to a tandem installation in order to disconnect the PAT during low flow operation, when the PAT would be absorbing instead of inputting power to the system.

The above demonstrates that, with a little imagination, a pump applied as hydraulic turbines has the potential to save substantial energy in numerous system applications. Each particular site must have a detailed evaluation to determine the most beneficial operating setup, to optimize the potential payback.

HOW TO OPTIMIZE THE OVERALL EFFICIENCY WHEN PUMPS ARE OPERATED AS BOTH PUMPS AND POWER RECOVERY TURBINES IN THE SAME SYSTEM

The objective of many PAT (pump as turbine) applications is to pump and retrieve water between two reservoirs at different elevations, as a means of energy storage and retrieval, to take advantage of lower energy rates at off peak periods, or to level out uneven energy generation.

In order to optimize the overall system efficiency for such an application, the different best efficiency (bep) flow points for pump and turbine operation, plus the different shapes of the pump and PAT head-capacity and efficiency performance curves, should be taken into consideration. The primary question for such operation then becomes: "What pump selection will allow the pump to operate on the desired system, so as to achieve a flow rate that will yield the optimum overall system efficiency (pump plus PAT)?"

Selection Factors

Such a pump selection, as shown in figure 4-50 (Ref 4-5) is complicated by the fact that the bep head and flow rate for a PAT is greater (by roughly the inverse of the pump bep efficiency) than the corresponding bep head and flow rate as a pump. In order to properly analyze this situation, the following factors must be considered and evaluated:

1. Determine the system head-capacity curve for pump operation. This includes the "static" (zero flow, elevation) head, plus the "friction" (velocity) head which increases as the square of the flow rate.
2. Make the initial pump selection and obtain the constant speed pump head-capacity curve. The maximum impeller diameter should typically be selected when the pump will also be used for PAT operation.
3. Plot the constant speed pump head-capacity curve against the system pump H-Q curve. The pump will operate at the intersection between the pump and pump system H-Q curves.
4. Determine the system head-capacity curve for turbine operation. This includes the "static" head, minus the "friction" head.
5. Establish the constant speed head-capacity curve for the pump operated in reverse as a turbine. The PAT bep can be approximated by dividing the pump bep head and capacity by the pump bep efficiency, with the curve shape being obtained from the dimensionless PAT H-Q curve shown in Figure 4-45, for constant speed operation. The pump manufacturer should be contacted for a more accurate/actual PAT curve, if one is available.
6. Plot the constant speed turbine (PAT) head-capacity curve against the system turbine H-Q curve. The PAT will operate at the intersection between the PAT and turbine system H-Q curves.
7. Obtain the constant speed pump efficiency vs. flow rate curve from the pump manufacturer.
8. Determine the pump efficiency for the above pump operating flow point.
9. Establish the PAT efficiency vs. flow rate curve. This can be approximated using the dimensionless plot presented in Figure 4-45, and assuming that the PAT BEP efficiency is equal to the pump BEP efficiency. It should be noted that, for VTP (vertical turbine pump) PAT operation, the BEP efficiency is normally about two points less than the pump BEP efficiency, due to the bowl vane angle mismatch with the impeller vanes, in reverse flow operation. Again, the pump manufacturer may be able to provide a more accurate PAT efficiency vs. flow rate, and head vs. flow rate curves. In all cases, at constant speed, the PAT efficiency will go to zero well before the zero flow point.
10. Acquire the PAT efficiency at the above turbine operating flow point.
11. Based on these two efficiency values (pump and PAT), determine the overall system energy storage and retrieval efficiency.
12. Repeat the above process for other pumps, to obtain the highest overall system efficiency. In one example, the best overall (pump x PAT) efficiency was obtained when the pump was selected to operate a little to the left of the pump BEP flow rate (between 90% to 95% of BEP). The optimum condition is dependent on the actual shape of the pump, PAT and system H-Q and efficiency curves.

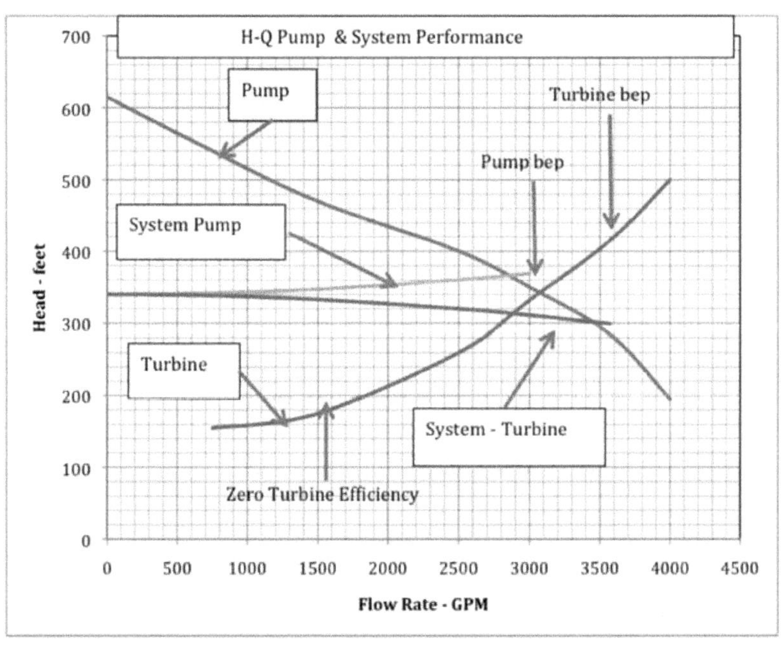

Figure 4-50: Pump, "Pump-as-Turbine," and Systems Head-Capacity constant speed performance

UNDERSTANDING PERIODIC SWITCHING OF PROCESS PUMPS AND PATS

Many process plants have "identical" centrifugal pumps (or Pumps as Turbines, "PATs") installed in a given service. Most of these were probably intended as spare equipment. They would be started up in the event the primary pump had to be serviced or repaired. In best practices plants, the two pumps are switched monthly. Switching is very important because it extends the bearing life of most pumps. For one, the lubricant is thus redistributed and corrosion damage is less likely. It should be noted that simply rotating the shaft one or two turns may not allow an oil ring (slinger ring) to throw oil into the bearing. Manual rotation is almost always inferior to alternate running of pumps.

Additionally, the extent of bearing damage due to vibration transmitted from an adjacent operating pump is cut in half. Operating each pump for about a month goes a long way to ensure that the rolling elements in the non-running pump do not remain in the same location for too long. More specifically, this reduces the severity and rate of incidents of a failure mode called ball indentation damage or "false brinelling." (The occasional concern that both pumps will wear out at the same time is unjustified and extremely unlikely).

As plants increase their throughput capacity, they often run two or more pumps simultaneously. Likewise, when pumping requirements are known to vary greatly it may be desirable to use several small pumps and stop one or more pumps when the throughput demand drops. The remaining pumps then operate closer to their respective best efficiency points (BEP). In new installations it is worth keeping in mind that, given a definite flowrate and head, several small pumps operating in parallel may allow increased pump speeds and may lower total initial pump cost. However, these pumps must have steep performance curves, typically with a 10% or greater head rise from operating point to shutoff.

Trouble-free parallel operation is possible only if the head vs. flow capacity curves (sometimes called H/Q curves) in Fig. 4-25 are relatively steep. A good rule of thumb would call for the head rise from operating point to shutoff off each pump to be 10 or more percent of the total head at BEP. New impellers are often available for existing pump casings and this upgrading may be easy to cost-justify.

Numerous references attest to the fact there is no such thing as two truly identical pumps. Each pump has its own H-Q performance curve because of its own unique internal roughness and wear or corrosion-affected clearances. There have been many instances of two pumps operating satisfactorily in parallel for years. Problems can occur suddenly after one pump has been overhauled or a new impeller has been installed. A thorough analysis of alternative solutions is always appropriate. These may include systems modifications and the use of a single, new, and more efficient pump instead of continuing to tolerate using two old, fundamentally weak pumps.

Chapter 5

Improved Pump Hydraulic Selection Extends Pump Life

INFLUENCE OF PUMP HYDRAULIC SELECTION

The major components of the life cycle cost of ownership are initial cost, installation cost, operating cost, and maintenance cost. In process plants it has been found that under many circumstances the cost of unscheduled maintenance is the most significant cost of ownership. Other chapters of this book deal primarily with mechanical means of improving reliability, while this chapter addresses pump life improvements that can be achieved through proper pump hydraulic selection. Further improvement could be realized if more attention were paid to key hydraulic factors, such as those associated with pump hydraulic forces (dynamic and static), cavitation, and wear within the pump (Ref. 5-1).

Improvements in MTBF will be limited in potential unless a holistic approach is used. Such an approach gives more attention to the best hydraulic fit to optimize reliability. The authors believe that there are four basic hydraulic selection factors that can have a significant effect on pump reliability. These are pump speed, percent of best efficiency flow, suction energy and NPSH margin ratio. These last two factors have further been combined into an NPSH margin reliability factor (NPSH-RF), which has been shown to be reasonably effective in predicting the reliability of high suction energy pumps.

The "Mean Time Between Repair" (MTBR) and Life Cycle Cost of most centrifugal pumps can be improved if slower pump speeds are used, and pumps are generally selected to operate in their preferred operating range (70% - 120% of BEP flow rate, Refs. 5-2 and 5-3). Further, the mean time between repair of high and very high suction energy pumps can be increased by keeping the NPSH margin ratio above certain minimum levels and/or by reducing the suction energy level. The easiest way to lower the suction energy and increase the NPSH margin of a pump application is by lowering the speed of the pump. This chapter deals with these issues.

Laboratory tests on three API end suction pumps, and maintenance data collected on 119 ANSI and split case pumps were used to validate the hydraulic selection reliability indicators.

OPERATING SPEED

Operating speed affects reliability through rubbing contact, such as seal faces, reduced bearing life through increased cycling, lubricant degradation and reduced viscosity due to increased temperature, and wetted component wear due to abrasives in the pumpage. Operating speed also increases the energy level of the pump, which can lead to cavitation damage.

Figure 5-1 compares the API-610 pump laboratory reliability predictor test results with the reliability trend line from actual MTBR data on 119 actual process pumps, as a function of the ratio of the actual to maximum rated pump speed (Ref. 5-3). The reliability factor for the field test data was based on zero pump repairs in a 48-month period, which was assumed to be equal to an MTBR of 72 months. Both curves show a marked increase in reliability with reduced speed.

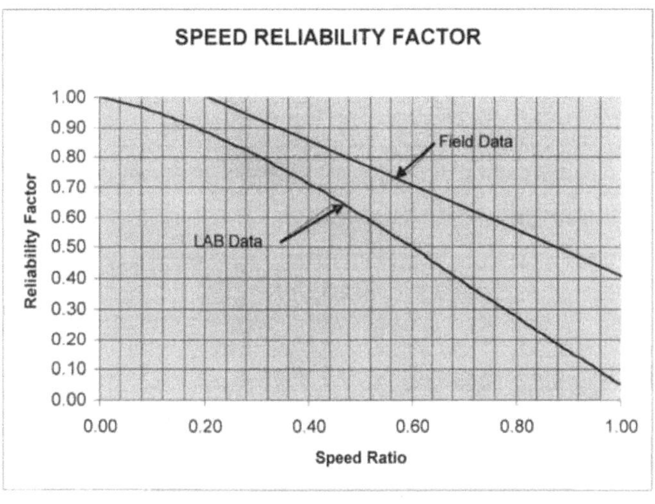

Figure 5-1: Pump speed reliability/life impact

ALLOWABLE OPERATING REGION

Design characteristics for both performance and service life are optimized around a rate of flow designated as the best efficiency point (BEP). At BEP, the hydraulic efficiency is maximum, and the liquid enters the impeller vanes, casing diffuser (discharge nozzle) or vaned diffuser in a shockless manner. Flow through the impeller and diffuser vanes (if so equipped) is uniform and free of separation, and is well controlled. The flow remains well controlled within a range of rates of flow designed as the preferred operating region (POR). Within this region, the service life of the pump will not be significantly affected by the hydraulic loads, vibration, or flow separation. The POR for most centrifugal pumps is between 70% and 120% of BEP.

A wider operating range is termed the allowable operating region (AOR). The AOR is that range of rates of flow recommended by the pump manufacturer over which the service life of a pump is not seriously compromised (Ref. 5-4). Service life within the AOR may be lower than within the POR. To use an analogy: While it may be possible to drive a standard automobile in first gear at speeds in excess of 25 mph (40 km/h), doing so for a long time will come at a price.

The following factors determine the AOR, with the degree of importance dependent on the pump type and specific design:

- Temperature rise
- Bearing life
- Shaft seal life
- Vibration
- Noise
- Internal mechanical contact
- Shaft fatigue failure
- Horsepower limit
- Liquid velocity in casing throat
- Thrust reversal in impeller
- NPSH margin
- Slope of the head-rate-of-flow curve
- Suction recirculation (Ref. 5-5)

The flow ratio (actual flow rate divided by BEP flow rate) affects reliability through the turbulence that is created in the casing and impeller as the pump is operated away from the best efficiency flow rate. As a result, hydraulic loads, which are transmitted to the shaft and bearings, increase and become unsteady. Also, the severity of these unsteady and often oscillating loads can reduce bearing and mechanical seal life.

Operation at reduced flow rates that put the pump into its recirculation mode can also lead to cavitation damage in high suction energy pumps.

Field and laboratory reliability data are presented in Figure 5-2, as a function of the flow ratio. Correlation between the field and laboratory data is good in the normal operating range, with the maximum reliability existing around 90 percent of the best efficiency flow rate.

SUCTION RECIRCULATION

Suction recirculation is a condition where the flow in the inlet portion of an impeller is separated from the vanes and ejected upstream, opposite to the direction of net flow entering the impeller, often well into the suction pipe (see Figure 5-3). This forms eddies and vortices within the impeller inlet, and pre-rotation of the liquid entering the pump. It occurs in all centrifugal pumps at reduced pump capacities, normally below the best efficiency rate of flow. In part, it helps a pump adjust to the lower throughput. Flow recirculation may, or may not, cause pump noise, vibration, erosion damage, and/or large forces on the impeller (Refs. 5-1 and 5-6). It can also cause the pump alignment to change, all of which tend to affect shaft seal and bearing life.

Flow Instabilities

Suction recirculation was first documented (in the 1950s) through investigations by noted pump expert Val Lobanoff. He demonstrated that the allowable pump operating region is smaller, and the minimum flow higher, with decreasing values of NPSHR, and subsequently increasing values of "Suction Specific Speed" (Nss), due to flow instabilities at low flow rates (from low system demand), in a number of different (often oversized)

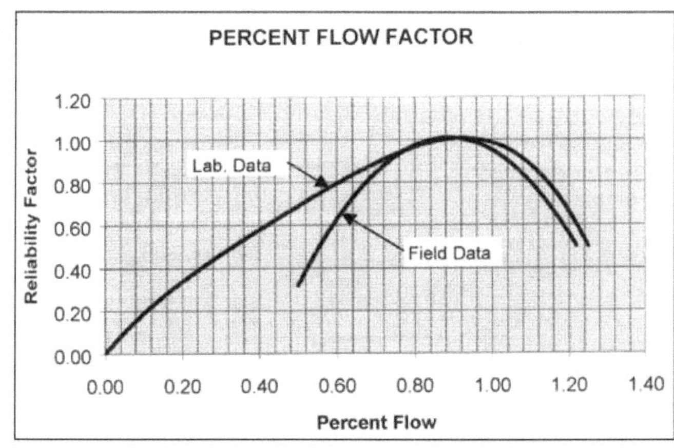

Figure 5-2: Percent flow factor

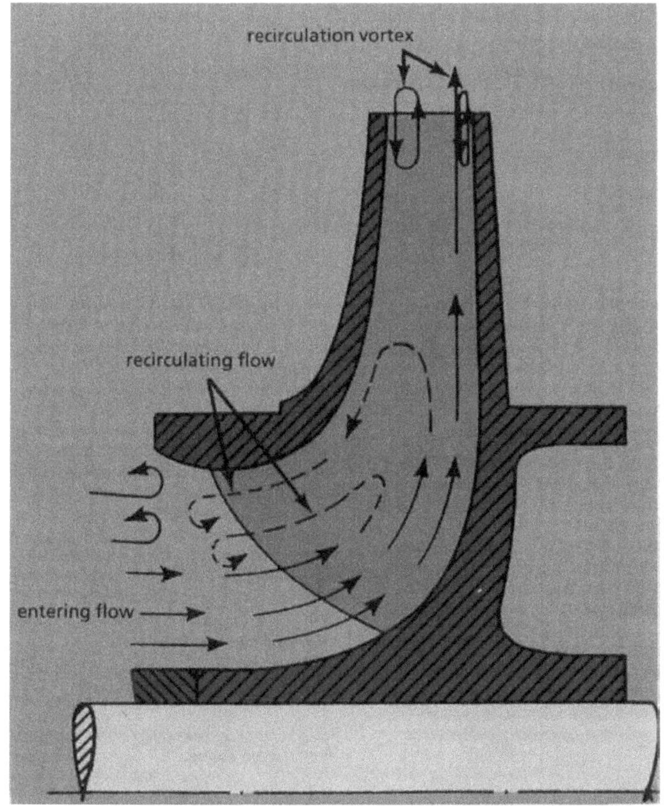

Figure 5-3: Suction recirculation

pumps. We now know that these flow instabilities are caused by suction recirculation. It should be noted that Nss (or "S" as it is sometimes referred to, see Equation 5-1), is a function of pump speed, the square root of the BEP flow rate, and inversely proportional to the NPSHR to the ¾ power.

In addition to suction specific speed, pump flow instabilities and the allowable operating region, are also impacted by the pump specific speed (Ns – see Figure 4-31). Figure 5-4, which was published in a recent article, presents this relationship, which recommends that higher specific speed pumps should be limited to distinct lower values of suction specific speeds (higher values of NPSHR), for "attainable and acceptable performance."

However, the above two criteria (Nss, and Nss plus Ns), that were developed to aid in determining the minimum allowable pump flow, are missing an additional important parameter, the impeller inlet tip speed (speed x diameter). This additional important parameter (inlet tip speed) is critical in determining if there is actually enough energy at the pump inlet to cause damage within the "flow instability" (suction recirculation) region. A pump operating at a very slow speed (which generates low suction energy) will not normally be damaged by cavitation or suction recirculation, even with high values of specific speed and/or suction specific speed. In addition, higher "inlet tip speed" is one of the reasons that, when the speed of a pump is increased, the reliability (MTBF) typically decreases (see Figure 5-1). It should be noted that suction specific speed and specific speed values are constant for a given pump geometry, regardless of the pump speed. This is why one of the authors (Mr. Budris) developed the concept of "Suction Energy," which includes both the pump suction specific speed and impeller inlet tip speed (speed x diameter). Once a pump is determined to have "High Suction Energy" (Nss x Inlet Tip Speed x specific gravity), the impact of the pump suction specific speed (Nss) and specific speed (Ns) on the minimum allowable flow (based on the start of suction recirculation) can then be established, as shown in Figure 5-5. This figure only applies to "High Suction Energy Pumps" since they have a high probability of being damaged by operation in the low flow suction recirculation flow region. The three curves show the effect of different specific speed levels on the start of suction recirculation, which increases with increasing

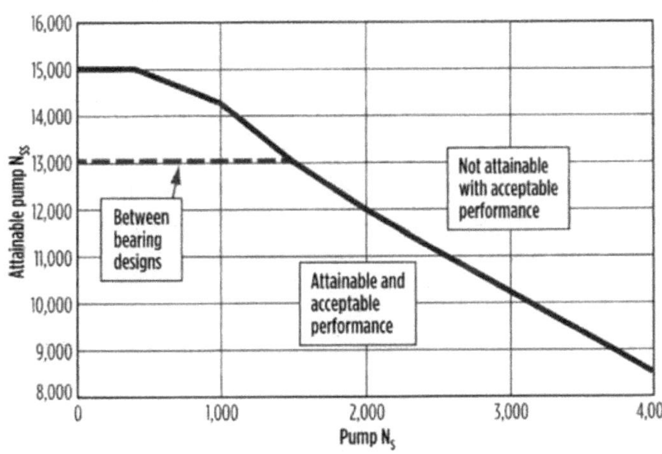

Figure 5-4: "Attainable and Acceptable" performance varies as a function of suction specific speed, Nss, on vertical axis, and pump specific speed, Ns, on horizontal axis. (Source: Bailey, J. and Bradshaw, S.; "Avoid Hidden Costs of Suction-Specific Speed in Pumping," *Hydrocarbon Processing*, May 2013)

specific speed and suction specific speed values. This then results in reduced allowable operating regions for higher values of suction specific speed and specific speed.

So the likelihood of pump damage in low flow suction recirculation (instable flow region) is a function of pump specific speed and suction energy (which includes the pump suction specific speed). In addition, the probability of damage in suction recirculation is also impacted by the NPSH margin to the pump, the nature of the flow provided by the suction piping, and to some extent the liquid being pumped.

Experience has shown that low suction energy pumps are not susceptible to damage from suction recirculation (Refs. 5-6 and 5-7). However, solids and/or corrosives can accelerate damage during suction recirculation, even with low suction energy applications.

MITIGATING SUCTION RECIRCULATION PROBLEMS

Other than avoiding the low flow suction recirculation zone all together, or selecting only low suction energy pumps, the normal first fix for a suction recirculation problem is to install an external by-pass line to bring the net pump flow rate above the start of suction recirculation.

If one of the above actions is not practical, there is a relatively simple pump modification which, in a number of cases, has been used quite successfully to reduce and even sometimes eliminate the unfavorable effects of suction recirculation. It consists of retrofitting pumps with a stationary casing ring, the apron of which extends inwardly of the impeller eye diameter (Figure 5-4a). (Ref. 5-12) If preferred, such rings can instead be rotating and mounted on the impeller. Such rings are commonly referred to as "bulk-head rings." This prevents the recirculation vortex from extending axially beyond the plane formed by the apron. Of course, since this does increase the required NPSH, the use of these bulk-head rings can only be resorted to if there is sufficient margin in the available NPSH. This was the fix used to solve a California pump recirculation pier damaging problem by one of the authors.

An additional interesting fix, a "stabilizer" can also alleviate this problem in certain cases. A stabilizer is a reduced diameter pipe, 4 to 12 inches (~100-300 mm) long, supported by equidistant struts, concentrically welded into the suction line (see Figure 5-4b) directly upstream of the pump suction nozzle. Occasionally, stabilizers are allowed to partially protrude into the pump casing. On certain pump sizes, these stabilizer pipes deliver stable performance over an extended flow range. Stabilizer performance prediction is largely empirical, which is probably why their existence is not widely known. Commercial versions are available as well.

On pumps without these bulk-head rings, or stabilizers, performance prediction is facilitated by graphics similar to Figure 5-5. In the absence of actual test results, Figure 5-5 can be used to approximate the start of suction recirculation for high suction energy pumps.

In essence, failure avoidance includes paying close attention to issues that center on hydraulic distress and avoiding low flow operation. High suction energy pumps should not be operated for extended periods of time in the suction recirculation region.

It should be noted that, the location of the pump material damage is an excellent diagnostic tool in iden-

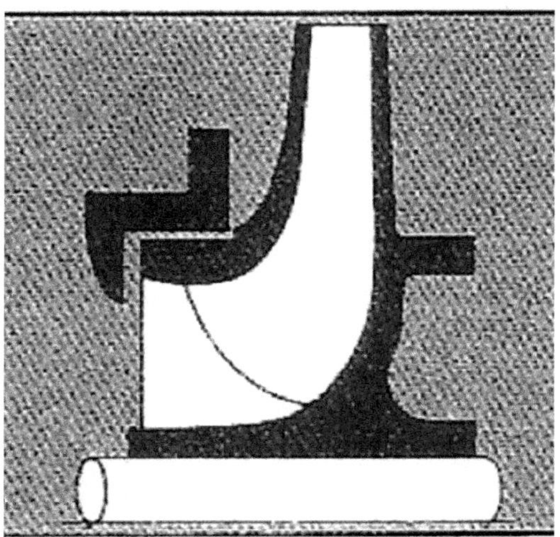

Figure 5-4a : Bulk-head ring construction

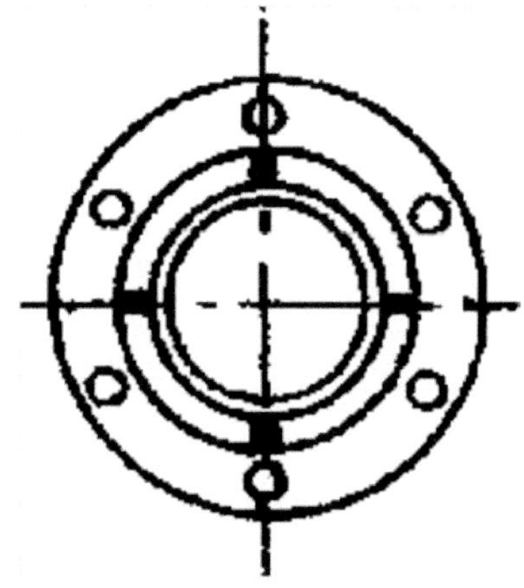

Figure 5-4b: Suction Recirculation "Stabilizer"

tifying whether the cause is classic cavitation or internal suction recirculation caused cavitation. If the damage is to the hidden (high pressure) side of the impeller vanes, and must be seen with the help of a small mirror, the cause is suction recirculation. Classic cavitation damage occurs on the visible (low pressure) side of the impeller vane, a short distance back from the leading edge.

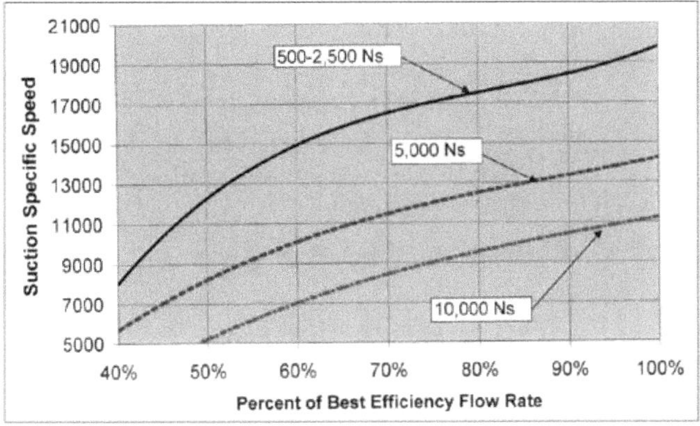

Figure 5-5: Minimum continuous stable flow

SUCTION ENERGY

Due to the very high NPSH margins required to completely suppress cavitation—normally 2 to 5 times the NPSHr of the pump, in the allowable operating range (AOR), we know that cavitation is sure to exist in a high percentage of pump applications. However, we also know that acceptable life is achieved in most installations, despite this cavitation. So how can we predict when cavitation is likely to cause problems? The amount of energy in a pumped fluid which flashes into vapor and then collapses back to a liquid in the high pressure areas of the impeller, determines the extent of noise and/or damage from cavitation,

Suction energy is another term for the liquid momentum in the suction eye of a pump impeller, which means that it is a function of the mass and velocity of the liquid in the inlet. Suction energy, as originally approximated in Ref. 5-6, is defined as follows:

$$\text{Suction energy (S.E.)} = D_e \times N \times S \times S.G. \qquad \text{Equation (5-1)}$$

Where:

D_e = Impeller eye diameter (inches)
N = Pump speed (rpm)
S = Suction specific speed (rpm × (gpm)$^{.5}$/(NPSHr)$^{.75}$
S.G. = Specific gravity of liquid pumped

If the impeller eye diameter is not available, suction energy can still be estimated by using the following approximations:

- End suction pump:
  D_e = Suction nozzle diameter × 0.9
- Horizontal split case and/or radial inlet pump:
  D_e = Suction nozzle diameter × 0.75

Since the suction energy numbers are quite large, the last six digits are normally dropped (S.E. × E6).

Ref. 5-6 also proposed distinct gating values for high and very high suction energy, for end suction, Figure 5-6, and radial suction, Figure 5-7 (also known as split case or double suction) pumps, based on the analysis of hundreds of pumps from several manufacturers.

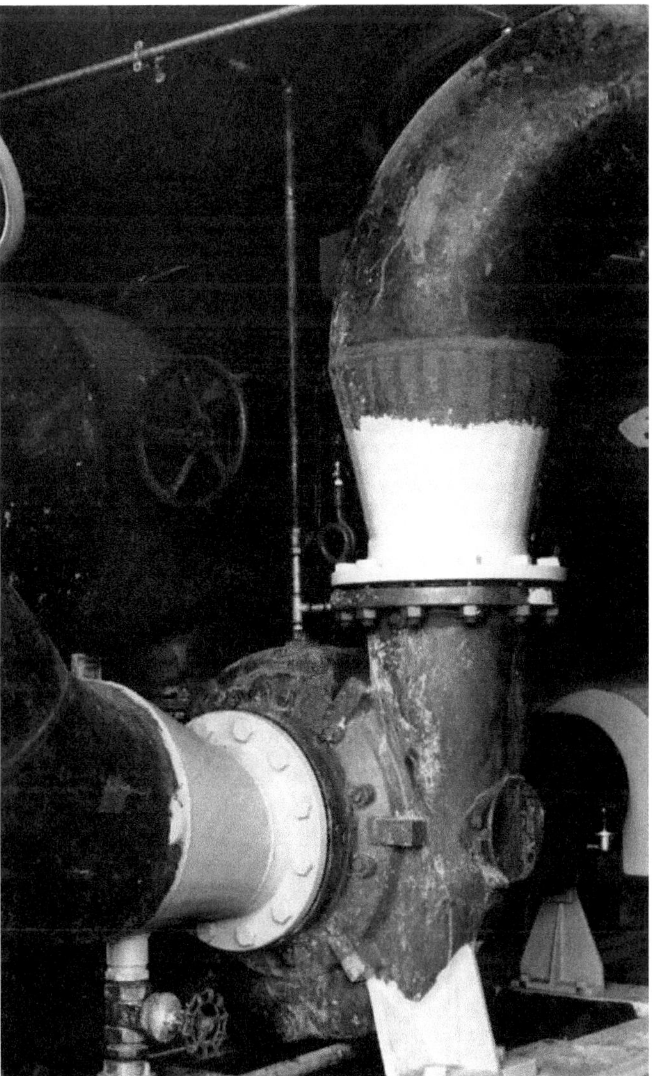

Figure 5-6: End suction pump (Source, KSB, Frankenthal, Germany)

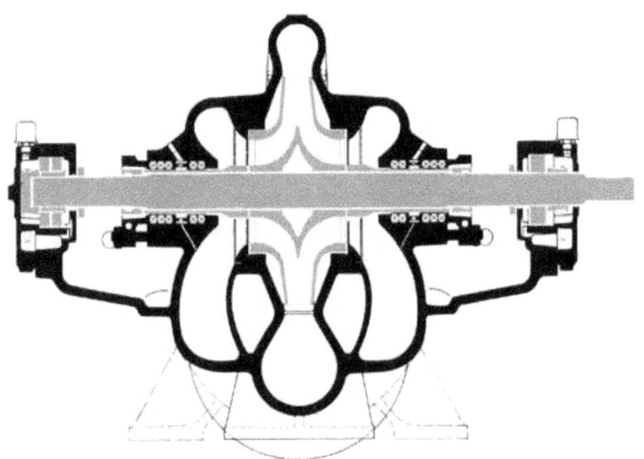

Figure 5-7: Single stage, axially split, double-entry volute casing pump (Source: Ruhrpumpen, Germany)

- Start of high suction energy:
  End suction pumps: S.E. = 160 × 10^6
  Split case & radial inlet pumps: S.E. = 120 × 10^6
- Start of very high suction energy:
  End suction pumps: S.E. = 240 × 10^6
  Split case/radial inlet pumps: S.E. = 180 × 10^6

The above definition of suction energy (Equation 5-1), and "high" and "very high" gating values are consistent with values presented in Ref. 5-12.

Suction Energy Gating Values

Based on the experience of hundreds of centrifugal pumps, the specific gating values for the start of "High Suction Energy" and "Very High Suction Energy," for various pump types, were established and are listed in Table 5-1.

In order to determine the likelihood of damage from cavitation, and/or suction recirculation, the calculated suction energy level (determined from the above formula) must be compared with the pump type gating value from Table 5-1. This table is based on field experience and the expected turbulence in the pump suction passage, due to the turns the inlet flow must navigate. This recognizes that not all pumps respond the same to the same suction energy level. To account for this, the suction energy ratio (SER) was established, and defined as the "Actual Suction Energy"/"Start of High Suction Energy Gating Value," which has proven to be a more accurate indicator of pump damage potential. From this, the specific pump suction energy severity levels have been developed, where Low Suction Energy is defined as SER values below 1.0, high suction energy has SER values of 1.0 to less than 1.5, and Very High Suction Energy has SER values of 1.5 and above.

It should be noted that some of the newer pump designs, developed with one of the computational fluid dynamics (CFD) computer programs, can operate with higher suction energy values, without developing damaging cavitation. Some split case/radial inlet pumps have superior impeller inlet designs, so that they can use the "end suction" gating values. Also, the shape of the leading edge of some impeller vanes has been shown to make the pumps less susceptible to cavitation damage. These improved cavitation resistant designs should, however, be verified by field experience.

Low Suction Energy

Pumps with levels of suction energy below the Table 5-1 "High Suction Energy" values are considered to have "Low Suction Energy." Low suction energy pumps are not prone to noise, vibration or damage from cavitation or recirculation, with just a few exceptions. When handling abrasives or a liquid that is corrosive to the pump materials of construction, even this low energy cavitation can be amplified to cause damage. Also, there could be detrimental effects on mechanical seals from the air or vapors which may be liberated from the liquid, during the formation of the cavitation bubbles, under very low NPSH margin conditions. Finally, there will be a small loss of the pump discharge head as the suction pressure approaches the pump NPSHR. A minimum NPSH margin ratio of 1.1 – 1.3 is typically sufficient for low suction energy pumps, not handling abrasives or corrosives.

High Suction Energy

Pumps with high suction energy and low NPSH margins, especially when operated in the suction recir-

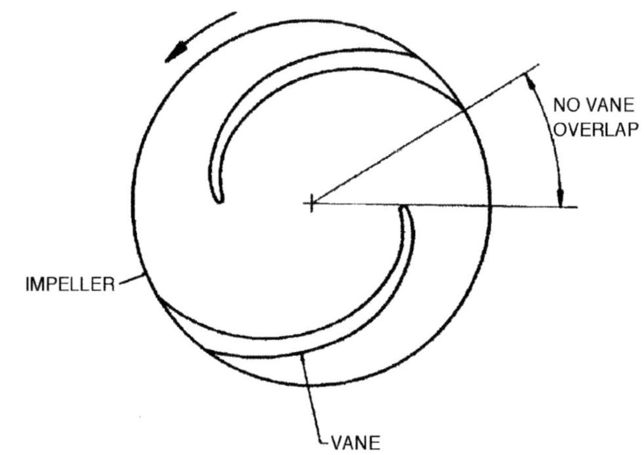

Figure 5-8: Impeller vane overlap

culation flow range (below the pump "Allowable Operating Region," AOR), may experience noise, vibration and/or minor cavitation erosion damage with impeller materials that have low cavitation resistance, such as cast iron (see Table 5-3). High suction energy starts at about 3,560 rpm in end suction pumps with 6" suction nozzles sizes, and split case pumps with 8" suction nozzles. At 1,780 rpm, high suction energy typically starts with 10" suction nozzle size end suction pumps, and 12" suction size split case pumps.

In addition to the impeller material, there are several other variables that can further mitigate cavitation damage under High Suction Energy conditions. The two major additional mitigating factors are: entrained gas (air) levels approaching one percent, such as found in paper stock and some sewage applications; and secondly: liquids that do not generate large cavitation bubbles (low thermodynamic liquids), such as hydrocarbons and very hot water. The most damaging liquid is water, at about 100°F to 120°F. High suction energy pumps, without one or more of the above mitigation factors, typically require minimum NPSH margin ratios of 1.5 to 2.0, within the AOR.

Very High Suction Energy

"Very High Suction Energy" starts at 1.5 times the start of "High Suction Energy." Pumps with very high suction energy and low NPSH margins, especially when operated in the suction recirculation flow range (below the pump AOR), may experience high vibration and erosion damage, even with cavitation resistant materials, such as stainless steel. Some reduction in damage, or extension in pump life, can be expected with the mitigating factors listed in the "High Suction Energy" section above. The typical recommended minimum NPSH margin ratio for very high suction energy pumps is 2.0 to 2.5, within the AOR.

"Very High" suction energy pumps should not be operated in suction recirculation for any extended period of time. However, if such operation is unavoidable, the minimum NPSH margin under these conditions should be at least 3.0.

Suction Energy Impact on Reliability

Figure 5-9 shows the impact of pump suction energy on pump reliability, and it is based on field data from close to one hundred ANSI and split case pumps. A reliability factor of 1.0 equates to no failures in 48 months, or a MTBR rate of 72 months. The trend is unquestionable, with higher suction energy pumps requiring the most repairs.

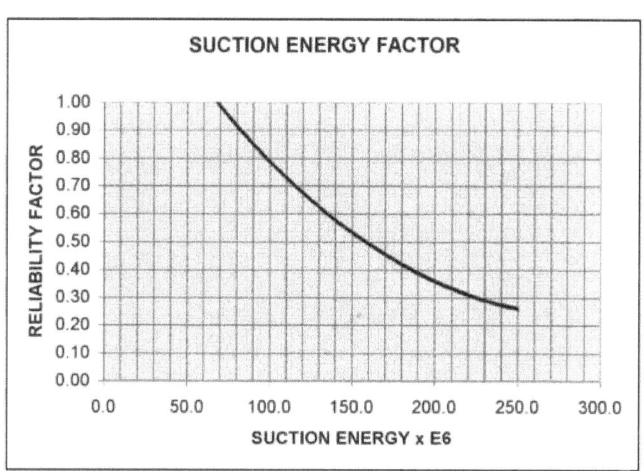

Figure 5-9: Suction energy factor

Example

To illustrate how to calculate suction energy, let us assume a split case pump with the following parameters:

Speed	=	1,780 rpm
Flow rate (BEP)	=	10,400 gpm (5,200 gpm per impeller eye)
NPSHr	=	35 ft.
S	=	1780 rpm × (5200 gpm)$^{0.5}$/ (35 ft)$^{0.75}$ = 8,920
Suction nozzle size	=	16 inch
D_e	=	16" × 0.75 = 12"
Specific Gravity	=	1.0 (cold water)
Suction Energy	=	12.00" × 1,780 rpm × 8,920 × 1.0
Suction Energy	=	190 × 10^6

This is a "very high suction energy" pump, since for split case pumps, high suction energy starts at 120 × 10^6 and very high suction energy starts at 180 × 10^6. Extreme care should be taken to ensure that very high suction energy pumps have good suction piping and adequate NPSH margin.

NPSH MARGIN

Having a system suction pressure at the entry to a pump in excess of the manufacturer's stated value of net positive suction head required (NPSHr) is no guarantee of satisfactory pump performance. You may need additional margin to suppress the cavitation that exists in a pump substantially above the published NPSHr value. Cavitation does exist above the true NPSHr, and may be noticeable at the NPSHr given on the vendor's proposal or even test curves. As a general comment, Hydraulic Institute standards allow vendors to quote as NPSHr a point, at which the differential pressure across a pump is reduced by 3% due to cavitation. On high energy pumps, cavitation may be both present and audible at the 3% deviation point.

More specifically, the NPSH margin ratio is defined as the NPSH *available* to the pump in a particular application, divided by the NPSH *required* by the pump. By Hydraulic Institute definition, the NPSHr of a pump is the NPSH that will cause the total head to be reduced by 3%, due to flow blockage from cavitation vapor in the impeller vanes. However, and to complicate matters, NPSHr is by no means the point at which cavitation starts. That level is referred to as incipient cavitation. It can take an NPSHa of from two to twenty times NPSHr to fully suppress cavitation within a pump, depending on pump design and flow ratio (percent BEP). It can take from 1.05 to 2.5 times the NPSHr value just to achieve the 100 percent head point (NPSH "required"—0%). The higher values are normally associated with high suction energy, high specific speed, pumps with large impeller inlet areas, or reduced flow operation in the region of suction recirculation. Collectively, References 5-6 through 5-8 give further insight into the matter.

Conventional Wisdom: *Pumps will not cavitate as long as NPSHa exceeds NPSHr by one ft (0.3 m) or more*

Fact: *For cavitation avoidance, NPSHa/NPSHr may have to be as high as 20!*

This means that a high percentage of pumps are operating with some degree of cavitation. It is the amount of energy associated with the collapse of the vapor bubbles that determines the degree of noise, vibration or damage from cavitation, if any.

Table 5-1: Pump types and suction energy categories

Pump type	*Start of "high suction energy"*	*Start of "Very high suction energy"*
2-Vane Sewage Pumps	100×10^6	150×10^6
Double Suction Pumps	120×10^6	180×10^6
End Suction Pumps	160×10^6	240×10^6
Vertical Turbine Pumps	200×10^6	300×10^6
Inducers	320×10^6	480×10^6

Table 5-2 summarizes the minimum NPSH margin ratio guidelines (NPSHa/NPSHr), which are applicable within the allowable operating region of the pump (Ref. 5-9).

High and very high suction energy pumps that operate with the minimum NPSH margin values recommended in the above table will normally have acceptable seal and bearing life, but not necessarily optimal. They may still be susceptible to elevated noise levels and erosive damage to the impeller. This can require more frequent impeller replacement than would otherwise be experienced, had the cavitation been totally eliminated. It will typically take a NPSHa of 4 to 5 times the 3% NPSHr of the pump to totally eliminate potentially damaging cavitation in a high suction energy pump.

Additional NPSH margin may be needed to cover uncertainties in the NPSHa to the pump or operating flow point (see Figure 5-10). If a pump runs farther out on the curve than expected (which is very common), the NPSHa of the system will be lower than expected and the NPSHr for the pump will be higher, thus giving a smaller (or possibly negative) actual NPSH margin. All pumping systems must be designed to have a positive margin throughout the full range of operation.

NPSHa margins of two to five feet are normally required (above those shown in Table 5-2) to account for these uncertainties in the actual NPSHr and NPSHa values. This added margin requirement could be even greater depending upon the severity of the conditions, especially if the pump is operating in suction recirculation. If the application is critical, a factory NPSH test should be requested (Refs. 5-8 and 5-9).

Figure 5-11 shows the effect of the NPSH margin ratio on pump reliability, based on close to one hundred operating pumps. It is evident from these data that the

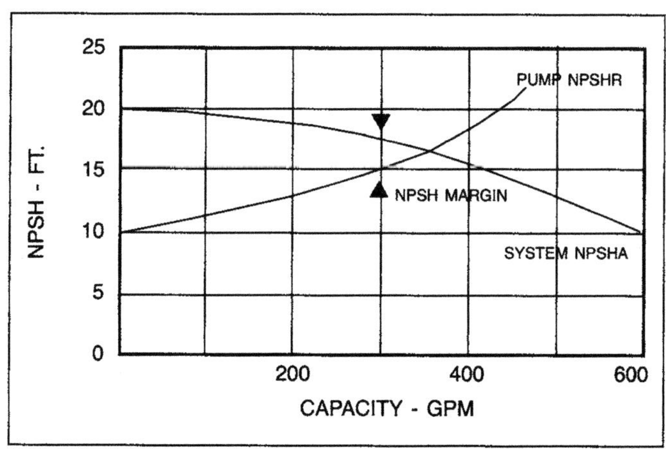

Figure 5-10: NPSH margin vs. flow rate

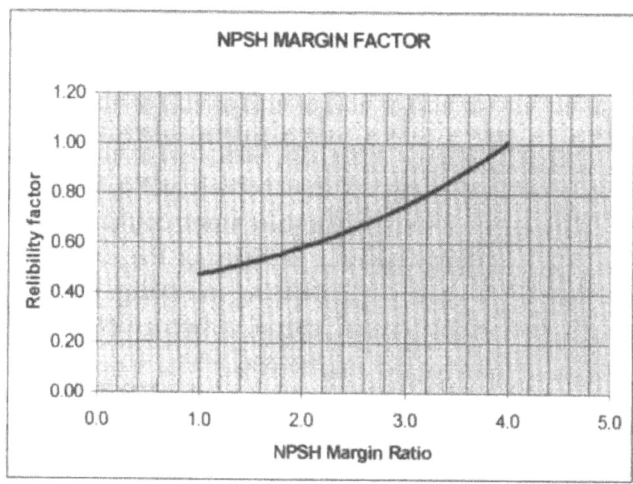

Figure 5-11: NPSH margin factor

NPSH margin ratio has a definite influence on pump reliability, especially for high and very high suction energy pumps. In fact, some cavitation usually exists below a ratio of 4.0.

Another factor to consider, when establishing the desired NPSH Margin for a pump, is that the actual field NPSH Margin may differ from the original calculated value, due to an assortment of uncertainties, such as:

- The accuracy of the initial pump NPSHR test. Even an NPSHR test performed to Hydraulic Institute standards has a tolerance.
- Any flow disturbance from the inlet pump piping
- The air content of the liquid.
- The liquid temperature.

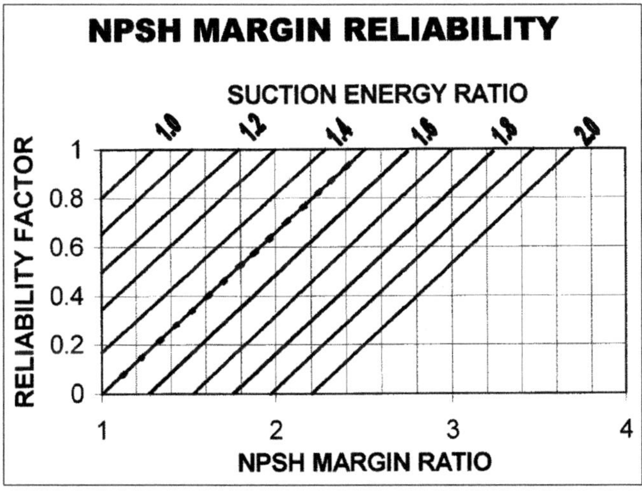

Figure 5-12: NPSH margin reliability factor

NPSH MARGIN RELIABILITY FACTOR

The NPSH margin reliability factor (Figure 5-12) was developed from field experience and guidelines from Ref. 5-10 (Table 5-2). It provides a reasonable approximation of the reliability that can be expected for aqueous liquids when various NPSH margin ratios are applied to high suction energy pumps of increasing energy levels (suction energy ratios)—see Figure 5-12. NPSH margin reliability factors are based on the fact that, above the gating suction energy values (start of high suction energy—see Table 5-1), cavitation becomes severe (Ref. 5-9 and 5-10). In other words, the greater the suction energy, the more important it is to suppress the residual cavitation that exists above the NPSHr, to prevent damage. This reliability factor is only applicable within the allowable operating flow region, above the start of suction recirculation. Much higher NPSH margin values are required in the region of suction recirculation, for high and very high suction energy pump applications.

Table 5-2: Minimum NPSH margin ratios recommended by the Hydraulic Institute

Suction Energy Level	*NPSH Margin Ratio*
Low	1.1 to 1.3
High	1.3 to 2.0
Very High	2.0 to 2.5

The diagonal "suction energy ratio" lines (in Figure 5-12) are lines of constant relative suction energy. The suction energy ratio can be obtained by using the actual calculated suction energy (Equation 5-1) divided by the gating "start of high suction energy" value for the pump type, per Table 5-1. For example a double suction pump with an actual suction energy value of 180 × 10^6 would have a suction energy ratio of 1.5 (180/120). The 1.5-line represents the start of very high suction energy. Pumps of this suction energy level require a minimum NPSH margin ratio of 2.5 for maximum reliability.

Figure 5-12 can also be used in place of Table 5-2 to determine the necessary NPSH available, for high suction energy pumps.

There are other factors that, although not generally affecting the suction energy of the pump, will affect the degree of cavitation erosion damage (and sometimes noise) within a pump, when sufficient NPSH margin is not provided above the NPSHr of the pump. These non-suction energy factors are:

1. *Impeller Material*
Table 5-3 provides probable relative life factors for the most common impeller materials under cavitation-erosion conditions. Steel and brass are the most frequently used material; hence, upgrading impeller materials in difficult applications can increase impeller life and pump operating time. Using a life cycle cost approach, the more expensive impeller materials may rapidly pay back the incremental cost over the more traditional materials. Repair costs and production loss expenses are frequently far greater than the cost of impeller material upgrades.

Table 5-3: Relative life of different impeller materials under cavitation-erosion conditions

Material	*Life Factor*
Aluminum Bronze	8.0
Titanium	6.0
Bronze	4.0
Stainless Steel	4.0
Monel	2.0
Cast Iron	1.5
Brass, gun metal	1.2
Mild Steel	1.0

2. *Cavitation resistant coatings*: A cavitation resistant coating could be a lower cost and better option than changing to a cavitation resistant material, such as stainless steel. Techniques available for cavitation damage repair and/or prevention include: weld overlays and inlays, plasma sprays, thermal sprays (such as Stellite 6), reinforced epoxy coatings, unreinforced polyurethane coatings, and ceramic coatings. Ideally, the material used to repair/prevent cavitation damage should be selected to minimize capital costs, while maximizing the service life, for the expected operating conditions. (Ref. 5-13)

Two such affordable polymer coatings are (ceramic) reinforced epoxies (which have been used widely) and unreinforced polyurethane (which have superior cavitation resistance). The more cost effective/cavitation resistant coatings appear to fall in the unreinforced polyurethane, elastomer family. Compared with a weld overlay technique, the advantages of epoxy and polyurethane compound coatings include: (1) significantly reduced labor costs, (2) avoidance of thermally-induced residual stresses in the repaired components, and (3) improved control of component contours through the use of templates.

However, whenever any coating is applied to the inlet of an impeller, care must be taken to insure that the coating thickness does not significantly increase the thickness of the leading edges of the impeller vanes, or markedly reduce the inlet throat area between the impeller vanes. Thicker vane leading edges and/or smaller inlet throat areas will increase the velocities at the impeller inlet, which increases the pump NPSHR and results in more cavitation. This additional cavitation can lead to increased damage, if the pump "Suction Energy" is high enough. Any such significant increase in cavitation could counter the benefits of the cavitation resistance coating. These repairs should also retain the original impeller vane shape to avoid negative impacts on the pump performance. A small change in impeller vane shape can have a relatively large impact on the head, capacity, efficiency and/or NPSHR of a pump. Generally the larger the pump the less it will be affected by the thickness of a coating.

3. *The gas content and thermodynamic properties of the liquid*: Small amounts of entrained gas (1/2 to one percent) cushion the forces from the collapsing vapor bubbles, and reduce the resulting noise, vibration and erosion damage. This is why high-energy pumps—even at low NPSH margins—handling paper stock that contains a large amount of entrained air, seldom experience cavitation damage. Indeed, small amounts of gas or nitrogen, somewhere between 1/2% and 2% by volume, are sometimes purposely introduced into pumps to suppress cavitation.

Warmer liquids entering a pump tend to contain less dissolved gas than cooler liquids and thus have less of the "cushioning" gas to release. Therefore, one often notices higher noise levels in pumps operating with higher liquid temperatures. On the other hand and by contrast, as temperature increases, the volume of vapor created from a given mass of liquid decreases as it vaporizes, which creates smaller vapor bubbles and decreased suction energy. This, in turn, tends to increase pump life. The net effect of these two opposing factors is that the maximum erosion damage occurs at around 120°F (49°C) for water, with the damage rate decreasing somewhat at lower, and markedly at higher temperatures, such as found in boiler feed service. This phenomenon also explains the high erosion rates found in low NPSH margin cooling tower services.

Most hydrocarbon liquids have relatively low vapor volume to liquid volume ratios, and are often mixtures of different hydrocarbon liquids with different vapor pressures. This means that, if the liquid should vaporize at or near the pump suction (impeller inlet), the volume of the resulting vapor does not choke the impeller inlet passages as severely as does water vapor during cavitation. There is thus a smaller drop in developed head for the same NPSH margin. Also, less energy is released when hydrocarbon vapor bubbles collapse. In essence, the velocity of the surrounding liquid micro-jet resulting from bubble implosion is less, and this means less damage occurs as a result of cavitation. It is, therefore, not as critical that cavitation be avoided, as might be the case with aqueous liquids. Also, lower NPSH margin and minimum continuous flow values may be used (see Figure 5-13). The KM factor is a multiplier that can be applied to reduce the onset of suction recirculation values obtained from Figure 5-5, for hydrocarbons.

4. *The corrosive properties of the liquid*: This can accelerate the damage from bubble implosions scouring away the corrosion deposits and coatings.

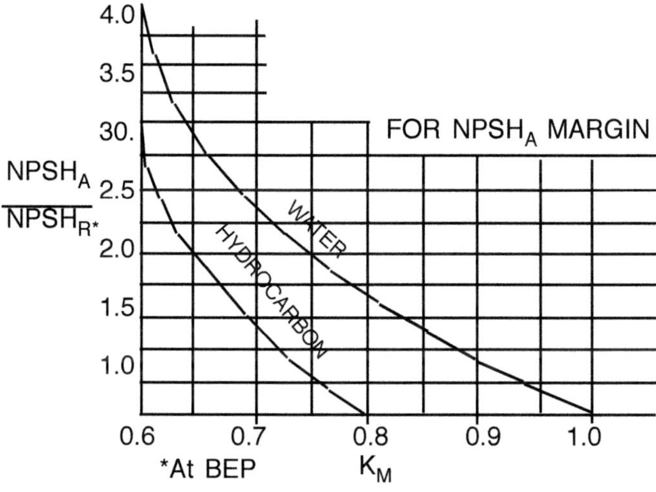

Figure 5-13: Minimum flow correction factor

5. *Solids/abrasives in the liquid*: Adding abrasives to the high implosive velocities from the collapsing vapor bubbles increases the wear rate.

6. *The duty cycle of the pump*: Cavitation damage is time related. The longer a pump runs under cavitation conditions, the greater the extent of damage. It is for this reason that fire pumps, which run intermittently, rarely suffer from cavitation damage.

TOTAL PUMP HYDRAULIC RELATIVE RELIABILITY

For comparison purposes, it is useful to combine several of the above hydraulic reliability factors in order to obtain the "total pump hydraulic relative reliability." This will allow various pump options, for a given application, to be evaluated on a common basis. The reliability factors to be combined are:

- Speed Reliability Factor (R_{sp})—Figure 5-1.
- Percent Flow Reliability Factor (R_{flow})—Figure 5-2
- NPSH Margin Reliability Factor (R_{npsh})—Figure 5-12.
- Material Reliability Factor (R_{mat})—Table 5-3 (only applies to high suction energy pumps).

The resulting formula for Total Hydraulic Relative Reliability (R_{total}) is:

$$R_{total} = R_{sp} \times R_{flow} \times R_{npsh} \times R_{mat}$$

As an example, we can use this formula to compare several pumps for a given application. Table 5-4 tabulates the results of this comparison.

- Option 1: Maximum speed, flow rate 75% of BEP, NPSH Margin Ratio = 1.3, Suction Energy Ratio = 1.5, and cast iron impeller.

- Option 2: VFD with speed ratio = 80%, flow rate 95% of BEP, NPSH Margin Ratio = 1.4, Suction Energy Ratio = 1.3, and stainless steel impeller.

Table 5-4: Total Hydraulic Reliability

Pump	*Speed*	*Flow*	*NPSH/SE*	*Matl.*	*Combined*
Opt. 1	0.40	0.95	0.20	1.5	**0.11**
Opt. 2	0.55	1.00	0.60	4.0	**1.32**
Opt. 3	0.78	0.98	0.83	4.0	**2.54**

- Option 3: Larger pump with speed ratio = 50%, flow rate 100% of BEP, NPSH Margin Ratio = 1.5, Suction Energy Ratio = 1.2, and stainless steel impeller

This example shows the large improvement in pump reliability that can be achieved from hydraulic upgrades, especially for high suction energy pumps. In this particular example, the larger pump *(Option 3)*, with a stainless steel impeller, would be the best choice. Its improved reliability may even pay for the added cost of the larger pump and stainless steel impeller.

PUMP RELIABILITY IMPACT FROM SYSTEM DESIGN

As was mentioned in Chapter 3, optimum pump performance also requires that proper suction/inlet piping practices be followed to ensure a steady uniform flow to the pump suction at the true required suction head. Poor suction piping can result in separation, swirl and turbulence at the pump inlet, which decreases the NPSHa to the pump and causes added cavitation. High suction energy pumps are obviously most sensitive to the effects of poor suction piping (Ref. 5-11).

Figure 5-14 shows test data on the effect of poor suction piping on a small, high suction energy, end suction pump, as it is throttled through the region of suction recirculation, 115—160 m^3/h (500 - 700 gpm). Suction pressure pulsation levels for normal piping (five diameters of straight pipe) are compared to the levels for suction piping with a short radius elbow mounted directly on the pump suction flange, and two short radius elbows at right angles, also mounted direct on the suction nozzle. The greater the flow disturbance entering the pump, the higher the pressure pulsations, noise and vibration, in the region of suction recirculation.

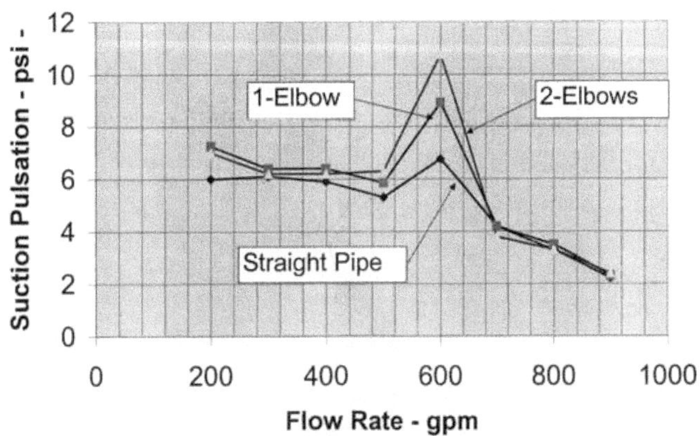

Figure 5-14: Cavitation vs. inlet configuration

PUMP SUMP INTAKE DESIGN

Pump performance and life cycle maintenance costs can be affected by the design of the pump intake structure and/or sump. A good design ensures that the following adverse flow phenomena are within the limits outlined in the Hydraulic Institute Standard (ANSI/Hi 9.8, Ref. 5-11):

- Submerged vortices
- Free-surface vortices
- Excessive pre-swirl of flow entering the pump
- Non-uniform spatial distribution of velocity at the impeller eye
- Excessive variations in velocity and swirl with time
- Entrained air or gas bubbles

The inlet bell diameter is a key element in the design of a sump. Favorable inflow to the pump or suction pipe bell requires control of various sump dimensions relative to the size of the bell. For example, the clearance from the bell to the sump floor and side walls and the distance to various upstream intake features is controlled in the Hydraulic Institute Standard by expressing such distances in multiples of the pump or inlet bell diameter. This reduces the probability that strong submerged vortices or excessive pre-swirl will occur. The recommended maximum inlet bell velocity is 5.5 ft/sec (1.7 m/s).

- The required minimum submergence to prevent detrimental free-surface vortices is also related to the inlet bell diameter (inlet velocity) and flow rate. The smaller the bell diameter and higher the flow rate, the greater the required minimum submergence. Examples of minimum required submergence values are shown in Table 5-5.

- The inlet structure may take several forms, such as rectangular, trench type, formed suction, suction tanks, unconfined or circular. The primary requirement of any structure is that it prevents cross-flows in the vicinity of the intake structure and creates asymmetric flow patterns approaching any of the pumps.

WATER HAMMER

Transient forces such as water hammer from sudden valve closure can produce large pressure spikes in

Table 5-5. Recommended minimum submergence of vertical pump inlet bells

Flow rate (gpm)	Recommended minimum submergence (inches)				
	Bell dia. for 2.0 ft/sec	Bell dia. for 3.0 ft/sec	Bell dia. for 5.5 ft/sec	Bell dia. for 8.0 ft/sec	Bell dia. for 9.0 ft/sec
200	15		19		25
1,000	25		31		40
2,000	33		40		49
4,000	43		50		60
6,000		51	57	64	
8,000		57	63	71	
10,000		62	68	76	
12,000		67	72	81	
14,000		72	76	85	
16,000		75	79	89	
18,000		79	82	92	
20,000		82	85	95	

the piping system, which can cause severe damage to the piping system and any attached equipment, such as pumps and valves.

When the flow of liquid is suddenly stopped, the liquid tries to continue in the same direction. In the area where the velocity change occurs, the liquid pressure increases dramatically, due to the momentum force. As it rebounds, it increases the pressure in the region near it and forms an acoustic pressure wave. This pressure wave travels down the pipe at the speed of sound in the liquid. If we assume the liquid is water, rigid pipe and ambient temperature, the wave velocity is 4,720 ft/sec (1,438.7 m/s). The acoustic wave will be reflected when it encounters an obstruction, such as a pump, fitting or valve.

The potential magnitude is dependent on the speed of valve closure and the liquid velocity in the pipe prior to the start of valve closure. If the time of the valve closure, in seconds, is greater than the total length of pipe (L), in feet, divided by 1,000, then momentum theory (Newton's second law of motion) applies.

$$F_m = (\rho/g)Q(V_1—V_2)$$

Where:

F_m = Momentum force—lbs_f
ρ = Fluid density - lbs/ft^3
g = Acceleration due to gravity - 32.2 ft/sec^2
Q = Rate of flow—ft^3/sec
V_1 = Initial velocity—ft/sec
V_2 = Final velocity—ft/sec

On the other hand, if the time to close the valve is less than L/1000, then the acoustic shock wave/elastic column theory applies. In that case,

$$P = \rho \times A \times V/(g \times 144\ in^2/ft^2)$$

Where:

P = Pressure rise — psi
ρ = Liquid density — lbs/ft^3
A = Velocity of sound in water - ft/sec
V = Velocity of the liquid in the pipe — ft/sec
g = Acceleration due to gravity — 32.2 ft/sec^2

As an example, if a pump is working at 200 psig (1.38 MPa), and the liquid is traveling at 15 ft/sec (4.6

m/s), the maximum recommended discharge velocity, the instantaneous pressure inside the casing would jump to 1,158 psig (7.98 MPa). This assumes the liquid is ambient water and the valve closure time is short enough for the elastic column theory to apply. Pump casings are not usually designed for this magnitude of pressure, especially if the casing is made of a brittle material, such as cast iron. Even if the casing is constructed of a more ductile material, the shock wave may still cause permanent deformation and ultimate failure.

The pump is not the only component that is affected by this phenomenon. Valves, sprinkler heads and pipe fittings are at risk of catastrophic damage. Pipe hangers and pump foundations can also be adversely affected by water hammer. PVC pipe and fittings are very susceptible to damage from water hammer.

Water hammer can be controlled through proper valve closure rates (with slow-closing valves), the addition of diaphragm tanks or similar accumulators to absorb the pressure surge, and relief valves to release the pressure.

Chapter 6

Improvements Leading to Pump Mechanical Maintenance Cost Reduction

Many "standard" ANSI and ISO-compliant pumps were designed decades ago when frequent repairs were accepted and plant maintenance departments were loaded with personnel. Unless selectively upgraded, a "standard" pump population will not allow 21st Century facilities to reach their true reliability and profitability potentials.

While primarily aimed at upgrading relatively inexpensive ANSI and ISO pumps, this chapter introduces a number of important issues that can't be overlooked in attempting to extend pump life. Pump sealing and bearing lubrication topics are primarily touched upon, although their overall importance is such that subsequent chapters will go into additional detail on these matters.

MECHANICAL MAINTENANCE LIFE CYCLE IMPACT

As mentioned in the introductory chapters, the cost of unscheduled maintenance is often the most significant cost of ownership, and failures of mechanical seals and bearings are among the major causes. In recent years, improved pump designs and the management of mechanical reliability issues have lead to significant increases in the pump mean time between failure (MTBF) rates for many process plants. Back in the 1970's and early 1980's, it was not unusual to find plant MTBF rates of 6 months to a year. Today, most reliability-focused plants have pump MTBF rates as high as 3 to 4 years. These improvements have been achieved by paying proper attention to pump components with the highest failure rates.

Then, in the late 1990's and early 2000's, there came to be plants with over 2,000 installed pumps in average sizes around 30 hp that enjoyed an MTBF of 8.6 years. These, as one might suspect, are plants that look at mechanical and process interactions. They fully understand that pumps are part of a system and that the system must be correctly designed, installed and operated if consistently high reliability is to be achieved. In addition and as further explained in Chapter 16 of this text, these plants are conducting periodic pump reliability reviews. As a consequence, the 8.6 year MTBF mentioned in the first edition has, in at least one documented instance, risen to a remarkable 11 years.

PUMP RELIABILITY REVIEWS START BEFORE PURCHASE

The best time for the first reliability review is before the time of purchase. This subject is given thorough treatment in Ref. 6-1. It is obvious that persons with a reliability engineering background and an acute awareness of how and why pumps fail are best equipped to conduct such reviews. This implies that these contributors should have an involvement in the initial pump selection process. Individually or as a team, those involved should consider the possible impact of a number of issues, including:

- The potential value of selecting pumps that cost more initially, but last much longer between repairs. The MTBF (Mean Time Between Failures) of a better pump may be one to four years longer than that of its non-upgraded counterpart.
- The published average values of avoided pump failures range from $2,600 to $12,000.
- One pump fire occurs per 1,000 failures. Having fewer pump failures means having fewer destructive pump fires. This 1970s statistic was reconfirmed and re-affirmed by 2009 statistics!

There are several critically important applications where buying on price alone is almost certain to ultimately cause costly failures. Included are the following:

— Applications with insufficient NPSH or low NPSH margin ratios (Ref. 6-2)
— High or very high suction energy services
— High specific speed pumps (Ref. 6-3)
— Feed and product pumps without which the plant will not run
— High pressure and high discharge energy

pumps
— Vertical turbine deep-well pumps

At a minimum, then, these are the six services where spending time and effort for pre-purchase reliability reviews makes the most economic sense. These reviews concentrate on typical problems encountered with centrifugal pumps and attempt to eliminate these problems before the pump ever reaches the field. Among the most important problems the reviews seek to avoid are:

- Pumps not meeting stated efficiency
- Lack of dimensional interchangeability
- Vendor's sales and/or coordination personnel being reassigned
- Seal problems (materials, flush plan, flush supplies, etc.)
- Casting voids (repair procedures, maximum allowable pressures, metallurgy, etc.)
- Lube application or bearing problems (Ref. 6-4)
- Alignment, lack of registration fit (rabbetting), base plate weakness, grout holes too small, base plates without mounting pads, ignorance of the merits of pre-grouted base plates
- Documentation: Manuals and drawings shipped too late
- Pumps that will not perform well when operating away from best efficiency point, i.e. prone to encounter internal recirculation (Refs. 6-3 and 6-5).

MEAN-TIME-BETWEEN-FAILURE CALCULATIONS

Although not perfect from a mathematician's or statistician's perspective, simplified calculations will give an indication of the extent to which improving one or two key pump components can improve overall pump MTBF (Ref. 6-6).

Say, for example, that there's agreement that the mechanical seal is the pump component with the shortest life, followed by the bearings, coupling, shaft, and sometimes impeller, in that order. The anticipated mean time between failure of a complete pump assembly can be approximated by summing the individual MTBF rates of the individual components, using the following formula:

$$1/MTBF = [(1/L_1)^2 + (1/L_2)^2 + (1/L_3)^2 + (1/L_4)^2]^{0.5} \quad \text{Eq. (6-1)}$$

In a 1980's study, the problem of mechanical seal life was investigated. An assessment was made of probable failure avoidance that would result if shaft deflections could be reduced. It was decided that limiting shaft deflection at the seal face to a maximum of 0.001 inch (0.025 mm) would probably increase seal life by 10%. It was similarly judged that a sizable increase in seal housing dimensions to allow the installation of the newest seal configurations would more than double the mean-time-between-failures of seals.

By means of such analyses, all of the components under consideration for upgrading were examined, the life estimates collected, and the latter used in mean-time-between-failure calculations.

In Equation (6-1), L_1, L_2, L_3 and L_4 represent the life, in years, of the component subject to failure. Using applicable data collected by a large petrochemical company in the 1980's, mean-times-between-failures, and estimated values for an upgraded medium-duty pump were calculated. The results are presented in Table 6-1. As an example, a standard construction ANSI B73.1 pump with a mechanical seal MTBF of 1.2, bearing MTBF of 3.0, coupling MTBF of 4.0 and shaft MTBF value of 15.0, resulted in a total pump MTBF of 1.07 years. By upgrading the seal and bearings, the total pump MTBF can be improved by 80% to 1.93.

Table 6-1 thus shows the influence of selectively upgrading either bearings or seals or both on the overall pump MTBF. Quite obviously, choosing a 2.4 year MTBF seal and a 6 year MTBF bearing (easily achieved by preventing lube oil contamination, as will be seen later), had a major impact on increasing the pump MTBF and—as-

Table 6-1: How selective component upgrading influences MTBF

ANSI Pump Upgrade Measure	Seal MTBF (yrs)	Bearing MTBF (yrs)	Coupling MTBF (yrs)	Shaft MTBF (yrs)	Composite Pump MTBF (yrs)
None, i.e. "Standard"	1.2	3.0	4.0	15.0	1.07
Seal and Bearings	2.4	6.0	4.0	15.0	1.93
Seal Housing Only	2.4	3.0	4.0	15.0	1.69
Bearing Environment	1.2	6.0	4.0	15.0	1.13

suming the upgrade cost is reasonable—may well be the best choice.

As has been noted, the failure of a *small* pump, based on year 2002 reports, costs $5,000 on average. This includes costs for material, parts, labor and overhead. Let us now assume that the MTBF for a particular pump is 12 months and that it could be extended to 18 months. This would result in a cost avoidance of $2,500/yr, which is greater than the premium one would pay for the upgraded medium-duty pump.

In addition, the probability of reduced power cost would, in many cases, further improve the payback. Recall also that the proper hydraulic selection of a pump can have a further marked positive impact on the life and operating efficiency of a pump. Audits of two large U.S. plants identified seemingly small pump and pumping system efficiency gains which added up to power-cost savings of hundreds of thousands of dollars per year.

Thus, the primary advantages of the upgraded medium-duty pump are extended operating life, higher operating efficiency and lower operating and maintenance costs.

Figure 6-1 provides a quick means to approcimate the annual pump repair cost based on the total MTBF and average repair cost. It can also be used to determine potential savings from upgrades.

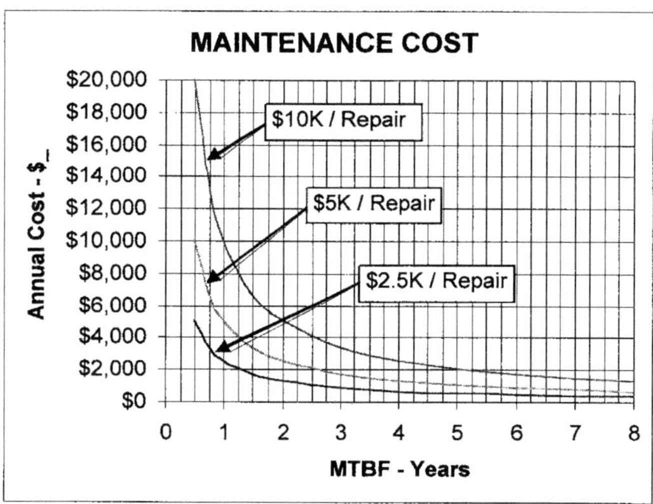

Figure 6-1: Pump MTBF vs. annual maintenance cost

Seal Environment

The number one cause of pump downtime is failure of the shaft seal. These failures are normally the result of an unfavorable seal environment such as improper heat dissipation (cooling), poor lubrication of seal faces, or seals operating in liquids containing solids, air or vapors. To achieve maximum reliability of a seal application, proper choices of seal housings (standard bore stuffing box, large bore, or large tapered bore seal chamber, Ref. 7) and seal environmental controls (CPI [Chemical Process Industry] and API seal flush plans) must be made.

Large Bore Seal Chamber

The original ANSI/AVS B73 specification was written around 1960. When originally written, packing was the predominant means of sealing the stuffing box.

Today the mix has changed and the vast majority of ANSI B73.1 pumps are provided with mechanical seals. This quasi "convertible" stuffing box, which can accommodate either packing or mechanical seals does not, however, have sufficient space for the diverse range of designs in mechanical seals, especially optimally dimensioned cartridge seals. This fact brought about the need for a second stuffing box cover design, called a seal chamber. It was introduced in the mid-80's, and has since been included in recent editions of the ASME/ANSI B73.1 standard. A seal chamber can be of a plain cylindrical (Figure 6-2) or one of several different taper bore (Figure 6-4) designs. There are variations of each and some of the many possible seal environments and available configurations are explained in Chapter 8.

Designed specifically for optimally dimensioned mechanical seals, large bores provide increased life of seals through improved lubrication and cooling of faces. The seal environment should be controlled through use of CPI or API flush plans. Seal chambers are often available with internal bypass to provide circulation of liquid to faces without using external flush. A properly dimensioned internal bypass taken off the back side of the impeller will reduce the solids concentration in the seal chamber. Large bore seal chambers are ideal for conventional or cartridge single mechanical seals

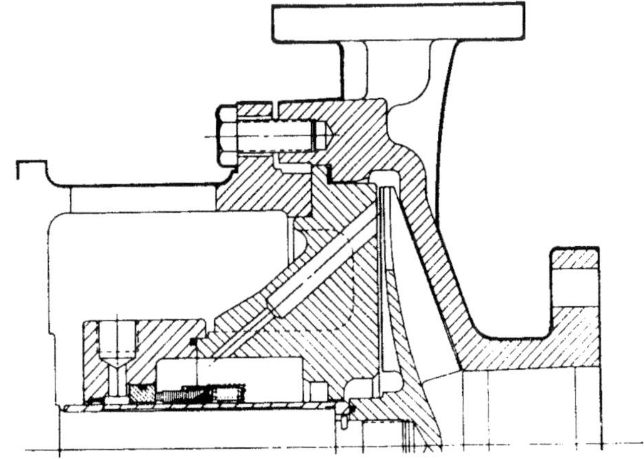

Figure 6-2: Large bore seal chamber with internal bypass (Source: ITT/Goulds, Seneca Falls, NY 13148

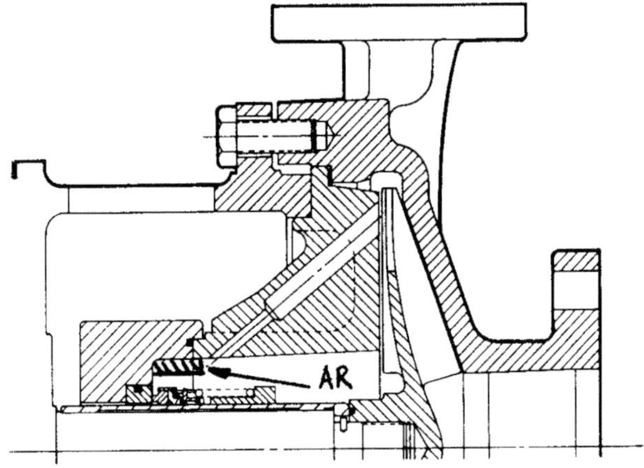

Figure 6-3: Large bore tapered seal chamber with vortex breaker baffles ("AR") (Source: ITT/Goulds, Seneca Falls, NY 13148

in conjunction with a flush and throat bushing at the bottom of the chamber. Large bore seal chambers are also excellent for conventional or cartridge double or tandem seals.

Table 6-2 compares the radial clearance (between the shaft sleeve and seal chamber bore) of the traditional old stuffing box (small bore) with the generally optional new seal chamber (big bore).

Table 6-2 and Figure 6-2 show the dramatic increase in mechanical seal space in the big bore seal chamber over the standard stuffing box cover. Virtually all standard O-ring mounted single and double mechanical seals will fit in the big bore seal chamber. This type of seal chamber can also be furnished with the internal bypass shown in Figure 6-2, but it is not normally used with outside mechanical seals.

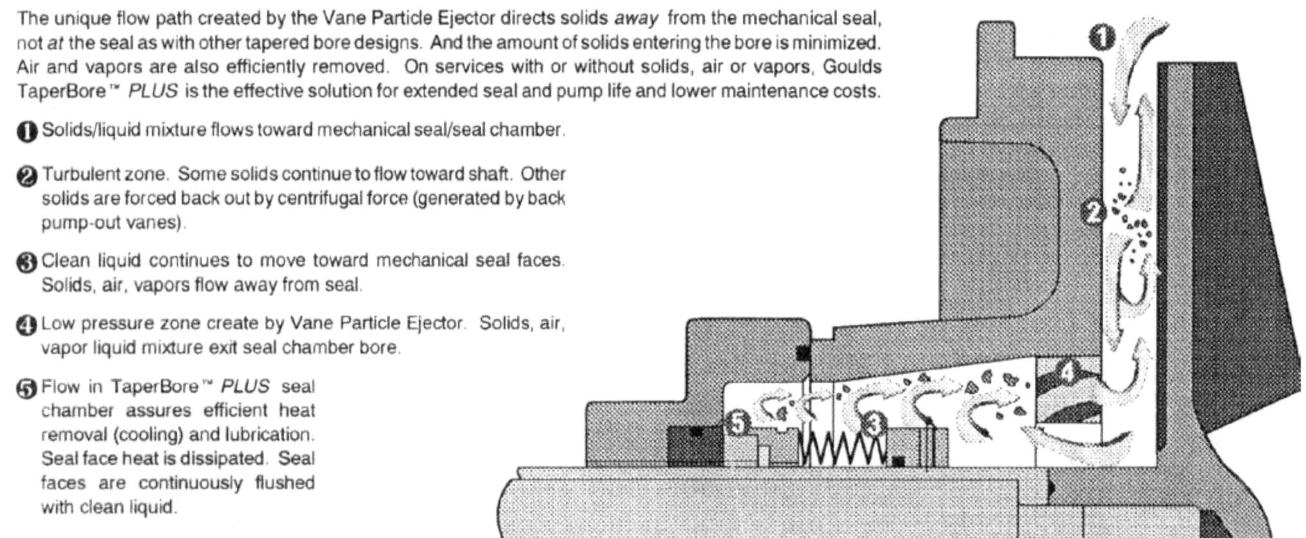

The unique flow path created by the Vane Particle Ejector directs solids away from the mechanical seal, not at the seal as with other tapered bore designs. And the amount of solids entering the bore is minimized. Air and vapors are also efficiently removed. On services with or without solids, air or vapors, Goulds TaperBore™ PLUS is the effective solution for extended seal and pump life and lower maintenance costs.
1 Solids/liquid mixture flows toward mechanical seal/seal chamber.
2 Turbulent zone. Some solids continue to flow toward shaft. Other solids are forced back out by centrifugal force (generated by back pump-out vanes).
3 Clean liquid continues to move toward mechanical seal faces. Solids, air, vapors flow away from seal.
4 Low pressure zone create by Vane Particle Ejector. Solids, air, vapor liquid mixture exit seal chamber bore.
5 Flow in TaperBore™ PLUS seal chamber assures efficient heat removal (cooling) and lubrication. Seal face heat is dissipated. Seal faces are continuously flushed with clean liquid.

Figure 6-4: Tapered Bore Plus™ seal chamber with vane particle ejector ring (Source: Goulds/ITT, Seneca Falls, NY 13148)

Table 6-2: Comparison between conventional and big bore seal chamber dimensions

	Radial clearance between shaft and seal chamber		
Frame Size	Small Bore	Big Bore	% Increase
AA - AB	5/16″	3/4″	140
A05 - A80	3/8″	7/8″	133
A90 - Al 20	7/16″	1″	128

Tapered Bore (Large Bore) Seal Chambers

Large bore seal chambers can be supplied with an internal taper. Tapered large bore seal chambers, Figure 6-3, have a minimum four-degree taper, open toward the pump impeller. This configuration favors circulation of liquid at seal faces without the use of external flush (see also Figure 6-2). It has been reasoned that tapered bore seal chambers offer advantages that include lower maintenance costs, elimination of tubing/piping, lower utility costs (associated with seal flushing) and extended seal reliability. The tapered bore seal chamber is commonly available with ANSI chemical pumps. API process pumps use conventional (cylindrical) large bore seal chambers.

Paper stock pumps use both conventional large bore and large tapered bore seal chambers.

However, as will be explained next, only tapered bore seal chambers with special flow modifier struts or vortex breaker baffles provide the expected reliability on services that contain solids, air or vapors. Without such vortex breaker baffles, labeled "AR" in Figure 6-3, there will be a cyclone separator effect whereby abrasive-containing pumpage tends to swirl around at great velocity and erode away the surrounding metal.

It should be noted that conventional (small bore) tapered bore seal chambers were often flawed. Developed in the mid 1980's, it was anticipated that solids and vapor would be centrifuged to the bore and then axially moved towards the impeller, where they would exit the seal chamber. However, once the solids made it into the box they could not escape. High pressure at the impeller vane tips and low pressure at the shaft made it difficult or impossible for solids and vapor to exit the confined space. Accumulation of solids resulted in severe erosion of the seal and pump parts. In addition, failures due to dry running of mechanical seals were often caused by vapors trapped in the seal chamber. This seal chamber type does provide good cooling of the seal faces with clean liquids.

Large bore tapered seal chambers with axial flow modifier baffles avoid these problems and provide better seal life when air or vapors are present in the liquid. The axial ribs dramatically change the flow profile, are effective at removing solids from the seal chamber, and prevent entrapment of vapors. Unfortunately, the new flow profile that allows the solids to escape also deflects solids towards the seal faces, which can cause failures. It is a good solution for vapor removal, but has solids handling and paper stock limitations of approximately one percent (1%) by weight. This seal chamber can be used with cartridge double seals.

Special Tapered Bore Seal Chamber with Vane Particle Ejector Ring

To eliminate seal failures on services containing vapors as well as solids, the flow pattern must direct solids away from the mechanical seal and purge air and vapors. Seal chamber designs that meet this objective incorporate a "vane particle ejector" (VPE) ring, Figure 6-4, at the seal chamber entrance. A low pressure zone created by the VPE creates a unique flow path that directs solids and vapor away from the mechanical seal. The back pump-out impeller vanes, in conjunction with the VPE, create a turbulent zone, which helps to minimize the amounts of solids entering the seal chamber bore. The VPE design extends the solids handling limit to 10% by weight, and paper stock to 5% by weight, to ensure that the seal faces are continuously flushed with clean liquid. No flush is required. The VPE design is not used with outside mechanical seals.

Tapered Bore Seal Chamber Comparison

The various large bore seal chambers (Figure 6-5) are fundamentally oversized, tapered bore seal chambers with flow modifiers. These could be baffles or—in "cyclone" chambers—two helical grooves cast into the seal chamber bore. The intent is to improve shaft sealing reliability. The baffles and/or helical grooves are key to the design; they act to maintain flow patterns within the seal chamber which both cool and lubricate the mechanical seal while preventing solids and vapor from collecting in the seal environment.

As an added benefit, the "cyclone" seal chamber flow modifiers can reduce, and sometimes eliminate, the need for auxiliary seal flush for a number of services containing up to ten percent (10%) solids by weight. This results in reduced installation cost and simplifies needed maintenance activities.

Table 6-3 provides a quick guide for selecting the right seal chamber/stuffing box for the application.

"ANSI PLUS®" MECHANICAL SEAL GLAND

According to ITT/Goulds, an ideal gland for big bore seal chambers should have the following features (see Figure 6-6):

1. Tangential flush connection—optimum versatility for standard flush, maximum circulation when utilizing a pumping ring.
2. Outside diameter pilot—dry, close tolerance fit to assure concentricity.
3. Metal to metal face alignment—O-ring should be utilized to seal the gland to the seal chamber, while the metal to metal fit assures the stationary seat will be perpendicular to the axis of rotation.
4. The gland should allow for the addition of a vent and drain chamber with a bushing if required.

SEALED POWER ENDS AND EXPANSION CHAMBERS

Now let us turn to improving the MTBF of the pump bearings. One way is to use a sealed power end;

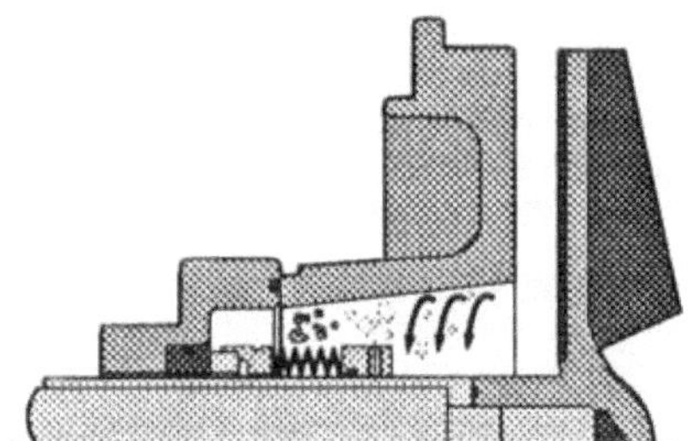

Conventional Tapered Bore Seal Chamber:
Mechanical Seals Fail When Solids or Vapors Are Present in Liquid

Many users have applied the conventional tapered bore seal chamber to improve seal life on services containing solids or vapors. Seals in this environment failed prematurely due to entrapped solids and vapors. Severe erosion of seal and pump parts, damaged seal faces and dry running were the result.

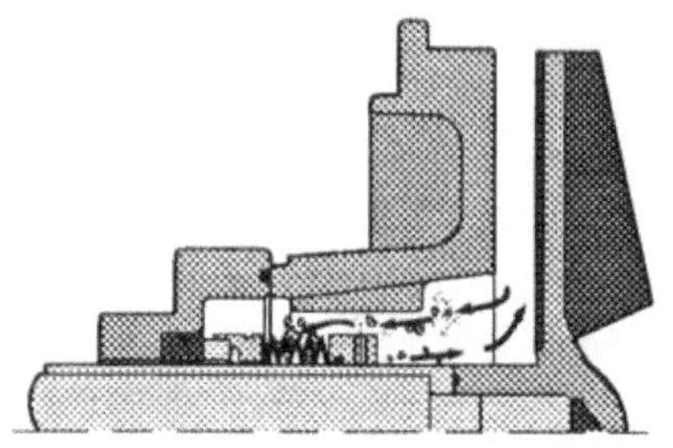

Modified Tapered Bore Seal Chamber with Axial Ribs:
Good for Services Containing Air, Minimum Solids

This type of seal chamber will provide better seal life when air or vapors are present in the liquid. The axial ribs prevent entrapment of vapors through improved flow in the chamber. Dry running failures are eliminated. In addition, solids less than 1% are not a problem.

The new flow pattern, however, still places the seal in the path of solids/liquid flow. The consequence on services with significant solids (greater than 1%) is solids packing the seal spring or bellows, solids impingement on seal faces and ultimate seal failure.

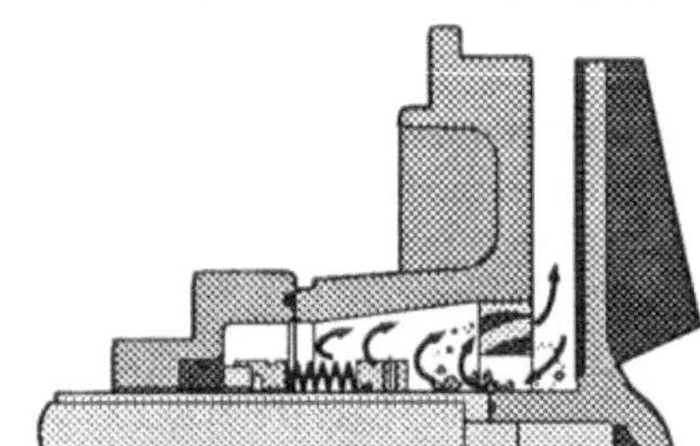

Goulds Standard TaperBore™ PLUS Seal Chamber:
The Best Solution for Services Containing Solids and Air or Vapors

To eliminate seal failures on services containing vapors as well as solids, the flow pattern must direct solids away from the mechanical seal, and purge air and vapors. Goulds Standard TaperBore™ *PLUS* completely reconfigures the flow in the seal chamber with the result that seal failures due to solids are eliminated. Air and vapors are efficiently removed eliminating dry run failures. Extended seal and pump life with lower maintenance costs are the results.

Figure 6-5: Tapered bore seal chambers (Source: Goulds/ITT, Seneca Falls, NY 13148)

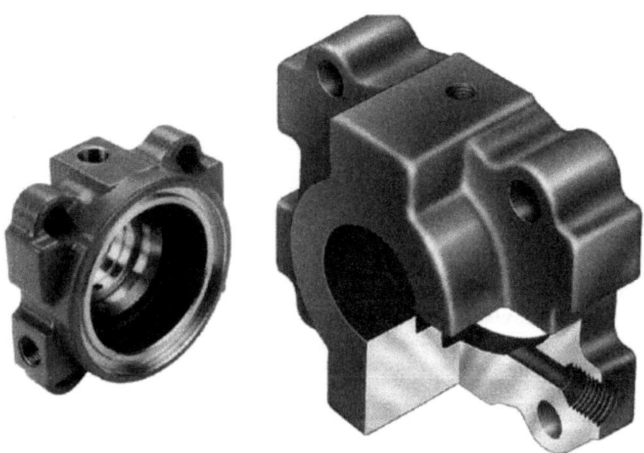

Figure 6-6: Optimized gland design for ANSI pumps (Source: ITT/Goulds, Seneca Falls, NY 13148)

alternatively, good sealing requires use of face-type housing seals. As will be seen later, it can be reasoned that wear-prone lip seals and even the many different types of rotating labyrinth seals are not achieving this near-hermetic containment. A sealed volume will undergo pressure changes in proportion to temperature changes. This was recognized in Charles' Law, dating back to the late 17th century. As regards pump bearing housings, it has been reasoned that the temperature-related expansion of a volume of gas floating above a liquid oil sump must be accommodated somehow.

Some sealed bearing housings are thus offered with a diaphragm-type expansion chamber (Figure 6-7 and Figure 6-8) to compensate for air volume changes within the frame caused by temperature differences. This small device, which incorporates an elastomeric diaphragm, is intended to facilitate making bearing housings completely enclosed systems. Expansion chambers are screwed into the housing vent opening. They accommodate the expansion and contraction of vapors in a sealed bearing housing without permitting moisture and other contaminants to enter. Carefully selected from a variety of either plain or fabric reinforced elastomers, the diaphragm will not fail prematurely in harsh chemical environments (Ref. 6-6).

However, expansion chambers rarely make either technical or economic sense on centrifugal pump bearing housings. If air can escape along the shaft due to existing air gaps, the expansion chamber serves no purpose at all. If the bearing housing is truly sealed and the constant level lubricator vented to the bearing hous-

Table 6-3: Typical seal chamber selection guide (Source: ITT/Goulds, Seneca Falls, NY 13148)

	Stuffing box cover—seal chamber type				
	Standard Box Cover	Cylindrical Large Chamber	Standard Tapered Bore	Axial Rib Tapered Bore	VPE & Cyclone Chambers
Service	Designed for Packing. Fits Mech. Seals	Enlarged Chamber. Use CPI Flush Plans	Enlarged Tapered Bore Chamber	Enlarged /Tapered Bore w/ Rib Chamber	Vane Particle Ejector & Spiral Groove
Ambient water with flush	A	A	A	A	A
Entrained air or vapor	C	B	B	A	A
Solids 0-10%, no flush	C	C	C	B	A
Solids up to 10%, with flush	B	A	B	A	A
Paper stock 0-5%, no flush	C	C	C	B	A
Paper stock 0-5%, with flush	B	A	A	A	A
Slurries 0-5%, no flush	C	C	C	B	A
High boiling point liquids, no flush	C	C	C	B	A
Self-venting and draining	C	C	B	A	A
Seal face heat removal	C	A	A	A	A

A - Ideally suited, B - Acceptable, C - Not recommended

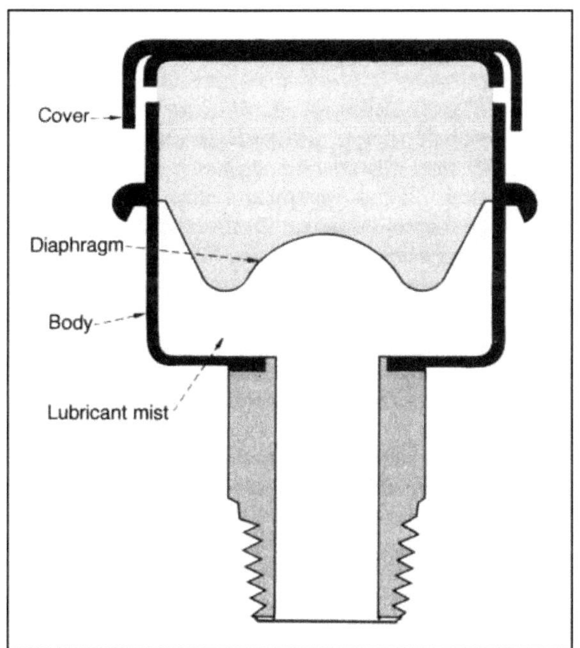

Figure 6-7: Diaphragm-type expansion chamber (Source: ITT/Goulds, Seneca Falls, NY 13148)

ing, the anticipated pressure increase would be minor. A quick and simple calculation using the fundamental thermodynamic relationship between pressure and temperature rise (e.g. $P_1T_2=P_2T_1$, using absolute pressures and absolute temperatures) would prove the point. It would show that for the actual and most probable temperature rise encountered in pump bearing housings, the accompanying pressure rise would be of no consequence as long as the lube application method is chosen in accordance with sound reliability principles. Again, this implies that the reliability-focused pump user would select balanced constant level lubricators (explained later in this text), whereas the risk-taking repair-focused user would continue to employ lubricators that allow oil levels to be contacted by ambient air.

In any event, the shaft is sealed with magnetic shaft seals (Figures 6-9, 6-10 and 6-11) at each end of the frame (Ref. 6-8). In each of these seals the rotary portion of the assembly is fastened to the shaft by an O-ring that performs both clamping and sealing functions. The stationary piece is O-ring mounted in an insulator, which is pressed into the pump housing or frame adapter. Magnetic bearing housing seals no longer use the traditional mechanical seal springs; they employ suitably positioned magnets instead. Early versions of magnetic seals are depicted in Figures 6-9 and 6-10 where one face is either fitted with rare earth rod magnets (Figure 6-9) or an aluminum-nickel-cobalt (Alnico) magnet (Figure 6-10) with the other face either being a magnetic material or encased in a magnetic material. An advanced version magnetic seal is shown in Figure 6-11.

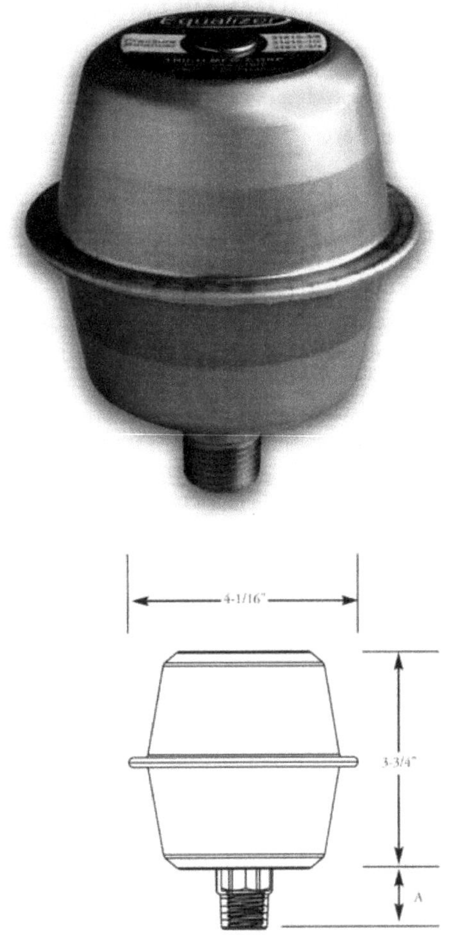

Figure 6-8: Diaphragm-type bearing housing expansion chamber (Source: TRICO Mfg. Co., Pewaukee, WI 53072)

On many sealed bearing housings, a sight glass is mounted in the side of the frame to ensure proper oil level (Figure 6-12, also Figure 9-24). Some manufacturers continue to supply these housings with constant level lubricators, although others have proven that with true hermetic sealing there is no longer the need to furnish make-up oil. Hence, they leave off the constant level lubricator and recommend level monitoring via sight glass observation. The user will be instructed to fill the unit to the proper oil level and this will be the only oil necessary until the normal oil change interval.

Although changing the oil periodically will still be required (since all lubricants will degrade over a period of time), keeping out contaminants will protect both oil and bearings. The recommended oil replacement intervals are therefore typically 2-4 years for "quasi-hermetically" sealed, synthetic lube-containing pump bearing housings, while 6-12 month intervals are more typical for conventional, open bearing housings using mineral oils. The frame is sometimes fitted with a magnetic oil plug to attract any magnetic wear particles during operation. Sealed power ends are designed to prevent the primary cause of premature failure, namely contamination of the oil.

Since the mid-1980s, "upgraded medium duty" pumps have often been specified by value-conscious, reliability-focused ANSI pump users. Typical of these are the features advertised as "standard" in the "ANSI Plus®" sealed power ends (Figure 6-13) of ITT/Goulds, a major pump manufacturer (for comments on sealed bearings frame see page 158):

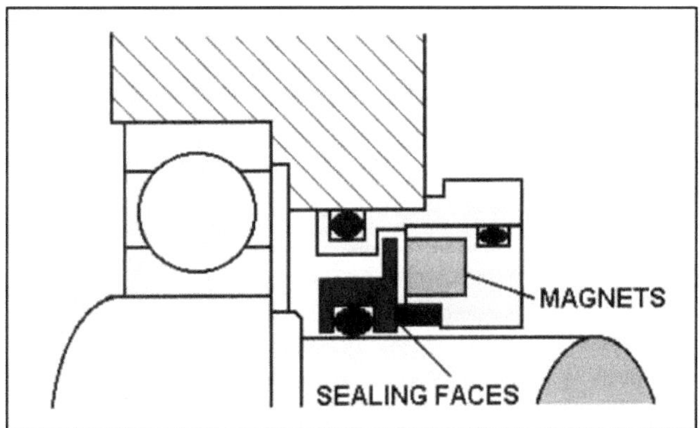

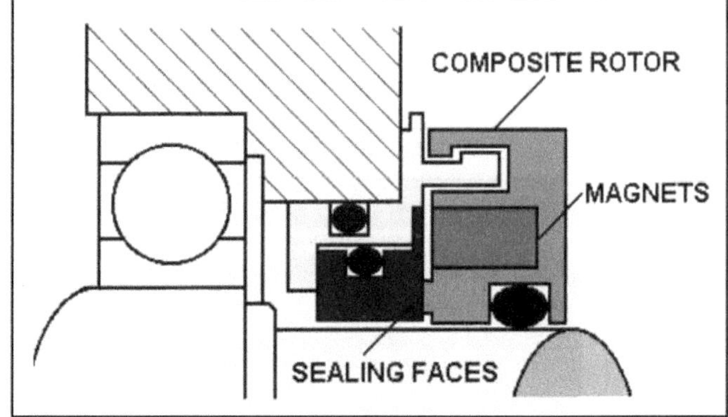

Figure 6-9: Cartridge-type magnetic shaft seal incorporating rare-earth bar magnets (Source: Isomag Corporation, Baton Rouge, Louisiana 70809)

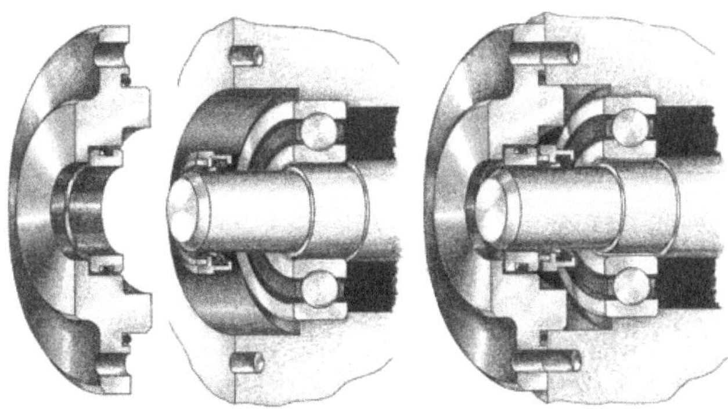

Figure 6-10: Magnetic shaft seal with Alnico face; shown here in the process of being assembled in three steps, left-to-right (Source: Magseal, Inc., West Barrington, Rhode Island)

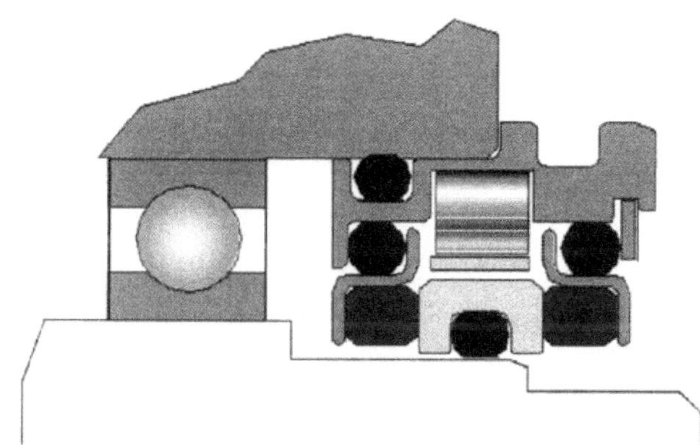

Figure 6-11: Dual-face magnetic bearing housing seal (Source: AESSEAL, plc, Rotherham, UK, and Rockford, TN)

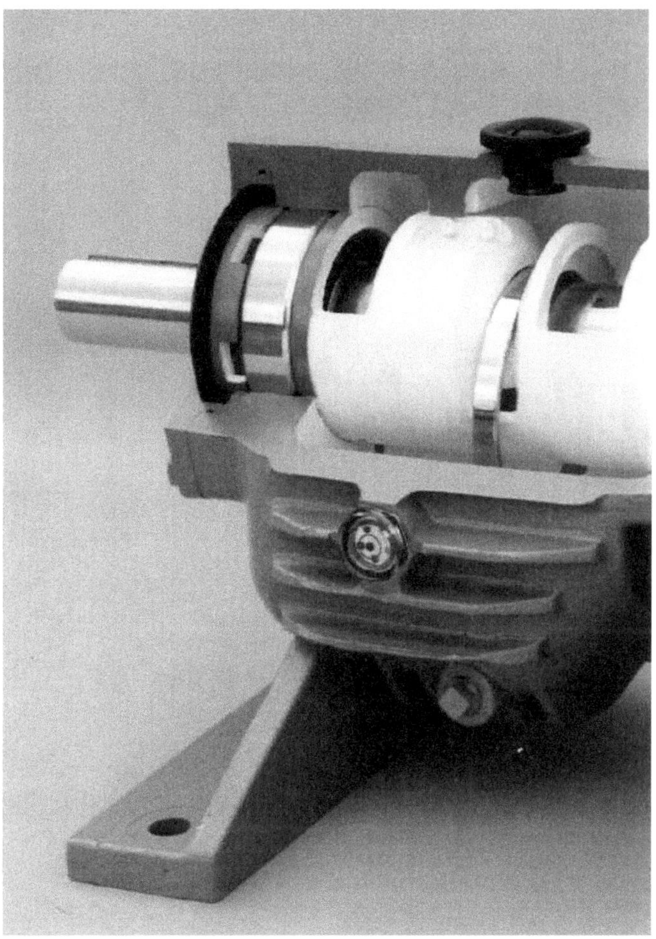

Figure 6-12: Sight glass assembly on a bearing housing For ease of observation, large diameter sight glasses should be specified by reliability-focused purchasers. (Source: Roto-Jet Pump®, Salt Lake City, UT 84110)

"ANSI Plus®" Feature	Replaces
1. Magnetic shaft seals	Lip seals
2. Expansion chamber	Breather
3. Magnetic oil plug	Oil plug
4. Sight window	Constant level lubricator

It should be noted that whenever the impeller is adjusted on a sealed power frame, it is possible for the shaft-contacting O-ring of a magnetic seal to travel with the shaft, which would risk opening the faces. This potential problem is avoided by selecting a cartridge-type magnetic seal, Figure 6-9, or the more recently developed variants of Figure 6-11, instead of certain non-cartridge configurations similar to Figure 6-10.

BEARING HOUSING SEALS IN THE LUBRICATION ENVIRONMENT

Most "traditional" bearing housing seals furnished with least expensive pumps are generally inadequate. ANSI pumps are usually furnished with elastomeric lip seals. When these seals are in good condition, they contact the shaft and contribute to friction drag and temperature rise in the bearing area. After 2,000 to 4,000 operating hours, they are generally worn to the point at which they no longer present an effective barrier against contaminant intrusion. Percent failure vs. hours to leakage for two types of lip seals is shown in Figure 6-14 (Ref. 6-7). It is noteworthy that the trace with the lesser life is for a dual-lip seal, whereas the trace with the longer life is for a single-lip seal. This counter-intuitive finding was attributed to lube-related issues.

EXTRA LARGE
OIL SUMP
Large oil capacity provides
optimum heat transfer for
cooler running bearings.

Figure 6-13: Upgraded medium duty pump frame components with sealed power end (Source: ITT/Goulds, Seneca Falls, NY 13148)

SHAFT/BEARINGS ASSEMBLY
Shaft designed for minimum deflection for long seal
and bearing life. Bearings sized for optimum life.
Duplex thrust bearings
optional.

Allows for viewing condition and level of
oil—critical for bearing life. Frame pre-
drilled for optional bottle oiler.

CONDITION
MONITORING SITES
Allow easy and consistent
monitoring of temperature and
vibration for preventive
maintenance. Optional
installation of sensors.

API-type pumps are generally furnished with non-contacting labyrinth seals or bearing isolators similar to the ones depicted in Figures 6-15 and 6-16. These rotating labyrinth seals, and especially the cleverly designed isolator of Figure 9-36, represent a commendable advancement over prior art. They have proven to give much better protection than either lip seals and standard stationary (API 5th Edition) labyrinth seals.

Yet, it has been stated that non-contacting rotating labyrinth configurations cannot possibly represent a fully effective barrier against the intrusion of atmospheric dust or moisture. Examining Figure 6-15, reliability professionals have reasoned that there is an interchange of air, as will be described later (Chapter 9, Figure 9-36).

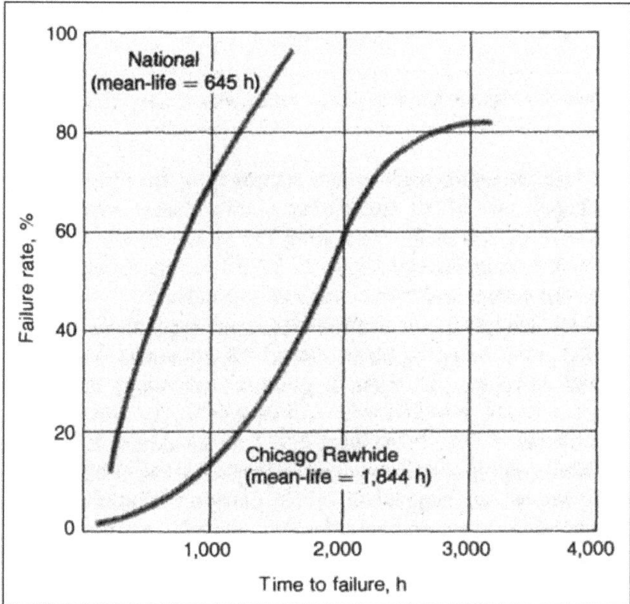

Figure 6-14: Percent failure vs. hours to leakage for two types of lip seals (Source: Bloch/Johnson, "Downtime prompts upgrading of centrifugal pumps," ***Chemical Engineering,*** **November 25, 1985)**

It should be noted, nevertheless, that all of the major ANSI pump manufacturers offer these seals, primarily because of user tradition. Also, the manufacturer of this bearing housing seal (Figure 6-15) has repeatedly affirmed the soundness of the design. This contention has been challenged by researchers who found degradation of the O-ring at lines of contact with the edges of O-ring grooves.

Isolators of the type depicted in Figure 6-16 have a fixed air gap that permits communication between bearing housing space and ambient environment. As the bearing housing undergoes temperature changes, air will flow in and out of the housing.

Airborne moisture, in particular, can cause unexpectedly high decreases in bearing fatigue life. Ref. 6-9 contains summaries on the deleterious effects of water in lube oil. It states that, in the 1960's, researchers had found the fatigue life at a water content of 0.002 percent reduced 48 percent and, at 6.0 percent water, found it to be reduced 83 percent. Others found a fatigue life reduction of 32-43 percent for squalene containing 0.01 percent water and a third group of researchers detected about an 80 percent drop with a moist air environment contacting dried mineral oil, as was shown earlier in Table 2-1. While the detailed mechanism for the reduction of fatigue life by water in a lubricant is not completely understood, it is thought to relate to aqueous corrosion. There is also much evidence that the water breaks down and liberates atomic hydrogen. This causes hydrogen embrittlement and increases the rate of cracking of the bearing material by a significant margin.

Expanding on our earlier discussion, a solution to the problem of sealing pump bearing housings can be found in aircraft and aerospace hydraulic pumps, which make extensive use of magnetic face seals. Earlier designs of this simple seal consist of two basic components, as shown in Figures 6-9 and 6-10:

1. A magnetized ring, having an optically flat sealing surface that is fixed in a stationary manner to the

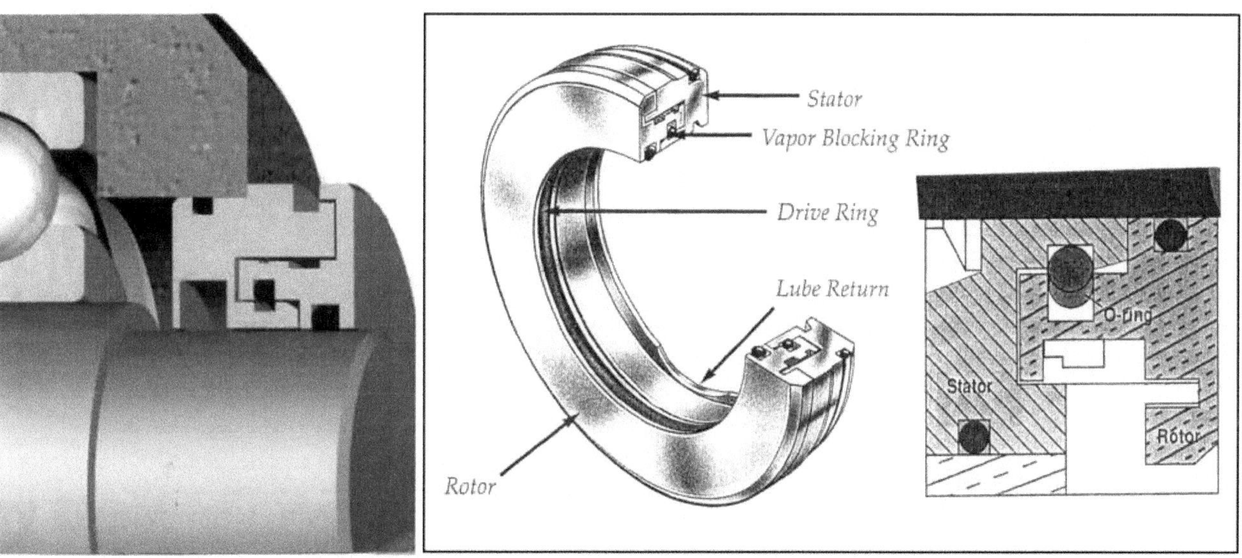

Figure 6-15: Bearing isolators (Source: INPRO/Seal Company, Rock Island, Illinois 61202)

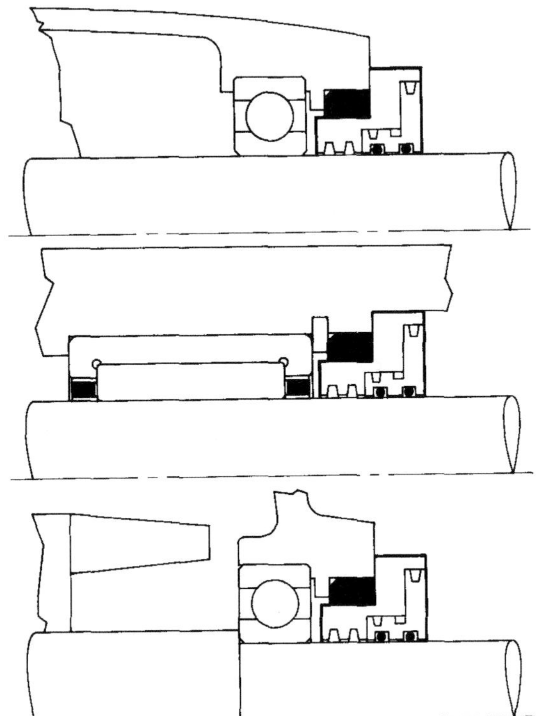

Figure 6-16: Bearing isolators with air gap separating rotor and stator rings

housing and sealed to the housing by means of a secondary O-ring.

2. A rotating ring having a sealing surface that is coupled to the shaft for rotation and sealed to the shaft with an O-ring.

The rotating ring, which is fabricated from a ferromagnetic stainless steel or is fitted with bar magnets, can be moved along the shaft. When no fluid pressure exists, the sealing surfaces are held together by the magnetic force, which is reliable and uniform, creating a positive seal with minimum friction between the sealing faces and ensuring the proper alignment of the surfaces through the equal distribution of pressure. Temperature rise and anticipated (low) leakage have been documented in Ref. 6-8.

DUAL-FACE MAGNETIC BEARING HOUSING SEALS

As was brought out before, it has long been recognized that lube oil contamination vastly reduces the life of pump bearings. This prompted the American Petroleum Institute (API) and pump users to seek out and recommend preventive measures, including devices such as rotating non-contacting labyrinth bearing housing seals ("bearing isolators") and contacting-face rotating magnetic seals. All of these measures have one primary objective: the extension of bearing life through reduction, and possibly even the virtual elimination, of lubricant degradation.

Various factory tests, the development of cost justifications and a thorough review of field experience have established the viability and effectiveness of cartridge-type magnetic dual-face bearing protectors. As of late 2009 and properly installed, these devices prevent external contaminants from degrading the lubricant in bearing housings for industrial machinery. The bearings of pumps, gears, mixers, small turbines, fans, blowers, star feeders, conveyor lines, rotary drum filters, and literally hundreds of other shaft-driven and bearing-supported types of equipment can be effectively sealed with dual (or "double") face magnetic seals. Figures 6-11 and 6-17 illustrate two different configurations of the dual-face (or double-face) product.

Features and Properties of Magnetically Energized Dual-face Cartridge Seals.

In pumps equipped with dual-face magnetic bearing housing seals, the lubricant is totally contained while the atmosphere is effectively excluded. Shafts are sealed with magnetic seals at each end of the frame. In general, the rotary portion of the seal is fastened to the shaft by an O-ring that performs both clamping and sealing functions. The opposing component is O-ring mounted in a stationary, which is then fitted into the

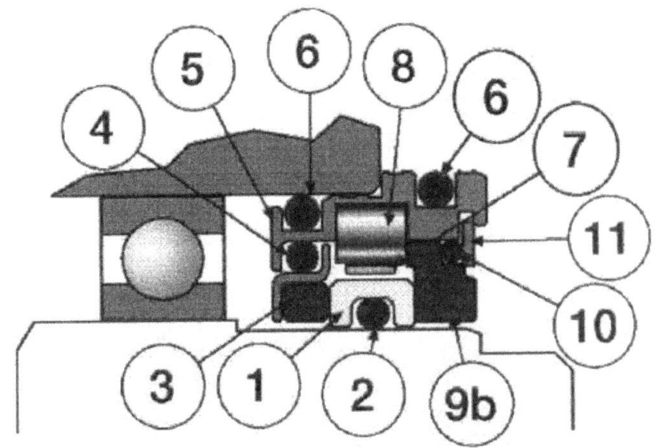

Figure 6-17: Dual-face magnetic seal and O-rings (2, 4, 6, 10), snap ring (11), rotating face (1), stationary faces (3, 9b), stationary magnets (8), magnet carrier (7), and outer body (5). (Source: AESSEAL, Inc., Rotherham, UK, and Rockford, TN)

pump housing or frame adapter.

Instead of springs to hold the faces together, one of the components of a modern dual-face magnetic bearing housing seal is fitted with a series of small rare earth rod magnets. The two opposing parts are either a high strength, corrosion-protected magnetic material or are made of a wear-resistant, low-friction face encased in a material attracted by the strong stationary rod magnets.

In dual-face magnetic seals two seal faces are magnetically held against their respective mating faces. However, the magnet materials often differ. Modern seals use stationary nickel-plated samarium-cobalt magnets, a tungsten-carbide rotary, an antimony carbon face on the side towards the housing-interior, and a bronze-filled Teflon face on the atmospheric side. While the single-face aerospace magnetic seal of Figure 6-10 dates back to the late 1940's, the new and improved dual-face magnetic seals have only been fully operational since about 2002.

Of the two types of dual-face magnetic seals shown in Figures 6-11 and 6-17, the one illustrated in Figure 6-17 has been independently third-party tested with regard to its electrostatic and electromagnetic properties to ensure safe operation in hazardous areas and potentially explosive atmospheres. This testing ensured ATEX compliance. ATEX is the European safety standard for explosive environments and test results, per Figure 6-18, are contained in an ATEX document which certifies compliance of these dual-face magnetic bearing protector seals for use in hazardous areas where a large number of Group II, Category 2 equipment is required. The temperature classification is dependent upon the specific application. The procedure and methodology contained in the ATEX document enable the purchaser to determine the safe use of the dual-face magnetic seals in a given application.

Limitations

The application options for dual-face magnetic bearing housing seals are almost limitless. They are presently used on many pump configurations, including horizontal and vertical pumps, rotary lobe, progressive cavity and gear pumps. Gear speed reducers and a wide variety of different machines found in pulp and paper mills, corn milling equipment, different pillow blocks and rotary valves have also been successfully sealed with these dual-face cartridge magnetic seals, which incorporate neither clips nor set screws. Their dimensional envelope fits many locations where lip seals were originally installed. Yet, their clamping locations avoid contacting the shaft surfaces worn off by lip seal contact.

The external faces of the modern dual-face magnetic seals shown in Figs. 6-11 and 6-17 were originally designed for dry running and will to do so without distress as long as the respective internal faces are kept cool. With marginal (oil splash) lubrication on the inner face of this seal and dry running conditions on the outer face, allowable peripheral shaft velocities were 22 m/s. Assuming a 3-inch (~76 mm) diameter shaft, it would rotate at slightly over 5500 rpm. At an 18°C (65° F) ambient, the resulting face temperature would stay within the stipulated ATEX limitation of 85°C (185°F), see Figure 6-18.

The internal face is made of antimony-infused carbon, selected for its desirable properties, including that of optimal heat dissipation. Additionally, and regardless of actual need, the lubricant application conditions in pump bearing housings are typically known to provide a small, or incidental, amount of oil to at least a segmented region of the inboard sealing faces. The inboard faces of the two dual-face magnetic seal models are designed to operate quite well with this incidental thin film lubrication.

Still, to safeguard against blatant misapplication, the manufacturer warns against dry running. As a matter of general policy, either continuous monitoring or other appropriate inspection and examination methods are advocated to ensure correct equipment oil levels. Fortunately, the bearing housings of properly designed pumps will always incorporate lube application methods that generate an oil fog that results in not only adequate bearing lubrication but also a thin coating of oil for the seal faces and only the complete loss of oil could cause

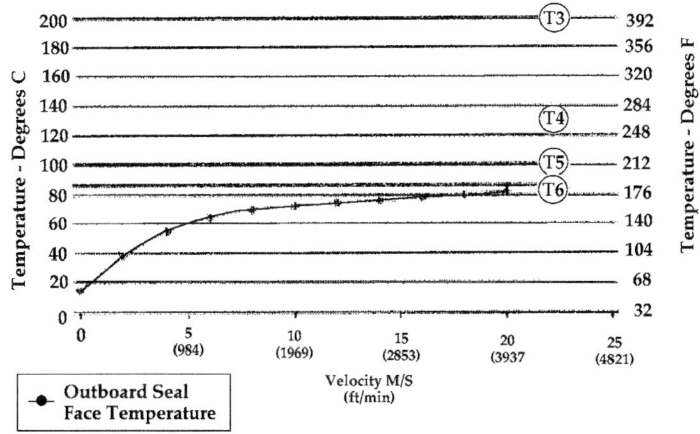

Figure 6-18: MagTecta II ATEX Temperature Graph (@18 deg. C/ 65 deg. F ambient for oil splash, i.e. marginal lubrication)

an unacceptable temperature increase at the inboard/outboard faces. Then again, that's an academic concern since deprivation of lube oil would invite catastrophic bearing failure in any case. For bearing housings in pure oil mist service, dual-face magnetic seals are provided with features that promote oil mist to coalesce on the seal faces. In that case, pre-lubrication of the faces will be required since, upon starting the equipment, oil mist will not instantly coalesce in sufficient amounts to form a separating film between seal faces.

In general, the manufacturers of different bearing housing seals are reluctant to publish leakage data on their respective devices and only occasionally will a test report find its way into the hands of consulting engineers. However, properly applied and installed, modern dual-face magnetic seals have always won in laboratory tests against every other bearing housing seal marketed in 2004. Better yet, many relevant case histories exist.

Case Histories and Cost Justification

There are a number of areas that merit being considered in justifying the incremental cost of dual-face magnetic bearing housing seals over the cost of simple lip seals or certain straight and rotating labyrinth seals.

Progressive, reliability-focused equipment users that seek to improve the profitability of their operations employ Life Cycle Costing, or LCC. The conscientious application of LCC concepts will help reliability-focused plants minimize waste. In the case of double-face magnetic bearing housing seals, LCC will often show dramatic savings in operating and maintenance costs.

In this instance, life cycle component cost comparisons investigate the total lifetime cost to purchase, install, operate and maintain equipment with, vs. equipment without, magnetic bearing housing seals. The comparison would have to include associated downtime, plus certain imputed values of having fewer failure events. Reliability-focused plants include here the *avoided cost* of plant fires occasionally brought on by catastrophic bearing failures and the implicit value of utilizing work force members that previously spent time on remedial tasks and are now free to pro-actively work on preventive tasks. A simplified mathematical expression for life cycle cost was already introduced on page 1 of this text; it could be used for a screening study of economic benefits.

Suppose, then, a facility had identified certain centrifugal pumps that suffered from disappointing bearing life and suppose further that these pumps were installed in a contamination-prone environment. It is not difficult to imagine bearings adjacent to steam quench injection points near mechanical seals, or bearings in mining pumps, or in areas experiencing sandstorms to be at special risk here. Moreover, given source data from Ref. 6-9 and 6-10, it is more than reasonable to anticipate a two-fold increase in bearing life with hermetically sealed bearing housings.

One of the simplest and most straightforward ways to assess the benefits of dual-face magnetic seals over sealing methods that allow an influx of atmospheric contaminants would be to look at a plant's pump failure frequencies and costs. The cost of a set of magnetic seals would be more than offset by the projected reduction in bearing-related pump failures. In virtually all cases so examined, upgrading to dual-face magnetic seals will show payback periods of less than six months.

First, a very simple payback calculation. Suppose a plant had centrifugal pumps with an average MTBF (mean-time-between-failures) of 2.5 years and suppose further that the average repair is costing the plant $6,400, including burden, overhead, field and shop labor, replacement parts, etc. The bearing housings are not hermetically sealed and there is clear evidence of lubricant contamination. Sets of magnetic seals cost $640 and hold the prospect of extending pump MTBF to 5 years. The plant would avoid a $6,400 repair and, over a five-year period, would realize a payback of $6,400/$640—a 10:1 ratio.

Along similar lines, more elaborate benefit-to-cost calculations could take many forms and one of these, labeled a simplified five-year benefit-to-cost calculation, is given below. It relates to a centrifugal pump that was originally equipped with lip seals and is now being upgraded to dual-face magnetic seals (Ref. 6-8).

Again, this calculation covers a five-year life for hermetic sealing with dual-face magnetic seals. As shown, an incremental expenditure of ($640-70) = $570 for dual-face magnetic seals would return ($16,138-$5,084) = $11,054 over a five year period. The payback would be approximately 19:1.

Compared against any other means of sealing, i.e. housing seals that would allow "breathing" (ambient air interchanges), dual-face magnetic seals win. The potential benefits might favor dual-face magnetic seals even more if a plant were to opt not to replace its lip seals every year. In other words, not replacing lip seals would likely result in more bearing replacements or even total pump overhauls and, in certain cases, unit downtime costs. Likewise, it is noteworthy that studies with lip seals on centrifugal pumps being replaced *twice* every year show cost breakdowns that again favor dual-face magnetic seals, and do so by greater margins.

Case History 1:
Axially split case centrifugal pumps in boiler feed service.

Although supplied by a well-known major manufacturer, five hot condensate pumps at a petrochemical plant had been furnished with lip seals at their respective bearings. These lip seals leaked immediately upon startup, which risked depleting the oil in the rather small volume (1 quart, or ~1 liter) oil sump.

Steam quench escaping from mechanical seals adjacent to the bearing housing seal area undoubtedly contributed to the disappointing performance of the factory-supplied elastomeric lip seals. Water intrusion caused rapid and repeated bearing failure, with estimated repair costs in the vicinity of $10,000 —a very reasonable per-event cost approximation for multi-stage pumps in this size range. Since three sealing locations are involved per pump (Figure 13-7, also Figure 13-9), and with the incremental cost of three dual-face magnetic seals remaining well below $1,000, avoiding even a single repair incident per year would equate to a benefit-to-cost ratio in excess of 10:1. Indeed, the facility advised payback in the vicinity of one month.

Case History 2:
Vertically oriented axially split pumps at a water works.

From Figures 6-19 through 6-21 it can be seen that the pillow block bearings are installed in close proximity to the product sealing areas of these medium-pressure, 8,300 gpm (1,925 cubic meter/hr) water pumps. Water intrusion and oil contamination required replacing the pillow blocks every six months. With dual-face magnetic seals subsequently operating flawlessly and no water intrusion or related distress being foreseen, the user claims that a $400 upgrade has avoided in excess of $10,000 worth of repairs. Investing in dual-face seals achieved a payback of two weeks.

There are many more case histories supporting the need to protect lube oil and bearings from external contaminants. Both moisture and particulates can be kept away by effectively sealing equipment bearing housings. Dual-face magnetic sealing represents an improvement and life-extension measure practiced by reliability-focused plants in many countries. The cost justification for these upgrades is easy to calculate and rapid paybacks are the norm in most instances.

	Lip Seal-Equipped	*vs.*	*Dual-face Magnetic Seals*
C_{ic}	(5 yrs)(2)($35)= $350		(2pc)($320) = $640
C_{in}	(0.08)($350) = $ 28		(0.08)($640) = $ 51
	[Procurement cost is ~8% of component cost. Assume no modifications needed to mount cartridge-type dual-face magnetic seals in pump bearing housing]		
C_e	= $ 0		= $ 0
	[Assumes no significant change in frictional energy magnetic seals vs. lip seals]		
C_o	= $186		= $ 45
	[Cost of mineral-type lube oil, replaced once per year, vs. synthetic lube, replaced once in 5 years. Includes labor for oil changes]		
C_m	= $ 10,625		= $2,125
	[Pumps being partially dismantled 5 times for lip seal installation, vs. once for magnetic seal installation. The alternative of NOT replacing lip seals yearly would incur bearing failures and more costly repair incidents]		
C_{dt}	= $ 0		= $ 0
	[Assumes a facility with redundant, or "installed spare" pumps]		
C_{env}	= $ 5		= $ 1
	[Assumes Kyoto Protocol values in a signatory country]		
C_d	= $ 0		= $ 0
	[Assume waste oil will be mixed with furnace feed; hence, no disposal cost]		
C_f	= $ 4,444		= $2,222
	[Incremental cost due to a $4,000,000 fire occurring once per 1,500 pump failures per year in refinery pumps. Assume one-third of these are bearing-related. Therefore, $4,000,000/1500X33.3%X5yrs = $4,444]		
C_v	= $ 500		= $ 0
	[Imputed loss due to not being able to assign technical work force to perform proactive or preventive tasks elsewhere]		
Five-Year [One-Year] Totals:	$ 16,138 [$3,228]		$ 5,084 [$1,017]

Figure 6-19: Axially split pump in vertical orientation at a municipal water facility in the United States

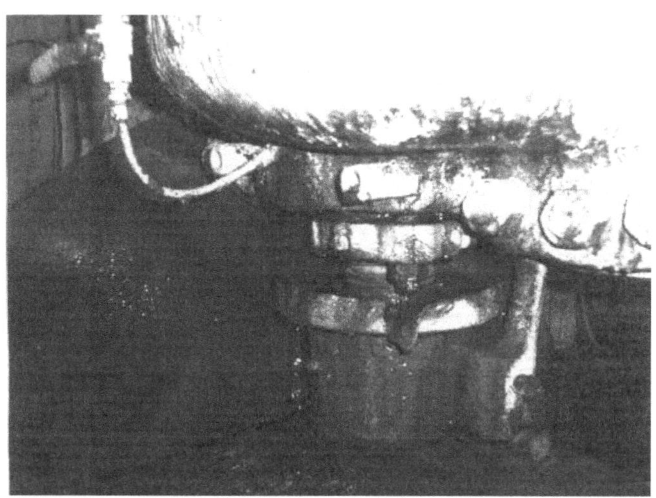

Figure 6-20: Lower bearing region of vertically oriented pump requires effective protection against water intrusion

Face-type magnetic seals are considered very reliable (Ref. 6-10). Also, according to an airworthiness certificate issued by the U.S. Federal Aviation Agency (FAA) and displayed in one vendor's sales literature, some early models have operated continuously for 40,000 hours without repair or adjustment. This life cycle was reached under conditions considerably more severe than those to which most petrochemical process pump bearing housings are typically exposed.

Figure 6-21: Dual-face magnetic seal installed near lower bearing in vertically oriented water pump at a water treatment facility

FEATURES OF UPGRADED CENTRIFUGAL PUMPS

The upgraded medium-duty bearing housing shown in Figure 6-22 incorporates the various bearing life improvement features that have so far been discussed:

1. A deep-groove Conrad-type bearing with loose internal clearance (C3)
2. A duplex, 40° angular contact, lightly pre-loaded back-to-back-mounted thrust bearing
3. A vent port that remains plugged for dry-sump oil-mist lubricated bearings and that can be fitted with an expansion chamber if deemed necessary by the user
4. A cartridge-type magnetic seal
5. A bearing housing end cover (Figure 9-11) made to serve as a directed oil-mist fitting (Ref. 6-6). A more detailed coverage of oil mist is given in Chapter 10.

If oil mist is not available at a given location, the conventional lubrication methods discussed later in Chapter 9 and involving a flinger disc instead of the vulnerable oil flinger, or "slinger" ring, would be the

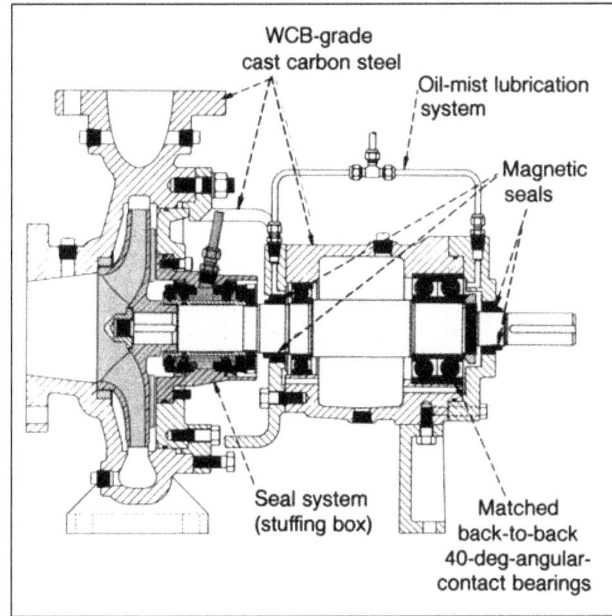

Figure 6-22: "Upgraded medium duty" pump (Source: Carver Pump Company, Muscatine, Iowa)

next best choice. As indicated in Figure 6-23, i.e. with the oil level reaching to the center of the lowermost ball, this choice is limited to pumps with operating speeds up to perhaps 1,800 rpm (Ref. 6-11). The two flinger discs of Figure 6-23 serve only to keep the oil well mixed. Intimate mixing prevents having a layer of hot oil floating at top.

Conventional Wisdom: *Oil rings are needed to feed lubricant into pump bearings.*

Fact: *In slow-speed applications with lube oil levels reaching the center of the lowermost bearing ball, oil rings (or flinger discs) are needed to maintain uniform oil sump temperatures.*

For pumps operating at speeds in excess of 1,800 rpm, and typically at DN (rpm times shaft diameter, in.) values exceeding 6,000 (see page 152), the oil level must very often be prevented from reaching the bearings so as not to cause churning and excessively high oil temperatures. In that instance, oil flinger rings or, preferably, larger diameter flinger discs (Ref. 6-12) would be necessary. Oil flinger rings, also shown later in Figures 7-12, 7-22, 9-16, 9-37, etc., are expected to either feed oil into the bearings or generate enough spray to somehow get adequate amounts of lubricant to the various bearings.

However, just to highlight the issue, proper lubrication of rolling element bearings in centrifugal pumps depends on such factors as bearing preload, cage inclination, oil cleanliness, viscosity and point of introduction of the lube oil. If oil rings are used as an application method, then shaft horizontality, oil ring design, depth of ring immersion and oil viscosity take on added importance. Unless these factors are understood and are taken into full account by pump manufacturers and users, bearing life may be erratic, or consistently (and even unacceptably) low. This is a very important subject and will be dealt with in Chapter 9 (Ref. 6-9).

ADVANCED LUBRICATION STRATEGIES

An advanced lubricant application strategy is specifically designed for owner/users that utilize oil mist lubrication systems in their plants. Here, standard shaft lip seals are replaced with stationary labyrinth seals, as required for API-610 compliance. Rotating labyrinth seals ("bearing isolators") would represent a further upgrade, and magnetic shaft seals, as listed in

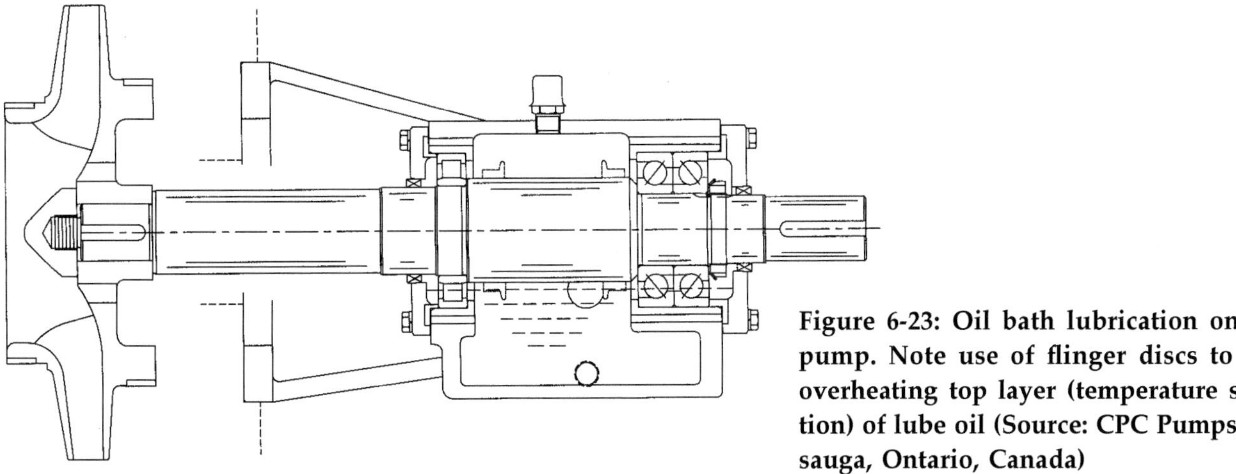

Figure 6-23: Oil bath lubrication on an API pump. Note use of flinger discs to prevent overheating top layer (temperature stratification) of lube oil (Source: CPC Pumps, Mississauga, Ontario, Canada)

the 9th and later editions of API-610, would represent the ultimate in hermetically sealing pump bearing housings. Used in conjunction with oil mist, well-designed rotating labyrinth seals have proven to be effective in keeping wash-down and contaminants out of the power end. With the oil mist creating a clean lubricant for the bearings and the mist venting out through the labyrinth seals, bearings can be expected to last, on the average, three to four times longer. This issue is given detailed attention in Chapters 9 and 10 of our text.

TELESCOPING SHAFT SEAL SLEEVE

One vendor has developed a special shaft sleeve package (Figure 6-24) to address the exacting requirements of mechanical seal setting. The key component is a telescoping sleeve. A short stub sleeve has the standard "hook" portion to properly locate the impeller. The mechanical seal mounts on a separate drive sleeve, which slides on the stub sleeve with an O-ring for sealing. This allows for adjustment of the mechanical seal after the pump has been assembled. A setscrew collar locks the sleeve in position and provides a positive drive. An important benefit of this sleeve arrangement is that it allows the use of a standard single inside mechanical seal. A customer's present seal inventory can be used without modification. It can also be used with a standard stuffing box, as well as with the big bore and taper bore stuffing boxes discussed earlier in this chapter.

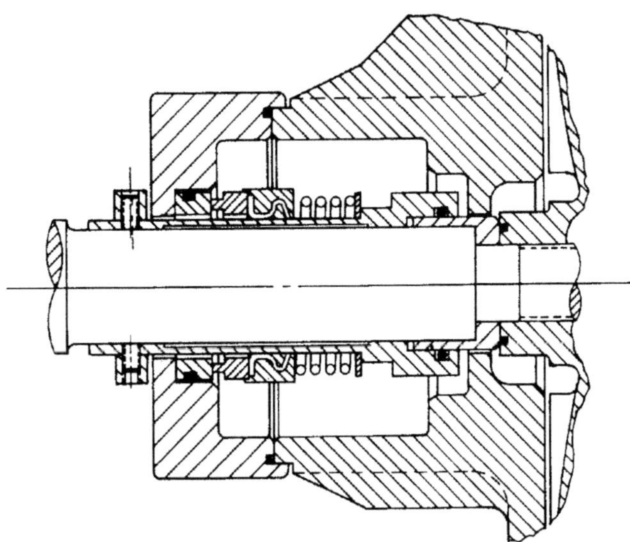

Figure 6-24: Telescoping shaft sleeves facilitate impeller adjustments (Source: ITT/Goulds, Seneca Falls, NY 13148)

Short of being a cartridge seal, this special seal/sleeve arrangement is designed for ease of installation and adjustment. The drive sleeve containing the rotary, stationary, gland and drive collar is installed on the shaft as a complete assembly. The rotary is pre-positioned on the drive sleeve, either against the sleeve shoulder as on an inexpensive single mechanical seal, or with a setscrew collar in a pre-determined position for shorter seals. The stub sleeve is then inserted on the shaft and under the end of the drive sleeve.

Seal setting is accomplished externally by setting the gap between the drive collar and the gland. A separate spacer piece is provided to precisely set this gap. When adjusting the impeller clearance, the spacer should be placed between the drive collar and the gland, and the drive collar setscrews loosened. After the impeller is adjusted, the setscrews are tightened, and the spacer removed.

DYNAMIC SHAFT SEAL—AN OCCASIONAL ALTERNATIVE TO THE MECHANICAL SEAL

On some tough pumping services like paper stock and slurries (note Figure 6-25, a paper stock pump with semi-open impeller and replaceable casing insert), mechanical seals require outside flush and constant, costly attention. Even then, unless superior seals and seal support systems are employed in conjunction with steeply tapered seal housings, seal failures are common, resulting in downtime. Dynamic seals, which use a repeller between the stuffing box and impeller, may eliminate the need for a mechanical seal in certain water-based services (Figure 6-26). There are, however, no reasonable incentives to ever apply dynamic seals in hydrocarbon, toxic, expensive or flammable services.

A repeller (sometimes called "expeller," Figure 6-27), works much like a small impeller. At start-up the repeller pumps liquid and solids from the stuffing box. The repeller soon becomes air-bound and an interface forms within the repeller vanes, sealing the pumped liquid. When the pump is shut down, packing (identified in Figure 6-25 and shown also in Figures 6-26 and 6-27), or some more advanced type of secondary seal (mechanical shutdown seal) prevents the pumpage from leaking into the ambient environment.

In pulp and paper plants, the advantages of a dynamic seal would include:

- External seal water is not required

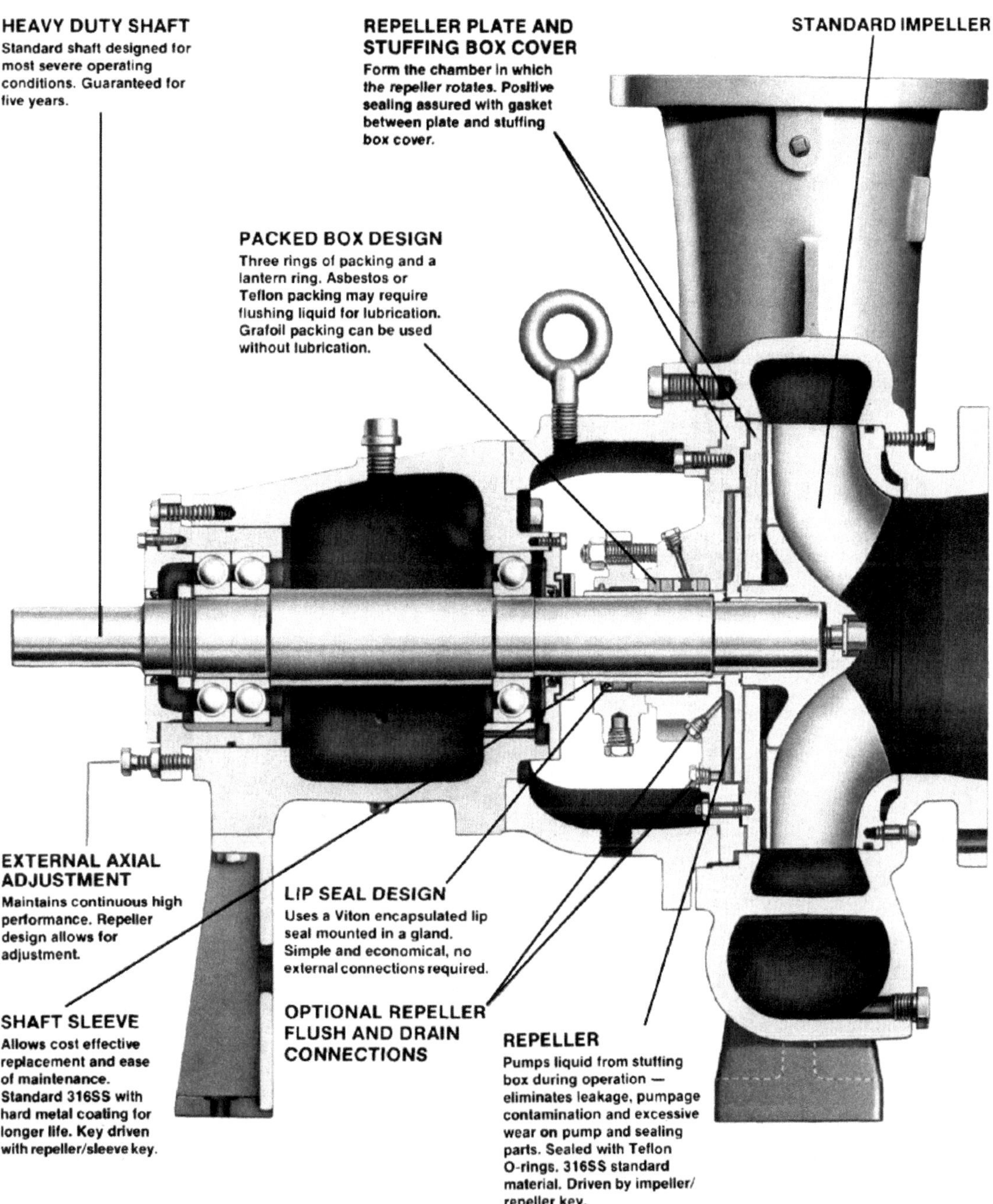
HEAVY DUTY SHAFT
Standard shaft designed for most severe operating conditions. Guaranteed for five years.
REPELLER PLATE AND STUFFING BOX COVER
Form the chamber in which the repeller rotates. Positive sealing assured with gasket between plate and stuffing box cover.
STANDARD IMPELLER
PACKED BOX DESIGN
Three rings of packing and a lantern ring. Asbestos or Teflon packing may require flushing liquid for lubrication. Grafoil packing can be used without lubrication.
EXTERNAL AXIAL ADJUSTMENT
Maintains continuous high performance. Repeller design allows for adjustment.
LIP SEAL DESIGN
Uses a Viton encapsulated lip seal mounted in a gland. Simple and economical, no external connections required.
SHAFT SLEEVE
Allows cost effective replacement and ease of maintenance. Standard 316SS with hard metal coating for longer life. Key driven with repeller/sleeve key.
OPTIONAL REPELLER FLUSH AND DRAIN CONNECTIONS
REPELLER
Pumps liquid from stuffing box during operation — eliminates leakage, pumpage contamination and excessive wear on pump and sealing parts. Sealed with Teflon O-rings. 316SS standard material. Driven by impeller/repeller key.

Figure 6-25: Paper stock pump with axial shaft adjustment provision, replaceable casing insert and repeller ("dynamic seal") arrangement (Source: ITT/Goulds, Seneca Falls, NY 13148)

- Elimination of pumpage contamination and product dilution
- No need to treat seal water
- Eliminating problems associated with piping from a remote source.

Among the disadvantages of a dynamic seal we note that:

- The repeller/expeller consumes a fair amount of power.
- The secondary seal does not seal as well as a mechanical seal when the pump is shut down. As

be used for hazardous liquids.

- Dynamic seals are limited to relatively low suction pressures.

CONSIDER COMPOSITE MATERIALS IN CENTRIFUGAL PUMP WEAR COMPONENTS

Recall that reliability improvement implies that both feasibility and cost-effectiveness of materials and component upgrading must be constantly on the mind of the reliability professional. Reliability-focused pump users view every repair incident as an opportunity to upgrade.

HP and other fluid processing industries have embraced the use of current generation composite materials in centrifugal pumps to increase efficiency, improve MTBR (mean time between repair), and reduce repair costs. Composite wear materials are included in API-610, 9th Edition, the latest centrifugal pump standard from the American Petroleum Institute.

One such material that has been used successfully by major refineries is Dupont Vespel® CR-6100, which is a PFA—a carbon fiber composite (Ref. 6-13). Its properties eliminate pump seizures and allow internal rotating-to-stationary part clearances to be reduced by 50% or more. CR-6100 has replaced traditional metal *and* previous generation composite materials in pump wear rings, throat bushings, line shaft bearings, inter-stage bushings, and pressure reducing bushings. Figures 6-28 and 6-29 show applications in horizontal and vertical pumps, respectively. For installation considerations refer to Chapter 15. Note that axial loads on thrust bearings may change when clearances are made smaller. Verify bearing loads are within allowable range.

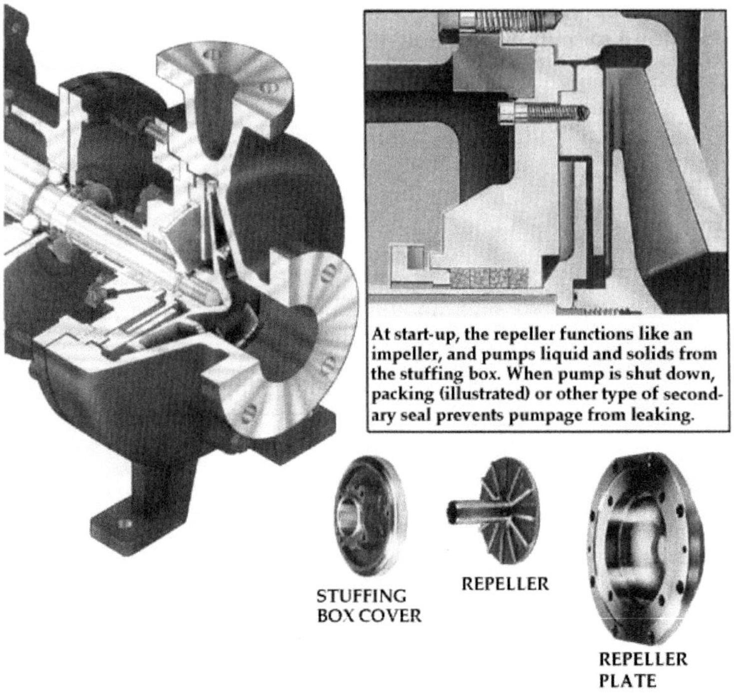

At start-up, the repeller functions like an impeller, and pumps liquid and solids from the stuffing box. When pump is shut down, packing (illustrated) or other type of secondary seal prevents pumpage from leaking.

STUFFING BOX COVER

REPELLER

REPELLER PLATE

Figure 6-26: Dynamic seal and repeller (Source: ITT/Goulds, Seneca Falls, NY 13148)
Note that terms "repeller" and "expeller" are used interchangeably.

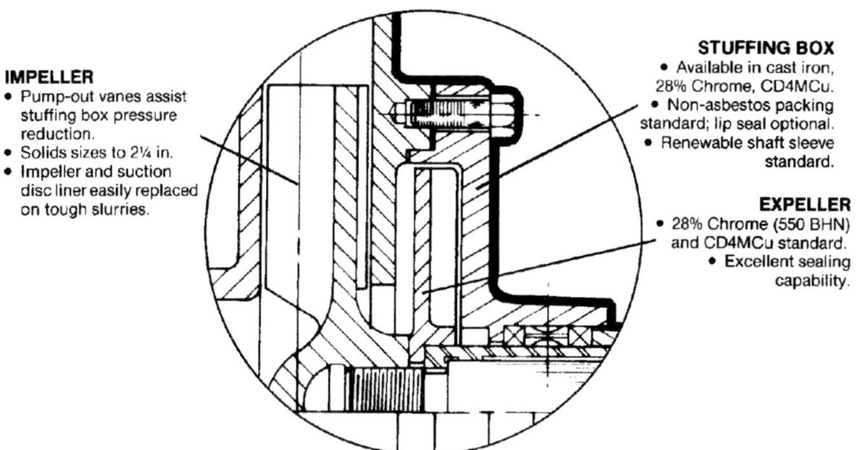
IMPELLER
• Pump-out vanes assist stuffing box pressure reduction.
• Solids sizes to 2¼ in.
• Impeller and suction disc liner easily replaced on tough slurries.
STUFFING BOX
• Available in cast iron, 28% Chrome, CD4MCu.
• Non-asbestos packing standard; lip seal optional.
• Renewable shaft sleeve standard.
EXPELLER
• 28% Chrome (550 BHN) and CD4MCu standard.
• Excellent sealing capability.

Figure 6-27: Packing components of a dynamic seal (Source: Morris Pump, Division of ITT/Goulds, Seneca Falls, NY 13148)

CR-6100 reduces the risk of pump seizures. Proper application of the high-performance material imparts dry-running capability and mitigates damage from wear ring contact. Pumps equipped with rings and bushings made from this carbon fiber material rarely—if ever—experience pump seizures during temporary periods of suction loss, off-design operation, slow-rolling, or start-up conditions. When the upset condition has been corrected, the pump can continue operating with no damage or loss of performance. Conversely, when metal wear components contact during operation, they generate heat, the materials gall (friction weld), and the pump seizes. This creates high-energy, dangerous failure modes, which can result in extensive equipment damage

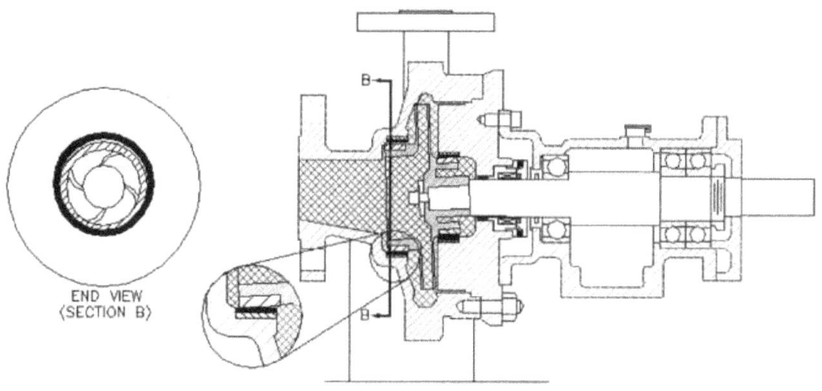

Figure 6-28: Vespel® CR-6100 applications in horizontal pump (Source: DuPont Engineering Polymers, Newark, Delaware. Also marketing publication VCR6100PUMP, 12/01/02)

and potential release of process fluid to atmosphere.

As usual, there are many ways to investigate the cost justification for upgrading to this proprietary PFA in pumps. Note how the justification in Table 6-4 incorporates the value of efficiency gains in a typical 75 kW centrifugal pump, where clearance was reduced by one-third (Table 6-4).

Essentially, then, a one-time incremental outlay of $1,000-$520 = $480 returns $1,890 per year for 7 years. The first year payback ratio is $ 1890/$480, almost 4:1. The seven-year payback is $ 13,230 / $ 480 = 27:1. However, several of the parameters listed in other case histories throughout this book could reasonably be used to arrive at even higher payback ratios. Among these are the imputed cost of avoided fire incidents and the value of reassigning freed-up workforce members to proactive tasks.

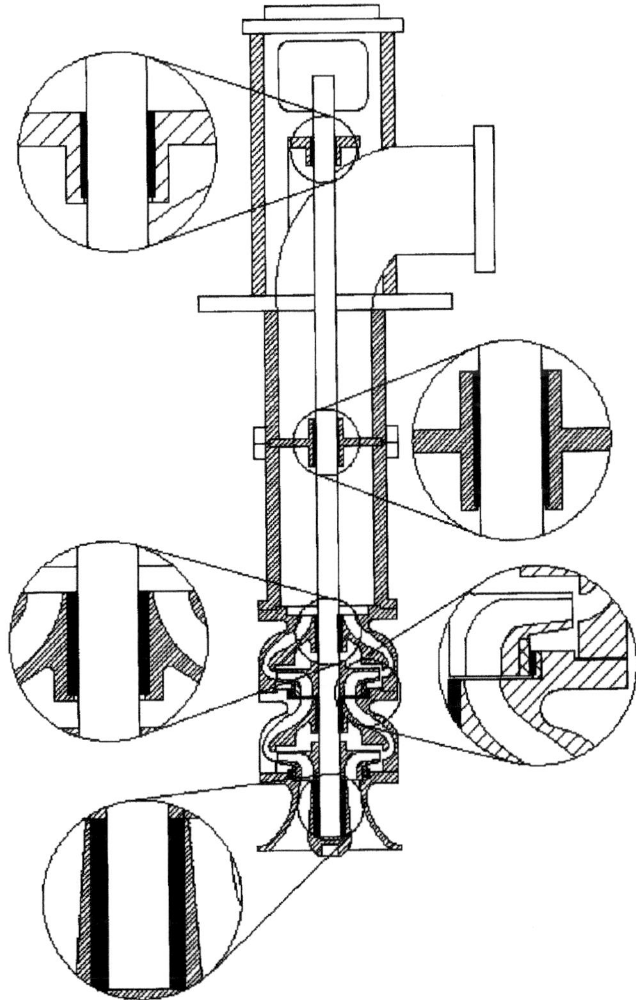

Figure 6-29: Vespel® CR-6100 applications in vertical pump (Source: DuPont Engineering Polymers, Newark, Delaware. Also marketing publication VCR6100PUMP, 12/01/02)

CR-6100 wear components do not gall or seize, and damage to expensive parts is eliminated. This reduces repair costs and mitigates safety and environmental incidents. Moreover, reducing wear ring clearance by 50% increases pump performance and reliability through increased efficiency, reduced vibration, and reduced NPSHr. The efficiency gain for a typical process pump is 4-5% when clearance is reduced by 50 % (Ref. 6-10). Minimized wear ring clearance also increases the hydraulic damping of the rotor, reducing vibration and shaft deflection during off-design operation. The lower vibration and reduced shaft deflection increase seal and bearing life, and help users achieve reliable emissions compliance. This clearance reduction also reduces the NPSHr on the order of 2-3 ft (0.6-1.0 m), which can eliminate cavitation in marginal installations (Ref. 6-14).

Users have had great success installing CR-6100 to achieve all of these benefits. One refinery installed this composite in wear rings and line shaft bearings to eliminate frequent seizures in 180°F (82°C) condensate return service. The condensate return pumps have subsequently been in service for 6 years without failure. Another user improved the efficiency and reliability of two gasoline shipping pumps by installing CR-6100 wear rings, inter-stage bushings, and throat bushings. The shipping pumps have been in service for 4 years without failure or loss of performance. Hundreds of other applications have benefited from composite wear

Table 6-4: Cost Justification for High-Performance Wear Parts

	Wear Rings, Cast Iron	Wear Rings, Vespel®
Cost, 2 Items	$ 520/5yrs = $ 104/yr	$1,000/7yrs = $ 143/yr
Efficiency Gain	Base	4% of 75 kW=3kW
Incremental $, (3kWx8,760hrs) x($0.06/kWh)	$ 1,576	Base
Repair Cost, per year	Basis $8,000/5yrs = $ 1,600/yr	Basis $8,000/7yrs = $ 1,143/yr
Total, per year	$ 3,176	$ 1,286
Savings, per year	Base	$ 1,890

components; these include light hydrocarbons, boiler feed water, ammonia, sour water, and sulfuric acid.

Another frequently used high performance composite for these applications is bearing grade Teflon® with carbon fiber filled PEEK (poly-ether-ether-ketone). This composite has exceptional chemical resistance, a relatively high temperature rating (480°F, or 249°C), improved dimensional stability and low coefficient of friction. Still, actual testing on pumps conducted in late 2003 showed that enthusiasm for Vespel CR-6100 is not unfounded. For follow-up and relevant information, reliability-focused readers may wish to contact www.industryuptime.com.

In any event, be sure to actively consider composite materials as your upgrade option wherever reliability improvements are sought. Also, don't overlook that the pressures now surrounding pump impellers could differ from those in your "old" installation. In other words, verify the new loads on a pump's thrust bearings are still in the acceptable range.

Chapter 7

Bearings in Centrifugal Pumps*

PUMP BEARING LOADS

Pump bearings support the hydraulic loads imposed on the impeller, the mass of impeller and shaft, and the loads due to the shaft coupling or belt drive. Pump bearings keep the shaft axial end movement and lateral deflection within acceptable limits for the impeller and shaft seal. The lateral deflection is most influenced by the shaft stiffness and bearing clearance.

The bearing loads consist of hydrostatic and momentum forces from the fluid, and mechanical unbalance forces from the pump rotor. The forces on the impeller are simplified into two components: axial load and radial load.

AXIAL LOAD

The axial hydraulic pressures acting on a single stage centrifugal pump are illustrated in Figure 7-1. The axial load is equal to the sum of three forces:

1) the hydrostatic force acting on the impeller front and back (hub) shrouds, due to the hydraulic pressures acting on the surface areas of the shrouds,

2) the momentum force due to the change in direction of the fluid flow through the impeller, and

3) the hydrostatic force due to the hydraulic pressure acting on the impeller (suction) opening and shaft. The hydrostatic forces dominate the impeller loading.

Both magnitude and direction of the axial force may change during the pump startup process, due to varying flow conditions in the side spaces between the impeller shrouds and casing walls. The changes in flow conditions and the resulting changes in pressure distributions on the impeller shrouds cause the axial load to fluctuate.

In single stage end suction pumps, the magnitude and direction of the net axial load is most influenced by the design of the impeller. Four typical impeller designs are illustrated in Figure 7-2. The semi-open impeller with pump-out vanes and the closed impeller with two wear rings and balance holes are most common in petrochemical and paper mill process applications.

In pumps with open and semi-open impellers, the axial load is normally directed towards the suction side owing to the pressure on the large area of the hub shroud. Closed pump impellers with wear rings can have near-balanced (zero) axial load or, more commonly, low axial load directed towards the suction. With increased suction pressures, the axial load can be directed to act opposite to the suction.

Impeller pump-out vanes and balance holes are employed to balance the axial load.

Pump-out vanes (also called back vanes) are small radial vanes on the hub shroud used to increase the velocity of the fluid between the hub shroud and the casing wall. This reduces the pressure of the fluid and

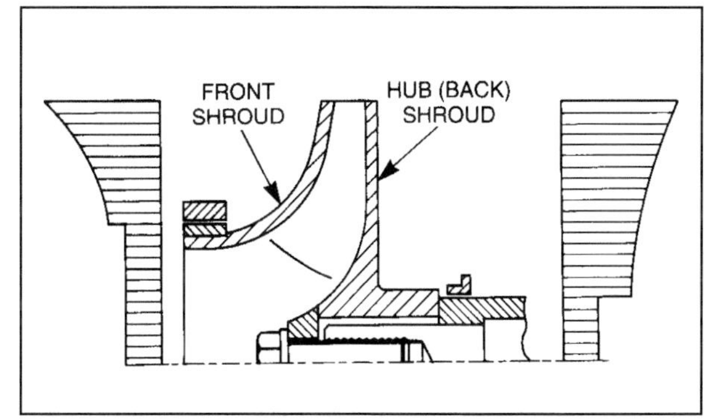

Figure 7-1: Axial hydraulic pressures acting on an impeller

*Courtesy of SKF USA Inc. Publication 100-955, Second Edition, by permission of SKF USA Inc., with minor updates provided by the authors. Please note that this material was derived from one of many manuals designed to provide specific application recommendations for SKF customers when used with the latest issue of SKF's General Catalog. It is not possible, in the limited space of this Pump Life Extension Handbook, to present all the information necessary to cover every application in detail. SKF application engineers should be contacted for specific bearing recommendations. The higher the technical demands of an application and the more limited the available experience, the more advisable it is to make use of SKF's application engineering service.

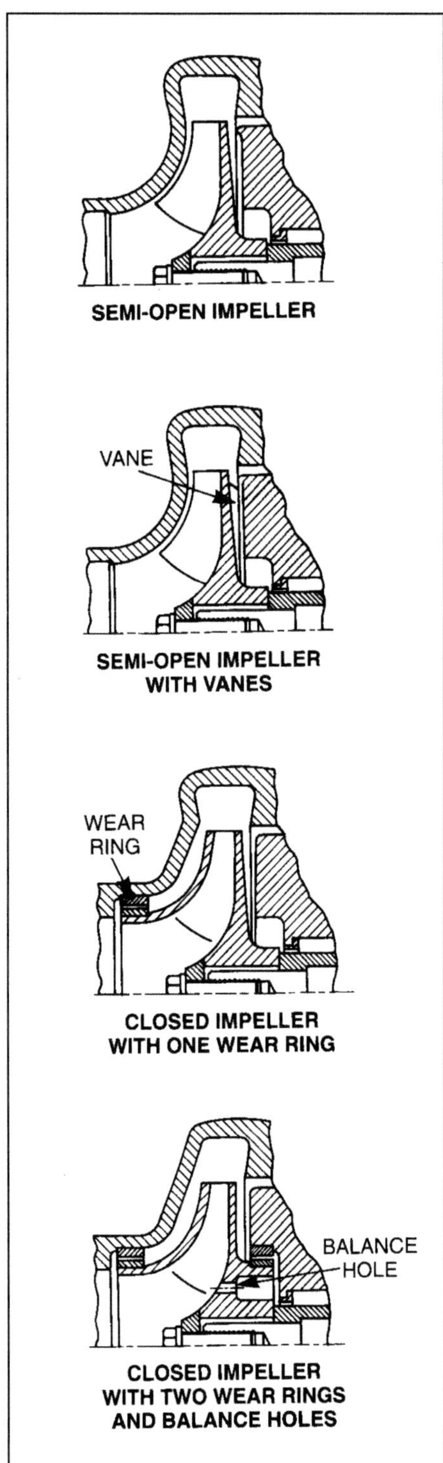

Figure 7-2: Typical impeller designs found in centrifugal pumps

results in reduced axial load on the impeller. The ability of pump-out vanes to reduce axial load is largely influenced by their proximity to, or clearance relative to, the back casing surface.

Balance holes are holes in the hub shroud used to equalize (balance) the pressure behind the impeller with that of the pump suction. Balance holes help to balance the two hydrostatic forces acting in opposite directions on the impeller shroud surfaces. Typical tests (Figure 7-3) illustrate the influence of these balance holes on pump axial load. The impeller without balance holes has greater axial load than the impeller with balance holes.

The magnitude and direction of the axial load can change from its design value if pump-out vane clearance changes due to wear, or is not set within tolerance, or if balance holes become plugged with debris. Pump-out vanes and balance holes reduce pump efficiency by several percentage points but may be necessary and unavoidable in certain pumps.

The axial load in double suction impeller pumps is balanced except for a possible imbalance in fluid flow through the two impeller halves. Such an imbalance might be caused by a piping configuration, proximity to pipe bends (elbows), or impeller-internal differences between seemingly mirror-image halves.

In multistage pumps, impellers are often arranged in tandem and back-to-back to balance the axial load.

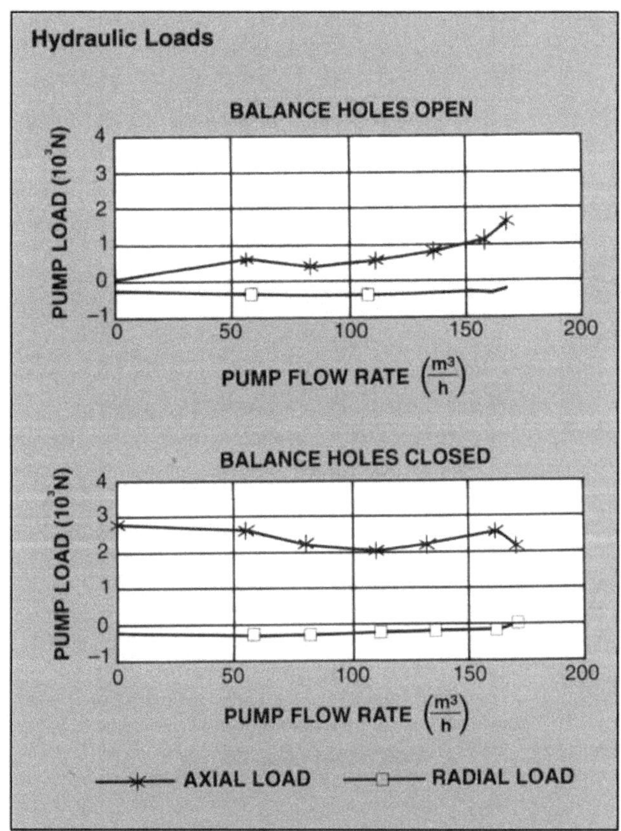

Figure 7-3: Influence of balance holes on pump axial load

RADIAL LOAD

The hydraulic radial load is due to the unequal velocity of the fluid flowing through the casing. The unequal fluid velocity results in a non-uniform distribution of pressure acting on the circumference of the impeller. The radial load is most influenced by the design of the pump casing.

The casing is designed to direct the fluid flow from the impeller into the discharge piping. In a theoretical situation at BEP, the volute casing has a uniform distribution of velocity and pressure around the impeller periphery, see Figure 7-4.

In a real volute at the BEP, the flow is most like that in the theoretical volute except at the cutwater, also called a "tongue." The tongue is, of course, needed for the volute construction, as illustrated in the real volute of Figure 7-4.

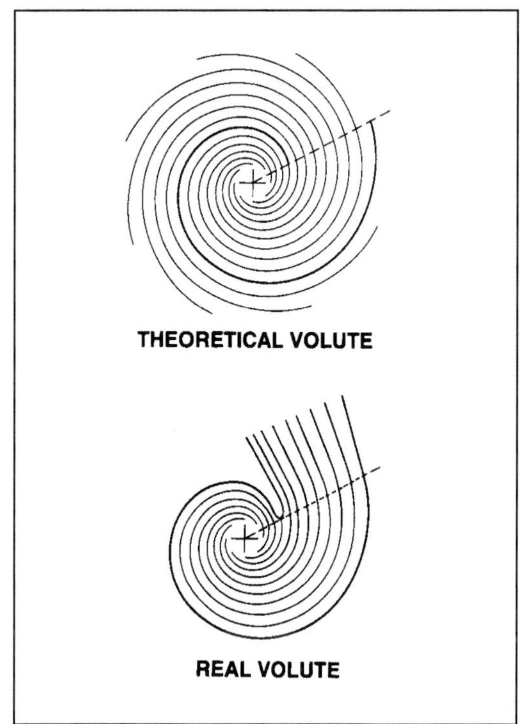

Figure 7-4: Fluid flow in pumps

The disturbance of flow at the cutwater causes a non-uniform pressure distribution on the circumference of the impeller resulting in a net radial load on the impeller. The radial load is minimum when the pump is operating at the BEP and is directed towards the cutwater. Radial load increases in magnitude and changes direction at flows greater than and less than BEP, see Figure 7-5.

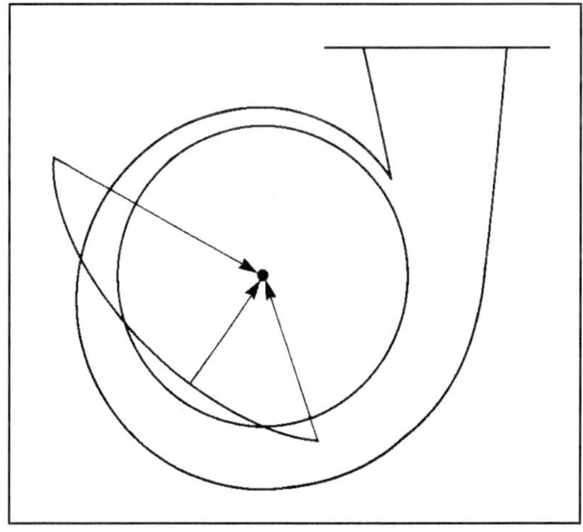

Figure 7-5: Radial load change magnitude and direction at flows greater than and less than BEP flows

Four typical casings are illustrated in Figure 7-6. The single volute casing is commonly used in small process pumps. The diffuser and circular volutes are also commonly used. Due to their diffuser vanes and geometry, they have a more uniform velocity distribution around the impeller and therefore have lower radial impeller loads. The radial load in a circular volute is minimum at pump shut-off, i.e. zero flow, and is maximum near the BEP.

Double volute casings are commonly used in larger pumps when this construction is possible. A double volute casing has two cutwaters that radially balance the two resulting and opposing hydraulic forces. This significantly reduces the hydraulic radial load on the impeller.

Fluctuating and unbalanced radial loads are superimposed on the steady radial load. Fluctuating loads are sometimes caused by the interaction of the impeller vanes passing the casing cutwater. The frequency of the fluctuating force is equal to the number of impeller vanes times the rotational speed. Unbalanced forces can either be due to unevenness in the flow through the passages of the impeller, or mechanical imbalance (mass unbalance).

The hydraulic loads are dependent on the type and size of impeller and casing, the pump operating conditions such as pump speed, fluid suction pressure, and the point of pump operation. The magnitude and direction of the hydraulic loads can change greatly with changes in these factors. In most instances, the lowest hydraulic loads exist only at the BEP.

Pump cavitation influences operation and consequently pump hydraulic loads. Bearing loads should be evaluated at the BEP condition and at the maximum and minimum rated conditions of the pump.

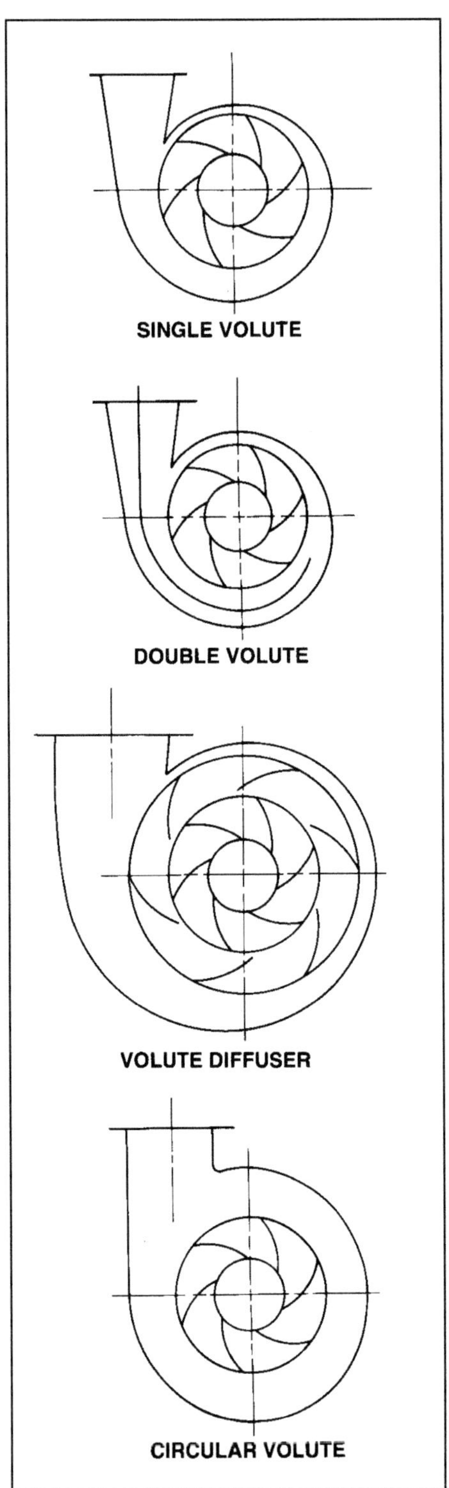

Figure 7-6: Casing configurations typically found in centrifugal pumps

Belt drives and flexible couplings also exert forces on the pump shaft. The force from a belt drive is greater than that from a flexible coupling. The forces resulting from flexible couplings are minimized with improved pump shaft and drive motor shaft alignment. Claims of misalignment having little or no influence on component life are simply false.

The magnitude and direction of hydraulic loads and the load from the belt or coupling drive are best obtained from the pump manufacturer or user. Figure 7-7 illustrates a typical bearing arrangement for an end suction centrifugal pump. From the figure, using the equations of engineering mechanics or SKF computer programs, the bearing reactions can be calculated.

BEARING TYPES USED IN CENTRIFUGAL PUMPS

Figure 7-8 illustrates the different rolling element bearings found in centrifugal pumps. The three most frequently used ball bearing types are the single row deep groove ball bearing, the double row angular contact ball bearing and the universally matchable single row angular contact ball bearing.

Table 7-1, provided by the MRC Division of SKF USA, expands on this illustration, giving relative load capacities in the radial and axial directions. Don't overlook the "matching" options implied on this table. For example, the split inner ring bearing (9000-series) is usually matched with a single-row angular contact ball bearing with the same angle (29° or 40°).

Ball bearings are most commonly used in small and medium sized pumps because of their high speed capability and low friction. Single row deep groove ball bearings and double row angular contact ball bearings are produced in Conrad (i.e. without filling slots) and filling slot type designs. For pump applications, Conrad bearings are preferred over the filling slot type bearing. Conrad double row angular contact ball bearings operate at lower temperatures than filling slot double row bearings in similar pump conditions and are not influenced by the filling slot.

Double row angular contact ball bearings with filling slots should be specially oriented so that the axial load does not pass through the filling slot side of the bearing. Since the late 1980's or early 1990's the API-610 Standard no longer allows filling slot bearings of any type.

Single row angular contact ball bearings of the BE design (40° contact angle) are used where high axial load capabilities are needed for greater pump operational reliability. Universally matchable single row angular contact ball bearings can be arranged as pairs to support loading in either axial direction.

The MRC Bearing Division of SKF has combined a 40° contact angle ball bearing with a 15° contact

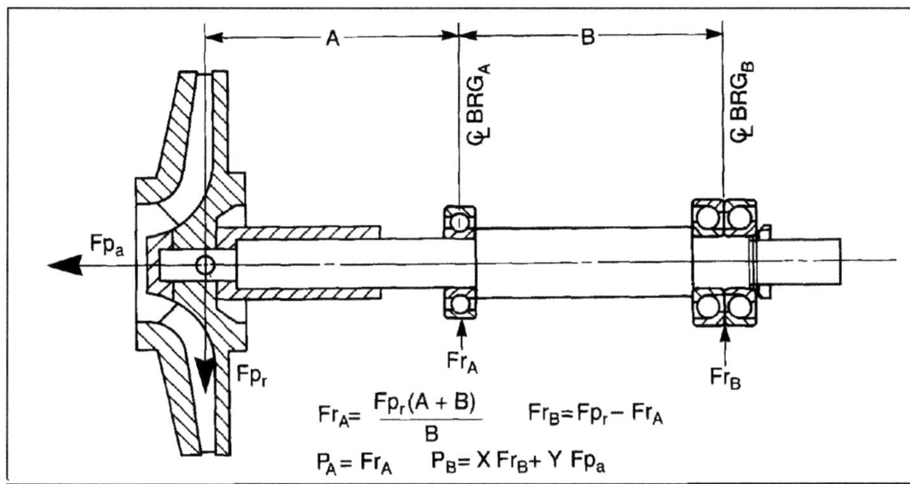

Figure 7-7: Load reactions at pump bearings

angle ball bearing in a bearing set call PumPac®*. (The PumPac® bearing set is described on page 175.)

Spherical, cylindrical, matched taper roller bearings and spherical roller thrust bearings are used in larger, slower speed pumps where the greatest bearing load carrying capacity is needed.

PUMP BEARING ARRANGEMENTS

The most common pump and pump bearing arrangements are shown in Figures 7-9 to 7-16.

Both the vertical inline pump, Figure 7-9, and the horizontal process pump, Figure 7-10, are used in light duty process services, including chemical and paper mill applications. The pump impellers are typically open or semi-open designs. The bearings of the vertical inline pump shown are grease lubricated and "sealed for life"—i.e. limited by the life of the trapped grease. The bearings are spring preloaded to control the endplay of the shaft.

Horizontal process pump bearings are most frequently oil bath lubricated. In some cases (as shown in Figure 7-10) the bearings supporting the axial load are mounted in a bearing housing, separate from the pump frame, to allow adjustment of the impeller in the casing. In these cases, the adjustable housing is shimmed with the frame to ensure good bearing alignment and to obtain the desired impeller-to-casing clearance.

Medium-duty, Figure 7-11, and heavy-duty, Figure 7-12, process pumps are used in refinery services where the highest reliability is required. The impellers are typically closed designs with one or more wear rings. Here, axial load is generally supported by universally matchable single row angular contact ball bearings. The bearings are most frequently oil-bath or oil-ring lubricated. (Lubricant application strategies are the subject of considerable discussion later in this text).

Two heavy-duty slurry pump arrangements are shown in Figures 7-13 and 7-14. Roller bearings are used to support the heavier loading common in these applications. Matched taper roller bearings with steep contact angles, arranged face-to-face or back-to-back, are well suited to support the combined axial and radial loads in these applications.

Spherical roller bearings are used in slurry pumps having very heavy loads. The radial loads are supported by spherical roller bearings. A spherical roller thrust bearing supports the axial load. It is spring preloaded to ensure that sufficient load is applied to the bearing conditions of axial load reversals during pump startup or shutdown, or under certain process upset conditions. This arrangement is most commonly oil-bath lubricated.

For vertical deep well pumps, the spherical roller thrust bearing, Figure 7-15, is a good choice. It easily accommodates the misalignment usually found in these and similar applications having long slender shafts.

Reliability concerns with certain mechanical shaft seals led to increased application of magnetic drive pumps, Figure 7-16. Here, impeller and shaft are supported by plain bearings lubricated by the pumped fluid. Rolling element bearings are used to support the drive shaft. Deep groove ball bearings are most commonly applied in these types of pumps. The bearings can be spring preloaded to limit shaft end movement and to maintain adequate load on the bearings. The spring preload prevents outer ring rotation in these often lightly loaded bearings.

*MRC® and PumPac® are registered trademarks of SKF USA Inc. They can be used when the pump axial load predominates in only one direction. This configuration is shown in Figs. 7-35 through 7-37.

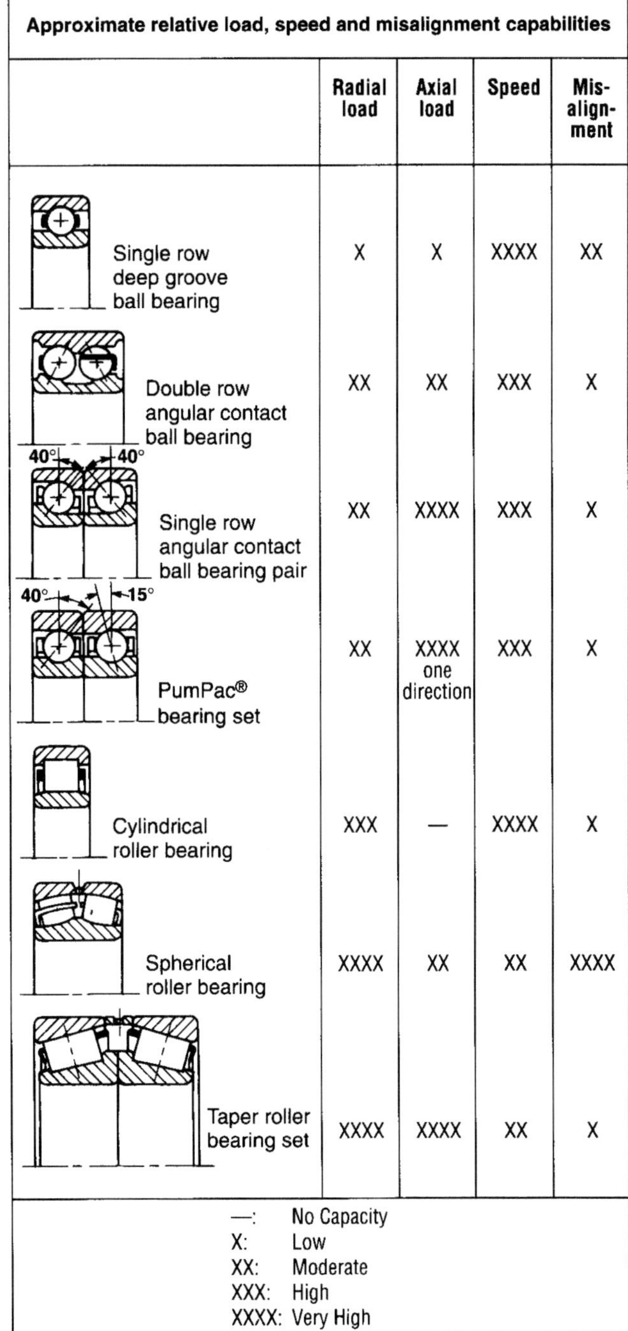

Approximate relative load, speed and misalignment capabilities				
	Radial load	Axial load	Speed	Mis-align-ment
Single row deep groove ball bearing	X	X	XXXX	XX
Double row angular contact ball bearing	XX	XX	XXX	X
Single row angular contact ball bearing pair (40°, 40°)	XX	XXXX	XXX	X
PumPac® bearing set (40°, 15°)	XX	XXXX one direction	XXX	X
Cylindrical roller bearing	XXX	—	XXXX	X
Spherical roller bearing	XXXX	XX	XX	XXXX
Taper roller bearing set	XXXX	XXXX	XX	X

—: No Capacity
X: Low
XX: Moderate
XXX: High
XXXX: Very High

Figure 7-8: Approximate relative load, speed and misalignment capabilities of pump bearing types

BEARING LIFE

When calculating the rating life of SKF bearings, it is recommended that the basic rating life L_{10h}, the adjusted rating life L_{10ah}, and the New Life Theory rating life L_{10aah} each be evaluated, provided sufficient information is available to satisfactorily evaluate the adjusted rating life and the New Life Theory.

The equations to use when calculating the bearing rating life are as follows:

$$L_{10h} = (C/P)^p\ (1{,}000{,}000/60\ n) \quad \text{Eq. 7-2}$$

$$L_{10ah} = a_{23}\ L_{10h} \quad \text{Eq. 7-3}$$

$$L_{10aah} = a_{SKF}\ L_{10h} \quad \text{Eq. 7-4}$$

where:

L_{10h} = basic rating life in operating hours
n = rotational speed, r/min
C = basic dynamic load rating, **N**
P = equivalent dynamic bearing load, **N**
p = exponent for the life equation
(= 3 for ball bearings)
(= 10/3 for roller bearings)
h = hours
L_{10ah} = adjusted rating life in operating hours
a_{23} = combined factor for material and lubrication, obtained from a diagram in the bearing manufacturer's General Catalog
L_{10aah} = adjusted rating life according to New Life Theory, in operating hours
a_{SKF} = life adjustment factor, based on New Life Theory

The ASME/ANSI B73.1 Standard (Ref. 7-1) for process pumps specifies that rolling bearings shall have rating lives greater than 17,500 hours at maximum load conditions and rated speed. The API-610 Standard (Ref. 7-2) for refinery service pumps specifies that rolling bearings shall have rating lives greater than 25,000 hours at rated pump conditions and not less than 16,000 hours at maximum load conditions at rated speed. Many supplemental standards invoked by users simply state that bearing L_{10} life shall exceed 40,000 hours. However, both ASME/ANSI B73.1 and API-610 Standards allow a basic rating life L_{10h}, or adjusted rating life L_{10ah} calculation.

The adjustment factor a_{23} and the adjusted rating life L_{10ah} are dependent on the viscosity of the lubricant at the bearing operating conditions. Figure 7-17 plots the a_{23} factor versus Kappa (**K**). Kappa is the ratio of the lubricant viscosity (v) at the operating conditions to the minimum required lubricant viscosity (v_1) at the operating conditions. The Kappa value should ideally be greater than 1.5.

The New Life Theory indicates that a rolling bearing can have infinite fatigue life provided the applied loads are below a fatigue limit, the bearing operates in a sufficiently clean environment, and was manufactured

Table 7-1: Bearing selection table giving relative load capacities in the radial and axial directions (Source: *TRW/MRC Engineer's Handbook*, 1982).

Description	Load Capacity	Relative Capacity
Type S — Single-Row Deep Groove Conrad Bearings. Ordinarily supplied with loose internal clearance. Other degrees of internal clearance may be necessary for special conditions. Outer and inner races have *full depth on each side*, with no filling notches. Assembly of the balls is made by eccentric displacement of the rings. This bearing type is the most universally used.	Has equal load-carrying capacities in either direction. Recommended for moderately heavy radial loads, thrust loads in either direction, or combination loads.	Type S
Type M — Single-Row, Maximum-Capacity Filling Notch Bearings. Ordinarily supplied with standard internal clearance. Other degrees of internal clearance may be necessary for special conditions. Sometimes designated as the "notched" type, because it has a filling notch on one side of the outer and inner rings a substantial distance away from the bottom of the raceway.	Can carry heavier radial loads than Type S and heavier combined radial and thrust loads. Recommended for a majority of heavily loaded bearing positions — heavy industrial machinery, farm equipment, tractors and trucks.	Type M
Type R — Single-Row, Angular-Contact Counterbored Outer Bearings. Supplied with loose internal clearance for normal applications. Other degrees of internal clearance may be necessary for special conditions. Type R bearing has a nearly *full circle of balls* giving maximum carrying capacity in combination with *continuous raceways*. Counterbored outer ring gives a full-depth shoulder on one side and a small shoulder on the other side. Non-separable type.	Higher radial load-carrying capacity than Type S, heavy thrust load in one direction only, or combined loads where radial load predominates and thrust is always against heavy shoulder of outer ring.	Type R

Relative Capacity	Description	Load Capacity
Angular Contact 7000	**7000 Series — Angular-Contact Bearings.** Counterbored outer ring, non-separable type with initial contact angle of 29°. Two-piece steel cage for normal use or one-piece non-metallic or solid bronze cage for high speed, high operating temperature or severe vibration applications.	Very high thrust load in one direction, combined radial and thrust load where thrust load predominates.
Angular Contact 7000-P	**7000-P — Angular-Contact Bearings, Heavy Duty Type.** Similar in design to 7000 Series Bearings but 40° contact angle. Ball complement and race groove depth designed for increased thrust capacity.	Capacity is 1.18 to 1.40 times that of 7000 Series, varies with individual sizes. Restricted to primarily **thrust** loads. Should not be used for radial loads only or combined radial and thrust loads where radial load is predominant.
Angular Contact 9000	**9000 Series — Angular-Contact, Split Inner Ring.** Designed with solid one-piece outer ring and two-piece inner ring, maximum ball complement and one-piece machined cage.	Construction allows bearing to carry greater thrust in either direction than Type S. May be used where there is substantial radial load providing there is always sufficient thrust load present.
Double-Row Angular-Contact 5000	**Double-Row — Angular-Contact, 5000 Series Bearings.** Filling notches on each side for inserting full quota of balls. Supplied with standard clearance but tighter or looser internal clearance can be supplied to suit special service conditions.	Will support very heavy radial loads, and equal heavy thrust loads in either direction, or heavy combined radial and thrust loads.
		Sizes 5200-5208, 5300-5305 furnished in conrad type with "inverted" contact angle. (With suffix SB)

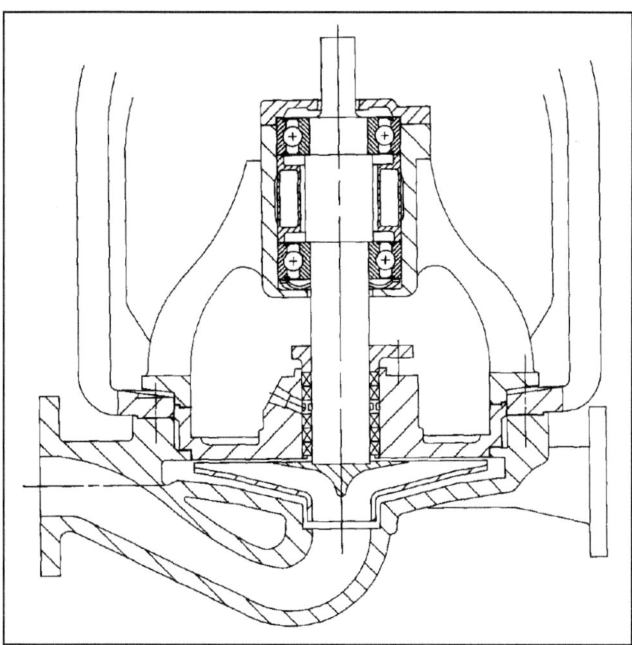

Figure 7-9: Vertical inline pump

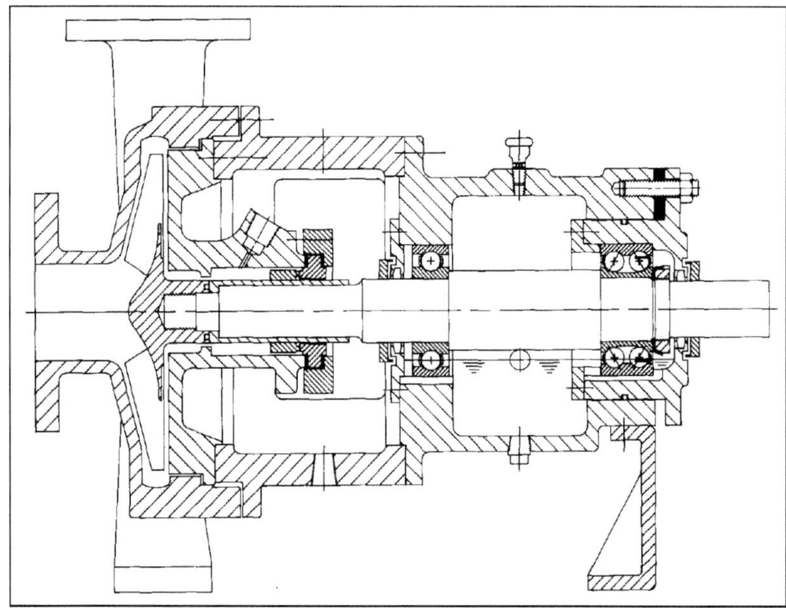

Figure 7-10: Typical horizontal process pump

to accurate tolerances. This is a consideration in some pump applications where hydrodynamic bearings are selected for infinite life.

The adjustment factor a_{SKF} for application of the New Life Theory is dependent on the viscosity of the lubricant at the operating conditions (v), the fatigue load limit of the bearing (P_u), and the contamination level in the application. To enable a systematic and consistent evaluation of the contamination level, SKF has developed a computer program for this purpose. Continued research on the quantification of contamination for use with the New Life Theory will undoubtedly lead to further refinement of this program. For evaluation of a bearing arrangement with the New Life Theory, SKF encourages reliability professionals to contact an SKF application engineer. However, we believe the data contained in our text are sufficiently accurate to yield the substantial life extension results desired by our readers.

In a few of the more unusual applications, pump bearing life can be extended by installing shielded or sealed grease-lubricated bearings within the sealed pump bearing housing. Recall, however, that the life expectancy of sealed bearings is closely tied to the serviceability of the grease charge. Moreover, using sealed bearings may make it necessary to limit the speed of the pump because of the additional seal friction.

The New Life Theory generally considers only solid particle contamination of the lubricant. Contamination of the lubricant by water and other fluids can also reduce the life of the bearings. The allowable free water content in mineral oil type lubricating oils generally ranges from 200 to 500 ppm by volume depending on the additives supplied in the lubricant. Preferably, the water content should not exceed 100 ppm, or about six drops of water per liter of oil! Some synthesized hydrocarbon oils (polyalphaolefins, or PAOs) have the same limits as mineral oils. There is a risk of reduced bearing life if the water content exceeds these values. Lubricant contamination by water is one of the most common reasons for bearing failures in pumps.

BEARING LUBRICATION

The lubricant separates the rolling elements and raceway contact surfaces and lubricates the sliding surfaces within the bearing. The lubricant also provides corrosion protection and cooling to the bearings. The principal parameter for the selection of a lubricant is viscosity, v.

Lubricating oils are identified by an ISO Viscosity Grade (VG) number. The VG number is the viscosity of the oil at 40°C (104°F). Common oil grades are shown in Figure 7-18. From this chart, the viscosity of an ISO Grade oil can be determined at the bearing operating

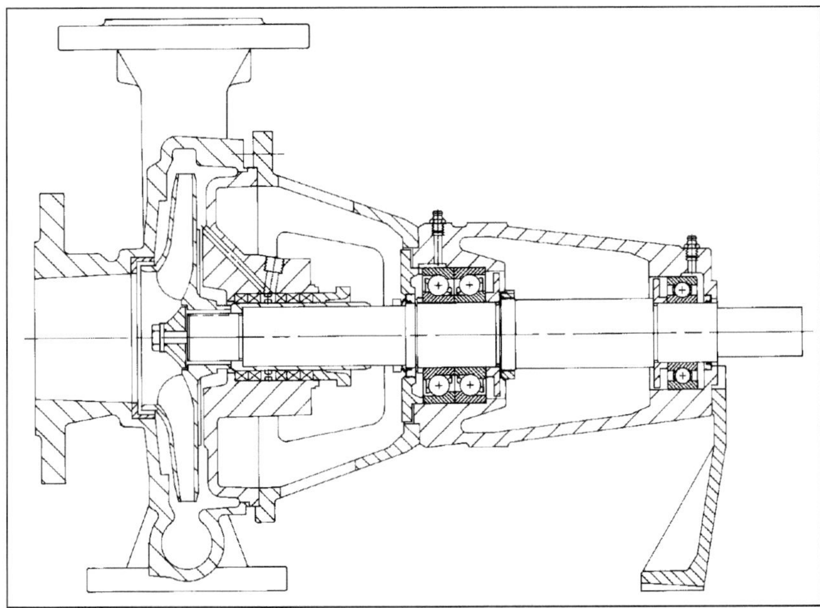

Figure 7-11: Grease-lubricated medium duty pump

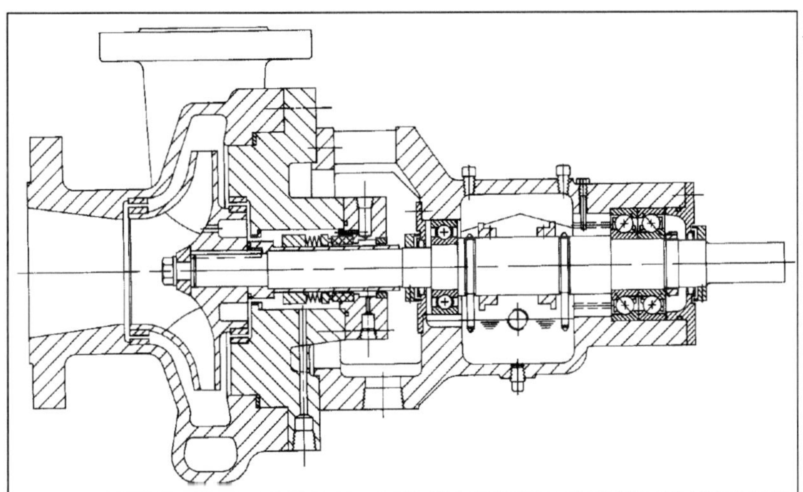

Figure 7-12: Typical heavy duty process pump

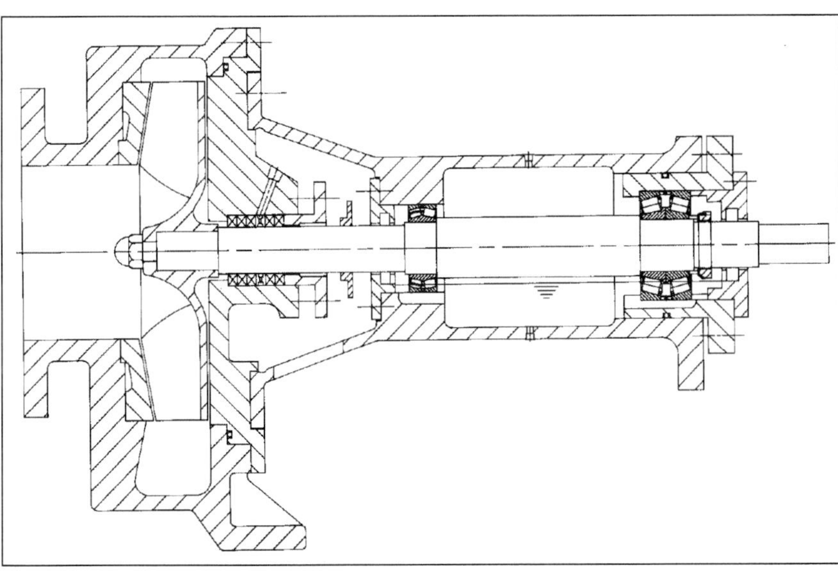

Figure 7-13: Slurry pump

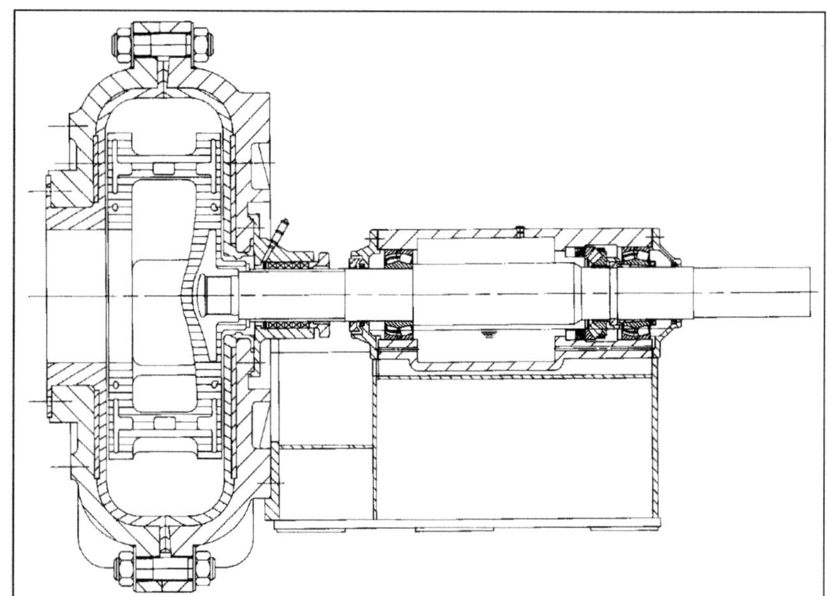

Figure 7-14: Rubber-lined heavy duty slurry pump

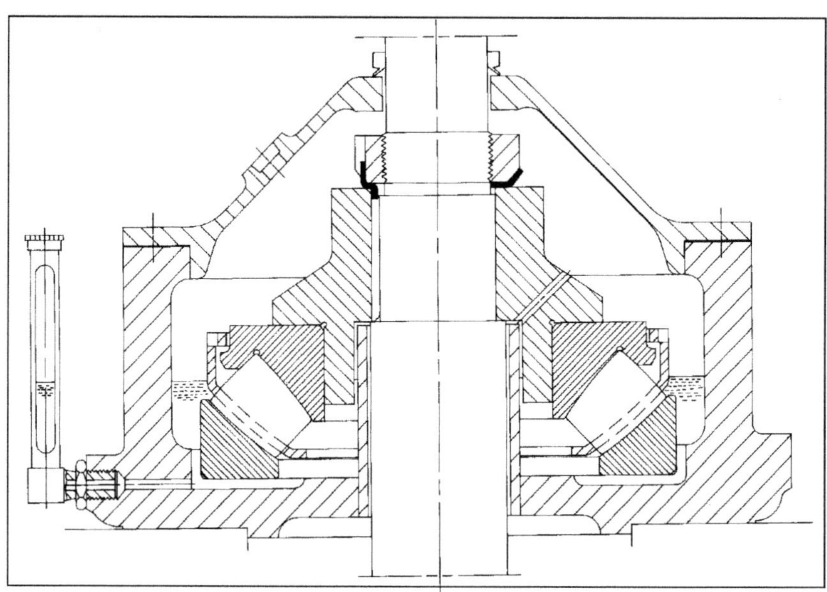

Figure 7-15: Bearing assembly for vertical deep-well pump

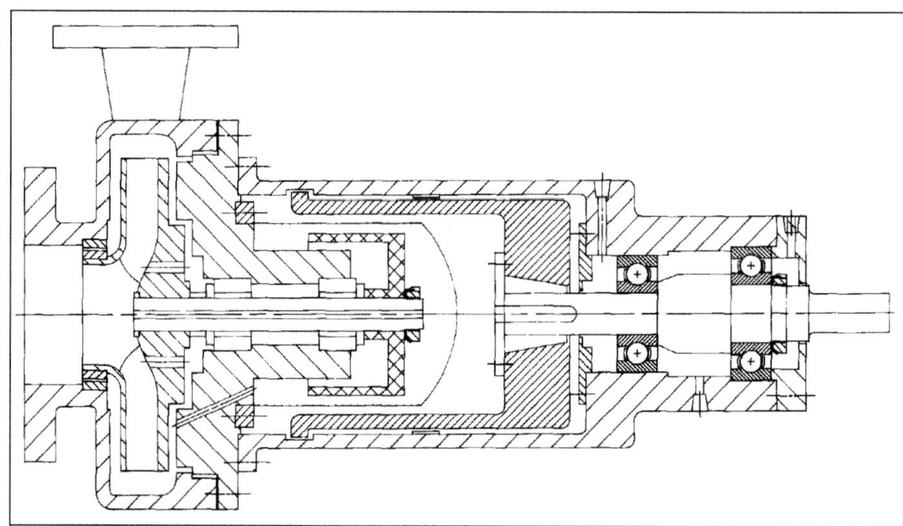

Figure 7-16: Magnetic drive pump

temperature. See also Figure 9-4.

Rolling bearing lubricant requirements depend on bearing size, dm, and operating speed n but little on bearing load. The minimum required lubricant viscosity v_1 needed at the bearing operating temperature is obtained from Figure 7-19. The actual lubricant selected for an application should provide greater viscosity v than the minimum required viscosity v_1 (i.e. Kappa, **K** > 1.0).

Table 7-2 provides general lubricant recommendations for bearings used in centrifugal pumps. These recommendations are valid for operating speeds between 50% and 100% of the bearing catalog speed rating. At lower speeds, higher viscosity grades should be considered and at higher speeds, lower viscosity grades should be considered. The viscosity ratio, Kappa, should be the guideline for evaluation of viscosity. Kappa > 1.5 is preferred. The lubricant viscosity should not be too high, since this would cause excessive bearing friction and heat. Note that Table 7-2 does *not* list VG 32 as a general recommendation for rolling element bearings in pumps. VG 32 *mineral* oils are too light for most typical applications and should only be used if a knowledgeable bearing application engineer makes a written recommendation. However, some pump manufacturers are mindful of the fact that slinger rings (oil rings, Figures 9-16, 9-21 and 9-22) have a greater tendency to malfunction with the higher viscosity ISO Grade 68 and Grade 100 lubricants. Thoughtful compromise is called for.

The frequency of oil change depends on the operating conditions and the quality of the lubricant. High quality mineral oils with a minimum Viscosity Index ("VI") of 95 are recommended. However, multi-grade oils and lubricants with detergents and viscosity

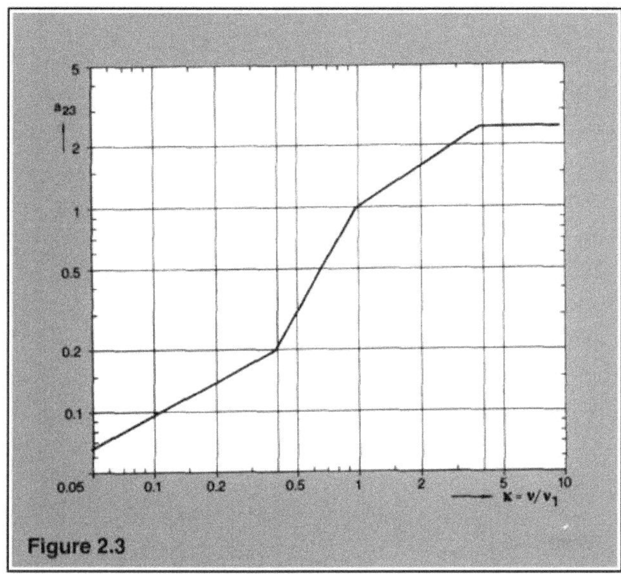

Figure 7-17: Combined adjustment factor a_{23} plotted against ratio of lube viscosity at operating conditions divided by minimum required lube viscosity at operating conditions

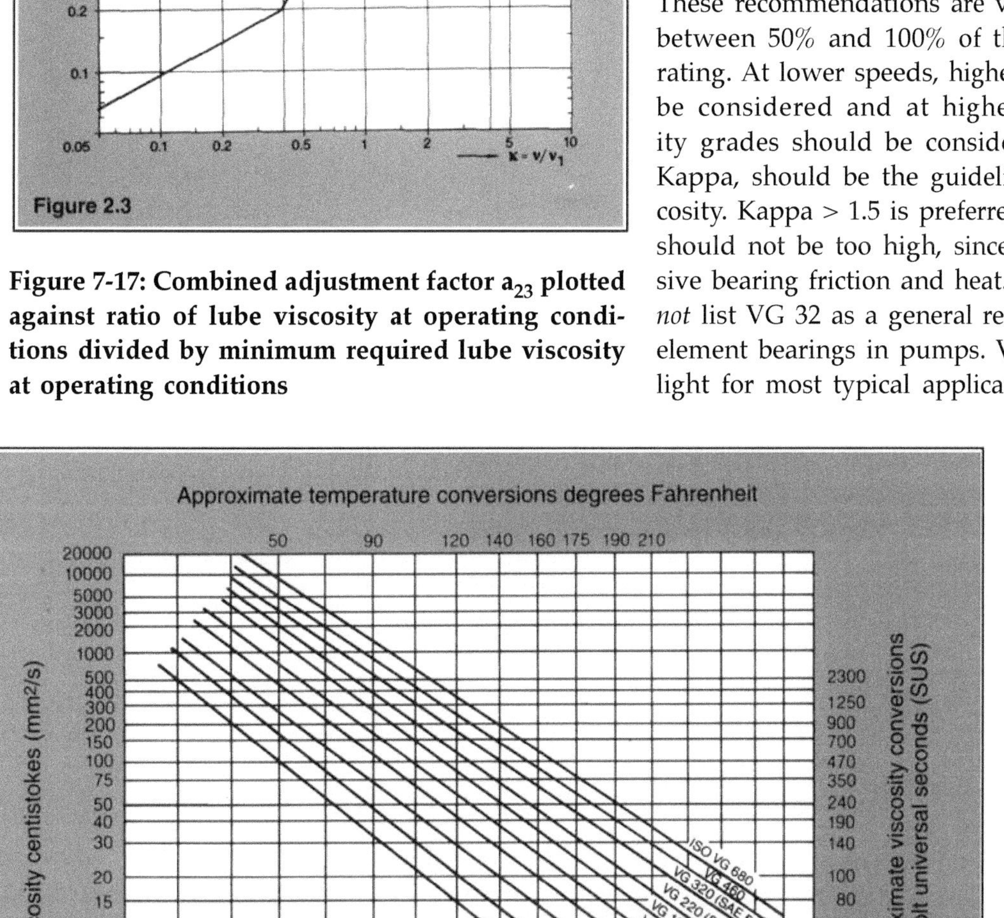

Figure 7-18: Viscosity-temperature relationships for different ISO-grade oils having a viscosity index (VI) of 95

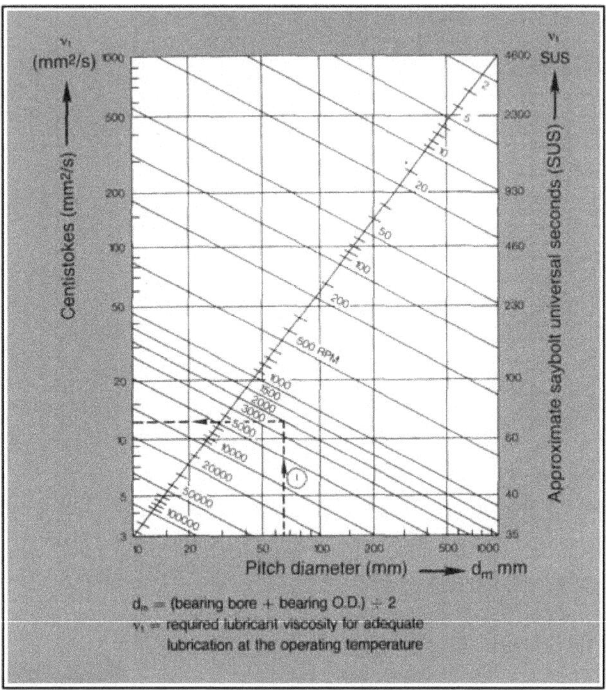

Figure 7-19: Minimum required lubricant viscosity needed at bearing operating temperature

Table 7-2: SKF recommendation of viscosity grades for rolling element bearings in pumps

	Recommended ISO Viscosity Grade	
Bearing operating temperature °C (°F)	Ball and cylindrical roller bearings	Other roller bearings
70(158)	VG46	VG68
80(176)	VG68	VG100
90(194)	VG100	VG150

improvers are not recommended. The use of additives other than those that were originally supplied by the oil manufacturer or formulator is strongly discouraged.

Mineral oils oxidize and should be replaced at three-month intervals if operated continuously at 100°C (212°F). Longer intervals between replacements are possible at lower operating temperatures. Synthetic oils are more resistant to deterioration from exposure to high temperature and may allow less frequent replacement. Lubricants may require more frequent replacement if contamination is present.

The most common methods of pump bearing lubrication are oil bath, oil ring, oil mist and grease. Circulating oil lubrication is also an option and will be cost-justified on larger pumps.

OIL-BATH LUBRICATION

In horizontally oriented applications and at low to moderate speeds, the oil bath level is set at the center of the bearing's lowest rolling element when the pump is idle (see Figure 7-20). At moderate to high speeds, allowing the oil to reach the center of the lowest rolling element often results in excessive heat generation. In those cases, the oil level must be lowered and a mechanical device (oil ring, flinger disc, auxiliary oil pump, etc.) must be used to apply the oil. Generalized rules of thumb recommend "oil flooding," i.e. allowing oil to reach the center of the lowermost bearing, up to a DN-value of about 6,000. Above DN=6,000, obtained by multiplying the shaft diameter (inches) times the shaft speed (rpm), it would be best to lower the oil level. A sight glass or window is needed to visually set the oil level in the bearing. The oil level observed in the sight glass will vary slightly when the shaft is rotating due to the splashing and flinging of the oil in the housing.

Bearing housings should allow the oil to freely flow into each side of the bearing. The housing should have a bypass opening beneath the bearings to allow the oil to flow freely. This opening and also an oil ring are depicted later, in Figure 7-22. The cross-section area of the opening can be estimated according to the following equation.

$$A = 0.2 \text{ to } 1.0\ (nd_m)^{0.5} \qquad \text{Eq. 7-5}$$

where:

A = bypass opening cross-section area, mm^2
n = rotational speed, r/min
d_m = $(d + D)/2$, mm

The small value from the above equation applies to ball bearings and the large value to spherical roller thrust bearings. Intermediate values can be used for other bearing types. If the bypass opening is not provided or not sufficiently large, the oil may not pass through the bearing. This is particularly true for bearings having steep contact angles (angular contact ball, taper roller and spherical roller thrust bearings) operating at high speeds. In these cases a pumping action caused by the bearing internal design may cause starvation of the bearing or flooding of the shaft sealing area.

Figure 7-21 shows an old style "constant level

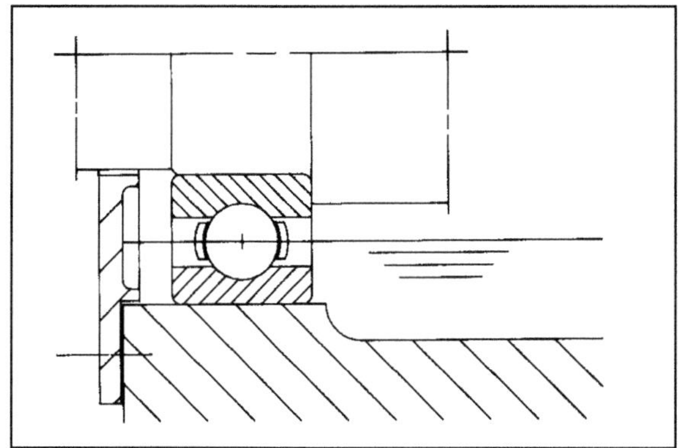

Figure 7-20: Oil bath lubrication

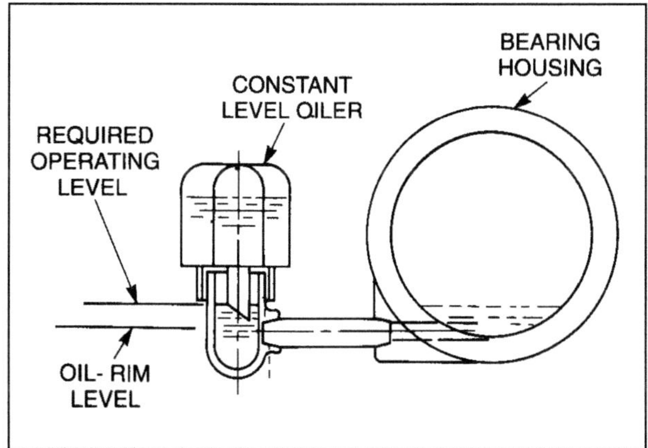

Figure 7-21: Elementary, unbalanced constant level lubricator

oiler," more correctly labeled a constant level lubricator. These lubricators comprise an oil reservoir mounted to the bearing housing that serves to replenish oil lost from the bearing housing. A sight glass is recommended along with these devices to enable the correct setting and examination of the lubricant level in the bearing housing.*

The recommended minimum oil volume V for each bearing in the housing is estimated from:

$$V = 0.02 \text{ to } 0.1\ D\ B \qquad \text{Eq. 7-6}$$

where:

V = oil volume per bearing, ml
D = bearing outside diameter, mm
B = bearing width, mm

For applications with a vertical shaft orientation, the oil level is set at or slightly above the vertical centerline of the bearing. Spherical roller bearings operating in a vertical oil bath should be completely submerged. For spherical roller thrust bearings, the oil level is set at 0.6 to 0.8 times the bearing housing washer height, C. Shaft sealing in these applications is best provided by a thin cylindrical sleeve inside the bearing inner ring support. This was shown earlier in Figure 7-15.

Horizontal oil bath lubrication represents the baseline of moderate bearing friction. The friction with other lubrication methods can be compared with that of oil bath lubrication. Vertical oil bath lubrication produces high friction if one or more bearings are fully submerged, possibly limiting the operating speed.

OIL RING LUBRICATION

The oil ring is suspended from the horizontal shaft into an oil bath below the bearings, as illustrated in Figure 7-22 and earlier in Figure 7-12. The rotation of the shaft and ring flings oil into the bearings and housing. The housing channels the oil to the bearings.

Oil rings are made of brass, steel, or high-performance elastomeric material and sit on the shaft. The inner diameter of an oil ring is generally 1.6 to 2.0 times the diameter of the shaft and the ring can be grooved for best oiling efficiency.

Some sliding may occur between the oil ring and the shaft causing wear and the shaft surface requires a fine finish to minimize this wear.

The large size of the bearing housing needed for

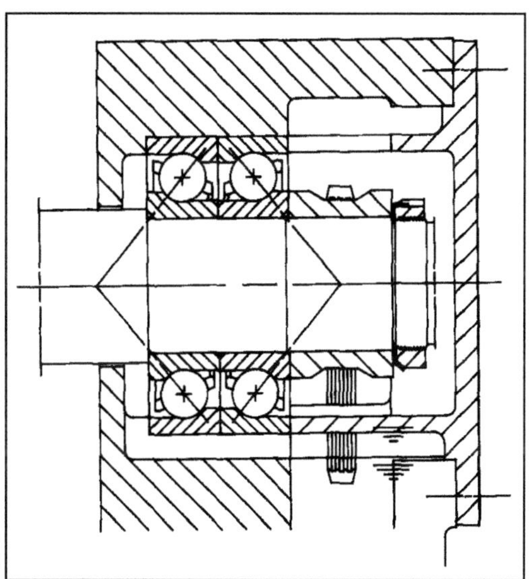

Figure 7-22: Oil ring with concentric grooves

*Note that the constant level lubricator illustrated here is no longer considered "state-of-art." More appropriate devices are discussed in Chapter 9 and shown in Figure 9-19 and associated illustrations.

the oil ring improves the heat transfer from the bearings and oil. Higher shaft speeds and lower viscosity lubricants are thus possible with oil ring lubrication because of the lower friction and better cooling.

However, oil rings suffer from a number of serious limitations that will be explained later in Chapter 9.

OIL MIST LUBRICATION OVERVIEW

Oil mist provides fine droplets of clean, fresh, and cool lubricant to the bearings. Contaminants are excluded from the bearings by the oil mist pressure inside the bearing housing. The mist can also be supplied to the bearings when the pump is idle for maximum bearing protection from contamination and condensation.

The oil mist may be introduced into the bearing housing (indirect mist) or directed at the bearing by a reclassifier fitting. In each case, the housing must be provided with a small vent, typically 3-mm (0.125 in.) diameter, so as to promote oil mist through flow. Directed oil mist (Figure 7-23) is highly recommended if the bearing nd_m* value is greater than 300,000 and/or if the bearing supports a high axial load.

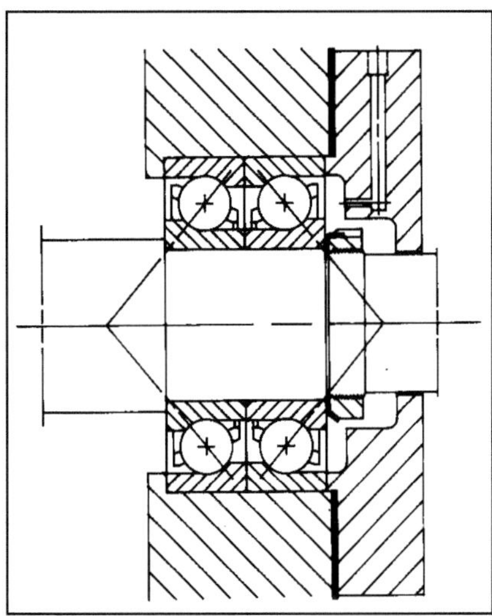

Figure 7-23: Oil mist lubrication using a directed oil-mist fitting

Synthetic or special de-waxed oils are often used for oil-mist lubrication. Paraffins in standard oils may clog the small oil-mist fittings. Ref. 7-3 through 7-5, or a specialist in oil-mist lubrication should be consulted for recommendations.

Bearings can be "purge" oil mist lubricated or "pure" oil-mist lubricated. Purge oil mist adds oil mist lubrication to bearings already lubricated with an oil bath. The "purge" oil-mist serves to prevent atmospheric contaminants from entering the bearing housing and, in special cases, safeguards against the possible loss of oil bath lubrication.

"Pure" oil mist lubrication, also called "dry sump," refers to oil mist application without an oil bath. The bearings are lubricated only by the clean mist lubricant and less likely exposed to contamination. Pure oil-mist lubrication has been shown to significantly improve bearing life (Ref. 7-4).

Although modern oil mist systems are highly unlikely to malfunction and it has been shown that systems interruptions of up to several hours can be tolerated by pump bearings, it is customary to safeguard the system with supervisory instrumentation and suitable alarms. To ensure satisfactory initial lubrication, it is important to pre-lubricate the bearings with a similar oil or to connect the bearings to the mist system for an extended time period before pump startup.

Environmental concerns may limit the use of old-style "open" oil mist lubrication. However, "closed" oil mist systems are in service at many locations worldwide. In closed systems, the bearing housings are fitted with magnetic or similar face-type shaft seals and oil mist collectors. These measures have been highly effective in either precluding or substantially limiting stray emissions to the environment. (See Chapter 10.)

Dry sump oil mist lubrication minimizes bearing friction. Correctly applied, oil-mist is far superior to oil bath lubrication of rolling element bearings. Chapter 10 treats this important topic in more detail.

GREASE LUBRICATION OVERVIEW

Lubricating greases are semi-liquid to solid dispersions of a soap-thickening agent in mineral or synthetic oil. The thickening agent is a "sponge" from which small amounts of the oil separate to lubricate the bearing.

Greases are selected for their consistency, mechanical stability, water resistance, base oil viscosity and temperature capability. Lithium soap thickened greases are good in all these respects and are recommended for general pump applications. However, electric motor drivers are best lubricated with polyurea-based greases.

Grease consistency is graded by the National Lubricating Grease Institute (NLGI). Consistency selection

*nd_m is the bearing speed n in r/min multiplied by the bearing mean diameter dm, in mm.

is based on the size and type of bearing used. NLGI 3 consistency greases are recommended for small-to-medium ball bearings, pumps operating with vertical shaft orientation and pumps having considerable vibration.

NLGI 2 consistency greases are recommended for roller bearings and medium to large ball bearings. NLGI 1 consistency greases are recommended for large bearings operating at low speeds.

The grease base oil viscosity is selected in a similar manner to that of lubricating oils. The viscosity v of the base oil at the bearing operating temperature should be greater than the minimum required lubricant viscosity v_1.

Greases of different thickener types and consistencies should not be mixed. Some thickeners are incompatible with other type thickeners. Mixing different greases can result in unacceptable consistency. With few exceptions, polyurea-thickened greases are generally incompatible with other metallic thickened greases, mineral oils, and preservatives.

The bearing and the adjacent housing cavity are generally filled 30% to 50% with grease at assembly. Excess grease is purged from the bearing into the housing cavity. The time period during which the grease can provide satisfactory lubrication (i.e. grease life) is dependent on the quality of the grease, operating conditions, and the effectiveness of the sealing to exclude contamination.

Bearing and grease manufacturers' catalogs and Chapter 9 of this text provide guidelines for the re-greasing interval and the quantity of grease to be added at re-greasing. The re-greasing interval (t_1) from the SKF USA general catalog is based on the use of lithium grease with mineral base oil at 70°C (158°F) temperature. Re-greasing intervals can be increased if the operating temperature is lower or if a premium quality grease is used. Re-greasing intervals must be shortened if the bearing temperature is higher than the rating basis.

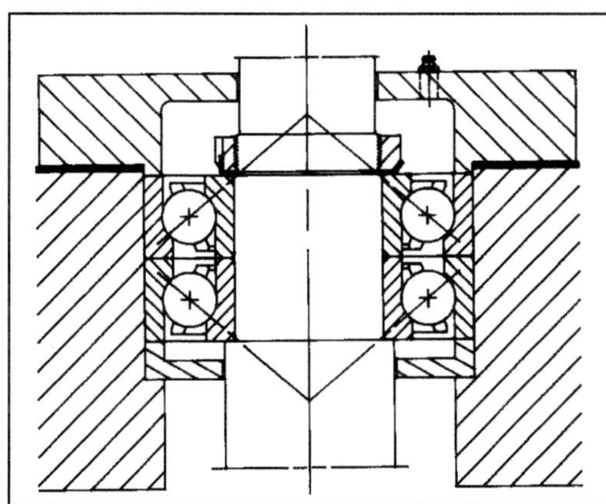

Figure 7-24: Grease lubrication applied to vertically oriented pump

Re-greasing intervals are to be reduced by half if the bearing is vertically oriented. It is best to provide a shelf beneath the bearing to help retain the grease. As illustrated in Figure 7-24, the shelf should have clearance with the shaft to allow excess grease to purge.

Excessive bearing temperatures may result if the bearing and the space around the bearing are completely packed with grease. The bearing housing should, therefore, be designed to purge excess grease from the bearing at start-up and while re-greasing.

BEARING TEMPERATURE

In general, the allowable operating temperature of a bearing is limited by the ability of the selected lubricant to satisfy the bearing's viscosity requirements (i.e. Kappa). Rolling bearings can achieve their rated life at high temperatures provided the lubrication is satisfactory, and other precautions such as the correct selection of internal clearance etc. are taken.

In some cases, a bearing operated from startup cannot achieve satisfactory low steady state temperature conditions due to incorrect bearing fitting (i.e. high initial internal preloading), or due to an excessively fast pump startup rate. These conditions can cause a thermal imbalance in the bearing, resulting in unintended bearing preload. This latter case is not unusual when the bearing housing is very cold due to low ambient conditions (cold climate or chemical process). The best solution in these instances is oil mist lubrication, or to control the bearing fitting (initial bearing clearance and shaft and housing fits) and slow the startup rate of the machinery to allow the establishment of thermal equilibrium. Machined brass cages (M or MA suffix) may be needed in these applications.

In order for bearings with polyamide cages to obtain their longest service life, the outer ring temperature should not exceed 100°C (212°F) in pump applications. Because of this limitation, API pump manufacturers and reliability-focused users do not allow the use of bearings with polyamide cages (TN9 or P suffix).

In some instances, the operating temperature is the limiting factor determining the suitability of a bearing application. Bearing operating temperature is dependent on the bearing type, size, operating conditions, and rate of heat transfer from the shaft, bearing housing, and foundation. Operating temperature is increased when heat is transferred to and from the bearings from external sources such as high temperature pump fluids and rubbing contact housing seals.

Bearing internal heat generation is the product of the rotational speed times the sum of the load-dependent and load-independent friction moments. The bearing load-dependent and load-independent friction moments can be calculated in accordance with the SKF USA general catalog. Recommended values of the lubrication factor f_0 for the pump bearing lubrication methods are shown in Table 7-3.

Values of f_0 for other bearing types can be found in SKF's general catalog. Bearing operating temperature and the viscosity of the lubricant can be estimated using SKF computer programs available to the user community. The values of f_0 given in Table 7-3 are recommended for use with this program. Higher bearing operating temperatures can be expected when rubbing contact shaft seals are used.

In the case of bearings operating in cold climates or with lubricants having very high viscosity, the radial load may need to be greater than the minimum required radial load estimated by the equations provided in the general catalog. Bearings having machined brass cages may also be necessary in these applications. In no case should a bearing be operated at temperatures lower than lubricant pour point.

Table 7-3: Lubrication factors for different lube application methods

Lubrication factor, f_0	*oil-mist*	*oil-ring*	*oil-bath(1)*	*grease*
Single row deep groove ball bearing	1	1.5	2	0.75-2
Double row angular contact ball bearing				
Conrad	2.3	3.3	4.3	2.7
with filling slot	3.4	5	6.5	4
Single row angular contact ball bearing pair	3.4	5	6.5	4

(1) - double value for vertical shaft orientation

Table 7-4: Bearing mounting tolerances

Shaft diameter, mm			Tolerance
ball bearings	cylindrical, metric taper roller bearings	spherical roller bearings	
≤ 18	—	—	j5
(18) to 100	≤40	≤40	k5
(100) to 140	(40) to 100	(40) to 65	m5
(140) to 200	(100) to 140	(65) to 100	m6

BEARING MOUNTING AND RADIAL CLEARANCE

Shaft Fits

The standard recommended shaft tolerances for ball and roller bearings in centrifugal pump applications supporting radial load or combined axial and radial loads are listed in Tables 7-4 through 7-6. These tolerances result in an interference fit between the bearing inner ring and shaft and are needed if the bearing supports a radial load.

The tolerances given in these two tables are recommended for bearings mounted on solid steel shafts. Heavier fits than normal, resulting in greater interference, may be necessary if the bearing is mounted on a hollow shaft or sleeve. Lighter fits using ISO j5 or h5 (k5 for large size bearings) tolerances may be necessary for bearings mounted on shafts made of stainless steel and having a large temperature difference between the bearing inner and outer ring. Stainless steels have lower conductivity than carbon steels and some stainless steels (AISI 316) have high coefficients of thermal expansion. High temperature in a bearing mounted with an excessive interference on a stainless steel shaft may cause unacceptably high stress in the bearing inner ring and unaccaptable internal clearance reductions. ISO j5 and h5 may also be used for bearings supporting pure axial loads.

The minimum required interference between the bearing inner ring and a solid shaft can be estimated from the following equation:

$$I = 0.08\ (d\ F/B)^{0.5} \qquad \text{Eq. 7-7}$$

where:

I = interference, micro-meters
d = bearing bore, mm
B = bearing width, mm
F = maximum radial load, N

An ISO j6 shaft tolerance, see Table 7-6(1), can be used for all types of bearings supporting only axial load. An ISO k5 tolerance is commonly used with paired universally matchable single row angular contact ball bearings supporting only axial load to control bearing internal clearance or preload.

Table 7-5. Bearing preload classes and respective clearances

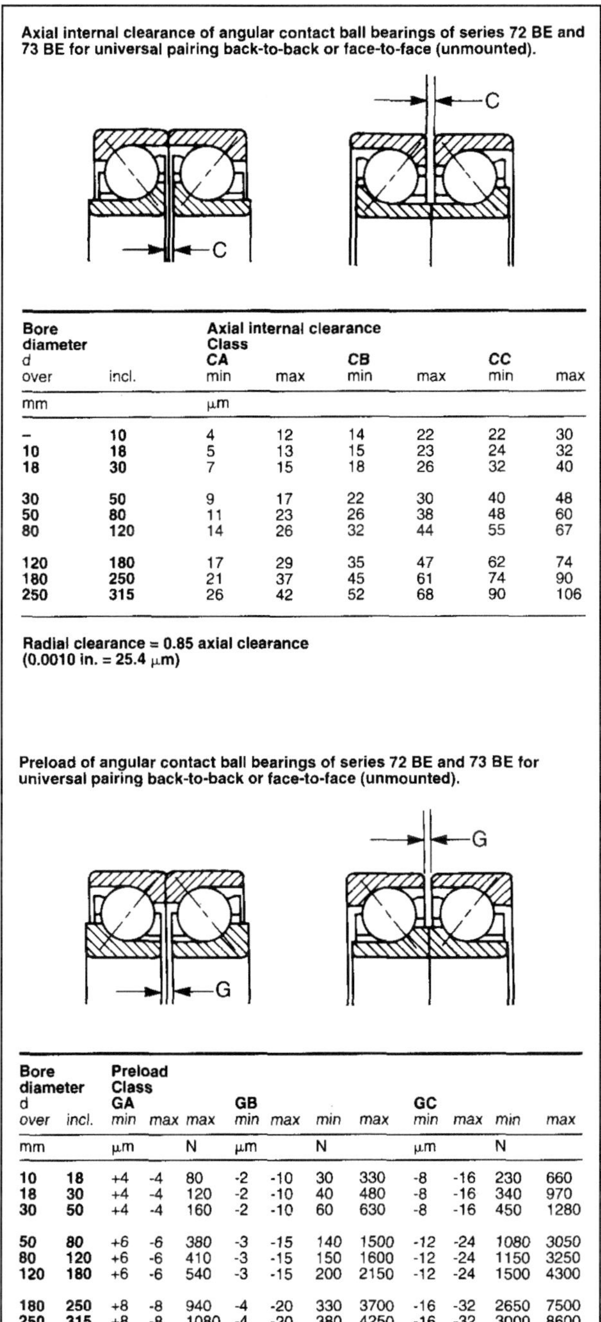

Axial internal clearance of angular contact ball bearings of series 72 BE and 73 BE for universal pairing back-to-back or face-to-face (unmounted).

Bore diameter d over	incl.	Axial internal clearance Class CA min	max	CB min	max	CC min	max
mm		µm					
–	10	4	12	14	22	22	30
10	18	5	13	15	23	24	32
18	30	7	15	18	26	32	40
30	50	9	17	22	30	40	48
50	80	11	23	26	38	48	60
80	120	14	26	32	44	55	67
120	180	17	29	35	47	62	74
180	250	21	37	45	61	74	90
250	315	26	42	52	68	90	106

Radial clearance = 0.85 axial clearance (0.0010 in. = 25.4 µm)

Preload of angular contact ball bearings of series 72 BE and 73 BE for universal pairing back-to-back or face-to-face (unmounted).

Bore diameter d over	incl.	Preload Class GA min	max	max	GB min	max	min	max	GC min	max	min	max
mm		µm		N	µm		N		µm		N	
10	18	+4	-4	80	-2	-10	30	330	-8	-16	230	660
18	30	+4	-4	120	-2	-10	40	480	-8	-16	340	970
30	50	+4	-4	160	-2	-10	60	630	-8	-16	450	1280
50	80	+6	-6	380	-3	-15	140	1500	-12	-24	1080	3050
80	120	+6	-6	410	-3	-15	150	1600	-12	-24	1150	3250
120	180	+6	-6	540	-3	-15	200	2150	-12	-24	1500	4300
180	250	+8	-8	940	-4	-20	330	3700	-16	-32	2650	7500
250	315	+8	-8	1080	-4	-20	380	4250	-16	-32	3000	8600

However, these tables assume that bearings with loose shaft fits, as identified by the letter "L" following the number indicating fit in micro-meters (1/1000mm), will be securely clamped against a shaft shoulder. Similarly, the tables assume that for shaft interference fits at or near the upper limit (letter "T" following the number indicating fit in micro-meters), the manufacturer or responsible reliability professional has ascertained the presence of ample and adequate lubrication.

Just so as not to leave any doubt: Simply because bearing-to-shaft interference fits are in the allowable range is absolutely no guarantee for long bearing life. High interferences will work as long as the lube application method ensures adequate heat removal. Rarely will oil rings deposit enough oil in bearings to ensure moderate temperature operation of high-interference bearings!

HOUSING FITS

The standard recommended housing tolerance for all bearing types is ISO H6. Dimensions can be obtained from Table 7-6(2). This tolerance results in a slight clearance between the bearing outer ring and housing and it allows for easy assembly and radial clearance for bearing expansion with increases in temperature. The risk of ring rotation is minimal with this tolerance. The ISO H7 tolerance is recommended for larger bearings.

A looser ISO G6 housing tolerance is recommended for larger bearings (d > 250 mm, or >10 in.) if a temperature difference greater than 10°C (18°F) exists between the bearing outer ring and the housing.

If the bearing is lightly loaded, it is recommended to spring preload the bearing outer ring. For radial bearings, the recommended spring preload is estimated from the following equation:

$$F = k\,d \qquad \text{Eq. 7-8}$$

where:

F = spring preload force, N
k = factor ranging from 5 to 10
d = bearing bore, mm

Bearing manufacturer's general catalogs provide guidelines for shaft and housing form, tolerance and surface finish. The housing material should have a minimum hardness in the range of 140-230 HB. A lower material hardness can result in housing wear at the bearing seat location.

The inner rings of double row ball bearings and paired universally matchable single row angular contact ball bearings arranged back-to-back should be clamped on the shaft with a lock nut and lock washer. The outer rings of these bearings can be loosely clamped or preferably provided with slight axial clearance (0 to 0.05 mm/0.002 in.) in the housing. Tight clamping can cause axial buckling of bearing outer rings and cannot be allowed in pump bearing housings. However, excessive axial looseness will cause axial shuttling and premature bearing failure.

Table 7-6. Shaft and housing fits (Source: NTN Corporation, Osaka, Japan)

Numeric value table of fitting for radial bearing of 0 class

(1) Fitting against shaft

Unit μm

Nominal bore diameter of bearing d (mm) over	incl.	Δ_{dmp} high	low	g5 bearing shaft	g6 bearing shaft	h5 bearing shaft	h6 bearing shaft	j5 bearing shaft	js5 bearing shaft	j6 bearing shaft	js6 bearing shaft	k5 bearing shaft	k6 bearing shaft	m5 bearing shaft	m6 bearing shaft	n6 bearing shaft	p6 bearing shaft	r6 bearing shaft
3	6	0	−8	4T~9L	4T~12L	8T~5L	8T~8L	11T~2L	10.5T~2.5L	14T~2L	12T~4L	14T~1T	17T~1T	17T~4T	20T~4T	24T~8T	28T~12T	–
6	10	0	−8	3T~11L	3T~14L	8T~6L	8T~9L	12T~2L	11T~3L	15T~2L	12.5T~4.5L	15T~1T	18T~1T	20T~6T	23T~6T	27T~10T	32T~15T	–
10	18	0	−8	2T~14L	2T~17L	8T~8L	8T~11L	13T~3L	12T~4L	16T~3L	13.5T~5.5L	17T~1T	20T~1T	23T~7T	26T~7T	31T~12T	37T~18T	–
18	30	0	−10	3T~16L	3T~20L	10T~9L	10T~13L	15T~4L	14.5T~4.5L	19T~4L	16.5T~6.5L	21T~2T	25T~2T	27T~8T	31T~8T	38T~15T	45T~22T	–
30	50	0	−12	3T~20L	3T~25L	12T~11L	12T~16L	18T~5L	17.5T~5.5L	23T~5L	20T~8L	25T~2T	30T~2T	32T~9T	37T~9T	45T~17T	54T~26T	–
50	80	0	−15	5T~23L	5T~29L	15T~13L	15T~19L	21T~7L	21.5T~6.5L	27T~7L	24.5T~9.5L	30T~2T	36T~2T	39T~11T	45T~11T	54T~20T	66T~32T	–
80	120	0	−20	8T~27L	8T~34L	20T~15L	20T~22L	26T~9L	27.5T~7.5L	33T~9L	31T~11L	38T~3T	45T~3T	48T~13T	55T~13T	65T~23T	79T~37T	–
120 140 160	140 160 180	0	−25	11T~32L	11T~39L	25T~18L	25T~25L	32T~11L	34T~9L	39T~11L	37.5T~12.5L	46T~3T	53T~3T	58T~15T	65T~15T	77T~27T	93T~43T	113T~63T 115T~65T 118T~68T
180 200 225	200 225 250	0	−30	15T~35L	15T~44L	30T~20L	30T~29L	37T~13L	40T~10L	46T~13L	44.5T~14.5T	54T~4T	63T~4T	67T~17T	76T~17T	90T~31T	109T~50T	136T~77T 139T~80T 143T~84T
250 280	280 315	0	−35	18T~40L	18T~49L	35T~23L	35T~32L	42T~16L	46.5T~11.5L	51T~16L	51T~16L	62T~4T	71T~4T	78T~20T	87T~20T	101T~34T	123T~56T	161T~94T 165T~98T
315 355	355 400	0	−40	22T~43L	22T~54L	40T~25L	40T~36L	47T~18L	52.5T~12.5L	58T~18L	58T~18L	69T~4T	80T~4T	86T~21T	97T~21T	113T~37T	138T~62T	184T~108T 190T~114T
400 450	450 500	0	−45	25T~47L	25T~60L	45T~27L	45T~40L	52T~20L	58.5T~13.5L	65T~20L	65T~20L	77T~5T	90T~4T	95T~23T	108T~23T	125T~40T	153T~68T	211T~126T 217T~132T

(2) Fitting against housing

Unit μm

Nominal outside diameter of bearing D (mm) over	incl.	Δ_{Dmp} high	low	G7 housing bearing	H6 housing bearing	H7 housing bearing	J6 housing bearing	J7 housing bearing	Js7 housing bearing	K6 housing bearing	K7 housing bearing	M7 housing bearing	N7 housing bearing	P7 housing bearing
6	10	0	−8	5L~28L	0~17L	0~23L	4T~13L	7T~16L	7.5T~15.5L	7T~10L	10T~13L	15T~8L	19T~4L	24T~1T
10	18	0	−8	6L~32L	0~19L	0~26L	5T~14L	8T~18L	9T~17L	9T~10L	12T~14L	18T~8L	23T~3L	29T~3T
18	30	0	−9	7L~37L	0~22L	0~30L	5T~17L	9T~21L	10.5T~19.5L	11T~11L	15T~15L	21T~9L	28T~2L	35T~5T
30	50	0	−11	9L~45L	0~27L	0~36L	6T~21L	11T~25L	12.5T~23.5L	13T~14L	18T~18L	25T~11L	33T~3L	42T~6T
50	80	0	−13	10L~53L	0~32L	0~43L	6T~26L	12T~31L	15T~28L	15T~17L	21T~22L	30T~13L	39T~4L	51T~8T
80	120	0	−15	12L~62L	0~37L	0~50L	6T~31L	13T~37L	17.5T~32.5L	18T~19L	25T~25L	35T~15L	45T~5L	59T~9T
120	150	0	−18	14L~72L	0~43L	0~58L	7T~36L	14T~44L	20T~38L	21T~22L	28T~30L	40T~18L	52T~6L	68T~10T
150	180	0	−25	14L~79L	0~50L	0~65L	7T~43L	14T~51L	20T~45L	21T~29L	28T~37L	40T~25L	52T~13L	68T~3T
180	250	0	−30	15L~91L	0~59L	0~76L	7T~52L	16T~60L	23T~53L	24T~35L	33T~43L	46T~30L	60T~16L	79T~3T
250	315	0	−35	17L~104L	0~67L	0~87L	7T~60L	16T~71L	26T~61L	27T~40L	36T~51L	52T~35L	66T~21L	88T~1T
315	400	0	−40	18L~115L	0~76L	0~97L	7T~69L	18T~79L	28.5T~68.5L	29T~47L	40T~57L	57T~40L	73T~24L	98T~1T
400	500	0	−45	20L~128L	0~85L	0~108L	7T~78L	20T~88L	31.5T~76.5L	32T~53L	45T~63L	63T~45L	80T~28L	108T~0

Conventional Wisdom: *Pump thrust bearing outer rings can be axially loose as much as 0.2 mm (0.008 in.).*

Fact: *Reliability-focused users will not permit thrust bearing outer rings to be more than 0.05 mm (0.002 in.) axially loose.*

For all bearing types, the axial clamping force on the bearing rings should not exceed one quarter of the basic static load rating of the individual bearing ($C_0/4$). In the case of the double row angular contact ball bearings, the clamping force should not exceed one eighth of the static load rating ($C_0/8$). The clamping forces must uniformly clamp the bearing rings without distortion.

The above recommendations for shaft and housing fits are in accordance with the ANSI/AFBMA Standard 7, a requirement for the API-610 Standard pumps.

INTERNAL RADIAL CLEARANCE

Bearing internal radial clearance greater than normal (C3 suffix) is recommended for bearings mounted with heavier than normal interference shaft fits and if high bearing temperatures are expected either from operation at high speed or from heat conducted to the bearing from an external source. Greater than normal (C3 suffix) internal radial clearance is recommended for radial bearings operating at speeds in excess of 70% of the General Catalog speed rating.

The API-610 Standard specifies that bearings, other than angular contact ball bearings, shall have a C3 internal radial clearance. Their internal clearances are thus somewhat larger than the designation Normal. However, angular contact ball bearings are sensitive to excessive clearance and special considerations apply to them.

BEARING HOUSING SEALING

Sealing of the shaft at the housing is important to exclude solid and liquid contaminant and to retain the lubricant. Radial lip seals and labyrinth seals (Figure 7-25) are among the many styles of bearing housing seals. In radial lip or "garter" seals, a synthetic rubber lip is spring-loaded to keep in contact with the shaft surface. Effective sealing depends on lubricant coating the seal and a well-finished shaft surface. Excessive seal friction can cause high temperatures and seal and shaft wear. Seal friction increases bearing operating temperature. The life of lip seals is usually short (2,000 to 4,000 hours).

Labyrinth seals are more effective in excluding contaminants and in retaining the lubricant. They cause no measurable friction and have a long life. Labyrinth seals provide natural venting for oil-mist lubrication and are not suitable for "closed" oil mist systems (see Chapter 10).

Rotating labyrinth seals are an improvement over the stationary version. However, only face seals—similar to mechanical seals, but narrower—hold the promise of hermetically sealing the bearing housing. See "magnetic seals," in Chapters 6 and 9 of this text; also note insert on Figure 7-25.

ANGULAR CONTACT BALL BEARINGS IN CENTRIFUGAL PUMPS

Single Row Angular Contact Ball Bearings

Single-row angular contact ball bearings are widely used in medium and heavy-duty centrifugal pumps, either as pure thrust bearings or for combined radial and axial loads. The most important features are high radial and axial load capacity combined with a high-speed rating. Single row angular contact ball bearings operate with a small clearance or a light preload, providing good positioning accuracy of the shaft.

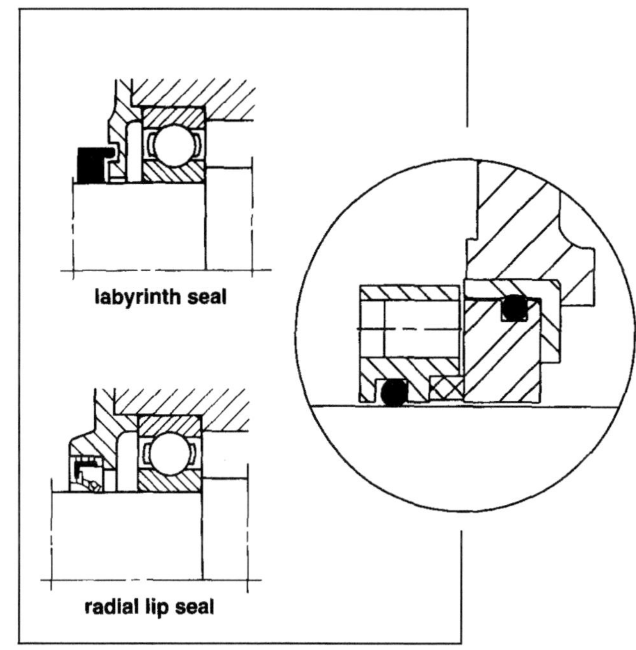

Figure 7-25: Elementary labyrinth seal, radial lip seal and magnetic face seal (insert)

Bearing Minimum Axial Load

For satisfactory operation, an angular contact ball bearing should carry a certain minimum axial load. At increased speed, centrifugal forces on the balls will cause a change in the contact angle between the inner and outer raceways (see Figure 7-26). These contact angle differences will cause sliding (skidding) which damages the raceways, balls and cage, and increases friction. This in turn increases the bearing temperature, reducing the effectiveness of the lubricant and the bearing life. Adequate axial load minimizes the risk of skidding.

Conventional Wisdom: *If two identical angular contact bearings are mounted back-to-back, the unloaded one will operate at a cooler temperature than the loaded one.*

Fact: *If the unloaded side is skidding, it will often generate destructively high temperatures.*

As bearing speed increases (nd_m* values in excess of 250,000), it is also necessary to apply greater axial load to minimize gyratory motion of the balls. Gyratory motion means ball spinning due to gyroscopic moment. Gyratory motion will increase ball sliding and bearing friction. The value of nd_m at which greater axial load is needed is influenced by the magnitude and direction of the applied loads, lubrication conditions, and construction of the bearing cage.

Insufficient load can also cause variation in the orbital speed of the balls. This will result in increased loads on the cage and possibly cause damage.

The minimum required axial load for angular contact ball bearings can be calculated from the following equation:

$$F_{a\,min} = A\,(n/1000)^2 \qquad \text{Eq. 7-9}$$

where:

$F_{a\,min}$ = minimum required axial load, N

A = minimum load factor

n = rotational speed, r/min

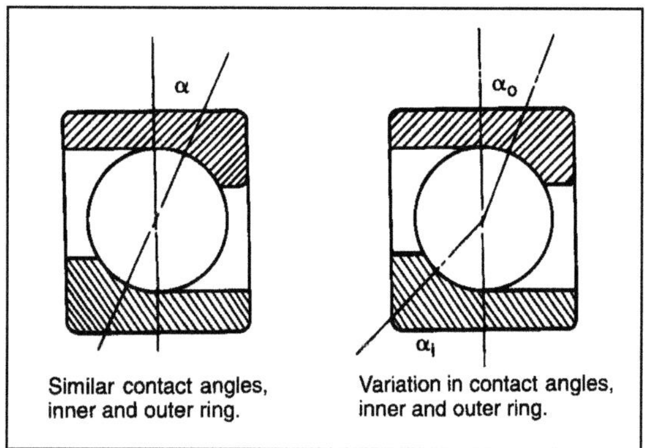

Similar contact angles, inner and outer ring. Variation in contact angles, inner and outer ring.

Figure 7-26: Variation in contact angles of inner vs. outer rings leads to detrimental skidding of rolling elements

The equation above is more accurate than the corresponding equation given in the SKF General Catalog.

Values of minimum axial load factor "A" for series 72 and 73 BE single row angular contact ball bearings are listed in Table 7-7.

Bearings having small contact angles are better suited for high speed, lightly loaded applications because of their lower axial load requirement.

During operation, the minimum required axial load in a bearing pair can be internally maintained by limiting the internal axial clearance. With a small

Table 7-7: SKF SRAC bearing minimum axial load factors

SKF single row angular contact ball bearing minimum axial load factor

	Minimum Load Factor A	
bore size	72 BE	73 BE
01	0.282	0.536
02	0.421	0.906
03	0.686	1.41
04	1.23	2.68
05	1.71	4.29
06	4.07	8.13
07	7.29	11.1
08	10.9	18.9
09	12.3	29.2
10	14.9	45.6
11	23.5	62.5
12	34.4	84.5
13	47.7	111
14	56.3	145
15	63.5	185
16	85.0	234
17	114	292
18	149	360
19	191	440
20	239	630
21	302	723
22	375	905

*nd_m is the bearing speed n in r/min multiplied by the bearing mean diameter dm in mm.

axial clearance, the balls are loaded by centrifugal force against the raceways with nearly equal inner and outer ring contact angles. As the axial clearance increases, so does the difference in the inner and outer ring contact angles. It should be noted that this promotes increased internal sliding and points out the need to carefully select the correct set of thrust bearings for a process pump.

The minimum axial load can also be maintained by spring preloading the bearings. However, in the case of pumps having vertical shaft orientation, the minimum required axial load may be satisfied by the weight of the shaft and pump impeller.

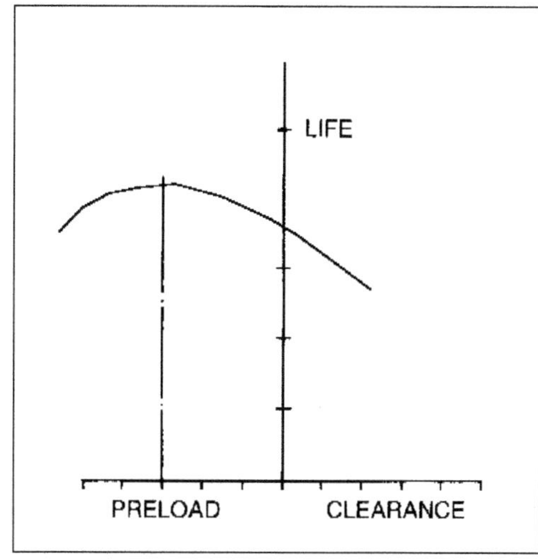

Figure 7-27: Light preloads increase bearing life; excessive internal clearances lower bearing life

BEARING PRELOAD

The five purposes of bearing preload are:

- avoidance of light load skidding
- control of contact angles
- improvement to internal load distribution
- increased bearing stiffness
- improved shaft positioning accuracy

Bearing preload can increase the fatigue life of a bearing by improving internal distribution of the applied external loads. This is illustrated in Figure 7-27. However, while excessive preload can reduce bearing fatigue life, the need for moderate preload often exists. In centrifugal pump applications, bearing preload is principally used to avoid light load skidding in angular contact ball bearings and to provide the shaft positioning accuracy needed by mechanical seals.

Preloaded bearings are more sensitive to misalignment and incorrect mounting than are bearings with clearance.

The load-deflection diagram for a pair of preloaded bearings rotating at 3600 r/min is shown in Figure 7-29. Under rotation, the preload force is increased, and the force in the inactive bearing does not fully reduce to zero due to centrifugal forces.

Figure 7-29 shows the static load-deflection diagram for two preloaded angular contact ball bearings. This diagram is typical of 40° bearings arranged either back-to-back or face-to-face. Preload P′ in this example is achieved by the elastic deflection of the bearings against one another. The initial deflection of the bearings due to the preloading is $delta_0$.

When an axial load is applied to the shaft, only one bearing supports this load. This bearing is denoted

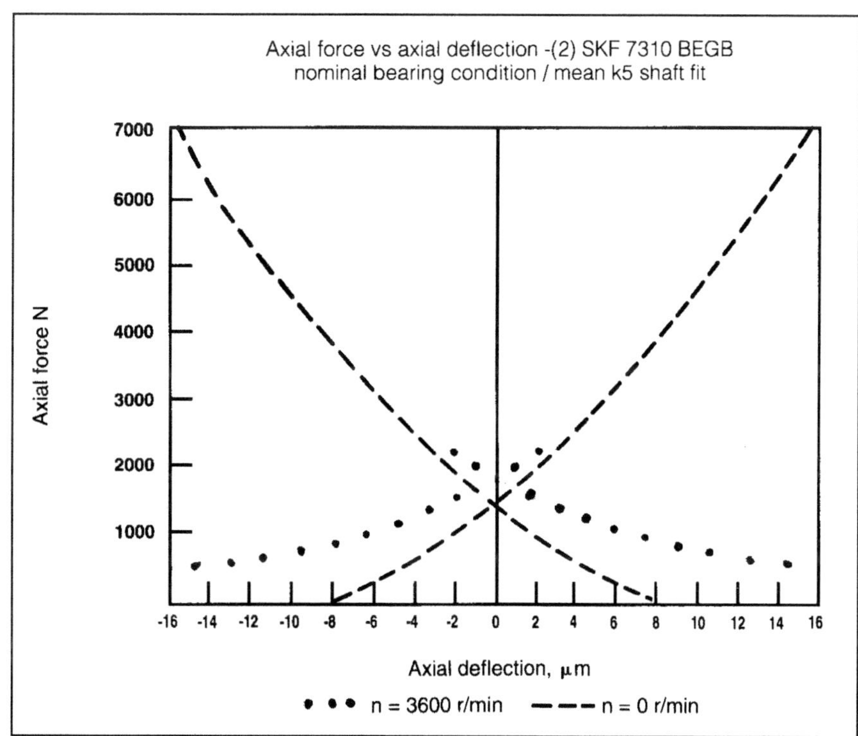

Figure 7-28: Axial force vs. axial deflection for a set of 7310 BEGB bearings

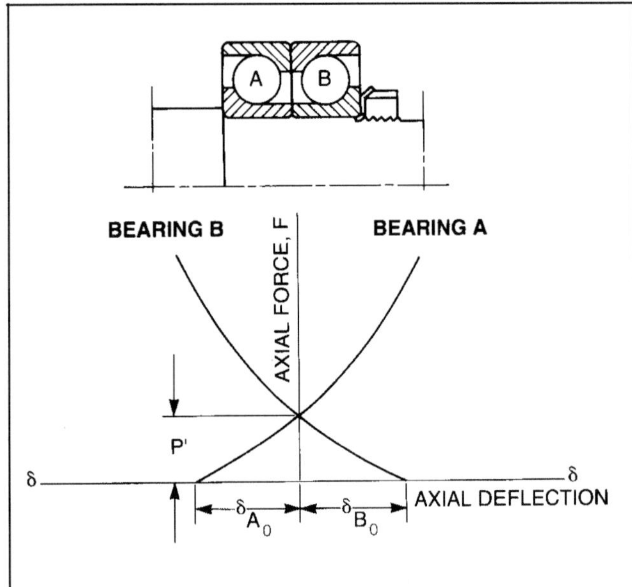

Figure 7-29: Static load deflection for two preloaded angular contact bearings

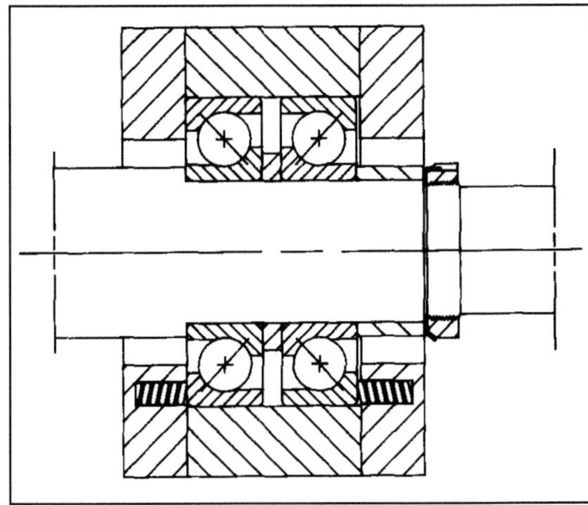

Figure 7-30: Angular contact bearing set, spring-loaded face-to-face arrangement

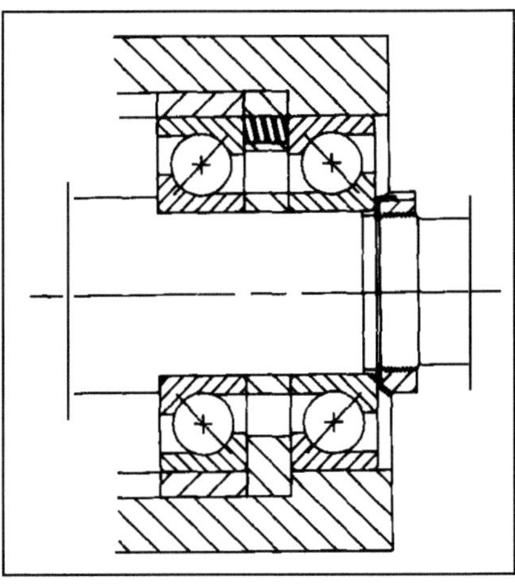

Figure 7-31: Angular contact bearing set, spring-loaded back-to-back arrangement

the "active" bearing. The deflection, delta, of the active bearing reduces the load (i.e. preload) in the adjacent "inactive" bearing.

At increased speeds (nd_m values 250,000 and greater), gyroscopic spinning of the balls will occur if the residual preload in the inactive bearing is less than the minimum required axial load, Fa min.

Bearing preload can also be applied by compression springs. The load from axial springs is constant and not affected by differences in the bearing mounting or temperature. Figures 7-30 and 7-31 illustrate two examples of bearings preloaded by springs.

UNIVERSALLY MATCHABLE SINGLE ROW ANGULAR CONTACT BALL BEARINGS

In the majority of pump applications, angular contact ball bearings of series 72 BE and 73 BE are mounted as pairs either in back-to-back or infrequently in face-to-face arrangements.

Bearing pairs can support combined axial and radial loads, and will assure accurate positioning of the shaft. A bearing pair will support an axial load equally in either axial direction. If the axial load of the pump is very heavy in one direction, the bearing pair can be arranged with a third bearing mounted in tandem. These arrangements are shown in Figure 7-32. It should be noted that lubricant application to back-to-back/tandem sets of bearings might merit special attention. Be aware that oil ring lubrication of heavily loaded sets of thrust bearings has often resulted in disappointing run lengths. Oil ring lube application to back-to-back/tandem sets will rarely succeed in high-load situations.

To be mounted in paired arrangements, the bearings must be manufactured for universal mounting. Different bearing manufacturers may use different suffixes. Standard SKF bearings available for universal matching have the CB or GA suffix, e.g. 7310 BECB or 7310 BEGA. The CB suffix denotes that the bearing is universally matchable, and that a pair of these bearings will have a certain axial clearance when mounted in any of the three arrangements shown in Figure 7-32. The GA suffix also denotes that the bearing is universally matchable, but a pair of these bearings will have a light preload when mounted in any of the three arrangements shown in

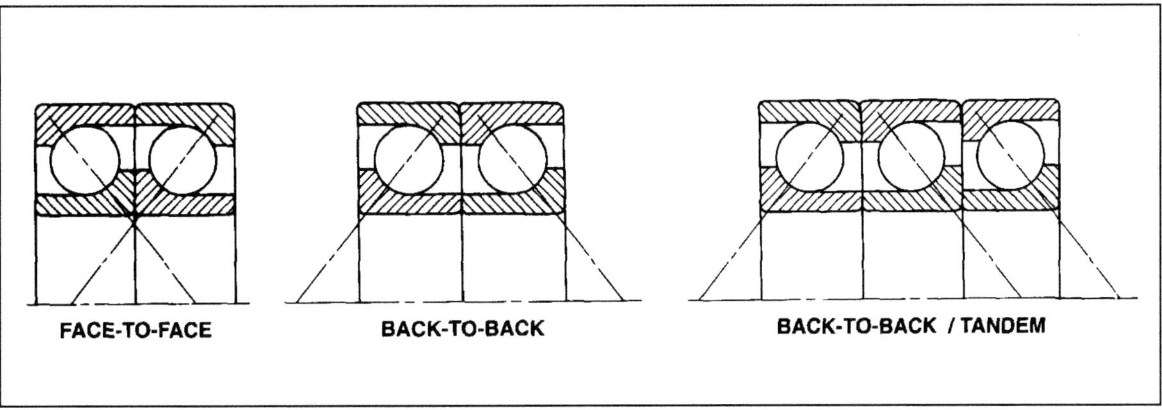

Figure 7-32: Matched sets of bearings for thrust take-up in centrifugal pumps

Figure 7-32.

SKF universally matchable bearings are also available with smaller or greater clearances (suffixes CA and CC, respectively) and with moderate and heavy preloads (suffixes GB and GC, respectively).

Greater axial clearance (CC suffix) bearings may be necessary for operation at high temperatures or with heavy interference shaft and/or housing fits. Preload (GA or GB suffix) may be necessary in bearings supporting predominantly axial load operating with light shaft and housing fits and operation at increased speeds (nd_m values approximately 250,000 and higher). More care must be used with preloaded bearings to ensure correct fit of the bearing rings on the shaft, in the housing and in the alignment of the shaft.

See Table 7-5 for the values of un-mounted axial clearance and preload for pairs of bearings. The initial bearing clearance or preload is assured when the bearing rings are axially clamped together. The initial clearance in a bearing pair is reduced or initial preload is increased by interference fits and if the shaft and inner ring operate with a higher temperature than the outer ring and housing. It should be noted that bearing inner rings generally operate at higher temperatures than their respective outer rings.

Conventional Wisdom: *Shaft interference fits obtained from standard tables apply equally to radial and angular contact thrust bearings*

Fact: *Shaft interference fits allowed for <u>sets</u> of angular contact thrust bearings (ACBBs) are <u>lower</u> than those recommended for individual ACBBs.*

SKF universally matchable bearings are produced with ISO class 6 (ANSI/AFBMA Class ABEC 3) tolerances as standard.

It is of extreme importance to note that single angular contact thrust bearings are *not to be used* where only radial loads are present. For two-direction thrust loads, use only paired bearings. Premature bearing failure should be expected if this requirement is not observed.

A back-to-back arrangement is recommended for most pump applications since clamping the inner rings controls the clearance of the pair, and no clamping of the outer rings is necessary.

The API-610 Standard specifies that the pump thrust loads shall be supported by two 40°, single row angular contact ball bearings, arranged back-to-back. The need for bearing axial clearance or preload is to be based on the requirement of the application.

It should be pointed out, however, that there are pump and load configurations where thrust loads should be supported by matched sets of 40°/15°, or 29°/29°, or even 15°/15° angular contact bearings. This is reflected in the supplementary example "Corporate Engineering" specification in Appendix 3 of this text.

The face-to-face arrangement is used when misalignment is unavoidable, such as in double suction pumps with slender shafts and housings bolted on the pump frame. The main advantage with the face-to-face arrangement is less sensitivity to misalignment, see Figure 7-33. However, there are disadvantage that arise from face-to-face orientation of preloaded angular contact bearings.

Recall that adjacent inner rings are configured such that the high inner ring contours touch each other (Figure 7-32). With outer rings typically clamped lightly against a bearing housing shoulder, neither inner nor outer rings are free to move axially. Visualize also that in most pumps, and especially in electric motors, bearing inner rings almost always operate at higher temperatures than bearing outer rings. The overall axial dimension of the clamped inner rings will therefore experience greater differential thermal growth than the overall axial

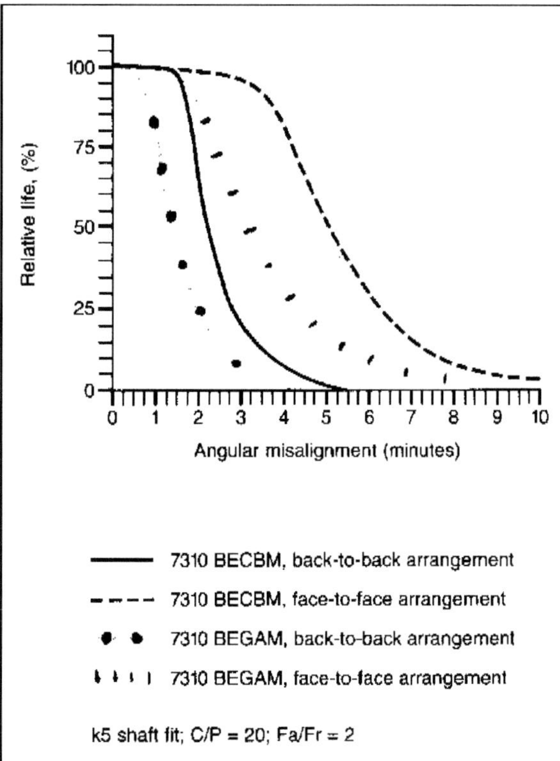

Figure 7-33: Bearing life is affected by bearing arrangement and angular misalignment

dimension of the bearing outer rings. This will now increase the preload. The resulting values could either go beyond the safe load capacity of the bearing, or could go beyond the heat removal capabilities of the often small amount of lubricant applied to the bearing.

If approved for use, and assuming proper functioning is desired (!), the outer rings of bearings arranged face-to-face must be securely clamped in the housing. The axial clamp force for bearings arranged face-to-face must be greater than the axial load supported by the bearings, but less than the limiting clamp load, $C_o/4$. The axial clamp force can be applied to the outer rings either by clamping of the housing cover or by springs. Spring loaded arrangements (Figures 7-30 and 7-31) better control the ring clamp load and are less sensitive to mounting tolerances, misalignment, and changes in bearing and pump temperatures.

CAGES

SKF's BE design bearings are produced with four optional cages, Figure 7-34: glass fiber reinforced polyamide 6.6 cages (P-suffix), pressed metallic cages of either steel (J-suffix) or brass (Y-suffix), and machined brass cages (M-suffix). As illustrated in Table 7-8, all cage types can be used for a large variety of pump applications. However, reliability-focused pump users will not allow the procurement and installation of bearings with polyamide cages. This stance is taken because cage degradation will not show up on the most widely used vibration data collectors and portable vibration analyzers and also because these cages have certain temperature limitations.

For bearings with polyamide cages (P-suffix) to obtain longest service life, the outer ring temperature should preferably not exceed 100°C (212°F). Operating continuously at higher temperatures could result in reduced service life of bearings with the polyamide cage. Obviously, special precautions are needed when using

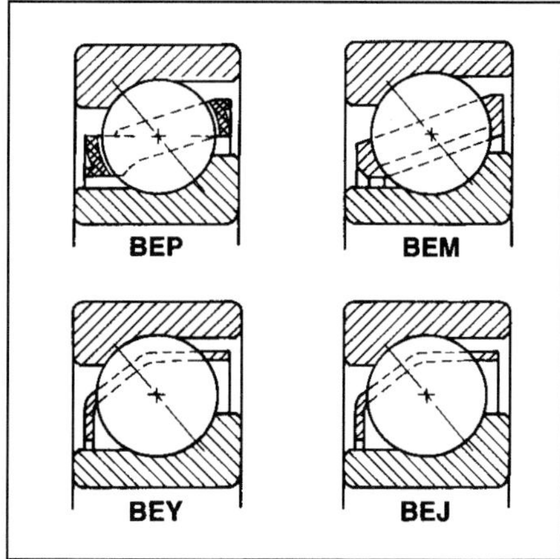

Figure 7-34: Four different cage materials identified by suffix:
P = fiberglass reinforced nylon
M = machined brass
Y = stamped brass
J = stamped steel

Table 7-8. SKF-recommended pump bearing executions

Shaft fit	k5		j5		h5
Housing fit	j6	H6, RF	J6	H6, RF	RF
nd_m value up to 250,000	BECBM BECBY		BECBJ BECBP		BECBM BEGAM BEGAY BEGAP BEGBY BEGBP
250,000 to 450,000		BECBM BEGAM		BEGAY BEGAP	BEGAM BEGAY BEGAP BEGBY BEGBP
450,000 to 650,000	BECBM	BEGAM			not recommended area

RF = Radially free. This means that there is a radial gap between the bearing outer ring and the housing.
For nd_m values lower than 450,000, housing fits J6, H6 may be replaced with J7, H7 respectively.
Applies to solid steel shafts/steel or cast iron housings.
Applies to bearing bore size 20 to 100 mm (including 100mm).
Applies to applications with inner ring temperature no more than 10°C warmer than outer ring temperature.
Circulating oil lubrication or other means for improved cooling may be necessary for control of the bearing operating temperature, in particular, at high speeds.
Consult SKF Application Engineering for details of these recommendations.

heat to mount these bearings.

Finally, reliability-focused users avoid, or even disallow bearing with riveted two-piece steel cages. Rivet heads are a "weak link" on some pump applications, if they pop off.

RECOMMENDED BEARING APPLICATIONS SUMMARIZED

Single row angular contact ball bearings with the BECBM suffix are recommended for centrifugal pump applications where the operating conditions are only generally known and other suitable clearance/preload classes and cage options cannot be satisfactorily evaluated.

- If the operating conditions and shaft and housing fits are known, Table 7-8 can be used for bearing selection.

- Bearings with larger clearances and having the polyamide cage (P-suffix), and to a lesser extent the stamped metallic cages (J, Y-suffix), are sensitive to operation at increased speeds (nd_m values approximately 250,000 and greater). With these cages, the axial clearance should be as small as possible at increased speeds. As can be seen from Table 7-8, these cages should not be used at nd_m values higher than 450,000.

- Machined brass cages (M-suffix) are recommended for centrifugal pump applications operating in heavy duty service conditions and requiring the highest reliability such as the API-610 refinery service pump.

- The complete suffix designation for the standard universally matchable bearings with the machined brass cage is, for example, BECBM or BEGAM, e.g. 7310 BECBM, 7310 BEGAM.

- The standard recommendation for most pump applications is the ISO k5 tolerance. This tolerance results in an interference fit between the bearing inner ring and the shaft. This fit is necessary if the bearing supports radial load.

- A lighter fit using ISO j5 or h5 shaft tolerances, may be necessary if the shaft is made of stainless steel and a large temperature difference between the inner and outer bearing rings in expected. These fits can also be used for steel shafts with pure axial loads.

- Size-on-size or minimum interference fits (0-0.005 mm, or 0-0.0002 in.) are recommended for back-to-back and face-to-face angular contact thrust bearing sets.

- The standard housing fit recommendation for bearings arranged back-to-back is the ISO H6 tolerance. This tolerance results in a slight clearance between the bearing and the housing. This allows for thermal expansion without risk of bearing outer ring rotation. For bearings arranged face-to-face, the standard recommended tolerance is J6.

- Bearings supporting axial loads only can be radially free in the housing, e.g. with a radial gap between the outer ring and the housing.

- Specific bearing cage and clearance/preload combinations are recommended depending on the shaft and housing fits and the shaft speed (Table 7-8).

MRC PUMPAC® BEARINGS

Proprietary series 8000 PumPac® bearings are manufactured by the MRC Division of SKF USA. They were developed and are used in heavy-duty applications where the thrust load is in one direction only. PumPac® bearings are matched sets of two angular contact ball bearings, one having a 40° contact angle and the other a 15° contact angle, Figure 7-35. The bearings are produced as standard with machined brass inner ring land-riding cages, and ISO class 6 (ANSI/AFBMA Class ABEC 3) tolerances.

PumPac® bearing sets must be mounted in the pump with the 40° bearing supporting the applied axial load as the active bearing. The outer rings of the two bearings are scribed together with a "V-arrow" (see Figure 7-36), and this "V-arrow" is to point in the direction of the applied axial load.

The matched, lightly loaded 15° angular contact bearing provides several advantages:

- — low sensitivity to gyratory motion and low requirement for axial load
- — lower sensitivity to mounting conditions resulting in lower mounted preload
- — greater initial preload deflection resulting in greater residual preload with applied axial loads
- — higher radial stiffness

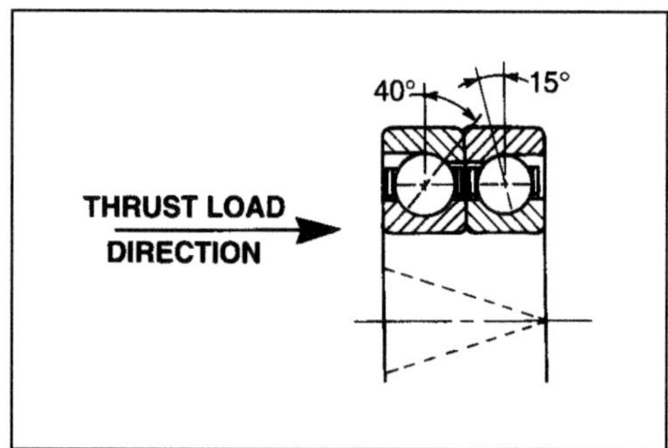

Figure 7-35: MRC's "PumPac®" thrust bearing set

Figure 7-36: Set of pump bearings showing V-arrow to be oriented in the direction of the applied axial load

The benefit of the greater initial preload deflection is best illustrated in a load-deflection diagram as shown in Figure 7-37. Greater axial load can be applied to a PumPac® bearing set compared with two 40° bearings before the residual load on the inactive 15° bearing is reduced to zero. A PumPac® bearing set can support greater axial load than two 40° bearings before fully unloading the 15° bearing.

The result of these features is reduced bearing operating temperature, documented in some cases to be as much as 10°C (18°F).

It must be noted that PumPac® bearing sets should not be used in pumps where the direction of axial load is unknown. Operation with axial load in the direction of the 15° bearing can result in bearing failure. These bearing sets are to be used in applications where the axial load is high in one direction and does not change direction during operation. PumPac® bearing sets can

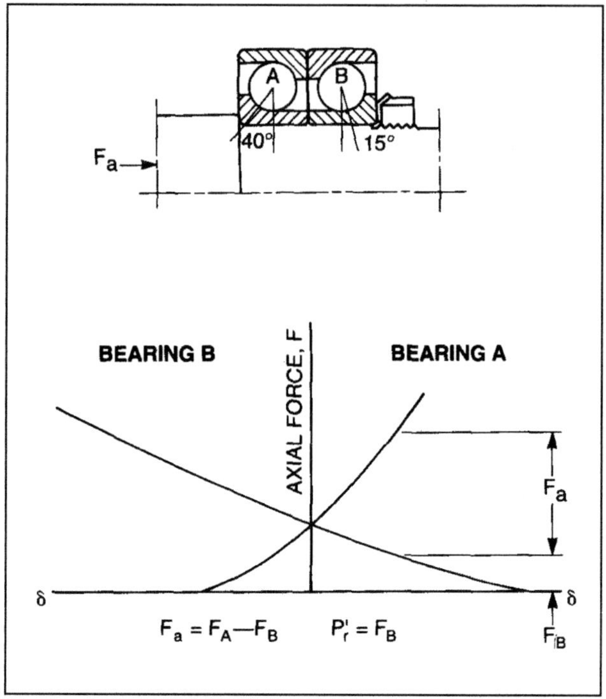

Figure 7-37: Load-deflection diagram for MRC "PumPac®" thrust bearing set

accept momentary reversals in axial load, such as those that occur during pump start-up and stoppage etc.

The inner ring land-riding cage of the PumPac® set requires special attention when grease lubricated. An initial charge of grease must be specifically injected between the cage and inner ring at assembly to ensure satisfactory lubrication. It has been said that PumPac® bearing sets with matched 40° and 15° bearings do not meet the requirements of the API-610 Standard since this standard "requires" the use of two 40° single row angular contact ball bearings for the thrust bearing. However, the API-610 Standard clearly states that its stipulations are

(a) guidelines, not law or mandatory regulatory ruling, and
(b) minimum requirements, leaving it to the user to ask for better, safer, longer lasting, or more reliable components

Hence, there are many occasions when experienced reliability engineers request that the requirements of API-610 Standard be waived. One such occasion is where PumPac® bearing sets have been demonstrated to outperform the matched traditional 40° angular contact bearing sets mentioned in API-610. Working with knowledgeable pump manufacturers and SKF application engineers will prove refreshing in this regard.

DOUBLE-ROW ANGULAR CONTACT BALL BEARINGS

Double-row "conventional" angular contact ball bearings (DRACBBs) are generally understood to have single inner and outer rings. Until the introduction of double row bearings with *dual inner rings* in the 1990's, the conventional DRACBBs were used extensively in medium duty centrifugal pumps. These bearings exhibit good load and speed capabilities and relative ease of mounting. Double row bearings have good radial and axial load capabilities.

Conventional SKF double-row angular contact ball bearings are produced in two designs: the European-made 32 and 33 series and the American-made 52, 53 and 54 series. Both are illustrated in Figure 7-38. With few exceptions, all series are available in Conrad (A-suffix) and high capacity filling slot (E-suffix) design and with seals, shields, and snap ring options. (It should be noted that reliability-focused pump users frown upon the use of bearings with filling notches and certain snap ring configurations, as will be discussed further in Chapter 9.)

The 32 and 52 series, and 33 and 53 series are dimensionally interchangeable. Their load and speed ratings are similar; however, the 32 and 33 A and E-series bearings have a 32° contact angle and are supplied with a polyamide (TN9 suffix) cage. The 32 and 33 series bearings without A or E-suffix are filling slot design only and have steel cages.

Series 52 and 53 A and E bearings have a 30° contact angle (except for the 52 A and E-series sizes 04, 05 and 06 which have a 25° contact angle) and are furnished with steel cages. The 54 A-series bearing is available only in Conrad design and without options of seals or shields.

The following notes are well worth noting when applying conventional double-row angular contact ball bearings. SKF reminds users that failure to follow recommendations may result in premature bearing failures:

Note 1: Differences in designation for seal, shield and snap ring locations exist for the European and American bearings. Interchanges should be made carefully.

Note 2: Double row angular contact bearings should be subject to sufficient radial load for satisfactory operation. The minimum required radial load for double-row bearings may be estimated from the following equation:

$$F_{rm} = k_r\,(\upsilon n/1000)^{2/3}\,(dm/100)^2 \qquad \text{Eq. 7-10}$$

where:

Frm = minimum radial load, N
k_r = minimum load factor
= 60 for bearings of series 52 A, 32 A
= 90 for bearings of series 52 E, 32 E
= 70 for bearings of series 53 A, 33 A
= 110 for bearings of series 53 E, 33 E
= 70 for bearings of series 54 A
υ = oil viscosity at operating temperature, mm^2/s

n = rotational speed, r/min.

d_m = mean diameter of bearing, 0.5 (d + D), mm

Note 3: Double row bearings should ideally not be subjected to axial load without radial load.

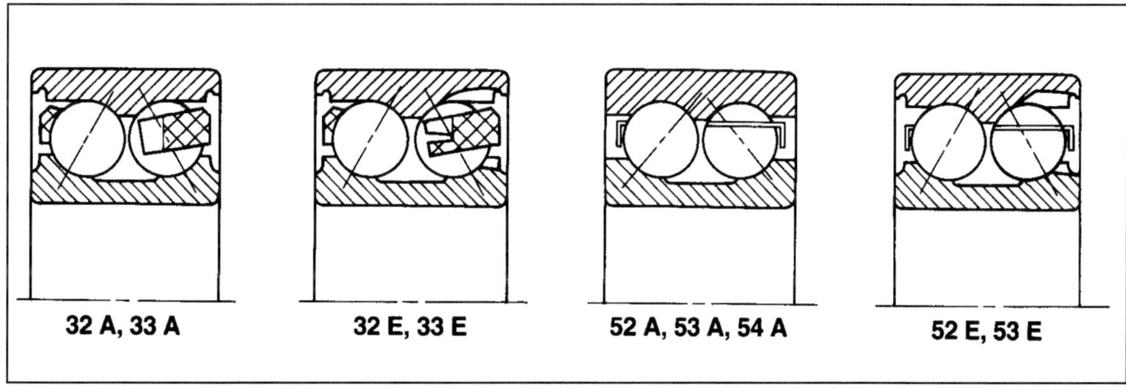

Figure 7-38: Double-row angular contact ball bearings used in ANSI pumps. Note that certain filling notch bearings are not recommended for reliability-focused applications

Note 4: The recommendations of shaft and housing fits for conventional double row bearings are the same as for other ball bearings in pump applications.

Note 5: Failure to follow the bearing manufacturer's engineering recommendations may result in premature bearing failures.

DOUBLE-ROW ANGULAR CONTACT BEARINGS WITH TWO INNER RINGS

Until the late 1990s, upgrading the double-row angular contact ball bearings (DRACBBs) in older API-610 Standard, 5th Edition and ANSI-style pumps to meet "ANSI-Plus" standards meant one of two things:

- The equipment owner could purchase sets of single-row angular contact ball bearings (SRACBBs), complete with a specially designed shaft and housing, at a premium price.
- Parties interested in upgrading could purchase sets of SRACBBs and then modify the existing shafts and housings in-house to accommodate them.

When it became evident that conventional double row angular contact ball bearings often represented the limiting factor in achieving extended pump life, there arose the need to devise a retrofit bearing with characteristics approaching those of the API 610-recommended 40° back-to-back orientation. This prompted SKF to develop DRACBBs with two inner rings.

Using a specially designed DRACBB with *two* inner rings (Figure 7-39) requires no labor-intensive retrofit work and presents an ideal upgrade for many API and ANSI/ISO pumps (Figure 7-40). The bearing employs a closely controlled axial clearance. This optimized clearance promotes load sharing between the two rows of balls—a design that reduces the possibility of skidding in the inactive ball set without the use of a preload.

Recall from our previous discussions that preloading a bearing can help prevent skidding, but may also have the undesired effects of generating excessive heat or contributing to poor bearing performance. While one ball set is supporting the axial load, the back-up set becomes inactive, supporting only a portion of the radial load. Without sufficient loading, the motion of the balls in the inactive set leads to skidding and heat generation.

The new SKF separable inner ring DRACBBs shaft and housing fits are identical to those for standard SRACBB and DRACBB with comparable sizes. ISO k5 is the recommended shaft tolerance for this bearing in most pump applications. This tolerance produces an interference fit between the bearing inner ring and the shaft. Interference fits are necessary for bearings supporting any radial load. A lighter fit using ISO j5 or h5 tolerances (see Table 7-6) may be necessary for bearings mounted on shafts made of stainless steel, or for bearings that have a large temperature differential between the inner and outer rings.

ISO H6 is the standard housing tolerance recommendation. H6 produces a slight clearance between the bearing outer ring and the housing, which facilitates easy assembly and provides radial clearance for bearings

Figure 7-39: SKF double-row angular contact ball bearing with separate inner rings is more skid-resistant than DRACBBs with single-piece inner ring

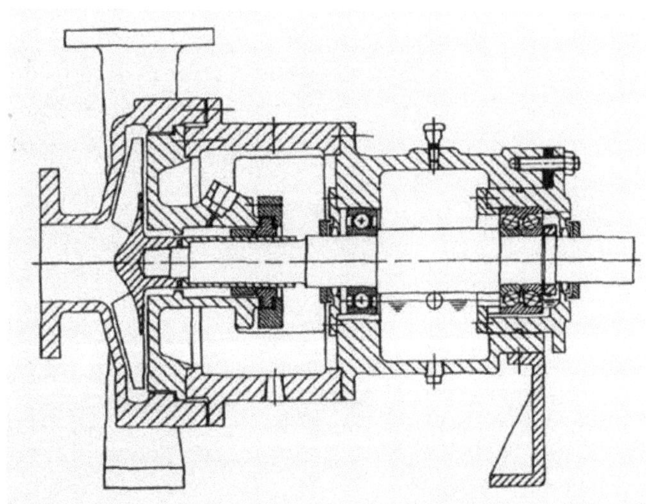

Figure 7-40: ANSI pump equipped with two-piece inner ring double-row angular contact ball bearing

when increased temperatures cause them to expand. H6 has minimal ring rotation risk.

Table 7-9 lists available dimensions and descriptive designations for the double-row, double inner ring angular contact ball bearing.

The mounting recommendations for DRACBBs with two inner rings differ from those applicable to the conventional double-row configuration. SKF recommends mounting the new DRACBB—they call it the "SKF Pump Bearing" with an induction heater. Following that, the craftsperson would

- Heat the entire bearing to approximately 110°C (230°F), but never higher than 125°C (257°F); higher temperatures might alter the structure of bearing materials and cause dimensional changes.
- Recall that the SKF Pump Bearing is designed with a two-piece inner ring—a feature that facilitates the use of a machined brass cage. This design requires the bearing to be mounted and held in place on the shaft with a locknut—an action that ensures positive clamping of the inner rings, both together and also against the shaft abutment.
- When the bearing is fully heated, place it on the shaft and immediately clamp it into position with a locknut.
- After the bearing has cooled and secured itself to the shaft, remove the locknut, add the lock washer, and then retighten the locknut hand tight plus an additional 1/8 to 1/4 turn.

As to the cost justification for DRACBBs, assume an incremental outlay of $40 per bearing plus $120 in

Table 7-9: Dimensions and nomenclature for SKF two-piece inner ring double-row angular contact ball bearings

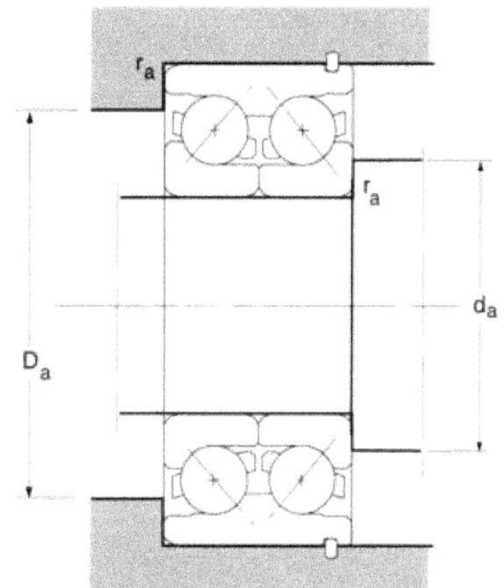

d Bore Dia.	d1	D1	r1,2 (min)	a	da (min)	Da (max)	ra (max)	e	Designation (complete bearing)
mm in	mm in	mm in	mm in	mm in	mm in	mm in	mm in		
40 1.5748	60.1 2.3661	79.5 3.1299	1.5 0.0591	71.1 2.7992	60 2.3622	80 3.1496	1.5 0.0591	1.14	3308 DNRCBM
45 1.7717	66.9 5.6339	88.4 3.4803	1.5 0.0591	79.2 3.1181	66 2.5984	89 3.5039	1.5 0.0591	1.14	3309 DNRCBM
50 1.9685	73.5 2.8937	98.2 3.8661	2.0 0.0787	88.2 3.4724	73 2.8740	99 3.8976	2.0 0.0787	1.14	3310 DNRCBM
55 2.1654	80.5 3.1693	107.8 4.2441	2.0 0.0787	96.6 3.8031	80 3.1496	108 4.2520	2.0 0.0787	1.14	3311 DNRCBM
65 2.5591	94.0 3.7008	126.8 4.9921	2.1 0.0827	114.0 4.4882	94 3.7008	127 5.0000	2.1 0.0827	1.14	3313 DNRCBM

conversion cost were to lead to a $8,000 repair avoidance on a large ANSI/ISO pump every 4 years (or $2,000 per year). In that case, $160/4 years = $40/year has returned $2,000/year. That's a rather attractive benefit-to-cost ratio of 2,000/40, or 50:1. Or, project an ultra-conservative scenario of spending $200 on conversion and extending a previous 1.5-year MTBR to a post-conversion MTBR of 3 years. In that case, avoiding even a $6,000 repair will still yield a solid 10:1 payback.

ROLLER BEARINGS IN CENTRIFUGAL PUMPS

Cylindrical Roller Bearings

SKF cylindrical roller bearings of EC design are used in many non-API centrifugal pumps for their high speed and high radial load capability. This manufacturer's EC design cylindrical roller bearings are produced with three optional cages:

- glass fiber reinforced "polyamide, 6.6" cages (P-suffix),
- stamped steel cages (J-suffix), and
- machined brass cages (M or MA-suffix)

The polyamide cage is standard in most sizes. Polyamide cages are used successfully in cylindrical roller bearings operating in many non-API centrifugal pumps. In order for bearings with polyamide cages to obtain longest service life, the outer ring temperature should not exceed 100°C (212°F).

Typically, either the NU or NUP types are utilized, see Figure 7-41. The NU type bearing is preferred because it can easily accommodate axial displacement due to heat expansion of the shaft. This feature makes it possible to use an interference housing fit. In pumps where impeller imbalance is unavoidable, an interference housing fit should be used to avoid bearing outer ring rotation.

The NUP type is used where it is desired for the bearing to function as a single unit much like single row ball bearings and spherical roller bearings. The NUP bearing is used in the free position and the loose inner ring flange is abutted against the shoulder of the shaft and retained on the bearing by the fitting of the inner ring. With this arrangement, oil lubrication is preferred. With grease lubrication, the housing fit should be G6 to assure axial displacement due to heat expansion.

Cylindrical roller bearings are somewhat sensitive to misalignment. The maximum allowable misalignment is three to four minutes, depending on the bearing series. For bearing housings machined in one setup, this is usually not a problem.

It should be noted that, for satisfactory operation, cylindrical roller bearings would have to be subjected to a given minimum radial load. The required minimum radial load to be applied to cylindrical roller bearings can be estimated from an applicable equation provided in SKF's General Catalog and is not duplicated in this text. Needless to say, not following the bearing manufacturer's engineering recommendations will often result in premature bearing failure.

Taper Roller Bearings

Taper roller bearings are used in pump applications to support high combined radial and axial loads and keep in check the axial play of the shaft. A limiting factor is the speed capability of taper roller bearings. The speed rating is governed by the sliding friction between the rollers and the inner ring flange. Taper roller bearings can be used singly at a bearing position or in matched pairs and are suitable for both oil and grease lubrication.

Matched pairs of single row taper roller bearings with preset axial clearances are used when the load carrying capacity of one bearing is insufficient and accurate axial guidance of the shaft is necessary.

The steep contact angle of SKF series 313 taper roller bearing makes them well suited for pump applications with high axial loads. The bearings are often arranged face-to-face (DF-suffix) or back-to-back (DB-suffix). The outer rings of the two bearings arranged face-to-face must be axially clamped in the housing to ensure correct operation. The axial clamp force must be greater than the applied axial load but less than the limiting clamp load, $C_o/4$, where C_o is the static load rating of one bearing.

If the clamp force is by necessity high, an outer ring spacer with greater stiffness may be needed for bearings arranged face-to-face to limit the deflection due to this clamping. Bearings having the stiffer spacer (DF003 suffix) are available.

Because of the steep contact angle in 313 series bearings, it is important to ensure free lubricant flow to each face of the bearing. The housing should have a bypass opening beneath the bearing to allow free lubricant flow into each face of the bearing and between the bearings through the holes provided in the outer ring spacer. Higher bearing temperatures will result if this is not possible.

For satisfactory performance, it is essential that taper roller bearings be subjected to a given minimum radial load. The minimum required radial load is estimated from:

$$Frm = 0.02\ C \qquad \text{Eq. 7-11}$$

where:

Frm = minimum radial load, N

C = basic dynamic load rating of the bearing or bearing pair, N

Finally we consider it worthy of note that SKF markets "Panloc" pre-adjustable non-locating bearing units that allow the internal clearance or preload to be adjusted to specific values. Although primarily developed for precision support of heavy printing machine rolls, pre-adjustable bearings have the potential of solving large pump bearing problems as well.

Spherical Roller Bearings

Spherical roller bearings and spherical bearings are used in centrifugal pump applications having heavy loads, These bearings are used for their high load capabilities and their ability to operate in misaligned conditions. Spherical roller bearings are produced as standard with steel cages (J suffix).

Spherical roller bearings are most commonly used in applications having low operating speeds. These bearings can be used at increased speeds provided the lubrication and cooling are satisfactory. Inductive pumps (small free-piston pumps operating on the induced current principle) can be used to supply the necessary quantity and quality of lubricant. See Chapter 9 for details.

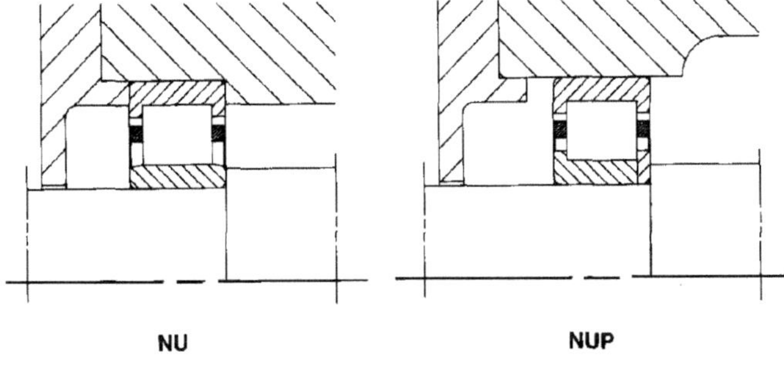

Figure 7-41: Cylindrical roller bearings used as radial bearings in many non-API pumps. Note that only the NU-type bearing can accommodate small axial shaft displacements.

Radial Spherical Roller Bearings

SKF spherical roller bearings of CC and E designs have very high radial load ratings and can operate with combined radial and axial load. The axial load capability of these bearings is sometimes limited by the bearing friction and resulting operating temperature. However, SKF bearings of the CC and E design have lower friction than many other spherical roller bearings. They are available with cylindrical bore and with tapered bore for adapter sleeve mounting on the shaft. The bearings can be oil or grease lubricated.

As before, for satisfactory operation spherical roller bearings must be subjected to a certain minimum load. The minimum load is necessary to maintain the motion of the rollers and can be estimated from the following equation:

$$P_o\ min = 10^{-4}\ C_o(vn/100)^{0.5} \qquad \text{Eq. 7-12}$$

where:

P_o min = minimum equivalent static load, N

C_0 = bearing static load rating, N

ν = lubricant viscosity, mm^2/s at operating temperature

n = rotational speed, r/min

The equivalent static load, P_o must be greater than the minimum equivalent static load P_o min. (The equivalent static load, P_o is calculated using the equation in the manufacturer's General Catalog.)

In vertical pump applications, the loading on spherical roller bearings can in some instances be relatively light, and rotating due to imbalances in the machine. This can cause rotation of the outer ring in the housing. Rotation of the outer ring can be prevented or minimized by pinning the outer ring with the housing or by using an O-ring mounted in a groove in the housing bore.

Spherical Roller Thrust Bearings

SKF spherical roller thrust bearings of the E design have very high axial load ratings and can operate with combined axial and radial loading. SKF spherical roller thrust bearings are most commonly oil lubricated but, in certain applications, can be grease lubricated.

Pump reliability professionals are again reminded that, for satisfactory performance, spherical roller thrust bearings *must* be subjected to a given minimum axial load. The required minimum axial load to be applied can be estimated by the following equation:

$$F_{am} = 1.8F_r + A(n/1000)^2 \qquad \text{Eq. 7-13}$$

where:

F_{am}	=	minimum axial load, N
F_r	=	applied radial load, N
A	=	minimum load factor, see bearing table in SKF's General Catalog
N	=	rotational speed, r/min

(If 1.8 Fr < 0.0005 C_o, then 0.0005 C_o should be used in the above equation instead of 1.8 Fr)

The axial load on the bearing must always be greater than that estimated by the above equation. If necessary, additional axial load must be applied to the bearing to satisfy the requirement for load. Compression springs are often used for this purpose.

The spherical roller thrust bearing is used in vertical pumps, as shown in Figure 7-42, to support combined axial and radial load. In some cases, the bearing is used with a separate ball bearing that supports radial load, or reversing axial load. In some cases, a thrust bushing is utilized to support reversing axial load. In heavy duty centrifugal pumps, deep-well pumps, and also in hydro-turbines, the spherical roller thrust bearing is sometimes used in combination with the spherical roller bearing which supports the radial load, and the thrust bearing supports the axial load. The thrust bearing is spring preloaded to ensure the minimum load requirements are satisfied when the pump's axial load reverses direction. Short-duration reverse axial loads are supported temporarily by the radial bearing.

Radial and thrust bearings are spaced so that their alignment centers coincide. They are fitted together in the housing with a small (1.0 mm/0.040 inch) overall axial clearance to allow for expansion. The thrust bearing is fitted with an ISO j6 tolerance on the shaft and with radial clearance (2.0 mm/0.080 in. minimum) in the housing. The radial bearing is fitted with the normal recommended shaft and housing fits. These arrangements are oil lubricated.

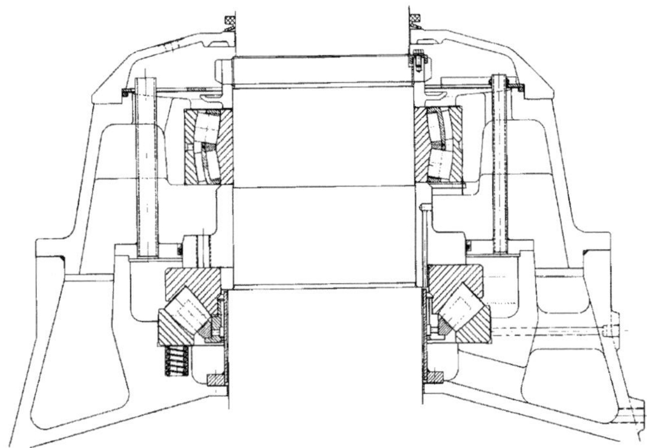

Figure 7-42: Spherical roller thrust bearings support axial and radial loads in large vertical pumps

SLIDING BEARINGS*

Sliding bearings are found in large to very large pumps. Typical configurations are shown in Figure 7-43. Life extension issues for these bearings are closely tied to material selection, bearing geometry, applied load and—above all—lubricant application and related oil characteristics.

Material selection is covered in the section on vertical pumps. Typical bearing diametral clearances are

C = [(0.001 in/inch of shaft journal diameter) + .002 in], or
C = [(0.001 mm/mm of journal diameter) + .05 mm]

A positive tolerance of perhaps 5% can be added to these guideline values.

Optimal bearing loads range from 100 to 250 psi (~690-1,724 kPa) on the projected area, i.e. (bore diameter) × (bearing length). Lubricant application ranges from the elementary and frequently no longer satisfactory oil ring to sophisticated continuous application of clean, temperature-controlled oils.

PIVOTED SHOE THRUST BEARINGS

Bearings transmit shaft loads to the foundation or pump support. Hydrodynamic bearings transmit (float) the load on a self-renewing film of lubricant. The bear-

Figure 7-43: "Plain" or "sleeve" bearing at right; combined sleeve/thrust land bearings, left and center illustrations. (Source: Waukesha Bearings Corporation, Waukesha, Wisconsin 53187)

*We are indebted to Kingsbury, Inc., for contributing much of the narrative and Figures 7-44 through 7-58.

ings may be solid for assembly over the end of the shaft, or split for assembly around the shaft. Figures 7-44 to 7-47 show some of the many hydrodynamic bearings produced by Kingsbury and other experienced manufacturers.

The hydrodynamic principle is based on theoretical and experimental investigations of cylindrical journal bearings. It can be shown that oil, because of its adhesion to the journal and its resistance to flow (viscosity), is dragged by the rotation of the journal so as to form a wedge-shaped film between the journal and journal bearing (Figure 7-48). This action sets up the pressure in the oil film which thereby supports the load (Figure 7-49).

This wedge-shaped film was shown by one of the earliest and most famous researchers, Osborne Reynolds, to be the absolutely essential feature of effective journal lubrication. Reynolds also showed that "if an extensive flat surface is rubbed over a slightly inclined surface, oil being present, there would be a pressure distribution with a maximum somewhere beyond the center in the direction of motion." This is represented in Figure 7-50.

Figure 7-44: Hydrodynamic equalizing pivoted shoe thrust bearing (Source: Kingsbury, Inc., Philadelphia, Pennsylvania 19154)

Figure 7-45: Hydrodynamic pivoted shoe journal bearing (Source: Kingsbury, Inc., Philadelphia, Pennsylvania 19154)

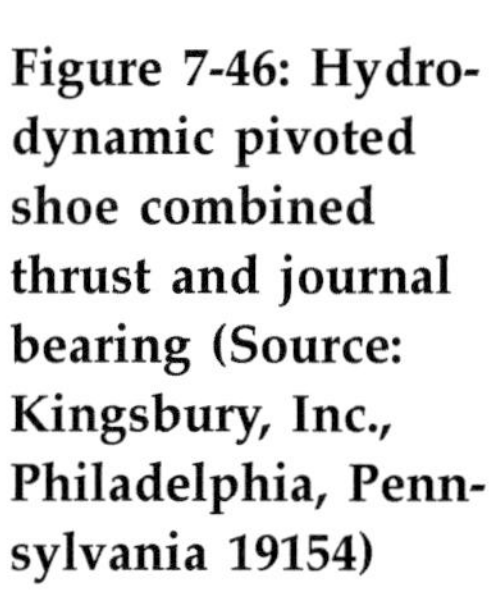

Figure 7-46: Hydrodynamic pivoted shoe combined thrust and journal bearing (Source: Kingsbury, Inc., Philadelphia, Pennsylvania 19154)

Pivoted Shoe

As with the plain cylindrical bearing, the pivoted shoe thrust and journal bearings rely on adhesion of the lubricant to provide the film with a self-renewing supply of oil.

Basic Pivoted Shoe Thrust and Journal Bearing Parts

A number of parts are associated with each bearing type:

- Rotating collar (journal). The collar transmits the thrust load from the rotating shaft to the thrust shoes through the lubricant film. It can be a separate part and attached to the shaft by a key and nut or shrink fit, or it may be an integral part of the shaft. The collar is called a runner in vertical machines. (In the radial direction, the shaft journal transmits the radial loads to the journal shoes through the lubricant film.) In hydrodynamic bearings, the fluid film is on the order of .025 mm (.001") thick. With this and the previous information referring to the hydrodynamic principle, two points can be realized:

    1. The stack-up of tolerances and misalignment in hydrodynamic bearings has to be conservatively less than .025 mm (.001"), or some means of adjustment has to be incorporated.

    2. The collar surfaces must be flat and smooth (and journal surface cylindrical and smooth) in comparison to the film thickness, but not so smooth as to inhibit the adhesion of the lubricant to the surface.

- Thrust shoe (journal shoe) assembly. The shoe (also called a pad, segment, or block) is loosely constrained so it is free to pivot. The shoe has three basic features: the babbitt, body and pivot, and so is usually referred to as an assembly.

- Babbitt. The babbitt is a high-tin material, metallurgically bonded to the body. As with the collar, the babbitt surface must be smooth and flat in comparison to the film thickness.

    The babbitt is a soft material (compared to

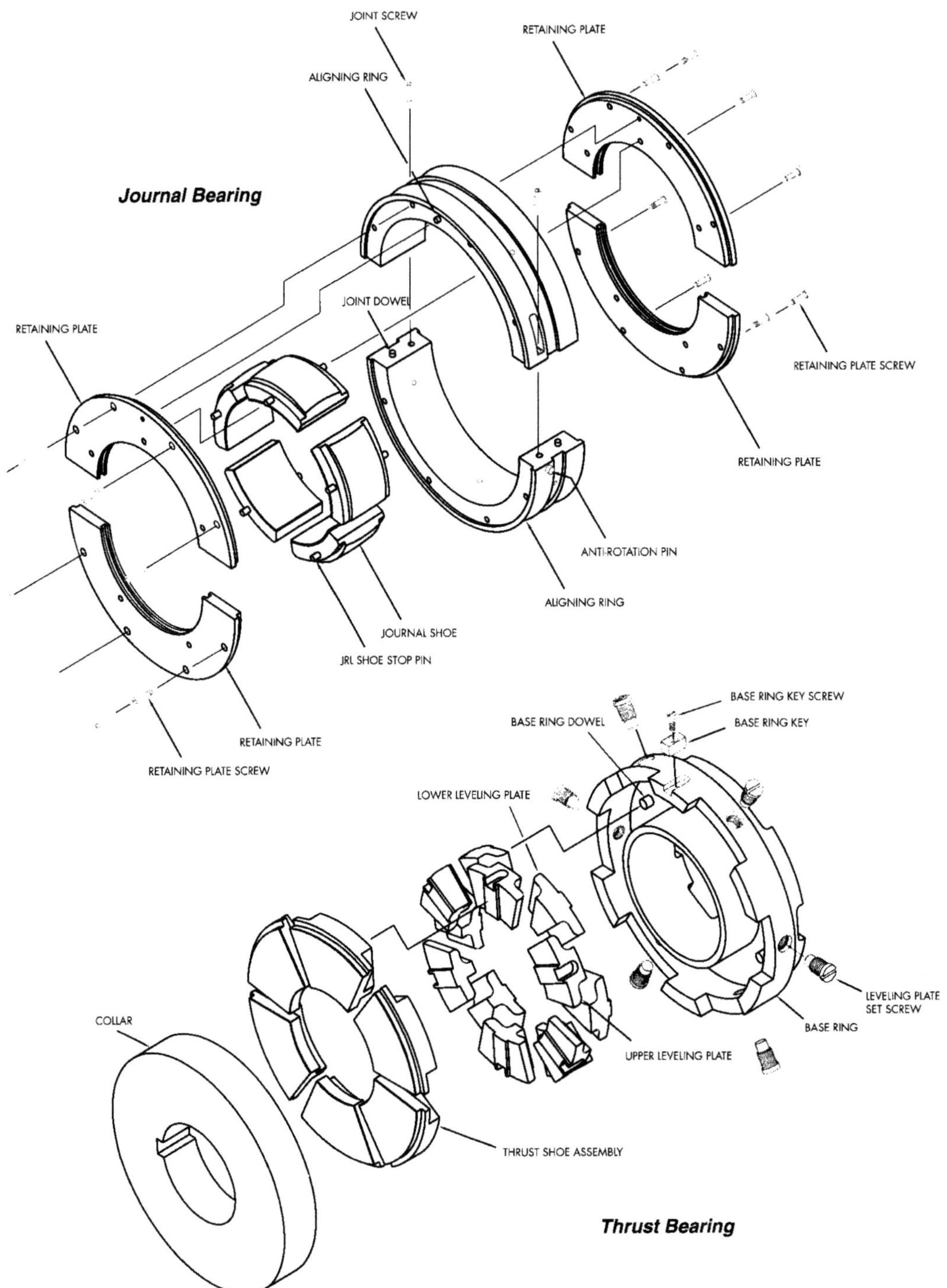

Figure 7-47: Combined thrust and journal bearing part schematic (Source: Kingsbury, Inc., Philadelphia, Pennsylvania 19154)

the shaft) which serves two functions: It traps and imbeds contaminants so that these particles do not heavily score or damage the shaft. It also protects the shaft from extensive damage should external conditions result in interruption of the film and the parts come in contact.

- Body. The shoe body is the supporting structure which holds the babbitt and allows freedom to pivot. The material is typically steel. Bronze is sometimes used (with or without babbitt), depending on the application. Chrome-copper is used to reduce babbitt temperature.

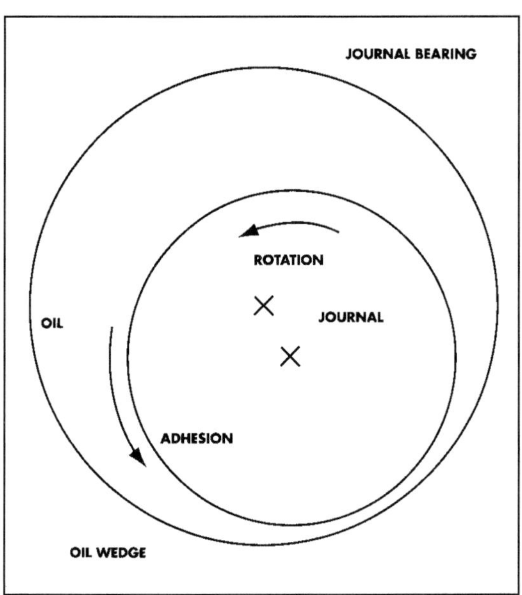

Figure 7-48: Oil wedge formation in journal bearings

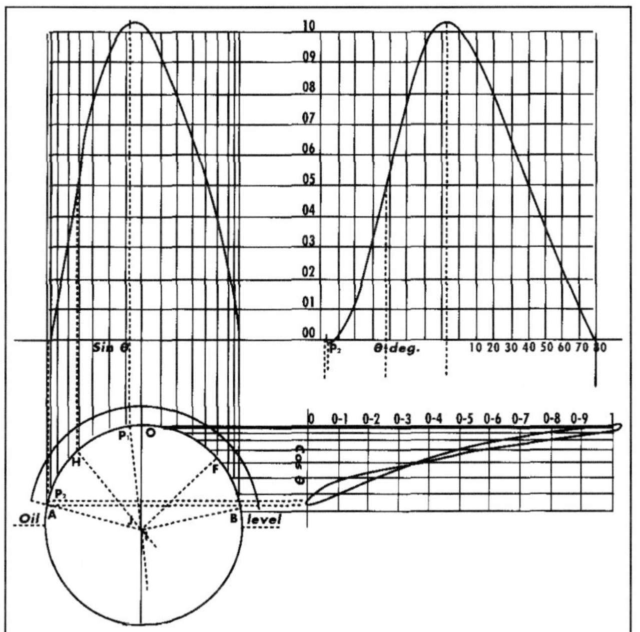

Figure 7-49: Oil film pressure distribution in hydrodynamic bearings

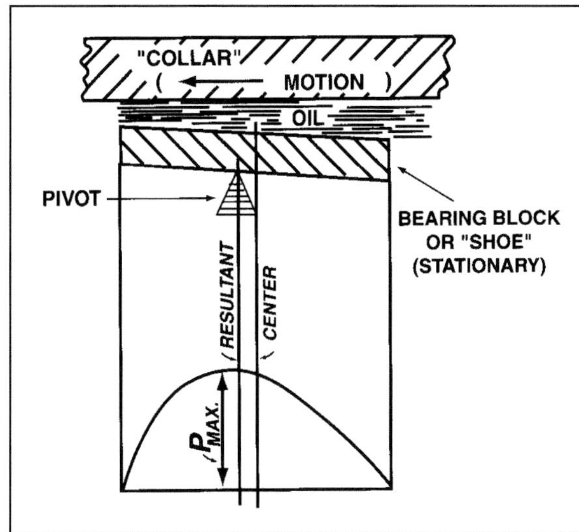

Figure 7-50: Maximum pressure exists beyond geometric center of pivoted shoe

- Pivot. The pivot allows the shoe to rotate and form a wedge. It may be integral with the shoe body, or be a separate insert. The pivot surface is spherical to allow 360 degree swiveling freedom.

- Base ring. The base ring loosely holds and constrains the shoes against rotating so as to allow freedom to pivot. It may have passages for the supply of lubricant, and contain features to adapt for misalignment and tolerance in the parts. The base ring (aligning ring) is keyed or doweled to the housing to prevent rotation of the bearing assembly.

- Leveling plates. The leveling plates (not applicable to journal bearings) are a series of levers designed to compensate for manufacturing tolerances by distributing the load more evenly between thrust shoes. The leveling plates also compensate for minor housing deflections or misalignment between the collar and the housing supporting wall.

- Lubricant. The lubricant is another important "element" of the bearing (Figure 7-51). The loads are transmitted from the shaft to the bearing through the lubricant which separates the parts and prevents metal to metal contact. The lubricant also serves to carry heat caused by friction out of the bearing.

PARAMETERS RELATED TO PIVOTED SHOE BEARINGS

Tolerance, alignment and equalization are key parameters affecting the operation of pivoted shoe bearings. In a machine, alignment and load distribution are not perfect because of manufacturing tolerances in the housing, shaft and bearing elements. There are three areas of concern:

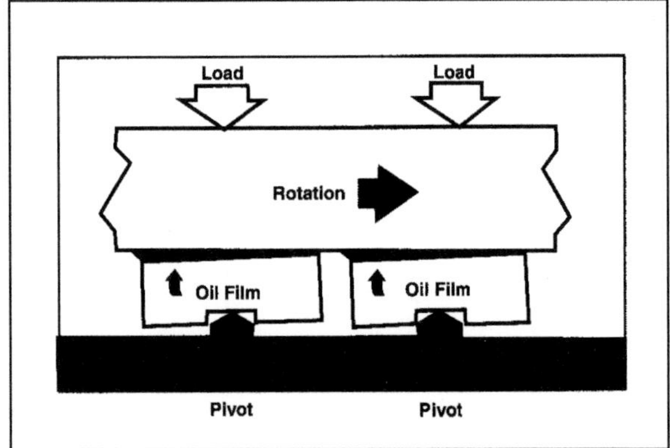

Figure 7-51: Hydrodynamic film formation in pivoting shoe bearings

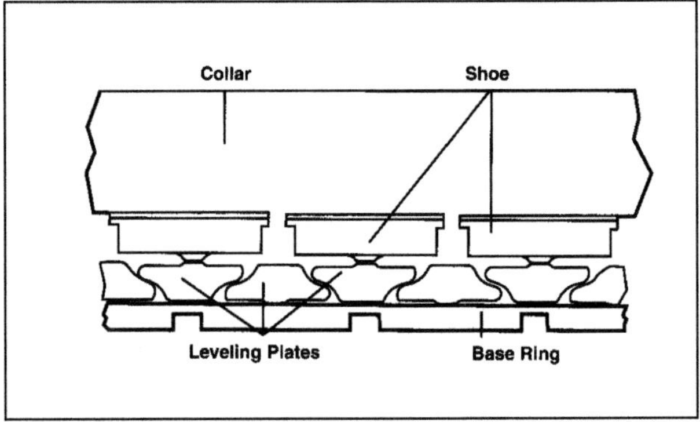

Figure 7-52: Thrust bearing load equalization and alignment

1. The squareness of the collar (and parallelism of the journal) to the axis of the shaft which is assembled to, or machined on the shaft.

2. The alignment of the shaft with the bearing and housing, which is a manufacturing tolerance stack-up of the bearing parts and the housing bores and faces.

3. The alignment of shafts between machines that are aligned and coupled together on site.

Misalignment of the shaft relative to bearing and housing, and between equipment shafts is considered static misalignment and can be adjusted at assembly if proper design features are incorporated. Other sources of misalignment, termed dynamic misalignment, are due to operating or changing conditions such as thermal housing distortion, shaft deflection from imposed loads, movement caused by thermal expansion, movement caused by settling of foundations, pipe strain, etc.

In pivoted shoe thrust bearings, static misalignment and manufacturing tolerances in the shoe height are accommodated by the leveling plates. Referring to Figure 7-52, the load transmitted by the rotating collar to any thrust shoe forces the shoe against the upper leveling plate behind it. If one shoe were slightly thicker than the others, the resulting higher film force would push the shoe down against the upper leveling plate. Each upper leveling plate is supported on one radial edge of each of two adjacent lower leveling plates. The lower leveling plates rock very slightly and raise the shoes on either side and so on around the ring. This feature also compensates for minor housing deflections or misalignment between the housing supporting wall and the collar face.

END-PLAY AND RADIAL CLEARANCE

End-play is the axial thrust bearing clearance which is the distance the shaft can move between opposing thrust bearings. For journal bearings the radial clearance is half the difference of the journal bearing bore and journal diameter. End-play and radial clearance are required to allow for misalignment, shoe movement, and thermal expansion of the parts. If set too tight, power is wasted. If too loose, the unloaded side shoes are too far from the shaft to develop a film pressure and can flutter, causing damage to the unloaded shoes. Filler plates and shim packs provide a means for setting end play and axial positioning of the rotating elements. Adjusting screws are also used to accomplish this function.

PRELOAD

Preload pertains mostly to journal bearings and is a measure of the curvature of the shoe to the clearance in the bearing. The shoe curvature is another parameter which affects the hydrodynamic film, allowing design variations in bearing stiffness and damping to control the dynamics of the machine. The geometry and definition are given in Figure 7-53.

LUBRICATION

For hydrodynamic bearings to operate safely and efficiently, a suitable lubricant must always be present at the collar and journal surfaces. The lubricant needs to be cooled to remove the heat generated from oil shear, before re-entering the bearing. It must also be warm

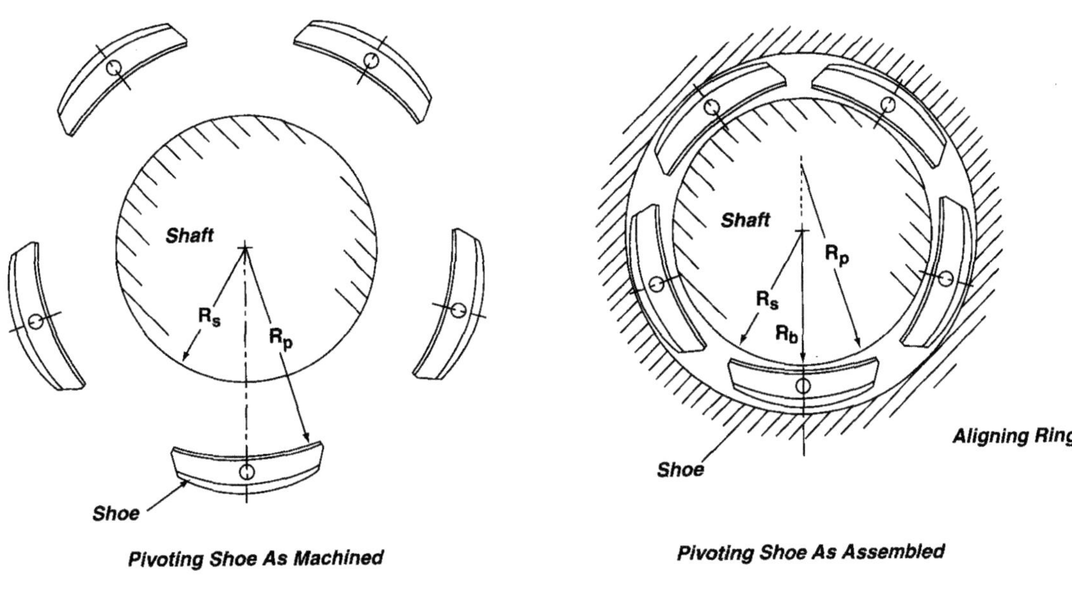

R_s = SHAFT RADIUS

R_p = SHOE MACHINED CURVATURE

R_b = BEARING ASSEMBLED RADIUS

C_p = SHOE MACHINED CLEARANCE = $R_b - R_s$

C_b = BEARING ASSEMBLED CLEARANCE = $R_b - R_s$

$$\text{Preload } M \equiv 1 - \frac{C_b}{C_p}$$

Figure 7-53: Component geometry and definition of preload in pivoted shoe bearings

enough to flow freely, and filtered so that the average particle size is less than the minimum film thickness.

Various methods are applied to provide lubricant to the bearing surfaces. The bearing cavities can be flooded with oil such as vertical bearings that sit in an oil bath. The bearings can also be provided with pressurized oil from an external lubricating system. The flow path of a horizontal, flooded pivoted shoe thrust bearing is shown in Figure 7-54.

For high-speed bearings, the frictional losses from oil shear and other parasitic losses begin to increase exponentially as the surface speed enters a turbulent regime. The amount of lubricant required increases proportionately. Industry trends for faster, larger machines necessitated the design of lower loss bearings. This has been incorporated by the introduction of other methods of lubrication.

Directed lubrication directs a spray of oil from a hole or nozzle directly onto the collar (journal) surface between the shoes. Rather than flooding the bearings, sufficient oil is applied to the moving surface allowing the bearing to run partially evacuated. Such a method of lubrication reduces parasitic churning losses around the collar and between the shoes.

In 1984 Kingsbury introduced its Leading Edge Groove (LEG) Thrust Bearings (Figure 7-55), another technology developed for high-tech machines, including feed pumps at power generating plants. In addition to reducing oil flow and power loss, LEG lubrication greatly reduces the metal temperature of the shoe surface. This effectively reduces oil flow requirements and power loss while improving the load capacity, safety and reliability of the equipment. Rather than flooding the bearing or wetting the surface, the LEG design introduces

Oil enters sump and is pumped through a filter and cooler.

Oil passes through inlet orifice which controls flow rate.

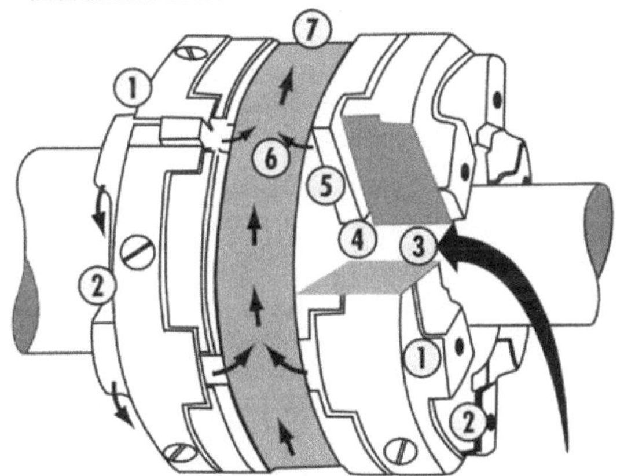

1 – Oil enters annulus in base ring.

2 – Oil passes through radial slots in back face of base ring.

3 – Oil flows through clearance between base ring bore and shaft.

4 – Oil flows to inner diameter of rotating thrust collar.

5 – Oil flows between shoes and into the films.

6 – At the collar rim, oil is thrown off into space around the collar.

7 – Oil exits tangentially through the discharge opening.

Figure 7-54: Typical oil flow path in a pivoted pad thrust bearing

cool oil directly into the oil film (Figure 7-56), insulating the shoe surface from hot oil that adheres to the shaft. The same technology is applied to pivoted shoe journal bearings (Figure 7-57).

Cooling System

A cooling system is required to remove the heat generated by friction in the oil. The housing may simply be air cooled if heat is low. Vertical bearings typically sit in an oil bath with cooling coils (Figure 7-58), but the oil can also be cooled by an external cooling system as typical in horizontal applications. The heat is removed by a suitable heat exchanger.

OPERATION AND MONITORING

Under operation, the capacity of hydrodynamic bearings is restricted by minimum oil film thickness and babbitt temperature. The critical limit for low-speed operation is minimum oil film thickness. In high-speed operation, babbitt temperature is usually the limiting criterion. Temperature, load, axial position, and vibration monitoring equipment are used to evaluate the operation of the machine so that problems may be identified and corrected before catastrophic failure. Of these, bearing health is commonly monitored through the use of temperature detectors. The temperature of the bearing varies significantly with operating conditions and also varies across and through the shoe. Therefore, for the measurement to be meaningful, the location of the detector must be known.

The recommended location for a detector is termed the "75/75 location" on a thrust shoe face, i.e. 75% of the arc length of the shoe in the direction of rotation and 75% of the radial width of the shoe measured from the ID to the OD. In a journal bearing, sensor location should be 75% of the arc length on the centerline of the shoes. This position represents the most critical area because it is the point where peak film pressures, minimum film thickness, and hot temperatures co-exist.

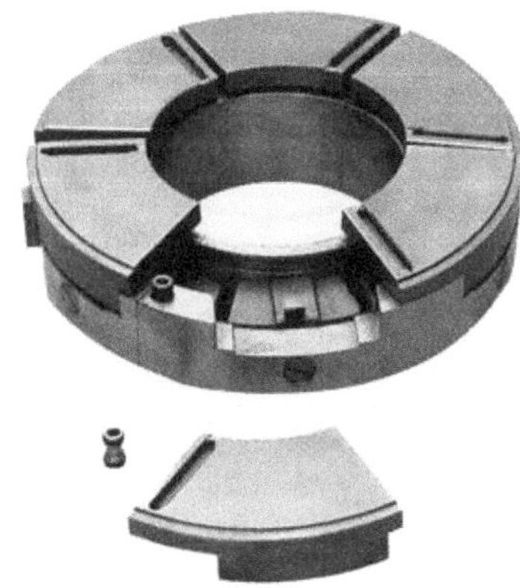

Figure 7-55: Kingsbury leading edge groove (LEG) thrust bearing

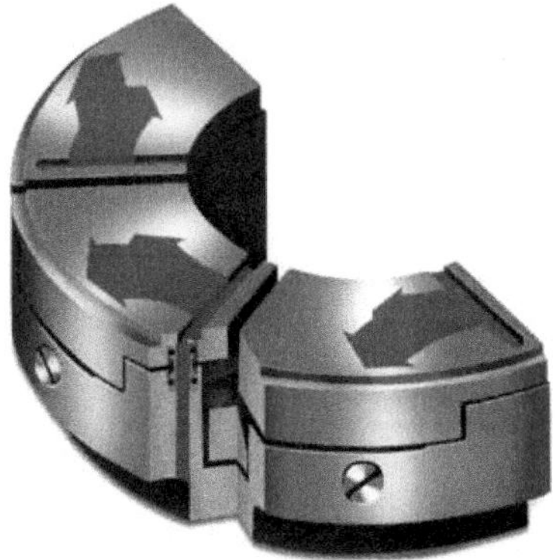

Figure 7-56: Oil flow path through a Kingsbury LEG thrust bearing

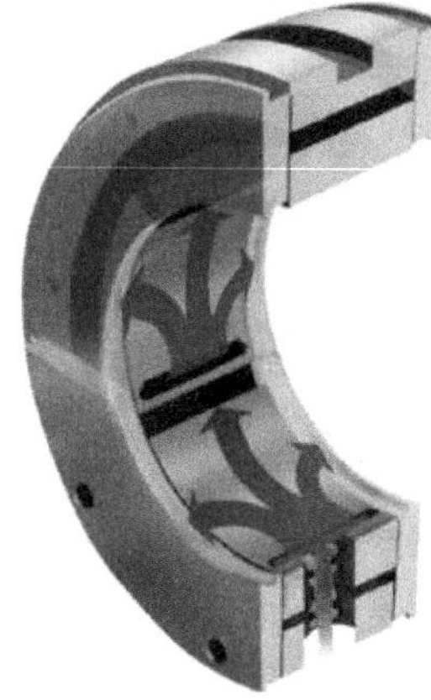

Figure 7-57: Oil flow path through a Kingsbury LEG journal bearing

Figure 7-58: Kingsbury vertical thrust bearing assembly with oil sump and cooling

BEARING HOUSING PROTECTION

Regardless of bearing style and type, bearing housings require provisions to prevent oil escape and dirt ingress. This is generally best accomplished with bearing housing protector seals. Dozen of different configurations are available in the market place; only three generic representations of bearing housing protector seals are shown in Figure 7-59.

The owner-purchaser must select bearing housing protector seals that are sturdy and cost-effective. Note that certain configurations risk O-ring degradation (top left) due to contact with sharp grooves. Others use a V-contoured ring (top right) that tends to increase frictional drag.

Look for optimized designs (bottom). In this particular design two O-rings are used to clamp the rotor to the shaft; another O-ring seals the stator against the housing bore. As the shaft turns and picks up speed, the two unclamped O-rings are activated. The smaller of these moves out--away from the shaft center, so-to-speak. This is due to centrifugal force action on the smaller of the two O-rings. The large diameter O-ring can then move axially a very small distance away from its contoured seat. When the machine is stopped, the smaller of the two O-rings moves back towards the shaft center, thereby pushing the large O-ring into sealing contact with its contoured seat. The dual O-ring setup acts as a valve and is highly effective.

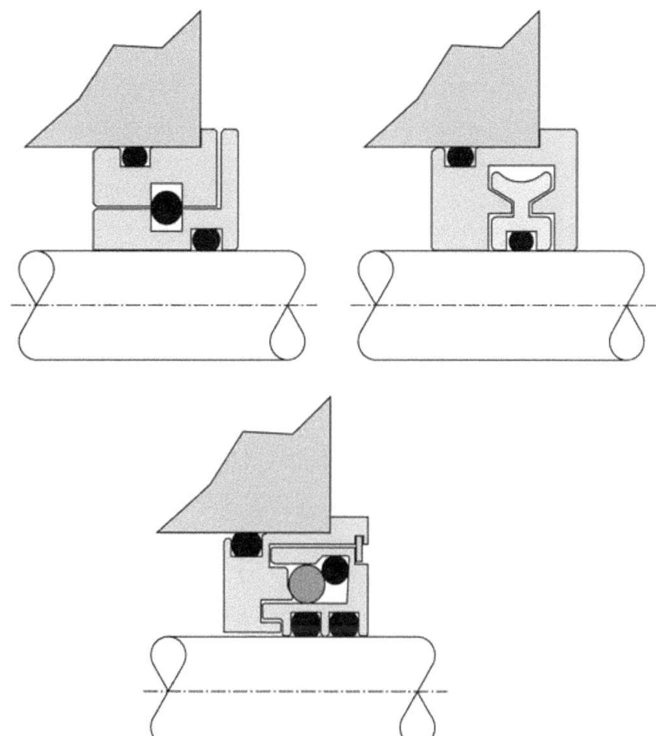

Figure 7-59: Generic representations of bearing housing protector seals. Note that certain configurations risk O-ring degradation (top left) due to contact with sharp grooves. Others use a V-contoured ring (top right) that tends to increase frictional drag. Look for optimized designs (bottom). Source: AESSEAL Inc., Rotherham, UK and Rockford, Tennessee.

Why Not Lip Seals?

Professional engineers often read, with great interest, articles on equipment upgrade opportunities. Because the effects of moisture contamination and the need for bearing protection are obvious subjects of importance, an editorial write-up on "Reducing Moisture Contamination in Bearing Lubrication" was judged of interest. It appeared in the February, 2006 issue of Sealing Technology (UK) and noted that "lip seals seem to be permanently denigrated when they are not used in an optimum configuration. The automotive industry uses cartridge arrangements ("cassettes"), to what appears to be good effect. They are widely used on trucks and buses, which now often have a half million-mile warranty. This must be in excess of 10,000 hours, so would be potentially quite adequate for many intermediate duty pumps." This editorial prompted a brief assessment of environmental and energy savings issues of the various bearing housing protection options available to pump users.

Environmental and Energy Savings Aspects of Bearing Protection

The rolling element bearings of literally billions of electric motors and hundreds of millions of automobiles, pumps and other industrial machines are protected against lubricant loss and contamination by lip seals. Lip seals have certainly served industry for more than a century in applications where the elastomeric component received ample lubrication and where the shaft surface velocities were moderate.

Although a 20 mm (0.78 inch) automobile drive shaft operating at a maximum speed of 2,000 rpm (2,093 mm/s, or 82.4 ips) would represent a rather strenuous application for a motor vehicle, it is far from the 12,250 mm/s, or 482 ips rubbing velocity of a 65 mm (~2.56 inch) / 3,600 rpm shaft in a centrifugal pump.

A rather universally accepted rule-of-thumb assumes that rubbing wear increases as the cube of the velocity ratio. Therefore, if a well-designed lip seal in an automobile had a life of 1,000,000 miles at 50 mph, this would equate to 20,000 operating hours on a set of lip seals. On the industrial equipment example and at a surface velocity 5.8 times greater, the wear life would be diminished by a factor of 200 and lip seals would last 100 hours--an unattractive choice by any measure. This, nevertheless, explains the 80% failure rate given for lip seal after 2,800 hours of operation at testing conditions mandated by a U.S. "MIL Spec," i.e., a specification followed by suppliers to the military.

Other Sealing Choices

It is reasonable, based on industrial experience, to assume two different scenarios for the two bearing housing seals illustrated in Figure 7-60. A lip seal is depicted on the upper portion of the shaft.

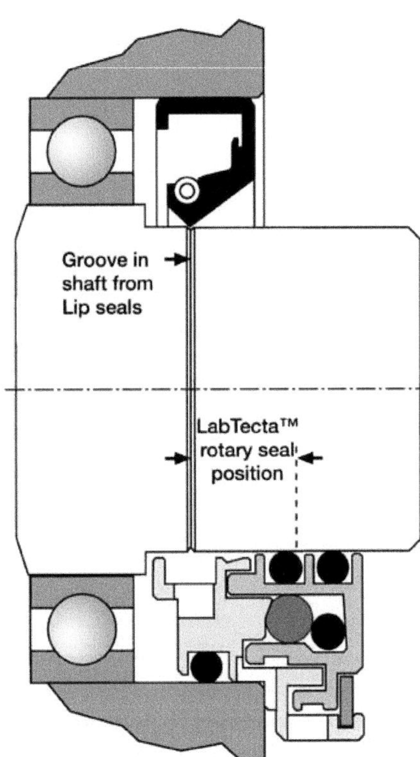

Figure 7-60: Comparison of elastomeric lip seal (top part of illustration) and modern rotating labyrinth seal (lower part)

Purely for the sake of illustration, we will assume the lip seal costs $5; the lower portion shows a modern rotating labyrinth seal and we choose to price it at $100.

Scenario 1—Machinery Bearing Housing Application

To avoid shaft fretting, moisture intrusion, and premature bearing failure (assuming labor and materials to remedy a bearing failure costs $6,000), we replace a $5.00 lip seal twice a year. Labor is accounted for at $500 per event and materials, per year, require an outlay of $1,010.

Alternatively, and purely for the sake of illustration, we make the decision to replace a $100 modern dynamic O-ring rotating labyrinth seal after just two years of operation (but hasten to add that its anticipated life is somewhere between 8 and 10 years). In the case of replacement after 2 years, labor is $250/yr and materials cost $50/yr. Our total is then $ 300 per year.

Scenario 2—Machinery Bearing Housing Application

This time, assume we run to failure. The labyrinth seal is allowed to degrade and a bearing fails after two years. Assume no production outage time, but the repair now costs the plant $6000, i.e. $3,000 per year

Energy Issues

As long as a lip seal is operationally effective and has not degraded to the point of shaft wear (top of Figure 7-60) or elastomer lip wear, it is reasonable to assume that 160 watts of frictional energy are consumed by an average lip seal. At $0.10/kWh, that equates to $140 per year.

If, in Scenario 1 above, precautionary lube oil replacements (oil changes) were performed and a lube oil charge and its environmentally acceptable disposal were factored in, the picture would shift even more in favor of modern dynamic O-ring rotating labyrinth seals--the type illustrated at the bottom of Figure 7-60.

Conclusions of Reliability-focused Pump Users Re Protector Seals

Lip seals have their place in disposable appliances; they also deserve consideration in some machines which, for unspecified reasons, must frequently be dismantled. However, engineers should always look at the full picture. While in no way claiming all lip seal applications are past their prime, there are now viable alternatives for a the reliability-focused and energy-conscious user community. In process pump applications, lip seals rarely measure up to the expectation of the overwhelming majority of intermediate duty pump users. The fact that lip seals are available in cassette configurations makes no difference.

Select the least vulnerable, longest life, rotating bearing housing protector seals for modern process pumps. A cost justification calculation will readily show the advantages of such protector seals.

Chapter 8

Mechanical Seal Selection and Application

DESIGN OF MECHANICAL SEALS

Mechanical seals have rapidly evolved from machined shaft shoulders contacting a stationary casing face (Figure 8-1a) through replaceable faces at shaft shoulders (Figure 8-1b) and gasket-backed inserts (Figure 8-1c) to the many modern mechanical seals available in the marketplace today. Considerable amounts of relevant data are contained in References 8-1 through 8-6, and Ref. 8-13. Also, since mechanical seals are vulnerable precision components, their preservation in "mothballed" pumps and recommended storage protection are discussed in Appendix 5.

Spring loading of the rotating seal ring and use of a dynamic (axially sliding) O-ring (Figure 8-2) constituted the next progression and, within a short period of time, pusher seals similar to the one shown in Figure 8-3 were commonly used in process pump applications. In pusher seals, one of the faces—usually the rotating one, but preferably the stationary one—is being pushed into the opposing face. These seals could now be unitized and mass-produced.

TYPES OF SEALS

There are two major mechanical seal groupings. In the first grouping—rotating flexure or "conventional" seals (Figures 8-2 and 8-3),—the spring-loaded face is part of the rotating shaft assembly. These seals are suitable for moderate speed applications and applications where shaft deflection is low. In the second grouping—stationary flexure or just plain "stationary" seals (Figures 8-4 and 8-5)—the spring-loaded face does not rotate. Stationary seals will accommodate greater shaft deflections and can operate at higher speeds than conventional seals.

Both conventional and stationary seals can be mounted either inside or outside the seal housing (stuffing box). When a seal is mounted inside the stuffing box of the pump, it is called an inside seal. Although inside seals are generally more difficult to install, the advantages of an inside seal will generally outweigh the disadvantages. Figures 8-2 through 8-4 represent just three of literally hundreds of types and styles of inside seals.

Among the advantages of inside seals we typically find that

- Cooling of the seal is facilitated by product flow through inlet ports in the stuffing box or gland.
- The rotary action of an inside seal helps to keep it clean. Centrifugal force makes it more difficult for suspended solids to enter in between the seal faces.

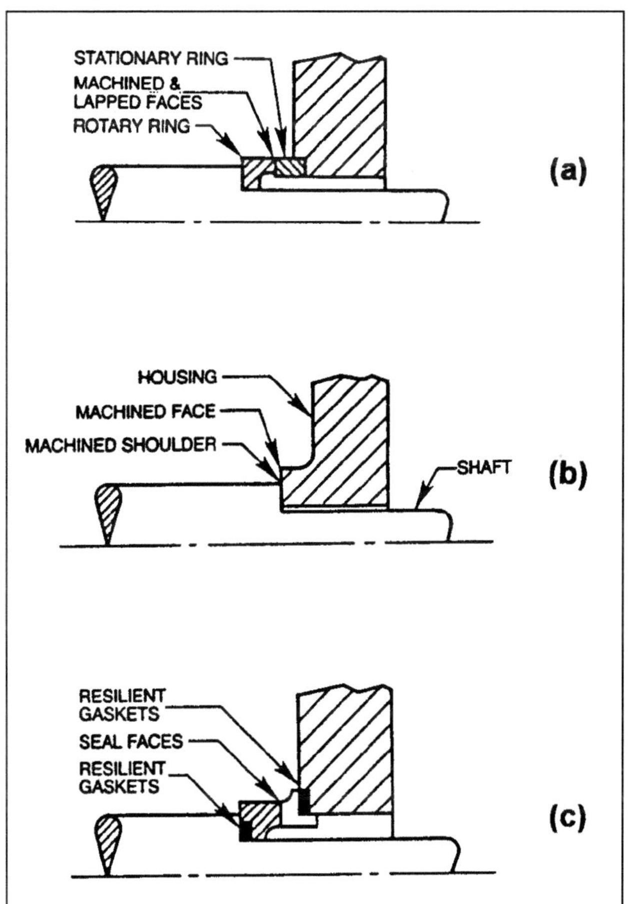

Figure 8-1: Mechanical seal evolution (Source: Ref. 8-1, Safematic, Muurame, Finland)

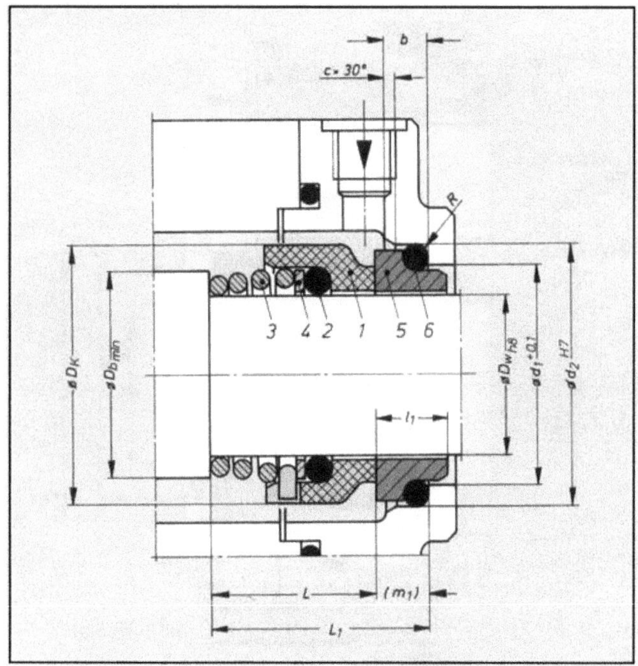

Figure 8-2: Elementary spring-loaded mechanical seal (Source: Ref. 8-2, Burgmann Seals America, Inc., Houston, Texas; also Dichtungswerke Feodor Burgmann, Wolfratshausen, Germany)

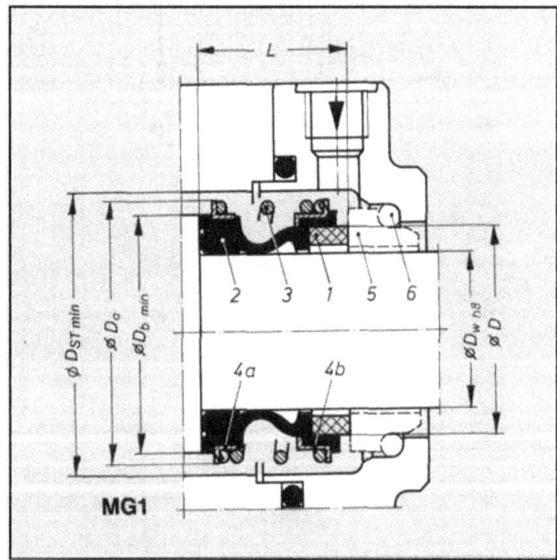

Figure 8-3: Pusher-type mechanical seal with rubber bellows (Source: Ref. 8-2)

- In case of seal leakage, inside seals are usually not prone to catastrophic leakage, because the strong seal gland is fitted with a close-clearance throttle bushing. Moreover, the hydraulic balancing forces tend to keep the seal faces closed.

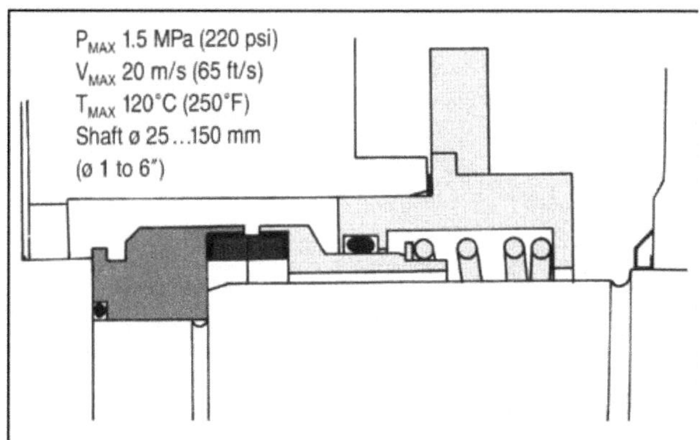

Figure 8-4: "Stationary" single seal (Source: Ref. 8-1)

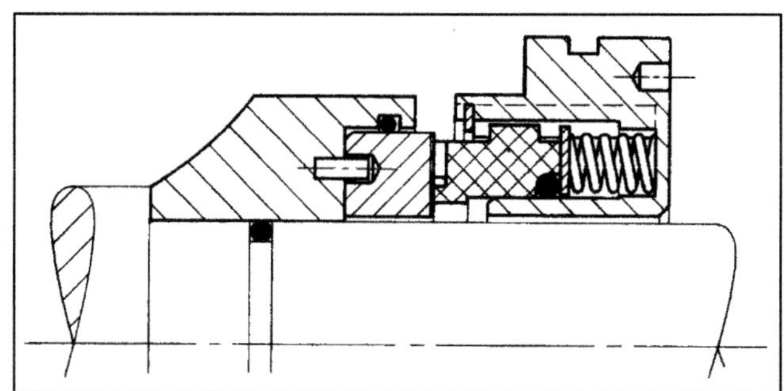

Figure 8-5: "Stationary" single seal with multiple pusher springs (Source: Ref. 8-3, Borg-Warner Corporation, Temecula, California)

- Environmental safety and support systems are easily fitted and attached to inside seals.

Outside Seals

An outside seal is located outboard of the pump stuffing box. Figure 8-6 depicts an outside-mounted elastomeric bellows seal.

Outside seal have the following advantages:

- easy installation
- can be inexpensively made from corrosion resistant materials
- suitable for services where it is necessary to quickly remove the seal for cleaning
- where stuffing boxes are shallow and inside seals cannot be used due to lack of axial or radial space
- where wear of the faces must be monitored and,
- where access to tightening the seal is difficult or practically impossible.

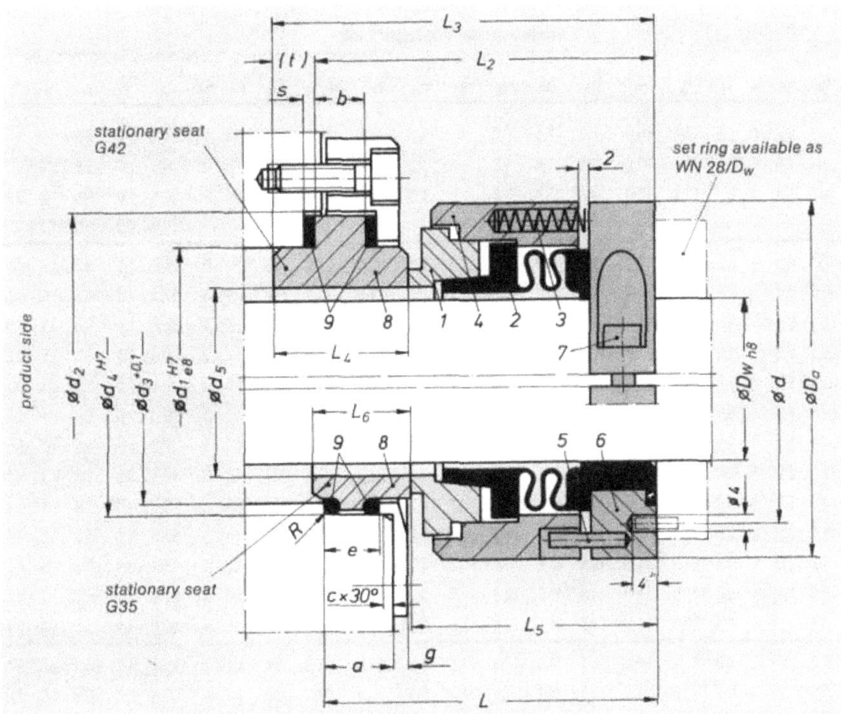

Figure 8-6: Outside mechanical seal, elastomer bellows, pusher-type (Source: Ref. 8-2)

Limitations on the design of outside seals are often significant and must be taken into account by safety and reliability-minded users. Specifically, due to lack of heat dissipation from below the seal faces, outside seals must be used in lower temperature, lower speed and lower pressure applications. (Pressures must be lower than on inside seals as the pressures are being exerted outward on seal parts rather than inward.)

Single Seals and Seal Torque Transmission

To re-state an earlier point, all mechanical seals can be either of "conventional" or "stationary" construction. Similarly, single seals can be mounted either inside or outside the stuffing box, although a user would want to be well aware of the vulnerabilities of outside-mounted seals. In general, these and virtually all other types and styles of mechanical seals consist of a rotary unit affixed to the shaft in some manner.

There are several practical ways to effect this torque transmission. While set screws, as shown in Figure 8-7/d, are prevalent due to their low cost, clamping rings (Figures 8-6, 8-7a/b, and 8-13), or drive boots (Figure 8-3) are preferred. For sets screws to hold, they have to "bite" into the shaft (Figure 8-7/d), which creates a rather undesirable burr or, alternatively, a depression will have to be machined into the shaft (Figure 8-7/c). Shrink discs and clamping rings are an identifying mark of many superior mechanical seals, although drive keys and snap locators (Figure 8-7f/h) also merit being ranked well ahead of setscrews.

Pressurized Dual Seals

These seals are arranged either facing away from each other or towards each other (back-to-back or face-to-face configurations). The most common of numerous available back-to-back geometries are similar to Figure 8-8. Both configurations are used in conjunction with a separate pressurized barrier fluid. In the back-to-back double mechanical seals illustrated in Figure 8-8, this bar-

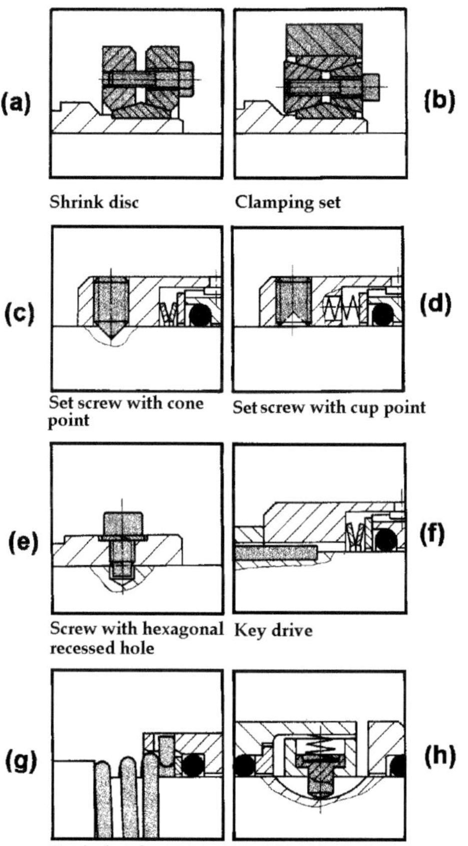

Figure 8-7: Torque transmission options for attaching seal rotating assemblies to pump shafts (Source: Ref. 8-2)

Liquids to be pumped which are toxic, explosible or otherwise endangering the environment require double mechanical seals.

According to the standard DIN 24 960 double mechanical seals are to be formed by installation of two single mechanical seals. For different working conditions the following back-to-back arangements are possible:

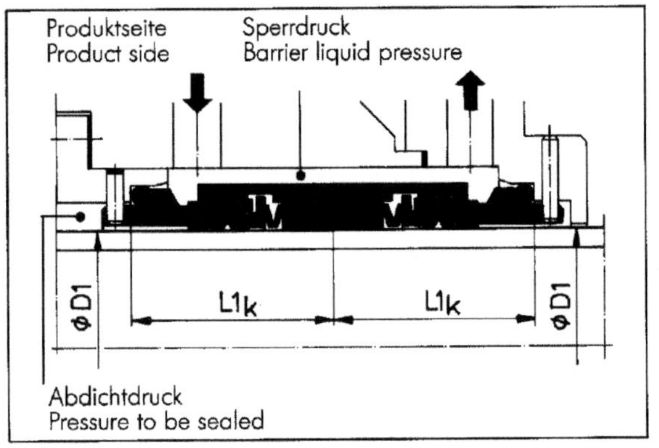

Form UU
(unbalanced/unbalanced)

Pressure to be sealed: ... 9 bar
Barrier liquid pressure: ...10 bar, min. 1 bar higher than the pressure to be sealed

Single mechanical seals:
Product side e.g. PACIFIC 600
Atmospheric side e.g. PACIFIC 600

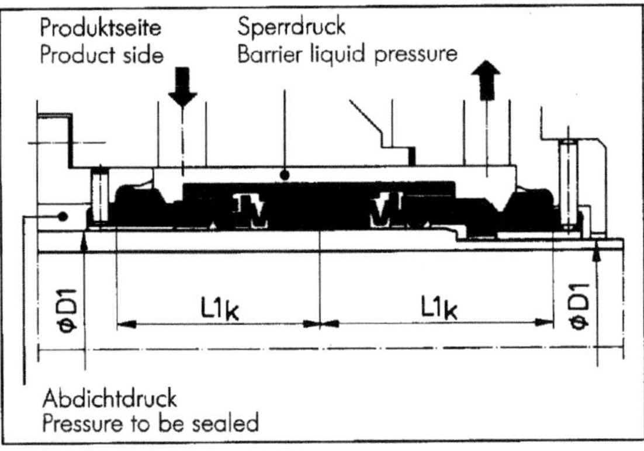

Form UB
(unbalanced/balanced)

Pressure to be sealed: ...49 bar
Barrier liquid pressure: ...50 bar, min. 1 bar, max. 10 bar higher than the pressure to be sealed

Single mechanical seals:
Product side e.g. PACIFIC 600
Atmospheric side e.g. PACIFIC 610

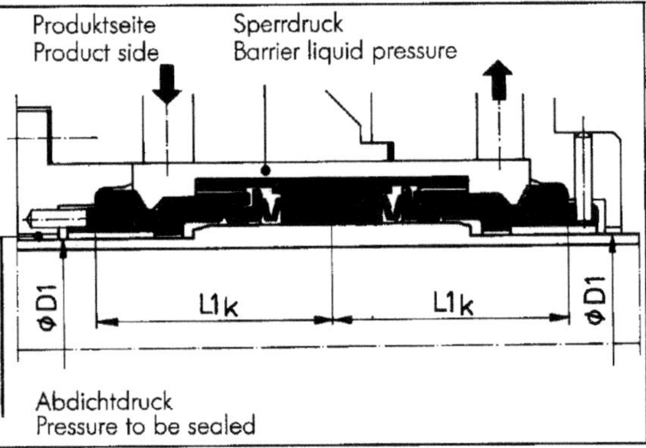

Form BB
(balanced/balanced)

Pressure to be sealed: ...49 bar
Barrier liquid pressure: ...50 bar, min. 1 bar higher than the pressure to be sealed

Single mechanical seals:
Product side e.g. PACIFIC 610
Atmospheric side e.g. PACIFIC 610

Figure 8-8: Pressurized dual mechanical seals, back-to-back oriented (Source: Ref. 8-4, Pacific-Wietz, Dortmund, Germany)

rier fluid is injected at a higher pressure between the two seals than the fluid being pumped. In pressurized dual ("tandem") seals, the barrier fluid is generally injected at a pressure below that of the stuffing box region.

The five principal advantages of all dual ("double") seals are:

(1) Increased protection against product leaking into the environment

(2) Barrier fluid level flow or pressure can be monitored to determine if either seal has failed

(3) The dual seal also works well when there is danger of gas pockets being formed in the seal chamber, as is the case with vertical pumps

(4) Very advantageous in applications with products that crystallize, freeze up or ignite when leaking to the surrounding atmosphere

(5) Heating or cooling can also be maintained via the barrier fluid called "buffer," if unpressurized).

In the double seal arrangement with the seals facing each other (Figure 8-9), the outside seal receives its lubricating and cooling liquid through the hole in the stationary seal ring. In this "face-to-face" dual seal configuration, the barrier fluid would be pressurized above normal product stuffing box pressure or near atmospheric pressure.

Tandem Seal
(Unpressurized Dual Seal)

The tandem seal arrangement (Figure 8-10) is the safest of all dual seal combinations. In effect, it is made up of two inside seals, both with the advantage of inside seals and double seals. In this arrangement a separate clean barrier fluid is sealed by the outside seal and process liquid is sealed by the inner seal. This eliminates one of the major drawbacks of certain conventional double seal arrangements, that of having process liquid on the underside of the primary seal, as is evident from Figure 8-9. The buffer fluid typically exists at a lower pressure than the process liquid. Leakage of the pumped fluid into the buffer fluid would manifest itself as either pressure or level increase in the buffer loop.

Seal Classification

Unbalanced Seals

Virtually all mechanical seals are available in either unbalanced (Figure 8-11) or balanced (Figure 8-12) ver-

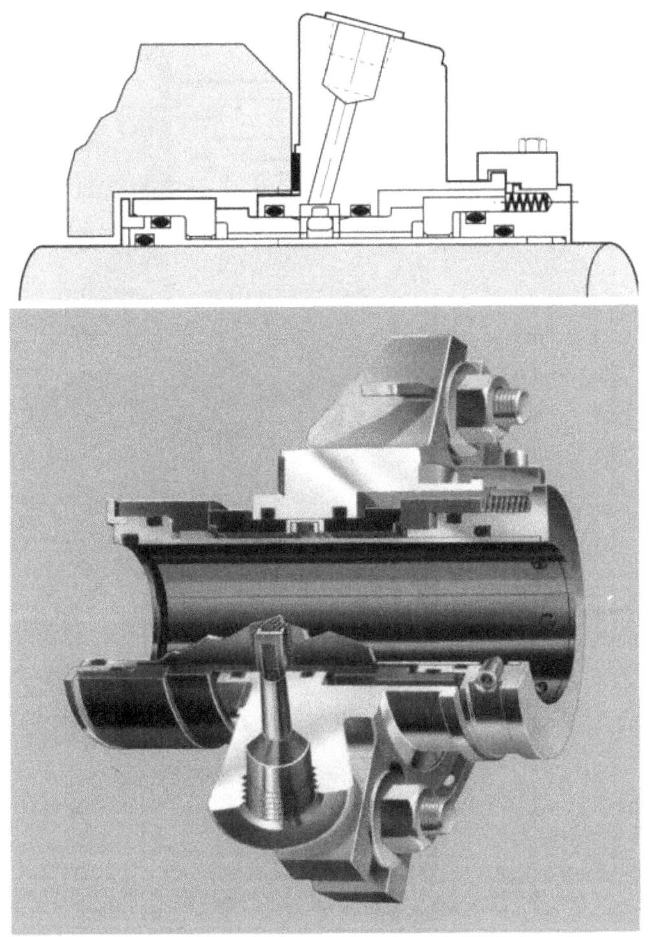

Figure 8-9: Double mechanical seal ("Chesterton 225 Dual Cartridge") oriented "face-to-face" (Source: Ref. 8-5, A.W. Chesterton Company, Stoneham, Massachusetts 02180)

sions. The term "unbalanced" is used when the stuffing box pressure times the area exposed to the pumped fluid (closing force), acting to close the seal faces, is greater than the average pressure between the seal faces (pressure gradient) times the area of contact between the faces. In other words, unbalanced mechanical seals exhibit net hydraulic closing forces which are generated by the actual pressures to be sealed.

For example, if there were a stuffing box pressure of 45 psig (310 kPa), the spring load would have to be added. Hence, the "face load" or closing force on the faces would be even higher than 45 psig times the face area. This, of course, limits the pressure sealing capacity of an unbalanced seal.

Unbalanced seals are often more stable than balanced seals when subjected to vibration, misalignment and cavitation. The disadvantage is their relatively low pressure limit. If the closing force exerted on the seal faces exceeds the pressure limit, the lubricating film between the faces is squeezed out and the highly loaded dry running seal fails (Ref. 8-7).

Balanced Seals

The balanced seal has the same opening (face) area as the unbalanced seal, but the closing area has been reduced in relation to the face area. Because force equals pressure times area, reducing the closing area reduces the closing force. Consequently, less heat is generated and the seal generally has a longer life. For example, if the stuffing box pressure were 200 psig (1,379 kPa), then the net closing force would be substantially reduced to perhaps 60 psig (414 kPa). To simplify the explanation, balancing a mechanical seal involves a small design change which reduces the hydraulic forces acting to close the seal faces. Balanced seals have higher pressure limits, lower seal face loading, and generate less heat. They are better able to handle liquids with low lubricity and high vapor pressures. This would include light hydrocarbons.

Because seal designs vary from manufacturer to manufacturer and from application to application, it is not possible to standardize on either configurations or materials that cover all conceivable services. Available basic designs have variations that were often developed so as to meet specific applications. Each seal design has its own strengths and weaknesses.

However, as will be seen later, it is indeed possible to select a certain seal model and use it for a very large number of applications and services in a given facility. This approach will inevitably involve the use of cartridge and/or cassette-type seal configurations.

Cartridge Seals

The cartridge design changes none of the functional components of the basic seal classifications. In a cartridge seal, all components are part of an assembled package. The term "containerized" has been used to describe the cartridge seal, which, incidentally, requires only the tightening of gland bolts, flush connections, and drive screws or clamping rings. There is no longer any need to scribe lines and to make critical installation measurements.

Cartridge seals are available in each of the basic types and classifications. Figure 8-9 is somewhat typical of many cartridge seals. The limiting factor in cartridge seal designs is the space available in a given pump stuffing box—or so it seems. A reliability-focused user will make an all-out effort to specify, redesign, modify or otherwise enlarge the stuffing box volume. Unless

Safety rules, temporary dry running, sealing pressure close to vapour pressure, odorous annoyance etc. often requires a sealing liquid (quench) for the primary mechanical seal.
The quench liquid mostly will be sealed by a second mechanical seal.
The following three tandem-arrangements are possible:

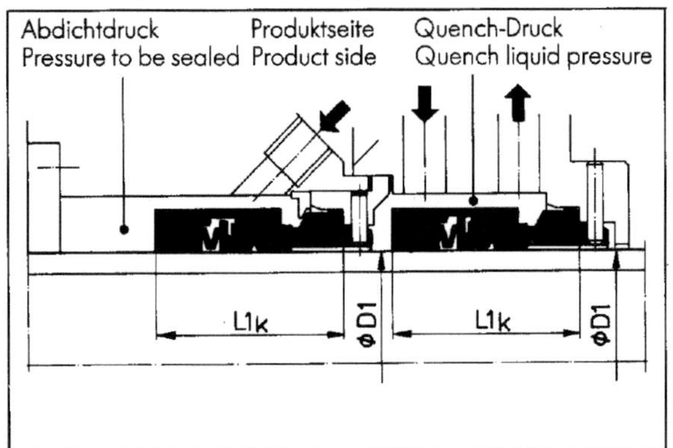

Pressure to be sealed: ...10 bar
Quench liquid pressure: lower than the pressure to be sealed

Single mechanical seals:
Product side e.g. PACIFIC 600
Atmospheric side e.g. PACIFIC 600

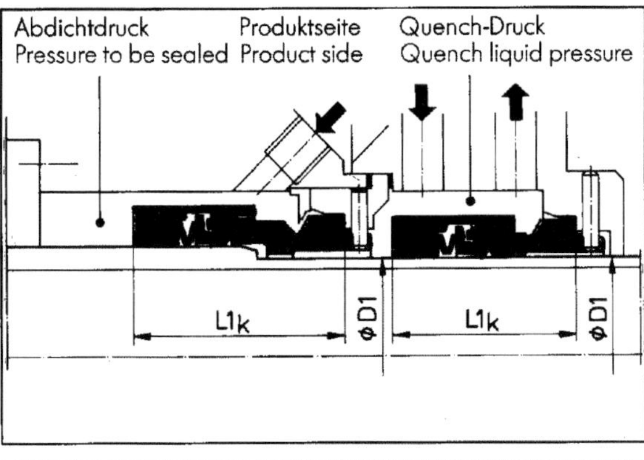

Pressure to be sealed: ...50 bar
Quench liquid pressure: lower than the pressure to be sealed, max. 10 bar

Single mechanical seals:
Product side e.g. PACIFIC 610
Atmospheric side e.g. PACIFIC 600

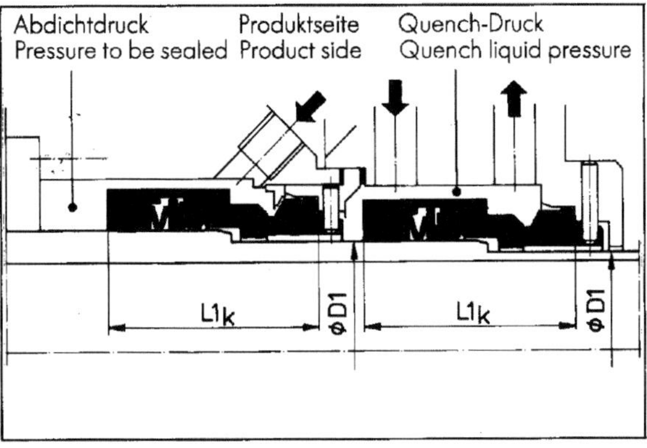

Pressure to be sealed: ...50 bar
Quench liquid pressure: lower than the pressure to be sealed

Single mechanical seals:
Product side e.g. PACIFIC 610
Atmospheric side e.g. PACIFIC 610

Figure 8-10: Unpressurized mechanical seal arrangements (Source: Ref. 8-4)

the user wishes to revert to packing—which reliability-focused users are not likely to do—there is no longer any reason to stay with the dimensional envelope that came into being around the year 1890, when rope packing was phased out in favor of standard-size braided packing.

Putting it in different words, suppose a rigorous selection procedure and reference check were to identify a sophisticated cartridge seal such as Burgmann's HR (Figure 8-13) as best suited for a slurry service (it generally is, by the way). Suppose further, that this seal does not fit into the confined stuffing box space that has been with us for well over one hundred years (a good assumption, because it won't). Before giving up on the HR and picking an inferior second choice, the reliability-focused user will examine his stuffing box enlargement options. Among these options, he would no doubt study if boring out is feasible. A reliability-focused user would probably prevail upon the pump manufacturer to furnish redesigned pump covers, or users would present their own redesign to a contract machine shop.

As mentioned above, the extent to which standardization will cover a certain range of services will vary from plant to plant. If judiciously tackled, it can indeed be done with great success. Demonstrated cost savings have materialized at sites where, as an example, the use of pressure-balanced, single and/or double acting, bi-rotational, conventional (Figure 8-14) or stationary (Figure 8-15) mechanical seals with non-product contacting multiple springs was found feasible for large numbers of pumps.

It should be noted that on the majority of cartridge seals, seal replacement will involve removal and shipment of the seal cartridge to the seal manufacturer for refurbishing, or replacement of the entire assembly. Appreciating these facts, the A.W. Chesterton Company began, in 2002, to market attractive mid-range replaceable seals, the "Streamline Cassette Cartridge," Figures 8-16 and 8-17. These cost-effective seals allow for the more expensive gland plate to stay with the pump and make it economically attractive to simply slip into place a replacement seal cassette. Defective seal internals are discarded rather than rebuilt, eliminating many of the administrative and inventory costs associated with seal repair. Of course, the overall cost-effectiveness of this discarding approach must be ascertained. Frequently

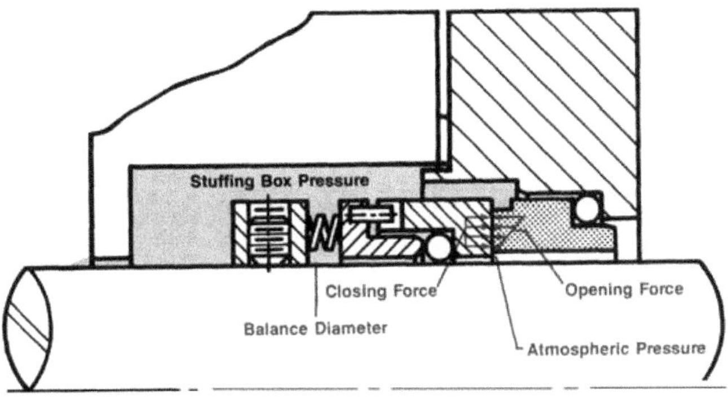

Figure 8-11: Inside unbalanced mechanical seal (Source: Ref. 8-6, Flowserve Corporation, Kalamazoo, Michigan)

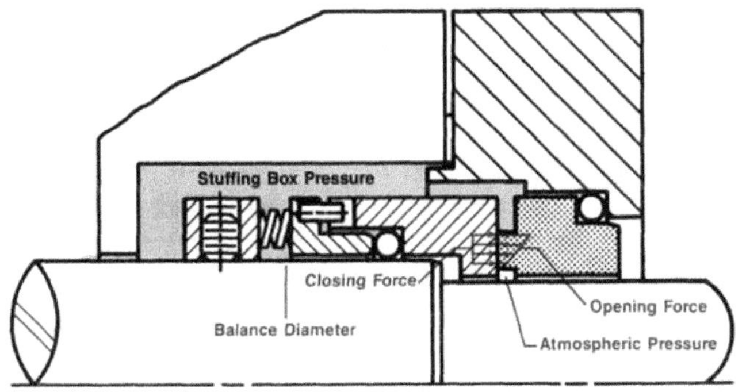

Figure 8-12: Inside balanced seal (Source: Ref. 8-6)

replacing even an inexpensive seal may cost a fortune when compared against upgrading a pump installation or taking remedial action by, among other things, opting for superior mechanical seals.

"U" Cup Seals

On U-cup seals (Figure 8-18) the secondary seal or "U" cup consists of PTFE, a suitable elastomer, or composite material. The large single spring does not drive the rotary element. It simply spreads the secondary seal and maintains face loading during running as well as standstill conditions. In general, U-cup seals are balanced designs. They are available for heavy-duty applications but will often require more axial and radial space than their multi-spring balanced seal counterparts.

"V" Ring Seals

"V" ring seals (Figure 8-19) require constant loading of the elastomeric "V" ring in order to seal. The open side of the "V" of the seal is on the pressurized fluid side. It should be noted that the flexing action of dynamic PTFE V-rings ("chevrons") will cause fretting on stainless steel shafts by removing the oxide layer and infusing it in the PTFE. These configurations rarely represent state-of-art solutions.

"O"-Ring Seals

O-ring seals incorporate a dynamic O-ring secondary seal and a static O-ring shaft seal, as shown earlier in Figure 8-2. However, the term "O-ring seal" is traditionally applied to seals where the springs are isolated from the pumped fluid by the O-ring seals and thus cannot become clogged unless leakage occurs across the seal face. Shown in the ISO-type pump of Figure 8-20, this type of seal is normally balanced within its own component parts.

Wedge Seals

This seal (Figure 8-21) is shown with a PTFE sliding O-ring—although PTFE wedges could also be used instead—mating with a wedge-shaped carbon seal face. The sliding O-ring or PTFE wedge makes contact with

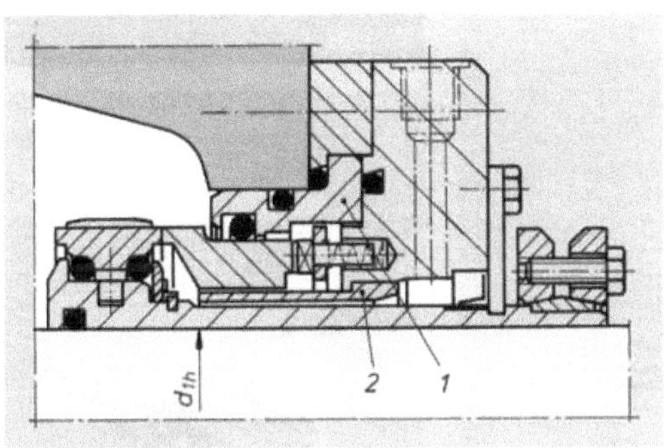

HR 2 ...

Cartridge-type single seal with guide sleeve (Item no. 2) for use with quench. Insert (Item no. 1) either metallic or SiC.

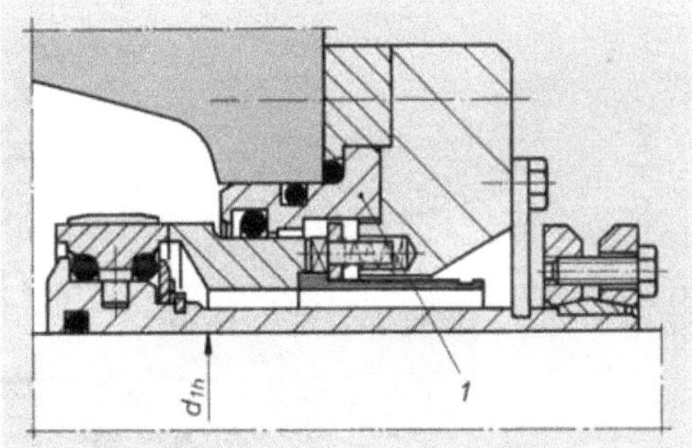

HR 3 ...

Cartridge-type single seal. Insert (Item no. 1) either metallic or SiC.

Figure 8-13: Sophisticated cartridge seals may not fit in confined stuffing box space (Source: Ref. 8-2 and 8-8 through 8-10)

Convertor II™ - Cartridge Seal Designed to Replace Packing

This cartridge seal is designed to replace two part component seals and conventional packing arrangements.

This seal also includes the following feature:

- Compact gland for use on applications with limited space

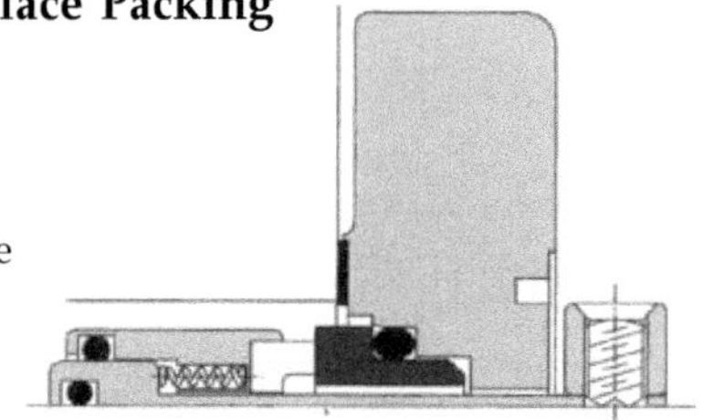

SCUSI™ - Short Cartridge Mechanical Seal

A short externally mounted cartridge seal, with flush and self aligning faces.

This seal also includes the following features:

- Available with flush port as standard for cooling/venting to maximize seal life
- Self aligning stationary face ensures perpendicular alignment of face to shaft axis, maximizing seal life
- Stationary face drive with contracting pins eliminates damage in stop-start applications and viscous fluids
- Flush port to increase seal life in arduous situations

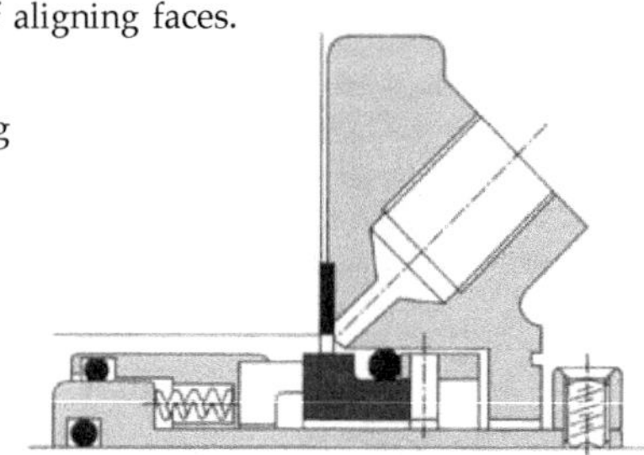

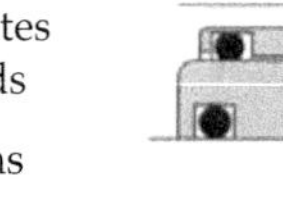

CURC™ - Cartridge Single Seal

The CURC™ is part of a modular range of seals designed to optimize self-aligning technology

This seal also includes the following features:

- Self aligning stationary face ensures perpendicular alignment of face to shaft axis, maximizing seal life
- Stationary face drive with contracting pins eliminates damage in stop-start applications and viscous fluids
- Quench, drain and flush ports for cooling/heating options to maximize seal life
- **Bi-Metal CURC™ option - maintains the features of the standard CURC™ but includes exotic alloy wetted components for use with corrosive chemicals.**

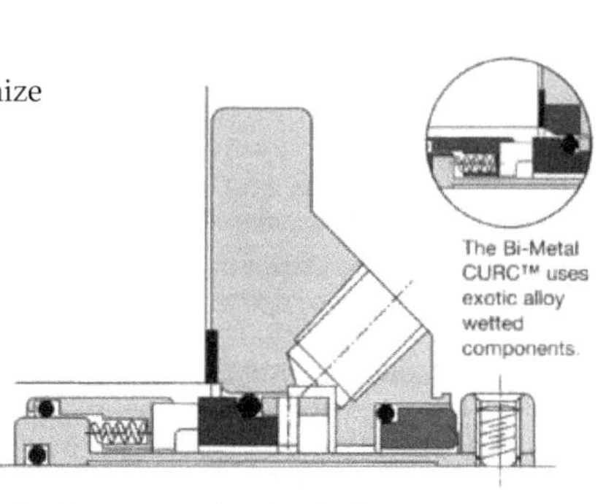

The Bi-Metal CURC™ uses exotic alloy wetted components.

Figure 8-14: Standardized "conventional arrangement" cartridge designs (Source: Ref. 13, AESSEAL, plc, Rotherham, UK, and Rockford, TN, USA)

both the shaft and carbon ring. It contains no close tolerance fits. Originally designed for either inside or low-pressure outside use, this configuration is infrequently represented in modern process plants.

Boot (Elastomer Bellows) Type Seals

Depicted earlier in Figure 8-3 and functionally similar to the elastomeric bellows seal of Figure 8-6, these seals employ a large single spring that maintains face contact. The rubber elastomer boot furnishes the drive—the turning of the unit with the shaft. Pump technicians must use sound installation practices to ensure that the elastomer grips the shaft without allowing slippage.

Bellows Seals

The welded metal bellows design (Figure 8-22) consists of thin convoluted discs that are joined by electron-beam welding on their outer peripheries and inside diameters. Each welded set of discs allows finite amount of axial travel. The more welded discs make up the seal, the greater the ability of the seal to accommodate face wear. This is an important point since certainly not all major mechanical seal manufacturers use 12 sets of discs or convolutions.

The welded sets of discs are usually made of a corrosion-resistant material, such as the Hastelloy® or 300 series stainless steels. There are no sliding elastomers in metal bellows seals. They are typically used at

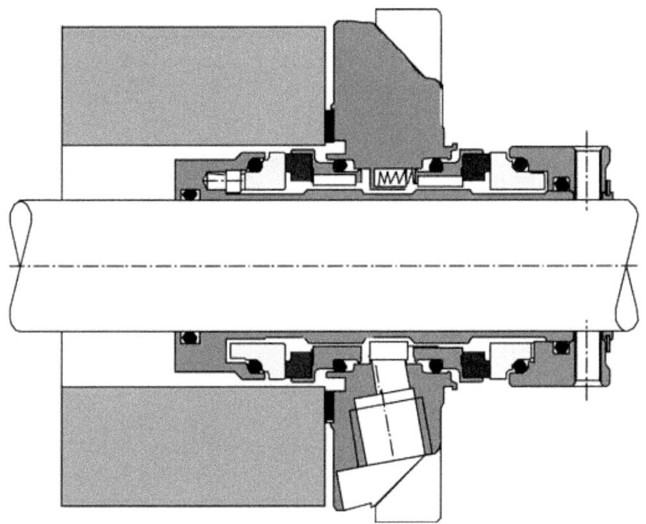

Figure 8-15: Modular cartridge seal (Ref. 8-13)

elevated temperatures when furnished with graphite or metal secondary seals. While bellows seals are balanced by design, the balance diameter will nevertheless change if the disc set axial length is changed for any reason. A good manufacturer will check this out and ascertain suitability for the application.

Re-sealing is as easy as popping a new
Cassette into place. A single pin
lines it up with the
reusable Gland.

Figure 8-17: Cassette-type seals are suitable for many straight-forward pumping services (Source: Ref. 8-5)

Construction Details

1. **Reusable End Plate** holds the inner sealing Cassette. Pop in a new Cassette for fast, easy, low cost re-sealing.
2. **Replaceable Inner Sealing Cassette** holds all the wearing parts in a simple to stock, easy to install package. Provides lower cost re-sealing, competitive with rebuilds, but with the reliability of new components.
3. **Hydraulically Balanced Faces** for low friction, reliable performance. O-ring mounted for better perpendicular alignment and isolation from shaft vibrations.
4. **Flush/Barrier Ports** are tapped into the *reusable End Plate, for optimum control of* the sealing environment. In the Single Seal, a single flush port is provided for cooling and cleaning. In the Double Seal, barrier fluid in and out ports are used to deliver circulating clean fluid to the faces.
5. **Springs** in both seal models are isolated from process fluid and any potential contaminants that could create hang-up.
6. **Single Screw Lock Ring** makes installation fast, easy and secure. Works in conjunction with the internal self-aligning features to *provide precise positioning and concentricity* for reliable performance.

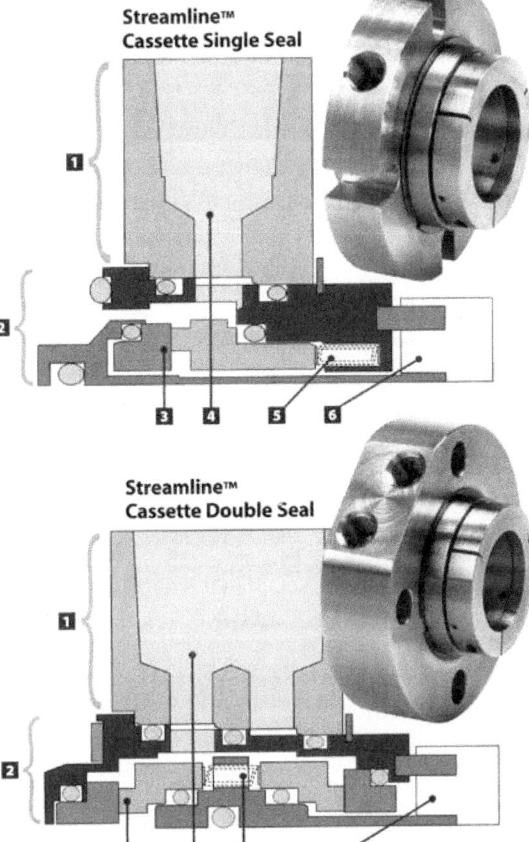

Figure 8-16: Cassette-type mid-range replaceable mechanical seals (Chesterton "Streamline," Ref. 8-5)

Certain construction features of bellows seals resist face opening. As is shown in Figure 8-23, leakage of certain fluids past the faces of pusher seals (upper three sketches) could impede movement of the spring-loaded face. This eventuality is reduced with the metal bellows seals illustrated in the lower two sketches. Note also the nomenclature insert that gives typical terms used for pusher seals and bellows seals.

A modern bellows seal is shown in Figure 8-24. The close-up shows bellows construction with 12 sets of convolutions.

Split Seals

Split seals (Figure 8-25) are used where equipment layout or pump configuration deprive mechanics of easy access. In other words, split seals may be considered in the relatively few instances where rapid installation and replacement

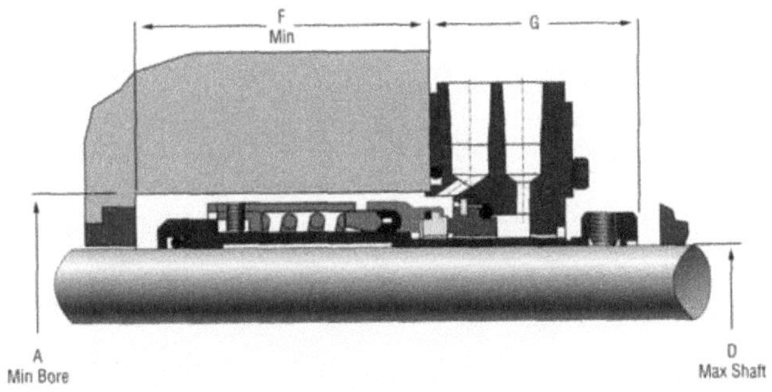

Figure 8-18: "U"-cup mechanical seal (Source: Ref. 8-6)

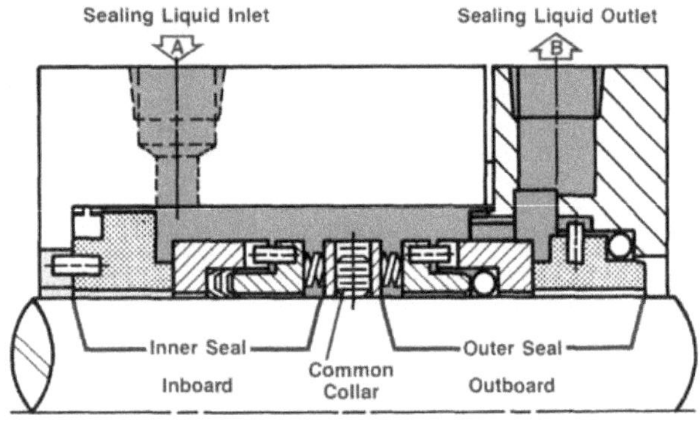

Figure 8-19: Double back-to-back seal with "V"-ring at inner seal-to-shaft contact (Source: Ref. 8-6)

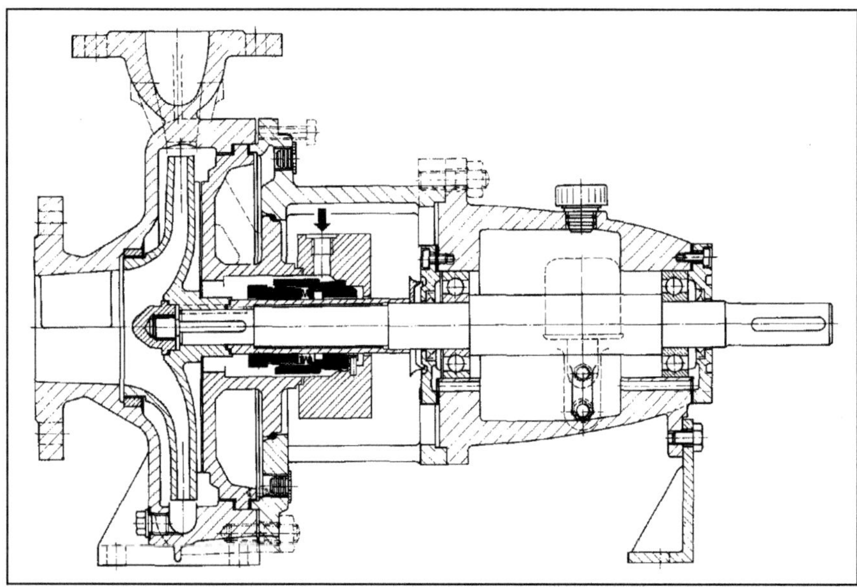

Figure 8-20: ISO-type pump with O-ring seal (Source: Ref. 8-4)

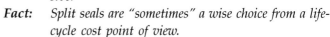

are of paramount importance. They will make neither economic nor life cycle cost sense elsewhere. Although split seals will occasionallly leak more than comparable non-split seals, all mechanical seals must have a small amount of leakage to separate the faces. However, this leakage will usually vaporize and not be visible to the naked eye. Without a few microns of face separation, most mechanical seals would have unacceptably short operating lives.

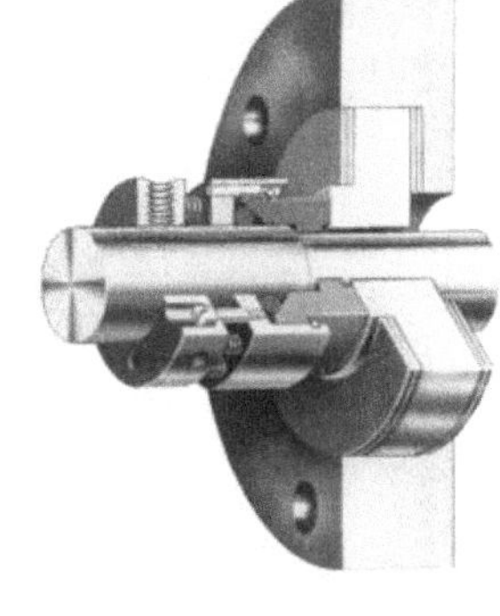

Figure 8-21: "Wedge" seal (Ref. 8-6)

Conventional Wisdom: *Split seals are "always" a wise choice, but expensive.*

Fact: *Split seals are "sometimes" a wise choice from a life-cycle cost point of view.*

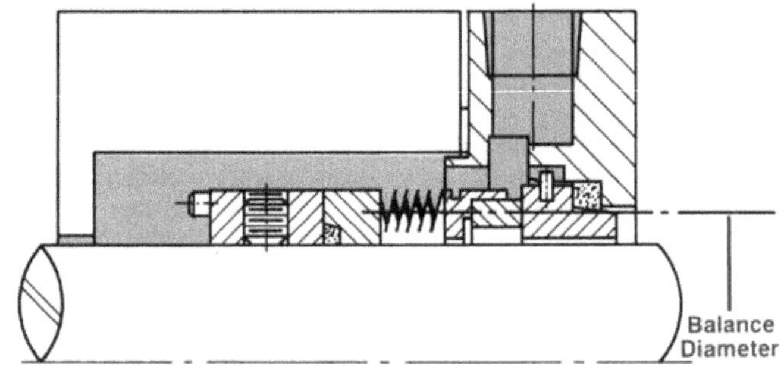

Figure 8-22: Metal bellows seal (Source: Ref. 8-6)

Gas and "Upstream Pumping" Technology*

Gas seals were first applied on compressors but are occasionally preferred for very specific liquid pumping services. Just as is the case with conventional mechanical seals, there are many different configurations to choose from. Likewise, gas seals for pumps are available from several competent seal manufacturers. They are similar in that well-controlled small amounts of a pressurized gas are introduced between the seal faces. The faces of dry gas seals are carefully etched or contoured so as to allow a minute amount of face separation to take place.

Figure 8-26 shows a rather compact design that uses an in-gland control system. This design maintains nitrogen (injection gas) pressure of 20 psig (138 kPa) greater than the opposing process liquid pressure. Fluid sealing takes place at the outside diameter of the seal faces. Upon shutdown, or loss of nitrogen gas, the seal reverts to a liquid-lubricated seal.

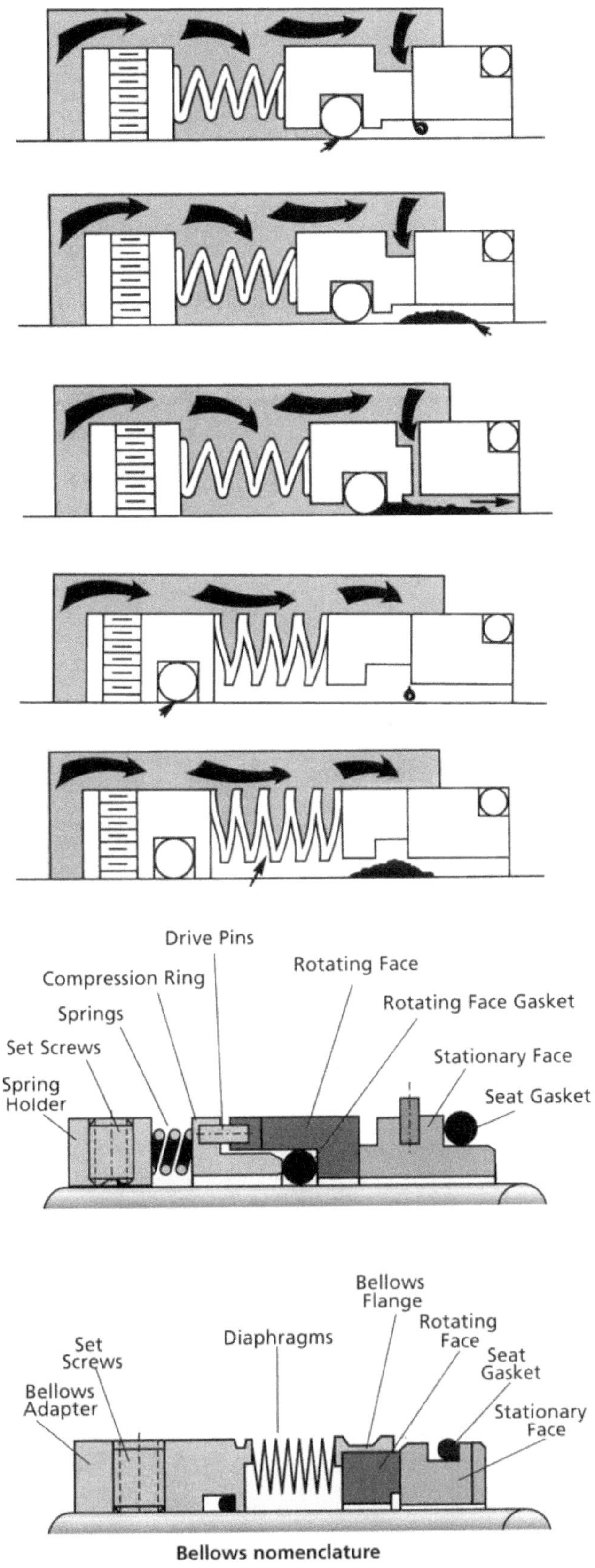

Figure 8-23: Pusher seals vs. bellows seals (Source: Ref. 8-6)

SEAL FACE LUBRICATION ENHANCEMENTS EMPLOYING LASER ETCHING TECHNOLOGY*

Laser-etching of seal faces represents technology available to many competent seal manufacturers. One manufacturer describes its LaserFace™ as "a unique seal interface technology that combines full fluid film lubrication with active leakage control." In its current stage of development LaserFace™ supports application where liquid based process fluids are sealed. Gas applications and applications with abrasive process fluids (i.e. slurries) are not yet supported (Refs. 11 and 12).

The laser-etched face groove is applicable within the customary geometrical interface dimensions of proven conventional seals. Its face pattern combines a pair of symmetrical precision-machined microgrooves (typically a few micrometers deep) each of which performs a different task (Figure 8-27).

The square "inlet-groove" is connected to the pressurized fluid area and through it, the process fluid can penetrate deeply into the sealing interface. The fluid within it is dragged in a tangential direction by the sliding counter face. When the fluid reaches the edge of the inlet groove it generates a strong hydrodynamic pressure that lifts the seal faces, thus stabilizing the sealing gap.

Full fluid films can be achieved at relatively low shaft speeds. This full hydrodynamic fluid face film promotes low friction and low wear performance. At this point, however, the radial leakage flow resistance of the sealing gap is vastly reduced and leakage levels would usually increase dramatically, as the Reynolds equation would suggest. With the laser-etched seal however, the semi-circular grooves, also known as "return-grooves," are designed to re-inject the excessive flow back into the pressurized fluid area. Additionally, the return grooves block radial fluid flow from the outer seal interface diameter. The efficiency of the return pumping mechanism results from the combination of shape, position and depth of the return groove.

The return groove is located within the sealing interface and is not connected to the pressurized fluid area, thus the groove collects the excessive fluid film and guides it along the trailing edges to the discharge end of the groove. The discharge end is very close to the high pressure edge of the sealing gap where the fluid pressure inside the groove is at a maximum and much higher than the sealed pressurized fluid (Figure 8-28).

*Courtesy of Neil Wallace and John Crane, Slough, UK

Materials of Construction

- Bellows: Alloy 718
- Rotating Face: Silicon Carbide
- Stationary Face: Carbon or Silicon Carbide
- Gasketing: Graphite
- Metal Parts: Rotating Assembly: 316 SS Stationary Assembly: Alloy 718, Low Expansion Alloy, 316 SS

Operating Parameters

- Maximum Pressure: Up to 300 psi (2070 kPa)
- Seal Chamber Temperature: -100°F to 800°F (-73°C to 427°C)
- Surface Speed: Up to 150 ft./sec. (46 m/sec.)
- Liquids: Hydrocarbons, Heat Transfer Fluids, Cryogenics
  Shaft Sizes: From 1 inch (25.4mm) to 5-1/8 inches (101.6mm)

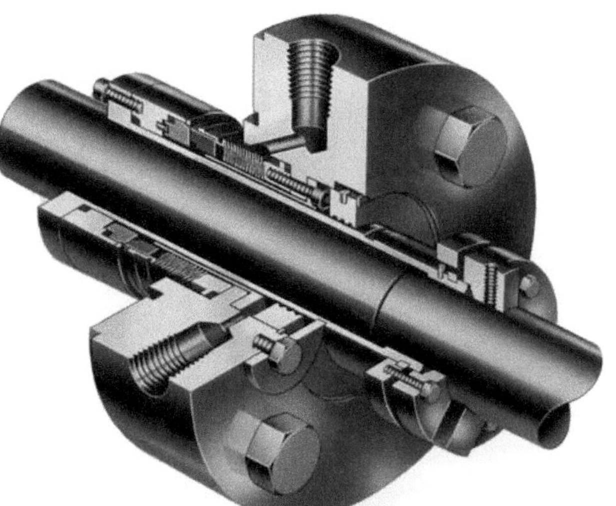

Figure 8-24: Modern bellows seal and close-up (Source: Ref. 8-6)

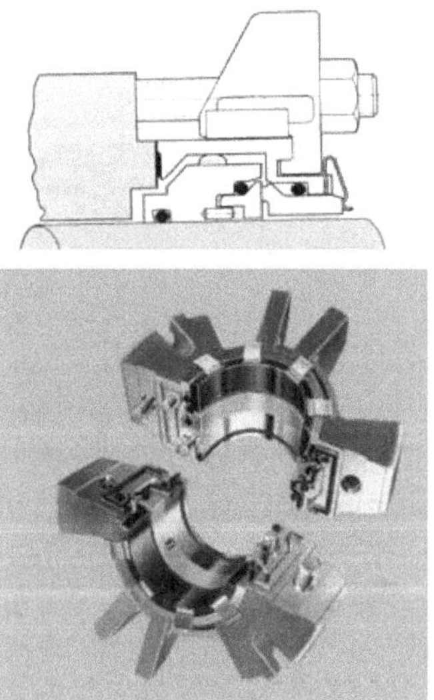

Figure 8-25: Split mechanical seal (Source: Ref: 8-5)

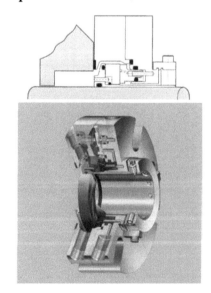

Figure 8-26: "Dry gas" seal (Source: Ref. 8-5)

Following the path of least resistance, a great deal of the fluid passes from the recess towards the pressurized fluid outside the sealing gap.

Through the combination of hydrodynamic grooves and fluid film re-circulation grooves, laser-etched faces combine the benefits of hydrodynamic lubrication and boundary lubrication. In other words, low friction and wear, but together with low static and dynamic leakages. Figure 8-29 supports this statement. In Figure 8-29, the two important performance measures for a mechanical seal, leakage and face friction, are compared for:

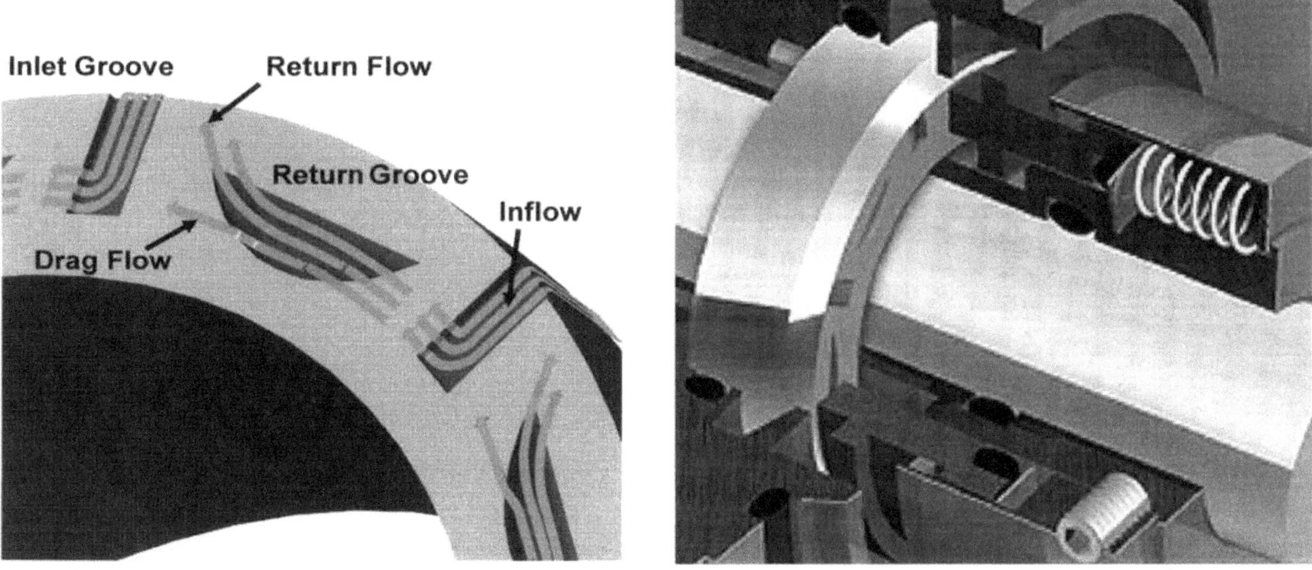

Figure 8-27: Seal Ring with (left) LaserFace™ groove, and mechanical seal with (right) LaserFace™ seal ring (Source: Ref. 8-16)

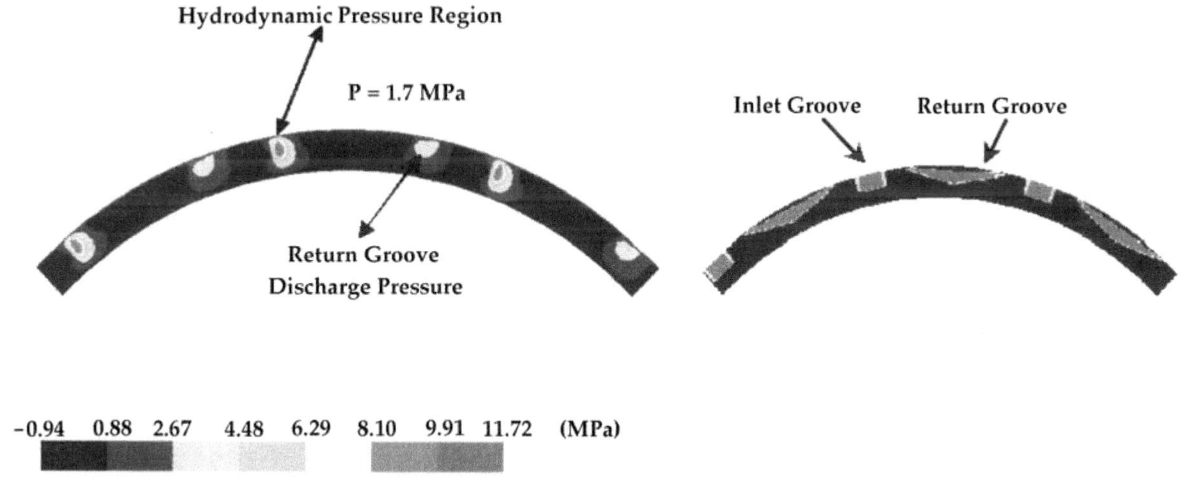

Figure 8-28: Typical interface pressure distribution of a LaserFace™ seal (computer model prediction)

- a standard plain face seal.
- the same seal with a conventional hydrodynamic face groove (inlet groove)
- a seal equipped with the laser-etched return groove.

The bar chart on the left illustrates the comparison using good lubricating fluids, the one on the right for a more difficult duty with flashing, poor lubricating liquids. The performance figures are presented in relative values where 100% is always the highest recorded value from the three seal configurations.

The standard seal operates for both process liquids with very low leakage but generates an appreciable amount of face heat due to asperity contact. When the seal is equipped with a simple hydrodynamic feature the operational gap is increased and asperity contact is prevented. The amount of face friction is a direct result of viscous shear in the seal fluid. This explains the much larger reduction in friction for the propane duty. However, the results also demonstrate that hydrodynamic face lubrication causes a significant increase in seal leakage. The laser-etched face seal on the other hand achieves full hydrodynamic interface lubrication at leakage levels as low as the plain face seal in the case of both liquids.

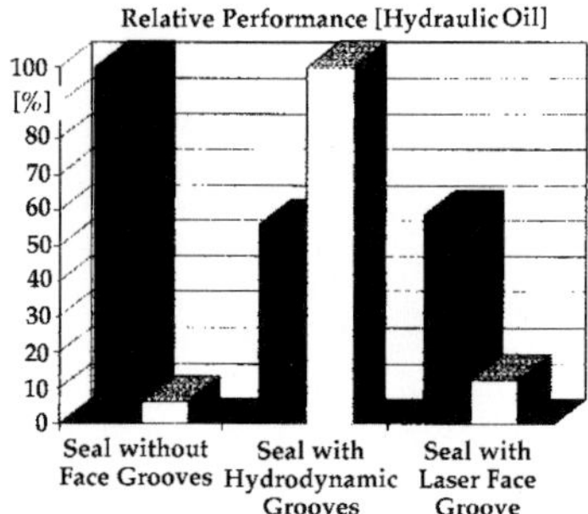

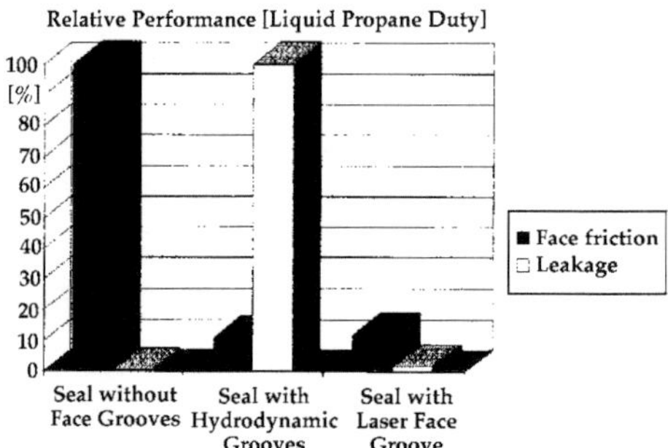

Figure 8-29: Mechanical seal performance comparison with conventional seal face technology and laser-etched faces (LaserFace™)

Case Study Involving Face Geometry

The benefits of laser-etching technology are best demonstrated in a comparison test of two identical seals, except that one is equipped with a LaserFace™ mating ring and the other with the conventional plain face mating ring. The test duty is liquid propane at 21 bar at a pumping temperature of 56°C and an API flush plan 11 (see Figures 8-34 through 8-46 for an overview of flush plans) set at 10 l/min flush flow. The flush temperature margin is only 6 degrees Kelvin to the point where vaporization occurs. API 682 would recommend a reduction in flush temperature of 14 degrees Kelvin to 42°C. This test duty was deliberately chosen as a more "praxis-near" example of a refinery duty where temperature margins at suction pressure are minimal and the option of cooling or increasing the box pressure is limited. Figure 8-30 shows the recorded pressure but, more interestingly, the temperatures recorded throughout the 200-hour test duration. The performance plot also indicates the vapor temperature of 62°C. The top graph shows the performance of the seal with laser-etched face and the bottom graph shows the performance without.

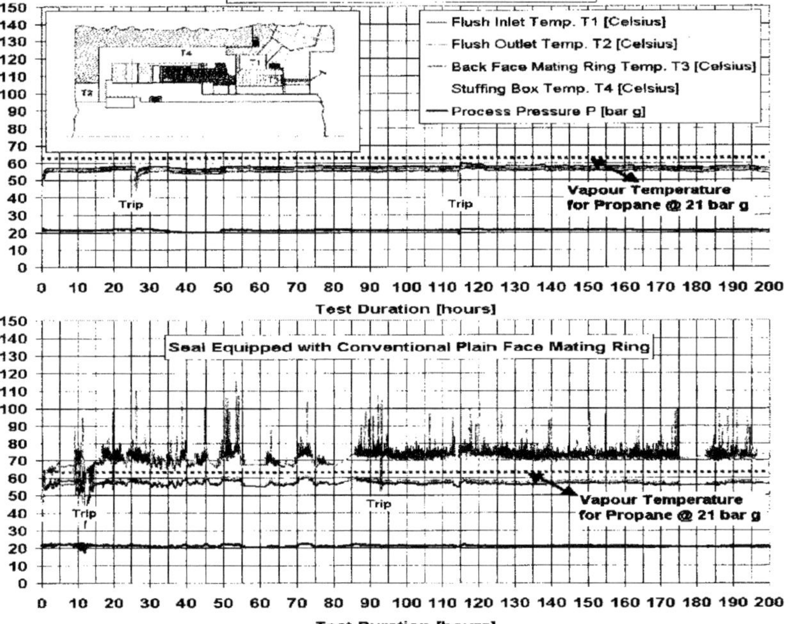

Figure 8-30: Test plot comparing a laser-etched (LaserFace™) vs. conventional mechannical seal

Taking leakage recordings throughout the test confirmed emission levels below 200 ppm for both seals.

The test readings clearly show the benefit of laser-etched seal faces. Although both seals show only a very small temperature rise between flush inlet T1 and outlet temperature T2, which would suggest good face lubrication and low friction performance, the actual performance of the seals is indeed significantly different. Measuring the temperature rise of the mating ring back-face with temperature probe T3 reveals a completely different picture of the true seal behavior. The conventional seal shows very high and erratic temperatures. The mean face temperature was above the vapor temperature throughout the test duration suggesting that the seal was subjected to dry running (operating in the boundary lubrication regime). In addition, the occasional very high temperature spikes indicate that the excessive component temperatures caused instant and complete flashing of the liquid around the seal, covering the components in produced vapor. At this point the heat transfer was dramatically reduced causing overheating of the mating ring until such time that the flush liquid could displace the "vapor blanket" that had been produced.

The laser-etched seal faces, however, show a completely different picture. The mating ring temperature T3 is constantly very low, almost identical to the flush inlet temperature. This suggests very low friction performance with no or negligible asperity contact. As a matter of fact, the heat generated by the laser-etched seal was so small that the cooling effect from the air around the ID of the mating ring reduced the back-face temperature by 0.5°C to the flush inlet temperature T1.

Analyzing the two seals with a proprietary computer program confirmed the empirical results. The program predicts a face temperature of 80°C (18°C above the vapor temperature) for the conventional seal with a back-face temperature of 71°C (mean T3 recorded at 72°C) assuming that liquid propane is present around the seal components. The computer analysis of the temporary 'flashing-off' condition also confirmed the recorded peak face temperatures of 130°C. The computer analysis confirmed the test readings and predicted that the seal would operate with full hydrodynamic interface lubrication. The average face temperature rise was estimated as only 0.2°C as a result of the low liquid shear from the interface fluid film.

Upstream Pumping (USP)*

In those applications where it is undesirable for the pumped fluid to act as the lubricating medium between the seal faces, the traditional approach has been to use a double seal arrangement supplied with a pressurized barrier fluid. The barrier fluid is pressurized from an external source to a pressure greater than seal chamber pressure; in this way a fluid film of the clean barrier fluid forms between the seal faces. This "contacting" seal technology is well established and its benefits are widely known. However, it is an expensive solution, often requiring a complex and costly support system, which itself requires maintenance.

A more recent technology to be developed is "active lift technology" or "upstream pumping." This active lift principle uses spiral grooving on the seal faces to produce the same result as the pressurized double seal, without the need for a complex pressurized seal support system.

Originally used in the development of "non-contacting" dry running gas seals for compressors, the spiral grooves form the inner portion of one of the seal faces. With rotation, the spiral grooves take the un-pressurized barrier fluid, and generate a pressure at the exit of the spirals. This pressure is greater than the pressure in the seal chamber. It is this pressure which forms the "active lift." The outer portion of the seal face, between the spiral exit and the process fluid, is a lapped region of the seal face known as the sealing dam. The pressure differential across this sealing dam determines the amount of fluid flow or "upstream pumping" which takes place from the barrier side to the process side. For a given design, as the process pressure increases, the rate of 'upstream pumping' reduces, whilst maintaining a "non-contacting", sealing gap between the seal faces.

Figure 8-31 shows, in simplified form, how USP technology varies from conventional dual seal arrangements.

Here the pressure distribution of the USP seal can be seen. The barrier pressure is a few psi above atmospheric pressure, but significantly below that of the process. (In a dual, pressurized "double" seal the barrier pressure is higher than the process pressure while in a dual non-pressurized "tandem" seal the barrier is normally at atmospheric pressure and depicted as the buffer fluid.)

The pumping action is achieved with the use of very shallow grooves on the hard face element of the USP seal. These generate significant pressure within the grooves, as shown in Figure 8-32.

Figure 8-33 indicates how the pressure is generated through the groove profile, from a groove inner radius of 54 mm the pressure increases from zero to approximately 55 bar at a radius of about 60 mm, which is the groove root. At this point the pressure decays back

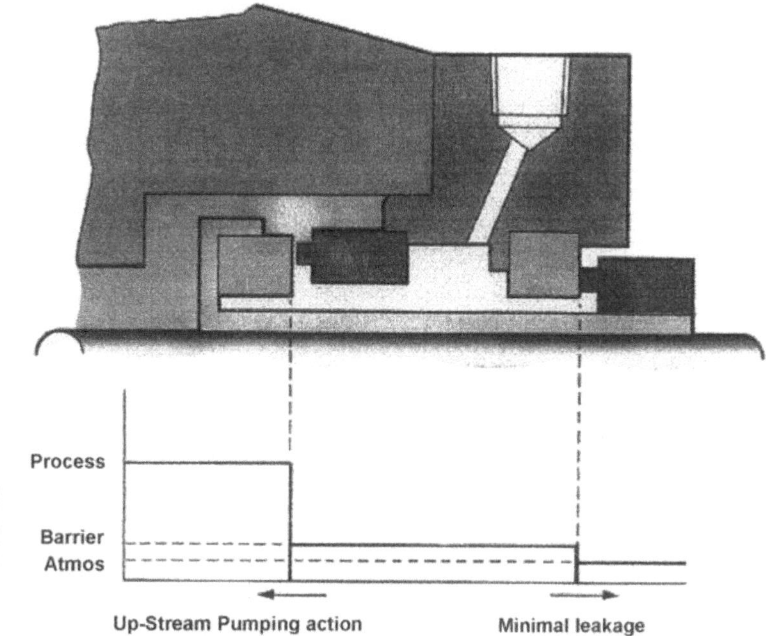

Figure 8-31: USP seal barrier pressure (Source: Ref. 8-16)

*Courtesy of Neil Wallace and John Crane, Slough, UK

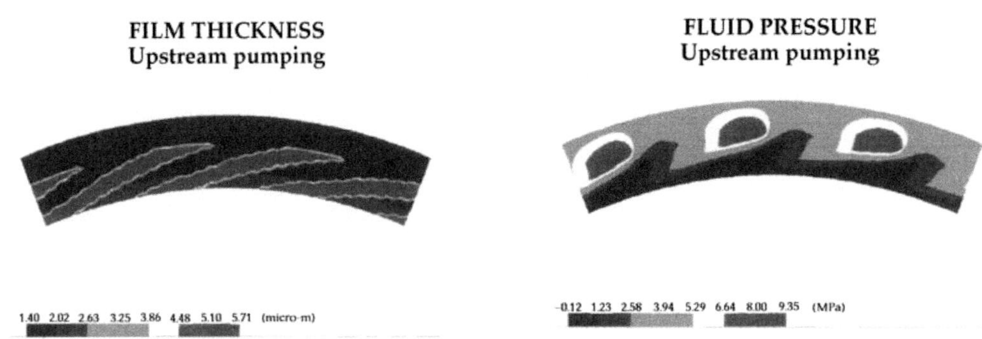

Figure 8-32: USP seal face pressure distribution (Source: Ref. 8-16)

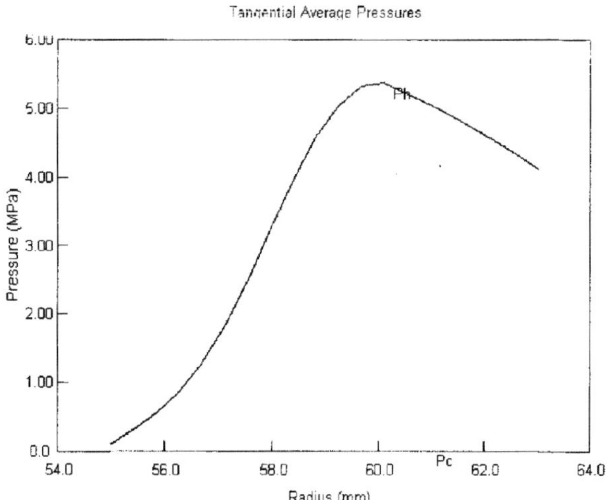

Figure 8-33: Pressure distribution in face grooves (Source: John Crane Company, Ref. 8-16)

across the un-grooved sealing dam to the process pressure at the outside of the seal, the 63 mm radius, which is about 40 bar.

This "active lift" technology therefore presents several advantages over the traditional double seal approach:

- The USP concept is "non-contacting" and therefore the usual PV limitations imposed by contacting seals and the resultant wear do not apply.
- The power consumed is significantly lower than a double or tandem seal arrangement.
- The positive flow of clean fluid into the seal chamber provides a cleaner sealing environment within the seal chamber.
- Seal leakage to atmosphere is significantly reduced when compared to a pressurized dual seal—where the outboard seal can often operate at considerable pressure.
- The USP concept allows a simple upgrade of single or multiple seal services, *where process changes have rendered the process fluid a poor seal lubricant.*
- In services where the process pressure is variable, or where pressure spikes are likely, USP technology constantly regulates against this varying pressure, maintaining a sealing gap at all times.

SEAL FLUSH AND ENVIRONMENTAL CONTROL PLANS

Mating mechanical seal faces must be separated from each other by a small, fluid-filled gap. Yet, seal leakage to the atmosphere must be minimized. For seals to perform as intended, fluid temperature, pressure, and overall properties must be right.

Seals will malfunction if:

- Low temperatures cause elastomers to stiffen
- High temperatures cause elastomers to soften and lose shape
- Temperature gradients cause thermal distortion of faces
- High temperatures cause pitting or heat checking of seal faces
- Face temperatures reach the point where fluid vaporizes
- Fluid pressures cause face distortion
- Fluid pressure forces faces into intimate contact
- Fluid pressure-temperature relationships reach the vaporization point
- A vacuum environment can cause "gassing out"—shrinkage, or even explosive disintegration—of certain elastomers

- Impurities, solids, abrasives or fibers get lodged in the seal faces
- Abrasive pumpage causes erosive wear
- Pumpage solidifies or crystallizes after migrating through seal faces
- Pumpage is of a kind that can plate out on the seal faces.

Upon leaking product through seal faces, some services in the pulp & paper and related industries may well encounter a combination of the above woes. Needless to say, toxic, flammable, corrosive and explosive fluids require environmental controls and auxiliary support systems that eliminate or at least reduce the significance of leakage events.

A number of flush plans have been catalogued by API and ISO Standards organizations. Some of these will utilize a barrier fluid system, in which case the selection of an appropriate or optimum fluid will be important. Keep in mind that the primary function of a barrier fluid is to provide cooling to the seal faces. Keep in mind also that water is a better coolant than oil, but water may need freeze protection and rust inhibitors. Certain synthetic lubricants are recommended for low and high temperature applications and are available from the Royal Purple Company Product Guide, "High Performance Lubricants." Knowledgeable formulators have thus produced pure, non-reactive synthetic fluids that provide superior protection for double and tandem mechanical seals.

An overview of the more frequently used flush plans is given in Figure 8-34 while application guidelines of several of these are highlighted and illustrated in Figures 8-35 to 8-59, courtesy of AESSEAL plc. Finally, Figure 8-53 illustrates inefficient uses of water. Steam is often applied to flush away leakage products and to thus preclude their accumulation in locations that could impede proper functioning of the seal (Ref. 13). See page 232 for smart water management.

RELIABILITY IMPROVEMENT IN MECHANICAL SEALS — ANALYSIS OF DESIGN CONSIDERATIONS

Very general application guidelines for inside seals are documented in Table 8-1. It should be noted that certain successful applications may well fall outside the limits given in this table. The manufacturer's experience should govern in those instances. However, an overview will be of benefit. If nothing else, it will illustrate the virtual impossibility of using a "one-type suits all" philosophy in many plants.

Table 8-1: Mechanical Seal Application Guide

PARAMETER	*UNBALANCED*	*BALANCED*
Below 400 deg. F (205°C)	Yes	Yes
Above 400 deg. F (205°C)		Yes
Below 50 psig (345 kPa)	Yes	Yes
Above 50 psig (345 kPa)		Yes
Below 3000 fpm (15.2 m/s)	Yes	Yes
Above 3000 fpm (15.2 m/s)		Yes

This overview is thus intended to show the seal user how mechanical shaft seal devices intended for high reliability will often differ from seals that were designed without reliability as the foremost consideration. It highlights such important topics as flow distribution enhancement in the seal face region, the merits of narrow face configurations, pumping screws and their influence on the simplification of buffer system, and other related topics.

Guidelines are included which enable the user to specify meaningful acceptance tests for mechanical seal systems. These guidelines will allow users to gauge the relevance of unusual material selections, seal geometries, face material impregnations and/or laser-textured surface treatments.

Seal Statistics Recap

While a more detailed treatment of seal statistics will be found in Chapter 16 of this text, the reader should keep in mind that more than one manufacturer is able to design and produce dependable mechanical seals. Indeed, most mechanical seals installed in modern rotating equipment are rightly considered to be reliable, low maintenance components. In all instances, however, seal life will be curtailed if the rest of the pump installation suffers from one or more of the pump or systems-related shortcomings that are the subject of this book. Two facts are nevertheless casting aspersions on the reputation of mechanical seals: Studies have shown that out of 100 pump failures in a typical petrochemical plant as many as 70 repair events are attributable to seal distress (Ref. 8-8), and mechanical seal-related failures can account for 50 cents of every maintenance dollar spent on rotating

Circulation, Flushing, Quenching

Please see VDMA 24297

Circulation

It is generally recommended that, for single acting mechanical seals a circulation pipe is laid between the pump discharge and the seal housing. A cross-section of G¼ is normally adequate.

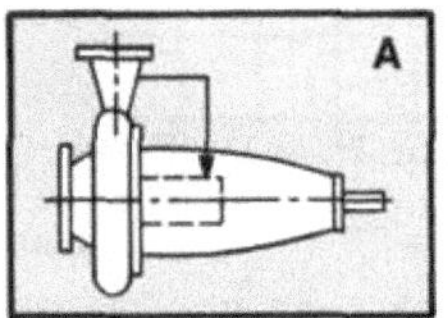

from pump discharge

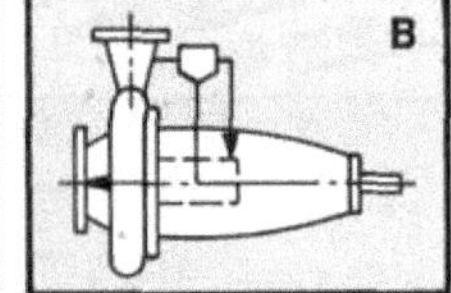

with cyclone separator

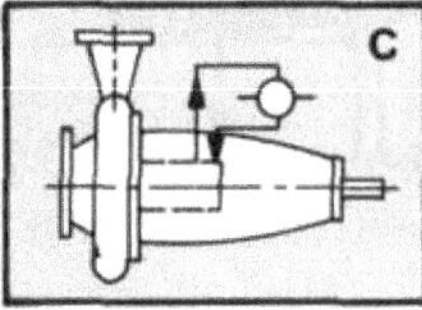

by pumping screw via cooler

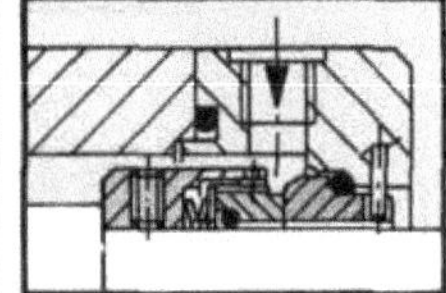

seal arrangement for A and B

Flushing

If, for heavily contaminated media, a double acting mechanical seal cannot be used, then a clean fluid is injected into the region around the sliding surfaces of a single acting mechanical seal.

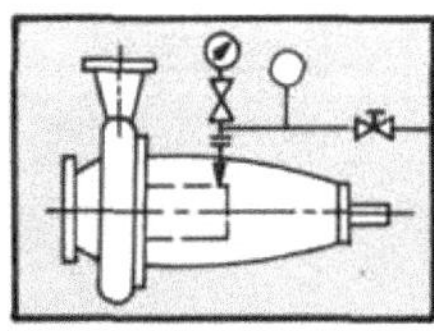

external connection

seal arrangement

Quenching

Quenching is recommended in the low temperature range (when danger of icing exists) or for media which tend to form deposits in contact with the atmosphere. Any leakage, occurring, is taken up by the quenching fluid and in the vacuum region dry running is avoided. The quenching pressure should not exceed 1 bar with conventional mounting.

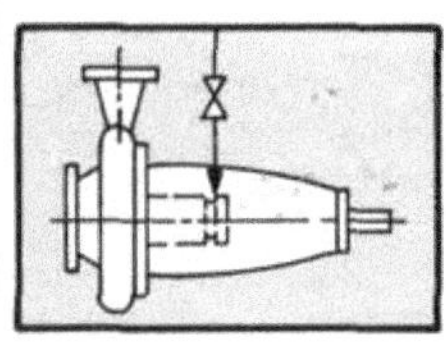

pressureless quench fluid

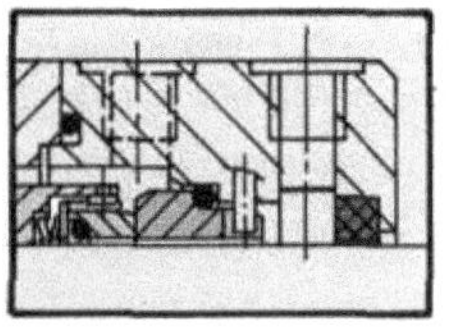

seal arrangement

Circulation Systems to API 610

Legend:

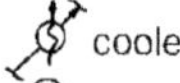 cooler

- (PI) pressure gage with block valve
- (TI) Dial thermometer
- (PS) Pressure switch, when specified including block valve
- cyclone separator
- (FI) flow indicator
- Y-type strainer
- Flow-regulating valve
- Block valve
- Check valve
- Orifice
- Sight flow indicator, when specified
- Valve A: inlet shutoff valve

  Valve B: Branch flow control valve

  Valve C: Outlet shutoff valve (optional)

Clean pumpage

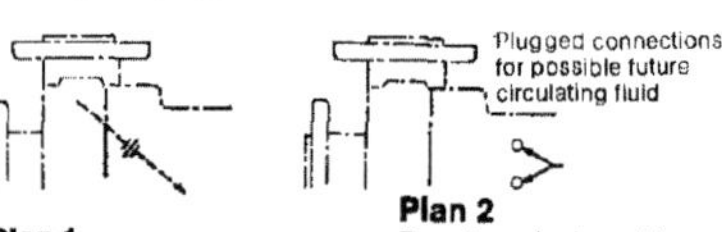

Plan 1
Integral (internal) recirculation from pump discharge to seal.

Plan 2
Dead-ended seal box *with no circulation of* flush fluid. Water-cooled box jacket and throat bushing required, unless otherwise specified.

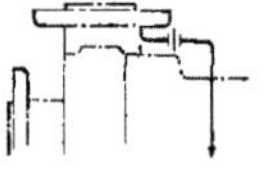

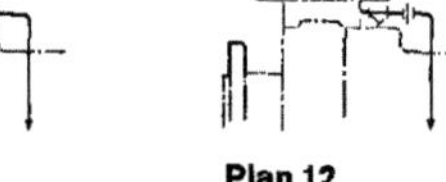

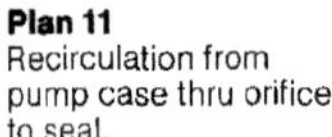

Plan 11
Recirculation from pump case thru orifice to seal.

Plan 12
Recirculation from pump case thru strainer and orifice to seal.

Plan 13
Recirculation from seal chamber thru orifice and back to pump suction.

Plan 21
Recirculation from pump case thru orifice and cooler to seal.

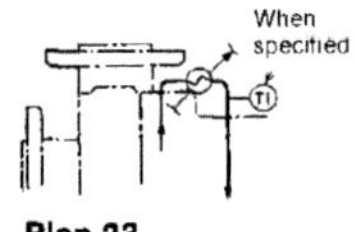

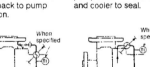

Plan 22
Recirculation from pump case thru strainer, orifice and cooler to seal.

Plan 23
Recirculation from seal with pumping ring thru cooler and back to seal.

Dirty or special pumpage

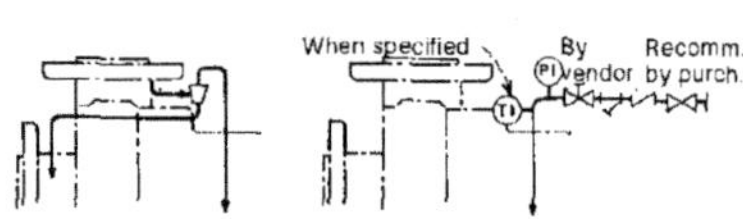

Plan 31
Recirculation from pump case thru cyclone separator.

Plan 32
Injection to seal from external source of clean cool fluid.

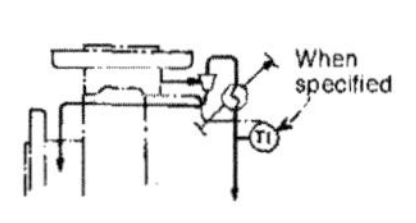

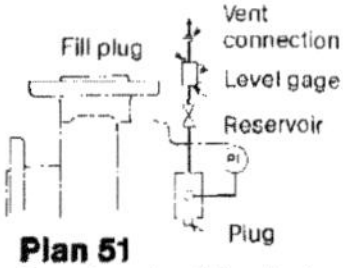

Plan 41
Recirculation from pump case thru cyclone and cooler to seal.

Plan 51
Dead ended blanket (usually methanol)

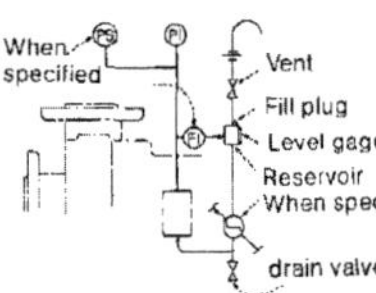

Plan 52
External fluid reservoir non pressurized; thermosyphon or forced circulation, as required.

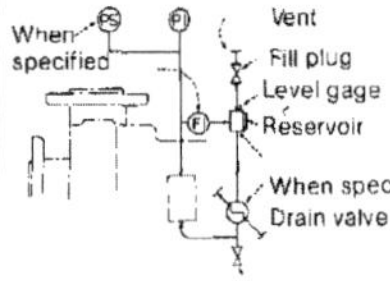

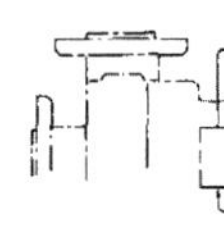

Plan 53
External fluid reservoir pressurized; thermosyphon or forced circulation, as required.

Plan 54
Circulation of clean fluid from an external system.

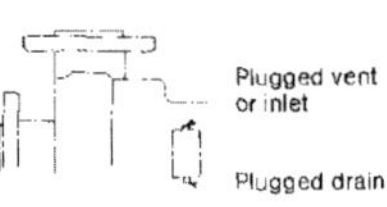

Plan 61
Tapped connections for purchaser's use.

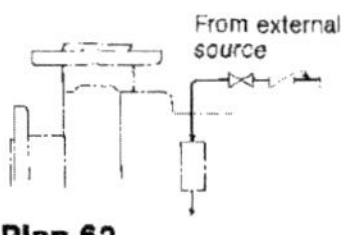

Plan 62
External fluid quench (steam, gas, water, other).

Figure 8-34: Overview of different seal flush plans (Source: Ref. 8-2)

PLAN 01 Description

Integrated (internal) product recirculation from pump discharge to seal chamber.

Features

1. Minimizes risk of freezing/polymerizing of fluid in flush piping plans exposed to atmosphere.
2. Removes heat from the seal chamber as well as acting as a vent connection in horizontal pumps.

Use

1. Recommended in clean fluids.
2. Recommended for fluids which thicken at ambient temperatures.

Caution

1. Ensure that the recirculation is sufficient for seal heat removal.

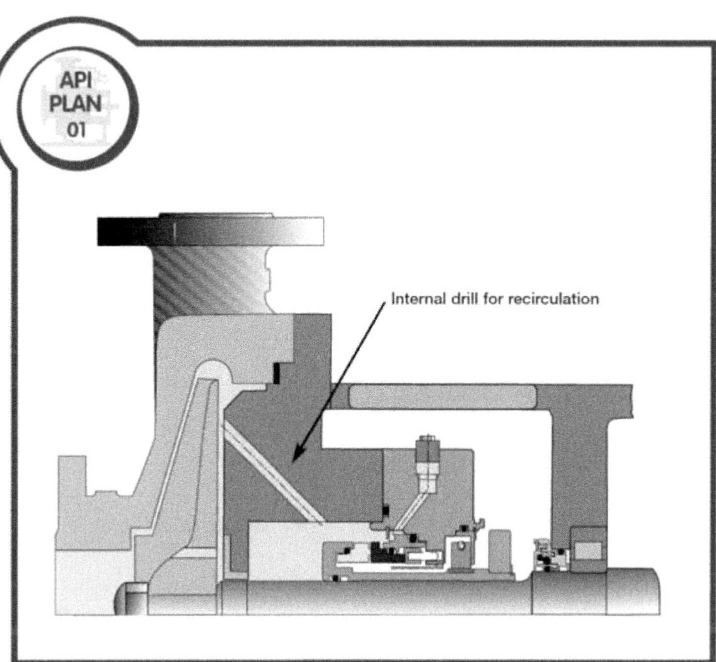

Figure 8-35: API Plan 01 shown with modern bearing protector seal (Source: Ref. 8-17)

PLAN 02 Description

Dead-ended seal chamber with no flush fluid circulation.

Features

1. Applicable to low seal chamber pressure and process temperature.
2. Can be used with tapered seal chambers, especially for slurries.
3. Normally is used along with a jacketed seal chamber.

Use

1. In cool clean fluids with high specific heat, such as water, in relatively low speed pumps.

Caution

1. To avoid flashing, process fluid temperature must be taken into consideration.
2. Avoid use without cooling/heating jacket (for cylindrical chambers).
3. Ensure top point vent in throat bushing (for cylindrical chambers in horizontal pumps).

Figure 8-36: API Plan 02 (Source: Ref. 8-17)

equipment in a typical refinery. Another study showed that the average seal-induced pump failure in the petrochemical industry results in repair costs ranging from $5,600 in a plant with an average size of 29 hp per pump (400 pumps) to $10,287 in a plant with an average size of 73 hp per pump (2,754 pumps). Although these costs include burden and overhead, they represent staggering figures (Ref. 12).

Experience shows that many of these failures could have been prevented if the selection process for the seal had been approached with more attention. Optimum seal selection practices rely to a great extent on the experience of qualified seal vendors. Since most applications of mechanical seals require seal systems, not just seals (Ref. 8-8), it is important that the full scope of operating data, specific machinery data and related information be

PLAN 11 Description

Product recirculation from pump discharge to seal through a flow control orifice.

Features

1. Prevents product from vaporizing by maintaining positive pressure above vapor pressure.
2. Becomes a self-venting plan for horizontal pumps.
3. Represents the default API plan for most single seals.

Use

1. In general, applications with clean non-polymerizing fluids with moderate temperatures.

Caution

1. Calculation of recirculation flow rate, heat removal and orifice size are required.
2. Orifice size should be at least 1/8" (3.2 mm).
3. Check the margin between discharge pressure and seal chamber pressure to ensure proper flow of fluid.
4. Do not use with media containing solids and abrasives.

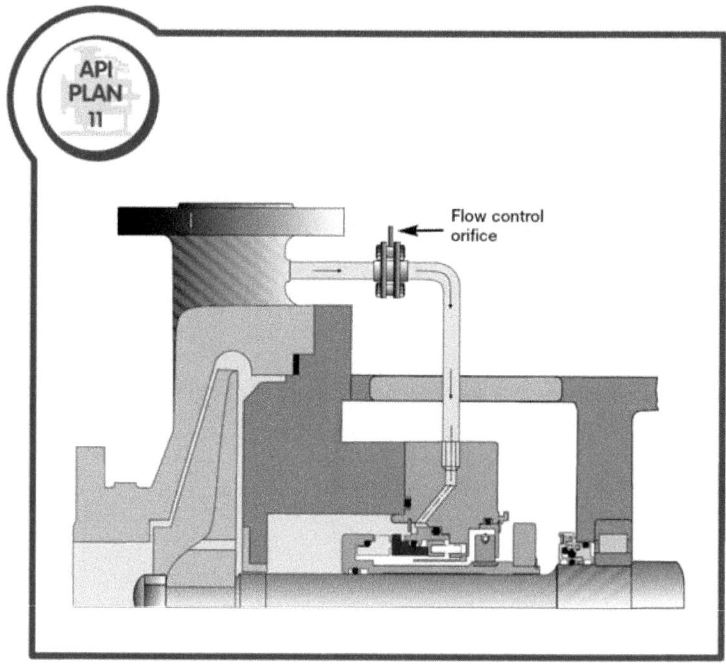

Figure 8-37: API Plan 11 (Source: Ref. 8-17)

PLAN 12 Description

Product recirculation from pump discharge through a small Y-strainer and a flow control orifice to seal chamber.

Features

1. Becomes a self-venting plan for horizontal pumps.
2. Can handle dirty liquids to some extent.

Use

1. Generally used in slightly dirty and non-polymerizing fluids.

Caution

1. Always ensure that orifice is placed behind the Y-strainer.
2. This plan is normally discouraged due to the relative unreliability of Y-strainer.
3. Calculation of recirculation flow often advisable.

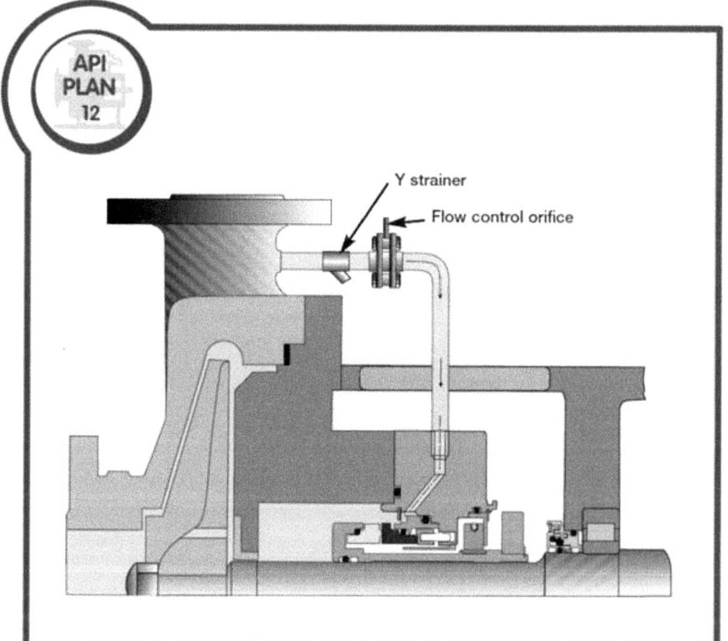

Figure 8-38: API Plan 12 (Source: Ref. 8-17)

made available to the seal manufacturer.

Equally important is the vendor-user interaction, that is to say the user must show more than superficial interest in product selection. Specifically, the seal user should be thoroughly familiar with the advantages and disadvantages of certain design features and should engage in a detailed comparison of physical configuration and material property of alternatives presented by the various bidders.

These then, are the key objectives of this overview:

- It should convince the user to take a far more active part in seal selection for vulnerable sealing services or applications where high reliability or fewer failures are to be achieved.

PLAN 13 Description

Product recirculation from seal chamber to pump suction via a flow control orifice.

Features

1. Provides continuous vent for vertical pumps.

Use

1. Wherever Plan 11 is not usable due to low-pressure margin between discharge and seal chamber pressure.
2. Used in vertical pumps.

Caution

1. Check margin between seal chamber pressure and suction pressure.
2. Orifice size should be at least 1/8″ (3.2 mm).

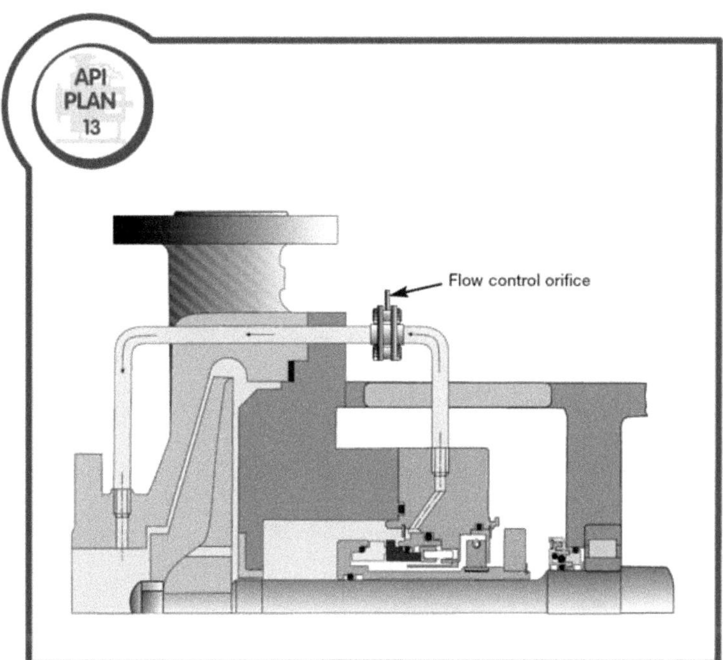

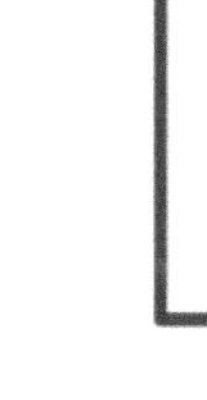

Figure 8-39: API Plan 13 (Source: Ref. 8-17)

PLAN 14 Description

Product recirculation from pump discharge to seal chamber through a flow control orifice and seal chamber back to suction through another flow control orifice.

Features

1. Ensures product recirculation as well as venting.
2. Reduces seal chamber pressure.

Use

1. Used in vertical pumps.
2. Used in light hydrocarbon services.

Caution

Check for pressure margin between discharge to seal chamber pressure and seal chamber to suction pressure.

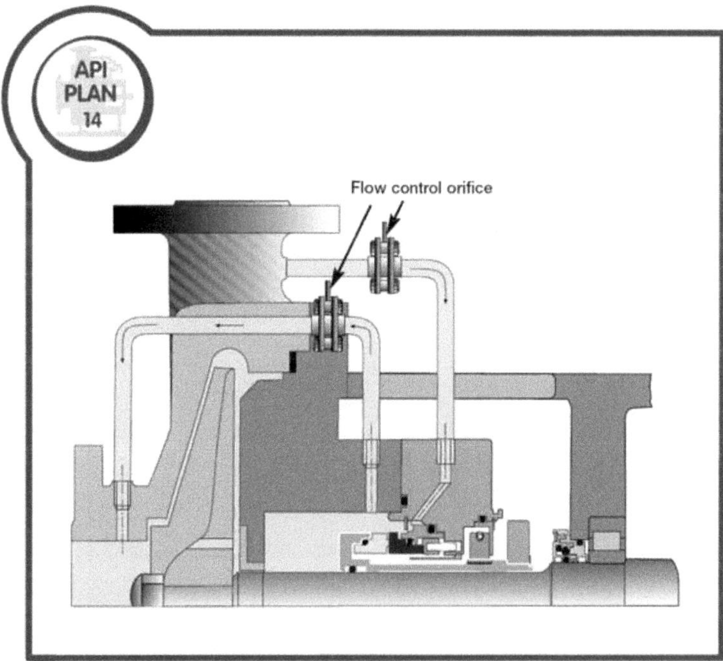

Figure 8-40: API Plan 14 (Source: Ref. 8-17)

- The user should become more familiar with the design features and concepts which distinguish seals of high potential reliability from seals with low potential reliability in critically important or difficult sealing applications.

- It should explain to the user why reliability-oriented seal manufacturers have elected to incorporate certain features in their design.

Once these objectives have been reached, the user should feel motivated to solicit detailed proposals for mechanical seals from several capable manufacturers. The vendor should be required to submit dimensionally accurate layout drawings for the proposed product, and the user should engage in a feature-by-feature comparison of the competing offers. Keeping in mind the high cost of seal-induced pump failures, the user would then assign a monetary value to the probable failure avoidance and ask whether or not the added cost for a superior seal is justified.

PLAN 21 Description

Product recirculation from discharge through flow control orifice and heat exchanger to seal chamber.

Features

1. Improves pressure margin over vapor pressure.
2. Improves temperature margin to meet secondary sealing element limits, to reduce coking or polymerizing and to improve lubricity.
3. Self venting plan.
4. Provides sufficient pressure difference to allow proper flow rate.

Use

1. For high temperature applications e.g. hot water application (temperature > 80°C), hot hydrocarbons etc.
2. In hot non-polymerizing fluids.

Caution

1. Always ensure that the cooler is placed downstream of the orifice.
2. Check pressure difference between discharge and seal chamber.
3. Cooler duty is high, leading to fouling on water side.
4. Potential plugging on process side if fluid viscosity increases rapidly.

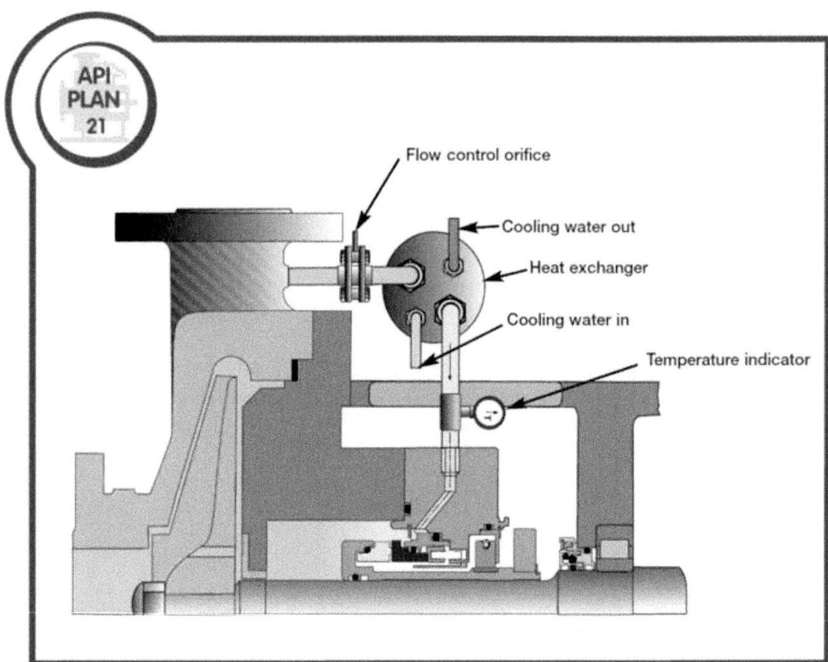

Figure 8-41: API Plan 21 (Source: Ref. 8-17)

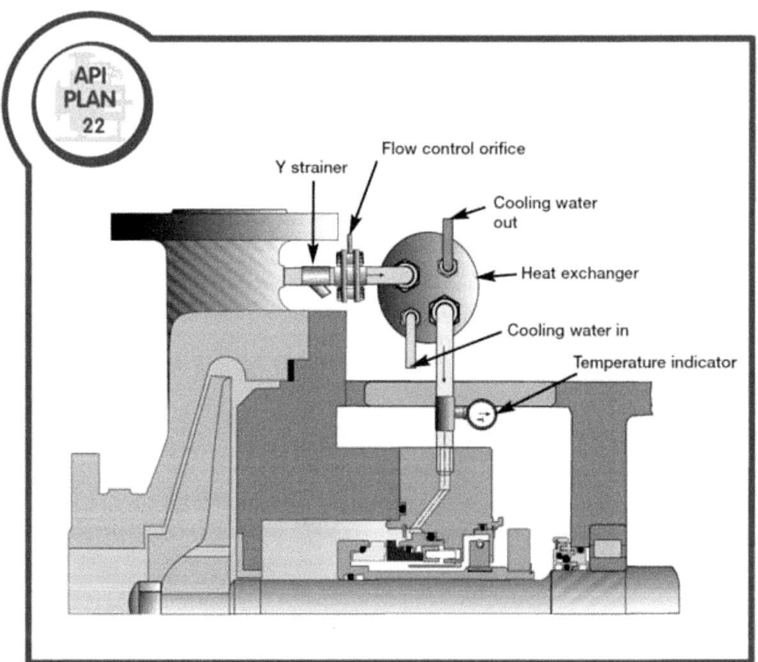

Figure 8-42: API Plan 22 (Source: Ref. 8-17)

PLAN 22 Description

Product recirculation from pump discharge through a Y-strainer, a flow control orifice and a heat exchanger to seal chamber.

Features

1. Improves pressure margin over vapor pressure.
2. Improves temperature margin to meet secondary sealing element limits, to reduce coking or polymerizing and to improve lubricity.
3. Self-venting plan.
4. Provides sufficient pressure difference to allow proper flow rate.

Use

1. For high temperature applications with slightly dirty liquid.

Caution

1. Always ensure that the cooler is placed downstream of the orifice.
2. Check pressure difference between discharge and seal chamber.
3. Cooler duty is high, leading to fouling on water side.
4. This plan is normally discouraged due to relative unreliability of Y-strainer.

PLAN 23 Description

Product recirculation from seal chamber to heat exchanger and back to seal chamber.

Features

1. Circulation is maintained by pumping ring.
2. In idle condition heat transfer is maintained by thermosiphon effect and in running condition by a pumping ring.
3. Lower product stabilization temperature is achieved.
4. Establishes required margin between fluid vapor pressure and seal chamber pressure.

Use

1. In hot and clean services e.g. in boiler feed water and hot hydrocarbon services.

Caution

1. Maintain minimum 0.5 m horizontal distance from seal chamber to heat exchanger.
2. Vent valve required at highest point of piping system.
3. Ensure that pump has a close clearance throat bushing.
4. Ensure that the seal outlet connection is in the top half of the gland.
5. Ensure that the cooler is mounted above the pump centerline.
6. Vent the system completely before start-up.

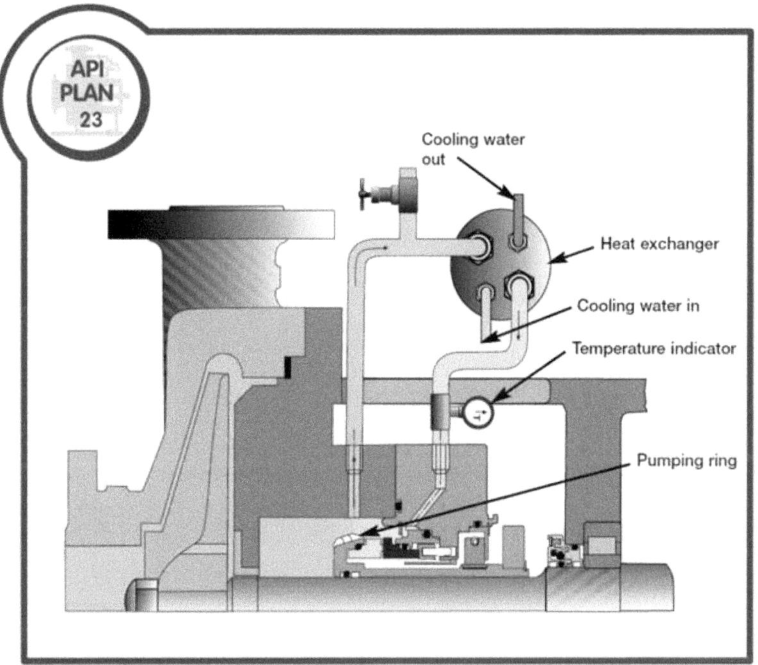

Figure 8-43: API Plan 23 (Source: Ref. 8-17)

PLAN 31 Description

Product recirculation from discharge through a cyclone separator, which directs clean fluid to the seal and routes solids back to the pump suction.

Features

1. Removes entrained solids from the side stream.
2. Particles from cyclone separator are returned to suction.

Use

1. Used in media with suspended solids (e.g., from mother liquor).

Caution

1. Pump throat bushing is recommended.
2. Ensure using only for services containing solids with a specific gravity twice or more than that of process fluid. May not be sufficiently effective unless there is a considerable difference in the specific gravities of the two.

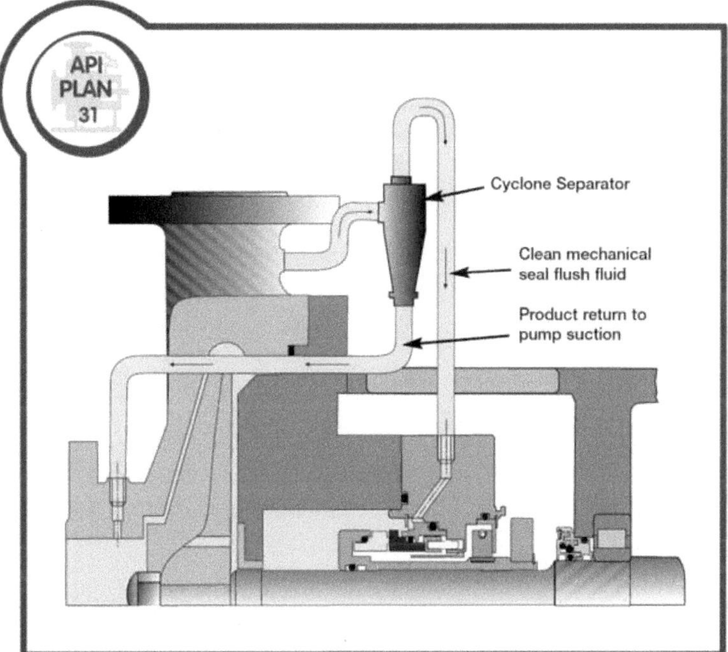

Figure 8-44: API Plan 31 (Source: Ref. 8-17)

PLAN 32 Description

Injection of clean or cool liquid from external source into the seal chamber.

Features

1. Reduces flashing or air intrusion across seal faces by providing a positive flush.
2. Maintains vapor pressure margin.
3. Always provided at a pressure greater than seal chamber pressure.
4. If maintained properly, it represents the best of all single seal plans (subject to acceptance of contamination).

Use

1. Dirty or contaminated fluids.
2. High temperature applications.
3. Polymerizing and oxidizing fluids.
4. Media with poor lubrication properties.

Caution

1. External source should be continuous and reliable at all times, even during start-up and shutdown.
2. Flush fluid must be compatible with process fluid due to product contamination.
3. Product degradation can occur.
4. Ensure use with close clearance throat bushing to maintain pressure in stuffing box and control the rate of contamination of pumped media.
5. Careful selection of flush fluid required to ensure that it does not vaporize on entering the seal chamber.

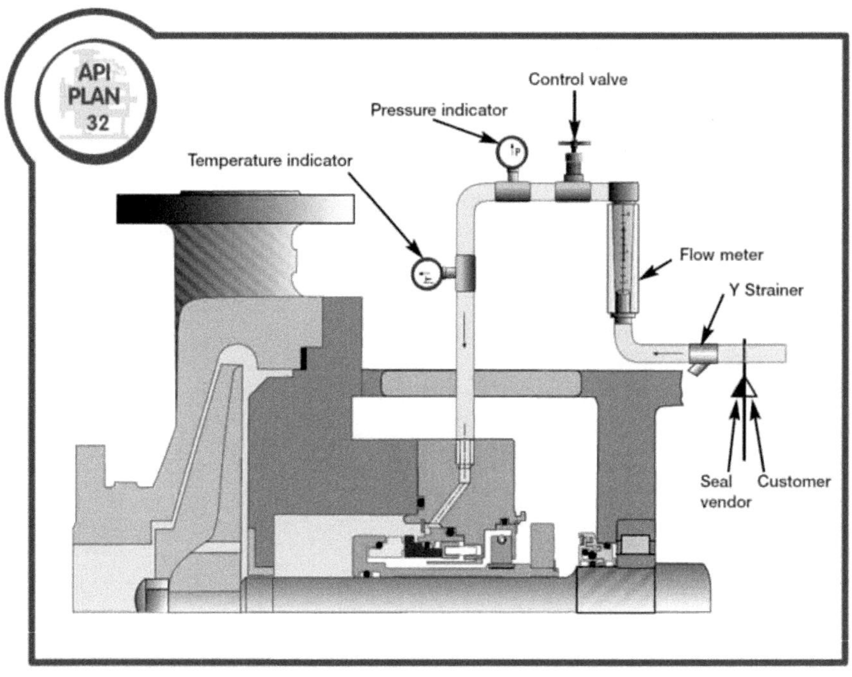

Figure 8-45: API Plan 32 (Source: Ref. 8-17)

PLAN 41 Description

Product recirculation from discharge through a cyclone separator and a heat exchanger to seal chamber.

Features

1. Improves pressure margin to vapor pressure.
2. Improves temperature margin to meet secondary sealing element limits, to reduce coking or polymerizing and to improve lubricity.
3. Removes entrained solids from side stream or mother liquor
4. Particles from cyclone separator are returned to suction.

Use

1. In hot services containing suspended solids.

Caution

1. Pump throat bushing is recommended.
2. Ensure use for services containing solids with specific gravity twice or more than that of process fluid.
3. Cooler duty is high, leading to fouling on water side.

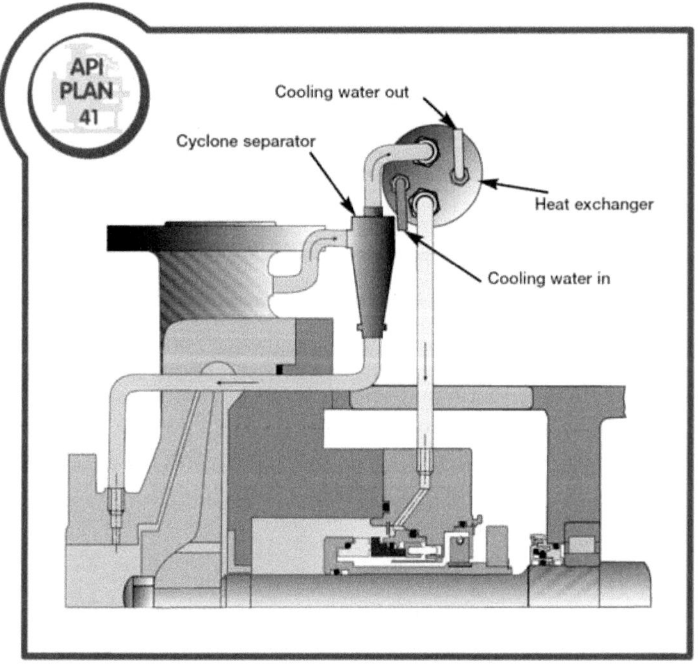

Figure 8-46: API Plan 41 (Source: Ref. 8-17)

Experience shows that in the overwhelming majority of repeat failure cases, a properly designed mechanical seal is worth many times its cost differential over the off-the-shelf "commodity" seal. We will first examine the various reliability enhancement features in detail. Thereafter, we plan to take a good look at some typical application examples, which incorporate one or more of these features.

BASIC CONSIDERATIONS FOR SEAL USERS AND MANUFACTURERS

Since by now it is assumed that the reader has a good working knowledge of mechanical seals, we can concentrate on relevant component considerations without first explaining how seals function. Here are some of the considerations that will lead to optimized seal selection (Ref. 8-8):

Design conditions must be within the range of feasibility. To establish a seal design for optimum operation at lowest long-term cost, the seal manufacturer must work within the following boundaries or requirements:

- high operating reliability
- appropriate interchangeability of components—low leakage
- satisfactory life expectancy—reasonable pricing

With the possible exception of dry gas seals, the selected mechanical seals must stay within the present limitations for these components; these are listed in Table 8-2.

Table 8-2

Shaft diameter:	5 to 500 mm	(~0.2 to 20 inches)
Pressure:	10^{-5} to 250 bar	(.008 mm Hg to 3,625 psig)
Temperature:	-200°C to 450°C	(-328°F to 842°F)
Face Speed:	100 m/s	(328 fps or 19,620 fpm)

Unless there is written proof and warranty data to back it up, it would not be reasonable to expect seals to have high life expectancy outside the application range of Table 8-2. Also, load conditions should not fall outside the boundaries of vendors' experience (Ref. 8-9). A convenient parameter for the load condition on a mechanical seal is the so-called p-v value, or pressure differential across one seal face multiplied by the face velocity experienced by this sealing surface. It is the p-v value that determines if a mechanical seal design should consist of balanced or unbalanced seals, whether these seals will have a narrow or more traditional wide face configuration, or if the arrangement is stationary or rotating—just to name a few significant features linked to this parameter. In addition, the p-v value has a major influence on the selection of materials of construction, particularly seal face materials (Ref. 10).

Table 8-3: Physical Properties of Seal Materials of Construction (mean values, per Ref.)

	Material	*Material*	*AISI*	*Hardness*	*0.2% yield strength (N/mm^2)*	*Modulus of Elasticity ($104N/mm^2$)*	*Coefficient of Expansion ($10^{-6}/K$)*	*Coeff of Thermal Cond (W/mK)*
Chrome Steel	14122	X35CrMo17	440C	2.250-2.750	600	21.3	10.5	29.3
Cr-Ni-Mo-Steel	14460	XSCrNiMo275	329	1.900—2.300	500	21.0	11.5	14.6
Cr-Ni-Mo-Steel	14571	X10CrNiMo Ti18	316	1.300—1.900	230	20.3	16.5	14.6
Hastelloy C	24602	NiMo16Cr	—	1.850	320	20	11.3	12.6
Monel K500	24375	NiCu30Al	—	1.300-3.150	280-420	18.25	13.6	17.6
Carpenter 20 Cb-3	—	X6NiCrMo CuNO3420	—	1.600-1.750	300	19.7	15.0	14.7
Inconel 625	24856	NiCr22 Mo9Nb		1.300-2.400	410-655	21.0	11.0	10.0

PLAN 51 Description

External reservoir providing a dead-ended blanket for fluid to the quench connection of the gland.

Features

1. No direct process leakage to atmosphere.
2. No need to maintain pressure system as in Plan 53A.

Use

1. Preferred for clean, non-polymerizing media with vapor pressure higher than buffer fluid pressure.

Caution

1. Keep pot vent continuously open; this is necessary to maintain buffer fluid pressure close to atmospheric pressure and vent the vapors to flare.
2. Should not be used in dirty or polymerizing products.
3. Never run the system with level in the sealant vessel being at low level as marked on the level gauge.
4. Vent the system properly before start-up.

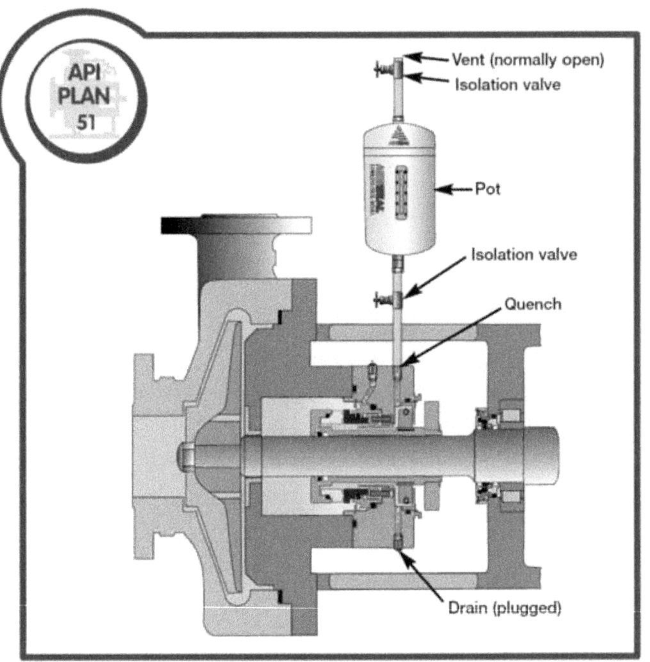

Figure 8-47: API Plan 51 (Source: Ref. 8-17)

PLAN 52 Description

Depressurized buffer fluid circulation in outboard seal of a dual seal configuration through a seal support system. Circulation is maintained by using pumping ring in running condition and by thermosiphon effect in stand-still condition.

Features

1. No process contamination.
2. No direct process leakage to atmosphere.
3. No need to maintain pressure system as in Plan 53A.

Use

1. For media where product dilution is not allowed, but leakage to atmosphere in diluted form may be allowed.
2. Preferred for clean, non-polymerizing media with vapor pressure higher than buffer fluid pressure (is also used for lower vapor pressure media).

Caution

1. Keep the sealant vessel vent continuously open; this is necessary to maintain buffer fluid pressure close to atmospheric pressure and vent the vapors to flare.
2. Should not be used in dirty or polymerizing products.
3. A restriction orifice is necessary in vent line to maintain back pressure in pot and facilitate quick release of vapors to flare.
4. Pressure switch setting should be done above minimum flare back pressure in order to avoid false alarms.

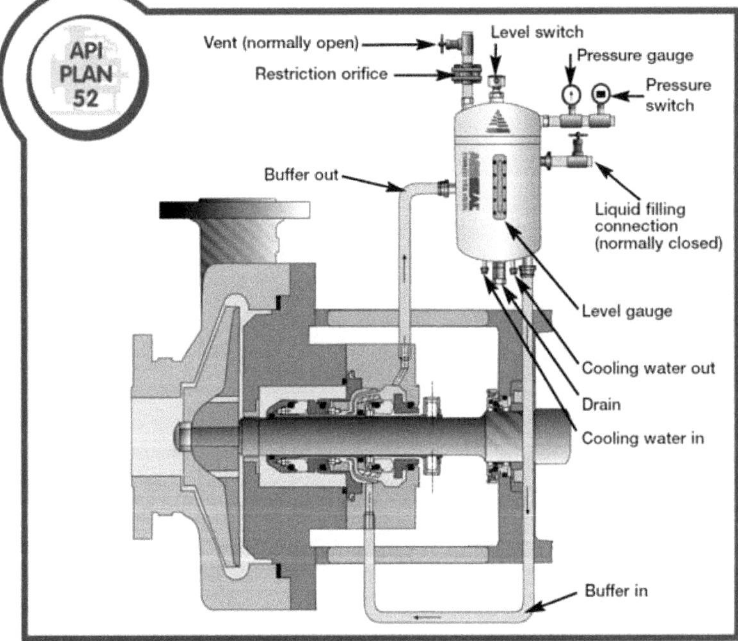

Figure 8-48: API Plan 52 (Source: Ref. 8-17)

5. Never run the system with level in the sealant vessel being at low level as marked on the level gauge.
6. Check for temperature difference in inlet and outlet lines to ensure that circulation is on.
7. Vent the system properly before start-up.

PLAN 53A Description

Pressurized barrier fluid circulation in outboard seal of dual seal configuration through a seal support system. Circulation is maintained by using pumping ring in running condition and with thermosiphon effect in standstill condition.

Features

1. In no case media leakage to atmosphere (provided the seal support system pressure is not lost).
2. Clean fluid film formation between the inboard seal faces gives better seal life.
3. Works as a Plan 52 arrangement if barrier fluid pressure is lost.

Use

1. Applications where no leakage to atmosphere can be tolerated e.g. hazardous, toxic, inflammable media.
2. For dirty, abrasive or polymerizing products where medium is unsuitable as a lubricant for inboard seal faces.

Caution

1. There will always be some leakage of barrier fluid in to the product. Check compatibility of barrier fluid with product.
2. Always ensure that the pressure source maintains higher pressure at the seal support system so that the process does not dilute the barrier fluid.
3. Vent the system properly before start-up.
4. In certain cases the inert gas can dissolve in the barrier media.
5. Product quality can deteriorate due to barrier fluid contamination.

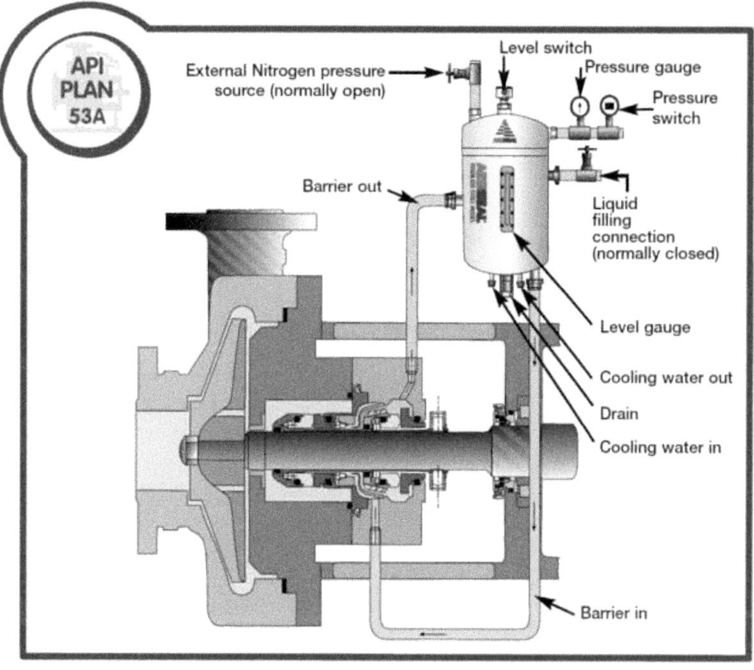

Figure 8-49: API Plan 53A (Source: Ref. 8-17)

Conventional Wisdom: *Seals from major manufacturers are always well-designed.*

Fact: *Major seal manufacturers have been known to offer "extrapolations of prior knowledge"—you are their laboratory! Investigate seal p-v and understand risk of buying products outside a given manufacturer's p-v experience. Always ascertain prior experience.*

Materials of construction utilized in a mechanical seal usually must resist corrosion. Since austenitic and ferritic materials are characterized by low thermal distortion and high modulus of elasticity, they adapt quite well to this requirement (Ref. 8-10). However, more sophisticated alloys can be selected for services requiring improved corrosion resistance. The application of such alloys in mechanical seals requires additional design work, since physical properties can be significantly different as opposed to standard materials. Typical physical properties of construction materials are listed in Table 8-2.

But, as most seal users know, sealing difficulties more often occur on the secondary, elastomeric sealing elements generally found on pusher seals. Where O-rings serve as secondary sealing elements, many failure incidents can be traced to the application of O-ring materials that are unsuitable for a given service. With O-ring manufacturers able to provide countless different compounds of O-ring materials, one or more will usually fit the requirements. Compromising by using the local distributor's standard selection may indeed be false economy. Developments such as PTFE (Teflon®)-wrapped elastomer materials combine necessary resilience and elasticity with good chemical and thermal resistance.

Teflon-wrapping may be the preferred solution in services with extensive temperature transients. Here, the utilization of solid PTFE may introduce vulnerabilities since the coefficient of thermal expansion of PTFE shows a rather extreme excursion at approximately 30°C (86°F). This leads to potential leakage at the surfaces contacted by the sealing element. Figure 8-60 illustrates PTFE-wrapped O-ring material and temperature-dependent expansion rates for Teflon, while temperature limits and Shore hardness values of frequently applied secondary seal elements are shown in Table 8-4.

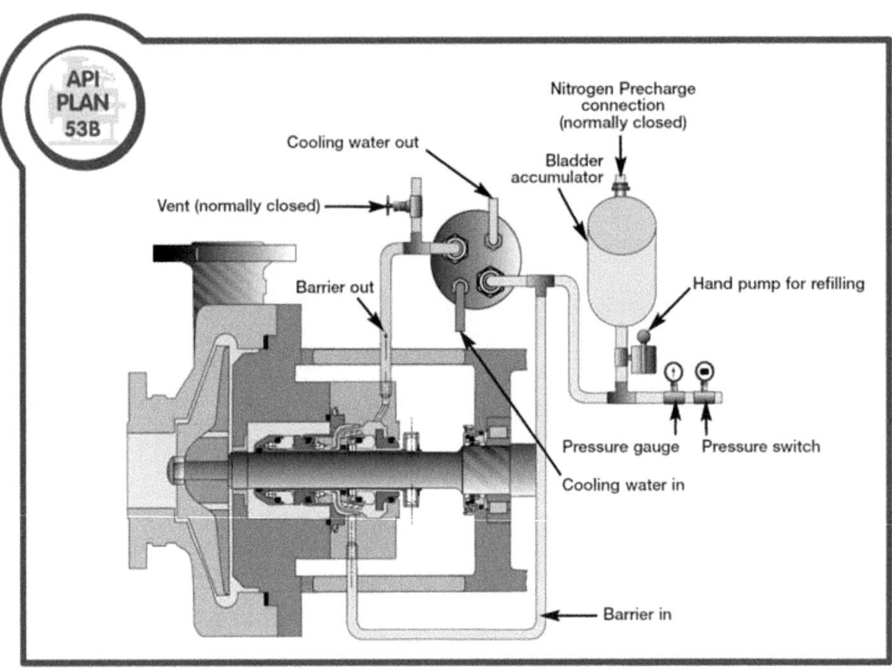

Figure 8-50: API Plan 53B (Source: Ref. 8-17)

PLAN 53B Description

Pressurized barrier fluid circulation in outboard seal of dual seal configuration. Circulation is maintained by using pumping ring in running condition and with thermosiphon effect in standstill condition. Pressure is maintained in the seal circuit by a bladder accumulator.

Features

1. Keeps barrier fluid and pressurized gas (inert gas) separate by using a bladder.
2. Heat is removed from the circulation system by an air-cooled or water-cooled heat exchanger.
3. Being a stand-alone system does not rely upon a central pressure source. Hence, much more reliable than a Plan 53A.
4. In no case media leakage to atmosphere.
5. Clean fluid film formation between the inboard seal faces gives better seal life.

Use

1. Applications where no leakage to atmosphere can be tolerated e.g. hazardous, toxic, inflammable media.
2. For dirty, abrasive or polymerizing products where medium is unsuitable as a lubricant for inboard seal faces.

Caution

1. There will always be some leakage of barrier fluid in to the product. Check compatibility of barrier fluid with product.
2. Low volume of barrier fluid in system; hence, heat dissipation is totally dependent on cooler efficiency.
3. Always recharge bladder to 0.9 times the working pressure.
4. Vent the system properly before start up.
5. Product quality can deteriorate due to barrier fluid contamination.
6. Cannot be used where seal chamber pressure varies. Use Plan 53C for such applications.

PLAN 53C Description

Pressurized barrier fluid circulation in outboard seal of dual seal configuration. Circulation is maintained by using pumping ring in running condition and with thermo-siphon effect in standstill condition. The pressure is maintained and fluctuations are compensated in the seal circuit by a piston-type accumulator.

Features

1. There will always be some leakage of barrier fluid in to the product. Check compatibility of barrier fluid with product.
2. Vent system properly before start-up.
3. Heat is removed from the circulation system by an air-cooled or water-cooled heat exchanger.
4. In no case media leakage to atmosphere.
5. Clean fluid film formation between the inboard seal faces gives better seal life.

Use

1. Applications where no leakage to atmosphere can be tolerated e.g. hazardous, toxic, inflammable media.
2. For dirty, abrasive or polymerizing products where medium is unsuitable as a lubricant for inboard seal faces.
3. Where pump pressure varies during operation will need an auto setting of barrier fluid pressure, thus maintaining the same differential throughout.

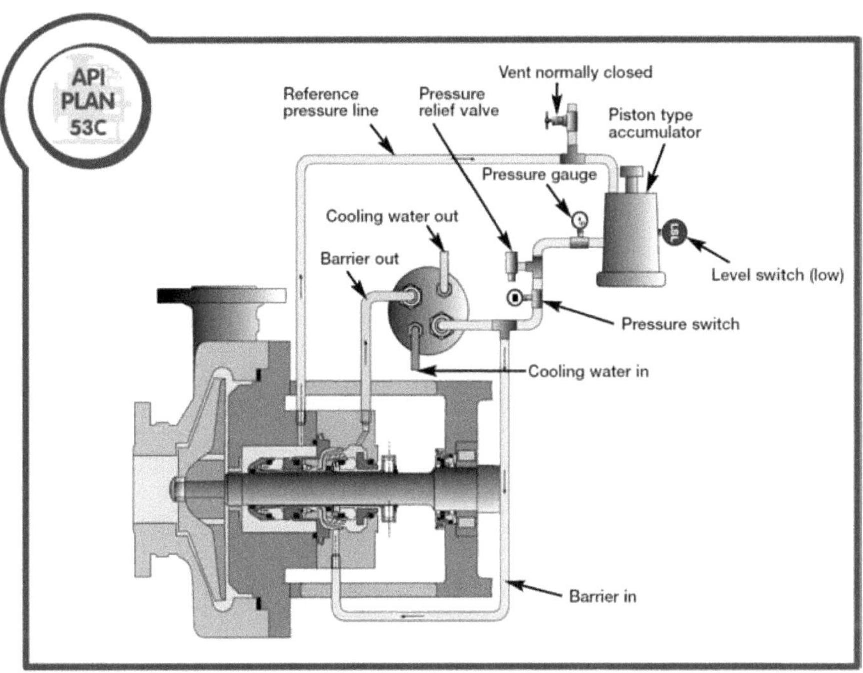

Figure 8-51: API Plan 53C (Source: Ref. 8-17)

Caution

1. Always connect reference pressure line from seal chamber to accumulator and keep it open.
2. There will always be some leakage of barrier fluid in to the product. Check compatibility of barrier fluid with product.
3. Vent the system properly before start-up.
4. Product quality can deteriorate due to barrier fluid contamination.

Table 8-4: Temperature limits and hardness of secondary sealing elements (Refs. 9 & 10)

Temperature limits and hardness of secondary seal components

Material	Temperature limit (°C)	Hardness (Shore)	
Nitrile-Butadiene - rubber (Buna N)	-55-+110	70-80	High resistance against swelling with oil, fuel and process gases. Low permeability to gas, resistance to aging and fatigue
Fluorosilicone	-70-+230	70	Excellent resistance against: hydrocarbons, oil. Preferred use if resistance against hydrocarbons and low temperature is required at the same time.
Fluorocarbon rubber (Viton)	-25-+190	75	High resistance against mineral oil, salt water, aliphatic and aromatic hydrocarbons, water <250°F and atmospheric conditions (e.g. ozone -solar radiation)
TTV, TTE, TTS (double - PTFE - coated)	-70-+280		Best chemical and thermal resistance against all aggressive liquids, elasticity ensured by use of Viton, silicone or ethylene-propylene core

PLAN 54 Description

Pressurized external barrier fluid circulation from a central pressure source or by a stand-alone pumping unit (e.g. AESSEAL® PUMPPAC™).

Features:

1. Ensures higher flow rate, better heat dissipation and positive circulation of barrier fluid.
2. If maintained properly, is the most reliable pressurized plan for dual seals as compared to Plan 53 A/B/C.
3. Can also be given as a stand-alone unit per pump.

Uses:

1. Applications where no leakage to atmosphere can be tolerated e.g. hazardous, toxic, inflammable.
2. For dirty, abrasives or polymerizing products where medium is unsuitable as a lubricant for inboard seal faces.
3. For media with high pressure and/or high temperature and/or high heat generation between faces.
4. Wherever Plan 53 A/B/C circulation is insufficient to dissipate heat.

Caution:

1. Carefully consider the reliability of barrier fluid source, if a central source is used.
2. Expensive system, proper engineering required.
3. Circulating system must be pressurized at least 1.5 bar higher than the pressure in the seal chamber.

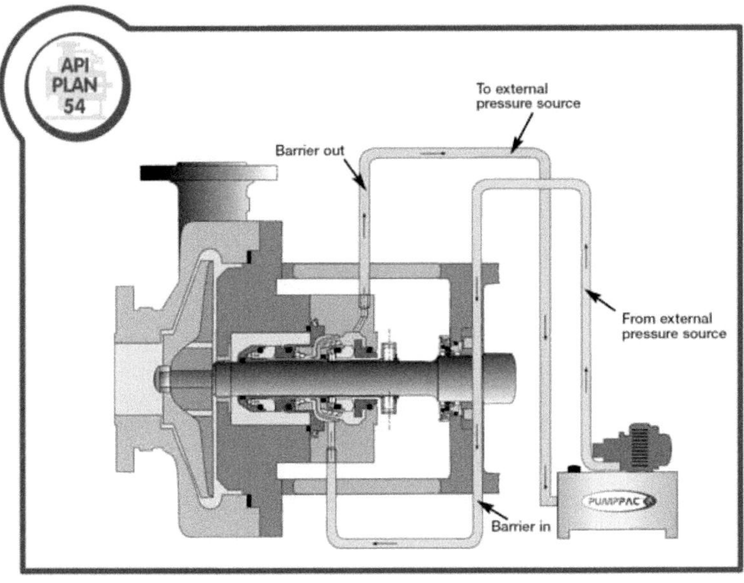

Figure 8-52: API Plan 54 (Source: Ref. 8-17)

4. Product contamination does occur. Barrier fluid selected should be compatible with the process fluid.
5. Always check filter/strainer in the system for any possible blockages.
7. Loss of pressure in system can lead to entire barrier liquid contamination.
8. Product quality can deteriorate due to barrier fluid contamination.

PLAN 62 Description

An external fluid stream is brought to the atmospheric side of the seal faces using quench and drain connections. Note that this is an inefficient use of water.

Features

1. The quench fluid acts as barrier in between atmosphere and process fluid.
2. The quench fluid reduces oxidation and cocking of product and also cools seal faces.
3. Flushes away undesirable material build up under seal faces.
4. Can be used with water, steam or an inert gas.

Use

1. In caustic or crystallizing fluids.
2. In oxidizing fluids or hot hydrocarbons.
3. Can be used to purge steam in hot applications, especially for stationary bellows to avoid coking.

Caution-

1. Ensure availability of continuous supply of low-pressure quench fluid limited to maximum 1 bar.
2. Use of throttle bushing on atmosphere side is mandatory.

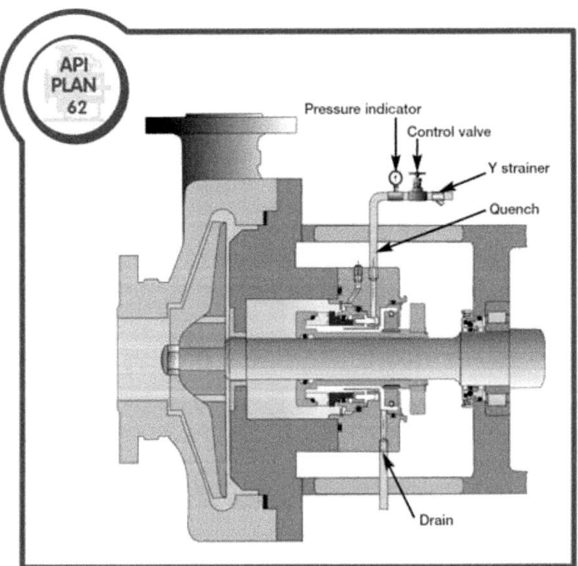

Figure 8-53: API Plan 62 (Source: Ref. 8-17)

3. Use proper bearing housing isolators to ensure that the quench fluid does not enter the bearings (see index for discussion of the advanced bearing housing isolators shown here).

Plan 65 Description

Leakage from seal faces is collected via the drain port and directed to a liquid collection system via a vessel equipped with a high-level alarm.

Features

1. The quench fluid acts as barrier in between atmosphere and process fluid.
2. The quench fluid reduces oxidation and coking of product and also cools seal faces.
3. Flushes away undesirable material buildup under seal faces.
4. Can be used with water, steam or an inert gas.

Use

1. In services where seal leakage is condensing.
2. Used for single seals.

Caution

1. Vent connection should always be plugged.
2. Orifice downstream of the level switch should be located in vertical piping leg to avoid accumulation of fluid in drain piping.
3. Shut down the pump as soon as high-level alarm is activated and attend the seal.

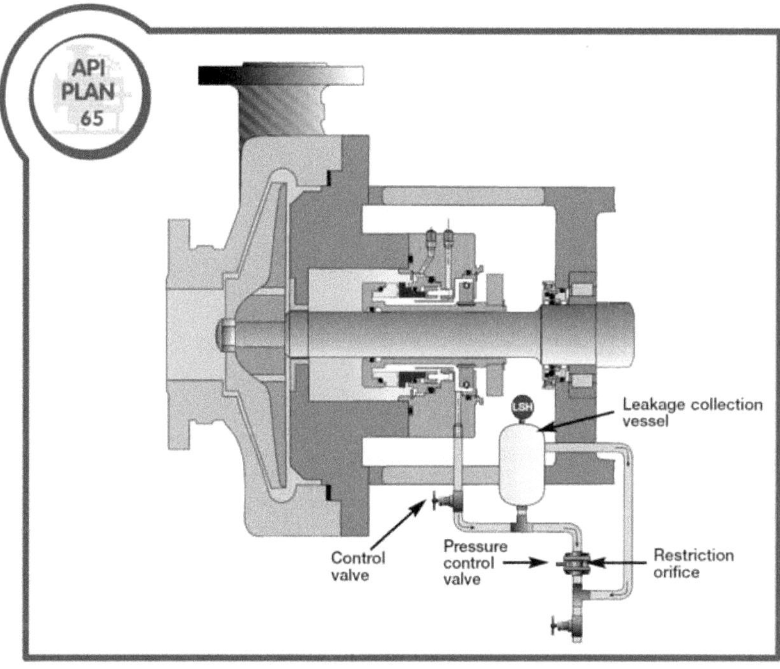

Figure 8-54: API Plan 65 (Source: Ref. 8-17)

PLAN 72 Description

Buffer gas is circulated in the containment seal chamber to sweep inner seal leakage away from outer seal to a collection system and/or dilute the leakage so that the emissions from the containment seal are reduced.

Features

1. Used in conjunction with API Plan 75 and/or 76.
2. Nitrogen provides cooling to seal faces.
3. Nitrogen blanket reduces the explosion hazard in high vapor pressure liquids.
4. This plan is used in conjunction with Plan 75 and 76.

Use

1. For flashing hydrocarbons

Caution

Always ensure that buffer gas pressure is less than seal chamber pressure.

2. Set the forward pressure regulator at min 0.4 bar above flare backpressure.

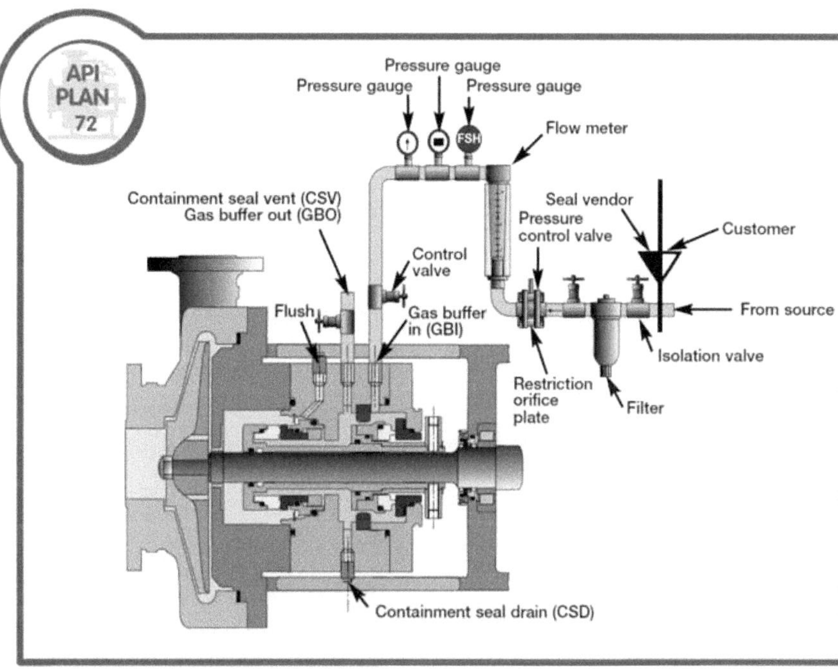

Figure 8-55: API Plan 72 (Source: Ref. 8-17)

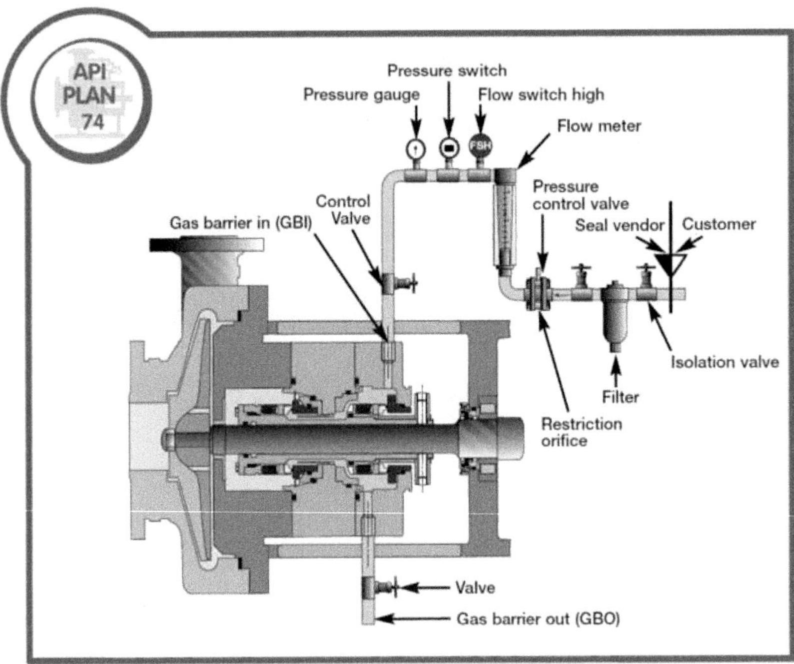

Figure 8-56: API Plan 74 (Source: Ref. 8-17)

PLAN 74 Description

Externally pressurized barrier gas through gas control system to a dual seal arrangement. An inert gas is used as a barrier gas.

Features

1. Media leakage to atmosphere is eliminated.
2. Achieves very high reliability. Solids or other materials that could lead to premature seal failure cannot enter the seal faces.

Use

This plan is intended for use in dual pressurized non-contacting gas seals.

1. Used in services which are not hot (within elastomer temperature limit) but which may contain toxic or hazardous material whose leakage to atmosphere cannot be tolerated.
2. In applications where solids or other materials are present in the sealing media.
3. Where process contamination is allowed but process liquid leakage to atmosphere is not allowed.

Caution

1. Always ensure barrier gas pressure is higher than seal chamber pressure.
2. Causes media contamination due to high-pressure nitrogen entering the pump.
3. Back pressure regulator should be set at least 1.7 bar higher than the seal chamber pressure.
4. Carefully consider the reliability of barrier pressure source, if central pressure is used.
5. Always check filter for any possible blockage
6. Do not use for sticking or polymerizing media.

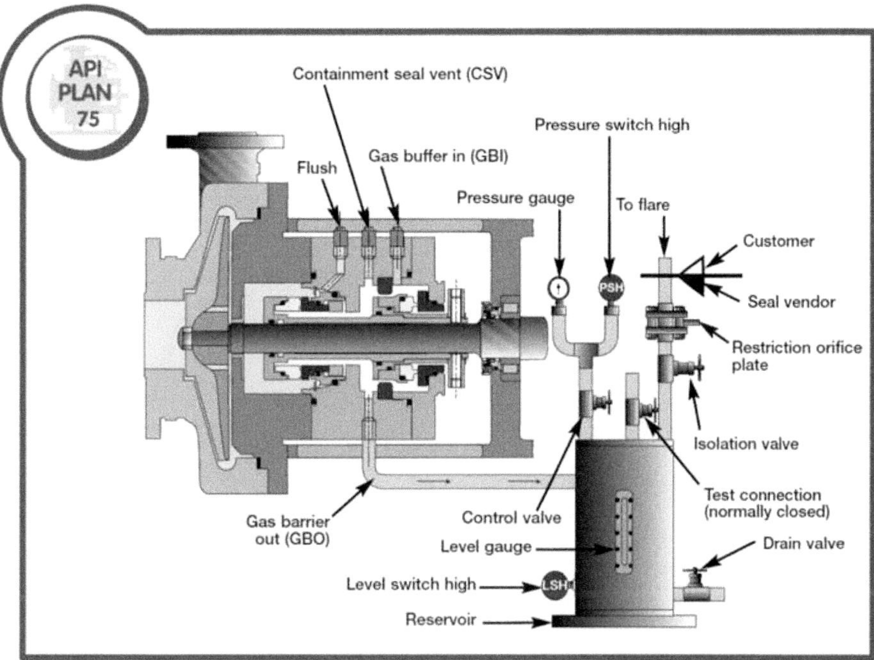

Figure 8-57: API Plan 75 (Source: Ref. 8-17)

PLAN 75 Description

Leakage of process liquid from inboard seal of a dual containment seal is directed to a liquid collector.

Features

1. Can be used with Plan 72 with buffer gas or with Plan 71 without buffer gas systems.
2. Collection can be redirected to process fluid by using separate pumping device.
3. Can also be used in single containment seal.
4. Test connection is provided to check the inner seal by closing the block isolation valve while pump is in operation and noting the time/pressure buildup relationship in the collector.

Use

1. In duties with condensing leakages.
2. With hazardous and/or toxic fluids.
3. May also be used for non-condensing leakages. In such cases, the collector can help in removing condensate from the vapor recovery system.

Caution

1. Ensure that collection system is located below the seal drain and with sloping pipelines.
2. Drain port should be at bottom of containment seal to allow the leakage to flow to the collection system.
3. Collection system should always be vented, releasing vapors of process liquid to vapor recovery system.
4. If present, valves should be accessible to operator relative to ground clearance and other obstructions.
5. A flow control orifice is required to create backpressure on collection system and to promote effective condensation of vapors.
6. Pressure switch should be set at a gauge pressure of 0.7 bar.

PLAN 76 Description

Vapor leakages from inboard seal of dual containment seal are directed to a vapor recovery system via a vent connection.

Features

1. Can be used with Plan 72 with buffer gas or with Plan 71 without buffer gas system.
2. Vapor leakage collection ensures zero to very low process emissions from outboard containment seal.

Use

1. For high vapor pressure fluids, light hydrocarbons
2. In hazardous or toxic media

Caution

1. Do not use for condensing media.
2. Ensure continuous vent to low pressure vapor recovery or flare system.
3. Tubing shall be 13 mm (1/2") minimum diameter and shall rise continuously from the CSV connection to the piping/instrumentation harness.
4. A flow control orifice is required to generate backpressure.
5. Ensure proper support to harness piping.
6. Ensure a low point drain in the piping loop.

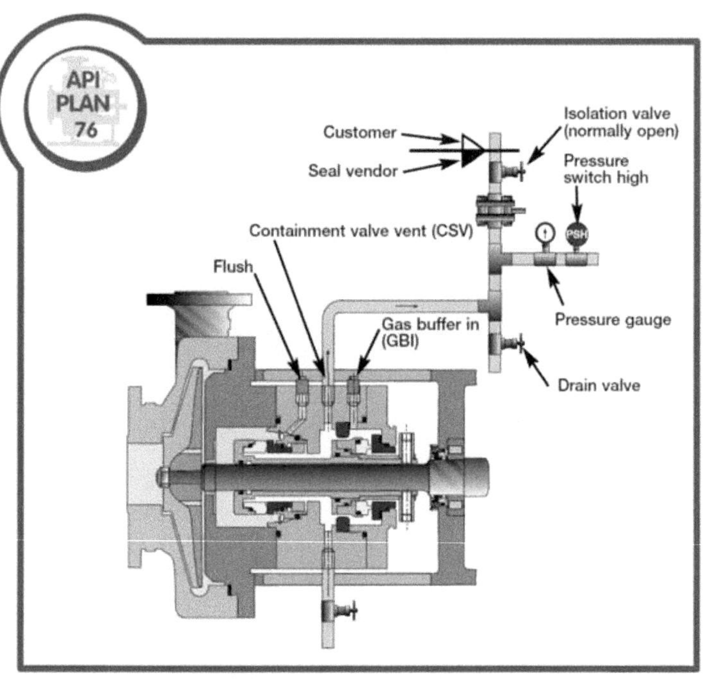

Figure 8-58: API Plan 76 (Source: Ref. 8-17)

PLAN 76M Description

This is a combination of Plan 75 and Plan 76. Leakage of process liquid from inboard seal of a dual containment seal is directed to a liquid collector. At the same time, any vapor leakages from inboard seal of dual containment seal are directed to a vapor recovery system via a vent connection.

Use

Whenever the vapor pressure of the medium is between minimum and maximum flare back pressures and so the leakage can either be condensing or non-condensing

Caution

Use only after verifying vapor pressure of fluid and limits of flare backpressure.

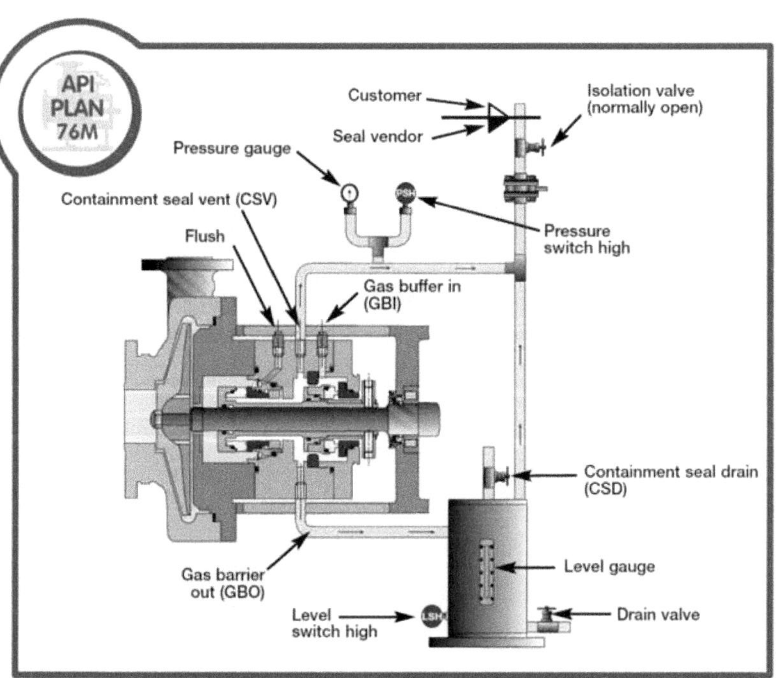

Figure 8-59: API Plan 76M (Source: Ref. 8-17)

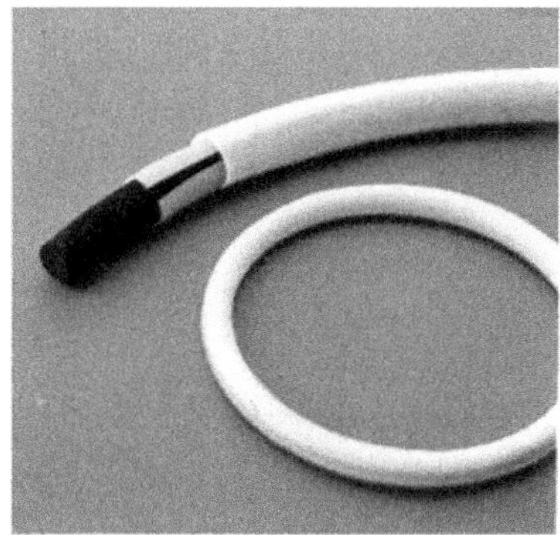

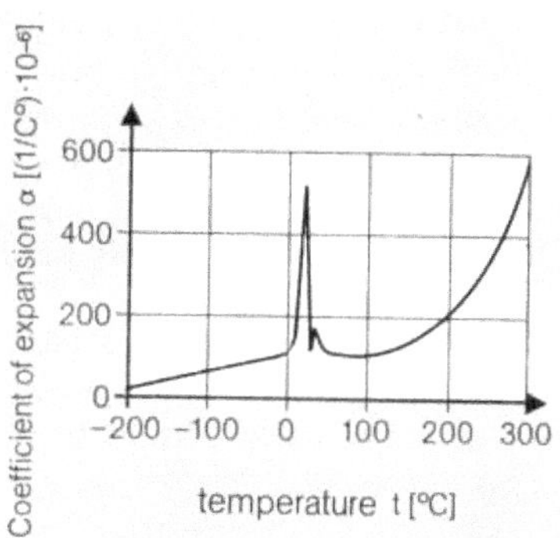

Figure 8-60: PTFE-wrapped O-ring and temperature-dependent expansion graph for PTFE (Teflon®)

O-rings in dynamic applications are expected to perform while undergoing movement in the axial and/or radial direction. They are typically installed with 10% to 15% radial compression. However, the compression set experienced by certain O-ring materials can have a significant negative impact on the performance of the O-rings.

Clearly then, O-ring selection influences the performance of a mechanical seal. This selection process should be entrusted to an experienced seal manufacturer, since only he will know whether or not certain materials adapt to existing design features. It can be inferred that on difficult sealing services, proper material selection must precede finalizing of seal geometries. That is to say, the seal must be engineered for the application. It is not acceptable to force a standardized seal geometry upon a difficult service or application.

The seal user must ask the vendor for relevant application charts. Be aware of the widely differing properties seemingly similar elastomer compounds may exhibit. Insist on the right one for vulnerable services.

Metal bellows can be used instead of O-rings in mechanical seals. Metal bellows, in fact, combine the function of a secondary sealing element with that of the necessary spring load. Traditionally, metal bellows seals were mainly used in high and low temperature services, or where the limitations of O-ring materials would not allow their application in a mechanical seal. In recent years metal bellows seals have also been widely used in services where O-rings might cause seal hang-up or shaft fretting. (8-23 showed hang-up.)

Aside from these advantageous features, metal bellows seals nevertheless have some significant limitations. Reliable long-term operation of bellows seals may be difficult to guarantee in high pressure, low specific gravity services because the hydraulic diameter on a metal bellows often varies as the pressure differential changes. Slip-stick motion due to partial dry running of seal faces in volatile hydrocarbons with low specific gravity has caused bellows failure due to torsional fatigue, thus posing a potential safety hazard.

Prior to installation, the resonant frequency of a metal bellows should be compared to the frequencies induced by the operating and resonance speeds of the rotating pumping machinery. It is thus rather important that seal users verify the manufacturer's experience before accepting metal bellows seals if offered for high pressure, low specific gravity services.

Advanced face material combinations, such as carbon graphite running against silicon carbide, are mandatory for mechanical seals operating at elevated p-v values. Corrosion and wear resistance, dry-running behavior, and temperature resistance determine face material selection.

Impregnations of carbon graphite (artificial carbon) have significant influence on the coefficient of thermal conductivity. Carbon graphite with antimony impregnation has a coefficient of thermal conductivity that is approximately 1.5 times that of resin impregnated carbon graphite. In some services, this could determine whether or not there will be partial dry running between two mating seal faces.

Hardness of face material and beneficial laser surface texturing (Ref. 8-11) translate directly into wear resistance. As of this writing, laser surface texturing is the latest and most promising face conversion treatment. It is suitable for virtually any seal face material and has proven to substantially reduce mechanical seal face temperature.

Reaction bonded silicon carbide with approximately 12% free silicon provides high hardness, good thermal conductivity, resistance to heat checking, and excellent dry-running behavior. Direct sintered silicon carbide has improved chemical resistance even if exposed to liquids with a pH value above 11, but also possesses an increased coefficient of friction. Increased friction will have a significant influence on the total power consumption of a seal system (Ref. 8-9).

The next-to-latest face material development is carbon graphite impregnated with silicon carbide. Both hardness and compressive strength of this face material are approximately three times higher than a comparable resin impregnated carbon graphite. The thermal conductivity, usually the weakest link in a face/housing combination, is five times higher than on an antimony impregnated carbon graphite. These unique physical values make the material a prime contender for application in high p-v, and high temperature services. An additional benefit is derived from the low coefficient of friction of this material. The resulting power savings will often compensate for the added cost. This is illustrated in the power consumption comparison of Figure 8-61, where the coefficient of friction of three face material combinations is plotted against a range of differential pressures (Ref. 8-11). Finally, the characteristic properties of the various materials are listed in Table 8-5.

It is again appropriate for users to ask if a given sealing service would benefit from some of these features. If yes, it will be important to choose from a source that can demonstrate proven experience and is willing to incorporate advantageous materials in mechanical seals.

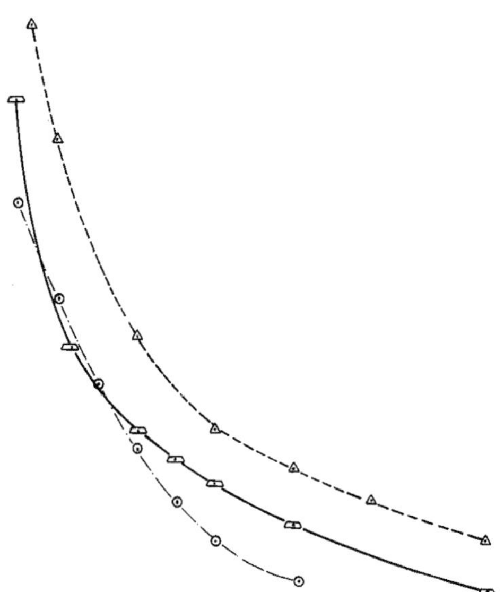

Figure 8-61: Coefficient of friction of various face material combinations at certain pressure differentials (Source: Ref. 8-2)

QUALITY CONTROL AND ACCEPTANCE TESTS

It is beyond the scope of this overview to give detailed specifications for quality control and testing of seals. Since it is evident that mechanical seal components are manufactured to very close tolerances, an extended quality assurance program that controls in-house manufacturing processes at strategically integrated locations will go far towards ensuring that quality is maintained at the established standards. Statistical and numerical quality control results should be compared against specifications given to the sub-supplier.

Generally, any mechanical seal should be tested prior to release for an industrial application. However, while leak testing is an API requirement, life cycle testing is not. Ideally, the seal vendor should be able to provide the seal user with data derived from operation of the seal on a test ring. This testing should have been done under conditions similar to those at the final installation site. Where this is not possible, seal face behavior under operating conditions should be simulated and studied by finite element analysis. It can thus be determined whether thermally and pressure-induced deflections will remain within acceptable limits to ensure long-term trouble-free seal operation.

Where acceptance tests are called for it would be prudent for seal purchaser and seal vendor to predefine expected or allowable deviations from norms in most or all of the following areas:

- leakage behavior of the mechanical seal during the hydraulic test and leakage behavior of the mechanical seal during the hydrodynamic test
- surface roughness and flatness of faces before and after test
- wear measurement of seal face after test; also power consumption of seal arrangement at various pressures
- temperatures, inlet versus outlet
- cooling water flow
- cooling water temperature, inlet versus outlet
- buffer fluid flow
- pumping device performance
- start/stop operation to evaluate torque transmission devices
- plotting radial and axial vibrations.

When in doubt, a reliability-focused user will specify testing, negotiate cost, and monitor seal test results.

Table 8-5: Physical properties of mechanical seal face materials (Ref. 8-9)

	Compressive strength (N/mm²)	Density (g/cm³)	Modulus of Elasticity (10^4N/mm²)	Coefficient of Expansion (10^{-6}/K)	Coefficient of Thermal Conductivity (W/mK)	Hardness (N/mm²)	Heat stress crack resistance factor (W/m)	Special suitability
I. Artificial carbons hard carbon, antimony impregnated	310	2.5	1.8	4.5	14	1.280	5.500	Good emergency running properties, high temperature resistance and compressive strength, therefore especially suitable for high pressure mechanical seals
II. Metals Chromium casting	1.200	7.5	20.5	9.5	20	3.200	3.900	Operational p<230 PSI (16 bar) v_g< 50 feet/sec, e.g. for secondary mechanical seals in tandem arrangement
III. Metal Carbides Tungsten carbide	5.000	15.0	60.0	5.2	79-82	15.000	12.650	High hardness and wear resistance
Silicon carbide	3.500	3.1	41.0	4.3	84	25.000	3.000	Harder than tungsten carbide, excellent chemical resistance, good emergency running properties and thermal conductivity
IV. Silicon impregnated artificial carbons SIC-C carbon silicon impregnated	800	2.65	1.4	4.5	70	3.100	12.000	Better hardness and wear resistance than carbon. Used in high temperature and high pressure mechanical seals

DESIGNING FOR DECREASED FAILURE RISK

When necessary for optimized safety and reliability, the more expensive engineered product makes eminent economic sense. Recall that in our introduction we highlighted seal failure frequencies and the cost of seal-induced pump repairs. Recall again our statistics: In 1983 it cost an average of over $5,000, before burden and overhead, to repair a centrifugal pump in U.S. Gulf Coast petrochemical plants (Ref. 8-12). If therefore, a better mechanical seal costs $1,000 more than the next choice and avoids even one repair, the additional investment is definitely recovered in the first few months of operation.

There are thus compelling reasons to analyze the design features of the various offers in efforts to identify the vulnerable executions and to give preference to those seal designs that avoid potential problems. The user must perform this analysis by comparing and evaluating the responses of several competing seal vendors. This can be done even if a user wishes to enter into a seal alliance, but the contract must be written with foresight and input from reliability professionals who know what's at stake here. (See pages 228-230.)

Conventional Wisdom: *Your single-source seal alliance partner must be your* *only* *source of seals.*

Fact: *Best-of-Class users have managed to structure the alliance such that the partner is both compelled and given incentives to furnish the best seal—even if it is being manufactured by his competitor.*

The bids must be conditioned with a view towards life cycle cost. In life cycle costing, the user considers the probable failure frequency, maintenance cost, or overall reliability of the machine train in which the competing seals would serve over an anticipated time period. This was done in many of the seals that are described next.

CASE STUDIES INVOLVING CONVERSION FROM AND TO DIFFERENT SEAL PLANS

The intrinsic value of different seal plans can best be explained by a few examples. Also, the technical and cost-based justifications for upgrading are always worth exploring.

Boiler circulation pumps: Conversion from API Plan 21 to Plan 23

Our first example illustrates the efficiency difference between two (process side) flush piping plans. Visualize that in fluid machinery operating with higher temperature media single seals often require cooling for reliable long-term service. This cooling may be needed to extend the temperature margin to vapor formation, or to meet the temperature limitations of certain secondary sealing elements (e.g. O-rings) that were chosen for their resistance to chemical attack. There may also be temperatures which, if exceeded, might promote undesirable coking or polymerization.

It should be noted that in such services two piping plans are commonly used by industry; they are referred to as API Plan 21 and API Plan 23.

In API Plan 21 (Figures 8-41 and 8-62), process fluid is diverted from the discharge of the pump, sent through a restriction orifice and a seal flush cooler, and then routed into the seal chamber. The seal flush cooler is removing heat from the process stream. However, if the production process demanded a hot fluid, such heat removal would benefit only the seal. In that instance, cooling the flush stream would reduce the overall process energy efficiency.

That is why API Plan 23 (Figures 8-43 and 8-63) often provides more efficient cooling for the mechanical seal. In API Plan 23, liquid is re-circulated from the seal chamber, moved through a cooler, and sent back to the seal chamber. Circulation is derived from a rotating pumping ring in the seal chamber, and the ring may be integral with the mechanical seal. Instead of constantly cooling a part of the process stream, only the contents of the seal chamber are cooled. This minimizes the load on the cooler, since the heat to be removed is only the heat generated by the seal faces plus the heat thermally conducted through the seal chamber casing.

A power plant application. A closely monitored application involved the boiler feed water pumps in a modern combined heat and power (CHP) plant at a paper-recycling mill. Operating at 160°C and a seal chamber pressure of 8 barg, this case study illustrates an experience with two different API flush plans. While each would be technically acceptable, one is clearly more economical.

The pumps had originally been fitted with a traditional 85 mm diameter seal and its faces received some cooling from the API Plan 21 configuration of Figure 8-62. But the service life of the seal was less than 12 months and this (for water) disappointing seal life was attributed to cooler fouling. A software package jointly developed by the cooler and mechanical seal suppliers

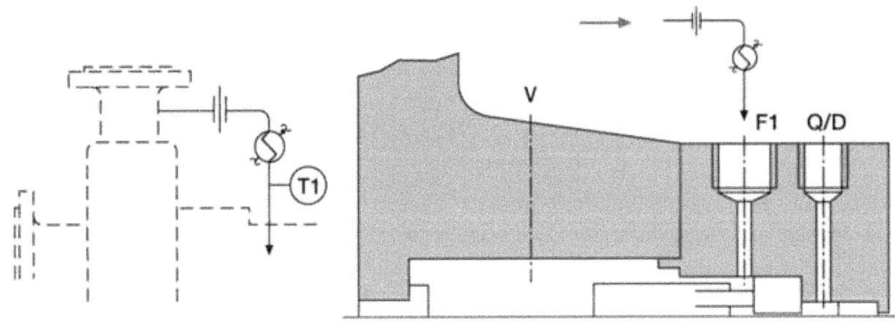

Figure 8-62: Recirculation from pump discharge through restriction orifice and cooler to seal chamber (API Plan 21)

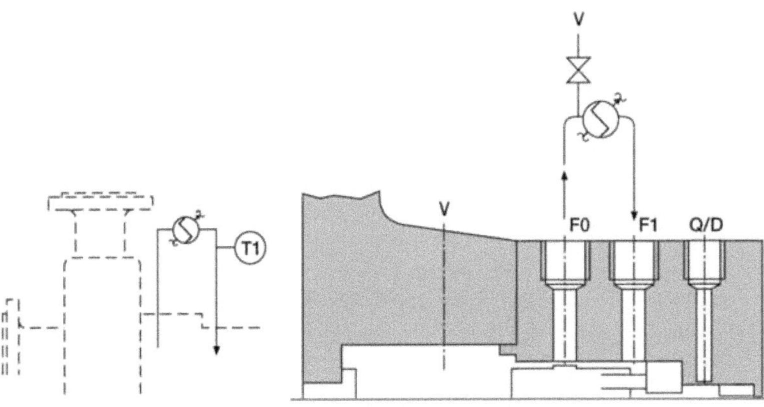

Figure 8-63: API Plan 23, recirculation from a pumping device in the seal chamber through a cooler and back to the seal chamber

was used to calculate the heat loads for both Plan 21 and Plan 23 operation. The Plan 21 simulations indicated that the seal would be operating at a seal chamber temperature of 108°C and with a cooler heat load in excess of 14 kW. Using the same basic operating parameters, a modern mechanical seal (Figure 8-64), equipped with the bi-directional pumping device shown to the right of Figure 8-64 and arranged per API Plan 23 (Figure 8-63) yielded significant efficiency improvements. The seal chamber temperature was lowered from 108°C to now only 47°C. The cooler heat load fell to 1.9 kW—less than 14% of the original Plan 21 system (Ref. 8-16).

The reduction in cooler load does not reduce the power absorbed by the pump itself, yet provides two potential savings. First, the overall energy demand to operate the boiler is reduced by about 12 kW. Second, the reduced load on the cooling water waste-heat removal system ultimately leads to lower maintenance costs.

A shipboard application. A similar exercise was undertaken on a shipboard application, which also demonstrated a reduction in cooler load. The MTBF of the original Plan 21 arrangement was less than 12 months, with heat exchanger fouling again listed as the cause of failure. A change to API Plan 23 was implemented and this conversion was started up in 2002. There had not been any failures as of 2007, which is when these results were first published.

It is noteworthy that API 682 3rd Edition (ISO 21049) explains "best practice" and offers an excellent tutorial on the reliability improvements that can be obtained with Plan 23. This industry standard states that Plan 23:

...is the plan of choice for all hot water services, particularly boiler feed water,

...is also desirable in many hydrocarbon services where it is necessary to cool the fluid to establish the required margin between fluid vapor pressure and seal chamber pressure.

...This duty is usually much less than that in a Plan 21. Lessening the duty is very desirable because it extends the life of the cooler.

...The industry has considerable negative experience with Plan 21 because of cooler plugging.

Industry adoption of Plan 23. Industry's adoption of the preferred and more efficient Plan 23 has been slow and Plan 21 was still widely used as of 2007. This is a "carryover tradition" because, decades ago, Plan 23 seals were more expensive. Pricing typically included the cost of add-on pumping rings and complex tangential porting in the seal chamber. However, modern machining techniques combined with innovative modular cartridge designs (Figure 8-64) have made the implementation of Plan 23 systems easier on both new and retrofit pumps. Costs have been reduced by fully integrating the pumping ring and ports into the seal cartridge. Moreover, the development of efficient bi-directional pumping rings (Figure 8-64) further simplified Plan 23 installation on between-bearing multistage centrifugal pumps.

For a petrochemical facility in the Middle East, in 2007, 10% of the applications had been originally specified as Plan 21. After demonstrating the energy savings potential, the Plan 21 specifications were changed to Plan 23. Note that whenever Plan 21 is specified for tradition's sake or because capital cost seems to be the overriding factor, this approach often merits a fundamental reassessment in light of the generally favorable operating and life-cycle costs of API Plan 23.

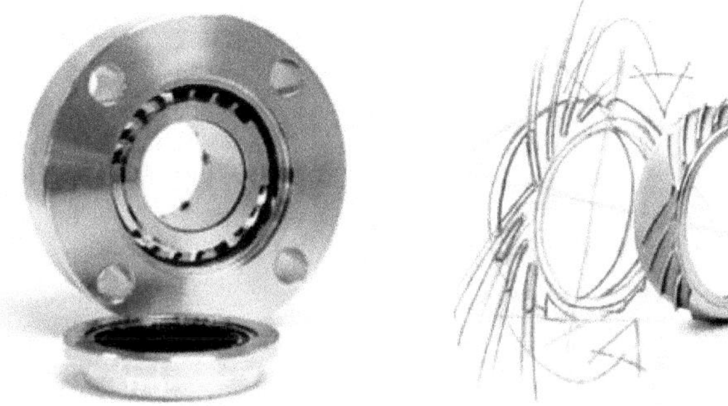

Figure 8-64: Modern cartridge seal with Plan 23 cooling arrangement. It incorporates the bi-directional pumping device on the right (Source: AESSEAL, Inc., Rotherham, UK, and Rockford, TN)

Conversion from API PLAN 32 to PLAN 53 on Evaporator Pumps

In this example energy efficiency was improved by converting from a single mechanical seal with external flush (Plan 32, Figures 8-45 and 8-65) to a double seal with a barrier fluid system (Plan 53A, Figures 8-49 and 8-66). This highly favorable application experience involved a Scotch whisky distillery where pot ale is being processed in a syrup evaporator that converts this byproduct into animal feed.

API Plan 32 and efficiency loss. There are three evaporator circulation pumps at this location; they had originally been fitted with single mechanical seals and a simple API Plan 32 (Figure 8-45). Here, clean fluid from an external source is used to flow through the seal chamber; the intent is to exclude solids or contaminants and also to reduce the temperature at the seal faces. It is also used to reduce flashing or air intrusion (in vacuum services) across the seal faces. The main driver for the use of an API Plan 32 system is its low initial cost.

A survey of the evaporator system revealed the inefficiency of this flush plan. Syrup was being fed to the evaporator at the rate of 150 liters per minute (l/m), but also being diluted by 2 l/m (per pump) of clean flush fluid. The additional total of 6 l/m of liquid into the process meant having to evaporate at least 4% more liquid. (An analogy would be walking 26 miles while disregarding the 25-mile route). In addition, the injected flush water came from a mountain stream source at 5°C. Consequently, more heat had to ultimately be added for the process to maintain evaporation temperature. In this instance, heating the cold flush water to the evaporation temperature required 19 kW, or 460 kWh per day. Fully evaporating the flush water required over 200 kW, or 4,800 kWh per day (Ref. 8-16).

Upgrading to Plan 53A (Figure 8-66) made considerable economic sense and was implemented on the three evaporator pumps at this distillery (Figure 8-67). Small pressurized tanks containing 10 liters of barrier fluid are installed at each pump; one is shown in the close-up of Figure 8-68. These tanks (and similar seal barrier fluid vessels) can be conveniently mounted on the fully adjustable stands, also shown in Figure 68.

The entire three-pump conversion described in this case history allowed syrup production to be increased from 88 to 98 tons/hr. This Scotch whisky distillery can now operate fewer hours while still meeting full demand.

Converting from Water Quench (PLAN 62) to API PLAN 53 in Water Treatment Service*

An ordinary water quench is similar to the now often superseded API Plan 62 (Fig 8-53). Water from an external source is passed between two sets of suitably configured mechanical seals before being discarded to a drain. The water quench is used to prevent solids accumulating on the atmospheric side of the mechanical seal, and can considerably improve seal reliability. Unfortunately, it can also be a rather obvious drain on precious resources (pun intended). See Ref. 8-18.

Quench Usage and Effluent. Open systems (Figure 8-69) can still be found in many industries and includes brewing, distilling, pulp and paper, corn milling, sugar

*Contributed by Richard Smith and Chris Booth of AESSEAL plc. Included in this acknowledgement are thanks to Messrs. Andrew Walker, (Machines Engineer, Sabic UK Petrochemicals), Rob Adam, (Sales Director AESSEAL pty), and Peace Katse Debswana, (Orapa Mines) for their assistance in preparing this case history segment. Also, items 8-14 through 8-19 in the listing of references—are acknowledged for relevance and helpfulness.

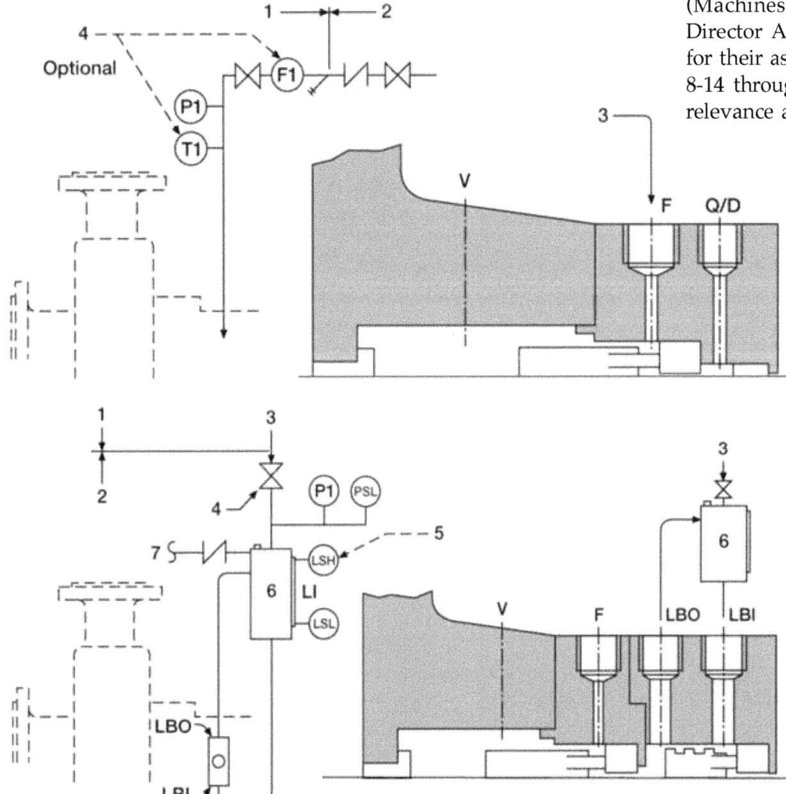

Figure 8-65: API Flush Plan 32; flush is injected into the seal chamber from an external source

Figure 8-66: API Flush Plan 53A, where a pressurized external barrier fluid reservoir supplies clean fluid to the seal chamber at a pressure greater than that of the process (pumpage). An internal pumping device per Figure 8-64 provides circulation

Fig 8-67: Plan 53A dual seal systems in use at a distillery in Scotland.

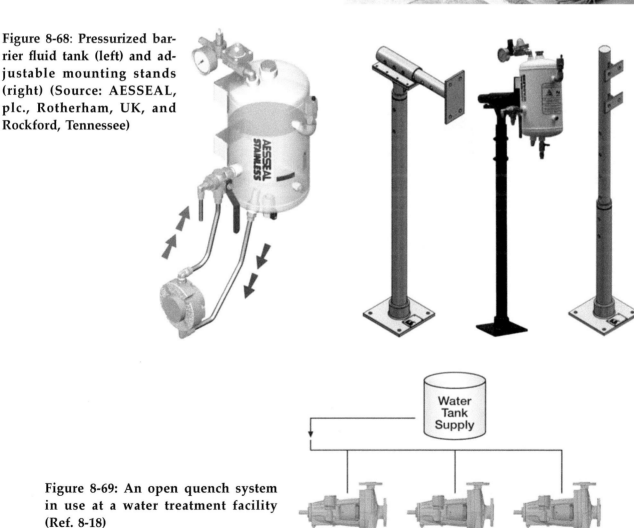

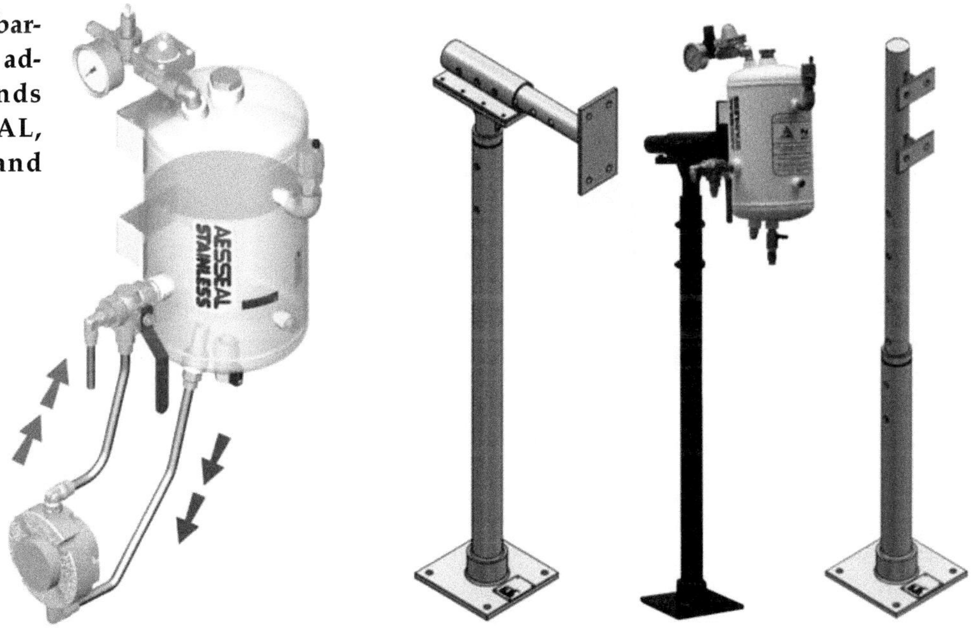

Figure 8-68: Pressurized barrier fluid tank (left) and adjustable mounting stands (right) (Source: AESSEAL, plc., Rotherham, UK, and Rockford, Tennessee)

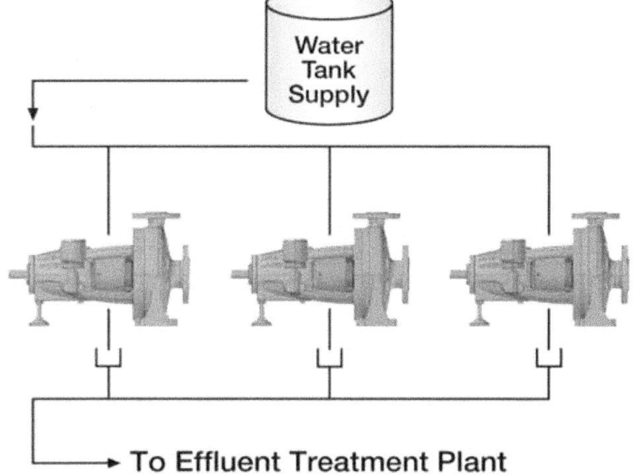

Figure 8-69: An open quench system in use at a water treatment facility (Ref. 8-18)

production, dairy processing and chemical manufacturing. At some installations quench flow rates are typically three l/m per seal installation and even larger flows have been reported. In addition to the cost of water, the cost and energy consumption of effluent treatment should not be neglected. Not to be overlooked in some parts of the world is the loss of goodwill incurred by wasteful water use practices. The main driver for this system is again its low initial cost. Yet, its ultimate cost should be studied and the wisdom of operating open systems must be periodically re-assessed in many situations.

Alternatives available. Plan 53A can eliminate effluent. A highly attractive alternative to the "open" system of Figure 8-69 is a barrier system similar to API Plan 53A. Such systems are self-contained (Figure 8-70). Perhaps even more important is the fact that they produce no effluent and, in many pulp & paper services, have conserved several hundred thousand gallons of water per pump per year. If the make-up water supply is connected so as to both pressurize and maintain the level in the tank, no manual intervention will be required and maintenance costs are reduced. An air space can serve as an expansion volume, minimizing the effect of temperature changes. In the event of interruption of make-up water supply, the air volume and a check valve maintain pressure until the supply is resumed.

Energy and CO_2 Emissions Savings

Converting a pump seal from open quench to a barrier system avoided discharging 1577 m^3 of water per year. This provides a power saving between 236 and 1352 kWh per year and, in the UK, equates to a CO_2 emission reduction between 118 and 676 kg.

It is intuitively evident that whenever an industrial plant generates its own power from product waste, the power saved in an effluent treatment process can be put to better uses. These include offsetting carbon production from other carbon generators. In any event, water management technology based on API Plan 54A sealing is well proven, and over 15,000 conversions to API Plan 54A have been made in the decade starting with 2000. Case histories are available from many different industry sectors. The mechanical seal industry has probably prevented many thousands of tons of CO_2 emissions as a result of these conversions (Ref. 8-19).

Conversions from Gland Packing to Mechanical Seals in Boiler Feed Pumps (Ref. 8-15)

This case history involved the conversion of a boiler feed pump from a traditional soft packing gland (Figure 8-71) to a Plan 23 mechanical seal arrangement.

Packing Glands. To help prevent rubbing friction causing damage to the packing, packing gland followers must be adjusted to allow some leakage and a slow drip rate of one drop per second is typical. The leakage obviously assists in cooling and lubricating the region of shaft contact. As packing and shaft wear progress, adjustment will be required to control the leakage rate and skilled maintenance labor is needed to do the job.

Multistage boiler feed pump with packing glands at both drive and non-drive ends had been installed at a petrochemical facility in the UK in the late 70's. This multistage configuration is typical of many applications around the world at chemical plants and refineries. Most of the feed water pumps installed in the 1970's were fitted with packed glands and many continued to run 40 years later with this outdated sealing technology. Leakage from the packed glands is a pure loss to the operation. In this example the feed water in the pump was at 121°C and losses through the packed gland would have to be made up with water from the treatment plant. An

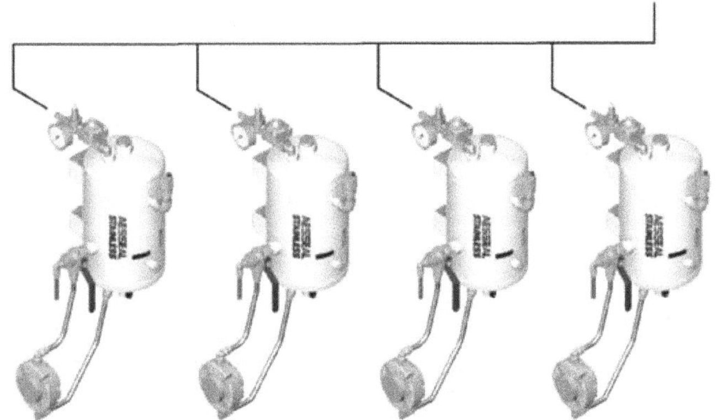

Figure 8-70: Alternative technology water management systems based on Plan 53A

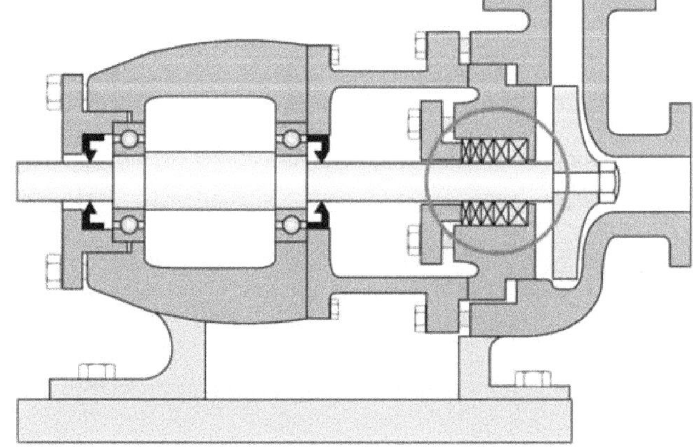

Figure 8-71: Packing used in a centrifugal pump

energy loss calculation was based on taking the make up water from 10°C to a feed water temperature of 121°C. In this instance the boiler was gas fired and the heat energy requirement could be translated into a net CO_2 contribution. Due to a reduction in available plant manpower, the gland follower adjustments required to compensate for packing wear were only made when the leakage was severe. As a result, the average leakage rate from the pumps was at least 1 l/m per gland (Figure 8-72).

Energy and emissions savings. With eight pumps (16 glands) leaking on average one quart (roughly one liter) per minute per gland, energy loss was calculated at 124 kW. The plant operated 24 hours per day, 365 days per year, causing an annual energy loss of 1,086,240 kWh. The potential savings in energy were purely based on heating requirements and did not include energy costs for water treatment, deaeration and pumping. It is also worth noting that the calculated energy savings did not take credit for the likely reduction in power absorbed by the pump; it provided for savings in the combustion process and boiler operating costs only.

Investigation of the combustion process by site personnel indicated that 0.0282 kg of CO_2 is emitted per liter of water heated. With losses of one liter per minute per gland (16 glands), there is a calculated saving of 237 tons of CO_2 pear year. For purposes of comparison, an average European high-efficiency diesel-fueled vehicle covering 20,000 km per year would emit 3.2 tons per year. Therefore, the savings equate to nearly 80 vehicles taken off the road. (See Refs. 8-15 through 8-19.)

Figure 8-72: Severe gland leakage from a boiler feed water pump

Extraction Slurry Pumps

It has been said that reliable and safe water supplies represent one of the greatest challenges facing the world in the 21st century (Ref. 5). With the use of flush water in pumping applications still a worldwide practice, our final case history highlights outstanding opportunities to conserve resources.

The Debswana mining operation consists of various diamond mines at Orapa in Botswana, a dry semi-desert area. Essential to the operation are slurry pumps pumping Kimberlite and iron silicate. These applications rank with the most abrasive in the slurry-pumping world and conversion to mechanical seals of a number of large slurry pumps with packed glands was of interest here.

Gland water requirements. The packed glands on the pumps needed flush water injection (Plan 32) into the process media, at a rate of 60 l/m to exclude slurry from the seal chamber. If the water supply failed, gland damage typically occurred within 30 seconds. Only 10% of the flush water is recoverable from the process. A typical train of five pumps consumes 134 million liters per year of water from underground sources. Due to constant extraction over many years, the water table was dropping and the resource was threatened. An official mandate called for reductions in water consumption by 10% immediately and 15% in the medium term. Pipelines from various alternative sources were considered. The most effective, and most extreme, was to build a 600 km pipeline from the Okovango Delta to Orapa at a cost of US$ 30 million. However, such a pipeline could have endangered sustaining water levels in the Delta, a World Heritage site.

Fitting dual seals with a self-contained external water circulation systems (Figure 8-73) to just 18 (initially) pumps reduced gland water consumption by a significant portion of the 10% total savings required across the site. A further 50 seals and systems have since been installed, covering all pumps that had previously consumed large volumes of gland service water.

Energy Savings. While the primary reason for the conversion was to tackle the issue of aquifer depletion, the secondary benefit of energy conservation should not be overlooked. This is typical of the need to use a systems approach; i.e., the complete original gland water system needed to be considered. The layout at

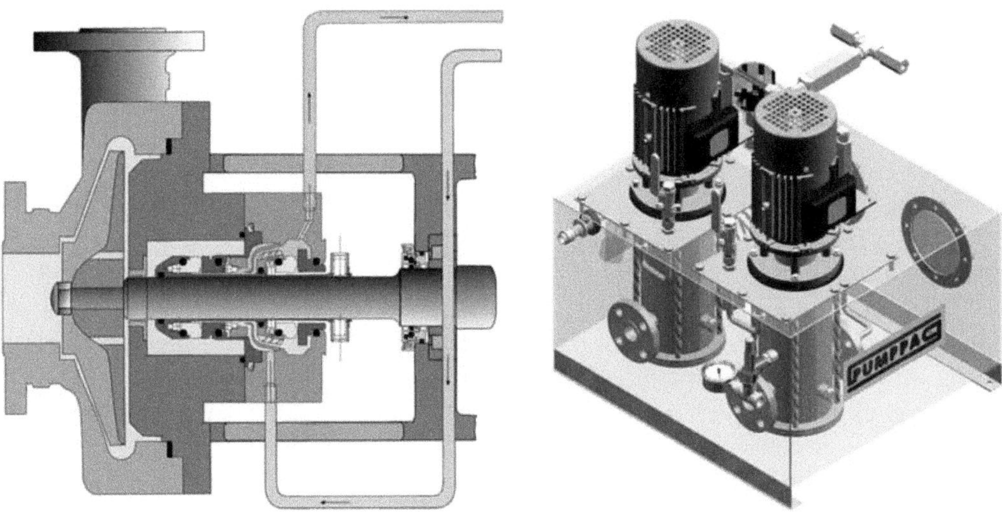

Figure 8-73: Dual seal arrangement (left) and self-contained external pressure source (right)

the Damtshaa mine (part of the Debswana operation) is typical of many of the mines in Orapa (Figure 8-74). Energy savings are realized because bore hole pumps are no longer required to lift the volume of gland water to a storage reservoir on the surface.

Typically, a bore hole is drilled to 250 meters with the water table at approximately 180 meters. Bore hole pumps are of the belt-driven helical rotor type, with an efficiency of approximately 75%. Taking suction from the reservoir, a multistage pump increases the pressure for injecting water into the sealing glands. The calculated energy consumption of this system is 394,084 kWh per year. With African power generation highly dependent on coal, for every kW saving there is an equivalent atmospheric saving of one kg of CO_2. (Ref. 8-19.)

The dual seal conversion utilized a simple tank re-circulation system, with a total energy consumption of less than 30,000 kWh per year. Hence, yearly energy savings of over 350,000 kWh were achieved.

Conclusions on the Importance of Seal Upgrading

Thoughtful selection of sealing support strategies and pump seal auxiliary piping systems deserves your time and attention. The energy indirectly consumed with traditional, non-optimized systems should not be ignored, and sole emphasis on minimum capital cost rarely results in best operational efficiency. Sizeable maintenance cost reductions are often possible when traditional gland packing arrangements are phased out and when modern mechanical sealing strategies are implemented instead. Both new and mature pump installations fitted with traditional high-maintenance

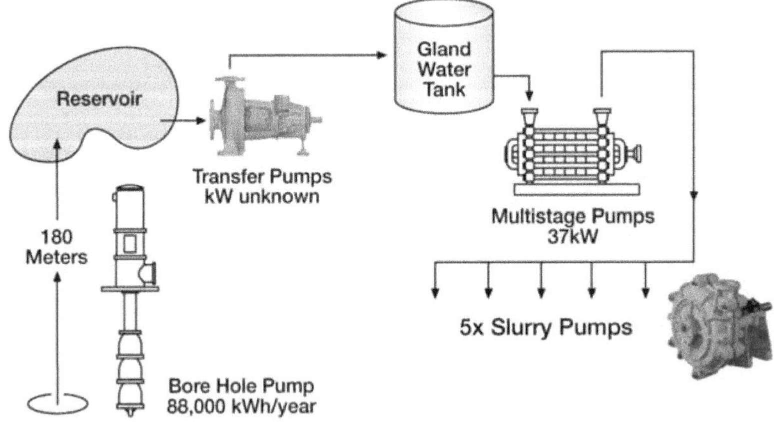

Figure 8-74: Original gland water system for pumps at Orapa Mines in Botswana.

packing arrangements or less-than-optimum seal flush plans merit detailed reviews in terms of plant efficiency improvement.

The mechanical seal industry has developed efficient and elegant solutions that provide owner-operators with higher equipment reliability and lower life-cycle costs. Sound solutions tend to add profits to the bottom line and can be shown to reduce environmental impact as well. These are win-win propositions well worth investigating.

ADDITIONAL EXAMPLES AND CASE HISTORIES*

Improved Seal Performance in Multiphase Pumping Applications

The oil and gas industry continues to increase the application of multiphase pumping as this technology itself develops and becomes more widely accepted as a realistic pumping solution. The aggressive nature of multiphase pumping can be characterized by:

- Variability of the pumped fluid
- The tendency to have water slugs, high gas volume fractions
- Significant solids content

This has led to recent advances in sealing technology being applied in efforts to offer the most reliable solutions. With multiphase pumping, the mechanical seal is challenged in several ways:

- The pumped fluid is usually a mixture of hydrocarbon liquid, water and gas in any proportion. Additionally, multiphase applications often include significantly high quantities of sand.
- The fluid properties can change both over short and long cycles. In positive displacement screw pumps seals may be required to operate for periods at near 100% gas, in effect running dry, unless alternative lubrication is provided.
- Fluid viscosity can be high and subject to significant variation with temperature. In some instances they may be reduced by the injection of a diluent.
- The pressure imposed on the seal can be highly variable and often unpredictable. Sudden pressure spikes and reversals are commonplace.
- Depending on the pump design, the seal is expected to accommodate significantly greater axial and angular misalignments than would normally be expected in a "standard" pump configuration. Axial and radial space available for the seal is often limited.
- Low, variable and high-speed operation, each presenting differing challenges for the seal. Screw pumps often operate in the range 300 to 3,600 rpm and heli-coaxial pumps in the range 1,800 to 6,000 rpm.

ENHANCING THE SEAL ENVIRONMENT WITH "SEAL PROTECTORS"

Spiral Trac® (Figure 8-75) and Seal Mate (Figure 8-76) are among the proprietary "Seal Protectors" devices that can enhance the seal environment. These inexpensive add-ons can reduce required flush quantities or prevent particulate from overwhelming the seal. There are, nevertheless, applications involving fluids at or near the vapor point where flashing concerns mandate caution. Also, different applications may require somewhat different versions or contours, and these are stocked by the manufacturer and its distributors.

These stationary devices are pressed into the pump backplate, just behind the impeller and essentially replace throat bushings. Fluid is drawn from the bore of the seal cavity by the operation of stationary vents or baffles. Fluid acceleration and the accompanying centrifuge effect separate abrasives for expulsion with the exiting flow. High operating flow assists in removing heat from the seal faces.

ENHANCING THE SEAL ENVIRONMENT BY USING ADVANCED FLOW INDUCERS AND BI-DIRECTIONAL INTERNAL PUMP-AROUND DEVICES

Dual seal applications and a number of API flushing plans (see Figs. 8-48 and 8-49) depend on thermal convection or thermosiphon effects to circulate the cooling fluid. Simple pumping rings and/or pumping screws as alluded to earlier are sometimes used to assist this circulation and thus heat removal from seal faces.

However, elementary pumping rings are generally inefficient. Similarly, and because shaft systems will nor remain concentric relative to the stationary parts, pumping screws must be manufactured and assembled with as much as 0.060 inch (1.5 mm) clearance. This, of course, affects the efficiency of the device. It can

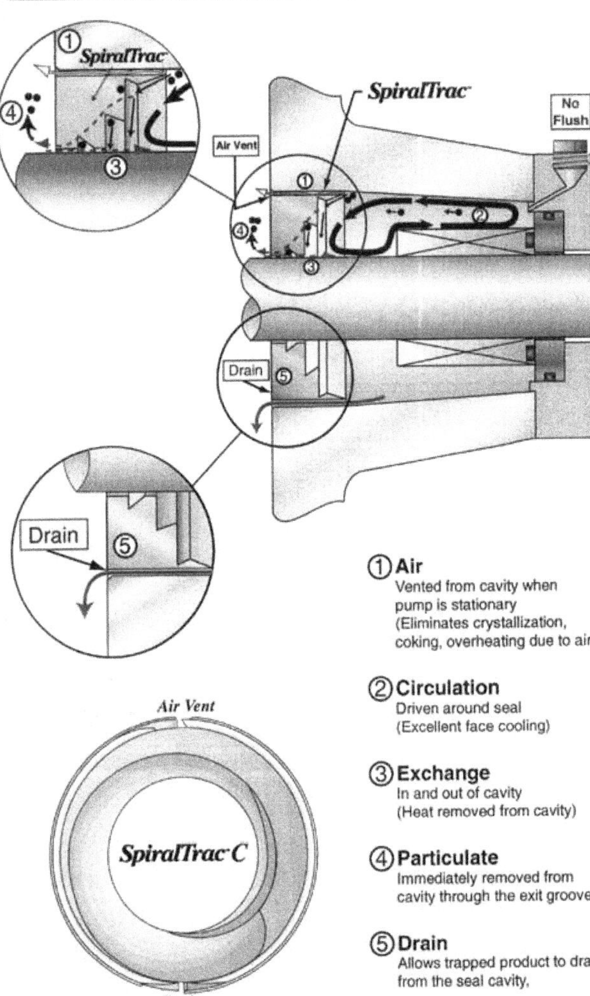
1 SpiralTrac
4
3
Air Vent
SpiralTrac
No Flush
1
2
4
3
Drain 5
Drain 5
Air Vent
SpiralTrac C
Drain
1 Air
Vented from cavity when pump is stationary (Eliminates crystallization, coking, overheating due to air)
2 Circulation
Driven around seal (Excellent face cooling)
3 Exchange
In and out of cavity (Heat removed from cavity)
4 Particulate
Immediately removed from cavity through the exit groove
5 Drain
Allows trapped product to drain from the seal cavity.

Figure 8-75: Spiral Trac® seal protector and how it operates (Source: Enviroseal, Waverley, Nova Scotia, Canada)

Figure 8-76 : "Seal Mate" ® solids exclusion device mounted in a typical stuffing box (Source: Enviroseal, Waverley, Nova Scotia, Canada)

be said that typical flow rates achieved with pumping rings are extremely low, with most of the energy of rotation going into the fluid as heat (Figure 8-77). Pumping screws are more efficient, but depend on a rather close gap between screw periphery and its opposing stationary bore. This close gap can be a serious liability in situations where shaft deflection deprives the stationary and rotating component parts from remaining perfectly concentric.

ENHANCING THE SEAL ENVIRONMENT BY USING ADVANCED FLOW INDUCERS AND BI-DIRECTIONAL INTERNAL PUMP-AROUND DEVICES

Dual seal applications and a number of API flushing plans (see Figs. 8-39 and 8-41) depend on thermal convection or thermosiphon effects to circulate the cooling fluid. Simple pumping rings and/or pumping screws as alluded to earlier are sometimes used to assist this circulation and thus heat removal from seal faces.

However, elementary pumping rings are generally inefficient. Similarly, and because shaft systems will not remain concentric relative to the stationary parts, pumping screws must be manufactured and assembled with as much as 0.060 inch (1.5 mm) clearance. This, of course, affects the efficiency of the device. It can be said that typical flow rates achieved with pumping rings are extremely low, with most of the energy of rotation going into the fluid as heat (Figure 8-77). Pumping screws are more efficient, but depend on a rather

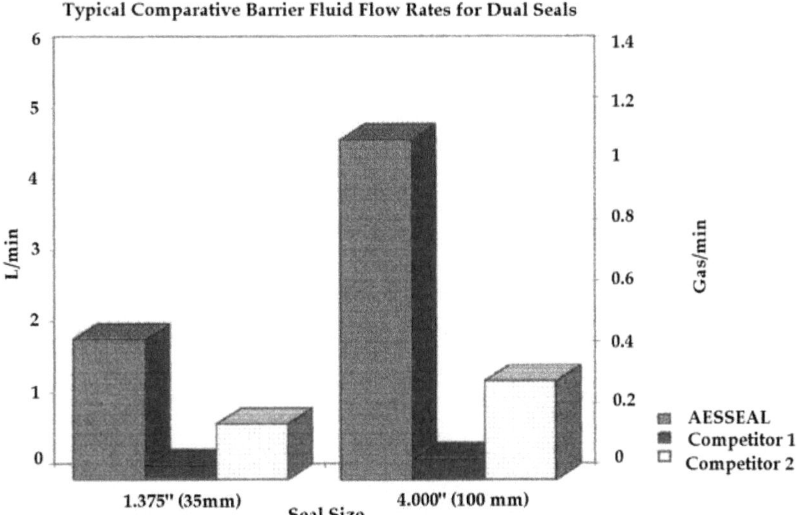

Figure 8-77: Typical comparative barrier fluid flow rates for dual seals (Source: AESSEAL plc)

close gap between screw periphery and its opposing stationary bore. This close gap can be a serious liability in situations where shaft deflection deprives the stationary and rotating component parts from remaining perfectly concentric.

As will be described later in this chapter, an innovative manufacturer of mechanical seals has developed a unique bi-directional barrier pumping design that imparts relatively high head and, especially, high flowrates to the barrier fluid. Optimized configuration of the separating wall between inlet and outlet flow is achieved through computational flow dynamics. The result is deposition of cool barrier fluid at the hottest and thus most important region of the mechanical seal.

It can be shown that seals incorporating these beneficial features are generally offered at prices that are virtually identical to those of conventional mechanical seals. In applications involving barrier fluid and where conventional mechanical seals do not incorporate the same capacity for heat removal, a bi-directional integral pumping design (Figures 8-64 and 8-85) represents a feature that extends seal life, reduces downtime risk, and benefits the maintenance budget. As in the case of many other zero-cost reliability enhancements, the payback can be deemed instantaneous and the benefit-to-cost ratio is implicitly in a range that far exceeds 10:1.

PHASING OUT PUMP PACKING

Statistics show that for every 1,000 pump failures in hydrocarbon processing plants there will be a fire. Although the incident shown in Figure 8-78 happened on a pump equipped with a mechanical seal, it is quite typical of the visible outcome of incidents involving leakage of flammable fluids.

Since roughly 60% of pump fires involve mechanical seal failures it has been assumed that the majority of mechanical seals are somehow flawed. However, this conclusion is quite incorrect; many pumps have been in service for years without seal failures. The entire experience points to the need to carefully select the right mechanical seal designs and to follow up by adhering to appropriate work processes and installation procedures.

Braided Packing or Mechanical Seals?

There is still the occasional assumption that critically important pumps will benefit from braided packing. Firewater pumps are mentioned in this context since they are expected to be available at all times and because, with water, a small amount of leakage is deemed tolerable. It might surprise that even here packing is not the first choice of experienced professionals. Best practice in modern firewater pumps has been (since about 1968) to use single-spring mechanical seals. As a matter of general practice, these mechanical seals are often backed up by a floating throttle bushing and a deflector guard.

Reliability professionals interested in "designing out" maintenance always check into the feasibility of systematically upgrading their equipment. In fact, the very best user companies mandate viewing every maintenance event as an opportunity to upgrade. For process pumps one of the many cost-effective upgrade

Figure 8-78: Scene of a refinery pump fire (Source: Release No. 2004-08-I-NM; U.S. Chemical Safety Board; http://www.csb.gov)

measures involves replacing braided packing with suitable mechanical seals.

In the 1960s, accurate statistics were kept (for insurance purposes) by a major multinational oil company. The statistics for firewater pumps showed that leaking packing tended to ruin bearings. Well-designed mechanical seals were then selected because they generally leak much less than packing and are considerably less likely to allow water spray to enter the adjacent bearing housing. Of course, we know that brittle mechanical seal faces might shatter when abused. However, seals that are properly designed, selected and installed are highly unlikely to shatter. Moreover, floating throttle bushings represent a "second line of defense" in firewater pumps.

Moves to phase-out braided packing were related to maintenance intensity. The text "Major Process Equipment Maintenance and Repair" (Ref. 8-20) gives a 9-step procedure dealing with packing maintenance and repair. While interested parties might wish to read the whole story, a short overview sequence includes the following:

- Distinguish between different packing styles and materials
- Preferably use a mandrel to properly wrap and cut new packing
- Use the right tool to remove old packing
- Remove corrosion products from shaft or sleeve surface
- Coat ring surfaces with an approved lubricant
- Insert one new ring at a time, then seat each with a proper tool (Figure 8-79)
- Install packing gland, initially only finger-tight
- Operate for run-in, initially allowing a stream of sealing water leakage
- Tighten sequentially to allow a leakage flow of 40-60 drops per minute. Then, make sure that leakage flow exists at all times. Not having such leakage could deprive the packing of lubrication. Without such lubrication, the packing risks overheating and failing.

The procedure will continue to be useful in certain inexpensive or old-style firewater pumps that have very limited packing box space. On those, conceivably, braided packing may still be a reasonable choice. However, AESSEAL and possibly other competent mechanical seal manufacturers have recently (2012) developed mechanical seal retrofit cartridges for these "tight fit" applications.

Suffice it to say that braided packing consumes

maintenance time and requires a measure of expertise which, with the passage of time, may no longer be readily available. The contention that packing rarely represents best practice also takes into account issues relating to periodic testing of all standby equipment. The question is sometimes if (a) switching the "A" and "B" pumps and running each for one month, or (b) turning on the standby pump once a month and then running it for 4-6 hours, is the better choice. When people argued--many decades ago--that plants might get away by testing only twice a year, responsible reliability professionals took the position that testing only twice a year would not be acceptable and monthly testing was needed. Depending on lubricant selection and lube application method, switching "A" and "B" every two or three months is considered Best Practice. This then keeps the bearings and packing lubricated. In installations with mechanical seals, this switching prevents seal faces from sticking.

Figure 8-79: Split bushing tool used for packing installation (Ref. 8-20)

In answer to the question whether braided packing or mechanical seals are preferred, we should opt for mechanical seals whenever possible. Leakage-prone braided packing tends to waste resources and will thus be avoided by risk-averse reliability practitioners.

IT'S WORTH IT TO SELECT BETTER SEALS

Selecting and utilizing better mechanical seals is without doubt the most cost-effective method of extending the mean-time-between-failure (MTBF) of centrifugal pumps in virtually all industries where pumps are used. A properly engineered seal will distinguish itself from the average product by such elements as material selection, component configuration, seal auxiliaries, etc. While it could be claimed that for a given service a certain enhancement feature results in only marginal improvement, the combined influence of these elements has been demonstrated to increase equipment reliability and to decrease maintenance expenditures in the long run by very substantial margins.

All too often, pricing pressures and cost considerations are prompting seal manufacturers to primarily offer standard, off-the-shelf, or as-per-prior-design mechanical seals. However, there are many services and applications where only an "engineered" seal will satisfy the safety and reliability requirements of modern process plants. It is thus important for the user to not only insist that engineered products are used where they make economic sense; the user must also be willing to pay for these superior products. Engineered seals may have special balance ratios, unconventional geometries and expensive materials of construction. Their anticipated face deflection is usually modeled on a computer and finite element analyses are not uncommon.

Assuming that the pumping system is correctly engineered and configurd, very few machinery component upgrade efforts will show higher, or more consistent returns on investment than thoroughly engineered mechanical seals. From the foregoing it should be evident that more reliable mechanical seals can add to plant profitability. But, while we wanted to describe how to spot the features which can make some seals more dependable than others, we must reemphasize a basic fact that was mentioned in out first chapter: Even the best product will fail if there are installation or assembly weaknesses.

Conventional Wisdom: *Better seals always make economic sense.*

Fact: *Better seals make economic sense if all seal auxiliary, control, pump base plate, piping, grouting, installation and assembly criteria have been observed to the letter!*

DUAL MECHANICAL SEAL DESIGNS AND API-682

Mechanical seal designers are being challenged when developing dual seals. Although originally aimed at the Hydrocarbon Processing Industry, the American Petroleum Institute's Seal Standard (API-682 / ISO 21049) is very widely used because it lists many universally applicable requirements. Among them is the intent to optimize seal face cooling, which must also mesh with the need to provide a seal that tolerates pressure reversal and abrasive or congealing fluids. Poor seal face cooling can certainly reduce equipment reliability. The seal standard, therefore, offers users a choice of three different configurations and, as can be expected, each version has advantages and disadvantages.

Although the stipulated cooling and other requirements can appear to be at odds with each other, innovative designs are now available that satisfy all of the diverse needs. Moreover, computational flow techniques (CFD) and testing have facilitated an understanding of uncompromising best available solutions. Also, field studies have fully validated the underlying design concepts.

Dual Seal Arrangements*

Dual seals are increasingly important in the hydrocarbon processing and many other industries. Plant hazard safety requirements, reductions in allowable fugitive emissions and the quest for increased equipment uptime are the main drivers. API- 682 / 3rd Edition (Ref. 8-21) collectively describes pressurized seal geometries as Arrangement 3. This arrangement comprises two seals per cartridge assembly and an externally supplied pressurized barrier fluid. This approach is used to provide a beneficial seal environment. The different configurations for Arrangement 3 are described by API-682 as:

- Face-to-back dual seal in which one mating ring is mounted between the two flexible elements and one flexible element is mounted between the two mating rings or seats (Figure 8-80 a).
- Back-to-back dual seal in which both of the flexible elements are mounted between the mating rings (Figure 8-80 b).
- Face-to-face dual seal in which both of the mating seal rings are mounted between the flexible elements (Figure 8-80 c).

However, the description is, in the opinion of the chapter segment contributor, incomplete. He points out the principal attribute of the face-to-back ("FB," Figure 8-80 a) configuration is locating the process fluid (pumpage) on the outer diameter of the inside set of seal faces. In the other two configurations the process liquid is on the inside diameter of the inside set of seal faces. Note that in this context an outside-facing set of faces would be contacted by the atmosphere.

Face-to-back arrangements are preferred (Ref. 8-22), although back-to-back and face-to-face orientations are offered as purchaser's options. In API-682, non-pressurized dual seals are called Arrangement 2 and face-to-back is the only arrangement option available in the standard (Ref. 8-23).

COMPARISON OF DIFFERENT DUAL SEAL ARRANGEMENTS

Advantages of Back-to-back and Face-to-face Seal Configurations

In the hydrocarbon processing industries, back-to-back and face-to-face configurations are widely represented (Ref. 8-24). They can potentially offer higher levels of performance—in large measure attributable to the cooling effect of barrier fluid flowing over both inner and outer seals. But there also are disadvantages.

Disadvantages of Back-to-back and Face-to-face Seal Configurations

The main shortcoming of back-to-back and face-to-face configurations is that the process fluid is on the inside diameter of the seal faces. Centrifugal force action tends to throw any entrained abrasive solids towards the seal faces, which increases the potential for damage. The "dead zone" formed by the small volume of process fluid underneath the inner seal creates susceptibility to trapped fluids congealing or solids accumulation. The secondary O-ring will tend to then move over this deposit-affected region of the sleeve (Figure 8-80a) and "hang up" is likely to be encountered.

The application of reverse balance can be more difficult with back-to-back designs. Upset conditions such as loss of barrier fluid pressure or increases in process pressure can adversely affect such configurations. Positive retention of the inner-seal mating ring can be difficult to accomplish since there will always be certain unavoidable dimensional constraints of associated hardware and seal chambers.

So, unless properly retained, thrust forces during pressure reversal may cause a ring to become dislodged. Also, with some designs, reverse pressure loading will

*Segment compiled and contributed by Richard Smith, AESSEAL plc, Rotherham, UK, who compiled this segment with valuable input from:
— The American Petroleum Institute (which granted permission to illustrations from API-682;
— Michael Munro, EEMUA Machinery Committee;
— Albany Pumps, Gloucestershire, England.
— Jaguar Automobile Manufacturing Company, Halewood, England.
— Chris Leeper, Thomas Broadbent & Sons, Ltd., Huddersfield, England.
— Dr. Chris Carmody, also Chris Booth BA, MBA, AESSEAL plc, Rotherham, Yorkshire, UK

apply a hydraulic force to the inner seal spring plate that then tends to open the seal faces.

Consequently, over the past two decades the chemical process industries have moved to face-to-back designs and overall seal reliability has improved as a result.

Advantages of Face-to-back Seal Configurations

Face-to-back configurations overcome virtually all the weaknesses of other designs with pumpage on the outside diameter of the seal faces. The preference for this configuration is noted by API-682 (Ref. 8-25) and is summarized below:

'The advantages of the series configuration are that abrasive contamination is centrifuged and has less effect on the inner seal.' A further note supports the case (Ref. 8-26). 'Liquid barrier seal designs arranged such that the process fluid is on the OD of the seal faces will help to minimize solids accumulation on the faces and minimize hang-up.'

Dual balance is easier to incorporate in this configuration and the seal O-ring can be located so as to permit it to move to either side of the groove (Ref. 8-27). This then supports a closing force regardless of the direction of pressure. Mating rings can be simply retained either positively or, with more modern designs, hydraulically (Figure 8-80b). Pressure reversal capability provides for greater safety; it increases the degree of tolerance for many process upset or loss of barrier fluid conditions. Again, API-682 reinforces this point in a note (Ref. 8-25):

'In the event of a loss of barrier fluid pressure, the seal will behave like an Arrangement 2'

Figures 8-81 and 8-82 illustrate O-ring retention and movement with pressure reversals.

Disadvantages of Conventional Face-to-back Configurations

Of major concern in face-to-back designs is cooling of the inner seal. Seal designers have typically approached the issue by mounting component seals with adaptive hardware (sleeve and gland) to form a cartridge. However, the barrier fluid flow path to the inner seal is compromised by a region of low or zero flow. The temperature in these stagnant areas will be elevated by heat soak and face-generated heat (Figure 8-81). The 2nd Edition of API-682 very eloquently provides a warning (Refs. 8-28 and 8-29):

'Restricted seal chamber dimensions and the resulting cartridge hardware construction can affect the ability of the barrier fluid flush to adequately cool the inner seal. Inadequate cooling of the inner seal can result in reduced seal reliability. Selection of a back-to-back or face-to-face configuration may resolve an inner seal cooling problem.'

Accordingly, the challenge for seal designers is to provide both a seal with optimized cooling and to simultaneously provide resistance to "hang up" and pressure reversal.

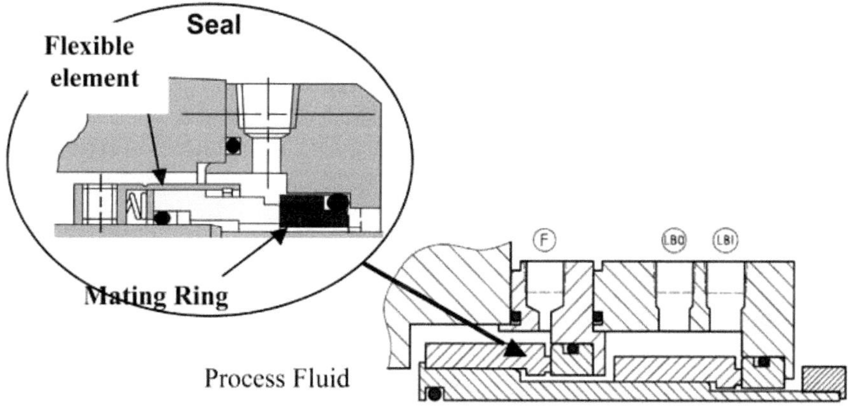

a) 3CW-FB, contacting wet seals in a face-to-back configuration

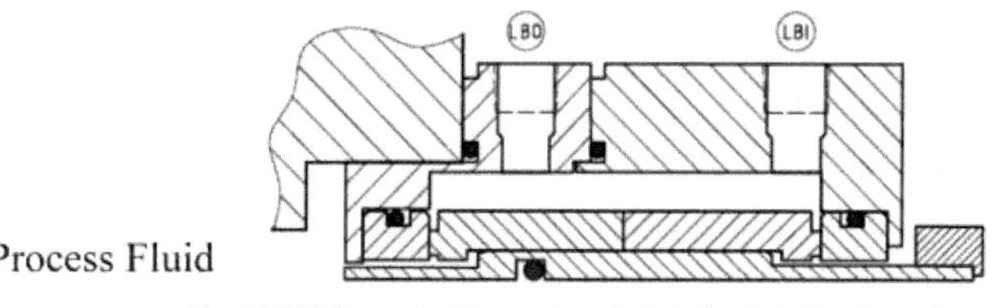

b) 3CW-BB, contacting wet seals in a back-to-back configuration

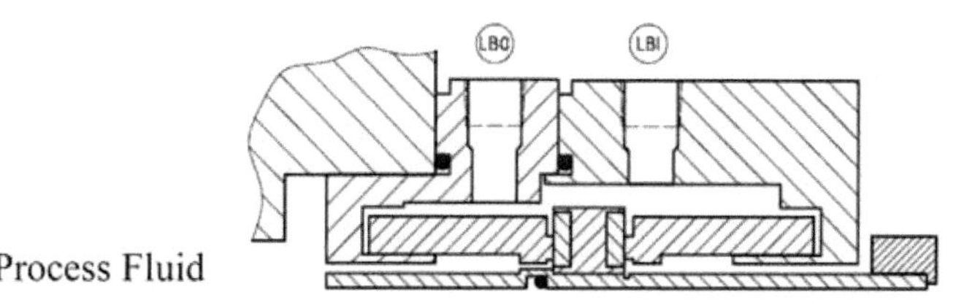

c) 3CW-FF, contacting wet seals in a face-to-face configuration

Figure 8-80: Three styles of dual face mechanical seals for pumps

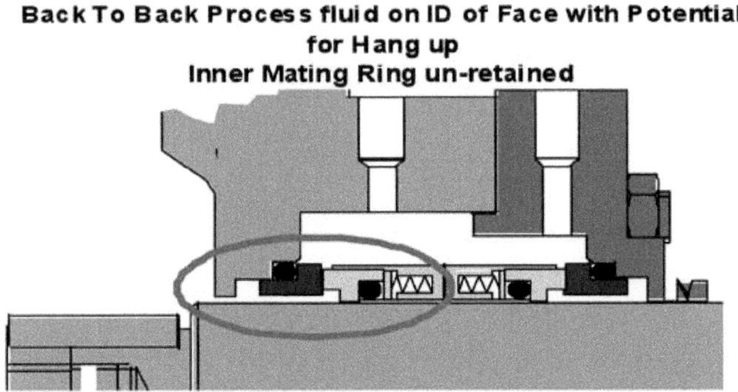

Figure 8-81: Explaining the potential for "hang-up"

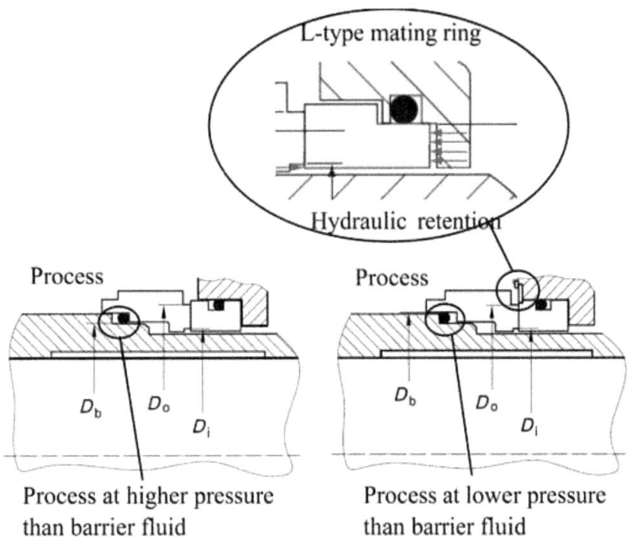

Figure 8-82: O-ring movement with pressure reversals

Deflector Baffle Developments

As can be seen in Figure 8-83 an ordinary face-to-back ("FB," upper right) seal would receive little or no cooling at the inner seal faces. That fact drove innovation in technologies to improve inner seal cooling. Superior cooling can now be achieved by incorporating high performance circulating devices and flow deflector baffle in these seals. The deflector baffle diverts the barrier fluid flow to the inner seal; this provides cooling to the inner seal faces and represents an elegant solution to the dual seal challenge.

Deflector baffles have long been employed on single high-temperature seals connected to quench stream. Restrictions in seal chamber dimensions and somewhat large cross-section of conventional dual balanced pusher seals have generally inhibited the more widespread use of deflector baffles. Except for bellows seals with their traditionally smaller (Figure 8-84) cross-sections, incorporating deflectors has generally been limited to "engineered specials" rather than off-the shelf designs. In applications where more space is available the deflector baffles can be optimally shaped and contoured. This contouring will guide the maximum amount of barrier fluid to the regions from which heat must be removed or where cooling is of greatest benefit.

Developments in Pusher Seal Technology

An alternative method to achieve seal balance is to place O-rings on the outside of the seal faces. This is now common practice in the chemical process industries. Making the sleeve serve also as the face holder is

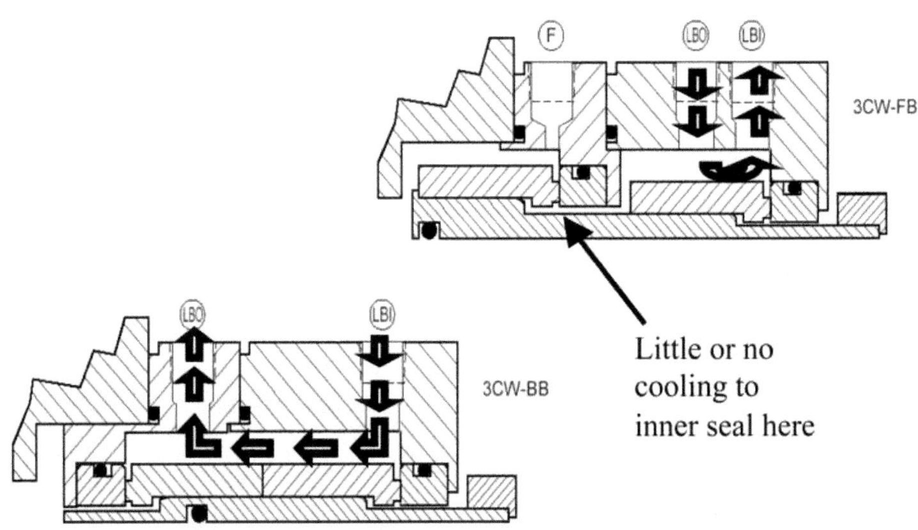

Figure 8-83: Barrier fluid cooling comparison face-to-back (upper right) versus back- to-back (lower left)

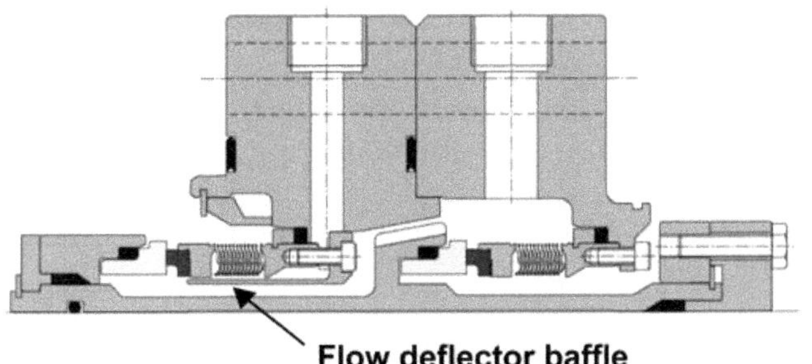

Figure 8-84: Face-to-face bellows seal with deflector baffle. This baffle guides the barrier fluid towards the seal faces (Ref. 8-13)

made possible by modern CNC machining techniques. These design and manufacturing techniques facilitate a more compact design; the techniques open up the inner seal envelope and provide a deflector baffle (Figure 8-85). Separation between barrier flow inlet and outlet causes the cooler barrier fluid to migrate towards the inner seal faces and is often critically important to promoting extended mechanical seal life.

The exact configuration of the deflector baffle shown schematically in Figure 8-85 was actually refined and optimized through the use of modern flow optimization techniques, typically called computational fluid dynamics (CFD). The before-versus-after results are shown in Figure 8-86.

Since it is, of course, desirable to maximize fluid flow in contact with the inboard seal faces, the end of the deflector was actually re-profiled to a triangular sharp edge. The analysis plotted in Figure 8-86 (right side) indicates there is now radial motion next to the extremity of the deflector. This re-circulation reduces the flow path and prevents some fluid close to the deflector nose from escaping before even reaching the seal faces. Compared with the original round shape, the triangular

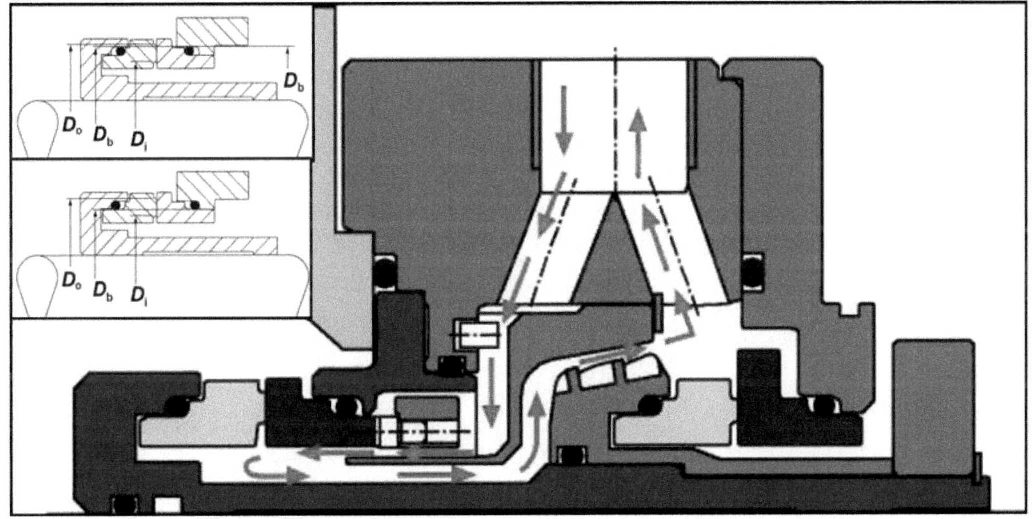

Figure 8-85: Flow of barrier fluid with deflector baffle. Flowrate increases are made possible by the tapered pumping devices shown here (Ref. 8-30)

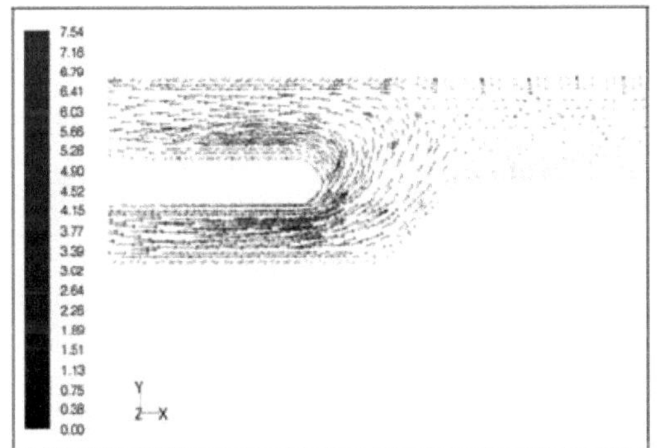
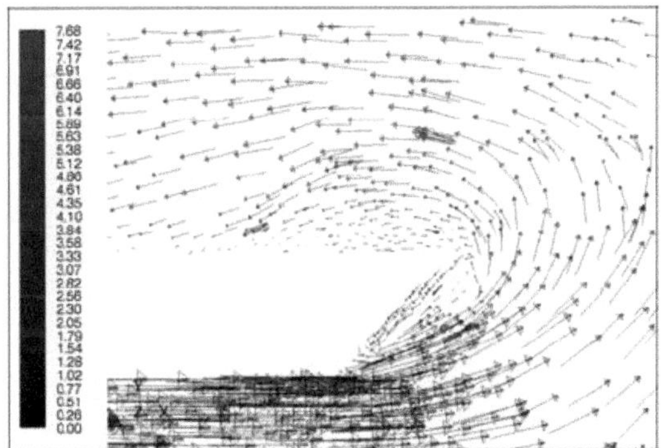

Figure 8-86: Deflector performance before flow optimization (left) and after making modifications for flow optimization (right)

sharp edge shape promotes vortex motion and redirects additional coolant flow to the seal faces.

Circulation Device Performance for API Plan 52 and 53 Systems

Dual wet seal barrier fluid circulation can be achieved by external means per API Plan 54 (Figure 8-54). It can also be achieved by internal circulating devices per API Plans 53 a, b, and c (Figures 8-49 through 8-51), and also per API Plan 52 (Figure 8-48, per Ref. 8-28). Effective cooling of the seal depends upon the efficiency of these devices. Internal devices are part of the seal cartridge; their performance can be affected by many factors and these are conveniently summarized in Table 8-6.

Traditional internal circulating devices (Figures 8-87a and 8-87b) fall into two groups, parallel slot (castellation, Figure 8-87a) and helical vane (Figure 8-87b). Parallel slot devices induce a radial flow and must be positioned adjacent to the barrier outlet orifice; they can be bi-directional only if used with radial ports. Helical vane devices are unidirectional, provide an axial flow, and are less dependent on port proximity.

Table 8-6: Factors that can affect circulation flow

Circulating or pumping device design
Direction of rotation
Seal size
Seal type
Shaft speed
Barrier fluid density
Barrier fluid viscosity
Barrier fluid containment vessel (or cooler)
Seal cavity flow path, concentricity, contour
Gland port orientation ports: top/bottom, tangential
Connecting pipe size & layout (bends/distances)
Fittings connections & roughness of pipe bore

However, the use of multi-axis CNC machine tools has given designers far more freedom to devise considerably more efficient arrangements. Figure 8-85 illustrated a modern bi-directional large clearance "taper vane" pumping ring that provides improved circulation.

The head versus flow performance of a 100 mm seal with the three different pumping devices in Figure 8-87 is plotted in Figure 8-88. The tapered vane device of Figure 8-87c excels with its higher head and higher flow capability (Ref. 8-30).

Practical Application and Limitations of Circulating Devices

Tangential porting arrangements can offer improved performance on all three types of devices and utilizing this feature on large between-bearing pumps is often straightforward. On smaller units port orientation can be more problematic since gland stud position and pump frame casting will often interfere. Regardless of pumping device chosen, its respective performance can be optimized by modifying the internal cavity. Suitably positioned cutwaters and eccentric or tapered bores merit attention. Due to gland plate machining costs these optimization practices tend to be found primarily on engineered API-610 compliant pumps.

Reductions in internal radial clearance can also improve device performance. However, there are safety implications and sufficient clearance is highly desirable to prevent contact between rotating and stationary components. While not normally occurring, such contact is possible under fault conditions. For many years API-682 had specified a minimum radial clearance of 1.5 mm (Ref. 8-31). But in their quest for increased device efficiency, some seal manufacturers lobbied for reduced clearances, although unduly small internal clearances tend to impede adequate thermosiphon action when the

Figure (a)

Figure (b)

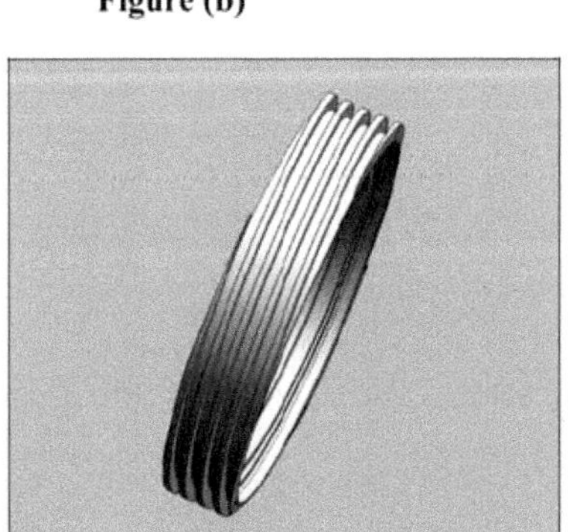

Figure (c)

Figure 8-87: Parallel slot pumping device (a), helical pumping device (b), and tapered vane bi-directional pumping ring (c)

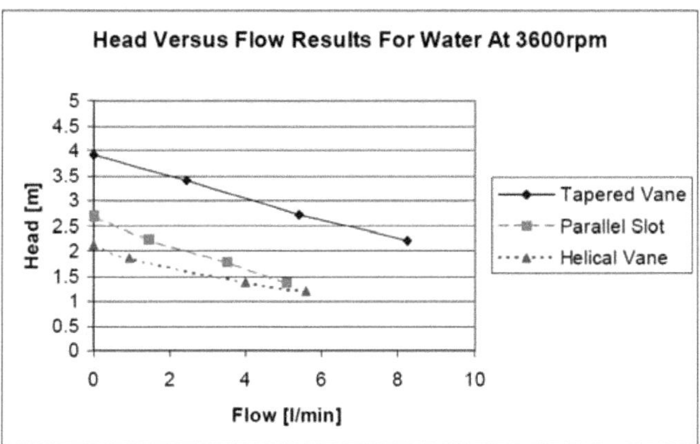

Figure 8-88: Head versus flow performance of a 100 mm seal with the three different pumping devices (shown earlier in Figure 8-87)

shaft is not rotating (Ref. 8-32). It should be noted that such fluid flow motion, i.e. thermosiphoning, is required to prevent overheating of the seal just after shutdown (when a pump will still be at full temperature) or during warm up.

The flow required in actual service depends largely on the amount of heat that must be removed from the seal by the barrier fluid system. Formulas for seal face temperature heat generation and heat soak from the process to the seal chamber are widely published (Ref. 8-13). On small higher temperature pumps the heat load on the seal barrier system will primarily be thermal soak and the cooling requirements are little affected by changes in shaft speed.

As with any pump impeller the performance of a circulating device is affected by its diameter and shaft rotational speed. Performance becomes critical to seal reliability on smaller pumps and on pumps operating at four-pole motor speeds or with variable speed drives. Circulating devices that can operate efficiently at lower speed on smaller shafts with large clearances and radial porting represent the widest potential application group. Bi-directional designs (Figures 8-64 and 8-87c) help eliminate installation errors on between-bearing pumps and can reduce spare parts inventory requirements.

The original taper vane bi-directional device of Figure 8-64 has been in successful commercial service since 1999. Notice that the vanes were straight. Still, research into vane profiles and vane angles has subsequently allowed flow increases of up to 40%. In fact, the vanes furnished after 2011 are contoured into a "swan's neck" shape, which prevents reverse flow on the back edge of the vane. The performance of this device with its now "swan-neck" vanes is virtually equal to that of a similarly configured unidirectional tapered vane. It has been tested at different shaft diameters and shaft speeds. Of interest are the relatively liberal radial clearances of 1.5 mm (0.060 in) between the rotor and stator. The design thus fully conforms to API-682 standards aimed at reducing the risk of seal-internal stationary components being inadvertently contacted by seal-internal rotating devices. Test results obtained with a 50 mm "swan-neck" tapered pumping device design (typical of many medium-size pumps) are readily available (Ref. 8-13).

MORE MECHANICAL SEAL CASE STUDIES

Also of interest are case studies where improved reliability has resulted from implementing some of the design elements discussed in this chapter segment.

Face-to-back Replacement of Back-to-back Configuration

There are many published examples that demonstrate improved reliability after replacing older back-to-back seals with modern by face-to-back geometries. Back-to-back seals had been installed in circulation pumps used on industrial laundry machinery manufactured in the U.K. by Thomas Broadbent and Sons, Ltd.

The duty for these ISO-compliant pumps would be considered "light" in most industries. They operate at 85°C with a seal chamber pressure of 3 bar and two seal sizes, 28 mm and 65 mm. The pumped fluid, however, is contaminated with fibers typically known as "lint". These fibers would enter the seal chamber and become packed near the inside diameter of the inner seal faces. The resulting seal face hang-up caused premature failure. After retrofitting dual seals with face-to-back orientation the pump MTBF (mean-time- between-failures) improved more than six-fold.

Replacement of Back-to-back with Advanced Face-to-back Technology Including Flow Deflector Baffle

A complex reactor circulation duty at a chemical plant in Wales, U.K., demanded that a twin-screw pump (made by Albany Engineering) be used and fitted with four dual seals. Conditions included process temperatures between 25-180°C, seal size 54 mm, seal chamber pressures ranging from slight vacuum to 3.5 bar, viscosities ranging fro 0.5cP to 5000cP, and shaft speeds ranging from 180 to 1500 rpm.

Light silicon oil was selected as the barrier fluid. This had proven problematic on other applications due to its poor lubricity and heat transfer properties at elevated temperatures. However, silicon oil had to be used for reasons of compatibility with the process fluid. An external pumping device Plan 54 configuration (Figure 8-52) was selected for these variable-speed driven pumps.

Several variants of back-to-back bellows cartridge seals were tested, but failures occurred typically after six months. These failures were primarily attributed to pressure reversal issues, as described earlier. Since then, face-to-back seals with deflector baffles have been successfully used at this facility; they replaced both traditional face-to-back and back-to-back designs in these screw pumps. Not only have the pumps achieved extended process runs in excess of two years, but the seal faces were still in pristine condition when examined well after two years.

Traditional Face-to-back vs. Advanced Face-to-back with Internal Pumping Plan 53 in an Automotive Paint Facility

Superior coating technologies are used in the automotive industry. Today, automobile body corrosion is virtually unheard of. One of the reasons for success is full immersion in primer dip tanks with an electrical charge applied. However, facilities were challenged by sealing issues in the electro-coat primer circulation pumps. The paint contains sub-micron abrasive particles, making it necessary to use hard-faced wear-resistant materials on the inner seal. The paint temperature has to be maintained near a process temperature of 25°C ±10°C. Upwards excursions in temperature are risky and could cause paint to rapidly congeal. Dual seals are used for this sealing duty and ultra-filtrate, essentially de-ionized water with other chemicals acting as a thinner, is used as a compatible barrier fluid.

End suction pumps with a 50 mm shaft are employed. At first, seal cooling and poor circulation device efficiency had prevented reliable operation of conventional face-to-back seals and API Plan 53a systems (Figure 8-49). The paint is tenacious and tends to overcome the barrier fluid film pressure between the inner seal faces. Heat generated by these faces causes the paint to congeal or polymerize. As the paint particles migrate, they cause the seal face gap to widen and, although the faces are undamaged, excessive leakage results. While it would seem logical to simply increase the barrier fluid pressure, this would increase temperature generation and the added heat load would merely accelerate problem development.

Reliable sealing has traditionally been achieved by moving pressurized barrier fluids at the high flow rates achievable with API Plan 54 (Figure 8-52). However, these relatively complex external barrier fluid circuits often require instrumentation ensuring that equal flow reaches each seal in the many pumps installed for parallel operation. Individual seal fault diagnostics can be difficult and cost-intensive with Plan 54 because a centralized pressure source provides external barrier fluid circulation to several pump sets (Ref. 8-33).

These concerns then prompted plant designers and operators to seek alternative solutions. Modern face-to-back cartridge seals with high efficiency circulation devices and barrier fluid flow separation baffles will give superior performance in such duties with API Plan 53a (Figure 8-49). API Plan 53a systems offer many advantages in terms of cost and simplicity; these systems have been successfully installed in five automotive plants in Europe and other continents. In fact, advanced face-to-back seals have, as of 2012, seen flawless operation for over seven years at one UK automotive plant.

Conclusion

Mechanical seal technology continues to develop and sound sealing principles are useful in all sectors of industry. Highly reliable face-to-back designs are now available with enhanced cooling features. These developments have further increased the application range for this advantageous dual-seal configuration. Well-proven developments combine API Plan 53a with tapered vane pumping devices. They certainly merit close consideration in many pumping services.

WHAT TO LOOK FOR IN STRATEGIC MECHANICAL SEAL PARTNERSHIPS

Since the late 1980's and early 1990's more and more mechanical seal users have entered into an alliance or partnership with a mechanical seal manufacturer. While it makes sense to involve the services of a knowledgeable company to tend to your needs, there are major differences in how some of these alliances are structured.

Not all mechanical seal supplier agreements benefit both parties equally, which is to say that we have often seen some rather one-sided ones. At an international pump users' conference in 2001, the suggestion that the "partner" should disclose pumping services where another manufacturer's seal would outperform the "partner's" product was met with derision by one, and silence by a second among the few world-scale mechanical seal manufacturers.

Nevertheless, there is at least one well-known major manufacturer that has experience offering reliability-focused users what they really need: Sealing products that represent best life-cycle cost value. And, of course, the application of sealing products that represent best long-term value is the only approach that makes sense, regardless of whether these products are

- Already in your plant and originally supplied by his competition
- Supplied by your partner company and installed in your plant at the next downtime event, or
- Supplied by the partner's competitors(!) and waiting to be installed at the earliest opportunity
- Standard products or—if needed—specially engineered mechanical seals.

Of equal or even greater importance, is the fact that a knowledgeable partner supplier will undoubtedly point out substantial and valuable pump improvement measures. The implementation of these measures will not only extend seal life, but will drive up equipment MTBR/MTBF as well. Valuable maintenance cost savings will result.

Objectives of Mutually Beneficial Agreements

For an agreement to meet the test of mutual benefit, achieving at least 90% of the seal life obtained elsewhere for a given service should be one of the stipulated goals. Remember, though: A good contract translates this goal into a legally binding clause.

Because energy efficiency, regulatory compliance and effective supply chain management are part of the life-cycle cost objectives, the use of (dry) gas seals, and perhaps even seals manufactured by the competition cannot be ruled out. Also, since the partners must share data on best available sealing technology, the supplier must compile and disclose this technology to the user. User and supplier will now have joint access to all available data and can compare their achievements against best available technology.

The supplier should benefit from the base fee paid and should, in addition, earn a success fee that gradually peaks at achieving the 90% mark mentioned above.

Additional bonuses may be negotiated and predefined as the 90% mark is exceeded.

Recall, again, that the contract must clearly deal with the issue of best life-cycle cost value for the user. This involves solid input from qualified reliability engineers. Purchasing departments that had no access to tangible life-cycle cost data have never been successful in defining the requisite contractual language.

Scope of Supply Must be Defined

The supplier should agree to furnish products and services including:

- Mechanical seal hardware (complete seals and components) and/or the repair of same for a given population of rotating equipment as defined in an appropriate tabulation that will have to be appended to the narrated agreement.
- Detailed and specific instructions, assessments, life-cycle cost impact calculations (including energy efficiency) and other recommendations, as needed for enhancing the reliability of all sealing systems predefined and covered by the agreement.
- Training of operators and maintenance workforces on a predefined time schedule. Unless otherwise agreed to, quarterly training is to be conducted.
- An on-site technician is to be supplied. This technician should spend a specified number of man-hours per year at the user's facility.
- Management of the supply chain of the inventory of complete seals, the rebuilding of seals (regardless of origin), without jeopardizing the availability of equipment within the population to meet the plant production requirements.

Inventory Buy-back

The supplier should purchase the current and useable mechanical seal inventory from the user subject to the following conditions:

- The inventory must be utilized on equipment currently in use and expected to be in use over the term of the contract.
- The inventory must be of current design and suitable for providing reliability commensurate with the goals set forth in this agreement.
- The inventory meeting the above criteria should be acquired by the supplier at an agreed discount on its present book value.
- The inventory should be paid for in the form of a credit memo to be used to offset the base fees to be paid to the supplier. The credit memo should not exceed a defined share of the base fee in any given month over the life of the contract.
- Any inventory purchased by the supplier that has not been used in the fulfillment of this contract should be repurchased by the plant owner at the termination or completion of this contract. Repurchase should be at the same price as paid by the supplier, plus carrying costs of the unused inventory at as high as 10% per year.

Obligations and Responsibilities of the Owner

The owner must provide the supplier with:

- Accurate data on the maintenance history and the parts consumption of the equipment population covered by this agreement. To the extent it is determined that insufficient history does not allow supplier to accurately estimate its cost of meeting obligations under the terms set forth in an agreement, the supplier should be allowed an adjustment of the fees to be paid to them, or delete the equipment from the population covered by the agreement.
- Timely implementation of recommended, and mutually agreed-to improvements necessary to achieve the goals and/or objectives of the agreement. If the improvements are not made in a timely and as agreed upon manner, the same remedies should be available to supplier as stated above.
- Suitable access to plant and working space is needed.
- Access for training, at least annually, during normal working hours and at no cost to the supplier. The training should be extended to such operations and maintenance personnel as have a direct impact on the reliability and performance of the equipment covered by the agreement.
- Adequately decontaminated equipment, or the knowledge and facilities allowing supplier to adequately decontaminate any equipment or device that is to be handled by supplier.
- Material safety data sheets and related documentation.
- Payment, net 30 days and with surcharges of 1.5% per month of unpaid balance thereafter.

Obligations and Responsibilities of the Supplier

The supplier should provide the products and services covered by this agreement, barring labor strife, catastrophes, shutdowns, etc., in which case equitable pro-rata separation of responsibilities would have to be made.

Success Oversight Committee and Designated Arbiter

- Each party should appoint at least two, but no more that three individuals, to a committee charged with overseeing the timely and successful implementation of the agreement at each plant site. This committee should convene periodically, but no less frequent than every 60 days.
- In plants with two or more sites, an executive committee should monitor the overall progress. The committee should have at least one member from each side with sufficient authority to resolve disputes and direct corrective steps to resolve deficiencies in their company's performance under the terms of this agreement.
- Disputes that cannot be resolved by the committee(s) should be submitted to an independent third-party arbiter who will be designated by mutual agreement at the inception of this agreement and whose fees should be paid equally by both supplier and owner.

Compensation for Products and Services

- The base fee should be set by mutual agreement based on prior yearly seal consumption (equipment MTBR), as well as historic expenditures for such related outlays as seal failure induced collateral damage, production losses, etc.
- In addition, the supplier should earn a mutually agreed-to success fee equivalent to 1/3 of the overall savings generated by the enhancements and to the extent that these exceed the predetermined goals of the program. Savings are defined as direct and measurable savings in labor, parts, energy, incremental production, etc.

Terms and Termination

- A typical agreement should cover a five-year period.
- Early termination of the contract without cause will cause monetary damage to both parties. Therefore, the party terminating without cause should agree to pay the other party damages equivalent to 12 months base fee and an annualized success (or, as the case may be, non-realized success fee based on the results achieved (or, as the case may be, forfeited) because of early termination.

This is a basic proposal for an equitable agreement that serves both parties. While it may not be considered complete, it should form the cornerstone of your partnership agreement with a seal supplier. A well thought-out agreement assists users in achieving pump life extensions and downtime avoidance.

REPETITIVE MECHANICAL SEAL FAILURES CAN CAUSE DISASTERS

Needless to say, pump life extension and downtime avoidance are one and the same. They are linked to the relationship between seal user and strategic partner. To what extent the incident described below could have been influenced by the right partnership structure is not known. It simply makes the point that root cause failure analysis should commence after the second failure.

Release No. 2004-08-I-NM, issued in October 2005 by the U.S. Chemical Safety and Hazard Investigation Board, addresses an incident at an oil refinery with a history of repeated pump failures. Located in the Southwestern United States, this facility's total of three primary, electric, and steam-driven spare iso-stripper recirculation pumps had 23 work orders submitted (see Table 8-7) for repair of seal-related problems or pump seizures in the one-year period prior to a fire and explosion. The catastrophic incident occurred during equipment disassembly on April 8, 2004 and caused over $13,000,000 in damage. At least six people were injured and production at this alkylation unit was shut down for months.

Because of the serious nature of the incident the U.S. Chemical Safety and Hazard Investigation Board [http://www.csb.gov] produced a detailed write-up which describes the sudden release of flammable liquid and subsequent fire and explosion that occurred. The type or make of seal used is not known to the authors.

Lessons for Those Willing to Learn

For many decades, truly reliability-focused organizations have avoided repeat seal failures by insisting on understanding and eliminating failure causes. They realize that collecting only the generalized failure descriptions of Table 8-7 would be analogous to a trucking company cataloguing repeated non-performance as "engine problems." More detail would be needed to implement sound remedial action.

The reliability-focused also realize that when a seal failure combines with one or more other deviations from the norm, disasters result. More specifically, reliability-focused users make it their business to know what fit-for-service mechanical seals are available. But, of course, these components have to be properly installed and will (usually) require a pump-around circuit and dual

Table 8-7

SEQUENCE / DATE <> PUMP	PROBLEM
01. April 17, 2003<> P-5A (Elec.)	Seal leak
02. May 9, 2003<> P-5B (Stm.)	Pump spraying from seal
03. May 23, 2003<> P-5A (Elec.)	Repair seal
04. June 9, 2003<> P-5B (Stm.)	Repair seal
05. June 9, 2003 <>P-5A (Elec.)	Repair seal
06. June 18, 2003 <>P-5A (Elec.)	Repair seal
07. June 20, 2003 <>P-5A (Elec.)	Replaced seal
08. July 31, 2003 <>P-5A (Elec.)	Replaced seal
09. August 22, 2003<> P-5B (Stm.)	Seal leak
10. August 25, 2003 <>P-5B (Stm.)	Replaced seal
11. September 26, 2003<> P-5B (Stm.)	Replaced seal
12. September 26, 2003<> P-5A (Elec.)	Replaced seal
13. October 14, 2003 <>P-5A (Elec.)	Seal leak
14. December 6, 2003<> P-5A (Elec.)	Replaced seal
15. December 9, 2003<> P-5B (Stm.)	Seal leak
16. December 9, 2003 <>P-5A (Elec.)	Seal leak
17. December 15, 2003<> P-5B (Stm.)	Replaced seal
18. December 15 2003<> P-5A (Elec.)	Seal leak
19. January 28, 2004<>P-5A (Elec.)	Seal leak
20. March 22, 2004<> P-5A (Elec.)	Seal leak
21. April 1, 2004<> P-5A (Elec.)	Pump seal leaking
22. April 3, 2004<> P-5A (Elec.)	Pump seal leaking
23. April 7, 2004<> P-5A (Elec.)	Repair pump seal

seals, generically depicted in Figure 8-89, that include a conservatively designed wide-clearance pumping ring.

We should remind ourselves of a key requirement of API 682/3rd Edition: Mechanical seals installed in refinery equipment should have a design life of 25,000 hours. Clearly, the mechanical seals in this particular installation fell far short of that requirement.

As to other lessons, there was an obvious breakdown in maintenance management at that refinery. It is hard to comprehend that an organization would tolerate 23 costly pump interventions without insisting on solid answers and remedies long before an inevitable catastrophic event occurs. Issues no doubt start with inadequate experience during the purchasing stage. After installation, it takes well trained, motivated mechanical engineers to correctly diagnose seal failures and find the true root causes. Unless the root cause is determined, there will be repeat failures. Don't employ the non-teachable and do develop an intense dislike for repeat failures. Be determined to reward managers and hourly employees who eradicate repeat pump failures.

For a certainty breakdown maintenance on machinery will lead to complete plant breakdowns, as occurred here. Identifying operating conditions and risky components that could contribute to equipment failure are critical ingredients of a sound mechanical integrity program.

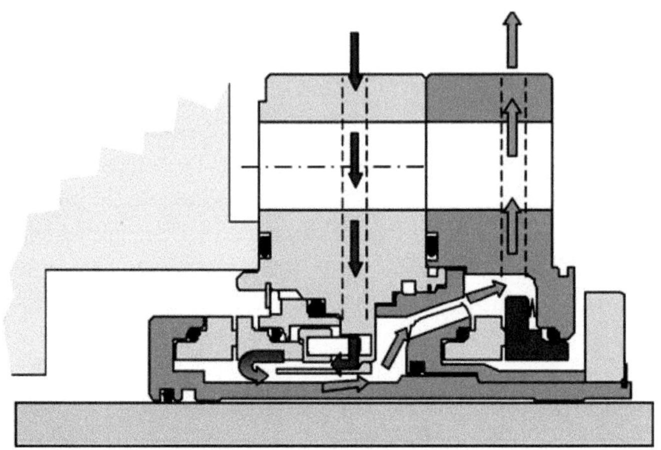

Figure 8-89: Modern dual seal with wide-clearance pumping ring and barrier fluid pump-around circuit. A dual seal and a Plan 53 system is thought to be the appropriate specification for this particular application.

Chapter 9

Improved Lubrication and Lubricant Application

EXTENDING PUMP LIFE BY UNDERSTANDING LUBE APPLICATION

After devoting many pages to bearing selection and application topics, why is there more information being presented here? The reasons are largely experience-based. Having spent literally thousands of hours analyzing pump failures and component distress, we know for a fact that lubricant *application* has not kept pace with *lube formulation* technology. It has been estimated that the life of roughly one-half of all pump bearings could be extended by better, or optimized lube application. More advantageous application methods are readily available and are, for the most part, rather inexpensive. We intend to explain them.

There is confusion as to when, where and why synthetic lubricants are justified in process pumps. We would like to clarify the issue.

Also, there exist costly misunderstandings regarding the applicability of "life-time" grease-lubricated bearings that will be dealt with in this chapter. We hope to provide authoritative answers to two important questions: what is an "optimized lubricant," and what constitutes an "optimized application method."

Generally, pumps with power input requirements in excess of 500 HP (~375 kW) use sliding bearings, although even a few 3,000 HP (~2,240 kW) pumps are known to do rather well with rolling element bearings. So, if your large pumps have rolling element bearings, don't automatically assume this to cause problems. Ask first if your rolling element bearings are being lubricated in optimum fashion. Table 9-1 may give some initial guidance.

Table 9-1 shows the possible influence of lubrication method on the fatigue life of rolling element bearings in pumps. It is extremely noteworthy that filtered jet-oil, circulating systems and pure oil-mist are ranked well above the traditional and widely used unbalanced constant-level lubricators.

It serves little purpose to investigate why pump manufacturers have been slow to offer improved lubrication. Reliability-focused readers are encouraged to review and absorb from the present and subsequent chapters relevant information on available lube application methods. If implemented, these superior methods can significantly extend bearing life. There can be no doubt that oil mist and jet-oil lubrication are worthy of consideration in many plant environments. It is fair to say that the proper application of jet-oil lube or dry sump oil mist will cure any rolling element bearing problem unless, of course, the root cause of the problem is not lube related.

It is reasonable to assume that the majority of centrifugal process pumps are small, say, less than 200 HP (150 kW). Therefore, most pumps encountered by reliability professionals are equipped with rolling element bearings. Avoiding or reducing rolling element bearing failures should thus be a prime topic whenever pump life extensions are being discussed. Close examination of Tables 9-2 and 9-3 may allow spotting performance

Table 9-1: Influence of Lubrication on Service Life of Pump Bearings

Influence of Lubrication on Service Life of Pump Bearings
(Condition: Lubricant Service Life Much Below Bearing Fatigue Life)

Ranking Order for Oil-Lubricated Rolling Element Bearings	*Ranking Order for Grease-Lubricated Rolling Element Bearings*
1. Jet-oil with filter	
	1.5 Automatic feed
2. Circulation with filter	
3. Oil-air; oil mist	
4. "Balanced" automatic lubricator, i.e. without ambient air contact	
5. Circulation without filter	
6. Sump, regular renewal	
	6.5 Regular regreasing of cleaned bearing
7. "Unbalanced" automatic lubricator, i.e. oil in contact with ambient air	
8. Sump, occasional renewal	
	8.5 Occasional renewal
	9.0 Occasional replenishment
	9.5 Lubrication for life

Table 9-2: Bearing types and characteristics (Source: MRC Division of SKF USA)

Bearing Type		Cage	Contact Angle	Performance Level	Characteristic: Radial Stiffness	Axial Stiffness	Speed	Radial Capacity	Thrust Capacity
S-Type Single Row Deep Groove	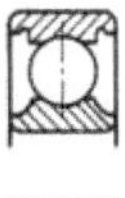	two-piece steel	Radial Type	Extremely High Very High High Moderate					
5000 Series Double Row		one-piece steel	30°	Extremely High Very High High Moderate					
5000 UPG Double Row Pump Bearing		ball-guided one piece machined brass	40°	Extremely High Very High High Moderate					
7000 Series Duplex	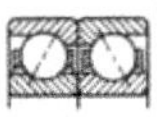	ball-guided two-piece steel	29°	Extremely High Very High High Moderate					
8000 Series PumPac Bearing Set	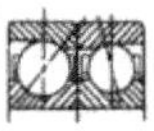	land-guided one-piece machined brass	40°/15°	Extremely High Very High High Moderate					
8000 AAB Series PumPac Bearing Set		land-guided one-piece machined brass	40°/15°	Extremely High Very High High Moderate					
8000 BB Series PumPac Diamond Bearing Set	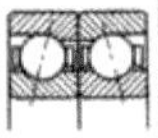	land-guided one-piece machined brass	15°/15°	Extremely High Very High High Moderate					
7000P & 7000PJ Duplex	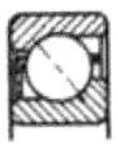	ball-guided one-piece machine or stamped-brass	40°	Extremely High Very High High Moderate					
97000U2 Duplex Tandem Pair Radial & Axial Looseness in Bearing System	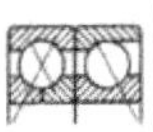	land-guided one-piece machined brass	29°	Extremely High Very High High Moderate					
97000UP2 Duplex Tandem Pair Radial & Axial Looseness in Bearing System	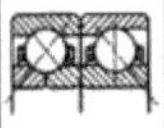	land-guided one-piece machined brass	40°	Extremely High Very High High Moderate					

Table 9-3: Bearing types and applications in pumps

Bearing Type		Cage	Contact Angle	Applications
S-Type Single Row Deep Groove		two-piece steel	Radial Type	Steady rest positions to accommodate radial load in centrifugal pumps. Most electric motors use this bearing to accommodate radial loads. Seldom used as a primary thrust bearing in centrifugal pumps.
5000 Series Double Row		one-piece steel	30°	Most often used in ANSI pumps with moderate thrust loads. 5000 series bearings can also be used in the radial position when the radial load is excessive for SRDG 200S and 300S series.
5000 UPG Double Row Pump Bearing		ball-guided one piece machined brass	40°	For moderate speed centrifugal pumps, used as thrust bearings where high thrust loads are expected. To upgrade your ANSI and API standard, fifth edition to meet ANSI + standards without the need to retrofit.
7000 Series Duplex		ball-guided two-piece steel	29°	For high speed centrifugal pumps. This bearing can run at higher speeds than the 7000P series and requires less thrust load to maintain proper traction forces between the ball and raceway surfaces.
8000 Series PumPac Bearing Set		land-guided one-piece machined brass	40°/15°	For centrifugal pumps with heavy thrust loads that are not reversing or reverse only momentarily. Very effective as a thrust bearing in high speed pumps when direction of thrust is known. Forgiving in an application when thrust loads and temperatures are not determined.
8000 AAB Series PumPac Bearing Set		land-guided one-piece machined brass	40°/15°	For pumps involving very heavy primary thrust loads consisting of a PumPac triplex set with two 40 degree bearings in tandem matched back-to-back with one 15 degree bearing.
8000 BB Series PumPac Diamond Bearing Set		land-guided one-piece machined brass	15°/15°	Balanced pumps, operating with light or no thrust loads at high speeds. Example: Double suction impeller pumps between bearing applications. Pumps with closed impellers, balance holes and pump-out vanes that result in light thrust loads.
7000P & 7000PJ Duplex		ball-guided one-piece machine or stamped-brass	40°	For moderate speed centrifugal pumps, used as thrust bearings where high thrust loads are expected. Temperatures, loads and shaft fits should be known to establish proper preload or clearance.
97000U2 Duplex Tandem Pair Radial & Axial Looseness in Bearing System		land-guided one-piece machined brass	29°	For vertical or other pumps when endplay is not a major concern. This bearing type will be specified for speeds higher than those accommodated by the 97000UP series. This product is often used in ethylene stirrer motor applications.
97000UP2 Duplex Tandem Pair Radial & Axial Looseness in Bearing System		land-guided one-piece machined brass	40°	For vertical pumps, other types of pumps and electric motor applications. For use where extremely high thrust load is possible. Often used in ethylene reactors as well as deep water pumps.

levels or characteristics of existing, vendor-supplied pump bearings that do not meet reasonable expectations. Perhaps some of these bearing types or configurations were less expensive when bought initially, but have since proved very costly after factoring-in long-term maintenance and downtime costs.

Although Chapter 7 dealt extensively with these issues, there is much more that needs to be stated and learned about lube types—both liquid and semi-solid—and their application. This chapter will thus serve as a catchall repository of items and information dealing with pump and driver lubrication. It will supplement and amplify Chapter 7 and must be viewed as just one more indispensable ingredient to the achievement of optimum pump life.

Rolling Element Bearing Life Expectancy

A rolling element bearing is a precision device and a marvel of engineering. It is unlikely that any other mass-produced item is machined to such close tolerances. While boundary dimensions are usually held to tenths of a thousandth of an inch or a few microns, rolling contact surfaces and geometries are maintained to even tighter tolerances. It is for this obvious reason that very little surface degeneration can be tolerated (Ref. 9-1).

The life of a rolling element bearing running under good operating conditions is usually limited by fatigue failure rather than by wear. Under optimum operating conditions, the fatigue life of a ball bearing is determined by the number of stress reversals and by the cube of the load causing these stresses. As examples, if the load on the ball bearing is doubled, the theoretical fatigue life is reduced to one-eighth. This is graphically shown in Figure 9-1. Also, if speed is doubled, the theoretical fatigue life is reduced to one-half.

For roller bearings, the number of stress reversals determines the fatigue life and the exponent 3.3 is used instead of 3 (Ref. 9-2). This means that, for roller bearings, doubling the load would result in the theoretical fatigue life being reduced to about one-tenth. Needless to say, solid contaminants and/or water further decrease the life of any rolling element bearing.

How bearings are rated, and how their respective lives are a function of bearing geometry, size, load, speed, lube type and lube contamination can be studied in the general catalogs of the major bearing manufacturers. It is thus well outside the scope of this text to duplicate these general catalogs. Suffice it to say, the authors know of process units in petrochemical plants where hundreds of grease-lubricated electric motor bearings are still in continuous service after 20 years, and pumping services where bearings have not been replaced in well over ten years. This chapter explains what reliability-focused users have known and practiced for many decades and what simple steps will positively lead to these attainable lives.

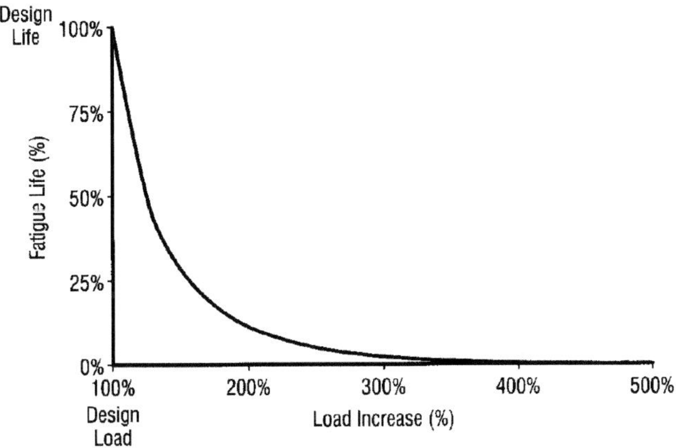

Figure 9-1: Effect of load on ball bearing fatigue life (Source: SKF USA, Inc., Kulpsville, Pennsylvania 19443)

Friction Torque

The friction torque in a rolling element bearing consists essentially of two components. One of these is a function of the bearing design and the load imposed on the bearing. The other is a function of the lubricant type, the quantity, and the speed of the bearing.

It has been found that the friction torque in a bearing is lowest with a quantity of the correct viscosity oil just sufficient to form a film between the contacting surfaces. This is just one of the reasons why oil mist (to be explained later) is a superior lubrication method. The friction will increase with greater quantity and/or higher viscosity of the oil. With more oil than just enough to separate the rolling elements, the friction torque will also increase with the speed.

Conventional Wisdom: *The more lubricant the better the bearing will perform*

Fact: *Excessive amounts of oil cause increased friction, heat buildup and lubricant degradation. Useful bearing life will suffer from too much oil.*

Function of the Lubricant

A bearing lubricant is necessary for a number of reasons. Here are some of them:

- To lubricate sliding contact between the cage and other parts of the bearing.
- To lubricate any contact between races and rolling elements in areas of slippage.
- To lubricate the sliding contact between the rollers and guiding elements in roller bearings.
- In some cases, to carry away the heat developed in the bearing.
- To protect the highly finished surfaces from corrosion.
- To provide a sealing barrier against foreign matter.

Oil vs. Grease

The ideal lubricant for rolling element bearings is oil. Grease, formed by combining oil with soap or non-soap thickeners, is simply a means of effecting greater utilization of the oil. In a grease, the thickener acts fundamentally as a carrier and not as a lubricant.

Although relatively few centrifugal pump bearings are grease lubricated, greases are in fact used for lubricating by far the largest overall number of rolling bearings. The extensive use of grease has been influenced by the possibilities of simpler housing designs, less maintenance, less difficulty with leakage, and better sealing against dirt. On the other hand, there are limitations to the use of grease. Where a lubricant must dissipate heat rapidly, grease should not be used. In many cases, associated machine elements that are oil lubricated dictate the use of oil for rolling element bearings. Listed below are some of the advantages and disadvantages of grease lubrication.

Advantages of Grease Lubrication

With grease lubrication, we find five principal advantages over lubricating with liquid oil:

1. Simpler housing designs are possible; piping is greatly reduced or eliminated.
2. Maintenance is greatly reduced since oil levels do not have to be maintained.
3. Being a solid when not under shear, grease forms an effective collar at bearing edges to help seal against dirt and water intrusion.
4. With grease lubrication, leakage is minimized where contamination of products must be avoided.
5. During start-up periods, the bearing is instantly lubricated whereas with pressure or splash oil systems, there can be a time interval during which the bearing may operate before oil flow reaches the bearing.

Disadvantages of Grease Lubrication

As with so many things, where there are advantages, there will also be disadvantages. Once both are understood, the task of selecting one over the other will become easier. Occasionally, the selection will vary depending on factors that may be unique to each application. Here, then, are the three main disadvantages of grease lubrication:

1. Extreme loads at low speed or moderate loads at high speed may create sufficient heat in the bearing to make grease lubrication unsatisfactory.
2. Oil may flush debris out of the bearing whereas grease will not.
3. The correct amount of lubricant is not as easily controlled as with oil.

Oil Characteristics are Important

The ability of any oil to meet the requirements of specific operating conditions depends upon certain physical and chemical properties. The pump reliability improvement specialist must be conversant with some of these. Also, since greases consist of 85% oil, physical and chemical properties may be equally important for both liquid oils and greases.

Viscosity

The single most important property of oil is viscosity. It is the relative resistance to flow. A high viscosity oil will flow less readily than a thinner, low viscosity oil.

Viscosity can be measured by any of a number of different instruments that are collectively called viscosimeters. In the United States, the viscosity of oils is often determined with a Saybolt Universal Viscosimeter. It simply measures the time in seconds required for 60 cc of oil to drain through a standard hole at some fixed temperature. The common temperatures for reporting viscosity are 100°F and 210°F (38°C and 99°C). Viscosities are now quoted in terms of centistokes (cSt), although years ago United States users were more accustomed to Saybolt Universal Seconds (SUS).

The correct operating viscosity for rolling element bearings is both speed and temperature dependent. While it would thus be appropriate to select a minimum viscosity for each application, inventory and maintenance constraints may make this impractical. For ball bearings and cylindrical roller bearings in centrifugal pumps, it is an acceptable compromise to select an oil which will have a viscosity of at least 13 cSt, or 70 SUS at operating temperature. An oil viscosity selection chart was given earlier (Figure 7-18) and is also shown in Figure 9-4, later in this chapter.

Viscosity Index ("VI")

All oils are more viscous when cold and become thinner when heated. However, some oils resist this change of viscosity more than others. Such oils are said to have a high viscosity index (VI). Viscosity index is most important in oils that must be used over a wide range of temperatures. Such oils should resist excessive changes in viscosity. A high VI is usually associated with good oxidation stability and VI can thus be used as a rough indication of such quality.

Pour Point

Any oil, when cooled, eventually reaches a temperature below which it will no longer flow. This temperature is said to be the pour point of the oil. At temperatures below its pour point an oil will not feed into the bearing and lubricant starvation may result. In selecting an oil for rolling element bearings, the pour point must be considered in relation to the operating temperature.

Flash and Fire Point

As oil is heated, the lighter fractions tend to become volatile and will flash off. With any oil, there is some temperature at which enough vapor is liberated to flash into momentary flame when ignition is applied. This temperature is called the flash point of the oil. At a somewhat higher temperature enough vapors are liberated to support continuous combustion. This is called the fire point of the oil. The flash and fire points are significant indications of the tendency of an oil to volatilize at high operating temperatures. High viscosity index ("high VI") oils generally have higher flash and fire points than lower VI oils of the same viscosity.

Oxidation Resistance

All petroleum oils are subject to oxidation by chemical reaction with the oxygen contained in air. This reaction results in the formation of acids, gum, sludge, and varnish residues which can reduce bearing clearances, plug oil lines and cause corrosion.

Some lubricating fluids are more resistant to this action than others. Oxidation resistance depends upon the fluid type, the methods and degree of refining used, and whether oxidation inhibitors are used.

There are many factors that contribute to the oxidation of oils and practically all of these are present in lubricating systems. These include temperature, agitation, and the effects of metals and various contaminants that increase the rate of oxidation.

Temperature is a primary accelerator of oxidation. It is a well known, but somewhat conservative assumption, that rates of chemical reaction double for every 18°F (10°C) increase in temperature. Below 140°F (60°C), the rate of oxidation of oil is rather slow. Above this temperature, however, the rate of oxidation of mineral oils increases to such an extent that it becomes an important factor in the life of the oil. Consequently, if mineral oil systems operate (conservatively) at temperatures above 140°F (60°C), more frequent oil changes would be appropriate. Figure 9-2 shows recommended oil replacement frequencies as only a function of oil sump or reservoir capacity. However, since it is intuitively evident that oil operating temperatures merit consideration, we draw attention to Figure 9-3. Here, the service life of a typical mineral oil is referred to its operating temperature.

The oxidation rate of oil is accelerated by metals, such as copper and copper-containing alloys, and to a much lesser extent by steel. In addition, contaminants such as water and dust also act as catalysts to promote oxidation of the oil. Note, however, that these guidelines are of academic importance on high-grade synthesized hydrocarbon lubricants, commonly called "synthetics." These lubricants have inherently greater stability and are often, theoretically, suitable for operating temperatures above those experienced in process pumps.

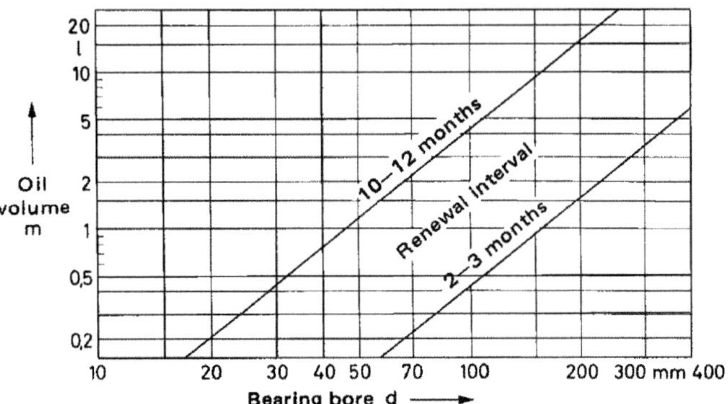

Figure 9-2: Recommended lubrication oil change frequency (Source: FAG, Stamford, Connecticut)

Emulsification

Generally, water and straight oils do not mix. However, when oil becomes dirty, the contaminating particles act as agents to promote emulsification. In rolling element bearing lubricating systems, emulsification is undesirable. Therefore, the oil should separate readily from any water present, which is to say it should have good demulsibility characteristics. Since these desirable characteristics are readily upset by the arbitrary addition of over-the-counter additives, a reliability-focused user company would not tamper with the premium lubricants supplied by responsible oil manufactures and formulators.

Conventional Wisdom: *A good additive offered by a renowned race car driver or other celebrity must en-*

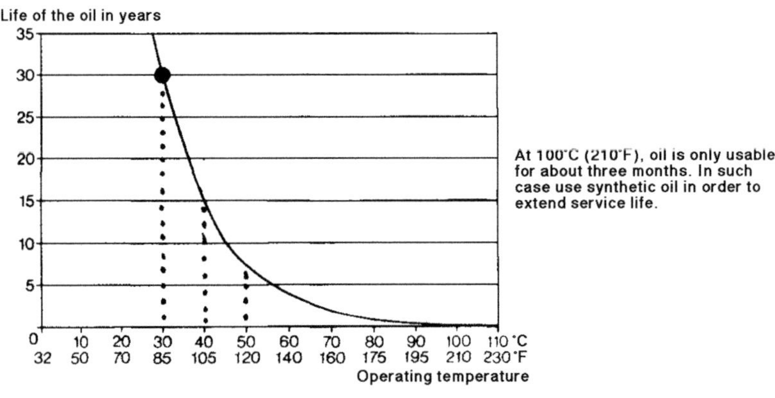

Figure 9-3: Service life of typical mineral oil as a function of oil operating temperature

hance equipment life.

Fact: *Premium grade lubricants are "in balance." Adding anything to a premium grade oil is either unnecessary or dangerous.*

Rust Prevention

Although straight petroleum oils have some rust protective properties, they cannot be depended upon to do an unfailing job of protecting rust-susceptible metallic surfaces. In many instances, water can displace the oil from the surfaces and cause rusting. Rust is particularly undesirable in a rolling element bearing because it can seriously abrade the bearing elements and areas pitted by rust will cause rough operation or failure of the bearing.

Additives

High-grade lubricating fluids are formulated to contain small amounts of special chemical materials called additives. Additives are used to increase the viscosity index, fortify oxidation resistance, improve rust protection, provide detergent properties, increase film strength, provide extreme pressure properties, and sometimes to lower the pour point. However, it is of great importance to again restate that adding other ingredients to a properly balanced oil formulation will upset the balance. It is thus always wise to leave additives selection to the knowledgeable oil manufacturer or lubricant formulator. Significant economic losses and safety concerns have been documented for locations that elected to disregard this sobering advice.

General Lubricant Considerations

The proper selection of a grease or oil is extremely important in high-speed bearing operation. Proper lubrication requires that an elastohydrodynamic oil film be established and maintained between the bearing rotating members. To ensure the build-up of an oil film of adequate load-carrying capacity, the lubricating oil must possess a viscosity adequate to withstand the given speed, load, and temperature conditions.

Liquid Oil Lubrication Generalized

Considering bearing lubrication with liquid oil, it may be said that the actual speed, load and temperature requirements of a particular bearing would probably be best satisfied by a different lubricant for each specific application. For the relatively unsophisticated rolling element bearing service found in the typical centrifugal pump, it is often possible to standardize and economize by stocking only a few lubricant grades. Maintaining a minimum base oil viscosity of 13 cSt (approximately 70 SUS) for pump bearings has long been the accepted norm for both users and bearing manufacturers. It can be applied to most types of ball and some roller bearing in pumps and electric motors. It is assumed, however, that bearings would operate near their published maximum rated speed, that naphthenic oils would be used, and that the viscosity be no lower than this value even at the maximum anticipated operating temperature of the bearings (Ref. 9-1).

Figure 9-4 shows how higher viscosity grade lubricants will permit higher bearing operating temperatures. ISO viscosity Grade 32 (147 SUS @ 100°F or 28.8-35.2 cSt @ 40°C) and ISO Grade 100 (557 SUS @ 100°F or 90-110 cSt @ 40°C) are among those shown on this chart. Entering the ordinate (vertical scale) at 13 cSt shows a safe allowable temperature of 146°F (63°C) for rolling element bearings with ISO Grade 32 lubrication. Switching to Grade 100 lubricant and requiring identical bearing life, the safe allowable temperature would be extended to 199°F (93°C). Although a change from Grade 32 to Grade 100 lube oil should cause the bearing operating temperature to reach some intermediate level, a higher oil viscosity would result and the bearing life would actually be extended. See also Figure 7-18.

Most ball and roller bearings can be operated satisfactorily at temperatures as high as 250°F from the metallurgy point of view. As mentioned above, the only concern would be the decreased oxidation resistance of common mineral oil lubricants, which might require more frequent oil changes. Whenever oil temperatures exceed 150°F (66°C), a reliability-focused user would favor the selection of synthesized hydrocarbon lubes, commonly and incorrectly called "synthetics." The various polyalphaolefin (PAO) and dibasic ester (diester) formulations, or also mixtures of the two, are ideally suited for pump bearing lubrication.

However, the "once-through" application of oil mist will also solve pump bearing elevated temperature problems. Specifically, oil mist lubricated anti-friction bearings are often, traditionally, served by ISO Grade 100 naphthenic oils, although ISO Grade 68 mineral lubricants and synthetic ISO Grade 32 oils would certainly be a more appropriate choice from energy conservation points of view.

In any event, where naphthenic oils are unavailable, or in locations with relatively low ambient conditions, PAO's and dibasic ester-based synthetic lubes have been very successfully applied. These synthetics will outperform mineral oils by lowering bearing operating temperatures, resisting oxidation, and eliminating

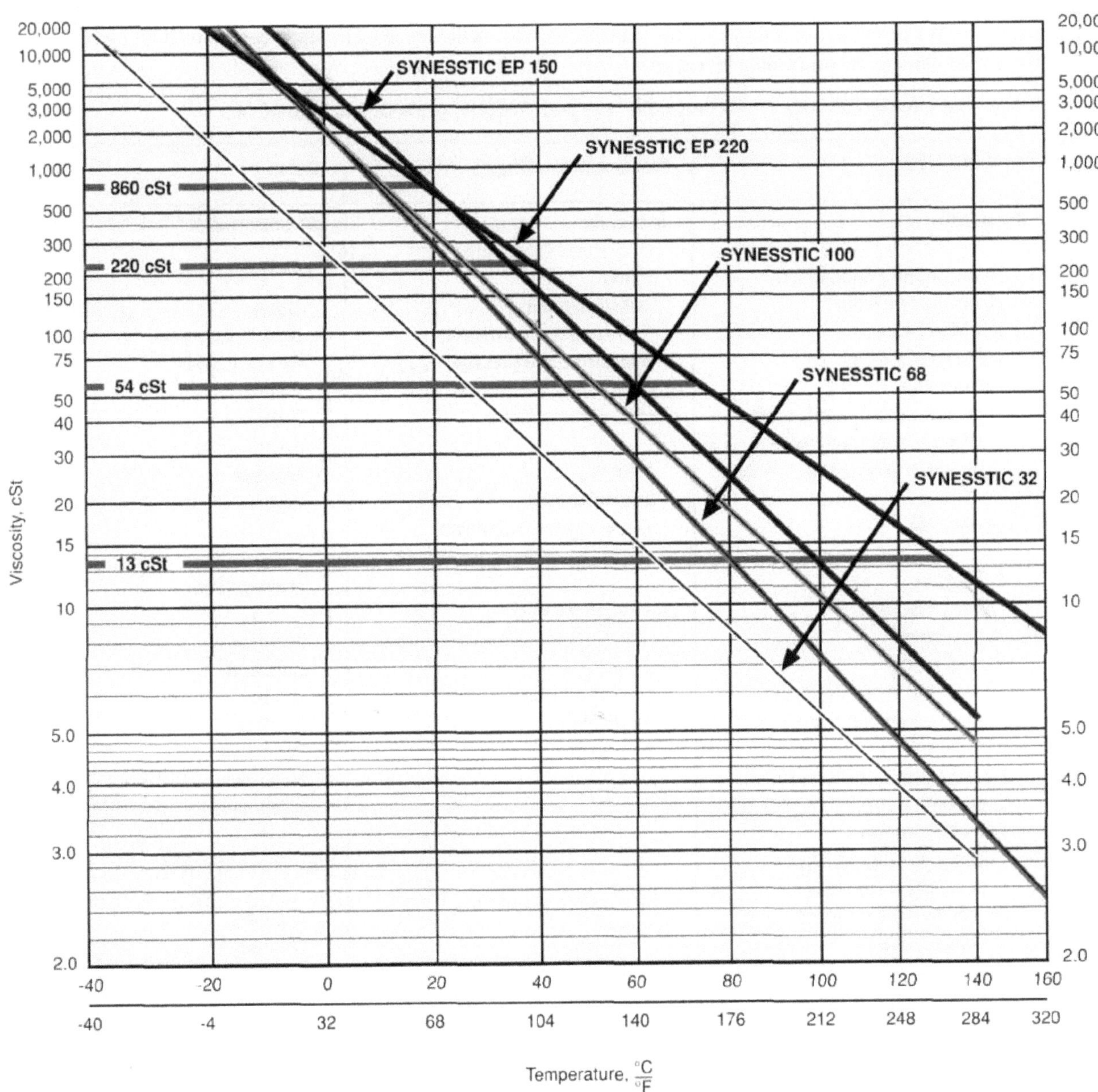

Figure 9-4: Temperature vs. viscosity relationships for typical synthetic lubricants used in different pump bearings (Synesstic 32 through 100) and certain gear speed reducers (EP-Grades 150 and 220). Note that the slopes of the various lines plotted here closely track the slopes of many temperature vs. viscosity lines for identical grades of mineral oils, Figure 7-18. (Source: ExxonMobil Lube Marketing Division, Houston, Texas)

the risk of wax plugging which has sometimes been experienced with naphthenic mineral oils at low temperatures. They also have greater film strength and film thickness than equivalent grade mineral oils—an added advantage.

SYNTHETIC LUBE OVERVIEW

What is a Synthetic Lubricant

Commonly termed "synthetic lubes" are actually synthesized hydrocarbons. They combine a synthetic base stock, or a mixture of base stocks. These carefully selected additives will influence lubricant viscosity, compatibility, volatility, low temperature properties such as seal swelling and many other performance parameters.

Synthetic lubricants can be blended with appropriately chosen mineral oils. While the choice of conventional and/or specialty additives is often compatibility and end-use dependent, the only synthesized hydrocarbons of interest to process pumps are either diester or PAO-based. These fluids are stable over the entire temperature range found in process pump bearings, i.e. from minus 40°F/40°C in the arctic, to maybe 120°C (248°F) in a few highly unusual refinery feed or transfer pump services.

The secret to high performance is in the additives that a competent formulator puts in the diester, or PAO, or combined base stock mixture. In many instances, these additives are necessary to obtain a particular level of performance. Also, since both of these base stock fluids can be combined with certain mineral oils, blends of diester and/or PAO's with mineral oils are feasible as well. The various proven PAO/diester blends contain synergistic additive systems identified with proprietary names (Synerlec, etc.). The synergism obtained in a competent additive blend combines all of the desirable performance properties plus the ability to ionically bond to bearing metals to reduce the coefficient of friction and greatly increase the oil film strength. The resulting tough, tenacious, slippery synthetic film makes equipment last longer, run cooler, quieter, smoother and more efficiently. After a short time in service, synergistic additive systems are in effect "micro-polishing" bearing surfaces, Figure 9-5. The result is demonstrably reduced severity of vibration (Figure 9-6), reduced friction, and minimized energy consumption.

For pumps, the most valuable synthetic lubricant types excel in high film strength and oxidation stability. However, while there are many high film strength synthetics on the market, these may not be appropriate for pumps. High film strength oils based on extreme pressure (EP) technology and intended for gear lubrication may typically incorporate additives such as sulfur, phosphorus and chlorine, which are corrosive at high temperatures and/or in moist environments. Sensitive to this fact, a reputable lubricant manufacturer would thus not offer an EP industrial oil with corrosive additives as a bearing lubricant for pumps, air compressors, steam turbines, high speed gear reducers and similar machinery.

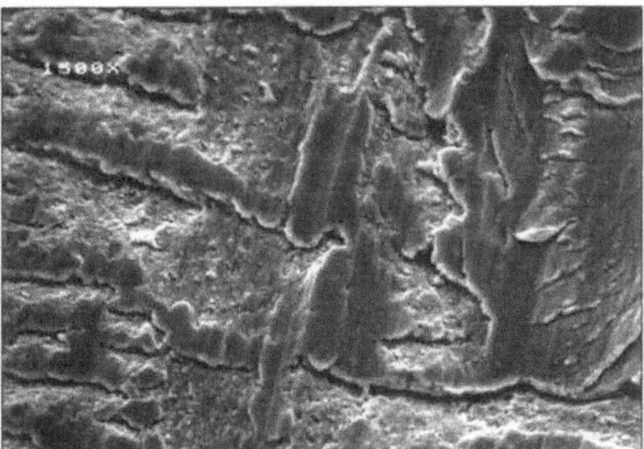

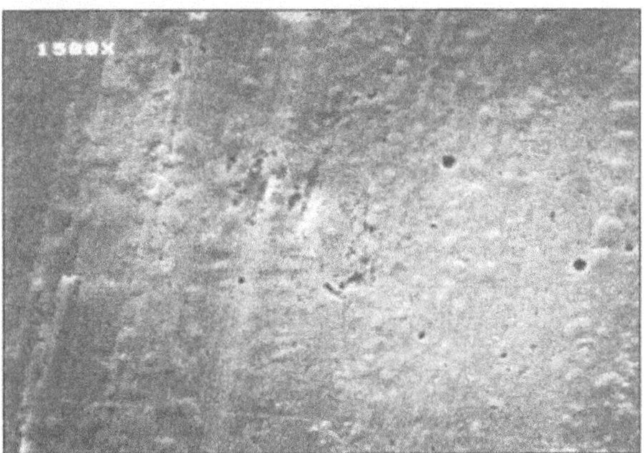

Figure 9-5: How synergistic additive systems "micro-polish" bearing surfaces (Source: Royal Purple, Ltd., Porter, Texas 77365)

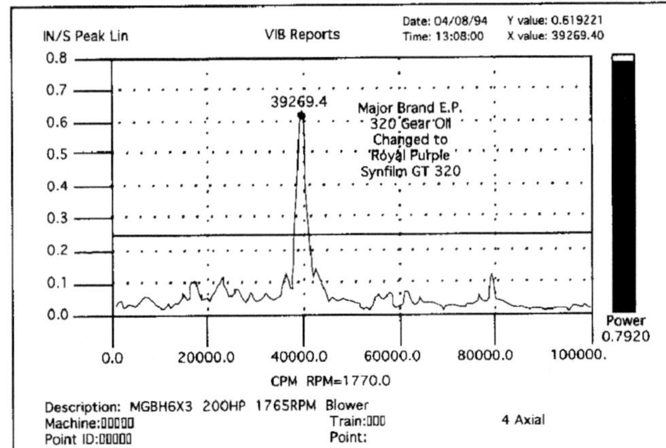

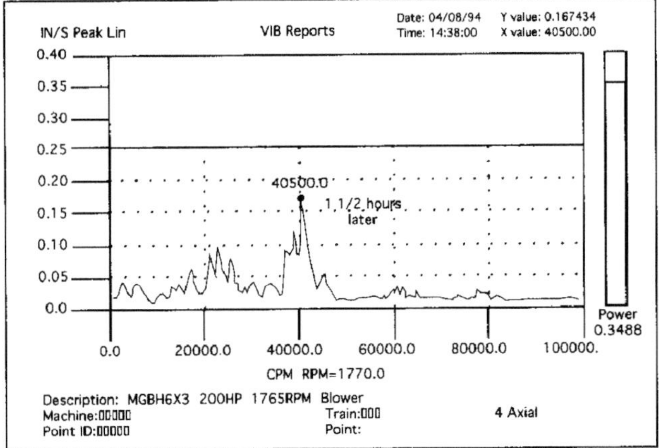

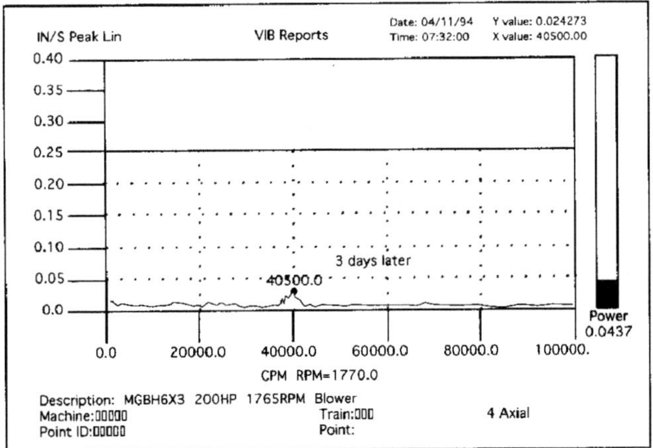

Figure 9-6: "Micro-polishing" reduces severity of vibration (Source: Royal Purple, Ltd., Porter, Texas 77365)

Conventional Wisdom: *EP oils are always advantageous.*
Fact: *EP additives are suitable only for well-defined applications and, unless used judiciously, may harm equipment performance.*

Synthetic Lubes for Troublesome Pumps

Because considerably more information would be available from Ref. 9-1, our text is limiting its coverage to issues dealing with synthetic lubes for pumps. Suffice it to say that some pump users have decided to use synthetic lubricants for their entire pump population. Others have made the decision to use synthetics only for the more troublesome pumps and consider these lubricants an indispensable component of conscientiously applied programs of "bad actor" pump management. "Bad actors" are pumps that experience a disproportionate number of failures—typically more than two failures per year.

A reliability-focused user facility will implement every reasonable and cost-effective upgrade feature. Since superior lubricants fit this description, these users will see to it that at least their "bad actors" receive synthetic lubricants. Needless to say, they will only allow carefully formulated synthetics from competent suppliers and will justify their use by pointing out the undisputed advantages of these fluids. An applicable case history involves a U.S. refinery (~230,000 bbls/day) with over 40 "bad actor" pumps initially lubricated by traditional mineral oils. After two years of upgraded operation using a suitable synthetic (but no other changes), only two of these pumps were still on the "bad actor" list.

There are six primary advantages of synthetics over comparable viscosity mineral oils:

- Improved thermal and oxidative stability
- More desirable viscosity-temperature characteristics
- Superior volatility characteristics
- Desirable frictional characteristics
- Better heat transfer properties, and
- Higher flash point and auto-ignition temperatures.

Experience clearly shows that these advantages result in the following economic benefits:

- Increased service life of the lubricant (typically four to eight times longer than mineral oils)
- Less lubricant consumed due to its low volatility
- Reduced deposit formation as a result of better high-temperature oxidation stability
- Increased wear protection resulting in less frequent maintenance
- Reduced energy consumption because of increased lubrication efficiency (lower friction), see Figure 9-7.
- Improved cold weather flow properties, especially in oil mist systems where certain moderate pour point mineral oils may cause droplets of highly

viscous or waxy particles to plug small reclassifier nozzles (Figure 9-8).

- Reduced fire hazard resulting in fewer pump fires and reduced insurance premiums
- Reduced bearing operating temperatures (Figure 9-9).
- Higher up-times and longer equipment life.

Case Histories Involving Synthetic Lubricants for Rolling Element ("Antifriction") Bearings

Our earlier discussions made the point that equipment reliability is obviously influenced by the quality of bearing lubrication. For good reasons and as will be emphasized throughout this text, the pursuit of quality lubrication must focus on application method, lube quantity, selecting the appropriate oil type and viscosity, properly storing and handling the lubricant, attending to bearing housing contamination issues, and implementing appropriate oil change intervals. A few case histories will also serve as helpful reminders.

It makes sense to summarize good lubrication practices as: choosing the right oil, taking proper care of it, and changing it on time. Yet, while good lubrication practices lead to improved equipment reliability by maximizing the performance of the oil selected, there are limitations. This is because in and of themselves, good practices cannot impart lubricating properties that the oil perhaps never possessed in the first place. At issue is thus the definition of the right oil, or appropriate oil type. Putting it another way, improvements in lubricant quality can only be achieved by selecting and utilizing oils with superior lubricating properties.

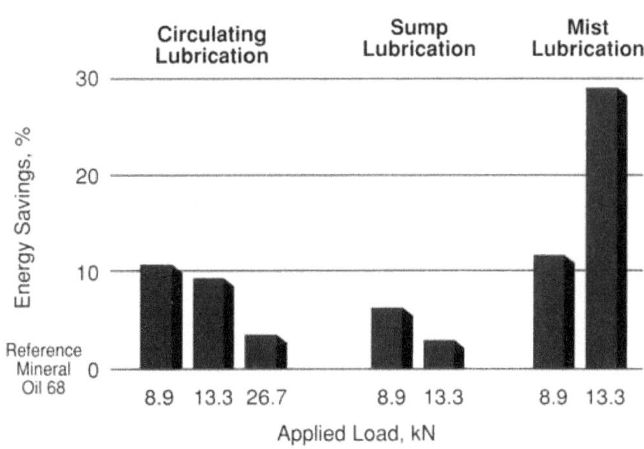

Figure 9-7: Typical energy savings with diester-based synthetics (Source: ExxonMobil, Houston, Texas 77001)

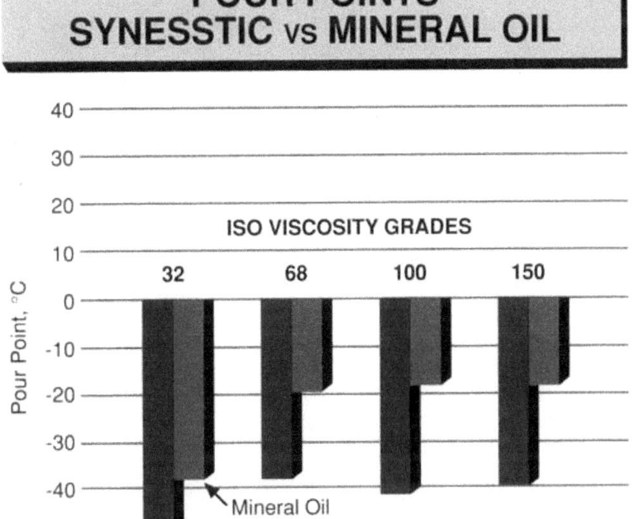

Figure 9-8: Improved pour points favor synthetics in cold ambients (Source: ExxonMobil, Houston, Texas 77001)

High Film Strength Synthetic Lubricants

Synthetic lubricants offer an obvious path for lube-related improvement. However, even among prominent synthetic lubricants, oil performance can vary greatly based on the amounts and types of additives in the oil. At least one company combines synthetic base oils with advanced additive chemistry so as to realize greater film strength. Numerous incidents have been documented where advanced lubrication technology has significantly improved pump reliability (Ref. 9-1, 9-11, 9-23). We will limit our coverage to just five examples in this topic category.

Hot Oil Pump Experience No. 1.

One chemical plant began to experience rolling element bearing failures in their 500°F hot oil pumps within 90 days of plant startup, despite the fact that the pumps were already being lubricated by a premium brand of synthetic oil. Root cause analysis determined that failures occurred because high temperatures had caused the synthetic oil to oxidize. All pumps underwent immediate oil changes with nine of the 18 hot oil pumps being converted to a superior film strength synthetic oil. Again, all

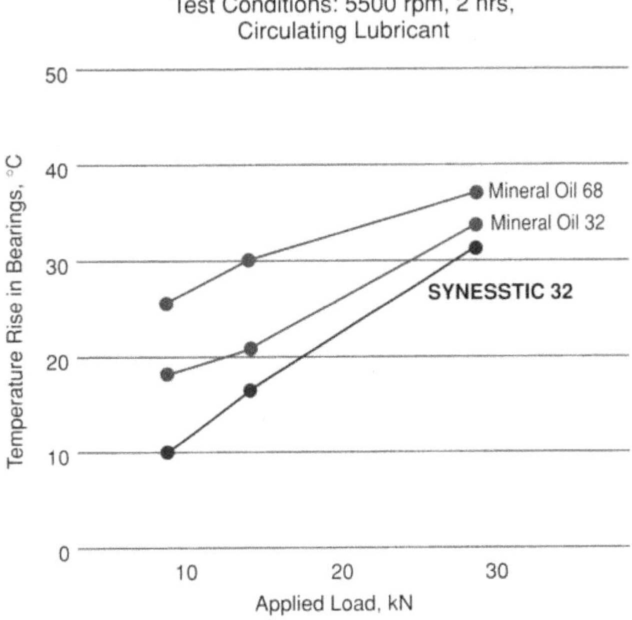

Figure 9-9: Quality synthetics result in lower bearing operating temperatures (Source: ExxonMobil, Houston, Texas 77001)

of the pumps using the original oil required an oil change within 90 days. The superior film strength oil, however, proved to be a lubricant upgrade that eliminated the bearing failures and enabled annual oil intervals to be established for all of the hot oil pumps.

A four-fold extension of oil exchange intervals results in a 75% reduction of oil usage after changing over to the high-grade synthetic lubricant. The reduction in consumption makes up for the fact that the synthetic lube costs perhaps four times as much as the mineral oil used before. Tangible savings accrue due to 200 man-hours of maintenance labor being reduced to only 50 man-hours. At $50 per man-hour, yearly savings are $7,500. Although intuitively evident, no additional credit was taken for the imputed value of reduced failure risk with superior film strength synthetic lubricants. Neither was credit taken for the gain due to pro-active use of the re-assigned workforce. In other words, the value of pro-actively employing 150 man-hours of freed-up manpower must logically be assumed to exceed $7,500 — or else the plant would not make any profit.

Hot Hydrocarbon Product Pump Experience No. 2

Another refinery was experiencing high vibrations and an audible noise from the inboard bearing of a critical, non-spared pump in one of its process units. The refinery was able to avoid a unit shutdown by draining the oil with the pump operating and replacing the quart (~ one liter) of ISO 32 synthetic oil already in the pump with a superior film strength ISO Grade 32 synthetic. The high vibrations disappeared (see Ref. 9-1 for a technical explanation), as did the audible noise and it was decided repairs to the pump were no longer necessary. The value of an avoided repair was estimated as $2,600 for a bearing change only, $12,300 for bearings and seals, and $52,000 for a complete pump overhaul. Additionally, unit downtime would have amounted to approximately $130,000 per day.

Whatever the differential cost of a quart (or liter) of high film strength synthetic, perhaps two or three dollars in 2004, it is simply insignificant compared to the value of a failure incident on critical, non-spared refinery pumps. Critically important and hot service pumps should, therefore, be lubricated with high film strength synthetic oils.

Hot Water Pumps—Experience No. 3

For years, a U.S. Gulf Coast chemical company had averaged two to three bearing failures every six weeks in their 30 hot water pumps. These pumps were lubricated by oil mist, using a premium brand synthetic oil. In an effort to improve pump reliability, the lubricant was changed to a greater film strength synthetic lube. In the three years since, only one pump failure has been reported and it was not lubrication related. While this may sound like a purely anecdotal report, we are including it here because it is quite representative of well over one hundred similar case histories that users have informally shared and reported over the past decade or so.

Disc Filter Shower and Bark Booster Pumps—Experience No. 4

A North American paper company experienced frequent difficulty with two 3,600 rpm pumps. These difficulties have been eliminated by changing the R&O mineral oil lubricants to synthetics with greater film strength. As is so often the case, the latter have the ability to avoid metal-to-metal contact and the resulting temperature reductions tell the story:

Temperature Readings on Disc Filter Shower Pump

	Before	*After*
Inboard Bearing	170°F	130°F
Outboard Bearing	185°F	160°F

Temperature Readings on Bark Booster Pump

	Before	*After*
Inboard Bearing	180°F	130°F
Outboard Bearing	170°F	114°F

With shower pump outages causing plant downtime, a pump repair incident cost the facility $35,600. A single instance of repair avoidance may make up for the incremental cost of supplying high film strength synthetics to an entire paper mill.

Similar temperature reductions were experienced on bark booster pumps. However, each booster pump failure was reported to cost only $3,600 because it did not cause a production stoppage.

Performance Evaluation on a Worthington/ Flowserve Pump Driven by a 450 HP Electric Motor—Experience No. 5

An experiment was conducted under controlled conditions in late 2008 by a world-scale multinational petrochemical corporation. It involved a relatively large pump bearing housing that held 2 quarts (~2 liters) of oil. A widely used "major" synthetic (ISO Grade 68) was replaced with a high-strength synthetic (Synfilm®) lubricant. Replacing the original charge of synthetic oil with Synfilm ISO Grade 68 reduced the housing temperature from previously 147 °F to now only 106°F and the power demand for this pump decreased by slightly over 4%. At a lubricant cost somewhere between $20 and $21, payback was achieved in less than five days of operation. We know that this is just one example in a long list of proven experiences. Collectively, these experiences illustrate and quantify why special formulations available from knowledgeable suppliers should not be overlooked by reliability-focused users.

These examples are but a few of the hundreds where lube selection was responsible for significant improvements in pump reliability. High bearing temperatures and vibration excursions related to elevated surface roughness of bearing metals can very often be cured by selecting and installing superior performing lubricants. Especially in problem pumps, upgrading to high-strength lubricants can improve equipment reliability in a manner unattainable by any other means.

Indeed, since oil changes are often feasible while pumps are on-line and running, using superior film strength synthetic lubricants often results in immediate payback. Virtually every cost justification calculation indicates unusually large benefits for employing these lubes on problem pumps and 10:1 payback in a single year is rather the norm.

Applications of Liquid Lubricants in Pumps

The amount of oil needed to maintain a satisfactory lubricant film in a rolling element bearing is extremely small. The minimum quantity required is a film averaging only a few micro-inches in thickness. Once this small amount has been supplied, make-up is required only to replace the losses due to vaporization, atomization, and seepage from the bearing surfaces (Ref. 9-1 and 9-7).

How small a quantity of oil is required can be realized when we consider that 1/1000 of a drop of oil, having a viscosity of 300 SUS at 100°F (38°C) can lubricate a 50 mm bore bearing running at 3,600 rpm for one hour. Although this small amount of oil can adequately lubricate a bearing, much more oil is needed to dissipate heat generated in high speed, or heavily loaded bearings.

Oil may be supplied to the rolling element bearings of various pumps in a number of ways. These include bath lubrication (Figure 9-10, upper left view shows sight glass), oil mist from an external supply (Figure 9-11), and wick feed—on rather small, sometimes high-speed, lightly-loaded and intermittently operated pumps (Figure 9-12). After passing through the bearing, the oil is returned to the sump.

Drip feed can occasionally be found on small, intermittently operated vertical pumps (Figure 9-13). Then there are circulating systems (Figure 9-14), jet-oil lubrication (Figure 9-15), and splash or spray from a loose slinger ring (Figure 9-16), or a shaft-mounted flinger disc (9-17). Each of these merits closer examination and will be discussed here.

By way of overview, we note that one of the oldest and simplest methods of oil lubrication consists of an oil bath through which the rolling elements will pass during a portion of each revolution. However, this "plowing through the oil" may cause the lubricant to heat up significantly and should be avoided on the great majority of process pumps. Where cooling is required in high speed and heavily loaded bearings, oil mist, oil jets and circulating systems should be considered. If necessary and with circulating systems, the oil can be passed through a heat exchanger before returning to the bearing.

It should be noted that bearing overheating might occur on many pumps operating at 3,000 or 3,600 rpm unless the oil level is lowered so as to preclude this plowing effect. In those instances, the oil level must be reduced to below a horizontal line tangent to the lowermost periphery of a rolling element at the 6-o'clock location of the bearing. It may then become necessary to use either oil rings or shaft-mounted flinger discs.

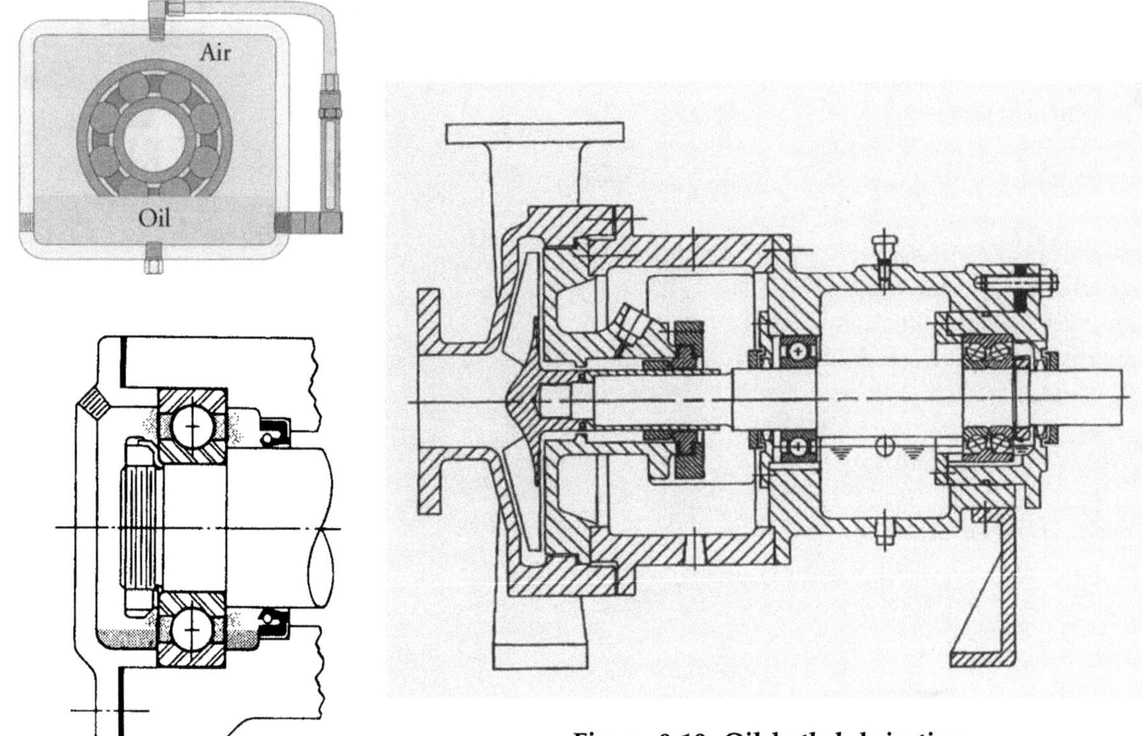

Figure 9-10: Oil bath lubrication

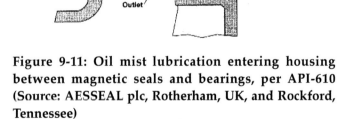

Figure 9-11: Oil mist lubrication entering housing between magnetic seals and bearings, per API-610 (Source: AESSEAL plc, Rotherham, UK, and Rockford, Tennessee)

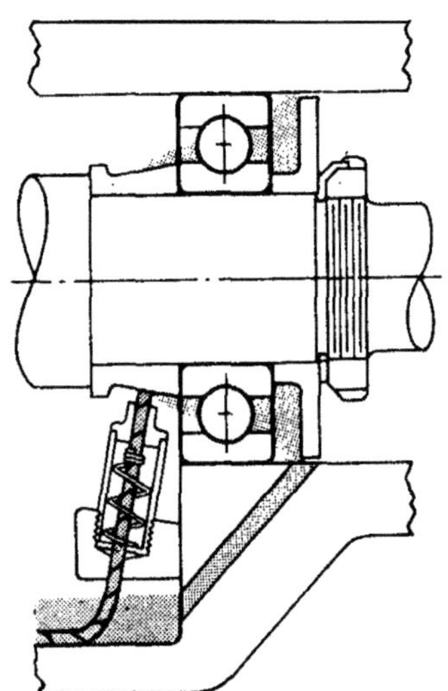

Figure 9-12: Oil wick lubrication for high speeds and light loads

However, caution is advised with oil rings, since these have clear shortcomings that disqualify them for use in high reliability pump applications. Oil rings are sensitive to dimensional concentricity, shaft orientation, shaft speed, depth of immersion, and oil viscosity. They tend to skip or hang up unless these five variables are in perfect balance. Skipping action inevitably results in wear. The wear debris contaminates the lubricant and is largely responsible for driving down bearing reliability and bearing life. Clearly then, flinger discs, jet-oil spray application, oil mist and circulating systems are given preference over oil rings by reliability-focused pump user companies. It should be noted that the bearing housing bore must be large enough to allow inser-

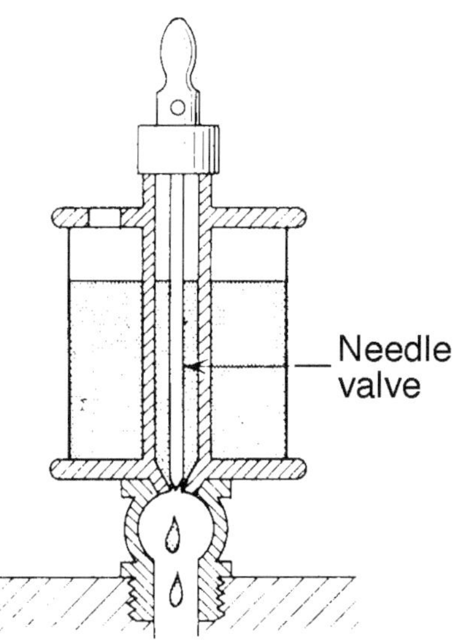

Figure 9-13: Drip feed lubricators for small vertical pumps (Source: Trico Mfg. Co., Pewaukee, Wisconsin 53072)

Figure 9-14: Crude transfer pump showing piping for circulating lube system (Source: Ruhrpumpen, Germany)

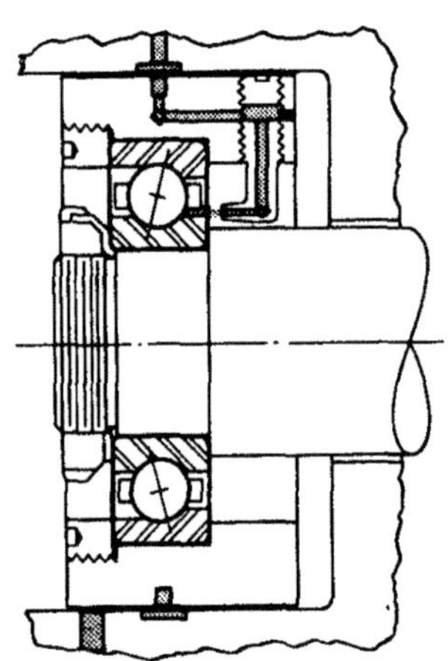

Figure 9-15: Jet-oil lubrication—most advantageous application method for rolling element bearings (Source: MRC Engineering Handbook)

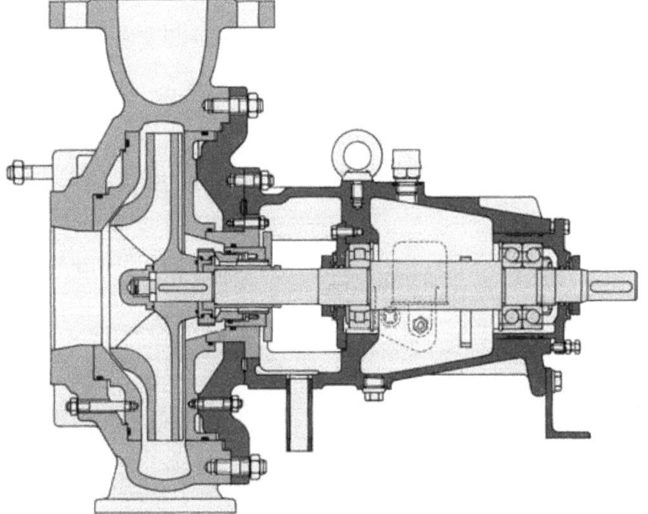

Figure 9-16: Lime suspension pump with typical loose oil ring ("slinger ring") lubrication (Source: Sulzer Company, Winterthur, Switzerland)

tion of flinger discs. This requirement can be met by mounting bearings in a cartridge (see Figure 7-10) or by manufacturing the discs from a flexible, elastomeric, material—usually Viton.

Oil Bath Lubrication

A simple oil bath method, shown earlier in Figure 9-10, is satisfactory for low speeds and loads. As a very general rule, most bearings in pumps operating at less than 1,800 rpm are suitable for oil bath lubrication. At least one rule-of-thumb states that for DN-values in excess of 6,000 (where D= shaft diameter, inches, and N= shaft rpm), the oil level should *not* be allowed to reach the bearing and that some means of "lifting" or "flinging" the oil should be employed. On pumps with DN-values of 6,000 and lower, the static oil level should be at the midpoint of the lowest ball or roller. A greater amount of oil can cause churning which results in abnormally high operating temperatures. In either case, systems with oil bath lubrication generally employ sight gages to facilitate inspection of the oil level.

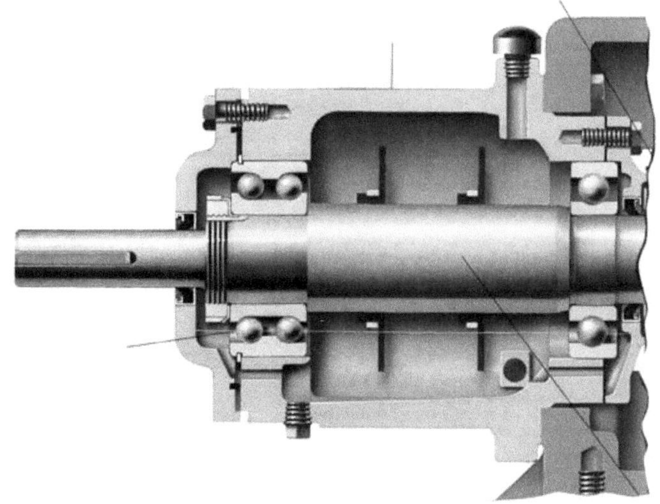

Figure 9-17: Flinger discs are used to prevent temperature-related lube oil stratification (Source: Worthington-Dresser, Harrison, New Jersey)

It is important to note that although on pumps with DN-values not exceeding 6,000 the oil bath will contact the bearings, either an oil ring (Figure 9-16) or a flinger disc (Figure 9-17) may have to be employed to prevent oil stratification. Oil stratification means that after passing through the bearing, a layer of hot oil would tend to float on the bulk of somewhat cooler and denser lubricant at the bottom of the sump.

Conventional Wisdom: *Oil slingers or flingers are not needed with oil bath lubrication.*

Fact: *Quiescent static sumps allow a hot layer of oil to float near the top. Flingers or similar means of agitation devices are needed to maintain uniform oil temperatures.*

Constant Level Lubricators

Constant level lubricators have been around for many decades. Unfortunately, they have also been misunderstood and misapplied for a very long time. There are essentially five points that are often overlooked and that merit being carefully considered by reliability-focused users:

1. The caulk or elastomeric sealing material at the lower portion of the transparent oil-containing bulb will not last forever. In outdoor installations subject to ambient temperature swings, the seal may crack and allow rainwater to contaminate the oil. This is an overlooked preventive maintenance item!
2. Unbalanced constant level lubricators allow the ambient environment to contact the lube oil level in the surge chamber of the device. This is a potentially serious contamination access route that explains why oil changes are necessary!
3. All constant level lubricators are direction-sensitive and will have to be installed on the tp-of-arrow side as shown in Figures 9-18(a) through 9-18(d). Installation on the wrong side of the bearing housing allows the oil level to undergo larger fluctuations between high and low oil levels before oil feed is being re-initiated.
4. The closed system oiler of Figure 9-18(d) is primarily intended for small ANSI/ISO pumps. Verify its satisfactory operation before you apply it on large pumps.
5. The user must select sturdy piping to connect API-style and similar heavy-duty pumps to constant level lubricators. While hydraulic tubing, Figure 9-19 (a), typically has more than adequate pressure rating, it may not adequately resist vibration-induced forces or inadvertent position-disturbing contact with workers or objects.

Figure 9-18, parts (a) and (b), shows an unbalanced or "vented" constant level arrangement for maintaining the correct oil level, while Figure 9-18, parts (c) and (d), depict arrangements that include a pressure-balanced constant level lubricator.

In the unbalanced constant level lubricator of Figure 9-18 (a) and (b), the oil level in the lubricator body—commonly called a surge chamber—is *always* contacted by the surrounding ambient air. Visualize a temperature increase causing a slight pressure increase inside the bearing housing. Such a pressure increase is a distinct possibility because of a drop of oil obstructing a small vent passage, or because lubricant effectively fills the narrow gap in a bearing housing seal. Bernoulli's law states that a housing-internal pressure increase causes a lowering of the oil level in the housing and will cause

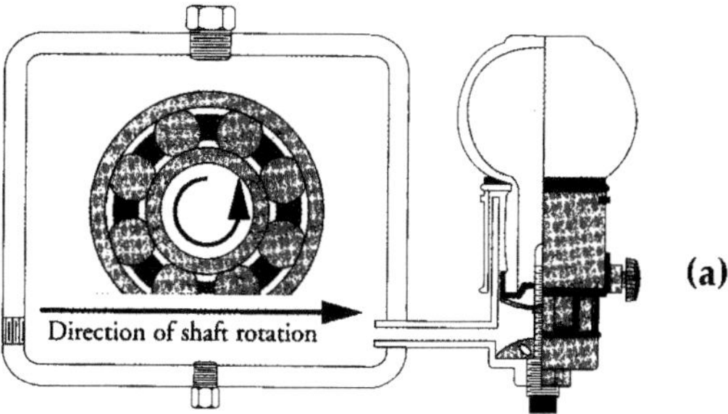

VENTED CONSTANT LEVEL OILER WITH SIDE CONNECTION

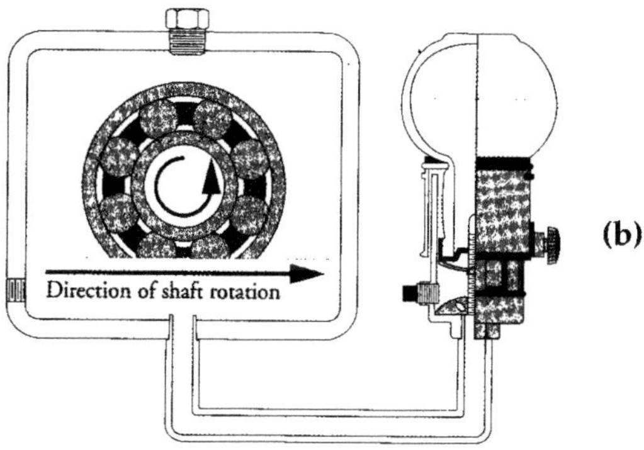

VENTED CONSTANT LEVEL OILER WITH BOTTOM CONNECTION

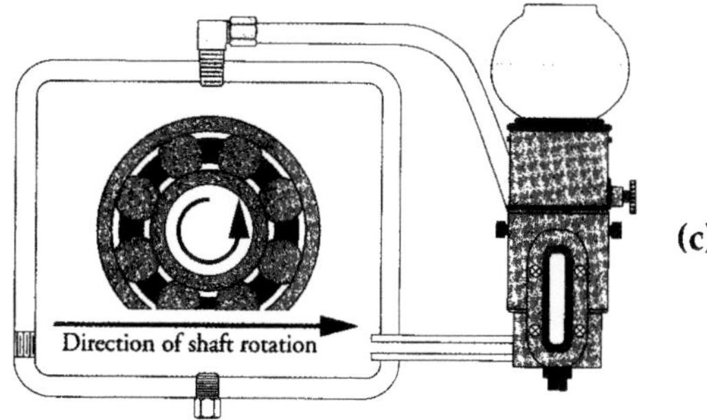

CLOSED SYSTEM OILER WITH PRESSURE BALANCING LINE

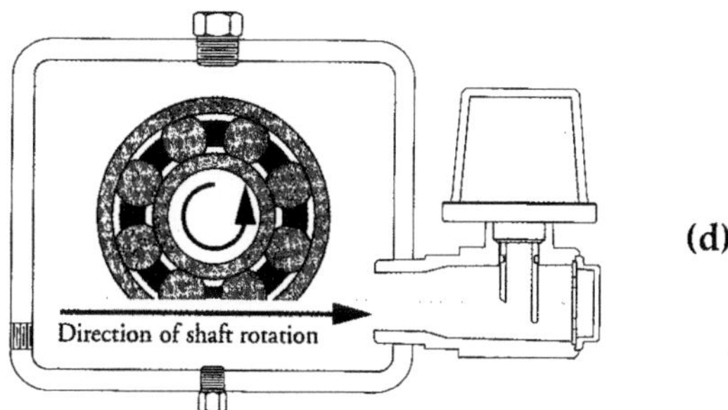

CLOSED SYSTEM OILER MOUNTED ON CENTER LINE OF DESIRED OIL LEVEL

Figure 9-18: Constant level lubricators (Source: Trico Mfg. Co., Pewaukee, Wisconsin 53072)

the displaced oil to be pushed into the surge chamber. The difference between the two oil levels will equal the pressure difference, in inches or millimeters, between the ambient atmosphere and the housing interior.

Conventional Wisdom: *"Unbalanced" constant level lubricators have served us well for close to a century, so why switch?*

Fact: *Since the late 1980's, environmental concerns have prompted use of smaller diameter vents and close-clearance housing seals. A small amount of oil collected at either location may act as a "seal" causing the bearing housing pressure to exceed atmospheric pressure.*

Realizing this possibility and having first-hand experience with "unexplained" lowering of oil levels in bearing housings with unbalanced constant level lubricators, reliability-focused pump users will only consider the pressure-balanced option, Figure 9-19. Here, an O-ring separates the atmosphere from the oil level in the lubricator surge chamber. Also, the space between the O-ring and the oil level in the surge chamber is vented back to the bearing housing interior space. Consequently, both oil levels are always exposed to the same pressure and thus are always at the same height. Figure 9-19 (b) shows one of the available configurations. The balance line shown in Figure 9-19 (c) can be connected to the top bearing housing location from which the customary vent has been removed. A generous range of height adjustments is available, Figure 9-19(d).

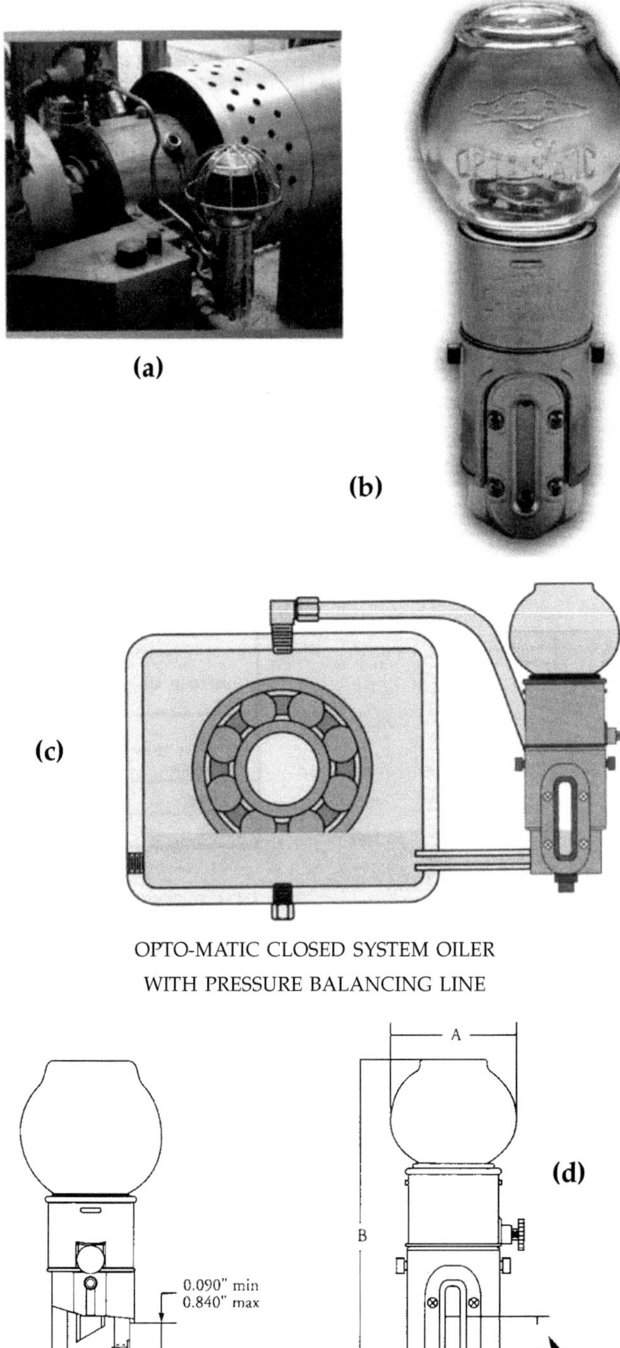

Figure 9-19: Optomatic "balanced" constant level lubricators (Source: Trico Mfg. Co., Pewaukee, Wisconsin 53072)

BALANCED CONSTANT LEVEL LUBRICATORS: HOW COST-EFFECTIVE?

Reliability enhancements are sometimes best retrofitted on an attrition basis. In this context, retrofitting on an attrition basis implies that every time a pump goes into the shop for repairs, the vulnerable old style, unbalanced, constant level lubricator (list price: $34) is being replaced by the fully balanced type. Note that modern, balanced constant level lubricators come with built-in sight glass level indicators (Figure 9-19). In early 2005, an 8-oz (~230 ml) capacity glass bulb model listed at a rather affordable $61.

But, suppose by the time one added incremental shop labor, each retrofit would costs $150. Suppose a mid-size refinery had 1,000 centrifugal pumps and, last year, had experienced 160 pump repairs. This year, one would retrofit 160 pumps and would spend (160)x($150) = $24,000. Assume now that in the following year, only 140 pumps would require repair and so the avoided cost would be calculated as (20 pumps)x($6,000 per repair) = $120,000. In that case, the payback would be 120,000/24,000 = 5:1. Over the (conservatively) projected 15-year life of the pump, an incremental investment of ($61-$34 = $27) would yield a payback well in excess of 10:1.

Another calculation approach accepted by some companies is even simpler. It assumes that a given pump will incur a $6,000 repair every five years with unbalanced, and every seven years with balanced constant level lubricators. The yearly repair cost would thus be $1,200 with unbalanced and $ 857 with balanced lubricators. The payback would be at least ($1,200-$857)/$27 = 12:1. In fact, assuming a 15-year life for the balanced constant lubricator and using its prorated yearly incremental cost of $27/15, the payback would exceed 100:1. Suffice it to say, a reliability-focused equipment owner would discontinue using unbalanced constant level lubricators and apply only pressure-balanced devices. Information on the deleterious lube oil contamination experienced by centrifugal pumps with unbalanced lubricators and vented bearing housings is found in Ref. 23. It allows us to see the rationale for conservatively estimating a 1.4-fold bearing life increase (i.e. seven-year MTBR vs. five-year mean-time-between-repairs, or MTBR) for pumps with balanced ("closed") lubricators vs. their predecessor unbalanced style lubricators.

Regardless of whether an unbalanced or pressure-balanced constant level lubricator is used, the preferred installation location for constant level lubricators is indicated by the direction of the arrows in Figure 9-18. Since the shaft in Figure 9-18a rotates counter-clockwise, the constant level lubricator should be on the right, its "up-arrow" side. "Up-arrow" location minimizes the oil level fluctuations in the bearing housing.

Conventional Wisdom: *Constant level lubricators will function the same way, regardless of location relative to direction of shaft rotation.*

Fact: *Locating the constant level lubricator on the "up-arrow" side minimizes oil level fluctuations.*

SPLASH LUBRICATION, OIL RINGS AND FLINGER DISCS

The term "splash lubrication" is rather all-encompassing and is extensively used on small reciprocating compressors and gear units. On pumps, splash lubrication refers to either an oil ring or a flinger disc dipping in the lubricant. Typical immersion values and diameter ratios for "slinger rings"—often just called oil rings—are shown in Figure 9-20. This illustration links depth of immersion to the 30-degree angular function. However, API-610 recommends that flingers or oil rings have an operating submergence of 3-6 mm (0.12-0.25 in) above the lower edge of a flinger or above the lower edge of the bore of an oil ring. Which of the two guidelines is more rigorous should be of academic interest to the reliability-focused user who will want to opt for less failure-prone means of oil application.

Why Oil Rings Are Not State-of-art Devices

As mentioned earlier in this text, unless the entire shaft system is truly horizontal, oil rings tend to move towards the lower end of the shaft. In doing so, an oil ring will make contact with a restraining surface and, as can be seen from Figure 9-16, this may be the interior wall of the bearing housing. The oil ring will slow down and abrade, or just skip around for a few seconds. To quote from Ref. 9-3:

> *"Oil ring stability (in tests) was variable and correlated strongly with oil viscosity and immersion depth. Generally, the oil ring operated erratically with pendulum motion while tracking backwards and forwards across the oil ring carrier."*

Abrasive wear can be spotted by examining the oil ring and noting that a previously slightly chamfered edge is now razor-sharp, or that an originally straw-colored lubricant has recently turned gray. Needless to

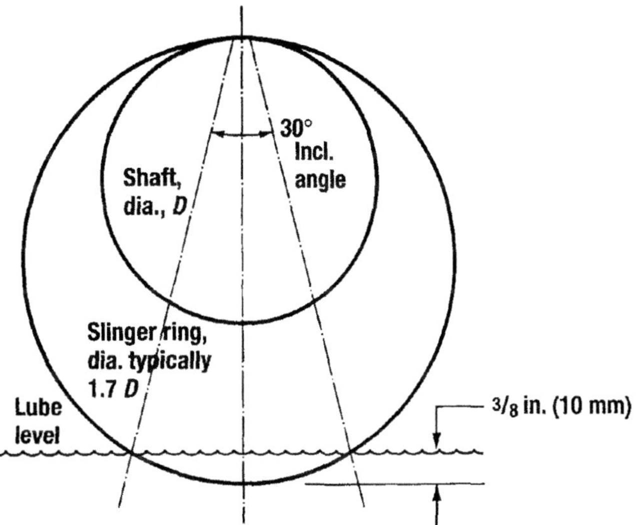

Figure 9-20: Typical oil ring ("slinger ring") dimensions of interest (Source: Hydrocarbon Processing, August 2002, p. 15)

say, the wear debris is suspended in the oil and bearing life is cut short.

In Ref. 9-4 (Wilcock & Booser, "Bearing Design and Application," First Edition, 1957), two world-renowned researchers comment on the need to have oil rings machined absolutely concentric. Based on their findings, oil ring concentricity tolerances of .002 inch (0.05 mm) maximum and surface finishes around 32 RMS should be considered mandatory.

Ref. 9-4 also cites surface velocity limits of from 3,500 to 4,000 fpm (~18-20 m/s) with water cooling, i.e. closely controlled oil temperature and viscosity. In the same reference text, Wilcock and Booser advise limiting velocities of 2,000 to 2,500 fpm (~10-13 m/s) without water cooling of the lubricant. Note that at 3,600 rpm, a 3-inch (76 mm) diameter shaft operates at a peripheral velocity of 2,827 fpm (~14.4 m/s). Another source, ExxonMobil Corporation's "Lube Marketing Course" text, suggest a DN value (inches of shaft diameter times rpm) of 6,000 as the threshold of instability for oil rings. With a DN value of 7,200, even a 2-inch shaft at 3,600 rpm would thus operate in the risky zone, whereas a 3-inch shaft operating at 1,800 rpm (hence, DN = 5,400) could use oil rings.

In the 1930's, adverse experiences with oil rings in large electric motors produced by Westinghouse (Philadelphia, Pennsylvania) prompted research which culminated in the recommendation to machine concentric grooves in the oil rings. Three or four of these grooves are typically taking up 60-70% of the axial width of the ring surface and their depth is approximately 0.04 inches (1 mm).

Grooved oil rings were shown in Figure 7-22. But grooved or ungrooved, when they "hang up" there's trouble (Figure 9-21). It should be pointed out that the peaks, slopes and separations of the plotted lines in Figure 9-22 are accurate only for a particular oil ring geometry and are valid only for operation at a specific depth of immersion in an oil of a given, constant, viscosity.

Note also that, regardless of the amount of oil delivered as a spray or oil fog, it must overcome the windage effect of the cage. In angular contact bearings (Figure 7-34) the cage acts as a tiny blower, trying to prevent the lubricant from migrating in the required direction, which is obviously left-to-right. In other words, with oil rings installed on pump shafts as illustrated earlier in Figure 9-16, neither the configuration nor the location of slinger ring and back-to-back mounted thrust bearings are favoring oil flow.

This is one of the reasons why lubricant flow in many pumps is either marginal or insufficient. Oil ring applications are anything but an exact science. It is well known that the speed of equipment has steadily moved up over the past decades. In addition, with ever- greater demands being made on equipment reliability, oil rings often are a very poor choice and are avoided by reliability-focused users.

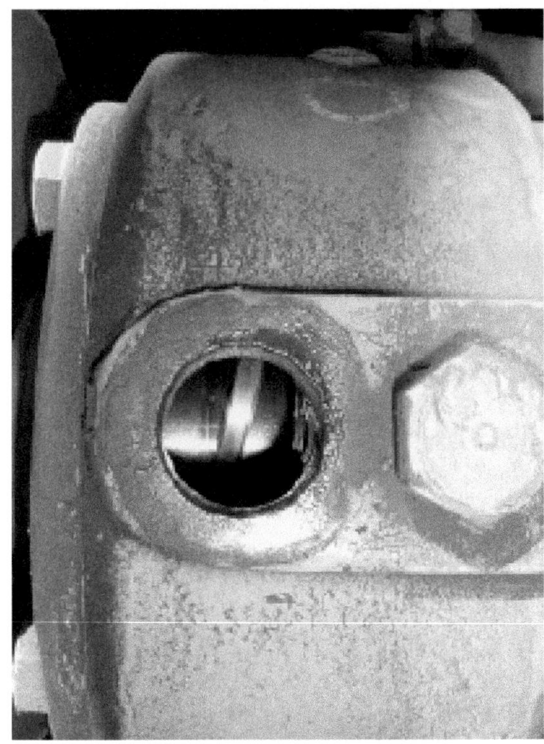

Figure 9-21: Oil rings (slinger rings) can hang up and become the source of bearing failure

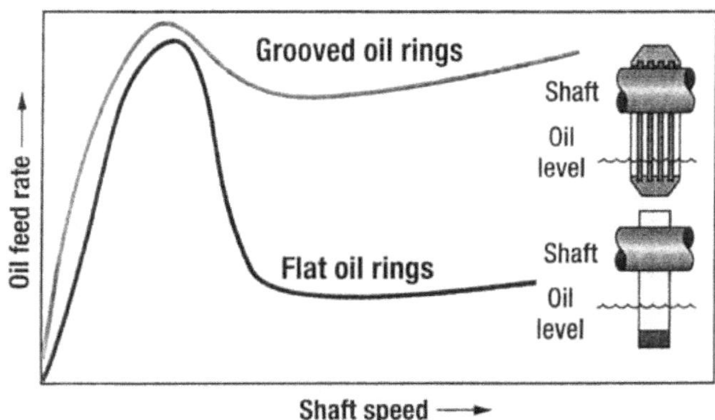

Figure 9-22: Performance comparison, grooved vs. ungrooved oil rings

Flinger Discs: A Much Better Choice than Oil Rings

Flinger discs securely mounted on the shaft as shown in Figure 9-23 are a much better choice than oil rings. Here, the Ingersoll-Dresser Company describes technical improvements made on Type JX radially split single stage process pumps and justifiably lists an "anti-friction oil thrower" among the demonstrated advantages. Knowing the issues of ring skipping and abrasion related to shaft horizontality, depth of immersion and lube viscosity, this reputable manufacturer states that properly mounted flinger discs eliminate the problems associated with oil rings.

Interestingly, the cheaper flinger disc substitute versions shown in Figure 9-24, a slip-on disc (left), or two roll pins (right), are considered unacceptable by API- 610. The relevant clause reads: "Oil flingers shall have mounting hubs to maintain concentricity and shall be positively secured to the shaft."

Flinger discs have been used in many successful pump designs for decades. So as to pick up a sufficient amount of oil the flinger disc has to be partially im-

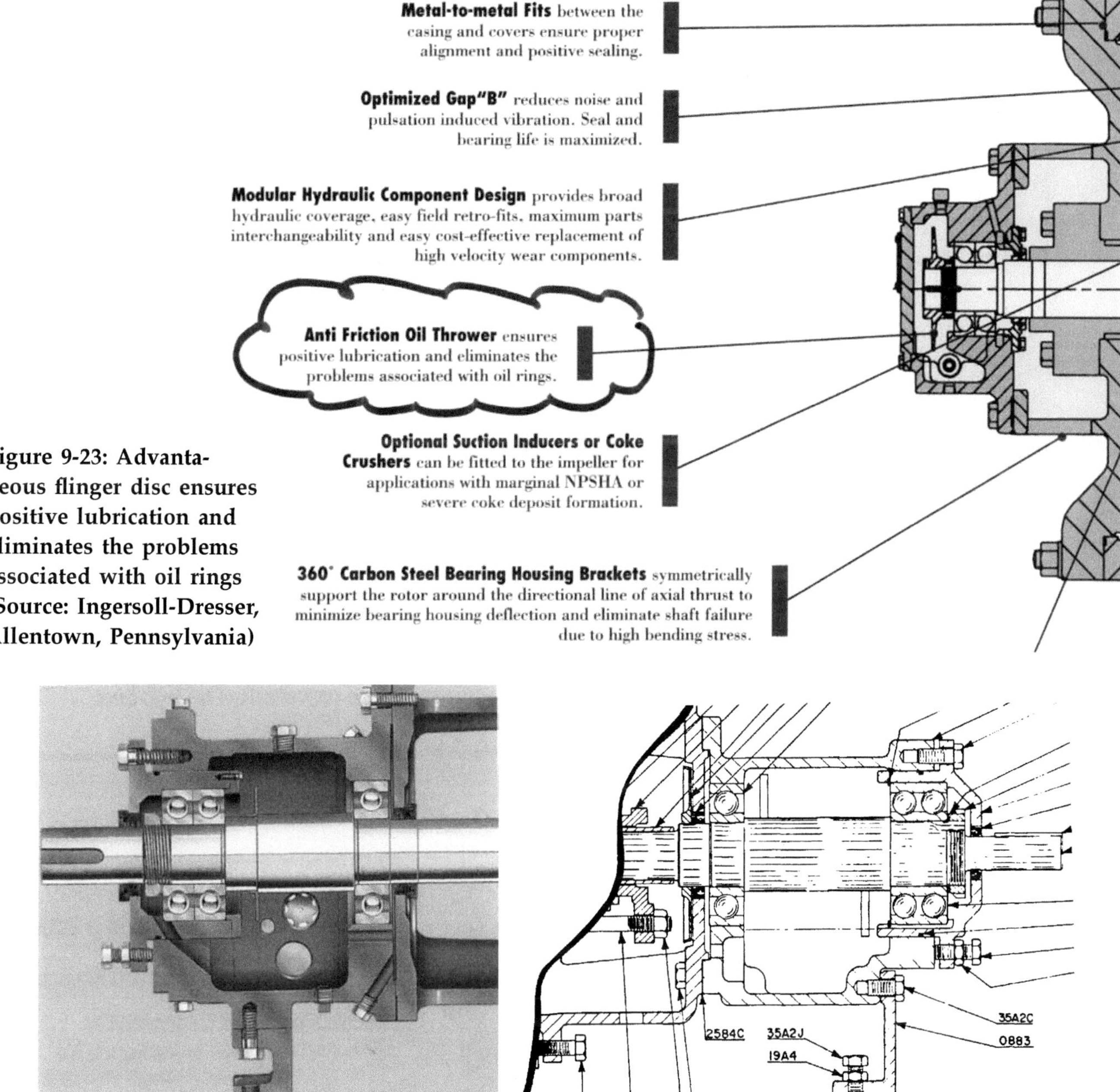

Metal-to-metal Fits between the casing and covers ensure proper alignment and positive sealing.
Optimized Gap"B" reduces noise and pulsation induced vibration. Seal and bearing life is maximized.
Modular Hydraulic Component Design provides broad hydraulic coverage, easy field retro-fits, maximum parts interchangeability and easy cost-effective replacement of high velocity wear components.
Anti Friction Oil Thrower ensures positive lubrication and eliminates the problems associated with oil rings.
Optional Suction Inducers or Coke Crushers can be fitted to the impeller for applications with marginal NPSHA or severe coke deposit formation.
360° Carbon Steel Bearing Housing Brackets symmetrically support the rotor around the directional line of axial thrust to minimize bearing housing deflection and eliminate shaft failure due to high bending stress.
2584C
35A2J
19A4
35A2C
0883

Figure 9-23: Advantageous flinger disc ensures positive lubrication and eliminates the problems associated with oil rings (Source: Ingersoll-Dresser, Allentown, Pennsylvania)

Figure 9-24: Poor substitutes for properly designed flinger discs: Slip-on (left) and twin roll pin versions (right)

mersed in the lubricant; in other words, it has to have an appropriate diameter. Note how, in Figure 9-25, the needed disc diameter is larger than the housing bore accommodating the smaller of the two bearings. To make assembling possible, the flexible disc(s) shown in Figure 9-25 were developed. Made of Viton® elastomeric material fused into a metal hub, they "fold up" when being inserted through a somewhat lesser diameter housing bore and then snap open again to full diameter. Molded to incorporate a series of concentric ring contours, flexible flinger discs can be trimmed to the correct size with a pair of scissors.

Figure 9-25: Shaft-mounted flexible flinger disc splashes lube oil into bearings (Source: TRICO Mfg. Corporation, Pewaukee, Wisconsin)

Although sometimes used to prevent temperature stratification of the oil, flinger discs serve as efficient oil spray producers. For reasons of strength and simplicity, solid steel flinger discs should be preferred over flexible discs. However, to accommodate the preferred solid steel flinger discs, bearings must be cartridge-mounted (Figures 9-24 and 9-26), in which case the effective bearing housing bore (i.e., the cartridge diameter) will have to be large enough for passage of a steel flinger disc of appropriate diameter. There are many situations where redesigning the bearing housing to accommodate solid steel flinger discs is easy to cost-justify.

In essence, oil rings are rarely (if ever) the most reliable choice of lubricant application. They often tend to skip around and even abrade unless the shaft system is truly horizontal, unless ring immersion in the lubricant is just right, and unless ring eccentricity, surface finish, and oil viscosity are within tolerance. Taken together, these five parameters are rarely found within close limits in actual operating plants.

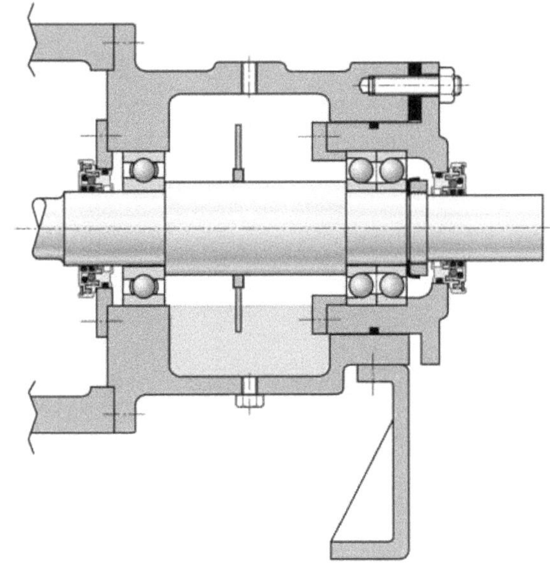

Figure 9-26: A bearing housing accommodating a cartridge on the outboard side. The bearing housing bore is slightly larger than the diameter of the steel flinger disc, making assembly possible

Oil rings in as-new condition incorporate a given width and are usually chamfered (Figure 9-27, left side of illustration). Examining and measuring these oil rings before and after a period of operation will often show significant wear and abrasion (Figure 9-27, right side of illustration).

Specifying and selecting pumps with flinger discs is considered prudent risk reduction by informed reliability-focused purchasers.

Figure 9-27: Oil rings in as-new ("wide and chamfered") condition on left, and abraded ("worn and narrow") condition on right side

Economic Value Explored

Upon close examination and with competent failure analysis, many observers have reached the conclusion that a large percentage of oil rings show signs of severe abrasion after one or two years of operation (Figure 9-27). It is well known that the resulting lube oil contamination is reflected in premature bearing failures. Based on these observations, it has been estimated that at least 5% of the centrifugal pumps installed in the average petrochemical plant suffer from oil ring deficiencies of sufficient magnitude to reduce bearing life from an assumed achievable 6 years, to typically only 3 years. Other pumps may experience oil ring degradation that reduces bearing life from five years to four years, and so forth. The issue is so intuitively evident that, to date, no one appears to have seen fit to spend research funds on scientific studies. Accordingly, empirical observations will have to suffice.

In any event and expanding on this conservative estimate, we might be dealing with a plant comprising 600 pumps. Suppose that of these, 18 "suspect" pumps were being repaired every three years to the tune of $6,000 per incident. This would require an expenditure of $36,000 per year. If, using flinger discs, the MTBR (mean-time-between-repairs) could be extended to six years, this expenditure would drop to $18,000 per year for the affected 3% of the plant's pump population. Needless to say, if you paid $50 per flinger disc, the 18 discs would have cost $900 and the investment would have had a payback of $18,000/$900 = 20:1. It is certainly no stretch to foresee greater savings and even better payback than demonstrated in this example.

Drip Feed Lubrication

Although not usually found on centrifugal pumps, this system is widely used for small and medium ball and roller bearings operating at moderate to high speeds where extensive cooling is not required. The oil, introduced through a filler-type sight feed lubricator, has a controllable flow rate that is determined by the operating temperature of the bearings. Recall that Figure 9-13 illustrates a typical design.

Forced Feed Circulation, or Pressure Lubrication

As alluded to earlier in Figure 9-14, this type of system uses a circulating pump and is particularly suitable for low to moderate speed, heavily loaded applications where the oil has a major function as a coolant in addition to lubrication. It should be noted that the bearings are not actually pressurized from an external source and that pressure is used only to get the lubricant to its destination. If necessary, the oil can be passed through a heat exchanger before arriving at the bearing. Entry and exit of the oil should be on opposite sides of the bearing. An adequate drainage system must be provided to prevent an excess accumulation of oil. Oil filters and magnetic drain plugs should be used to minimize contamination.

Conventional Wisdom: *With pressure-lubed bearings, the housings are pressurized.*

Fact: *Not so in most applications! Pressurization is used only for the purpose of delivering the right amount of oil to the bearing. In most cases the bearing housing is not under pressure.*

Jet-Oil Lubrication

Jet-oil lubrication, shown earlier in Figure 9-15, is considered the best possible method of applying lubricant to rolling element bearings. It is unexcelled for high speeds and heavy loads. The oil jet is directed at the space between the outside diameter of the bearing inner ring and the bore of the cage. For extremely high speeds, means for scavenging the oil should be provided on both sides of the bearing. The oil system may be used to assure free axial flotation of the bearing cartridge in the housing on a thin pressurized oil film. If axial flotation is desired, a clearance of 0.001 in (0.025mm) between the housing and the cartridge is recommended.

In applications with large, heavily loaded, high-speed bearings operating at high temperatures, it may be necessary to use high velocity oil jets. In such cases the use of several jets on both sides of the bearing provides more uniform cooling and minimizes the danger of complete lubrication loss from plugging. The jet stream should be directed at the opening between the cage bore and inner ring O.D., as was illustrated in Figure 9-15. Adequate scavenging drains must be provided to prevent churning of excess oil after the oil has passed through the bearing. In special cases, scavenging may be required on both sides of the bearing.

At extremely high speeds, the bearing tends to reject the entry of sufficient oil to provide adequate cooling and lubrication with conventional oil jet and flood systems. Figure 9-28 shows an under-race lubrication system with a bearing having a split inner ring with oil slots. This method ensures positive entrance of oil into the bearing to provide lubrication and cooling of the inner ring and may well become a standard on the high-performance pump of the future.

However, a pressurization source is required for

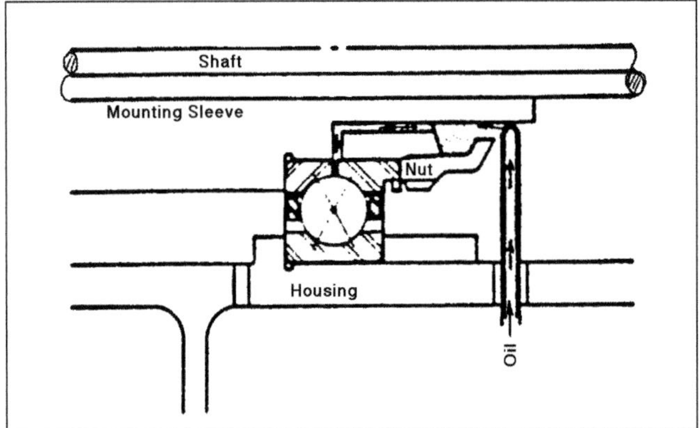

Figure 9-28: Under-race jet-oil lubrication (Source: MRC Engineering Handbook)

the oil and, while circulating oil systems can economically provide this pressurization for large pumps, the economics do not favor full-fledged circulating systems for smaller pumps. This is where inductive pumps merit consideration.

Small Stand-alone Pumps used for Jet-oil Delivery

Until recently only large-scale oil mist systems were economically attractive, although smaller units are now available for a variety of stand-alone applications (see HP In Reliability, June, 1999). Where the overall economics or unavailability of compressed air preclude oil mist, jet-oil spray systems are the best solution. Including a novel inductive pump (Figure 9-29) in oil spray units makes the most recent systems inexpensive, virtually maintenance-free, and thus highly attractive.

Inductive pumps use electromagnetic force to drive a completely encapsulated internal piston, creating positive piston displacement within a sleeved cylinder. By using one-way check valves, both ends of the piston can be used for simultaneous suction and pumping. Since each stroke displaces a fixed volume, any increment of this volume can be delivered with a high degree of accuracy.

One of several pump sizes obtainable from Inductive Pump Corporation (www.inductivepump.com), the Model 1.5 can easily pressurize lubricant taken from the bottom of the sump to appropriate spray pressures. Lubricant rates are adjustable from 30 ml/min to 1.5 gpm (~ 6 l/min). Weighing a scant 14 lbs (~6 kg), the 3.25 × 9.5 × 4 inch (approximately 83 × 241 × 100 mm) unit can be combined with an automotive spin-on filter and an industrial spray nozzle connected to a length of flexible tubing. Taking suction from the bottom of the bearing housing oil sump, a simple inductive pump represents a highly effective pump-around system, Figure 9-30.

Figure 9-29: Inductive pump (Source: www.inductivepump.com)

It is only fair to point out that suitably-sized conventional pumps, especially those that are pre-packaged with a small motor direct-coupled to the pump, merit consideration as well. One such type, made by the Dodge Manufacturing Company, has recently been introduced to the equipment retrofit and upgrade

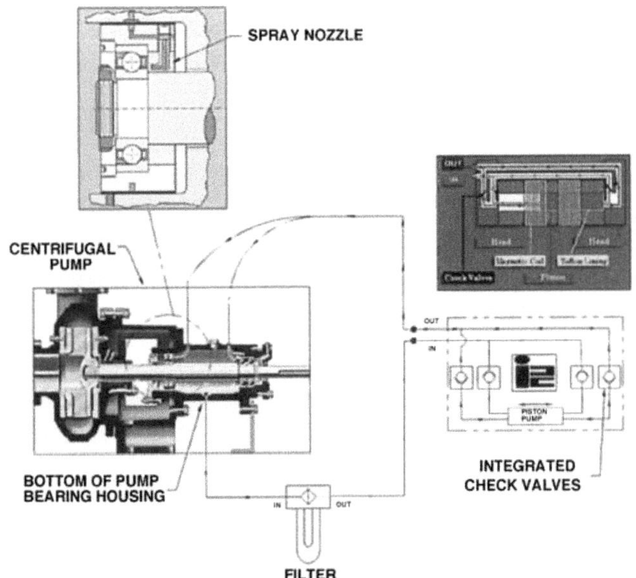

Figure 9-30: Schematic illustrating inductive pump used to provide clean, pressurized lubricant to jet-oil nozzles in paper stock pump (Source: www.inductivepump.com)

markets. It has the pump-driver assembly mounted on a small lube oil reservoir and it, too, should be considered in view of the often attractive anticipated payback.

Payback Anticipated for Stand-alone Pumps with Jet-oil Delivery.

Suppose a user had to deal with a situation where large pillow block bearings support the shaft of an overhung blower impeller. This may well be a situation where the oil ring option described earlier in this text might be judged a reliability risk and where objections might be raised to a conventional continuous lubrication oil system on purely economic grounds. Using an estimated incremental cost basis, we see fit to construct the matrix on page 241 to justify an attractive solution that uses an inductive pump.

It can thus be shown that a small stand-alone packaged circulating "mini-oil system" based on conventional or inductive pumps would win out on economic grounds and would be technically superior to an oil application strategy depending on oil rings.

Oil Mist Lubrication

Used extensively for pump lubrication by best-of-class companies, oil mist lubrication offers all the inherent benefits of centralized lubrication systems. Such benefits include reduced labor costs and reduced contamination of the lubricant—but at a much lower capital outlay than is required for centralized grease or circulating oil systems. Moreover, the benefits include the stand-still protection of installed pumps and electric motors against corrosion and dirt ingestion—a highly desirable feature in modern plants. Oil mist is a suspension of droplets of oil, one to three microns in size and is used primarily to lubricate rolling element bearings. It could also be characterized as an "oil fog" or an aerosol (Ref. 9-5, 9-6, 9-8).

It can thus be shown that a small stand-alone packaged circulating "mini-oil system" based on conventional or inductive pumps would win out on economic grounds and would be technically superior to an oil application strategy depending on oil rings.

Oil mist was originally developed in the 1930's in Europe to lubricate high-speed spindles because grease and liquid oil could not be effectively used. The oil in grease would not provide adequate lubrication and liquid oil generated too much heat. Typically, oil mist will cause bearings to operate up to 20°F (12°C) cooler. However, numerous examples exist where temperature reductions exceeded these figures (Ref. 9-6).

	Oil Ring Lube	*Inductive or Small Circulating Pump*	*Conventional Circulating System*
Cost of Blower:	$80,400	$84,300	$ 124,000
Preventive Maint. Cost/Year:	$4,200	$1,100	$2,000
Anticipated failures per Year:	0.25	0.1	0.05
Imputed per-year cost of a $60,000 failure	$15,000	$6,000	$3,000
Yearly cost outlay	$19,200	7,100	$6,000
Total payback over a 20-year equipment life	Base	(19,200-7,100)(20) = $ 240,000	(19,200-6,000)(20) = $264,000
Ratio:	Base	(240,000/84,300-80,400) = 61:1	(264,000/124,000-80,400) = 6:1

Oil mist technology, of course, is an extension of the liquid oil application family. However, it merits considerable discussion that includes economic justification and checklist-type information. Additional information can be found in Chapter 10, which deals with both oil mist and oil mist preservation.

WHY AND HOW TO PROTECT BEARINGS AGAINST LUBE CONTAMINATION-RELATED FAILURES

The fundamental reasons for favoring hermetic sealing devices for bearing housings can be derived from basic physics (Amonton's Law) and the technical literature. As regards Amonton's Law we will recall that, upon cooling, the density of a gas mixture will increase. For a closed volume, a pressure reduction would result. However, if this cooling takes place in a bearing housing and a path or opening exists to the external environment, ambient air will be induced to flow into the housing. Should the air contained in the bearing housing heat up again later, the reverse would take place and warm air would be expelled into the surrounding atmosphere until the pressures are equalized.

It is certainly appropriate to anticipate that bearing housings that no longer have access to ambient air environments will preclude oil contamination from external sources. But, how clean is the oil? The various bearing manufacturers use such *qualitative* terms as ultra-clean, very clean, clean, normal, contaminated, and heavily contaminated to describe oil cleanness. However, *quan-*

Table 9-4: Effect of oil moisture content on the life of rolling element bearings (Source: Royal Purple, Ltd., Porter, Texas 77365)

Effect of Moisture Content on Machine Life

Life Extension Factor (LEF)

Current Moisture Level (PPM)

A \ B	2x	3x	4x	5x	6x	7x	8x	9x	10x
50,000	12,500	6,500	4,500	3,125	2,500	2,000	1,500	1,000	782
25,000	6,250	3,250	2,250	1,563	1,250	1,000	750	500	391
10,000	2,500	1,300	900	625	500	400	300	200	156
5,000	1,250	650	450	313	250	200	150	100	78
2,500	625	325	225	156	125	100	75	50	39
1,000	250	130	90	63	50	40	30	20	16
500	125	65	45	31	25	20	15	10	8
250	63	33	23	16	13	10	8	5	4
100	25	13	9	6	5	4	3	2	2

Table 9-5: ISO 4406:99 oil cleanliness levels and reporting routines (Sources: Diagnostics, Inc., and Royal Purple, Ltd., Porter, Texas 77365)

ISO 4406:99 Cleanliness Level Standards

Range Number	Number of Particles per ml	
	More Than	Up to and Including
24	80,000	160,000
23	40,000	80,000
22	20,000	40,000
21	10,000	20,000
20	5,000	10,000
19	2,500	5,000
18	1,300	2,500
17	640	1,300
16	320	640
15	160	320
14	80	160
13	40	80
12	20	40
11	10	20
10	5	10
9	2.5	5
8	1.3	2.5
7	.64	1.3
6	.32	.64

Oil cleanliness levels may be reported using either a two- or three-digit rating. The cleanliness level of an oil with a particle count of:

2000 ≥ 4 microns
500 ≥ 6 microns
60 ≥ 14 microns
would be reported as 18/16/13.

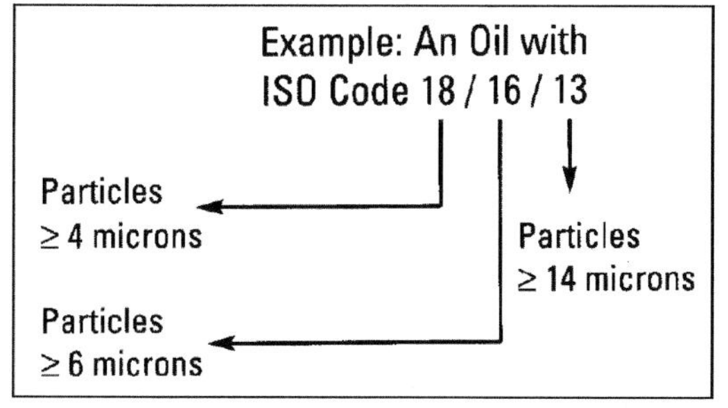

titative data are needed for purposes of cost justification or life-cycle-cost studies. We find these numbers in technical papers, articles and books that deal with the quantitative effects of contaminated lube oil. Among the many references, only five technical data sources will be mentioned here:

In Ref. 9-1, we find plots (provided by Dr. Richard Brodzinski, BP Oil, Kwinana, Western Australia) that show bearing lives with oil contamination "normal" to be less than one-half of the lives of bearings with oil labeled "clean."

Tables 9-4 through 9-6, derived from marketing literature of Royal Purple, Ltd., Porter, Texas, (Ref. 9-1), assess and quantify the benefit of clean oil by assigning a life extension factor. Using the example of an ISO 4406:99 cleanliness level of initially 22/19 and bringing it up to a new level of 14/11, Royal Purple's experience shows bearing life extended by a factor of four.

Oil analysis experts at Tulsa, Oklahoma-based NORIA Corporation (Ref. 9-3) consider an ISO 4406:99 cleanliness level of 23/20/17 typical for pumps. A level of 16/13/10 would be seen as "world class," and >28/25/22 as "evidence of serious neglect." For a cleanliness improvement from NORIA's "typical" to NORIA's "world class" and after converting its three-range numbers to the equivalent two-range ISO numbers 20/17 and 13/10, Reference 9-2 would again give a bearing life extension factor of four.

In Eschmann, Hasbargen and Weigand's 1985 text "Ball and Roller Bearings" Ref. 9-15, pg. 183), European bearing manufacturer FAG emphasizes that the severity

Table 9-6: Effect of oil cleanliness level on the life of rolling element bearings in pumps (Sources: Diagnostics, Inc., and Royal Purple, Ltd.)

Effect of Fluid Cleanliness on Rolling Contact Bearing Life

Life Extension Factor (LEF)

Current Machine Cleanliness (ISO 4406) A \ B	2x	3x	4x	5x	6x	7x	8x	9x	10x
26/23	22/19	20/17	18/15	17/14	16/13	15/12	15/12	14/11	14/11
25/22	21/18	19/16	17/14	16/13	15/12	14/11	14/11	13/10	13/10
24/21	20/17	18/15	17/14	16/13	15/12	14/11	13/10	13/10	12/9
23/20	19/16	17/14	15/12	14/11	13/10	13/10	12/9	11/8	11/8
22/19	18/15	16/13	14/11	13/10	12/9	11/8	11/8	—	—
21/18	17/14	15/12	13/10	12/9	11/8	11/8	—	—	—
20/17	16/13	14/11	13/10	11/8	—	—	—	—	—
19/16	15/12	13/10	11/8	—	—	—	—	—	—
18/15	14/11	12/9	—	—	—	—	—	—	—
17/14	13/10	11/8	—	—	—	—	—	—	—
16/13	12/9	—	—	—	—	—	—	—	—
15/12	11/8	—	—	—	—	—	—	—	—
14/11	11/8[1]	—	—	—	—	—	—	—	—
13/10	11/8[1]	—	—	—	—	—	—	—	—
12/9	11/8[2]	—	—	—	—	—	—	—	—

[1] Life Extension Factor 1.5 [2] Life Extension Factor 1.3

of the undesirable end effects of contamination depends on the ratio of operating viscosity of a lubricant divided by its rated viscosity. While there obviously could be an almost infinite number of combinations in the amount of contamination and ratios of viscosity, ratios of 0.5 to perhaps 1.0 are thought rather typical. Using 0.5 for this ratio, and plotting from the mid-point of a zone labeled "contaminants in lubricant" to the mid-point of a zone labeled "high degree of cleanliness in the lubricating gap," one would find a four-fold increase in bearing life for the cleaner oil. At a viscosity ratio of 2, the projected bearing life increase traversing from "contaminated" to "clean" would be approximately seven-fold. It should be noted that we are not here considering "ultra-clean" oil, since it would be unrealistic to find this degree of cleanliness in field-installed process pump bearing housings.

The most authoritative data on the effects of lubricant contamination might perhaps be gleaned from the General Catalog of one of the world's leading bearing manufacturers, SKF (Ref. 9-2). For the example shown in their catalog, SKF applied its New Life Theory to an oil-lubricated 45 mm radial bearing running at constant load and speed. Under ultra-clean conditions (nc = 1), this example bearing was calculated to reach 15,250 operating hours. The SKF catalog text goes on to explain that, if the example were to be calculated for contaminated conditions such that nc =0.02, bearing life would be only 287 operating hours.

Reasonable engineering judgment considers fully sealed bearing housings fully capable of maintaining "clean" oil conditions. Opinions to the contrary are without foundation and do not coincide with field experience. As an example, Ref. 9-24, published in 1996, reaches the seemingly startling conclusion that "the type of bearing housing closure device (labyrinth, lip seal, or magnetic seal) shows no significant correlation with either particulate or water contamination levels."

However, this conclusion is easily explained by the fact that the bearing housings involved in the survey were all fitted with vented filler caps and were thus allowed to "breathe." In this context, and as if to corroborate Amonton's Law, "breathing" means that temperature expansion of the air/oil mixture floating above the oil levels causes the gas mixture to be expelled through the vented filler cap into the ambient environment. Upon cooling of the air/oil mixture, ambient air is again drawn into the bearing housings. This is very obviously an ongoing in-out process of real consequence for plants located in unfavorable environments.

As Ref. 24 states, five out of seven sample plants were located in the Houston, Texas, area which, in the past, has been known for some of the worst industrial air pollution in the USA. Moreover, it is known that many, if not most, of the tested pumps employed oil ring lubrication. Oil rings are sensitive to shaft horizontality and, when "sliding downhill" often contact housing-internal stationary parts. This has been shown to cause ring wear and generally serious oil contamination (Ref. 9-25).

The oil cleanliness condition of any bearing housing interior that is accessed by the surrounding ambient air might, at best, be labeled "normal." All rotating labyrinth seals have an open gap that allows communication between the housing and ambient environment. While no definitive quantifiers are given that describe a refinery ambient, most users are aware that considerable amounts of particulates and moisture exist in the ambient air of industrial regions. A simple test situation may help in visualizing the issue.

Say, a new automobile is being washed and polished, and then left somewhere in the open near an industrial plant. Three days later, a person takes a clean paper towel and swipes it over the hood of the car. Not surprisingly, the paper towel will no longer be clean. We can certainly envision that a considerable amount of dust is likely to find its way into bearing housings that continually "breathe" because of temperature expansion and contraction of the air that fills the space above the oil level.

The severity of contamination can also be seen from the same study, Ref. 9-24, which had paradoxically concluded that the amount of contamination found in pump bearing housings is independent of the type of bearing housing seal employed. It found that bearing housings (with open vents) have particle contamination levels at least 10 times greater than recommended levels. Moreover, a staggering 54 percent of the more than 150 samples taken from industrial pumps contained contamination levels more than 100 times greater than recommended!

Recall that bearing manufacturers are using the terms "clean" and "normal" to describe the degree of lubricating oil contamination. It can be reasoned that bearing housings with closed vents and equipped with face-type seals, or bearing protectors that employ mechanical seal principles and technology are able to keep the lubricant "clean," whereas rotating labyrinth seals designed with an air gap will allow the lubricant to degrade to the point of "normal" contamination. As we consider all of the above, it is simply reasonable to accept the premise that bearings lubricated with "clean" oils will live at least twice as long as bearings with oil in "normal" condition of cleanness. Therefore, this doubling of bearing life is often used in the most conservative and simplest cost justification calculations, as will be seen later.

Protecting Bearings against Lube Contamination-related Failures*

According to the Barden Corporation, fewer than 10% of all ball bearings run long enough to succumb to normal fatigue failure (Ref. 9-7). Most bearings fail at an early age because of static overload, wear, corrosion, lubricant failure, contamination, or overheating. SKF USA provided the tabulation in Figure 9-31, emphasizing that lubrication-related failures account for 54% of all bearing failures. Experience shows that a very large portion of this 54% failure slice must be attributed to lubricant contamination. As shown in our introductory chapter and Figures I-1 and I-2, a major oil refinery reached very similar conclusions.

Then also, while rolling element bearings have been around for well over 100 years and are being used by the millions each year, costly misunderstandings persist. Not everyone knows that lightly loaded bearings are as likely to fail as are heavily loaded ones. Skidding, the inability of a rolling element to stay in rolling contact at all times because of too light a load, can cause bearing components to experience abrasive wear.

*For a discussion of "black oil" and its elusive root causes, see Bloch, Heinz P., *Practical Lubrication for Industrial Facilities*, (2009), The Fairmont Press, Lilburn, GA 30047 (ISBN 0-88173-579-5)

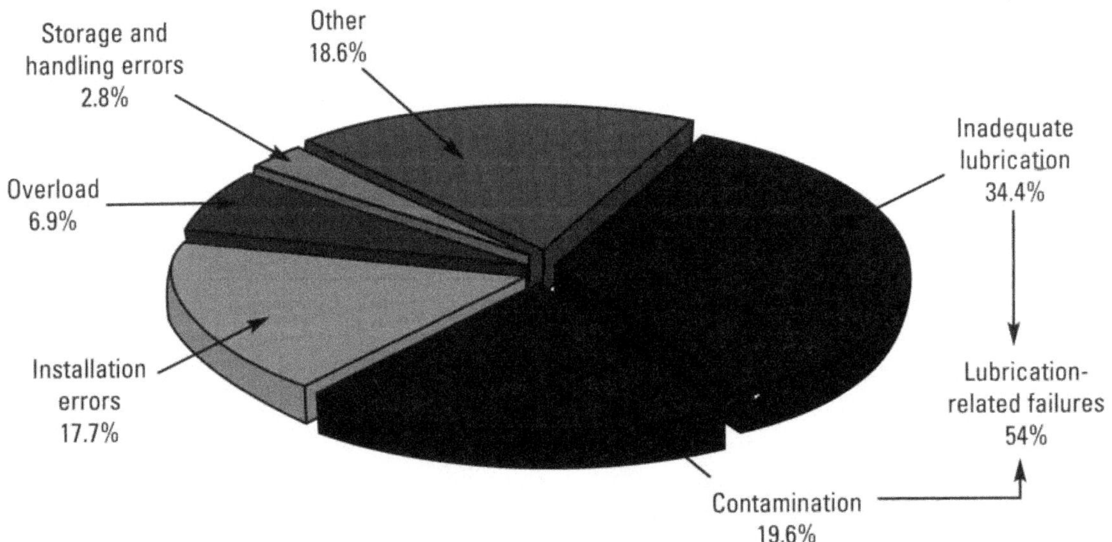

Figure 9-31: Causes of bearing failures (Source: SKF USA, Kulpsville, Pennsylvania 19443)

Even in small concentrations, wear particles will cause extremely serious oil contamination. Skidding can be largely eliminated by ensuring the bearing will always be loaded. In certain pumps, sets of 15° angular contact bearings live longer than sets of 40°/40° bearings!

Actual operations have shown that better bearing specification practices will avert the majority of static overload problems. Here, the reader may wish to refer back for overview and general guidance to Tables 9-2 and 9-3, or even to Chapter 7. Other problems caused by wear, corrosion, lubricant failure, oil contamination and overheating can be prevented by the proper selection, application and preservation of lubricants. Oil viscosity is of concern and higher-viscosity lubricants are generally preferred (Ref. 9-9). Yet, an excessive viscosity film could increase pump bearing friction losses to the extent that total pump power draw goes up by as much as 4%. Moreover, such an oil has been known to cause excessive oil ring drag, depriving certain pumps of adequate lubrication.

Pump Lube Contamination and its Origins

Lubricant contamination by solid particles and moisture is often overlooked and must be addressed by reliability-focused users. While this concern points the way to sealing the bearing housing against the intrusion of airborne dirt and moisture, let's first understand where the contaminants originate and how detrimental these impurities really are.

Except for the occasional dislocated oil ring and forced spray of water or sandblasting action directed against pumps, both moisture and airborne dirt enter bearing housings by the breathing effect of the air space that exists above the lubricating oil. Atmospheric air contacts not only the oil level in most bearing housings, but also the oil in the surge chamber of unbalanced constant level lubricators.

As the volume of air inside the bearing housing warms up and expands, some air is expelled along the shaft protrusion and through the housing vent into the atmosphere. When at some later time the housing temperature declines, the remaining air contracts. Unclean atmospheric air now re-enters the housing along the same two pathways (Refs. 9-9 and 9-10).

The detrimental effects of moisture contamination can be gleaned from Table 9-4. If, for example, the current moisture level is 500 ppm and a suitable bearing housing seal would protect against moisture intrusion and thus limit the moisture level to 45 ppm, bearing life would probably be extended by a factor of 4. Recognizing the advantage of using lubricants with absolutely minimum moisture contamination prompted the development of OEM-supply (original equipment manufacturer) as well as add-on ("aftermarket") desiccant cartridges. These cartridges can be threaded into the vent ports generally located at the top of pump bearing housings. Once saturated with water vapor, the desiccant charge typically changes color, signaling the need for replacement. Along those lines, electronic moisture detection and annunciation devices are available and have found use in the United States (Refs. 9-9 and 9-10. Also: Trico Manufacturing Company, Pewaukee, Wisconsin 53072).

Another path to a clean and moisture-free bearing environment would be to look at the cost justification for installing bearing housing seals of the type (see Chapter 6) that would essentially preclude the ingress and exit of air.

Quite obviously, bearing protection must not be limited to the exclusion of moisture only; beneficial results are certain to result from reduced particle contamination of bearing lubricants. Clean oil greatly extends the life of bearings and pumps (Ref. 9-11). Many oils are full of small particles in the range from 2 to 30 micron. The eye cannot see these particles; they consist mainly of fibers, silica (dirt) and metals. The amounts and sizes of particles can be measured with a laser particle counter and then quantified using the ISO 4406:99 Cleanliness Rating, Table 9-5, which reports on the particle content of one ml of oil—approximately the volume of an eye dropper (1 milliliter = 1 cc).

It is of interest to note that the cleanliness levels requested by major entities and corporations vary. ISO 12/9 and 13/10 are listed for bearing manufacturers FAG and SKF, ISO 14/12/10 is mentioned in STLE publications and the CRC Lubrication Handbook, and ISO 15/12 and 16/12 are recommended by turbine manufacturers Siemens, Westinghouse and General Electric.

The probable bearing life extension resulting from cleaner lubricants can be obtained from Table 9-6. If, in the highlighted example, the ISO 4406:99 cleanliness level would be improved from 20/17 to 13/10, the rolling element bearing life would probably be quadrupled—a 300% increase. Again, there are compelling reasons to apply cost-effective means of excluding contaminants from pump bearing housings.

Contamination Control through Bearing Housing Seals

While many pumps are today still equipped with large open vents and short-lived, wear-prone lip seals (Figure 9-32), reliability-focused users have long since favored reducing the vent size. Many have even eliminated vents altogether and opted to install rotating labyrinth seals with expulsion ports, Figure 9-33. Rotating labyrinth seals, sometimes called "bearing isolators," seem to work best when the housing vent is plugged. To quote from one well-known manufacturer's marketing literature (Ref. 9-12):

> *"If the housing vent is left open, the slight vacuum created by the contaminant expulsion elements will induce the flow of airborne dust, dirt, vapors and everything available in the immediate environment through the bearing enclosure not unlike an oil-bath vacuum cleaner. This action is constant and the amount of induced debris build-up can be significant."*

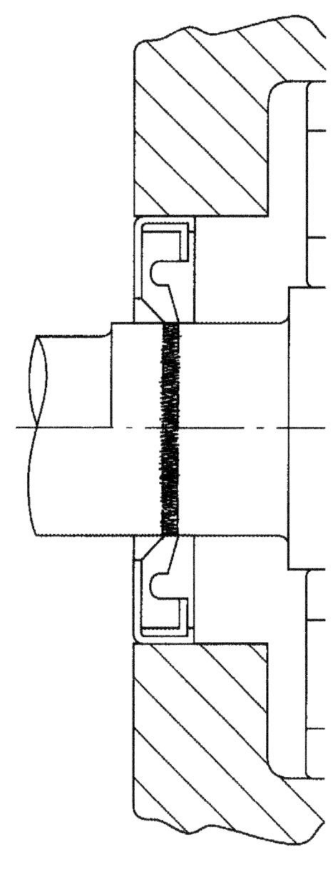

Figure 9-32: Lip-seals are wear-prone and can damage shafts

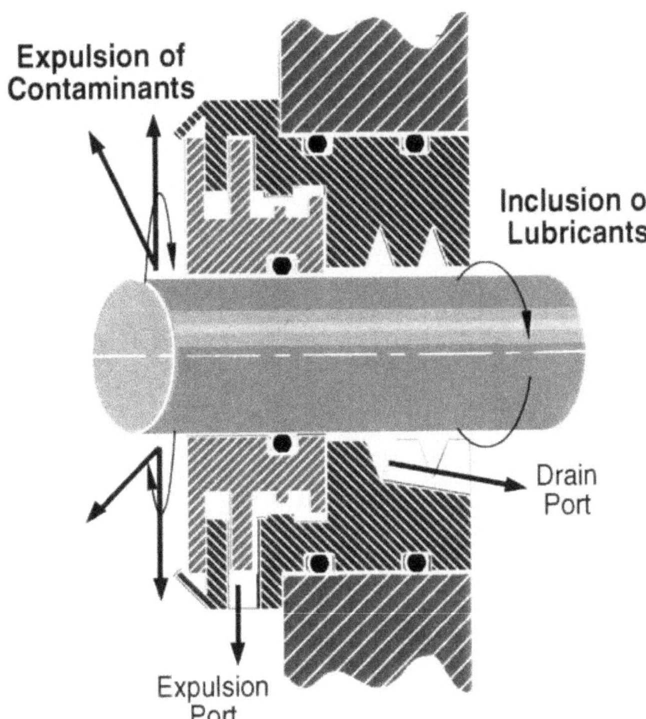

Figure 9-33: Typical rotating labyrinth-type bearing housing seal (Source: Garlock Sealing Technologies, Palmyra, NY 14522)

Other users have begun to protect pump bearing housing interiors by installing magnetically activated face seals, Figure 9-34, also 6-11 and 6-17. Both rotating labyrinth and magnetic face seals take up less axial space than even the narrowest available mechanical seals, one of which is shown in Figure 9-35.

One of the most recently developed bearing housing seals, the repairable IP66-rated "LabTecta®" model (Figure 9-36), is a vast improvement over ordinary non-rotating labyrinths and lip seals. (See also Appendix 6.) Some corporate equipment purchase and repair standards require this type of "bearing isolator" for new and rebuilt electric motors. However, even this device cannot be considered a totally effective housing seal for oil-lubricated pump bearings since it is unable to prevent the interchange of vapors inside the bearing housing with the surrounding ambient environment. Let's examine why this is so.

The rotor of this bearing housing seal is clamped to the pump shaft by the friction fit of two O-rings. Two different O-rings are located in the rotor bore groove. Designed to lift off slightly on running pumps, they are intended to re-seal the bearing housing after the pump is stopped. Applying Amonton's Law, some have reasoned that if the O-rings lift off during operation, some of the air, which during equipment operation exists in the housing at higher-than-ambient temperature, will escape into the environment. When the pump is stopped, the residual air cools and a slight vacuum is produced. If the O-rings lift off after restarting, ambient air at higher pressure gets into the bearing housing. If the rings do not lift off, then there is no gap between the rotor's O-rings and an opposing stator surface. In that case, the rotating O-rings could undergo wear and again not perform as a "nothing in/nothing out" hermetic seal. However, the O-rings are field-replaceable—a big advantage!

It can be shown that for true sealing, a face seal will have to be selected. With even the narrowest conventional mechanical face seals taking up too much axial space, several manufacturers have opted to use small bar magnets (Figures 7-25 and 9-34) to provide the closing force. Although thousands of magnetic seals have been in widespread use in aerospace applications since the mid-1940's, rigorous experimentation and analysis of their industrial counterparts had

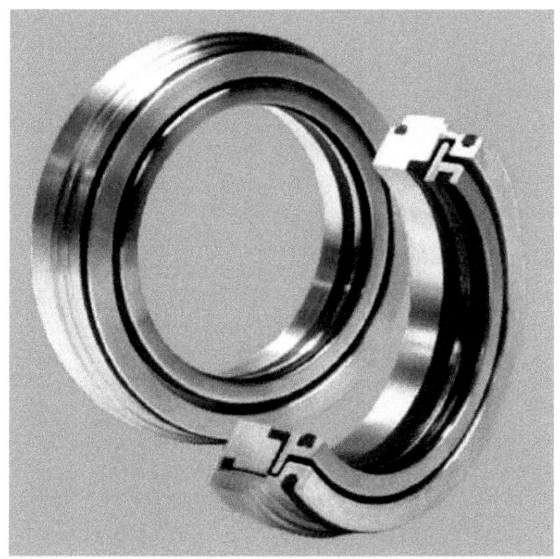

Figure 9-34: Narrow-width magnetic face seals can completely isolate bearing housing interior from atmosphere and external contamination (Source: Isomag Corporation, Baton Rouge, Louisiana 70809)

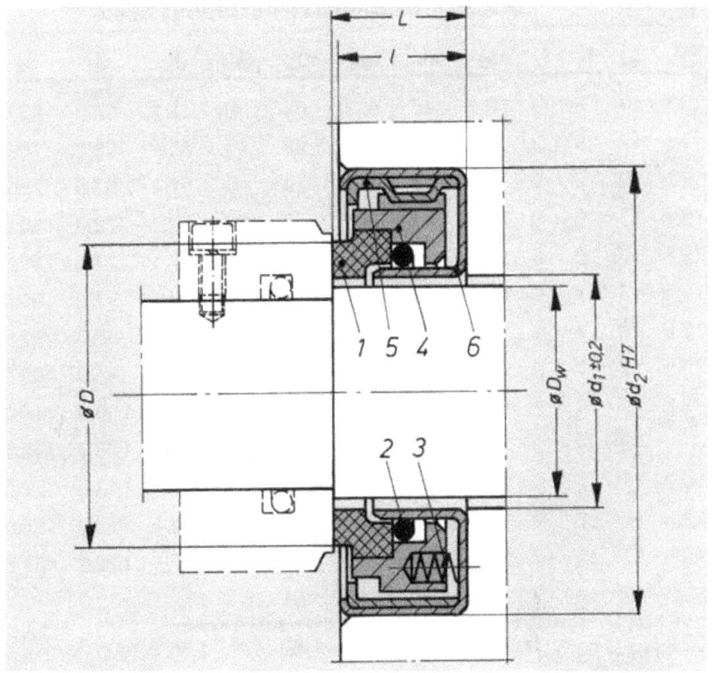
L
l
1 5 4 6
2 3
øD
øDw
ød1±0,2
ød2 H7

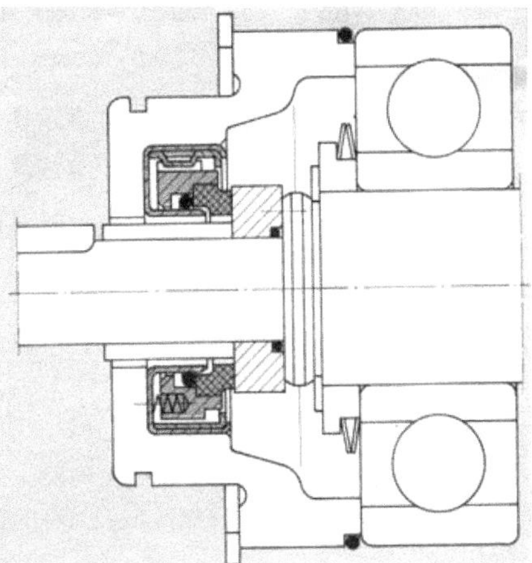

Figure 9-35: Narrow conventional mechanical seal for pump bearing housing (Source: Burgmann Seals America, Houston, Texas 77041)

to wait. In 2001/2002 research was performed at Texas A&M University by Leonardo Urbiola Soto under the tutelage of Prof. Dr. Fred Kettleborough. The essential components of Texas A&M's test setup are shown in Figure 9-37.

In his master's thesis (Ref. 9-13), Urbiola reported that the steady-state torque required to make the magnetic seal disc rotate is a function of the oil level, but remains a fraction of the torque needed to make the (adjacent) ball bearing balls roll. This fraction ranges from a low of 0.13 to a high of 0.83. He also noted that the differential temperature between the seal contact surfaces and an adjacent ball bearing outer race is independent of oil pressure and averaged 12°C (22°F). To quote from his summary:

> "A magnetic seal leaks less than any other type of seal and operates with a low coefficient of friction, torque, electric power consumption and heat generation. Experimental data evidences this device as an affordable way of sealing ball bearing housings..."*

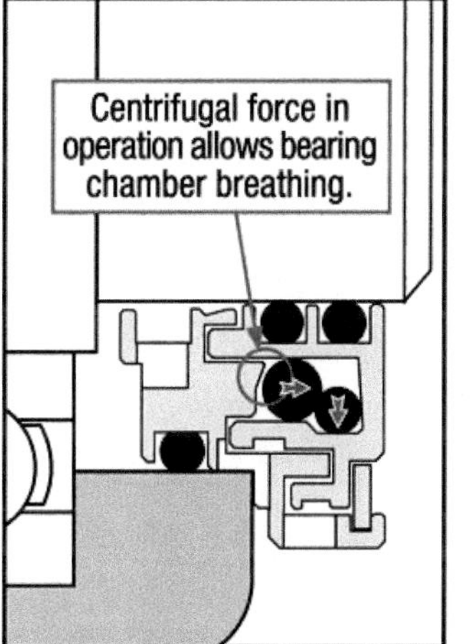
Centrifugal force in operation allows bearing chamber breathing.

Axial force at standstill creates a positive seal.

Figure 9-36: Advanced "non-contacting" bearing housing seal (Source: AESSEAL, plc, Rotherham, UK, and Rockford, Tennessee

---

*Although unplanned, the experiments at Texas A&M corroborated and validated two of our earlier statements:

—that incorrectly machined oil rings will quickly cause bearing damage (page 210), and

—that operating a 65 mm bearing at 3,600 rpm with lube oil reaching the center of the lowest bearing ball will cause oil temperatures to increase significantly.

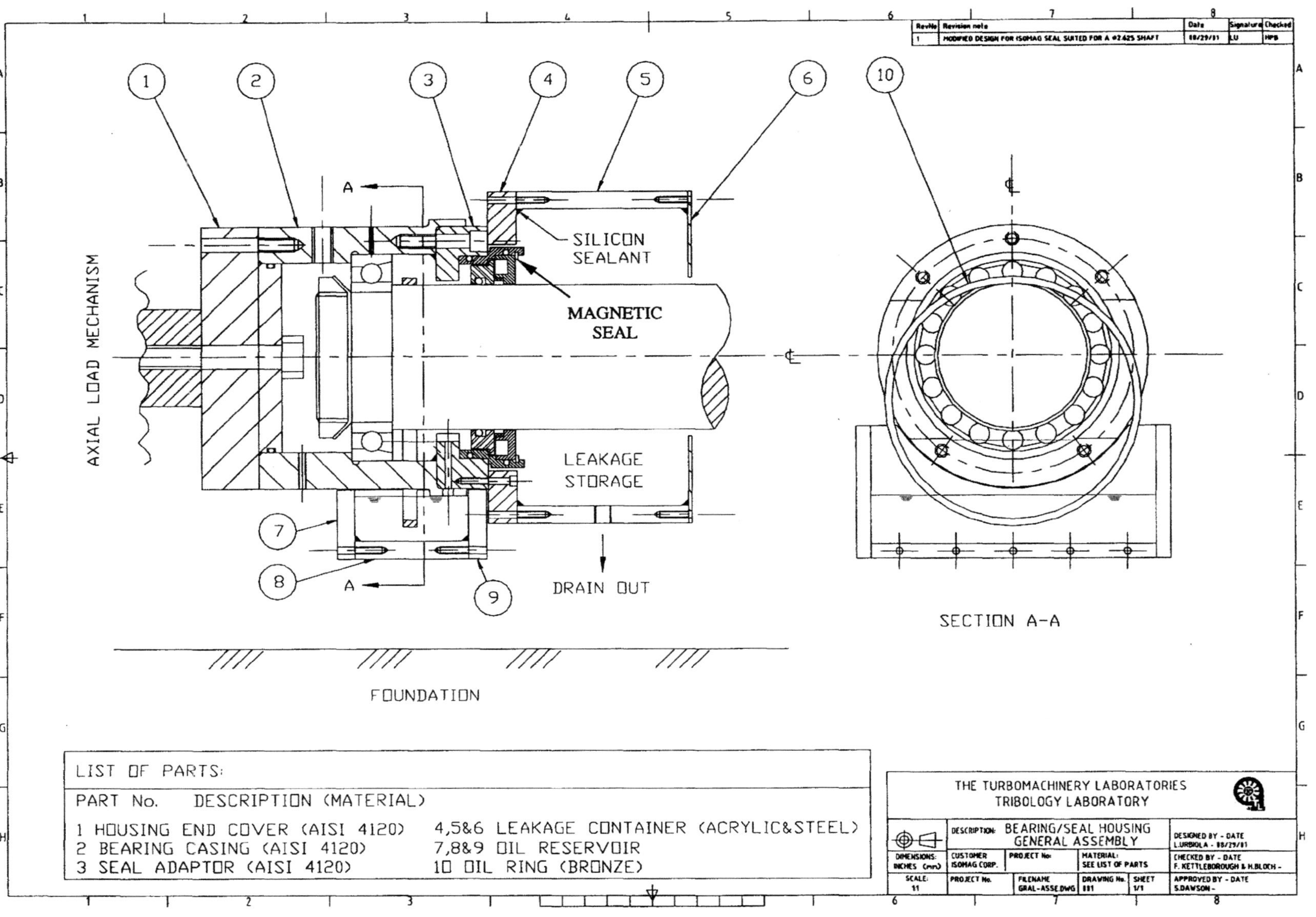

Figure 9-37: Magnetic seal test setup at Texas A&M University, 2001/2002 (Source: Urbiola Soto, Leonardo, Master's Thesis "Experimental Investigation on Rotating Magnetic Seals.")

Getting Back to Bearing Life Issues

Unlike API-610 pump bearings, which petrochemical companies often specify for an L_{10} life of 40,000 hours, ANSI pump bearings are selected on the basis of an expected 24,000-hour life. Nominally, this means that 90% of ANSI pump bearings should still be serviceable after approximately three years of continuous operation. However, the failure statistics quoted in Ref. 9-7 and elsewhere indicate that conventionally lubricated ANSI pump bearings do not even approach this longevity. Lack of lubrication, wrong lubrications, water and dirt in the oil, and oil-ring debris in the oil sump all cause bearing life expectancies to be substantially less. It must be assumed that similar findings by other major users of ANSI pumps prompted the search for "life-time lubricated" rolling element bearings which we had alluded to earlier, but which nevertheless have their own particular vulnerabilities and make economic sense only at low DN values.

Problem incidents caused by dirt and water have been substantially reduced by oil mist lubrication. However, serious failure risk can also be introduced by certain specification practices, including some that are perhaps implied in the various editions of API-610. Without going into the many possible factors which could influence bearing life, a number of items must be considered in the mechanical design of reliable centrifugal pumps. First among these would be that deep-groove Conrad-type radial bearings with loose internal clearance (C3) are more tolerant of off-design pump operation than bearings with standard internal clearance. Also, it should be recognized that centrifugal pumps which undergo significant bearing temperature excursions are prone to cage failures. Phenolic cages are typically suited for operation below 220°F (105°C) only.

New cage materials, such as polyamide 6.6, provide a higher temperature limit and excellent lubricity even at slightly higher temperatures. For process pumps, however, metallic cages are the component of choice. There are two reasons why reliability-focused users disallow non-metallic cages and insist on machined brass or other metals. The first reason is because pump repair shops are often unable to guarantee heating and assembly procedures that will not overheat the bearing as it is being placed on the pump shaft. The second reason is linked to the fact that elongation or similar deformation of the ball separators (cages) will not be detected by conventional and typically practiced vibration detection methods. Plastic cages thus have been known to fail without warning.

The pump designer and pump user must also come to grips with questions relating to preload values and contact angles. The API-610 requirement to utilize duplex 40° angular contact angle thrust bearings was prompted by the desire to obtain maximum load capacity from a given size bearing (Ref. 9-14). Similarly, the requirement of installing these bearings back-to-back with a light axial preload was aimed at reducing axial shuttling of rotors to prevent brinelling of contact surfaces (raceways) and to prevent ball skidding.

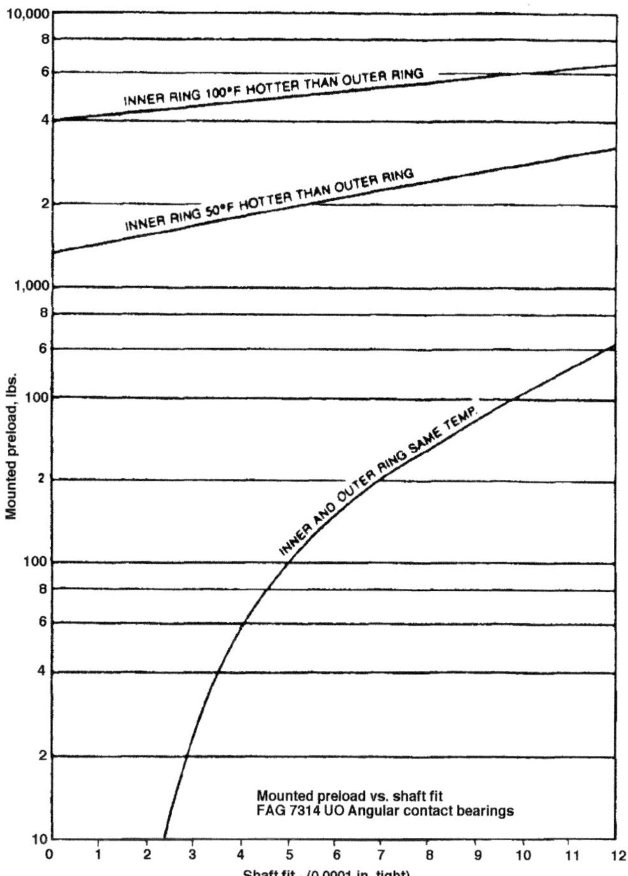

Figure 9-38: Mounted preload vs. shaft fit for a 70 mm angular contact thrust bearing (Source: FAG Bearing Corporation, Stamford, Connecticut 06904)

Research by the FAG Bearing Corporation demonstrated that under high thrust conditions 40° angular contact bearings generate more heat than thrust bearings with less angularity (Ref. 9-15). However, preloading adds to the heat load, and using an interference fit between shaft and inner ring compounds the problem.

In this regard, Figure 9-38 will prove very enlightening. It shows that for a given bearing (FAG 7314, 70 mm bore diameter) a shaft interference fit of 0.0003 in. will produce an almost insignificant preload of approximately 22 lbs (~10 kgf), whereas an interference fit of 0.0007 in. would result in a mounted preload of 200 lbs

(~91 kgf).

However, a much more significant preload would result from temperature differences between inner and outer bearing rings. Such differences often exist in pumps, since some heat migrates from high-temperature pumpage along the shaft, and also because many pump designs incorporate cooling provisions that tend to artificially cool the outer ring and will thus prevent it from expanding.

By far the worst scenario would be for a pump operator to apply a cooling water stream from a fire hose. It is baffling to still see this practice today, in the age of high tech, space travel, and information explosion.

Conventional Wisdom: *If it feels too hot, pour water on a hot bearing housing.*

Fact: *Doing so is almost certain to accelerate the demise of the bearing. The outer ring will shrink more than the inner ring, causing excessive preload and failure.*

Lubricant Flow Direction Influenced by Bearing Orientation

There are compelling reasons to mount pump "duplex" angular contact thrust bearings back-to-back, as was shown earlier in Figure 7-31 and again in Figures 9-16 and 9-21. In process pumps, it must be assumed that axial thrust will often act in both directions. This is why, typically, two opposing thrust bearings are installed. Experience also shows that the bearing inner rings are hotter than the outer rings, and Figure 9-38 alluded to this possibility. Therefore, pump thrust bearings are almost always installed back-to-back so as to allow the load contours of the bearing inner rings to thermally "grow away" from ball contact.

If thrust bearings were mounted face-to-face (Figure 7-30), the load contours of the bearing inner rings would thermally "grow into" more severe, and potentially excessive, contact between balls and bearing raceways.

Although the foregoing serves as a reminder that duplex angular contact pump thrust bearings should generally be mounted back-to-back, our earlier discussion showed that this manner of orientation is not optimized for oil flow. As can be gleaned from Figure 9-39, this fact did not escape the manufacturers of high reliability pumps. Aiming for the least vulnerable thrust bearing execution prompted one major manufacturer to utilize a flinger disc that tosses the oil into the bearing housing wall. From there, the lubricant runs into a sloped trough or gallery. The left branch of the trough leads to the radial (roller) bearing while the right branch guides the lube oil into a spacer ring separating the bearing outer rings.

With oil thus being deposited at the thrust (ball) bearing cage location closest to the shaft and centrifugal force obviously acting away from the shaft, lubricant will flow through both thrust bearings before exiting at each end. A second spacer ring clamped between the bearing inner rings facilitates making preload adjustments. The periphery of the flinger disc dips into the lube oil level; however, the lube oil level is generally maintained well below the center of the lowermost ball. This reduces oil churning and friction-induced heat-up of lube oil and bearings. Needless to say, unless lubricant application methods take into account all of the above, bearing life and reliability may be severely impaired.

Pump users should also realize that double row "filler notch" bearings are considerably more vulnerable in pump thrust applications than other bearing types and should not be used. Similarly, ball bearings are sensitive to misalignment and must be properly mounted to eliminate this cause of failure. Misalignment must be no greater than 0.001" per inch (1 mm/m) of shaft length.

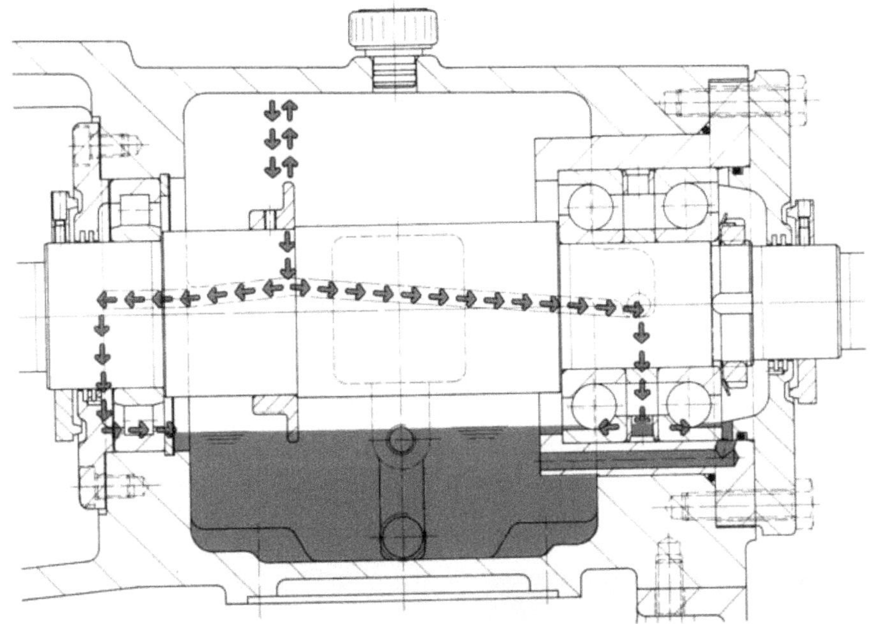

Figure 9-39: Optimized lube application between two back-to-back mounted angular contact bearings (Source: KSB AG, Germany)

Bearings operating in a misaligned condition are subject to failure regardless of cage type, although riveted cages seem particularly prone to rivet head fatigue in misaligned condition. This is another reason why machined brass or bronze cages are strongly recommended for pump applications.

Grease Lubrication Summary

While grease lubrication of pump bearings is not very common in the average U.S. process plant environment, entire pulp and paper plants in Scandinavia and other overseas locations use grease lubrication on thousands of pumps. However, grease lubed bearings are the accepted standard for the majority of electric motors throughout the world. And so, regardless of what equipment type is involved, certain facts are worthy of consideration and a few precautions are in order.

The cost difference between superior greases and "minimum acceptable" greases is relatively small. Process plants and other facilities wishing to be reliability-focused and to avoid bearing repairs will find it economically attractive to buy only premium greases. Also, while there may be an optimum grease selection for electric motor bearings, this grease would not serve well in grease lubricated couplings, or open gearing, or steam turbine linkages and so forth. In other words, a reliability-focused pump owner will not even consider using the same "all-purpose" grease product for every application requiring grease.

It serves no useful purpose to request grease properties that are not actually needed in a given application. A good example would be EP (extreme pressure) additives. While of importance in sliding gear engagements, these additives would in no way improve the life of electric motor bearings. Electric motor bearings are almost always lightly loaded.

General Properties Reviewed

Soft, long-fibered type greases, or excessively heavy oils, will result in increased churning friction at higher speeds, causing bearing overheating due to the high shear rate of these lubricants. Excessive amounts of lubricant will also create high temperatures.

Using oils of adequate film strength, but light viscosity, or using channeling or semi-channeling greases has the benefit of substantially reducing the heat-generating effects of lubricants. The advantages of these greases rest in their ability to "channel" or be pushed aside by the rotating ball or roller elements of a bearing and to lie essentially dormant in the side cavities of the bearing or housing reservoir.

Channeling greases normally are "short-fibered" and have a buttery consistency that imparts a low shear rate to the lubricant. This low temperature aids an operating bearing to establish temperature equilibrium, even if a lubricant is applied having a slightly higher viscosity than the application demands. Higher fluid friction increases the temperature of the lubricant until the viscosity is reduced to the proper level. It should be noted, however, that short-fibered greases might lead to "false brinelling" damage in applications subject to vibration without equipment rotation.

Greases are generally applied where oils cannot be used, e.g. in places where sealing does not exist or is inadequate, or in dirty environments and inaccessible locations. Greases are also used in places where oil dripping or splashing cannot be tolerated and where "sealed-for-life" lubrication is desired. It should be recognized, however, that "sealed for life" and "lubricated for life" are terms that simply refer to bearings that cannot be re-lubricated and which will fail once the grease has been used up, or is no longer serviceable due to oxidation of the oil. This explains why, at the beginning of this chapter, Table 9-1 ranked "Lubrication for Life" last in tabulating the influence of lubrication on service life.

Greases are fine dispersions of oil-insoluble thickening agents—usually soap—in a fluid lubricant such as a mineral oil. When a bearing lubricated with grease starts to move, the grease structure (created by the thickening agent) is affected by the shearing action, and lubrication is provided by the oil component of the grease. As the bearing slows to a stop, the grease regains its semi-solid structure. In non-moving parts of the bearing, this structure does not change.

The type and amount of the thickener, additives used, the oil, and the way in which the grease is made, can all influence grease properties. The chosen base-oil viscosity generally matches that for a fluid lubricant used for the same service—low-viscosity oil for light loads, fast speeds and low temperatures, and high-viscosity oils for differing conditions. The thickener will determine grease properties such as water resistance, high-temperature limit, resistance to permanent structural breakdown, "stay-put" properties, and cost.

Greases are classified on the basis of soap (or thickener) type, soap content, dropping point, base oil viscosity and consistency. Consistency is mainly a measure of the sealing properties, dispensability, and texture of a grease. Once the grease is in a bearing, consistency has little effect on performance. But despite this, greases are widely described primarily on the basis of consistency

(Ref. 9-1). They come in an endless array of formulations and with many different soaps. Suffice it to say that we will only describe some of the most common types.

Sodium-soap greases are occasionally used on small pump bearings because of their low torque resistance, excellent high-temperature performance and ability to absorb moisture in damp locations. Since all sodium soaps are easily washed out by water sprays, they should not be employed where splashes of water are expected.

Lithium-soap greases are generally water resistant and corrosion inhibiting, and have good mechanical and oxidation stability. Many automobile manufacturers specify such grease—often with additives to give wide protection against problems caused by shipment, motorist neglect, and now popular extended lubrication intervals. Widely used in centralized lubrication systems, these versatile greases are also favored in both sliding and rolling element bearings.

Simple calcium-soap greases resist water-washout, are non-corrosive to most metals, work well in both grease cups and centralized lubrication systems, and are low-cost lubricants. They are, depending on manufacturer and ingredients, limited to services cooler than 160°F to 200°F (~71°-93°C).

Complex calcium-soap greases, wisely applied, can provide multi-purpose lubrication at a fraction of the cost of a lithium-soap grease; however, misapplication of these greases will likely cause more difficulty than the same error committed with lithium greases. Special-purpose greases are available for food processing (both the thickener and oil are nontoxic), fine textile manufacture (light colors for non-staining, or adhesive grades to avoid sling-off), rust prevention and other special services.

Aluminum complex greases have outstanding EP (extreme pressure) capabilities and excellent water resistance to both emulsion and water washout. They can be pumped at low temperatures, are stable at high temperatures, and have excellent oxidation stability. They have solved seemingly insurmountable problems in high-speed electric train wheel lubrication and other tough applications (Ref. 9-16). To obtain maximum benefit from aluminum complex greases, work with a competent provider and purge existing grease fill, as per vendor's instructions.

Premium-grade polyureas ("EM" electric motor polyureas) are often considered the best choice for electric motor bearings. They incorporate a synthesized hydrocarbon base oil with high temperature capability, excellent rust inhibition properties, low friction, and other desirable attributes. Recent studies confirmed that the re-lubrication intervals for polyureas exceed those of lithium greases by at least a factors of two, and often by a factor of four.

Application Limits for Greases

Bearings and bearing lubricants are subject to four prime operating influences: speed, load, temperature, and environmental factors. The optimal continuous operating speeds for ball and roller type bearings—as related to lubrication—are functions of what is termed the DN factor. To establish the DN factor for a particular bearing, the bore of the bearing (in millimeters) is multiplied by the revolutions per minute, i.e.:

$$75 \text{ mm} \times 1000 \text{ rpm} = 75{,}000 \text{ DN value}$$

Rule-of-thumb application limits for premium-grade mineral oil-type greases in continuously operating bearings have been established to range from 100,000 to 150,000 DN for most spherical roller type bearings, and 200,000 to 300,000 DN values for most conventional ball bearings. Higher DN limits can sometimes be achieved for both ball and roller type bearings, but require close consultation with the bearing manufacturer. When operating at DN values higher than those indicated above, use either special greases incorporating good channeling characteristics, or use circulating oil.

Grease Re-lubrication Intervals

Typical grease re-lubrication intervals have been published by motor, bearing, and grease manufacturers. However, no two recommendations are alike and many do not take into account ambient or operating temperature effects. The SKF guidelines given in Figure 9-40 are based on engineering analyses and tests. They have been extensively validated by industry experience and have for many years been used by reliability-focused user companies all over the world.

In Figure 9-40 then, the re-lubrication intervals for normal operating conditions can be read off as a function of bearing speed "n" and bore diameter "d" of a certain bearing type. The diagram is valid for bearings on horizontal shafts in stationary machines under normal loads. It applies to good quality lithium base greases at a temperature not exceeding 158°F (70°C). SKF recommend that, to take account of the accelerated aging of the grease with increasing temperature, the intervals obtained from the diagram be halved for every 15°C (27°F) increase in bearing temperature above 158°F (70°C). However, the user must keep in mind that the maximum operating temperature for the grease—as

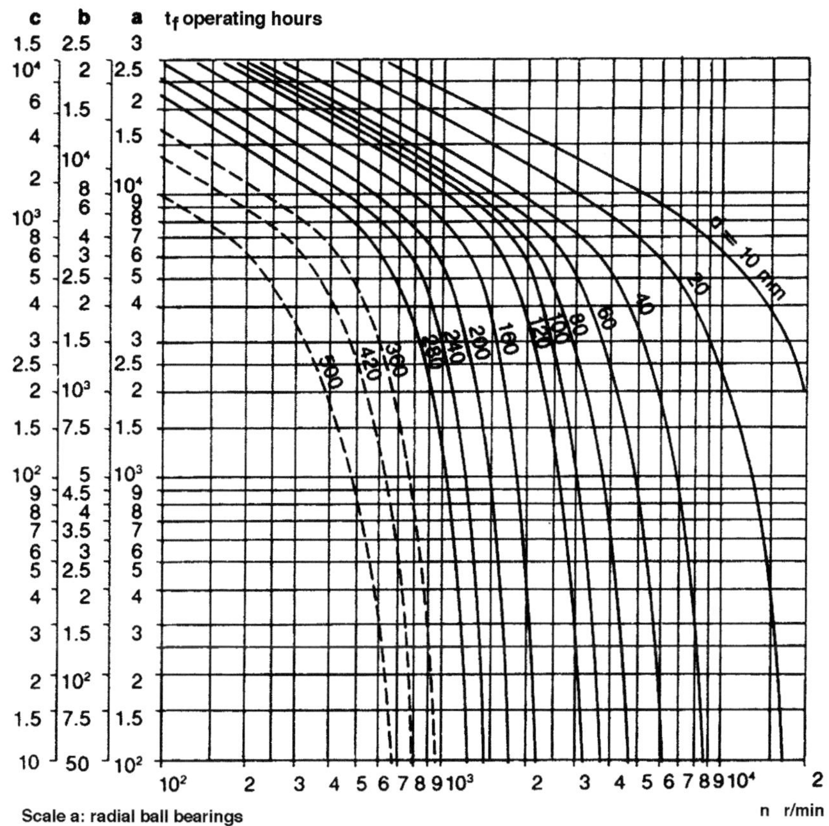

Scale a: radial ball bearings

Scale b: cylindrical roller bearings, needle roller bearings

Scale c: epherical roller bearings, taper roller bearings, thrust ball bearings;
full complement of cylindrical roller bearings ($0.2t_f$);
crossed cylindrical roller bearings with cage ($0.3t_f$);
cylindrical roller thrust bearings, needle roller thrust bearings, spherical roller thrust bearings ($0.5t_f$)

Figure 9-40: Grease relubrication intervals (Source: SKF USA, Kulpsville, Pennsylvania 19443)

stipulated by the grease manufacturer—should not be exceeded.

The intervals may be extended at temperatures lower than 158°F (70°C), but as operating temperatures decrease the grease will bleed oil less readily; therefore, at low temperatures, an extension of the intervals by more than two times is not recommended. It is not advisable to use re-lubrication intervals in excess of 30,000 hours. For bearings on vertical shafts the interval obtained from Figure 9-40 should be halved.

For large roller bearings having a bore diameter of 300 mm and above, the high specific loads in the bearing mean that adequate lubrication will be obtained only if the bearing is more frequently re-lubricated than indicated by Figure 9-40, and the lines are therefore broken. It is recommended in such cases that continuous lubrication be practiced for technical and economic reasons. The grease quantity to be supplied for continuous lubrication can be obtained from Equation 9-1 for applications where conditions are otherwise normal, i.e. where external heat is not applied.

$$Gk = (0.3\ldots\ 0.5)\ D\ B \times 10^{-4} \qquad \text{(Equation 9-1)}$$

where

Gk = grease quantity to be continuously supplied, grams/hr

D = bearing outside diameter, mm

B = total bearing width (or height, for thrust bearings), mm

While the above quantities refer to continuous grease lubrication, it should be noted that the quantities used for periodic replenishment are substantially different. By adding small amounts of fresh grease at regular intervals, the used grease in the bearing arrangement will only be partially replaced. Suitable quantities to be added can be obtained from Equation 9-2:

$$Gp = 0.005\ DB \qquad \text{(Equation 9-2)}$$

where

Gp = grease quantity to be added when replenishing, grams

D = bearing outside diameter, mm

B = total bearing width (axial length of bore), mm

Alternatively, the user may use Equation 9-3 as a periodic grease replenishment guideline

$$Gv = 0.2\ W \times d^2 \qquad \text{(Equation 9-3)}$$

where

Gv = grease quantity to be added when replenishing, cubic inches
W = width of bearing, inches
d = bearing bore diameter (or shaft size), inches

A rather different, much more general approach to grease re-lubrication of rolling element bearings is taken by FAG Bearing Company of Stamford, Connecticut. As shown in Figure 9-41, the user would divide equipment operating speed n by the manufacturer's stipulated limiting speed n_g and enter this value on the abscissa. From the intersect of a vertical line with the appropriate curve, the recommended re-lubrication interval can be found to the left, on the ordinate.

Finally, we want to introduce a grease manufacturer's guidelines for electric motor re-greasing, Table 9-7. It represents an effort to make adjustments for motor size, severity of service, and grease type. Ronex-MP is typical of good multipurpose greases, whereas Unirex-N2 is a premium product intended for electric motors only. More recently, Unirex-N2 has been replaced by "Polyrex EM," a superior electric motor bearing grease.

Grease Better than Oil?

From the above, we note that grease lubrication is sometimes simpler, or more convenient than liquid oil applied to rolling element bearings. A rather elementary definition of grease would call it a composite of around 85% oil and perhaps 13% soap, with the remainder made up of additives. However, since it is the oil and not the soap that does the lubricating, it can be said that the quality of lubricating with grease cannot possibly exceed the quality of properly applied liquid oil.

We do not have to be reminded that bearing failures often occur in the electric motors that drive pumps. There are many reasons for these failures and the use of better bearings is not in and of itself a solution to the problem. Again, some statistics will be of interest.

In the petrochemical industry, approximately 60 percent of all electric motor difficulties are thought to originate with bearing troubles. One plant, which had computerized its failure records, showed bearing problems in 70 percent of all repair events. This figure climbs to 80 percent in household appliances with "life-time" lubrication. If a bearing defect is allowed to progress to the point of failure, far more costly motor rewinding and extensive downtime often results. Improvements in bearing life should not be difficult to justify under these circumstances, especially since it has been established that most incidents of bearing distress are caused by lubrication deficiencies, including, of course problems brought on by very pervasive faulty lubrication procedures.

There is some disagreement among electric motor manufacturers as to the best bearing arrangement for horizontal-type, grease-lubricated, ball bearing motors. There is disagreement also on the best technique for replenishing the grease supply in the bearing cartridge. If the user of these motors wanted to follow the recommendations of all these manufacturers for their specific motors, he would have to stock, or have available, ball bearings in a given size with no shields (i.e. open, Figures

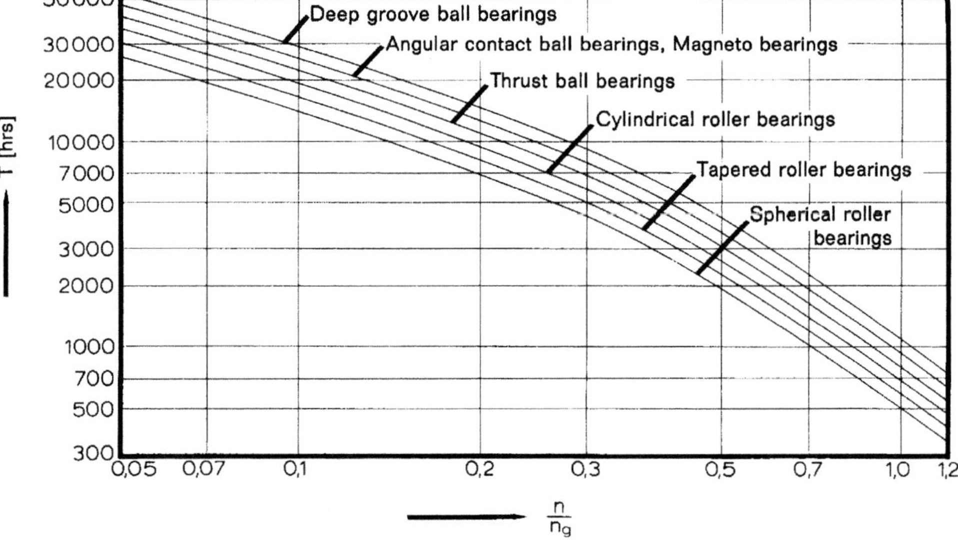

Figure 9-41: Grease re-lubrication intervals recommended by FAG Bearing Corporation (Stamford, Connecticut, 06904).

Table 9-7: Maximum grease re-lubrication interval for electric motors recommended by Exxon in 1980

Maximum Relubrication Intervals for Motors Lubricated with Ronex-MP and Unirex-N2 (Months)

Motor Size, Horsepower	1/4 – 7 1/2		10 – 40		50 – 150		Over 150	
Type of Grease	Ronex	Unirex	Ronex	Unirex	Ronex	Unirex	Ronex	Unirex
Type of Service								
I. Easy, infrequent operation (1hr/day). Valves, door openers, and portable tools.	60	120	60	84	48	48	12	12
II. Standard, 1 or 2-shift operation, Machine tools, air-conditioners, conveyors, refrigeration equipment, laundry and textile machinery, woodowrking machinery, light-duty compressors and pumps.	60	84	48	48	12	18	6	6
III. Severe, continuous running (24 hr/day). Motors, fans, motor-generator sets, coal and other mining machinery, steel-mill machinery and processing equipment.	36	48	12	18	6	9	2	3
IV. Very severe. Dirty, wet, or corrosive environment, vibrating applications, high ambient temperatures (over 40°C, 100°F), hot pumps and fans.	DO NOT USE	9	DO NOT USE	4	DO NOT USE	3	NOT	2

(1) Relubrication interval for Class F motors in Service Types III and IV should not exceed 12 months.

9-42 and 9-43, with single-shields, Figure 9-44, and double-shields, Figure 9-45 (Ref. 9-17). The user would also be asked to train personnel in the often-different re-lubrication techniques specified for different makes of motors. The confusion thus created in the mind of maintenance personnel may indeed bring about a less than satisfactory method of maintaining expensive, important equipment.

Users, too, disagree on such matters as lubrication method, bearing type, and re-lubrication frequency in seemingly similar plants. A 1980 study of twelve petrochemical facilities showed that lubrication practices for electric motors varied from having no program at all, to the opposite—and certainly laudable—continuous application of oil mist. Four plants stated they had no lubrication program for motors and ran motors to failure. These plants specified sealed bearings for motors. Two plants were apparently trying oil mist on some motors.

Another plant had all (which is to say several thousand) electric motors with anti-friction bearings lubricated by oil mist. The facility using oil mist lubrication was able to point to nine years of highly satisfactory experience as of 1983. They reaffirmed their complete

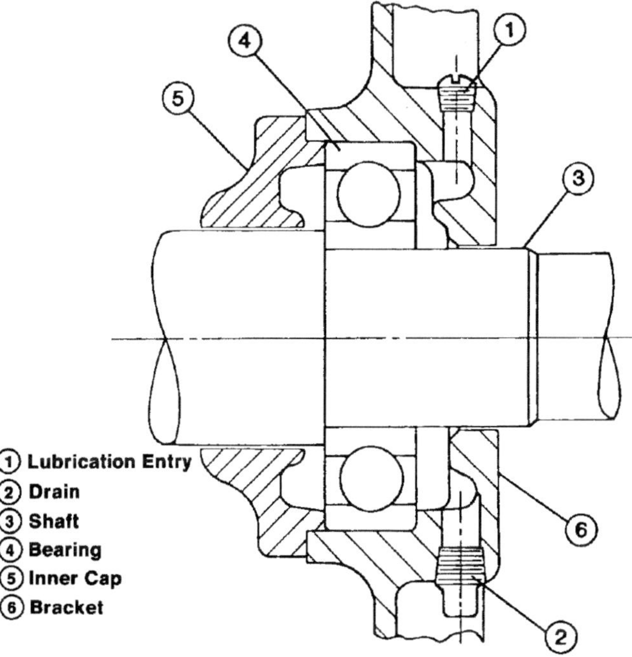

Figure 9-42: High load and/or high speed bearing supplied without shield

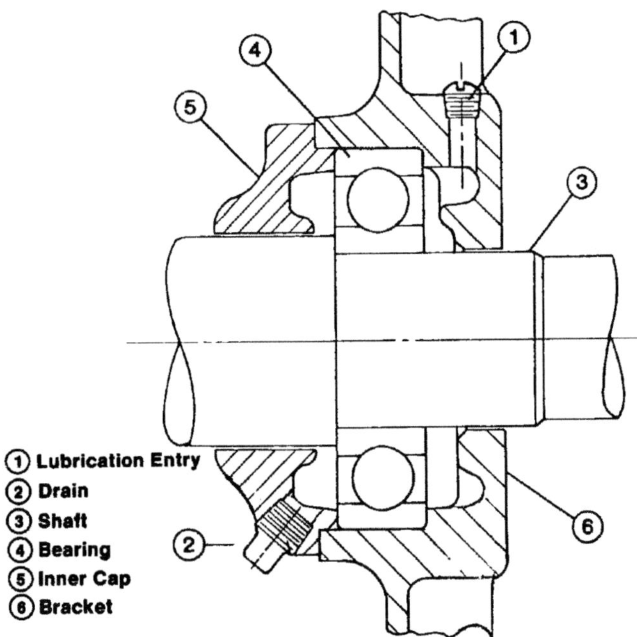

Figure 9-43: Open bearing with cross-flow grease lubrication

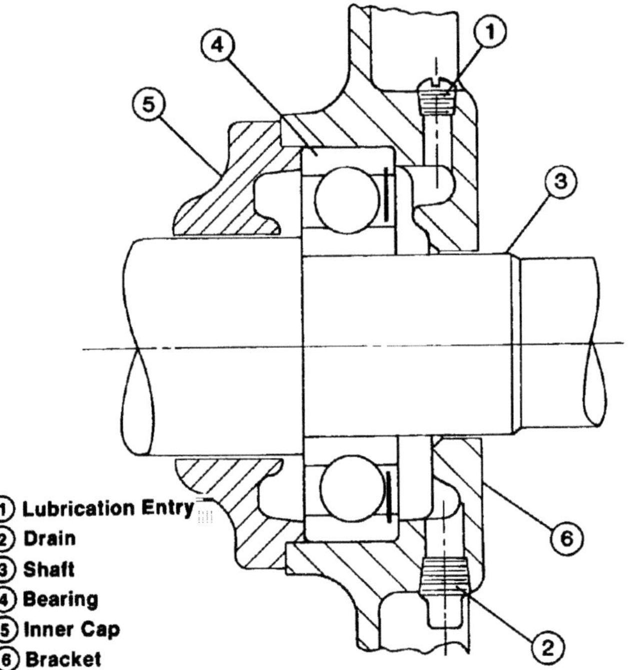

Figure 9-44: Single-shield motor bearing with shield facing the grease cavity

success in 1992 and are as satisfied as ever in 2003. In 1992, six of the twelve plants used a formal program to lubricate oiled and greased bearings. As of the early to mid 2000's, no reliability-focused plant was found without a conscientious and well-managed pump and driver lubrication program. Many used plant-wide oil mist lubrication systems.

However, before we explain what we consider optimized lubrication for pump drive electric motor bearings, it might be helpful to review the most frequently used motor bearing housing configurations and lubrication arrangements. Also, it is of interest to note the many persistent misunderstandings that deserve to be rectified and explained (Ref. 9-18).

How Grease-lubricated Bearings Function

Open Bearings

High-load and/or high peripheral speed bearings are often supplied without shields to allow cooler operating temperature and longer life. Two such bearing were illustrated in Figures 9-42 and 9-43. It is important to note that these motor bearing housings (and also the other types of motor bearings dealt with in this text) are shown with inner caps, item 5. Inexpensive "throw-away" motors might be found with inner caps lacking. Our entire premise here is that the user is dealing with the electric motors that incorporate item 5.

If grease inlet and outlet ports are located on the same side, bearings are commonly referred to as "conventional-flow grease lubricated." If grease inlet and outlet ports are located at opposite sides, we refer to "cross-flow lubrication." Both shielded and sealed bearings can be modified to become open bearings by simply removing shields or seals.

A shielded, grease-lubricated ball bearing (Figure 9-44) can be compared to a centrifugal pump having the ball-and-cage assembly as its impeller, and having as its impeller eye the annular space between the stationary shield and the rotating inner ring. From our earlier chapter on bearings, recall that shields act primarily as metering orifices and that make-up grease bleeds across the annular gap and into the bearing from an adjacent grease cavity which is part of the bearing housing. Using proper methods and controlled application, shielded bearings can thus be re-lubricated, whereas sealed bearings cannot.

Single-shielded Bearings

A large petrochemical complex in the U.S. Gulf Coast area considers the regular single-shield bearing with the shield facing the grease supply, as shown in Figure 9-44, to be the best arrangement. Their experience indicates this straightforward arrangement will extend bearing life. It will also permit using an extremely simple lubrication and re-lubrication technique, if so

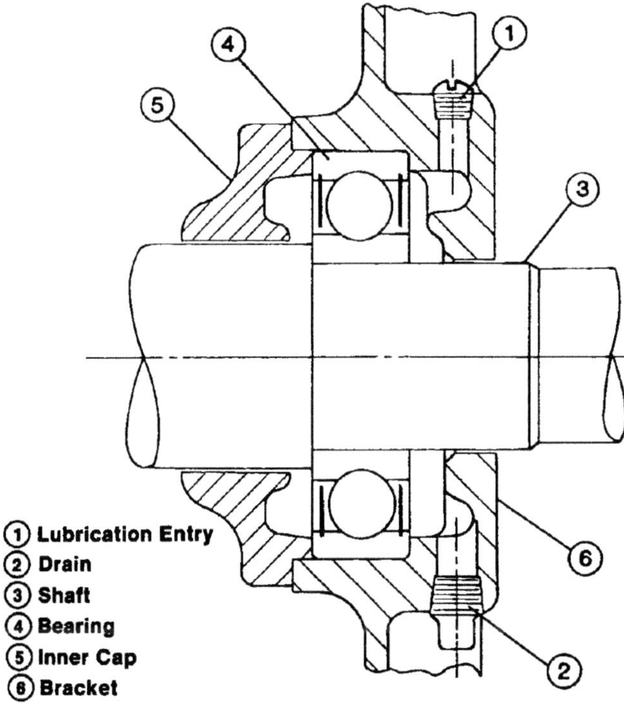

Figure 9-45: Double-shielded bearing

installed. This technique makes it unnecessary to know the volume of grease already in the bearing cartridge. The shield is important in that it acts as a baffle against agitation.

Since the shield-to-inner-ring annulus serves as a metering device to control grease flow, overheating from excess grease is much less likely to occur. Also, the risk of premature ball bearing failures caused by contaminated grease is reduced. Further, warehouse inventories of ball bearings can be reduced to one type of bearing configuration for the great bulk of existing grease-lubricated ball bearing requirements. For other services where an open bearing is a "must," as in some high-speed applications or flush-through arrangements, the shield can be removed in the field.

Conventional Wisdom: *Shielded bearings cannot be re-lubricated.*
Fact: *Using proper procedures, shielded bearings can be re-lubricated.*

As mentioned above, with the shielded type of bearing, grease—or the oil constituent in the grease—may readily enter the bearing, but dirt is restricted by the close-fitting shields. Bearings of the sealed design will not permit entry of new grease, whereas with shielded bearings grease will be drawn in through capillary action as the bearing cage assembly rotates. The grease will then be discharged by centrifugal force into the ball track of the outer race. Since there is no shield on the backside of this bearing, the excess grease can escape into the inner bearing cap of the motor bearing housing.

Double-shielded Bearings

Some motor manufacturers still subscribe to a different approach, having decided in favor of double-shielded bearings. These are usually arranged as shown in Figure 9-45 and function no differently than the single-shielded version of Figure 9-44. The housings serve as a lubricant reservoir and are filled with grease. By regulating the flow of grease into the bearing, the shields act to prevent excessive amounts from being forced into the bearing. A grease retainer labyrinth is designed to prevent grease from reaching the motor windings on the inner side of the bearing.

On motors furnished with this bearing configuration and mounting arrangement, it is not necessary to pack the housing next to the bearing full of grease for proper bearing lubrication. However, packing with grease helps to prevent dirt and moisture from entering. Oil from this grease reservoir can and does, over a long period, enter the bearing to revitalize the grease within the shields. Grease in the housing outside the stationary shields is not agitated or churned by the rotation of the bearing and consequently, is less subject to oxidation. Furthermore, if foreign matter is present, the fact that the grease in the chamber is not being churned reduces the probability of the debris contacting the rolling elements of the bearing.

Along those lines, occasional claims that shielded bearings are *not* regreasable deserve to be labeled fundamentally flawed. The annular gap between the shield and bearing inner ring allows bleeding of grease at just the right amount to ensure long and satisfactory bearing operation. So-called "tests" done while regreasing and showing that the grease cannot be forced into the bearing are highly misleading in that they are overlooking the design intent. For shielded bearings, the design intent is to allow slow bleeding, not pressure-based instantaneous grease replenishment.

Double-shielded Bearings with
Grease Metering Plate

On many motors furnished with grease-lubricated double-shielded bearings, the bearing housings are not usually provided with a drain plug. When grease is added and the housing becomes filled without applying excessive pressure, some grease will be forced into the bearing. Some of these bearings come with an additional shield, typically called a metering plate, and shown in Figure 9-46. Any surplus grease will be squeezed out along the close clearance between the shaft and the outer cap because the resistance of this path is less than the resistance presented by the bearing shields, metering plate and—on open, non-shielded bearings—the labyrinth seal. However, excessive pressure applied with a grease gun may again deflect the adjacent shield and cause it to contact the bearing rotating elements. This creates frictional heat and, in some instances, abrasive metal wear that are sure to terminate bearing life. In that case, the user would have been better off leaving things alone or following best practices, below.

Best Practices for the Expulsion of Spent Grease

Closest to "Best Practices" are plants that consistently manage to remove their grease drain plug before attempting to inject the new grease. There are several facilities in the Middle East that do just that. Their electric motor bearings fail at a rate of 14 per 1,000 motors per year. Drain plugs are also being removed prior to grease replenishment at a petrochemical unit located in the South Texas area of the United States. Their electric motor bearings are being replaced at a rate of 18 per 1,000 motors per year.

The very best statistic from an installation that did not remove their motor bearing drain plugs was 156 per 1,000 motors per year; the worst report came from a plant that replaced 900 sets of bearings on their 900 electric motors each year. Not allowing the spent grease to be expelled inevitably causes much of the old product to be forced into the bearing. This, and the heat build-up from the excess grease, would explain the poor performance of plants tolerating grease injection into spaces that are not configured to first expel the spent grease.

True best practices plants have eliminated human forgetfulness by discarding the drain plug and leaving the drain passage open at all times. They have installed at or near the bottom of the motor bearing housing the setup illustrated in Figure 9-47. Typically consisting of a length of 3/8th-inch threaded pipe followed by a pipe elbow and a short pipe nipple, this "escape route" is always filled with a "plug" of residual spent grease. Upon re-greasing through a conventional application fitting located at the top of the bearing housing, the semi-solid "drain plug" will be expelled and a new plug will take its place.

The use of pressure relief fittings instead of the permanently open drain should be discouraged. Experi-

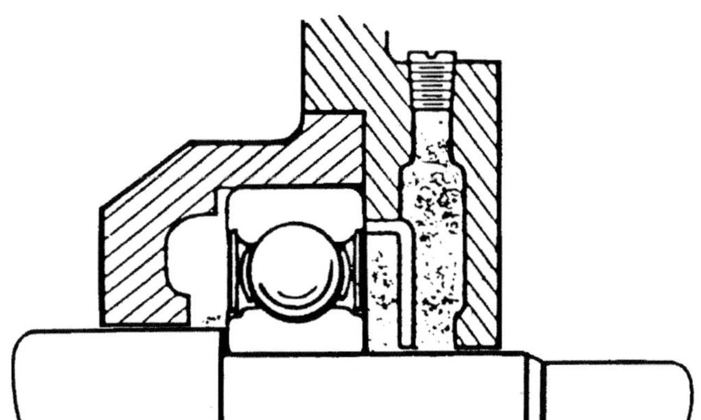

Figure 9-46: Double-shielded bearing with grease metering plate facing grease reservoir

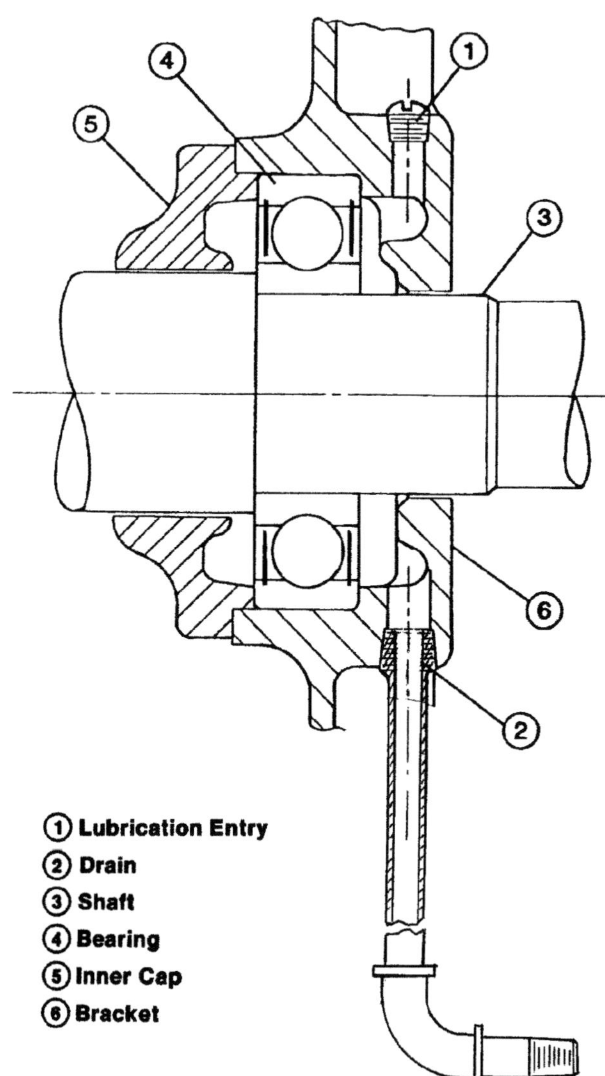

Figure 9-47: Best practices grease expulsion through open drain pipe

ence shows that these fittings have often refused to open at pressures below those that cause unacceptable shield deflection. Many European manufacturers of electric motors have therefore designed bearing housing end caps with a labyrinth configuration that doubles as an "escape valve," Figure 9-48.

Procedures for Re-greasing Electric Motor Bearings

Electric motor bearings should be re-lubricated with greases that are compatible with the original charge. It should be noted that some of the polyurea greases often used by the motor manufacturers may be incompatible with the rather well represented lithium-base greases that have traditionally been chosen for their favorable all-around properties. However, since the mid 1990's premium-grade "EM"-polyureas have rightly been considered superior for a very wide range of electric motor lube applications in reliability-focused plants. See Table 9-8 for grease compatibility information.

Grease replenishment should be undertaken only after the drain plug has been removed. The preferred solution would be to install permanent escape piping per Figure 9-47, or to utilize the labyrinth escape approach depicted in Figure 9-48.

Here, then, are experience-based procedures for grease replenishment of the various pump drive motor bearings.

Single-shielded Bearings

To take advantage of single-shielded arrangements in electric motors, a large Phillips Petroleum facility developed and validated three simple procedural recommendations:

1. Install a single-shield ball bearing with the shield facing the grease supply in motors having the grease fill-and-drain ports *on that same side* of the bearing. Add a finger full of grease to the ball track from the open back side of the bearing during assembly.

2. After assembly, the balance of the initial lubrication of this single-shielded bearing should be done with the motor idle. Unless a permanently open pipe is used as a drain passage, remove the drain plug. With a grease gun or high volume grease pump, fill the grease reservoir until fresh grease emerges from the drain. The fill and drain plugs should then be reinstalled and the motor is ready for service.

It is essential that this initial lubrication would not be attempted while the motor is running. It was observed that to do so would cause, by pumping action, a continuing flow of grease through the shield annulus until the overflow space in the inner cartridge cap is full. Grease would then flow down the shaft and into

Table 9-8: Grease compatibilities published by Royal Purple, Ltd.

General Grease Compatibility Chart	Aluminum Complex	Barium	Calcium	Calcium 12-Hydroxy	Calcium Complex	Clay	Lithium	Lithium 12-Hydroxy	Lithium Complex	Polyurea	Silicone
Aluminum Complex	X	I	I	C	I	I	I	I	C	I	I
Barium	I	X	I	C	I	I	I	I	I	I	I
Calcium	X	I	X	C	I	I	I	I	C	I	I
Calcium 12-Hydroxy	C	C	C	X	B	C	C	C	C	I	I
Calcium Complex	I	I	I	B	X	I	I	I	C	C	I
Clay	I	I	C	C	I	X	I	I	I	I	I
Lithium	I	I	C	C	I	I	X	C	C	I	I
Lithium 12-Hydroxy	I	I	B	C	I	I	C	X	C	I	I
Lithium Complex	C	I	C	C	C	I	C	C	X	I	I
Polyurea	I	I	I	I	C	I	I	I	I	X	I
Silicone	I	I	I	I	I	I	I	I	I	I	X

C = Compatible

B = Borderline Compatible:
Typically results in a light softening or hardening of the NLGI Grade and a lowering of the dropping point of the mixture of grease.

I = Incompatible:
Typically results in a softening or hardening of greater than 1 1/2 the NLGI grade, a shift in the dropping point, and a possible reaction of additives or base oils.

Source: E.H. Meyers' paper entitled "Incompatibility of Greases" 49th Annual NLGI Meeting

Royal Purple Ultra-Performance® Grease Compatibility Chart	Aluminum Complex	Barium	Calcium	Calcium 12-Hydroxy	Calcium Complex	Clay	Lithium	Lithium 12-Hydroxy	Lithium Complex	Polyurea	Silicone
Operating Temps. <225°F	C	C	C	C	C	I	C	C	C	C	I
Operating Temps. 225-350°F	C	B	B	C	B	I	B	B	C	B	I
Operating Temps. >350°F	C	I	I	C	I	I	I	I	C	I	I

Note: Ultra-Performance® greases are more stable. This chart is generated from independent lab testing and field experience. Actual compatibility results may vary. It is recommended that bearings be purged of old grease per OEM instructions to ensure proper lubrication and performance.

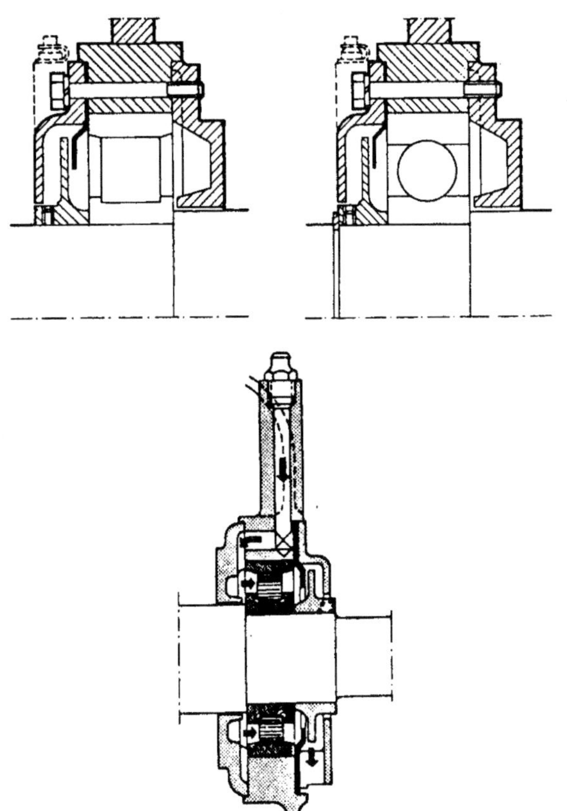

Figure 9-48: Labyrinth grease drain passages popular in European-built electric motors

the winding of the motor where it would often cause heat-related winding damage. This unwanted flow in the direction of the motor interior would take place before the grease can emerge at the drain.

Although bearing manufacturers have occasionally expressed other preferences, re-lubrication may be done while the motor is either running or idle. Plant tests showed that fresh grease takes a wedge-like path straight through the old grease, around the shaft, and into the ball track. Thus, the overflow of grease into the inner reservoir space is quite small even after several re-lubrications. Potentially damaging grease is thus kept from the stator winding.

Further, since the ball and cage assembly of this arrangement does not have to force its way through a solid fill of grease, bearing heating is kept to a minimum. In fact, it was observed that a maximum temperature rise of only 20°F (~12°C) occurred 20 minutes after the grease reservoir was filled. It returned to a 5°F (3°C) rise two hours later. In contrast, the double-shield arrangement caused a temperature rise of over 100°F (56°C). At 90°F (32°C) ambient temperature the resulting temperature was 190°F (88°C) and maintained this 100°F (56°C) rise for over a week.

Double-shielded Bearings

One of several well-proven procedures for re-lubricating ball bearings consists of the following two steps:

- Hand-pack (completely fill) the cavity adjacent to the bearing. Use the necessary precautions to prevent contaminating this grease before the motor is assembled.
- After assembly, lubricate *stationary motor* until a full ring of grease appears around the shaft at the relief opening in the bracket.

If under-lubricated after installation, the double-shielded bearing is thought to last longer than an open (non-shielded) bearing given the same treatment. This is because a certain amount of grease is retained within the shields (plus grease remaining in the housing from its initial filling).

If over-greased after installation, the double-shielded bearing can often be expected to operate satisfactorily without overheating. That said, it must be ascertained that over-lubrication was not caused by simultaneously over-pressurizing the grease reservoir. Over-pressurizing the to the point of causing contact between shields and rolling elements would invite frictional heat and abrasive wear damage to both grease and bearing. Plain over-lubrication would still allow excess grease to escape through the clearance between the shield and inner race. Hence, the grease in the housing adjacent to the bearing is not being churned, agitated and overheated.

It is not necessary to disassemble motors at the end of fixed periods to grease bearings. Bearing shields do not require replacement.

Double-shielded ball bearings should not be flushed for cleaning. If water and dirt are known to be present inside the shields of a bearing because of a flood or other unusual circumstances, the bearing should be removed from service. All leading ball-bearing manufacturers are providing reconditioning service at a nominal cost when bearings are returned to their factories. As an aside, reconditioned ball bearings are generally less prone to fail than are brand new bearings. This is because grinding marks and other asperities are now burnished to the point where smoother running and less heat generation are likely.

Re-lubrication of Cylindrical Roller Bearings

- Hand pack bearing before assembly
- Proceed as outlined in (1) and (2) for double shielded ball bearings.

Open Bearings

Open bearings have neither shield(s) nor seals and generate less frictional heat than shielded or sealed bearings. Also, they are typically used in higher peripheral speed applications than their shielded or sealed counterparts. Note that motors with open, conventionally greased bearings are generally lubricated with slightly different procedures for drive-end and opposite end bearings.

To Lubricate Drive-end Bearings, Proceed as Follows:

1. Re-lubrication with the shaft stationary is recommended. If possible, the motor should be warm.
2. Wipe off grease fitting or, if grease entry passage is provided with a plug, remove plug and replace with grease fitting.
3. Unless permanently open drain pipe assembly is provided, remove large drain plug at bottom of bearing housing.
4. Using a low pressure, hand operated grease gun, pump in the recommended amount of grease, or use an amount equal to 1/4 of bearing bore volume.
5. If purging of system is desired, continue pumping until new grease appears either around the shaft or at the drain opening. Stop after new grease appears.
6. On large motors provisions have usually been made to remove the outer cap for inspection and cleaning. Remove both rows of cap bolts. Remove, inspect and clean cap. Replace cap, being careful to prevent dirt from getting into bearing cavity.
7. After re-lubrication allow motor to run for fifteen minutes before replacing drain plugs.
8. If the motor has a special grease relief fitting (caution: relief fittings have tendency to "freeze up," as mentioned earlier!), pump in the recommended volume of grease or until a one inch long string of grease appears in any one of the relief passages. Replace plugs.
9. Wipe away any excess grease that has exited at the grease relief port.

Lubrication Procedure for Bearing Opposite Drive End:

- If bearing hub is accessible, as in drip-proof motors, follow the same procedure as for the drive-end bearing.
- For fan-cooled motors, note the amount of grease used to lubricate shaft-end bearing and use the same amount for opposite-end bearing.

Motor bearings arranged with housing provisions as shown earlier in Figure 9-43, with grease inlet and outlet ports on opposite sides, are called cross-flow lubricated. Re-greasing is accomplished with the motor running. The following procedure should be observed:

1. Start motor and allow to operate until motor temperature is obtained.

2. Inboard bearing (coupling end)
   a. Remove grease inlet plug or fitting.
   b. Remove outlet plug. Some motor designs are equipped with excess grease cups located directly below the bearing. Remove the cups and clean out the old grease.
   c. Remove hardened grease from the inlet and outlet ports with a clean probe. Nylon or Delrin® cable wraps make useful probes.
   d. Inspect the grease removed from the inlet port. If rust or other abrasives are observed, do not grease the bearing. Tag motor for overhaul.
   e. Bearing housing with outlet ports:
      (1) Insert probe in the outlet port to a depth equivalent to the bottom balls of the bearing.
      (2) Replace grease fitting and add grease slowly with a hand gun. Count strokes of gun as grease is added.
      (3) Stop pumping when the probe in the outlet port begins to move. This indicates that the grease cavity is full.
   f. Bearing housings with excess grease cups:
      (1) Replace grease fitting and add grease slowly with a hand gun. Count strokes of gun as grease is added.
      (2) Stop pumping when grease appears in the excess grease cup. This indicates that the grease cavity is full.

2. Outboard bearing (fan end):
   a. Follow inboard bearing procedure, provided the outlet grease ports or excess grease cups are accessible,
   a. If grease outlet port or excess grease cup is not accessible, add 2/3 of the amount of grease required for the inboard bearing.
   c. Leave grease outlet ports open—do not replace the plugs. Excess grease will be expelled through the port.
   d. If bearings are equipped with excess grease cups, replace the cups. Excess grease will expel into the cups.

"Automatic" Grease Lubrication

Table 9-1 had clearly listed automatic grease replenishment as one of the leading contenders for extending bearing life. Plant-wide automatic grease lubrication systems are popular in Scandinavian pulp and paper mills. The experience of some mills dates back to the 1970's and some plants with many hundreds of installed pumps have reported zero bearing failures in time periods often spanning more than a year. This application method is thoroughly explained in Refs. 9-19 through 9-22 and a brief overview will have to suffice here.

The systems are configured as shown in Figure 9-49. Modular in design and easily expandable, they consist of a single or multi-channel control center (Item 1), one or more pumping stations (Item 2), appropriate supply lines (Item 3), tubing (4), which links a remote shut-off valve (5) and lubrication dosing modules (6) and also interconnects dosing modules and points to be lubricated. Different size dosing modules are used to optimally serve bearings of varying configurations and dimensions. The dosing modules are individually adjustable to provide an exact amount of lubricant, thus avoiding over-lubrication. A pressure sensing switch (Item 7) completes the installation and pressure is applied only during the feed stroke of the system.

A similar, but vastly scaled down approach is used in the many different types and models of single point automatic lubricators, or SPALs, available in North America. The estimated two or three dozen available configurations can be categorized into refillable and non-refillable models, or models that keep the grease under pressure at all times and models that apply pressure only during the feed stroke, i.e. intermittently. To date, any high performance motor bearing grease subjected to long-term pressurization will undergo separation into its oil and soap components. Among the very few greases that will not "come apart" when pressurized for long times we typically find coupling greases. However, coupling greases would perform very poorly in rolling element bearings expected to perform reliably in process machinery. They should not be used in pumps and electric motors.

In any event, Figure 9-50 is introduced here to depict some of the more widely available SPALs that are primarily marketed in the Americas. Marketing and pricing considerations may have prompted some manufacturers to produce pressure by either springs or chemical propellants (Figure 9-51). An awareness of grease behavior may have prompted some manufacturers to offer certain SPALs with electronic timing, others with microchips, and even one with a motor-driven agitator auger (Figure 9-52). Many SPAL models can feed into a

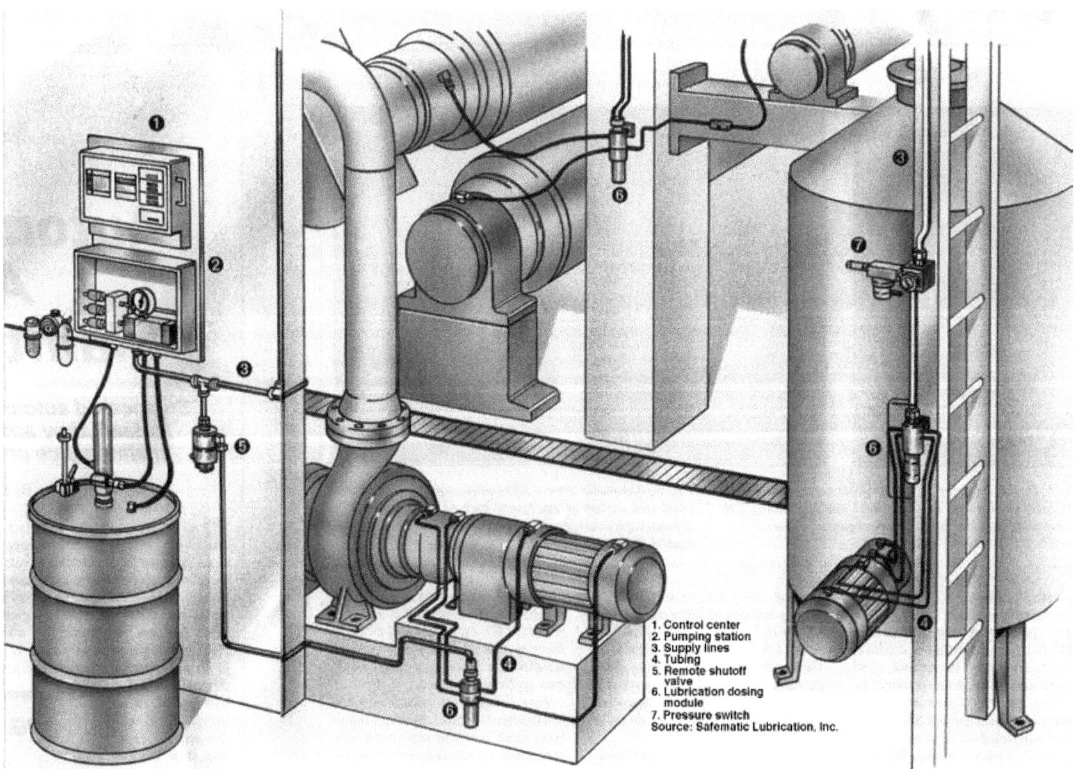

Figure 9-49: Automatic grease lubrication system (Source: Safematic Oy, Muurame, Finland)

small distribution manifold with outlet tubing connected to four or more separate, or remotely located, points to be lubricated. The device would thus become a small self-contained multi-point automatic lubricator.

Conventional Wisdom: *Single Point Automatic Lubricators are an inexpensive means of ascertaining proper lubrication.* ***Fact:*** *Unless the right grease is applied with the correct pressure and at the appropriate feed rate, SPALs may not be a good lubrication choice.*

While each model or variant thus comes with its own claims and features, we know of no best practices plants or profitable high-reliability performers that are large-scale users of SPALs for either pumps or pump drive motors. Therefore, before deciding to apply SPALs on a plant-wide basis it would be wise to thoroughly investigate the costs and benefits of each and every one of the many contenders discussed so far. Until such a rather detailed study is undertaken, competent engineers would be hard pressed to choose between oil mist and fully automated grease systems, or conventional oil lubrication and manual re-greasing, and perhaps even the plant-wide use of well-engineered SPAL units. But, regardless of which route is being investigated or considered, never forget to ask where the spent lubricant will reappear and how one plans to dispose of it.

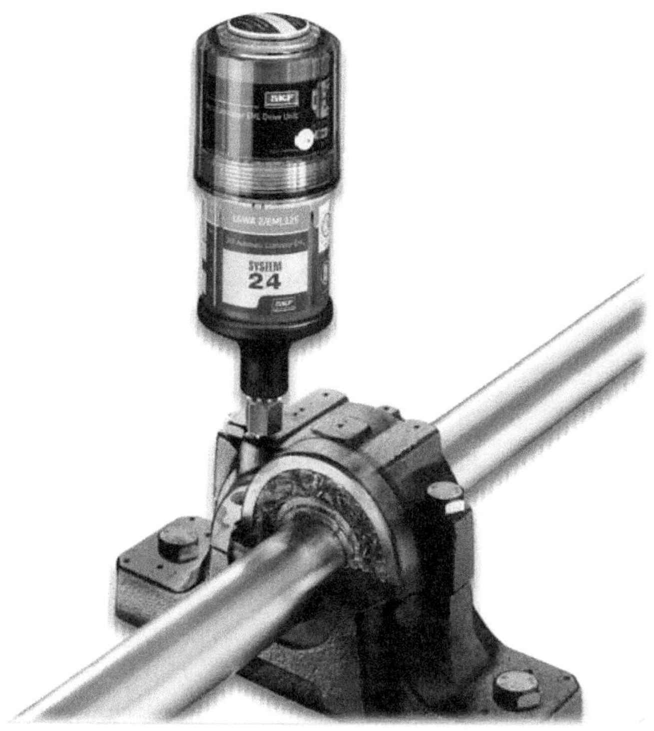

Figure 9-50: Single-point automatic lubricator. (Source: SKF USA Inc., Kulpsville, PA 19443)

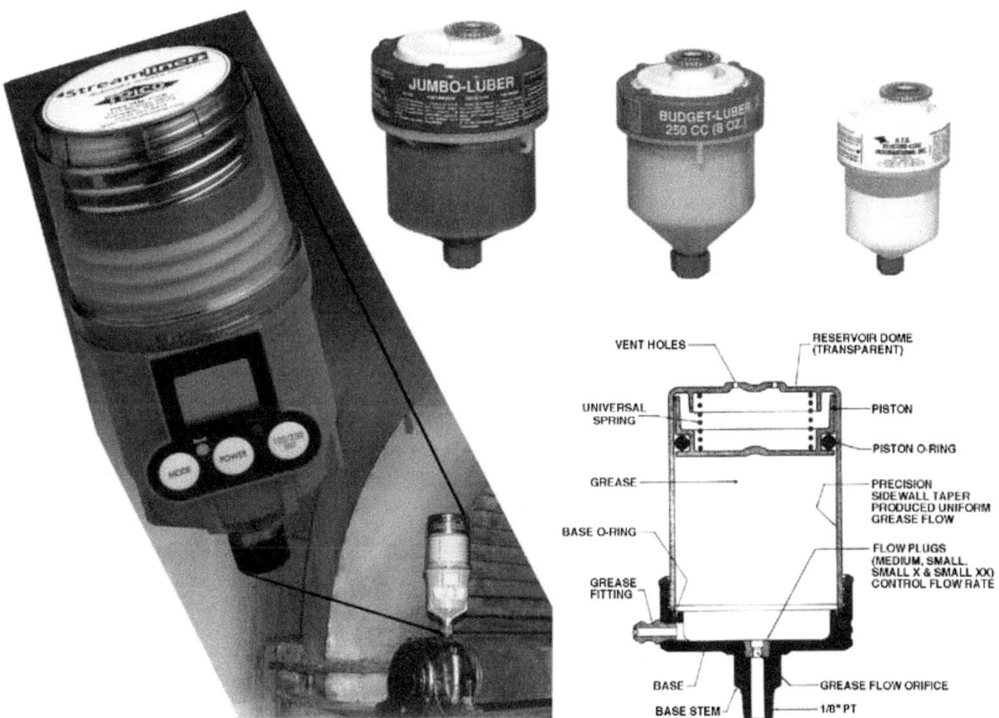

Figure 9-51: Spring and electrically activated single-point automatic lubricators (Trico Mfg. Corp.)

Life-time Lubricated, "Sealed" Bearings

Lifetime lubricated bearings are often misapplied and are rarely suitable for continuous operation in process pumps and their motor drivers. As shown in Table 9-1, "lubrication for life" ranks last in order of preference. Lubed-for-life bearings incorporate close-fitting seals in place of, or in addition to shields. They are customarily found in low horsepower motors or in appliances that operate intermittently. But, it is well known that close-fitting seals can cause high frictional heat that leads to accelerated oxidation of the grease charge. Once the seals have worn and are loosely fitting they will no longer be effective in excluding atmospheric air and moisture. Again, rapid grease deterioration may result.

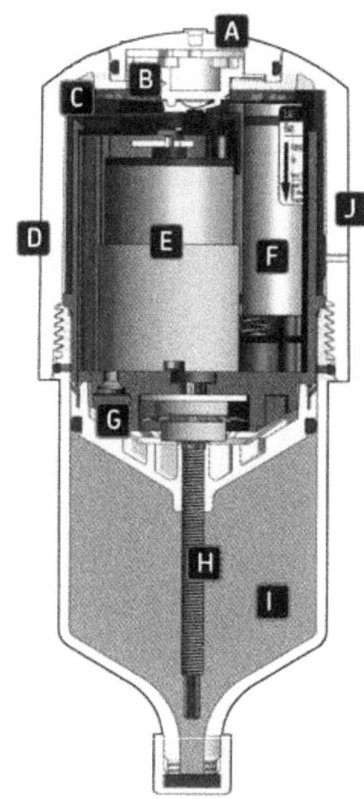

A B **On/off knob and time setting dial**
Enables easy activation and dial setting

C **LED status indicators**
Helps verify operating status

D **Drive cover**
Easily removable, seals and helps prevent ingress of dirt and moisture

E **Electric motor and gearbox**
Helps enable constant discharge pressure

F **Battery pack**

G **Piston**
Special piston shape helps ensure optimum emptying of lubricator

H **Spindle**
Rotates to drive piston, enabling lubricant to be dispensed

I **Lubricator canister**
Filled with high quality SKF lubricant

J **Anti-vacuum membrane**
Helps prevent vacuum forming

Figure 9-52: Positive displacement, auger-type automatic single-point lubricator (Source: SKF USA, Inc., Kulpsville, PA 19443)

Nevertheless, at least one large petrochemical company in the United States deviated from customary practice. This facility claimed satisfaction with sealed ball bearings in small centrifugal pumps as long as bearing operating temperatures remained below 150°C (302°F) and speed factors DN (mm bearing bore times revolutions per minute) did not exceed 300,000.

Notwithstanding the above 1980's claims, reliability-focused users in the decade beginning with 2001 have ruled out the use of lubed-for-life bearings. Let there be no doubt that reliability-focused installations expect "life" to last more than three years in the typical plant environment. This is one of the reasons why knowledgeable bearing manufacturers advise against the use of sealed bearings larger than size 306 (30 mm bore) at speeds exceeding 3,600 rpm. Sealed bearings are thus generally excluded from 3,600 rpm motors of 10 or more horsepower (>7.5 kW).

In line with this recommendation, a DN value of 108,000 was quoted by an experienced bearing application engineer as the economic, although not technically required, upper limit for "life-time-lubrication." This assumes an average machine operating continuously in a moderate ambient environment. Obviously, if a device operates intermittently in a benign environment, grease lubricated rolling element bearings may be acceptable at DN-values in excess of 108,000. As an example, this would be the case in a motor-operated control valve, or certain pumps used in batch processing facilities. Conversely, bearings with DN values ranging from 108,000 to 300,000 are almost universally given periodic re-lubrication.

LUBRICANTS, BASE STOCKS AND PERFORMANCE TRADE-OFFS

With the possible exception of a few experimental pumps equipped with gas bearings and perhaps a hundred large pumps with magnetic bearings, oil-based lubricants are essential in process pumps. In any applications using either oil-containing greases or liquid oil lubricants, proper oil choices are needed to dependably operate these pumps.

Every so often, a lube marketer will advocate "standardized lubricants." However, the same type or lubricant formulation cannot optimally serve all kinds of different machines, or even different pumps. Some of the issues were highlighted earlier in this chapter in conjunction with grease selection. Of course, there are many properties that are critically important for particular machines and without the right lubricant formulation equipment reliability will be impaired.

Both mineral oils and synthetic oils compete in the market place. Mineral oils will (usually) perform satisfactorily in an industrial process machine. As of 2011 or 2012, certain new and advantageous mineral oil formulations are recommended for many (but not all) process pump applications. For satisfactory performance overall, oil viscosity, lube application method and oil cleanliness must conform to best available practices. While one can also not neglect these (and other) parameters in synthetic lubricants, the various parameters (occasionally called "inspections" in tribologists' terms) nevertheless cover a wider range of applications. Some inspections/parameters available in good synthetics permit wider latitudes of deviation. As an example, with synthetic lubricants there will be lower frictional heat generated than with performance-equivalent mineral oils. Also, synthetic lubes will typically perform in wider ranges of ambient environments (Figure 9-53).

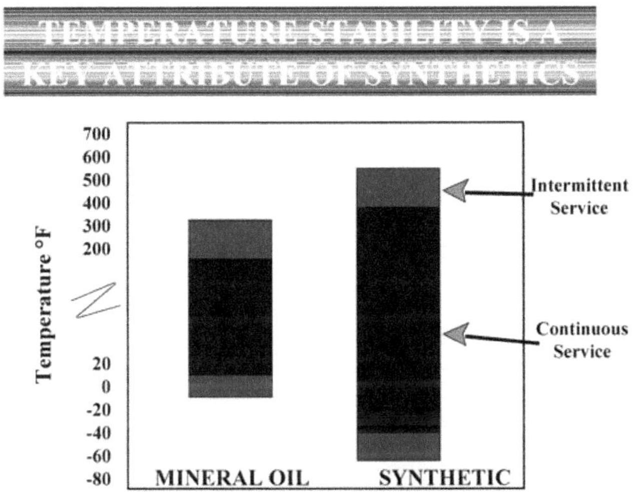

Figure 9-53: Temperature stability of mineral oils versus synthetic lubricants

As a general rule, whenever the requisite equipment operating conditions drift or deviate from their ideal ranges or conditions, consideration should be given to "SH"—synthesized hydrocarbons. Customarily these—collectively called—"synthetic lubricants" will then serve better than mineral oils. Their higher initial cost can be easily cost-justified under less-than-ideal or highly stressed operating conditions. And while good synthetic lubricants can cost 10 times (usually 4 or 5 times) as much as mineral oils, they almost always allow extended operation. Four- to six-fold longer-than-traditional drain intervals are rather typical, assuming the lubricant is kept clean. Looking at the full life-cycle cost picture, there is much evidence that the judicious application of synthesized hydrocarbon lubricants has helped Best-of-Class process plants achieve their enviable profitability and reliability leadership position. Yet, as with everything else in life, there will be trade-offs for different lubricants. In fluid machines, these trade-offs must be considered by the reliability technician, engineer or lubrication professional.

Rapid payback is achieved whenever SH lubricants are used in failure-prone equipment and equipment that is judged critically important. In essence, it should be very easy to cost-justify SH on pumps and other equipment where repeat failures have been experienced. Conscious efforts directed towards avoidance of repeat bearing failures in pumps place emphasis on well-formulated SH fluids. These superior lubes should be used at the slightest hint of increased vibration excursions. It has been postulated that the SH will fill the micro-cracks in the bearing races. Experience shows that SH will often reduce the severity with which rolling elements strike discontinuities in bearing races. Lessened impact severity can then reduce noise and vibration.

Base Stocks and Additives of Interest

Many different base stocks make up the SH family; however, the overwhelming majority of modern process plants are best served by, and make primary use of, fluids with base stocks belonging to the subgroups PAO (poly-alpha-olefin) and diester (actually, dibasic ester). Superior SH formulations for process pumps and other machines with rolling element bearings should exhibit good "tenacity," i.e., the lubricant should cling to the bearing, minimize friction (and, therefore, bearing temperature). The lubricant should also have reasonable pressure capability, high cleanliness, and good demulsibility (Ref. 9-1). In short, a superior lubricant must perform under less-than-optimum conditions for extended time periods, must keep machine systems, sumps, and/or reservoirs clean, and should separate rapidly from water.

PAOs are used more frequently than diesters and mixtures of the two are occasionally found very beneficial. Either base stock is of real practical and economic value in the modern plant environment, but all modern lubricants need additives. Reaching for superior or merely "acceptable" performance, a blending plant, or lube manufacturer/formulator selects the most appropriate additive package. While many top-tier oil companies have the know-how to provide both base stock and the right additive package, not every lube supplier is interested in selling products in small quantities. That fact, plus the degree of competence of a lube manufacturer's service and marketing personnel, explains market penetration and customer satisfaction. Dealing with a reputable supplier and tapping into the supplier's application know-how is the preferred choice of reliability-focused users (Ref. 9-26).

Top suppliers can often prove the advantages of full synthetic lubricants. The same top suppliers can also point out where modern lubricant formulations based on mineral oils can be completely satisfactory and where SH lubes are simply not needed. Rely on these suppliers for sound advice. Consider entering into a mutually rewarding networking relationship with their best application engineers. Tapping into the expertise of lube suppliers with extensive application experience is of great value in failure avoidance and troubleshooting. Dealing with those suppliers will always be worth the incremental cost (Ref. 9-27).

LUBRICANTS, TESTS AND PROPERTIES

Advertising is part of any marketing or business strategy. Literature is available that points to synthesized hydrocarbon lubricants or "SH" fluids (from even big-name brands) occasionally exhibiting lower performance than fluids marketed by their smaller competition. Speaking of lubes for machines, not all that's better is better for all.

As a general rule, a few years ago SH fluids were found to be superior to mineral oils. Certain performance differences may have been rooted in the fact that smaller formulators, at times, had superior additives know-how. It was then also possible that SH lubricants from small "niche-market" formulators represented best overall value because of undisclosed proprietary additive know-how. Try not to draw the wrong conclusions from old data. Competitors sometimes change their marketing approaches and their product offers. Also, in recent years the "majors" have linked up with independent distributors able of catering to the needs of both low- and high-volume customers. In any event, users are advised to periodically reassess their knowledge; it will always be wise to expend efforts to stay informed. One might even find that certain modern mineral oil formulations (Figure 9-54) are as good as needed and that using SH fluids is "overkill" in some situations.

Because lubricant producers are making progress, we may have to re-learn or adjust our understanding on the subject of superior additive know-how. There have been instances when products use additives that look good at one temperature but are not thermally or shear stable. Gear oils are a case in point. Some gear oils truly can smooth out surface pits and scratches. If there is an overload condition, or possibly damage from moisture-induced pitting corrosion, some gear oils can restore the surface of the gear. However, aggressive EP additives should be used sparingly since they can affect tooth profile and produce deposits.

Some Lubricants Disappoint due to Insufficient Foam Control

Foaming in an industrial oil system is a serious service condition that may interfere with satisfactory performance and may even lead to mechanical damage.

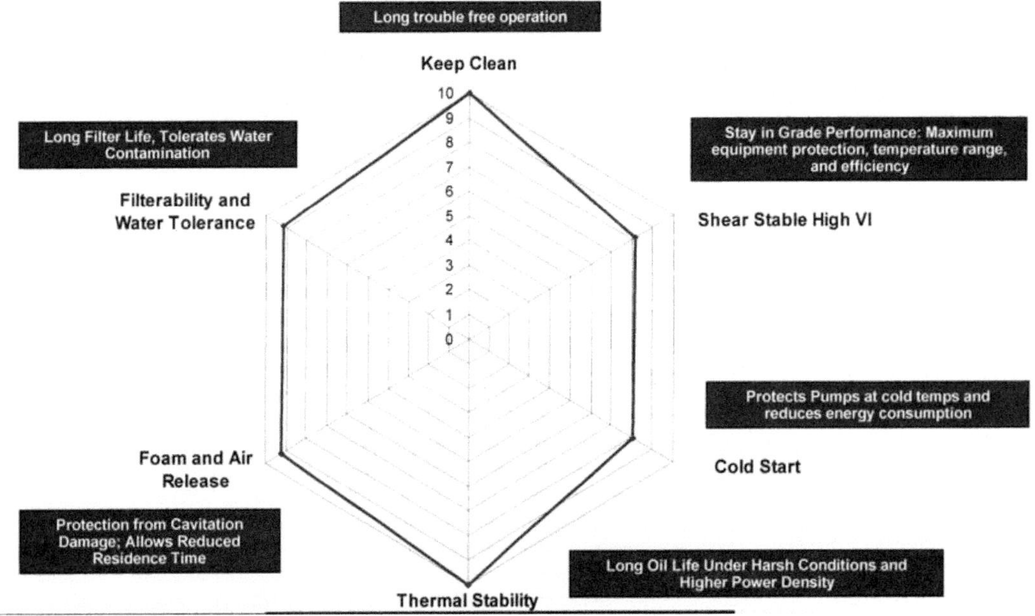

Figure 9-54: Spider diagram for an advanced mineral oil (Source: ExxonMobil Lubricants & Specialties, Fairfax, VA)

Insufficient foam control can disqualify oils that are otherwise satisfactory. Antifoaming additives may be used in oils; however, many such additives tend to increase the air entrainment tendencies of a lubricant. Aided by competent suppliers, a reliability-focused user should seek to balance these two undesirable characteristics.

When examining foaming tendency, be aware that this is a significant concern in gear applications. On the other hand, some SH lubricant with not-so-perfect foaming behavior may test excellent in Falex performance (Ref. 9-1), which is a measure of extreme pressure behavior. Be aware also of the demulsifying properties and look for centrifuged water, or "cuff." In some SH lubricants, these can exceed the limits for EP (extreme pressure) gear oils stipulated in two important industry benchmarks. The two benchmarks are commonly known as standards issued by AGMA (American Gear Manufacturers Association) and USS, the United States Steel company. An FZG (Forschungsstelle für Zahnräder und Getriebebau) test is more representative of gears in service, although it is also more expensive to run.

Be mindful of the fact that some oils could exhibit insufficient viscosity and might test below allowable limits for a particular ISO Grade lubricant. A sample may still meet allowable limits of AGMA (American Gear Manufacturers' Association), USS and other specifications, but its FZG rating could fall below the minimum necessary to protect the equipment. Perhaps the EP additive system in the oil is a borate. Borates are widely used in various structures for anti-wear (AW) and extreme pressure (EP) performance. One good feature of the borate is that it typically produces an anti-scuff film of several times the thickness of sulfur-phosphorus. It may also improve the anti-pitting (surface fatigue) performance for the same reason.

However, one risk commonly associated with borates is the influence of water contamination. It has been theorized that the submicron-size borate spheres adhere to bearing and gear surfaces by electromechanical precipitation. When the dielectric properties change, or the oil becomes more conductive due to water contamination, the film "plates out" and protection is disrupted or lost altogether. If an application risks water intrusion steps must be taken to exclude water. In general, SH lubricants have better demulsibility, better air separation, better viscometrics and better load carrying ability than their hydrocarbon counterparts. They are often used in applications that are subject to contamination because they give extra assurance of lubrication even in the presence of water. But the message here is that not all synthesized hydrocarbon lubricants will perform well in the presence of water. Again, what is done to prevent water intrusion into equipment bearing housings in, say, process pumps? Using advanced bearing housing protector seals (Figure 9-36) should be the first line of defense in pumps, gears, and other machines.

Failure to Measure up in Wear Control and Demulsibility

A few comments are helpful when geared pump sets are involved. Gear oils benefit from additives that allow sliding under the extreme pressures exerted on mating parts. If a lubricant would fall significantly below average in the Falex test, this might be a concern to the user. Was the sample perhaps low in sulfur, boron, zinc and phosphorous? The average EP gear oil would be expected to have moderate amounts of these ingredients. And again, the latest wisdom is to focus on FZG rather than Falex or Timken.

If a lube sample has excellent foaming tendencies and also exhibits very good Falex EP properties, it may still score below average in wear performance if it allows only an insufficient Timken load. That may well disqualify the oil for some uses.

As to demulsibility, recall that water promotes the rusting of ferrous metal parts and accelerates oxidation of the oil. For effective removal of the water, the oil must have good demulsibility characteristics. When tested for oil-water demulsibility, an oil should meet AGMA, USS, and perhaps other industry specifications (Ref. 9-28).

What to Expect from Top-Performing Lubricants

It is reasonable to expect as-tested or as-advertised performance from the lubricants we select for important process machinery. Comparisons among competing offers are of interest and may require explanations.

Assume you noted that both TAN (Total Acid Number) and the elemental concentrations in a lubricant allowed you to conclude you were dealing with a high-load sulfur-phosphorus EP lubricant. Chlorine found in such lubes might well be a typical byproduct of the synthesis of sulfur and phosphorus during formulation, the concentration varying proportionally. Chlorinated hydrocarbons such as chlorinated paraffins are always a cause for concern, since derivative compounds (among them hydrochloric acid), are frequently corrosive.

While chlorinated paraffins are rarely encountered by pump engineers, a brief note might still be of interest. Lubrication engineers can see a possible use for chlorinated paraffin; these have historically been used in metalworking applications, i.e., broaching. Their activation temperature is lower than sulfur/phosphorus. If one had a very slow moving, extremely heavily loaded gear, there would be boundary lubrication, but the very slow speeds would not reach the activation temperature of S/P. Chances are that a chlorinated product would be a good choice and would fit the type of specialty formulation that could be developed by a niche blender.

But modern additive formulating technology is effective at mitigating the risk by forming more stable halogenated molecules. Since sulfur-phosphorus additives have been useful through years of service in a variety of applications, it may not be reasonable to reject an oil with a chlorine concentration as low as 0.01% (100 ppm). Translation: In spite of traditional concerns with oils containing chlorine, the oil may outperform its competitors.

An experienced formulator may perhaps substitute antimony for the traditional zinc (or copper and other metals) as the transition element. The result could be exceptional wear performance without compromising much of the antifoam and demulsifying characteristics we might seek in an attractive SH lubricant. Nevertheless, the level of zinc or other metals does not indicate the oil's anti-wear capability. It is the phosphorus, forming a phosphorus glass, which provides the protection. That is why additives such as TCP (tricresyl-phosphate) are so effective.

What Some Additives Are Achieving

As alluded to earlier in this segment of our text, the formulators/marketers of certain lubes often ascribe special performance to their products. Rarely, if ever, does a company disclose its exact recipes. One company explains that their proprietary additive technology forms a tough, ionic, slippery, chemical film on all metal surfaces (Ref. 9-29). This film is said to help separate metal surfaces in three ways:

1. It provides a thicker oil film, i.e., adds to the oil film thickness created by the oil viscosity alone. An increase in oil film thickness is proven to increase bearing life in direct proportion to the percentage increase.

2. It creates a tougher oil film, i.e. the probability of oil film rupture is greatly reduced. Higher load carrying capacities are thus achieved and factors of safety are increased.

3. It promotes and induces micro-polishing of contacting metal surfaces. When opposing asperities (microscopic projections or irregularities on metal surfaces) breach the oil film of the more conventional oils, surface-initiated fatigue failures will often result.

It would appear that some of the constituents of this proprietary technology bridge, or fill-in the asperities and create smoother surfaces, which then easily separate with a high film strength oil film. It has been theorized that antimony is the "magic ingredient" here, but we may never know if that is true.

A similar additive technology, from the same formulator-producer, lists the earlier performance mechanisms and adds EP protection against boundary lubrication conditions typically caused by heavy load, shock loads, and low operating velocities. Boundary conditions exist when a full-fluid lubricant film is not developed between two rubbing surfaces.

As we can readily surmise, additive know-how is of great importance and does not remain static. To be realistic, job priorities may not enable pump engineers to become lube experts in the fullest sense of the word. Likewise, pump reliability specialists should resist the opposite extreme, i.e. "hands off." Retain an interest in the matter and do not let a facility simply buy from the lowest bidder. The right approach is for reliability professionals to stay informed of lube-related developments. Knowledgeable lubricant vendors and their application engineering staff should rank very high among our technology providers.

LUBRICANTS AND THE IMPORTANCE OF CLEAN PRODUCTS

Oil cleanliness deserves our utmost attention. Suppose the as-supplied water content of a certain lubricant was nearly one percent, which would be rather unusual and would really be unacceptable for new oils. But this high initial contamination has happened, on occasion. Not unlike particle contamination, it may have had more to do with packaging and handling than manufacturing and blending operations. We can deduce that low water contamination should be made part of any rigorous purchase specification. In any event, water contamination can contribute to the lower wear and extreme pressure (EP) performance of a lubricant sample.

From the above, we should learn to pay attention not only to initial water content, but oil cleanliness in general. Clean oil greatly extends the life of bearings and equipment. Most oils are not clean. They may not look dirty to the unaided eye, but will often contain a profusion of particles in the 2-30 micron (0.0001-0.0012 inch) range of diameters. The particles may consist of fibrous, metallic, and sand-like contaminants. With water as the catalyst, these contaminants form sludge, and sludge is bad for machines in modern process plants. Fortunately, advanced bearing housing protector seals are available to limit ingress of airborne contaminants (Ref. 9-26, also Fig. 9-36).

The merits of some SH fluids have been reported in the two previous parts of our three-part series of articles. As to shipping containers, the SH fluids are perhaps packaged in new, clean plastic drums and other plastic containers that ensure oil cleanliness. However, some lube manufacturers will not use plastics because they tend to sag and deform greatly in hot climates. Regardless of packaging, your lubricants deserve to be filtered to ISO 4406 cleanliness level 14/13/11, which is explained in Tables 9-4 through 9-6 and in the marketing bulletins of virtually all prominent suppliers (Ref. 9-27). It used to be that this degree of cleanliness was often much better than conventional lubricants in steel drums or lubes shipped by bulk delivery truck, but that is not always the case today. Both world-scale and small niche manufacturers can provide superior cleanliness in all kinds of containers; so, "check it out." The merits of clean oil are intuitively evident. Bearing and hydraulic equipment manufacturers believe that buying ultra-clean oils will increase bearing and hydraulic component life from 300 to 500 percent over oils with barely passing cleanliness values (Refs. 9-28 and 9-29).

Bottom Line: Not all SH Lubricants Perform Alike

The reader may realize that some of the oils discussed, or to which we have alluded in this text, are produced and marketed by large multinational oil companies. Of course, these companies manufacture many lubricants that will do exceedingly well in many applications. Indeed, as a practicing engineer, this author is quick to point out there might even be some machines where it would not matter which of the various SH lubricants was chosen. But advanced mineral oil formulations should also be considered and oil mist comes to mind. In oil mist systems on the U.S. Gulf Coast, highly advanced mineral oils will be just as good, if not better, than some SH fluids. But before making an informed choice, take time to look at their respective spider diagrams (Figure 9-54) and ask relevant questions of the various competitors. Actual field experience should be of real interest here.

Of course, there will always be some machines which greatly benefit from using a truly superior lubricant. And superior lubricants can quite obviously contribute to moving equipment owners from "average" into the "Best-of-Class," high profitability category. The trick will be to sort out which are these superior lubricants and to apply them in the many instances where economics favor this choice.

It would be inappropriate to tell you which lubricants are always superior. Few things deserve the attribute "always," and a vendor's diligence (or lack thereof) in 2000 may no longer be the same in 2013. Still, PAO or PAG are the standard of reliability-focused users in worm gears, PAO for temperatures above a certain temperature (we might use 170°F as a guide), or PAO for temperatures below a certain threshold.

Explore your options and do not base your opinions on hearsay. See what the "majors" are capable of doing today. Let them explain data obtained through rigorous testing and examine the case histories they can provide. Link up with and favor vendor-formulators willing to share information. Many will be pleased to transfer their application know-how into extending the life of your machines. Compared to equipment failures and downtime expenses, good lubricants are a bargain. Be prepared to pay for lube application engineering expertise. This kind of technology sharing should be considered the most desirable and perhaps most valuable contribution anyone can make in an industrial partnership involving rotating equipment.

Chapter 10

Oil Mist Lubrication and Storage Protection

Oil mist lubrication, as shown schematically in Figure 10-1, is a centralized system which utilizes the energy of compressed air to produce a continuous feed of atomized lubricating oil (Ref. 10-1). This atomized lube oil, or "oil mist," is conveyed by the carrier air—usually dry air from the plant instrument compressor or dry plant air—to multiple points through a low pressure distribution system. The system typically operates at a header pipe pressure of about 20 inches H_2O (~5 kPa). The oil mist then passes through a reclassifier nozzle before entering the point to be lubricated. This reclassifier nozzle establishes the oil mist stream as either a mist, spray, or condensate, depending on the application of the system (Ref. 10-2). At its destination, the oil mist provides either only a protective, or both a protective and lubricating environment (Ref. 10-3).

Reclassifier nozzles can either be located at the mist manifold ("distribution block") located at the end of the branch or at the stainless steel tube-male pipe connector of Figure 10-1. Mist manifolds are shown in Figures 10-1 through 10-3. A typical "mist" reclassifier fitting is shown in Figure 10-4, where it essentially provides a protective blanket for a gear speed reducing unit (Ref. 10-4). Here, the reclassifier is located at the stainless steel tube-male pipe connector.

Not shown in Figure 10-1 are a mist density detector, filters associated with both the instrument-quality air supply and a bulk oil storage vessel, and snap drains (or snap valves) at the ends of the various pipe branches. Properly designed oil mist consoles will include air and oil heaters. Maintaining constant mixing temperatures ensures that the proper air-to-oil ratio is maintained. Once the mist has been produced, it will flow equally well in Canadian and South American ambient environments. Neither the galvanized steel header pipes nor the branch piping (or "drops") shown in Figures 10-1 and 10-2 require heat-retaining insulation.

Generally speaking, a mist stream is ideal for high-speed bearing lubrication and for the preservation and protection of bearings in non-operating pumps and electric motors. A spray may be needed for low-speed bearings, whereas "condensed" or, to be more precise, "coalesced," droplets of lubricant are primarily suitable for sliding guideways, and application of viscous lubes to certain gear configurations and a limited spectrum of other equipment. Few, if any, condensing fittings are ever used on pumps.

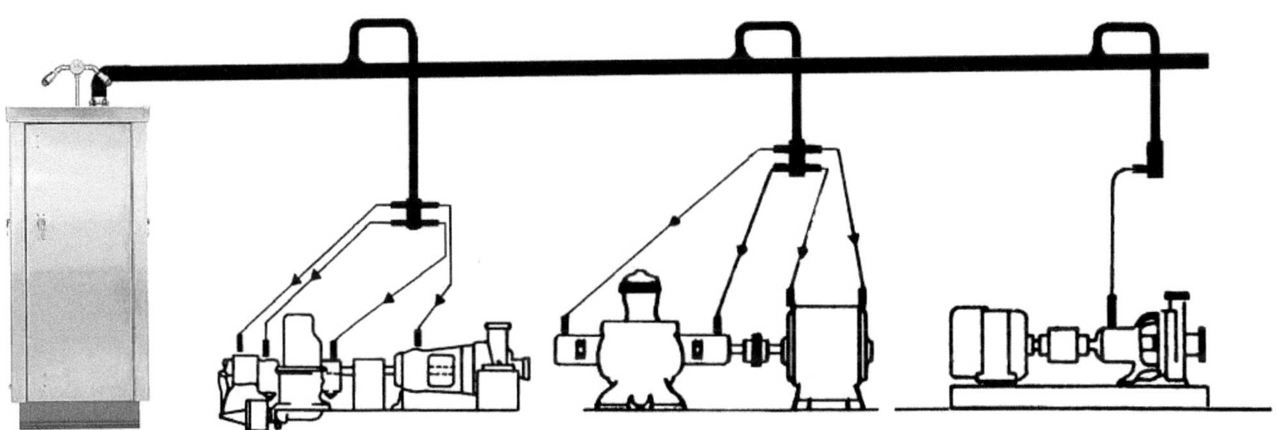

Figure 10-1: Oil mist lubrication system schematic (Source: Lubrication Systems Company, Houston, Texas 77024)

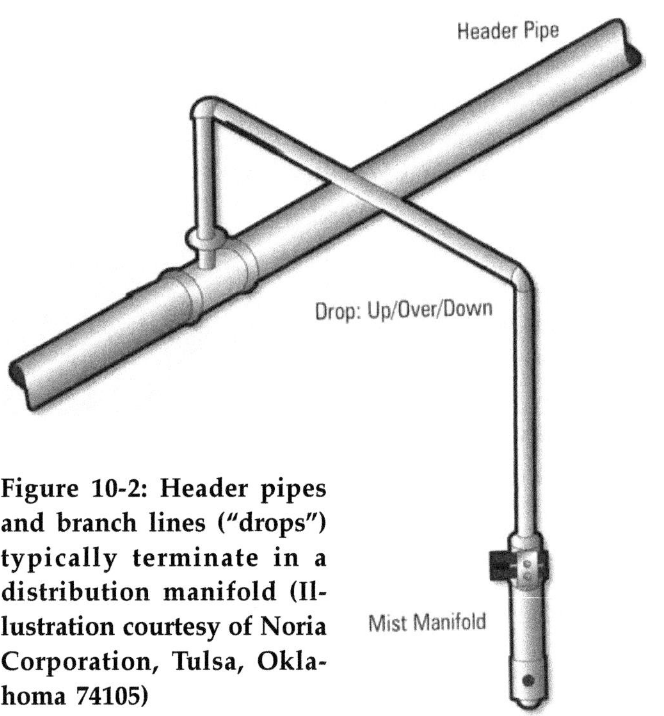

Header Pipe
Drop: Up/Over/Down
Mist Manifold

Figure 10-2: Header pipes and branch lines ("drops") typically terminate in a distribution manifold (Illustration courtesy of Noria Corporation, Tulsa, Oklahoma 74105)

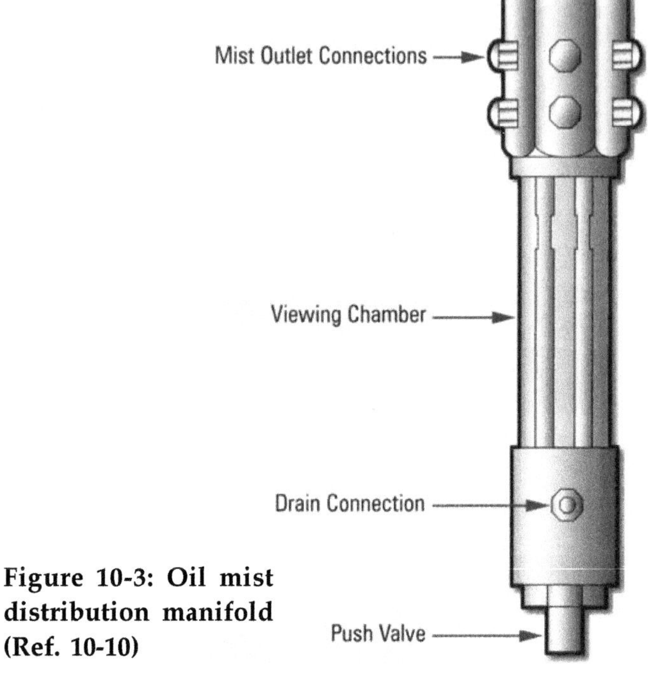

Mist Outlet Connections
Viewing Chamber
Drain Connection
Push Valve

Figure 10-3: Oil mist distribution manifold (Ref. 10-10)

SYSTEM USAGE

Oil mist lubrication is a mature, straight-forward, well-understood technology. Since the mid-1950's, the oil mist lubrication concept has been accepted as a proven and cost-effective means of providing lubrication for centrifugal pumps (Ref. 10-5). Typical large-scale applications include not only numerous different types of pumps, but also electric motors, gears, chains, and horizontal shaft bearings, such as on steam turbines and steel rolling mill equipment.

Oil mist units come in different sizes and with capacities that vary by orders of magnitude. Small units serve perhaps only one or two critically important pumps, consuming no more than 2.1 gallons (~8 liters) of oil per pump per year. Medium-size units often serve 15 or 20 pump sets. Large, centrally located oil mist consoles (Figure 10-5) may serve the combined lubrication requirements of 100 pump sets (Ref. 10-6). Both the small and large units comprise built-in heaters and pressure switches.

The size of an oil mist unit can also be expressed as "bearing-inches," abbreviated "B.I." A 10,000 B.I. unit would have the capacity to generate the oil mist lubrication requirements of 10,000 bearings located on one-inch, or 5,000 bearings located on two-inch, or 2,000 bearings on five-inch shafts, and so forth. Incidentally, this sizing

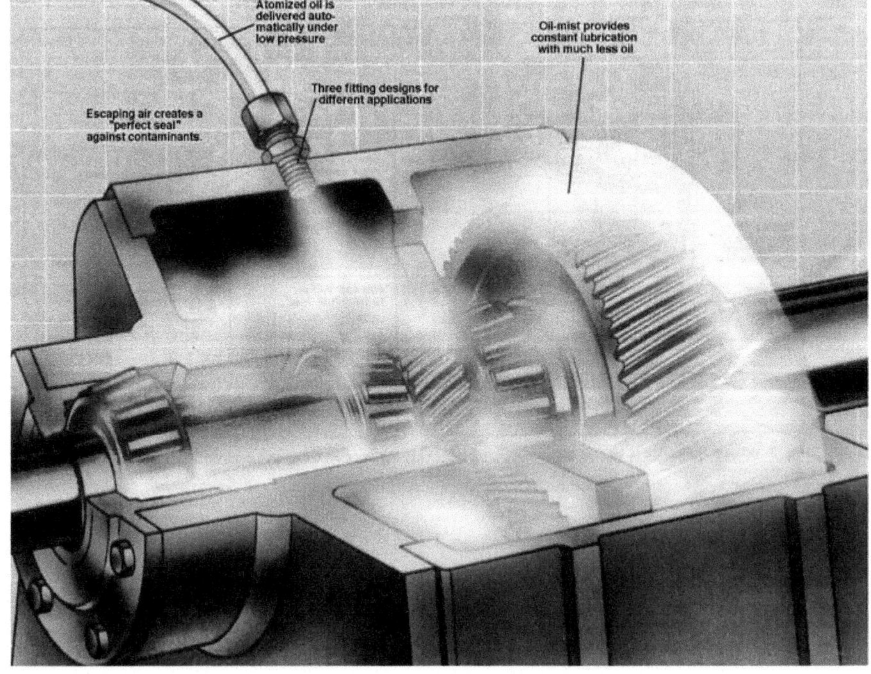

Atomized oil is delivered automatically under low pressure
Oil-mist provides constant lubrication with much less oil
Three fitting designs for different applications
Escaping air creates a "perfect seal" against contaminants.

Figure 10-4: Typical mist reclassifier fitting supplying "protective blanketing" to a gear speed reducing unit (Illustration courtesy of Noria Corporation, Tulsa, Oklahoma 74105)

method explains that oil mist consumption is largely a function of bearing size, not bearing speed. However, heavily loaded bearings do require more oil mist than lightly loaded bearings.

Figure 10-5: Oil mist console suitable for lubricating 60 or more pumps—LubriMate suitable for lubricating 1 or 2 pumps. Both are furnished with built-in heaters and supervisory instrumentation. (Source: Lubrication Systems Company, Houston, Texas 77024)

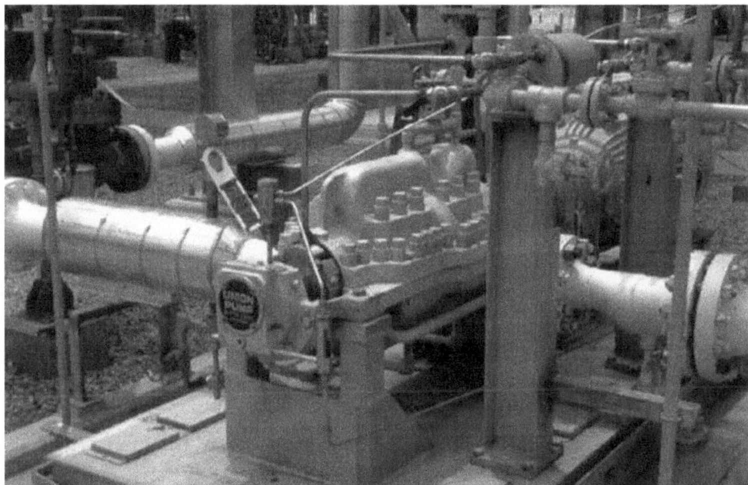

Figure 10-6: Oil mist provides purge to multistage pump and dry-sump mist to electric drive motor (Source: Lubrication Systems Company, Houston, Texas 77024)

Plant-wide oil mist systems allow the oil mist to travel as far as 600 ft (~200m) from a centrally located oil mist console to the most distant pump. Dry air, usually instrument air, is supplied at 55 psig (379 kPa) pressure and subsequently reduced to about 20 inches of H_2O (~5 kPa). Headers are sized for an oil mist velocity of approximately 20 fps (~7 m/s). Experience shows that higher velocities would cause too many of the atomized lube oil globules to collide with other globules. The resulting droplets would then be too heavy to remain in suspension; they would tend to drop to the bottom of the header, leaving only "lean" mist to reach the various points to be lubricated.

A typical optimum header size would consist of two-inch nominal galvanized steel pipe. This size is generally large enough to keep mist velocities low, and rigid enough to limit deflection between pipe supports. Headers are installed at a slight slope (Ref. 10-7). Any liquid oil not kept in suspension by the carrier air will thus be able to flow back into the console and its associated lube oil reservoir (Ref. 10-8).

"PURGE MIST," OR "WET SUMP" OIL MIST

The actual method of applying oil mist to a given piece of equipment is governed to a large extent by the type of bearing used. For pumps equipped with sleeve bearings, oil mist alone is not considered an effective means of lubrication because relatively large quantities of oil and considerable heat removal would be required. However, in pumps with sleeve bearings (Figure 10-6), oil mist is used effectively as a bearing housing purge and, to a limited extent, as fresh oil make-up to the reservoir.

An effective purge must exclude atmospheric contaminants from the bearing environment. The purge mist must, therefore, exist at a higher-than-ambient pressure. Either a balanced oil sight assembly (Figure 10-7, left side), or a balanced constant level oiler (Figure 10-7, right side) is required for oil mist purged bearing housings (Ref. 10-8). Balanced constant level lubricators, also shown (see Figure 9-19) and discussed in the preceding chapter, are required so as not to force the lubricant from the higher pressure region (the housing interior) to the region of ambient pressure. Over the years, numerous costly bearing failures have occurred because the user neglected to use balanced constant level lubricators as needed. The lower

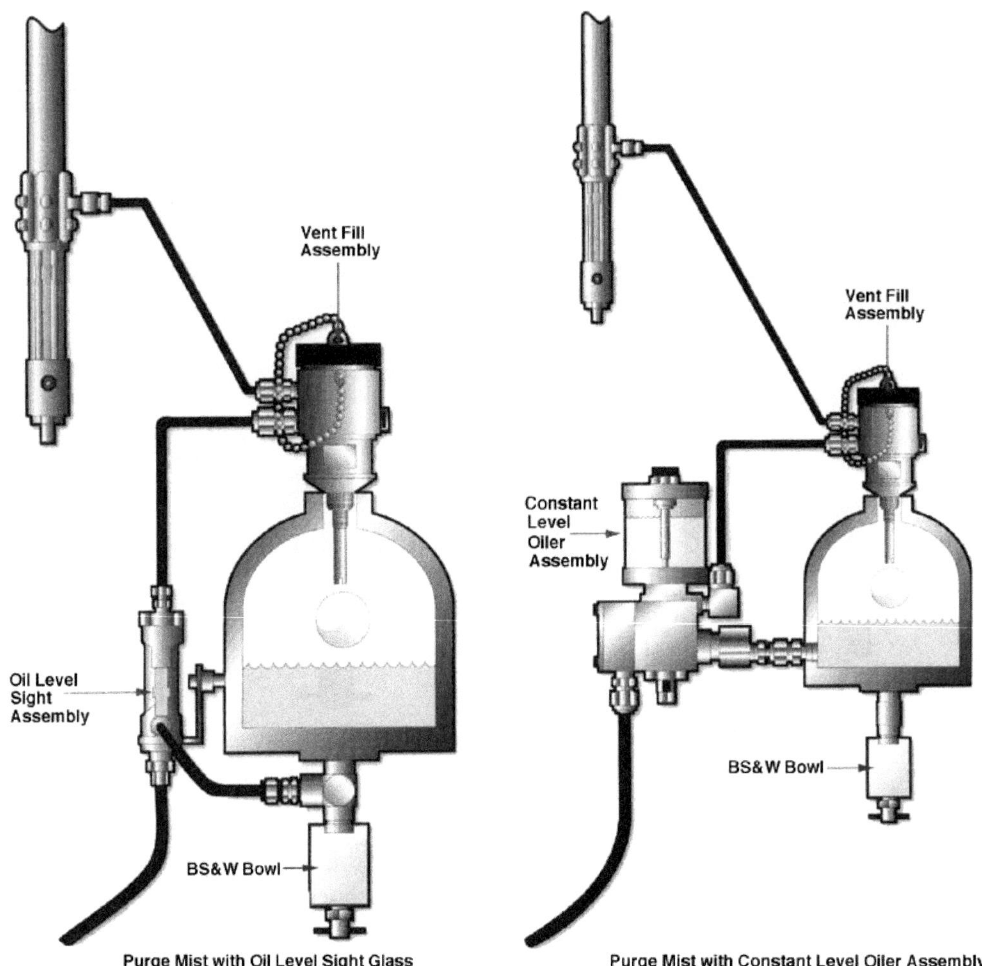

Figure 10-7: Cross-section views of purge mist applications (Illustration courtesy of Noria Corporation, Tulsa, Oklahoma 74105)

side-port of the vent-fill assembly shown in Figure 10-7 connects the vapor space of the bearing housing to the surge chamber of the constant level lubricator (Ref. 10-9). The needed pressure balance is thus achieved.

DRY SUMP OIL MIST IS ADVANTAGEOUS

Rolling element bearings, on the other hand, are ideally suited for dry-sump lubrication. With dry sump oil mist, the need for a lubricating oil sump is eliminated and oil rings or flinger discs are removed. If the equipment shaft is arranged horizontally, the lower portion of the bearing outer race serves as a mini-oil sump. The bearing is lubricated directly by a continuous supply of fresh oil collecting on the bearing components. Turbulence generated by bearing rotation causes oil particles suspended in the air stream to coalesce on the rolling elements as the mist passes through the bearings and exits to atmosphere.

The pumps illustrated in Figures 10-8 through 10-10 are dry-sump lubricated. Since no liquid oil sump exists in dry-sump lubricated pumps, there are no constant level lubricators. This differentiation is highlighted in Figure 10-11. In the wet sump case, left illustration, a closed (non-vented) oil sight bottle is used with the oil level sight assembly to visually observe oil contamination. In the dry sump case, right illustration, the oil mist will coalesce on the bearings and overflow into the (vented) oil sight, or oil collection bottle. However, a keen observer might note that, on dry-sump lubricated bearing housings, the mist should be routed through the bearings as shown earlier in Figures 9-11 and 10-8. Routed as shown on Figure 10-11, the mist might tend to flow directly from the top of the oil sight bottle. In that instance and to make up for the bypassing mist, the pump owner would have to "overfeed" or over-lubricate this particular bearing

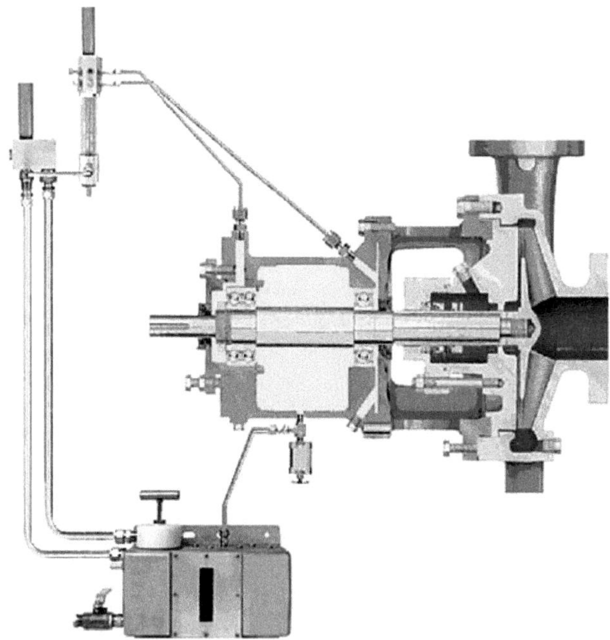

Figure 10-8: Single-stage overhung pump using two reclassifier locations for dry- sump oil mist lubrication as described in API 610 8th Edition (Source: Lubrication Ssystems Company, Houston, Texas 77024)

Figure 10-9: Single-stage overhung pump in hot service is pure dry-sump oil mist lubricated (Source: Lubrication Systems Company, Houston, Texas 77024)

Figure 10-10: API-style back pull-out pumps lubricated with dry sump pure oil mist (Source: Lubrication Systems Company, Houston, Texas 77024)

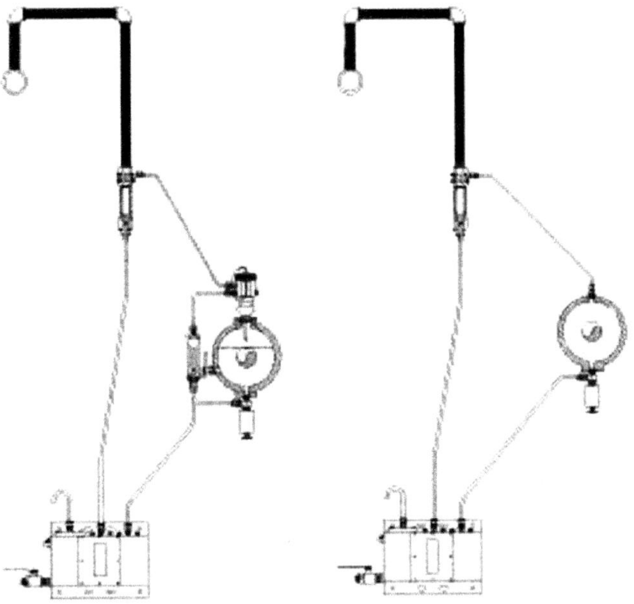

Figure 10-11: Pump bearings lubricated by (left) purge mist/wet sump, and (right) dry-sump/pure mist (Source: Lubrication Systems Company, Houston, Texas 77024)

housing or use a directional reclassifier to force the oil mist into the bearings!

The dry sump oil mist technique offers four principal advantages:

- Bearing wear particles are not recycled back through the bearing, but are washed off instead
- The need for periodic oil changes is eliminated
- There are no oil rings whose wear debris might jeopardize bearing life
- Long-term oil breakdown, oil sludge formation, and oil contamination are no longer factors affecting bearing life
- The ingress of atmospheric moisture into pump or motor bearings is no longer possible and even the bearings of standby equipment are properly preserved

Recall that without oil mist application, daily solar heating and nightly cooling cause air in the bearing housing to expand and contract. This allows humid, often dusty air, to be drawn into the housing with each thermal cycle. As was demonstrated earlier in this text, the effects of moisture condensation on rolling element bearings are extremely detrimental. Water, or condensed water vapor, is largely responsible for few bearings ever achieving their catalog-rated design life in a conventionally lubricated plant environment.

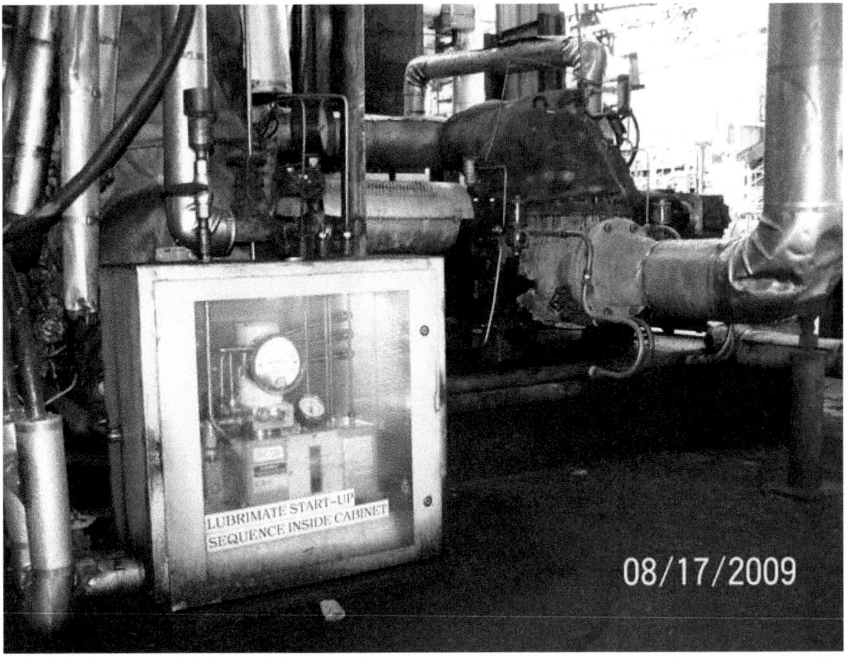

Figure 10-12: Pump and turbine combination served by a small modular oil mist unit (LubriMate®). Purge mist is used in the turbine bearing housings, pure oil mist is applied to all pump bearings (Source: Lubrication Systems Company, Houston, Texas 77024)

MODERN PLANT-WIDE SYSTEMS ENHANCE SAFETY AND RELIABILITY

It has been established that loss of mist to a pump or motor is not likely to cause an immediate and catastrophic bearing failure. Tests by various oil mist users have proven that bearings operating within their load and temperature limits can continue to operate without problems for periods in excess of eight hours. Furthermore, experience with properly maintained oil mist systems has demonstrated outstanding service factors. Because there are no moving parts in the basic oil mist console and mist generator components, and because the system pressure is very low, oil mist ranks near the top of the most reliable lubrication methods. Proper lubrication system operation can be interlocked with pump operation or an alarm system, assuring adequate lubrication.

It should even be possible to use an oil mist system as a plant-wide fire monitor. Branch piping from oil mist headers could be led to locations in the proximity of vulnerable bearing housings. Each branch pipe could be capped with a low melting point alloy. Elevated temperature would cause the fusible alloy to melt, causing the header pressure to drop and the low pressure alarm to be activated. Manual intervention would isolate mist flow to the stricken machine. Meanwhile, the oil mist system could continue to serve non-affected machines without interruption.

The savings due to lower preventive maintenance labor requirements, equipment repair cost avoidance, and reductions in unscheduled production outage events have been very significant and cannot possibly be overlooked by a responsible manager or cost-conscious manufacturing facility. Oil mist systems have become extremely reliable and can be used not only to lubricate operating equipment, but to preserve stand-by, or totally deactivated ("mothballed") equipment as well, as will be shown later.

Moreover, oil mist can be applied in a variety of ways. The small modular unit of Figure 10-12 serves both the pump and its turbine driver. Pure oil mist ("dry sump") is applied to this asphalt pump which operates at slightly over 650°F (at or near auto-ignition). The steam turbine bearings are lubricated by liquid oil and oil mist fills the housing volume above the "wet sump" oil level, making it a "purge mist" application. Installed adjacent to this pump is its standby spare; it is served by a second LubriMate® unit. An insulated and heated enclosure offers protection from extreme cold experienced by this refinery during winter months. Although located in mid-U.S. just south of the Canadian border, the oil mist headers need no such protection.

As of 2010, both units had seen highly satisfactory service for well over two years. Only routine maintenance was performed on a semi-annual basis.

ENVIRONMENTAL CONCERNS ADDRESSED

As oil mist lubrication has become more widely used, some questions have arisen about whether it contributes to pollution, is a health hazard, or is combustible. Studies have been made and have concluded that oil mist is none of the above. With 200,000 volume parts of air per volume of oil, the mixture is far too lean to be explosive.

U.S. regulatory agencies do not consider oil mist lubrication an air pollutant. They concluded that oil mist is not a vapor and therefore is not a contributor to air pollution. Generally, lubricating fluids are not a volatile organic compound (VOC). California Rule #1173, with the title "Fugitive Emissions of Volatile Organic Compounds" (VOC), specifically exempts lubricating fluids.

Still, modern plant-wide oil mist systems are closed-loop. Traveling through the rotating bearing will have caused the atomized oil to coalesce; the spent mist will have become "lean." Now mostly air, the lean mixture is drawn into a lower pressure return header (see Figure 10-13) and, finally, to an oil knock-out drum, or reclamation unit, Figure 10-14. The reclassified oil is filtered and reused. For small oil mist systems or

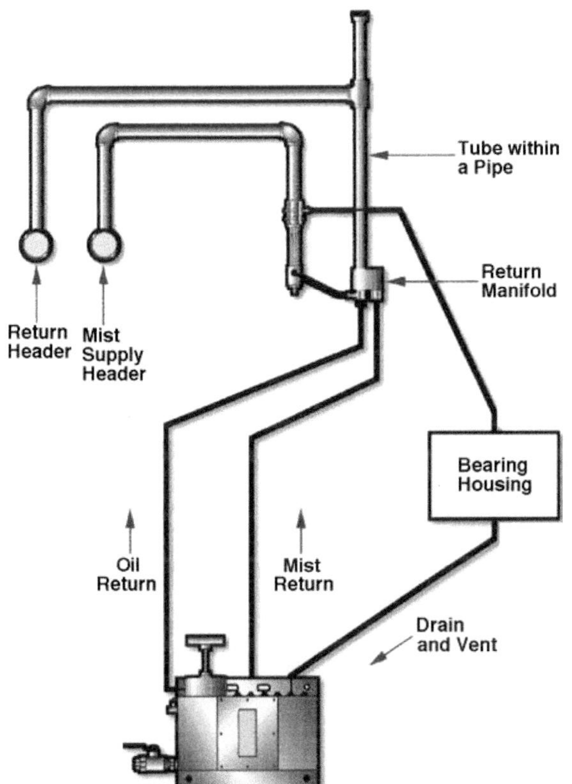

Figure 10-13: Layout of header system and associated oil mist tubing (Source: Lubrication Systems Company, Houston, Texas 77024)

Figure 10-14: Medium-size oil mist console (left) and collecting tank are part of a mid-size closed-loop oil mist system (Source: Lubrication Systems Company, Houston, Texas 77024)

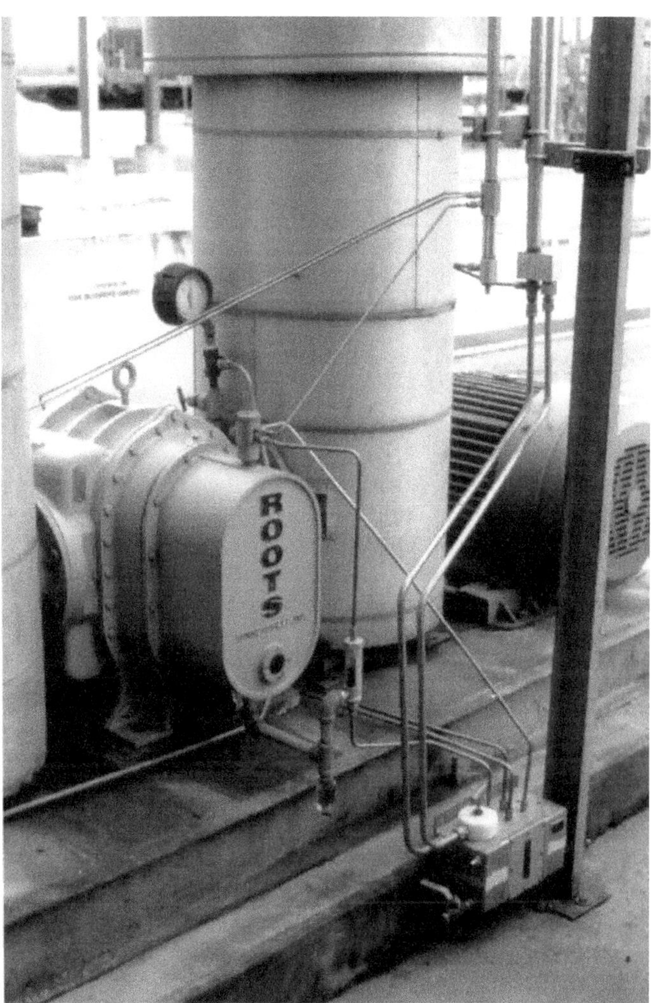

Figure 10-15: Collecting tank (lower right) associated with wet-sump lubricated equipment (Source: Lubrication Systems Company, Houston, Texas 77024)

installations with as few as a single pump where the economics perhaps are not favoring closed systems, specially designed collection containers (Figure 10-15) may be substituted. Liquid oil collecting in these containers could simply be pumped back to either the mist supply or mist return header. Since both headers are installed with a gentle downward slope, the liquid oil would flow back to the console reservoir.

It should be noted that closed loops can be easily retrofitted to existing systems. The main additions necessary are return headers and an oil reclamation unit. The significant benefit of oil reclamation is that the oil may be recycled. At a cost of $18-$25 per gallon for synthetics, closed loop systems can have an enhanced pay-out.

What—If Anything—
Can Shut Down an Oil Mist System?

There are only two events that can completely shut down an oil mist system. The first is the loss of instrument air, in which case a unit process would probably have to be shut down for a host of safety-related reasons. So, in that case the pumps would be shut down anyway and lubrication will certainly not be an issue. The lubrication provided by the oil mist system will fully support the equipment as it coasts to a stop.

The second possibility is the loss of oil supply to the oil mist generator (OMG). That possibility was estimated three or more orders of magnitude less likely than the inadvertent loss of oil level, or grease contamination, or oil ring malfunction with any of the far more vulnerable default methods of pump and driver lubrication. With oil mist systems, loss of electric power does not impede the generation of oil mist. Any electricity connection exists only for monitoring and maintaining an optimum temperature balance (oil versus air) in the misting chamber reservoir. An incorrect temperature would be annunciated in the plant's operations control center.

There is an interesting potential side benefit mentioned in one of the many references. A capped plastic oil mist tubing line placed above an oil mist-lubricated machine will melt in the event of a fire at (or near) the equipment. Oil mist will then flow through the opening and the pressure in the oil mist header will decay. Declining header pressure would trigger an alarm or similar malfunction indicator in the control center. Think of it: The oil mist piping, while itself filled with a non-explosive mixture, can also perform duty as a fire alarm system!

Typical oil consumption rates for average-size oil mist systems are about 1 gallon (3.85 liters) per 24 hours. The main misting chambers in the large OMG console units range from 5 to 9 gallons (19 to 35 liters), which provides several days of operation. In addition to the misting chamber there is usually a 75 to 110 gallon (289 to 423 liter) bulk oil supply tank supporting the OMG. There are thus several months of oil supply available. Low level alarms are provided on both main misting chamber and bulk oil tank. Therefore, the probability of running out of oil is exceedingly remote.

With the now increasing use of "closed loop" oil mist systems, the initial charge of oil may last for years. In closed loop systems up to 95% of the oil that goes out into the system can be recovered. That fact, plus highly favorable experience with oil mist lubricated electric motors, can easily shift most cost justification calculations into the favorable column.

ECONOMICS OF DRY SUMP OIL MIST LUBRICATION

The economics of plant-wide oil mist may differ from location to location (Ref. 10-10). Plants that have wisely decided to maximize dry sump applications and to include electric motor drivers have seen higher returns on their investment than plants that overlooked this potential. The service and maintenance requirements of modern oil mist systems are extremely low and must be factored in favorably.

Reliable failure statistics of primary importance in calculating the economic justification of oil mist. Also, a number of references contain outright failure data:

1. *Oil Refinery, Netherlands Antilles—Electric Motor Bearings*
   Annual bearing failure rate, grease lube: 15.6%
   Annual bearing failure rate after conversion to dry sump oil mist: 0.9%

2. *US Gulf Coast Chemical Plant —Pump Bearings*
   Prorated failure events for conventionally lube control sample: 100/year
   Prorated failure events for wet-sump lubricated control sample: 38/year
   Prorated failure events for dry-sump lubricated control sample: 10/year

3. Discounted Cash Flow returns, system serving 80 pumps and electric motors: 19%

4. US Gulf Coast Chemical Processing Unit — Pump Bearings
   Before conversion to oil mist (prorated failures): 100/year
   After conversion (wet/dry sump): 25/year

5. US Gulf Coast Phenol 1 vs. Phenol Unit 2— Electric Motor Bearings
   Unit 2, 35 motors, number of failures with grease lubrication: 10/year
   Unit 2, 35 motors, failures after conversion to dry sump oil mist: 1/year

Pump bearing failure reductions in petrochemical plants have been demonstrated to range from 75 to 90 percent (Ref. 10-1). A straightforward 1990s vintage screening approach is represented in Table 10-1, "Justification Example and Worksheet." Work with an oil mist vendor to update the cost data.

SELECTING THE CORRECT OIL

In the United States and South America, thousands of refinery pumps and their electric motor drivers have been lubricated with oil mist, many of them since the mid-1960's. Moderate bearing temperatures are typically achieved with this reliable once-through lubrication method. ISO Grade 68 or Grade 100 mineral oils are prevalent in moderate climate zones, although Grade 100 is often "overkill" that will create somewhat higher than expected bearing friction and, hence, power waste.

Note, also, that certain mineral oil base stocks (primarily naphthenic base stocks) promote wax platelet formation at low temperatures. These platelets might then plug the small bore passages in some reclassifier fittings. It is thus recommended to use ISO Grade 32 or 68 *synthetic* lubes at locations that experience ambients below 40°F (~5°C).

Oil mist has clear and substantial benefits over conventional lubrication systems, as many users around the world can attest. Not only have pumps used this advantageous lube oil application method, but so have electric motors.

Figure 10-16 shows a cross-section of the bearing housing of a Siemens TEFC motor. These electrical machines have for decades been available with rolling element bearings that could be lubricated with either grease or oil mist.

Indeed, there are few if any technical, environmental, safety or health reasons that would preclude the use of oil mist lubrication for rolling element bearings. As was discussed earlier, modern oil mist systems are available to suit many different objectives, sizing criteria, etc., (Table 10-2). Large systems are sometimes designed with very economic backup systems.

OIL MIST FOR ELECTRIC MOTORS

In the mid-1960s oil mist—a mixture of 200,000 volume parts of clean and dry plant or instrument air with one part of lubricating oil—was found to provide ideal lubrication for thousands of rolling element bearings in electric motors. Since then, this lubrication method has gained further acceptance at reliability-focused process plants in the United States and overseas.

However, while striving to be more reliable, the majority of process plants have been slow to abandon their traditional, and costly, repair-focus. Questions and concerns relating to oil mist that were answered decades ago are again surfacing in the minds of people who seem to have no access to a body of literature that addressed and explained matters literally decades ago.

A brief overview of considerations that, to this day, have allowed oil mist lubrication to maximize the reliability and energy efficiency of electric motors is, therefore, of value.

In particular, a general update on oil mist lubrication and favorable efficiency should be of interest here.

Electric Motors and Mechanical Efficiency

Reliability-focused readers know quite a bit about oil mist and many use plant-wide oil mist systems for their process pumps and electric motor drivers. In fact, some major grass-roots olefins plants used oil mist on motors with rolling element bearings as small as one hp (0.75 kW) and as large as 1250 hp (925 kW) as early as 1975. What's less known is that oil mist, together with the right viscosity synthetic lubricant, will save considerable energy.

In 1980, a research laboratory in New Jersey determined that a readily available synthetic lubricant, having a viscosity of 32 cSt at a temperature of 40°C (98°F), offered long-term contact surface protection for process pumps and their electric motor drivers (see Chapter 9). Although "only" 32 cSt, the protective effect of this *synthetic* lubricant was found to be equivalent to that of a base line *mineral oil* with the higher viscosity of 68 cSt. The same good wear protection could not be achieved with a reduced viscosity mineral oil. Thorough evaluation also showed the lower viscosity synthetic lubricant providing energy savings of 2-3% (even 5%) of the normal power draw of the electric motor used in these tests.

Table 10-1: Oil mist justification example and worksheet (1 of 3)

JUSTIFICATION EXAMPLE AND WORKSHEET

	EXAMPLE	ACTUAL UNIT
		(Name) ______
	40 Pumps	Pumps ______
	1 Turbine	Turbines ______
	45 Motors	Motors ______

PUMP CONVERSION INCENTIVE

• Annual Maintenance Cost Savings:	EXAMPLE	ACTUAL UNIT
— Pump population	1. 40	1. ______
— Annual bearing failures due to lubrication (10% of line 1)	2. 4	2. ______
— Reduction with LubriMist	3. 90%[1]	3. ______
— Total cost of a pump/bearing repair	4. $4,500[2]	4. ______
— Maintenance cost savings Line 2 × Line 3 × Line 4	5. $16,200	5. $
— Number of pumps in hot service	6. 5	6. ______
— Cooling water cost reduction ($300 per pump per year [3]) Line 6 × $300	7. $1,500	7. $
— Eliminate cooling water line maintenance ($200 per line per year [3]) Line 6 × $200	8. $1,000	8. $

MOTOR CONVERSION INCENTIVE

• Annual Maintenance Cost Savings:	EXAMPLE	ACTUAL UNIT
— Motor population	9. 45	9. ______
— Annual bearing failures due to lubrication (10% of line 9)	10. 4.5	10. ______
— Reduction with LubriMist	11. 90%[1]	11. ______
— Total cost of motor/bearing repair	12. $1,200[2]	12. ______
— Maintenance cost savings Line 10 × Line 11 × Line 12	13. $4,860	13. $

STEAM TURBINE CONVERSION INCENTIVE

• Annual Maintenance Cost Savings:	EXAMPLE	ACTUAL UNIT
— Turbine population	14. 1	14. ______
— Outtages per year due to lubrication (once every 4 years)	15. 0.25	15. ______
— Reduction with LubriMist	16. 60%[4]	16. ______
— Total cost of a turbine/bearing repair	17. $5,800[4]	17. ______
— Maintenance cost savings Line 15 × Line 16 × Line 17	18. $870	18. $

(Continued)

Table 10-1: Oil mist justification example and worksheet (2 of 3)

	EXAMPLE	ACTUAL UNIT
MANPOWER CREDITS		
• Elimination of oil sump changes:		
— Changes per pump per year	19. 2	19.
— Man hours per change	20. 1	20.
— Number of sumps	21. 42	21.
— Annual manpower credits + wage rate at $18/hour + benefits at 30%		
($18/hr plus 30%) × Line 19 × Line 20 × Line 21	22. $1,965	22. $
• Eliminate routine oil sump filling:		
— Minutes per shift	23. 15	23.
— Annual manpower credits		
Line 23 ÷ 60 × 3 shifts × 365 × ($18/hr plus 30%)	24. $6,405	24. $
OTHER COST INCENTIVE AREAS		
• Annual reduction in lube oil consumption:		
— Normal pump consumption of lube oil	25. 18.5 gal./yr.	25.
— Oil consumption reduction with LubriMist	26. 40%[5]	26.
— Lube oil savings		
Line 1 × Line 25 × Line 26 × $4/gal)	27. $1,184	27. $
• Annual fire loss reduction credits:		
— Pumps in low flash point service	28. 10	28.
— Fire loss reduction credit per pump[6]	29. $25	29.
— Fire loss credit		
Line 28 × Line 29	30. $250	30. $
• Annual energy savings:		
— Power loss in pump reduced by 0.34 kw[7]		
— Cost of electrical energy	31. 5¢/kwh	31.
— Per cent time pumps operate	32. 50%	32.
— Energy credits		
Line 1 × 0.34 × Line 31 × Line 32 × 24 × 365	33. $2,978	33. $
• Production outtage: — Add as appropriate	34. TBD	34. $
SYSTEM ANNUAL OPERATING COSTS		
• Utilities -air:		
— SCFM of air per drop	35. 0.36	35.
— Number of drops	36. 46	36.
— Cost per 1000 SCFM of air	37. 0.32 cents	37.
— Total cost of air		
Line 35 × Line 36 × Line 37 × 60 × 24 × 365 ÷ 1000	38. $28	38. $
• Utilities -electricity:		
— "EXP" console require 1.4 kw		
— Total cost of electricity		
Line 31 × 1.4 × 24 × 365	39. $613	39. $

(Continued)

Table 10-1: Oil mist justification example and worksheet (3 of 3)

	EXAMPLE	ACTUAL UNIT
• Routine maintenance at recommended six month intervals:		
— Parts: oil and air filters, reservoir gasket and bulbs		
Yearly cost	40. $50	40. $
— Labor: 4 man hours per inspection		
4 × ($18 + 30%) × 2	41. $187	41. $

NET ANNUAL CREDITS

	EXAMPLE	ACTUAL UNIT
• Total all red boxes	42. $37,212	42. $
• Total all blue boxes	43. $878	43. $
• Net annual credits		
Line 42 minus Line 43	44. $36,334	44. $

LUBRIMIST SYSTEM TURN KEY INSTALLATION COST

	EXAMPLE	ACTUAL UNIT
• Average cost per slab, up to six points per slab	45. $1,000	45.
• Number of separate slabs		
— pump/motor pairs	46. 40	46.
— separate motors	47. 5	47.
— turbines/other	48. 1	48.
• Installation cost		
Line 45× (Line 46 × 47 × 48)	49. $46,000	49. $

PROFITABILITY MEASURES

	EXAMPLE	ACTUAL UNIT
• Payback period:		
— Installation cost (Line 49) divided by net annual credits (Line 44)	50. 1.27 years	50. years

• Pre-tax rate of return:
— Based on ten-year life
— Tax effects not included
— Use following table to estimate:

Payback period-years	0.50	0.54	0.62	0.77	0.83	1.00	1.10	1.23	1.40	1.67	1.96	2.41	3.09	3.57	4.19
Rate of return %	200	180	160	140	120	100	90	80	70	60	50	40	30	25	20

Footnotes:

1.) Towne, C.A.,"Practicle Experience with Oil Mist Lubrication", *ASLE Preprint*, No. 82-AM-4C-1, pg. 7 (May 1982).
2.) Bloch, H.P., *Oil Mist Lubrication Handbook*, Gulf Publishing Company, Houston, Texas, 1987 pg. 160.
3.) *Cost data from a refinery that chooses to remain unnamed.*
4.) Bloch, pg. 161.
5.) Bloch, H.P., "Large Scale Application of Pure Oil Mist Lubrication in Petrochemical Plants", *ASME Publication*, 80-C2/Lub 25, pg. 6 (August 1980).
6.) Bloch, H.P., *Oil Mist Lubrication Handbook*, pg. 160.
7.) Morrison, F.R., Zielinski, J., James, R., "Effects of Synthetic Industrial Fluids on Ball Bearing Performance". *ASME Publication*, 80-Pet 3, pgs. 9, 10 (November 1980).

Quantifying the Energy Savings Potential

Using both oil mist and synthetic lubes makes economic sense. According to tests conducted by a bearing manufacturer, the frictional losses in rolling element bearings can be reduced as much as 37% (Ref. 10-13). The results of these tests are summarized in Table 10-3 and Figure 10-18.

On the 65 mm bearings typically used in 15 hp (~11.3 kW) process pumps and electric motors, 0.11 kW could be saved (Figure 10-17). While the small absolute value of 0.11 kW per bearing tends to make the savings appear insignificant, petrochemical process pump rotors are typically supported by a double row radial ball bearing and two angular contact ball thrust bearings. These then represent a total of 4.8 test-equivalent bearings [(4 x .7) + (2 x 1) = 2.8 + 2 = 4.8]. Therefore, the total savings available from an actual motor-driven pump set are 4.8 times the single test bearing energy savings of 0.11 kW; that equals 0.53 kW or 4.7% of 11.3 kW.

Assuming that the average pump operates 90 percent of the time and rounding off the numbers this difference amounts to energy savings of 4,180 kWh per year. At $0.10 per kWh, yearly savings of $418 could be expected. By using synthetics on conventionally lubricated equipment, oil replacement schedules are typically extended four-fold; the extended drain intervals more than compensate for the higher cost of synthetic lubricants.

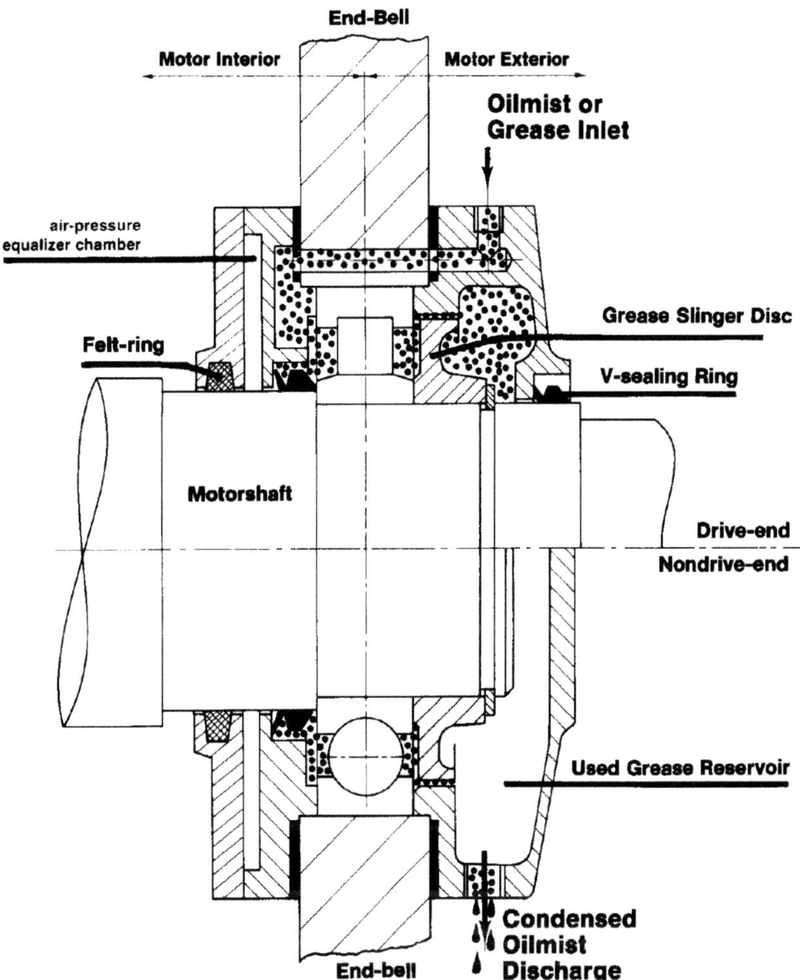

Figure 10-16: Dry sump oil mist path through large electric motor bearing (Source: Siemens A.G., Erlangen, Germany)

With oil mist the bearings run cooler and last longer than those typically lubricated by conventional oil sumps. While open oil mist systems typically consume 12-22 liters (3.1-5.7 gallons) per pump set per year, closed oil mist systems consume no more than 10% of these yearly amounts. Again, at these extremely low make-up or consumption rates and compared to the cost of mineral oils, the incremental cost of synthetic lubricants is relatively insignificant.

Considering annual energy saving per 15 hp pump and driver set to be worth $418, we realize that these savings should be multiplied by the number of pumps actually operating in large refineries—850 to 1,200. Again using $0.10 per kWh, annual savings in the vicinity of $ 450,000 would not be unusual. A detailed calculation can be found in Ref.10-13; it will prove the point:

- Total pump hp installed at the plant = 15 hp x 1,000 = 15,000 hp
- Total pump kW installed at the plant = 15,000 x .746 = 11,190 kW
- Total consumption kWh per year, considering 90% of 8,760 h/yr = 8,760 h x .90 = 7,884 h/yr; then 7,884 x 11,190 kW = 88,220,000 kWh/yr
- Total U.S. Dollar value of yearly energy consumed, assuming $0.10/kWh: = $ 8,822,000

Total energy savings for 1,000 average-sized pump sets would thus equal 0.047 x $ 8,822,000 = $414,600. That is an amount that should not be overlooked.

Electric motor lubrication by pure oil mist ("dry sump" oil mist) is highly advantageous. It saves both

Table 10-2: Oil Mist Generator Comparison Table (Source: Lubrication Systems Company, Houston, Texas 77024)

Model	Controls and Monitoring	Mist Density Monitor	Oil Supply	Design and Materials of Construction	Electrical Supply	Auxiliary Back-up Unit	Third Party Approvals and Electrical Classification	Application Comments
Consoles for centralized oil mist systems:								
IVT	<u>Microprocessor</u> • Monitoring and control of a full range of system operating conditions including ambient temperature. • Provides extensive troubleshooting guidelines and a comprehensive data/event log. • Alphanumeric display panel, individual LED condition lights plus analog gauges.	Integral and standard.	• Bulk supply integral to the unit. • No separate tank or drum required.	• Fully enclosed cabinet with access doors on three sides. • All SS and no brass containing components.	115 or 230 Volt, Single Phase.	Integral and semi-automatic start-up.	• UL and DEMKO. • NEC - Class 1, Div 2, Groups B/C/D. • IEC - Zone2 Ex nC IIC T4 • Unit being evaluated to ATEX standards.	• For large-scale, fully automated systems in process units where maximum control/monitoring, interface with DCS and reliability are required. • Up to 1000 BI capacity. • Cold and hot weather packages available.
JD	<u>Microprocessor</u> • Monitors and/or controls mist pressure, oil/air temperature, oil level and regulated air pressure. • Provides troubleshooting guidelines and a data/event log. • Alphanumeric display panel plus analog gauges.	Integral monitor is optional.	• Separate bulk supply required.	• Pedestal mounted, SS, single-door cabinet. • All SS reservoir and no brass containing components.	115 or 230 Volt, Single Phase.	Optional manual system available.	• Base unit UL approved for NEC – Class 1, Div 2, Groups B/C/D. • Unit being evaluated to ATEX standards.	• For intermediate to large size systems where microprocessor control/monitoring and feedback to DCS is desired. • Up to 1000 BI capacity.
JR	<u>Electromechanical</u> • Switches that signal alarm condition for high/low mist pressure, high/low oil level and low oil temperature. • Analogue gauges.	Stand-alone monitor optional.	• Separate bulk supply required.	• Pedestal-mounted, single-door aluminum cabinet. • Painted carbon steel reservoir is standard.	115 or 230 Volt, Single Phase.	Optional manual system available.	• Self certified to Class1, Div 2, Groups B/C/D.	• For intermediate size systems where auto-fill is desired and minimum external DCS interface is required. • Up to 500 BI capacity.
VFP	<u>Electromechanical</u> • Switches that signal alarm condition for high/low mist pressure and low oil level. • Analogue gauges.	Stand-alone monitor optional.	• Must be manually filled.	• Panel mounted unit constructed of painted carbon steel.	115 of 230 Volt, Single Phase.	Optional manual system available.	• Self certified to Class 1, Div 2, Group D.	• For systems with Very Few lubrication Points and where manual fill is acceptable. • Up to 500 BI capacity.
Stand-alone oil mist unit:								
LubriMate®	• Mist pressure gauge and oil level viewing window in the reservoir. • No electronic or mechanical monitoring devices.	Stand-alone monitor optional.	• Closed-loop so routine refilling not required.	• Cast aluminum reservoir.	None required.	Not available.	• None required, as there are no electrical devices.	• Ideal for use with isolated equipment. • Not for multi-equipment system use.

<u>Notes:</u>

1. It is impossible to provide complete comparison between and among the units with one table. Please consult product brochures and operating manuals for more details on the design features of the units.
2. Each of the above system generators can be incorporated into an LSC designed closed-loop oil mist system that is based on technology covered by US patent 5,318,152 and other international patents.
3. The incorporation of a microprocessor into an oil mist system for the purpose of both control and monitoring of the mist process is covered by US patent 5,125,480 and other international patents.
4. Each of the above centralized system consoles have external operating status lights and a dry form C common alarm relay contact.

energy and labor costs. Sealing and mist drainage are well understood and have been thoroughly explained (Refs. 10-2 and 10-3).

Table 10-3: How changes in lubes and application methods affect power losses

Change	*Δ Power loss per bearing*	*Total reduction*
Sump: MIN 68 to SYN 32	0.017	6%
Mist: MIN 68 to SYN 32	0.022	8%
Sump MIN 68 to Mist MIN 68	0.080	29%
Sump SYN 32 to Mist SYN 32	0.085	31%
Sump MIN 68 to Mist SYN 32	0.11	38%

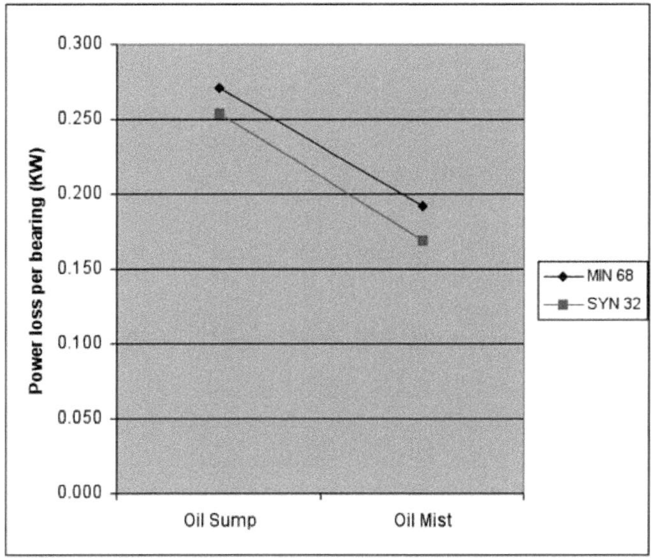

Figure 10-17: Power loss plot for the ball bearing tests reported in Ref. 10-13. The two different oils have different viscosities, but their protective properties are identical. Oil mist reduces power losses, as do carefully selected synthetic lubes (Ref. 10-14)

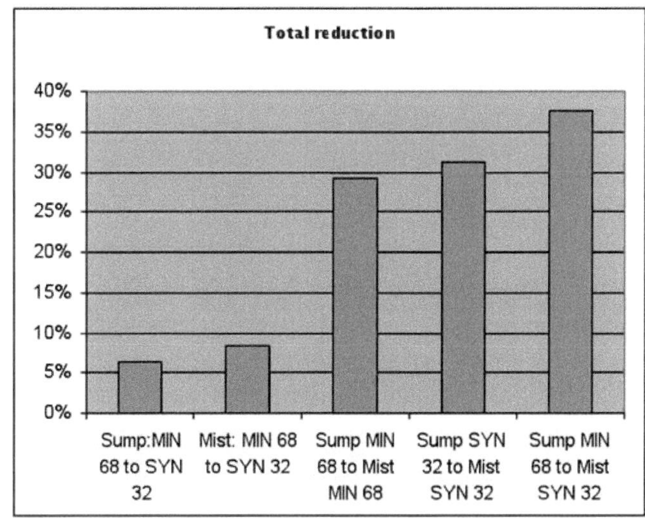

Figure 10-18: Percentage savings in frictional energy when changing lubricant type and method of lubricant application (Ref. 10-13)

Wide Application Range Documented

Although major grass-roots olefins plants commenced using oil mist on motors as small as one HP (0.75 kW) in 1975, the prevailing practice among reliability-focused users is to apply oil mist on horizontal motors sized 15 HP and larger, and vertical motors of 3 HP and larger, fitted with rolling element bearings.

As called for in the recommendations of API-610 Standard for pumps in the petrochemical and refining industries, the oil mist is routed through the bearings instead of past the bearings. This recommendation will work equally well for electric motor rolling element bearings. The resulting diagonal through-flow route (see Figure 9-11) guarantees adequate lubrication, whereas oil mist entering and exiting on the same side might allow some of the mist to leave without first wetting the rolling elements. Through-flow is thus one of the keys to a successful installation. As shown in Figure 10-16, Siemens A.G. is among the proponents of oil mist lubrication for rolling element bearings in electric motors and has published technical bulletins showing oil mist applied in the through-flow mode as a superior technique for equipment ranging in size from 18 to 3,000 kW.

The required volume of oil mist is often translated into bearing-inches, or "BI's." A bearing-inch is the volume of oil mist needed to satisfy the demands of a row of rolling elements in a one-inch (~25 mm) bore diameter bearing. One BI assumes a rate of mist containing 0.01 fl. oz., or 0.3 ml, of oil per hour. Certain other factors may have to be considered to determine the needed oil mist flow. These are known to oil mist providers and bearing manufacturers. The various factors are also well-documented in numerous references and are readily summarized as:

a. *Type of bearing*. The different internal geometries of different types of contact (point contact at ball bearings and linear contacts at roller bearings), amount of sliding contacts (between rolling elements and raceways, cages, flanges or guide rings), angle of contact between rolling elements and raceways, and prevailing load on rolling elements. The most common bearing types in electrical motors are deep groove ball bearings, cylindrical roller bearings and angular contact ball bearings.

b. *Number of rows of rolling elements*. Multiple row bearing or paired bearing arrangements require a simple multiplier to quantify the volume of mist flow.

c. *Size of the bearings*, related to the shaft diameter—inherent in the expression "bearing-inches."

d. *The rotating speed*. The influence of the rotating speed should not be considered as a linear function. It can be linear for a certain intermediate speed range, but at lower and higher speeds the oil requirements in the contact regions may behave differently.

e. *Bearing load conditions* (preload, minimum or even less than minimum load, heavy axial loads, etc.)

f. *Cage design*. Different cage designs might affect mist flow in different ways. It has been reasoned that stamped (pressed) metal cages, polyamide cages, or machined metal cages might produce different degrees of turbulence. Will different rates of turbulence cause different amounts of oil to "plate out" on the various bearing components? Is this a valid concern for electric motor bearings? Let's explore and explain.

Through-flow Represents "Best Technology"

It can be stated without reservation that through-flow oil mist, as explained earlier in Figure 9-11, addresses the above concerns and will accommodate all of the arrangements listed above. Major legacy electric motor manufacturers, such as Reliance Electric (Cleveland, Ohio), were aware of this fact and successors continue to supply excellent products. Representing "Best Technology," their bearing housings are arranged for through-flow.

Experience shows that no further investigation is needed for bearings in the operating speed and size ranges encountered by motors driving process pumps. As of 2013, more than 26,000 oil mist lubricated electric motors continue to operate flawlessly in reliability-focused facilities. Moreover, a survey of these plants confirmed that their procurement specifications for new installations and replacement motors require oil mist lubrication in sizes 15 HP and larger.

For the past 40 years, empirical data have been employed to screen the applicability of oil mist. The influences of bearing size, speed, and load have been recognized in an oil mist applicability formula, limiting the parameter "DNL" to values below 10E9, or 1,000,000,000. Here, D= bearing bore, mm; N= inner ring rpm; and L= load, lbs. An 80 mm electric motor bearing operating at 3,600 rpm and a load of 600 lbs, would thus have a DNL of 172,000,000—less than 18% of the allowable threshold value.

In the 1960's, it was customary to apply oil mist near the center of the bearing housing, letting the excess mist vent to the atmosphere after passing through the bearings, Figure 10-19. More recently and in accordance with the recommendations of API-610 8th and later editions of this standard governing centrifugal pumps in the petrochemical and refining industries, the oil mist is routed *through* the bearings. Here, the oil mist enters at a convenient location between the bearing housing protector (bearing isolator, or end seal) and bearing, as was shown in Figures 2-20, 9-11 and 10-20. The metering orifice (reclassifier) may or may not be located as shown in this illustration, although locating it close to the bearing is considered advantageous. This through-flow is recommended wherever feasible, regardless of whether the machine is an electric motor or a pump.

It is worth mentioning that with the growing use of bearing housing *face* seals reliability-focused users and oil mist providers have become quite conscious of the need to ensure the flow of oil mist *through* pump bearings. User-operators installing these new *face seals* on pumps, motors, and other equipment that have one mist inlet point in the top of the housing, Figure 10-19, need to realize that this mist application worked quite well when the housing seals "leaked" and allowed mist to flow through the bearing. However, after installation of an effective bearing housing seal, not all the mist will probably reach the outer race of the thrust bearing unless cross-flow (Figures 2-20, 9-11 and 10-20) is established. Without cross-flow, whereby the oil mist is introduced on one side and vented on the opposite side of the bearing, inadequate lubrication may exist and bearings may fail quickly. The indiscriminate installation of effective bearing housing seals without thinking through the surrounding issues can thus lead to problems. The leading service providers are trying to be proactive with customers on this, and this text is certainly aimed at imparting the beneficial knowledge.

It must again be emphasized that, while originally intended for centrifugal pumps, the recommendations of API-610 have worked equally well for electric motors with rolling element bearings. The resulting *diagonal*

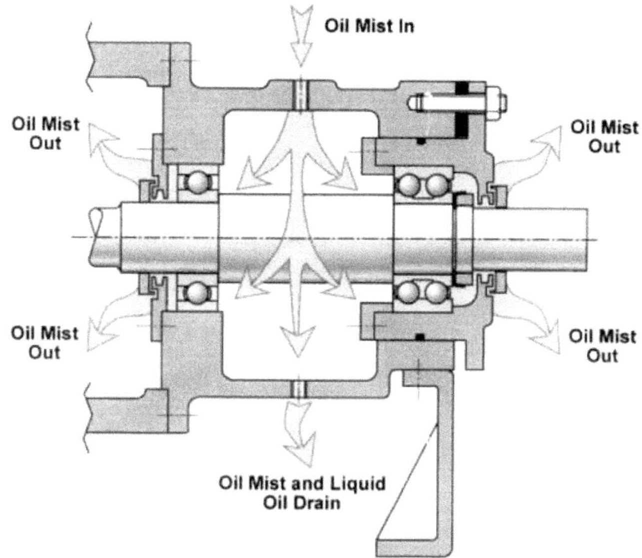

Figure 10-19: Old-style application of oil mist stresses environment

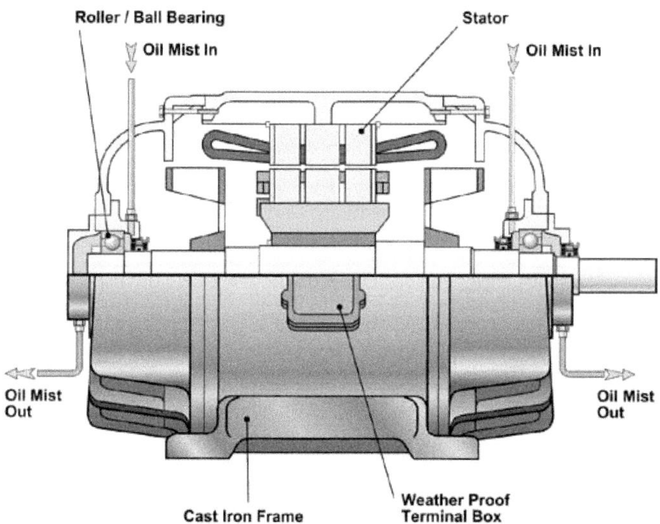

Figure 10-20: Through-flow application on electric motor in compliance with API-610

through-flow route shown in Figure 10-20 guarantees adequate lubrication, whereas oil mist entering and exiting per Figure 10-19 might

(a) cause some of the oil mist to leave at the drain port, without first wetting the rolling elements (Ref. 10-1)

(b) inadvertently be kept away from the rolling elements due to windage, or fan effects, generated by certain inclined bearing cage configurations

It is acknowledged that a few "business-as-usual" oil mist users continue to be satisfied with the routing of Figure 10-16. Nevertheless, it can be shown that highly-loaded bearings and bearings operating at high speeds *must* use the API-recommended routing of Figure 10-20 (or 2-20 and 9-11). A risk-averse user thus recognizes through-flow as one of the key ingredients of successful oil mist implementations.

It must be re-emphasized and again stated without reservation that *through-flow* oil mist addresses the above concerns and will accommodate all the lubrication requirements of electric motors furnished with rolling element bearings. Electric motor bearing housings from all competent vendors are arranged for *through-flow.*

Sealing and Drainage Issues

Although oil mist will not attack or degrade the winding insulation found on electric motors made since the mid-1960's, mist entry and related sealing issues must be understood and merit discussion.

Regardless of motor type, i.e. TEFC, X-Proof or WP II, cable terminations should never be made with conventional electrician's tape. The adhesive in this tape will last but a few days and become tacky to the point of unraveling. Instead of inferior products, competent motor manufacturers use a modified silicone system ("Radix") that is highly resistant to oil mist. Radix has consistently outperformed the many other "almost equivalent" systems.

Similarly, and while it must always be pointed out that oil mist is neither a flammable nor explosive mixture, it would be prudent not to allow a visible plume of mist to escape from the junction box cover. The wire passage from the motor interior to the junction box should, therefore, be sealed with 3M Scotch-Cast Two-Part Epoxy potting compound to exclude oil mist from the junction box.

Finally, it is always good practice to verify that *all* electric motors have a suitably dimensioned escape passage at the bottom of the motor casing. Intended to drain accumulated moisture condensation, the weep hole (or a plug purposely provided with a transversely drilled weep hole) will allow liquefied or atomized oil mist to escape. However, competent motor manufacturers have fitted either an appropriate explosion-proof rated vent or a suitably routed weep hole passage at the bottom of the motor casing or lower edge of the end cover. Intended to drain accumulated moisture condensation, the vent or weep hole passage will allow liquefied or atomized oil mist to escape. Although code issues should be addressed, the suitability of oil mist for Class 1, Group C and D locations was specifically re-affirmed by a highly competent U.S. manufacturer of electric motors in July of 2004.

Conventional Wisdom: *Oil mist lube for electric motors is messy, could cause winding failures, and is not economically justified.*

Fact: *Using well-proven application techniques and closed systems, oil mist lubrication has proven to be both clean and capable of significantly extending motor bearing life. Factoring in demonstrable energy savings and maintenance cost avoidance, it is cost-justified in many applications.*

TEFC vs. WP II Construction

On TEFC (totally enclosed, fan-cooled) motors, there are documented events of liquid oil filling the motor housing to the point of contact with the spinning rotor. Conventional wisdom to the contrary, there were no detrimental effects, and the motor could have run indefinitely! TEFC motors are suitable for oil mist lubrication by simply routing the oil mist *through* the bearing, as has been explained in Figure 9-11, Refs. 10-2, 10-3 and numerous other references, including API-610. No special internal sealing provisions are needed.

On weather-protected (WP II) motors, merely adding oil mist has often been done and has generally worked surprisingly well. In this instance, however, it was found important to lead the oil mist vent tubing away from regions influenced by the motor fan. Still, weather-protected (WP II) electric motors do receive additional attention from reliability-focused users and knowledgeable motor manufacturers.

Air is constantly being forced through the windings and an oil film deposited on the windings could invite dirt accumulation to become objectionable. To reduce the risk of dirt accumulation, suitable means of sealing should be provided between the motor bearings and the motor interior. Since V-rings and other elastomeric contact seals are subject to wear, low-friction face seals are considered technically superior. The axial closing force on these seals could be provided either by springs or small permanent magnets.

As is so often the case, the user has to make choices. Low friction axial seals—face seals, as discussed in Chapter 6 of this text—are offered by Isomag and the traditional mechanical seal manufacturers AESSEAL, (Figure 6-11), Burgmann, Chesterton, John Crane, Flowserve and others. More specifically and as of early 2005, a second-generation dual-face magnetic seal has been designed and marketed by AESSEAL, plc, for oil mist-lubricated equipment, including electric motors, Figure 10-21. Rotating labyrinth seals with radial lift-off dynamic O-rings and—very recently developed—repairable rotating labyrinth seal designs with two *axially contacting* O-rings (Figure 9-36) are available as well.

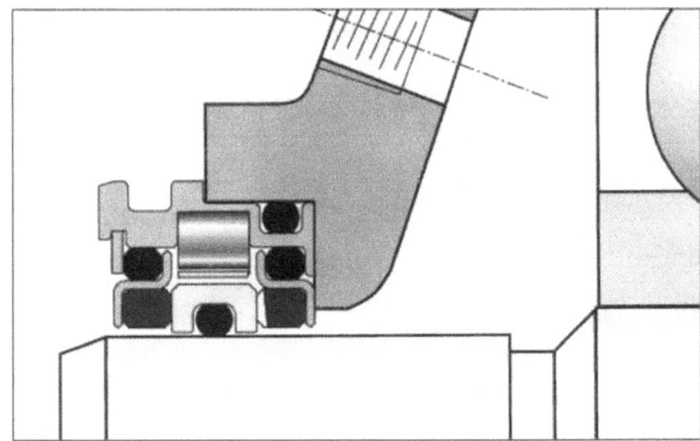

Figure 10-21: Dual-face magnetic bearing housing seal with component features that allow oil mist to coalesce (Source: AESSEAL, plc, Rotherham, U.K., and Rockford, Tennessee)

Some of these may require machining of the cap, but long motor life and the avoidance of maintenance costs will make up for the added expense. Nevertheless, double V-rings using Nitrile or Viton elastomeric material should not be ruled out since they are considerably less expensive than face seals. Note that an elastomeric ring and a felt seal are shown in Figure 10-16, although we confirm that, based on three decades of field experience, these provisions are really not needed on TEFC motors.

Sealing to Avoid Stray Mist Stressing the Environment

Even when still accepted by prevailing environmental regulations (e.g. OSHA or EPA), the regulatory and "good neighbor climate" will sooner or later force industry to curtail stray oil mist emissions. Of equal importance and to set the record straight, it must be noted that state-of-art oil mist systems are fully closed, i.e. are configured so as *not* to allow any mist to escape. In the late 1980's, one of the authors collaborated with a California-based engineering contractor in the implementation of two plant-wide systems in Kentucky. As of 2005, these systems continued to operate flawlessly and had even been expanded. The owner company has added another fully closed system at its refinery in Minnesota and is proceeding to convert existing, open systems, to closed systems.

It should be noted that combining effective seals and a closed oil mist lubrication system is a proven solution. It not only eliminates virtually all stray mist and oil leakage, but makes possible the recovery, subsequent purification, and re-use of perhaps 97% of the oil. These recovery rates make the use of more expensive, superior quality synthetic lubricants economically attractive.

Needless to say, closed systems and oil mist-lubricated electric motors give reliability-focused users several important advantages:

- Compliance with actual and future environmental regulations
- Convincing proof that oil mist lubrication systems exist that will not contaminate the environment
- The technical and economic justification to apply high-performance synthetic oils.

Optimized Energy Efficiency

To capture energy efficiency credits, lubricants with suitably low viscosity must be used in combination with the correct volume of mist. Moreover, and as mentioned above, low-friction seals are desired on WP II motors.

PAO and diester (or dibasic ester) lubricants embody most of the properties needed for extended bearing life and greatest operating efficiency. These oils excel in the areas of bearing temperature and friction energy reduction. It is not difficult to show relatively rapid returns on investment for these lubricants, providing, of course, the system is closed, and the lubricant re-used after filtration.

Again, very significant increases in bearing life and overall electric motor reliability have been repeatedly documented over the past four decades.

Needed: The Right Bearing and a Correct Installation Procedure

Regardless of equipment type or size, oil mist cannot eliminate basic bearing problems, it can only provide one of the best and most reliable means of lubricant application. Bearings must be:

- Adequate for the application, i.e. deep groove ball bearings for coupled drives, cylindrical roller bearing to support high radial loads in certain belt drives, or angular contact ball bearings to support the axial (constant) loads in vertical motor applications
- Incorporating correct bearing-internal clearances
- Mounted with correct shaft and housing fits
- Carefully and correctly handled, using tools that will avoid damage
- Correctly assembled and fitted to the motor caps, carefully avoiding misalignment or skewing
- Part of a correctly installed motor, avoiding shaft misalignment and soft foot, or bearing damage incurred while mounting either the coupling or drive pulley
- Subjected to a vibration spectrum analysis. This will indicate the lubrication condition as regards lubricating film, bearing condition (possible bearing damage) and general equipment condition, including misalignment, lack of support (soft foot), unbalance, etc.

Additional Considerations for Converting Electric Motors Already in Use

When converting operating motors from grease lubrication to oil mist lubrication, consider the following measures in addition to the above:

1. Perform a complete vibration analysis. This will confirm pre-existing bearing distress and will indicate if such work as re-alignment, base plate stiffening, is needed to avert incipient bearing failure.

2. Measure the actual efficiency of the motor. If the motor is inefficient, consider replacing it with a modern high efficiency motor, using oil mist lubrication in line with the above recommendations. This will allow capture of all benefits and will result in greatly enhanced return on investment.

3. Last, but not least, evaluate if the capacity of the motor is the most suitable for the application. "Most suitable" typically implies driven loads that represent 75% to 95 % of nominal motor capacity. The result: Operation at best efficiency. Note that converting an overloaded, hot-running electric motor to oil mist lubrication will lead to marginal improvement at best.

ELECTRIC MOTOR CODES AND PRACTICES

Although explosion-proof motors have been successfully and safely lubricated with pure oil mist since 1970, questions are occasionally raised whether explosion-proof (XP) electric motors are suitable for this demonstrated reliable and cost-effective mode of lubrication.

The selection, operation, and even maintenance of industrial equipment in the developed countries are often influenced by industrial standards, regulatory agencies and certain applicable codes. Major companies superimpose their own design standards, specifications and best practices. It can be shown beyond any doubt that many of these "Company Best Practices" reflect advanced thinking that is often years ahead of current regulatory edicts. But, for a variety of reasons, even time-tested and thoroughly well thought-out practices come under occasional scrutiny. In the case of oil mist applied to explosion-proof electric motors and because oil mist is not an explosive mixture, such scrutiny has never been prompted by any safety incidents.

It thus appears that both scrutiny and acceptability of dry sump oil mist on explosion-proof motors relate to third-party* approval and the original equipment manufacturer's certification of the motor. As mentioned earlier, users have for three or more decades provided *all* except their explosion-proof electric motors with a small (3-5 mm diameter) weep hole and have only given XP-motor drains closer attention. Some of the more recently produced XP-motors are furnished with either an explosion-proof rated vent or a suitably routed weep hole passage at the bottom of the motor casing or lower edge of the end cover. Intended to drain accumulated moisture condensation, the vent or weep hole passage will allow liquefied or atomized oil mist to escape. Note, however, that explosion-proof motors are still "explosion-proof" with this weep hole passage. At least one motor manufacturer tack-welds an explosion-proof "XP-breather drain" to the motor brackets.

So, what's the problem? It seems that, not being familiar with dry sump oil mist lubrication, some motor manufacturers and third-party* validation providers have taken the position that explosion proof-motors lose this "XP-Listing" once *any modifications whatsoever* are made to the motor. That's regrettable in instances where the modification actually makes the motor less likely to fail.

Dealing with Explosion-proof Motors

As the name implies, explosion- proof motors are intended for use in hazardous areas. The majority of hazardous areas in hydrocarbon processing facilities are designated as Class 1, Division 2, Groups B, C and D.

The Class 1 area designation indicates that either a flammable liquid or vapor or both are present. (Class 2 designations are reserved for areas where combustible metal, carbon fines or other combustible dusts such as grain flour or plastic are present).

The "Division" label is used to better describe the probability of flammable gases or vapors being present in a Class 1 or Class 2 location.

- Division 1 is intended for locations where ignitable concentrations of flammable gases or vapors can either exist under normal operating conditions, or might be present while the equipment is undergoing repair or maintenance
- Division 2 defines the area or location where the flammable liquids or vapors are possibly present and/or

  (1) normally confined within closed containers or closed systems and are present only in case of accidental rupture or breakdown of such containers, or in case of abnormal operation of equipment, or
  (2) where ignitable concentrations are normally prevented by positive ventilation, or
  (3) an area *adjacent* to a Class 1, Division 1, location.

The "Group" designation has four subgroups, or gas groups—appropriately called Groups A, B, C and D. Determining the proper group classification for flammable gases and vapors requires monitoring and describing explosion pressures and maximum safe clearances between parts of a clamped joint under certain prescribed conditions whereby a test gas is mixed with air and ignited. The test values obtained for a reference gas are compared with the gas or gases of interest; these must now be tested under the same conditions. Gases having similar explosion pressures are grouped together. However, Groups C and D contain the majority of flammable gases and vapors.

Group A only contains acetylene, while Group B generally contains hydrogen and other hydrogen-rich gas mixtures, plus a few other flammable gases.

An important concession is made by the National Electrical Code (NEC) for equipment used in Division 2 areas, where flammable gases are normally not present, i.e. a refinery or petrochemical plant under normal conditions. If they meet stipulated criteria, the NEC allows the use of certain types of devices and materials that may not be listed by third-party, or "listing" agencies. For instance, these exceptions to the NEC's general code requirements permit general-purpose enclosures if the electrical current interrupting contacts are:

---

*The authors and The Fairmont Press are expressly disclaiming knowledge or criticism of the particular listing agencies that either allow or disallow oil mist for explosion-proof electric motors.

Much of the material for this segment of our text was contributed by, and/or reflects the world-wide application experience of major oil mist users and Lubrication Systems Company, Houston, Texas.

1. Immersed in oil, or
2. Enclosed within a chamber that is hermetically sealed against the entrance of gases or vapors; or
3. Situated in non-incendive circuits; or
4. Part of a "listed" non-incendive component; or
5. Are without make-and-break or sliding contacts

Except for the above exclusions, the National Electrical Code/NFPA 70 ("NEC") requires that all electrical apparatus installed in classified (hazardous) areas must be approved for use in the specified Class and Group where it is to be used. Once an electrical apparatus is described as "explosion-proof," it is implied that the device has been evaluated and approved for use in a particular Class and Group. The evaluation or approval agency was earlier called a "third party." In the United States, third parties include Underwriters Laboratories (UL), Factory Mutual (FM) and others*. Once an apparatus or device has been evaluated and approved for a particular Class or Group, it is labeled "listed" by the agency.

In most Class 1 Division 2 hazardous areas, the electric motors are not, and do not need to be "explosion-proof." The overwhelming majority are non-arcing induction motors that meet the requirements of the applicable and allowed exceptions. These non-explosion proof motors can be adapted for dry sump oil mist lubrication by simply connecting oil mist supplies and vents to the existing connections on the motor. Because these motors are non-arcing and an explosion-proof housing is not needed for Division 2 service, the case drain fitting can be removed and a drain can be installed without in any way affecting the suitability of the motor for Division 2 service.

ON THE SAFETY OF EXPLOSION-PROOF MOTORS FOR CLASS 1 DIVISION 1 SERVICE

Regrettably, some listing agencies* seem to believe that oil mist applied to the bearings makes the motor different from what was originally approved. Not understanding oil mist, they take the position that by in any way adapting plugs and drain fittings to oil mist application, mist venting and mist draining, the safe clearance requirements between clamped components used in the original design requirement may have been changed. Therefore, they consider the approval listing void and claim the motor is no longer suitable for use in Class 1 Division 1 service. In view of this stance taken by third parties, even a major provider of oil mist systems in the United States does not allow its employees to make on-site modifications to convert or connect an explosion-proof motor to oil mist.

However, a number of clarifications are in order here. First and foremost is the fact that explosion proof motors were successfully converted to oil mist lubrication by undisputed best-of-class petrochemical companies in the late 1960's and have since given safe and reliable service. These forward-looking companies, for whom safety is of utmost importance, correctly reasoned that all electric motors, regardless of classification, were assembled and are being operated in an ambient environment. They are thus always filled with ambient air; certainly none of these explosion-proof motors are provided with mechanical seals that would positively prevent an interchange with, or communication between, motor-internal air and the surrounding ambient air. Should an explosive gas mixture prevail in the vicinity of such motors, there would now exist the possibility of the motor ingesting this explosive gas mixture. If, on the other hand, this motor were filled with the demonstrably non-explosive oil mist at slightly higher-than-atmospheric pressure, the probability of the motor becoming filled with an explosive gas mixture would be greatly reduced. In other words, knowledgeable user companies have long recognized that *any* oil mist lubricated motor operating in a Class 1 Division 1 environment is safer than a conventionally lubricated electric motor operating in the same environment.

It may also be argued that item 2, and possibly one or two other items cited as exclusionary by NEC, allow the user to reason that oil mist existing at a pressure higher than atmospheric complies fully with the spirit of the listed exclusions.

Feedback from Motor Manufacturers

A major U.S. manufacturer provides oil mist lubrication for explosion-proof motors in Class 1, Groups C and D. In this vendor's design, the bearing is located outside of the Class 1 shaft opening of the motor and also outside of the Class 1 joint of the inner bearing cap-end shield. The design provides lip seals outside the bearing and, optionally, between the inner bearing cap and bearing. This manufacturer has obtained third-party listing for oil mist lubricated motors in Class 1 Groups C and D and is asking that one of its sales offices be contacted regarding specific applications. The tack-welding of its XP breather drain to an existing Class 1 motor to accept oil mist lubrication would either be done as part of the initial order for the motor, or as a modification by one of this manufacturer's motor service shops.

A number of motor manufacturers have taken the time and effort to understand oil mist lubrication and have worked with knowledgeable users. Then again, there are perhaps other manufacturers that have not taken the time to study the matter and, regrettably, will show a lack of responsiveness. Some simply do not wish any involvement in the often difficult and expensive third-party approval process.

Summary and Guidance for TEFC Electric Motors

1. On operating or field conversions apply oil mist per Figure 10-16 or 10-20. The "inner" bearing housing seals are not mandatory because the motor case is furnished with a weep hole to drain off accumulated moisture. However, these inner seals could be retrofitted when the motor is in the shop for any reason.
2. When specifying new motors intended for oil mist lubrication, exercise whatever internal sealing option is deemed desirable
3. On in-shop conversions, install internal seals to reduce dust accumulation

Summary and Guidance for Explosion-proof Electric Motors

1. Specify oil mist lubrication on data sheets when purchasing new explosion-proof motors
2. Before applying oil mist to an existing explosion-proof motor, remove the motor from service and send it to an authorized motor repair facility to make the necessary XP-breather drain modifications. The repair facility should be recognized by the original motor manufacturer and, if re-listing is desired, be able to certify the modified motor with the required third party for the applicable hazardous listing.

LONG-TERM STORAGE OF PUMPS AND MOTORS USING OIL MIST

Operating machinery must be protected from the elements. Painting, plating, sheltering, use of corrosion-resistant materials of construction and many other means are available to achieve the desired protection.

A similar set of protection requirements applies to both not yet commissioned or temporarily deactivated equipment. The analysis discussed herein refers to that "inactive" machinery category. The means or procedures chosen for the preservation or corrosion inhibiting of fully assembled, but inactive pumps will logically depend on the type of equipment, expected length of inactivity, geographic and environmental factors, and the amount of time allocated to restore the equipment to service.

The basic and primary requirement of storage preservation is exclusion of water from those metal parts that would form corrosion products ("rust") that could find their way into bearings and seals. A secondary requirement might be the exclusion of sand or similar abrasives from close-tolerance bearing or sealing surfaces. All or any of the storage preservation strategies must aim at satisfying these requirements.

Machinery preservation during pre-erection storage or long-term deactivation (mothballing) will have an effect on machinery infant mortality at the startup of a plant or process unit. Many times, machinery arrives at the plant site long before it is ready to be installed at its permanent location. Unless the equipment is properly preserved, scheduled commissioning dates may be jeopardized, or the risk of premature failure is increased.

Long-term storage preservation by nitrogen purging is well known in the industry. Generally, this method of excluding moisture is used for small components, such as hydraulic governors or large components, such as turbomachinery rotors kept in metal containers. Nitrogen consumption is governed by the rate of outward leakage of this inert gas and may be kept at a low, highly economical rate if the container is tightly sealed. Alternatively, the container could be furnished with an orificed vent to promote through-flow of nitrogen at very low pressure. This is called nitrogen sweep (Ref. 10-11).

However, the preservation of field-installed inactive pumps and their drivers should be our primary concern. Here, purging can be applied as either the more hazardous (suffocation risk!) and expensive nitrogen blanketing or, the subject of this segment of our text, as a moderate-cost oil mist environment.

General Setup

An oil mist generator, like the one normally used to lubricate rotating equipment, will be used to generate a preserving mist. Since none of the equipment is rotating, a basic unit without all the alarms and backups will often suffice. It is recommended that air and oil heaters be used to ensure mixing effectiveness and maintaining the correct air/oil ratio. These heaters are mandatory if ambient temperatures during the period of storage drop below 50°F(10°C). A typical turbine oil (ISO Grade 32) can be used in the mist generator lube reservoir to provide oil mist at an approximate header pressure of 20″ of water column (~5 kPa).

A temporary outdoor storage yard is shown in Figure 10-22. The eight plastic tubing connectors in Figure 10-23 connect the oil mist manifold to the pump and motor locations to be protected by oil mist. A temporary storage yard is also depicted in Figure 10-24, where all of the equipment is left in the shipping crates and connected to the oil mist system using a preservation packaging kit.

Storage Site Preparation

A few common-sense considerations will assist in defining long-term storage measures:

1. Choose a site that has good drainage and is located out of the main stream of traffic. This will reduce possible mechanical damage from trucks, fork lifts, cherry pickers and cars. A covered fenced storage area is preferred for the convenience of personnel, but not needed by the stored pumps and motors.
2. Position stored equipment on cribbing (pallets) if storage site is not paved or concreted. Arrange equipment in an orderly fashion with access for lifting equipment.
3. Install temporary overhead supports for the oil mist headers as per Figure 10-22. Piping for the oil mist headers should be Schedule 40 screwed galvanized steel with minimum size of 1-1/2.″ All piping should be blown clean with steam prior to assembly to remove dirt and metal chips. All screwed joints are to be coated with Teflon sealant (no Teflon tape) prior to assembly to prevent oil mist leakage.
4. Install laterals (3/4″ min.) from top of mist header at each piece of equipment. Attach a distribution manifold to each lateral. Each distributor manifold typically has eight connection points in which to attach the 1/4″ tubing. This should be sufficient to mist most driver and pump combinations.

Pump and Driver Preparations

A typical connection sequence would include:

1. Connect 1/4″ plastic or copper tubing from distributor manifold (with reclassifier fitting attached) to pump bearing housings. (See Figure 10-23).
2. Connect 1/4″ plastic or copper tubing from distributor manifold (incorporating reclassifier fittings) to pump and turbine suction flanges. If wooden, plastic or metal flanges are used, drill 1/4″ hole through the suction flange protector to permit the

Figure 10-22: Oil mist protected outdoor equipment storage in a hot and humid U.S. Gulf Coast environment.

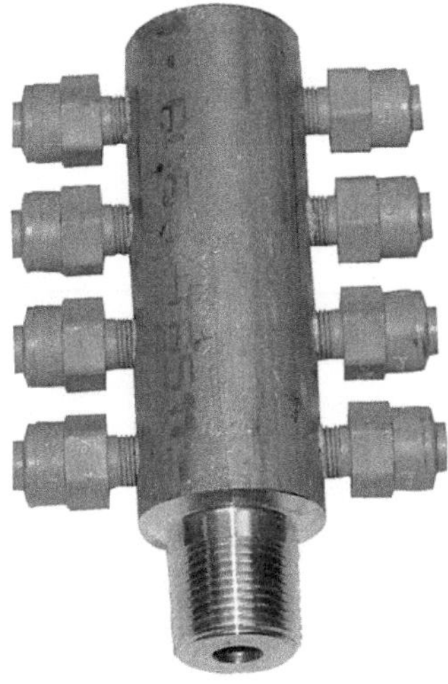

Figure 10-23: Preservation distribution manifold with built-in orifices and quick-connect plastic tube connectors (Source: Lubrication Systems Company, Housing, Texas 77024)

insertion of an adaptor fitting. Once adaptor fitting and tubing are inserted in the suction flange, seal the hole with duct tape. This prevents moisture and dirt from entering the mechanical seal and wetted area of the pump by maintaining a positive pressure of oil mist. No vent holes are required because of normal leakage around flange protectors.

3. Electric motors modified for oil mist lubrication should be stored with oil mist flowing through the smallest size built-in orifices to each bearing cavity. Electric motor and steam turbine hookups are shown in Figures 10-25 and 10-26. Oil mist is venting from the steam turbine governor in the foreground of Figure 10-26.

4. Coat all exposed machine surfaces with an asphalt base preservative similar to ExxonMobil's "Rust-Ban 373." Re-coat exposed machine surfaces every 6 months if needed. Preservative may be spray or brush applied.

   Rotate pump and driver shafts 1/4 revolution each month to prevent "brinnelling" (bearing ball indentation damage) of anti-friction bearings and perhaps even slight bowing (sagging) of heavily loaded pump shafts.

Oil Mist Generator Maintenance

1. Check weekly that air supply is dry
2. Refill mist generator oil reservoir weekly.
3. Perform weekly checks of air and oil heaters on mist generator
4. Check oil mist header pressure daily. (Verify ~ 20" or 5kPa of H_2O).

Equipment preserved by oil mist blanketing can be stored for years with minimum maintenance and cost. Figures 10-27 and 10-28, dating from the mid-1970s, shows the thoughtfulness and professionalism that have brought us to modern oil mist preservation. It is assumed that the internal surfaces of equipment stored in crates will have been coated with a light film of a preservative oil and that the various equipment-internal volumes are kept at slightly more than atmospheric (ambient) pressure. Oil mist through-flow is being achieved by providing a venting of the equipment casings blanketed with this oil mist environment.

Whenever possible, the equipment purchase documents should state that oil mist will be used as a long-term storage means. This might allow pump vendors to select or pre-define the most convenient oil mist inlet and vent locations.

Figure 10-24: Crated oil mist-protected outdoor equipment storage in a hot and humid Gulf Coast environment. Oil mist header, distribution manifolds and oil mist generator are shown to the right of picture. (Source: Lubrication Systems Company, Houston, Texas 77024)

Figure 10-25: Cutaway of crate showing preservation packaging kit for electric motor-pump package for storage protection outdoors (Source: Lubrication Systems Company, Houston, Texas 77024)

Figure 10-26 Oil mist tubing connections for outdoor storage of large lube oil skid and small steam turbine (Source: Lubrication Systems Company, Houston, Texas 77024)

Figure 10-27: Temporary storage highlighting oil mist tubing connections from distribution manifold to reclassifier fittings (1970s)

Figure 10-28: Oil mist tubing connections for outdoor storage of large electric motor and small steam turbine.

Purposefully Designed Lubricant Transfer Containers Save Money

A final point: Many bearings fail because unclean containers contaminate the lube oil as it is being transferred from storage drums to pump bearing housings. Reliability-focused pump users will only use properly designed plastic containers, Figures 10-29 through 10-32 for their lube replenishing and oil transfer tasks. Each of these containers will cost only a fraction of the cost of a single bearing failure. This is one product for which the payback is virtually measured in days.

That said, reliability-focused equipment maintenance and operating technicians will no longer permit rusty cans, zinc-plated (galvanized) buckets and discarded or unclean plastic bottles to be used in any area where lube oil contamination risks costly failures. It should be noted that cost justifications are easy to obtain.

Nozzles twist close to reduce contamination
Spout valve regulates oil flow and reduces oil contamination due to rain, sand, and dust entering through open spouts
Lids available in eight colors – yellow, red, blue, green, purple, orange, black, and dark green
HYDRAULIC OIL
GEAR OIL
ENGINE OIL
CONTENTS
Color coded labels provide easy identification of lubricant type
All lids and drums are interchangeable
Large carry handle, which doubles up as a hook so drums can be hung up out of the way
High heat and chemical resistance with ultra-violet and anti-static additives
Convoluted drum profile and large inlet hole allows for easy grip when filling
See through graduated drums make liquid levels easy to see and measure
TRICO
Oil Safe

Figure 10-29: Features of a high-grade lube oil transfer container (Source: Oil Safe® Systems Pty. Ltd., also Trico Mfg. Company and PdM USA)

Figure 10-30: Variations of high-grade lube oil transfer containers suit every reasonable requirement (Source: Oil Safe® Systems Pty. Ltd., also Trico Mfg. Company and PdM USA)

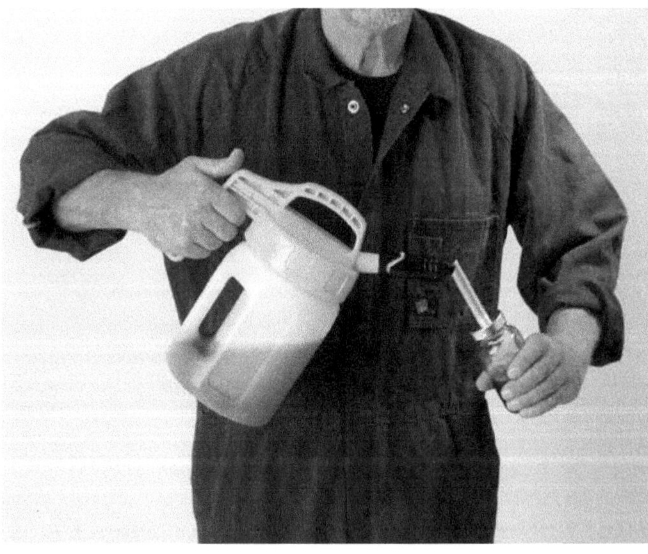

Figure 10-31: Using a high-grade lube oil transfer container to refill a pump constant level lubricator bottle (Source: Oil Safe® Systems Pty. Ltd.)

Special attention might be directed at Ref. 10-12, where the Linden, New Jersey plant of the General Motors Corporation employed the most rigorous accounting steps. General Motors reached the conclusion that investing in lube program upgrades by using properly designed, color-coded plastic transfer containers of suitable size and configuration made economic sense. In fact, this General Motors facility had achieved a two-month payback and 738% return on the money and effort investment.

In a similar case history dating back to the 1990s, a paper mill estimated that an expenditure of $6,000 for 100 transfer containers had resulted in $240,000 worth of downtime avoidance within three years of program initiation. Their three-year payback had reached 40:1.

Figure 10-32: Contamination-prone lube oil transfer containers (left) being replaced by high-grade special containers on right (Source: Oil Safe® Systems Pty. Ltd., also Trico Mfg. Company and PdM USA)

Chapter 11

Coupling Selection Guidelines

COUPLING FUNCTION

A coupling is used wherever there is a need to connect a prime mover to a piece of driven machinery, such as a pump. The primary function of a coupling is to transmit rotary motion and torque from one piece of equipment to another. Couplings may also perform other (secondary) functions, such as accommodating misalignment between shafts, transmitting (or *not* transmitting) axial thrust loads from one machine to another, permitting axial adjustment of shafts to compensate for machinery wear, and maintaining precise alignment between connected shafts.

Couplings can be divided into two primary groups, rigid and flexible. Rigid couplings are used to connect machine shafts (and shaft sections), where it is desired to maintain shafts in precise alignment, transmit torque and (frequently) also transmitting axial thrust. Rigid sleeve-type couplings (Figure 11-1, circled area) are mainly used on vertical turbine pumps. They, and the rigid flange-type couplings of Figure 11-2, used when a separate pump thrust bearing is not provided, cannot accommodate misalignment between shafts; therefore, precise alignment of the driving (normally an electric motor) and driven (pump) shafts is essential. This alignment should be checked prior to startup.

Flexible couplings accomplish the primary function of any coupling, transmitting the driving torque between the driver and driven machine. In addition, they perform a secondary function of accommodating a set amount of misalignment between the driving and driven shafts. Most horizontal, base plate mounted (long-coupled), pumps use flexible couplings.

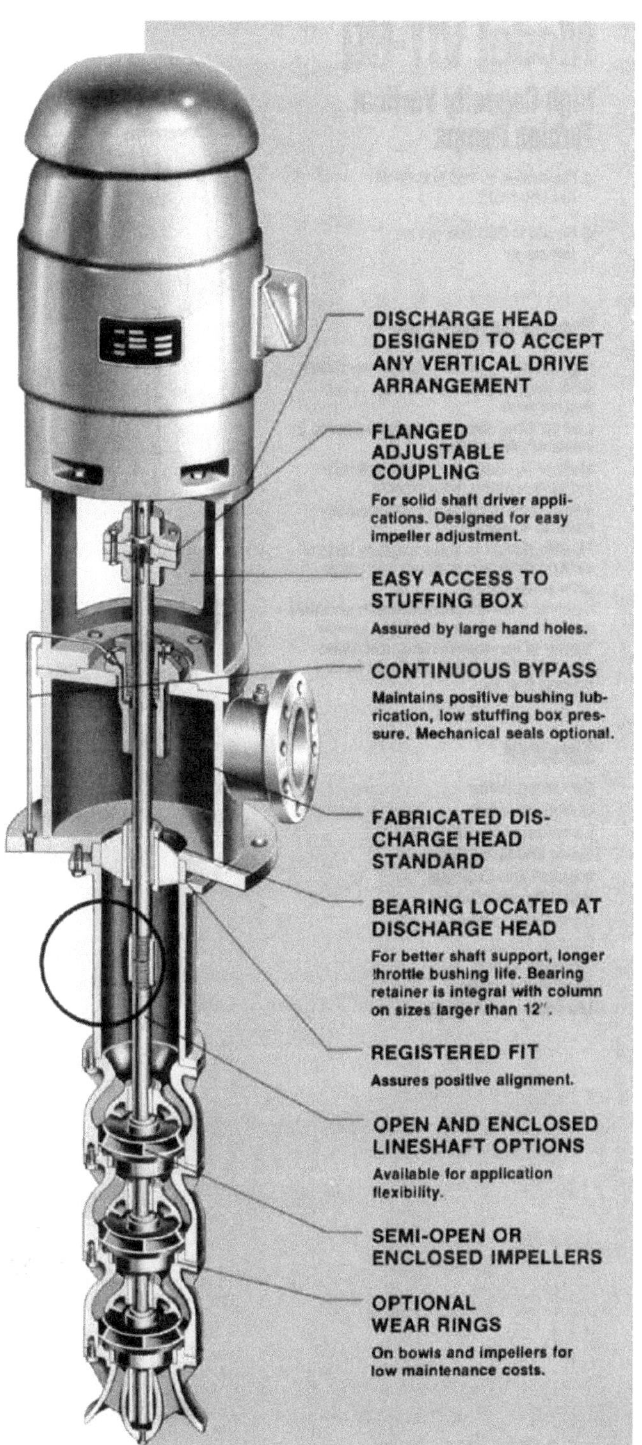
DISCHARGE HEAD DESIGNED TO ACCEPT ANY VERTICAL DRIVE ARRANGEMENT
FLANGED ADJUSTABLE COUPLING
For solid shaft driver applications. Designed for easy impeller adjustment.
EASY ACCESS TO STUFFING BOX
Assured by large hand holes.
CONTINUOUS BYPASS
Maintains positive bushing lubrication, low stuffing box pressure. Mechanical seals optional.
FABRICATED DISCHARGE HEAD STANDARD
BEARING LOCATED AT DISCHARGE HEAD
For better shaft support, longer throttle bushing life. Bearing retainer is integral with column on sizes larger than 12".
REGISTERED FIT
Assures positive alignment.
OPEN AND ENCLOSED LINESHAFT OPTIONS
Available for application flexibility.
SEMI-OPEN OR ENCLOSED IMPELLERS
OPTIONAL WEAR RINGS
On bowls and impellers for low maintenance costs.

Figure 11-1: Vertical turbine pump (VTP) and principal components. Note threaded sleeve coupling, circled (Source: ITT/Goulds, Seneca Falls, NY 13148)

Conventional Wisdom: *Couplings will accommodate the misalignment quoted by the manufacturer.*

Fact: *The quoted misalignment often refers to static capabilities. Once running, vibration and bearing overload may occur unless the shafts are aligned more carefully.*

Another secondary function provided on most process pump couplings is a separate "spacer" section that allows the diver and driven shaft ends to be positioned a specified distance apart (see Figure 11-3). The coupling spacer on horizontal pumps can then be removed to allow sufficient axial clearance to replace a mechanical seal without moving the pump or motor, or to allow room for the pump "back-pull-out" element to be replaced without moving the motor (Ref. 11-1). Also, since allowable shaft misalignment is normally expressed in fractions of degrees of arc, a longer spacer at, say 1/3rd of a degree will accommodate more parallel shaft offset than a shorter spacer or couplings without spacers. Most couplings can be provided in spacer or non-spacer configurations.

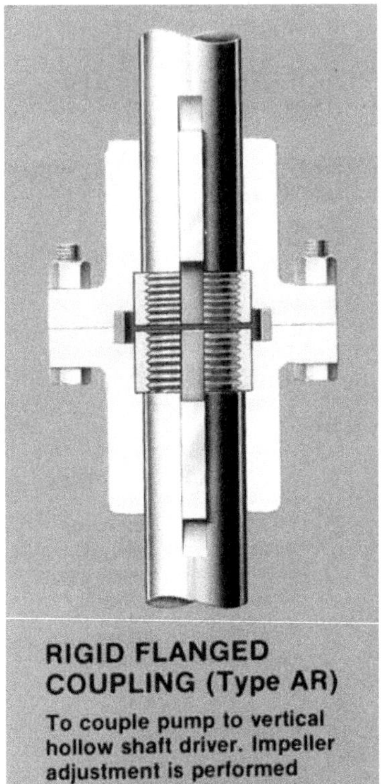

RIGID FLANGED COUPLING (Type AR)
To couple pump to vertical hollow shaft driver. Impeller adjustment is performed on adjusting nut located on top of motor.

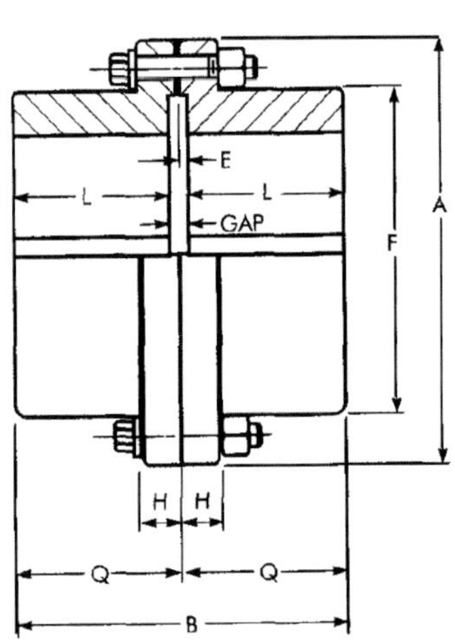

Figure 11-2: (Left)—rigid flanged vertical pump shaft coupling arrangement (Source: ITT/Goulds, Seneca Falls, NY 13148) (Right)—rigid flanged horizontal pump coupling (Source: The Falk Corporation, Milwaukee, WI 53208)

RIGID COUPLINGS

Rigid couplings are found in certain vertical in-line pumps and at several locations on vertical turbine pumps where individual shaft sections must be joined together. These couplings do not compensate for any misalignment between the mating shaft ends. Therefore, failure to ensure precise alignment of the connecting shafts will result in increased vibration, bearing wear, and shortened mechanical seal life (if the pump is so equipped).

Solid shaft drivers require one of the rigid coupling types shown in Figures 11-2 (left) or 11-3 to connect driver and driven shafts.

a. For short set pumps, a rigid flanged coupling (Figure 11-3) may be used, if axial rotor adjustment is not required. This coupling cannot be used with open impellers

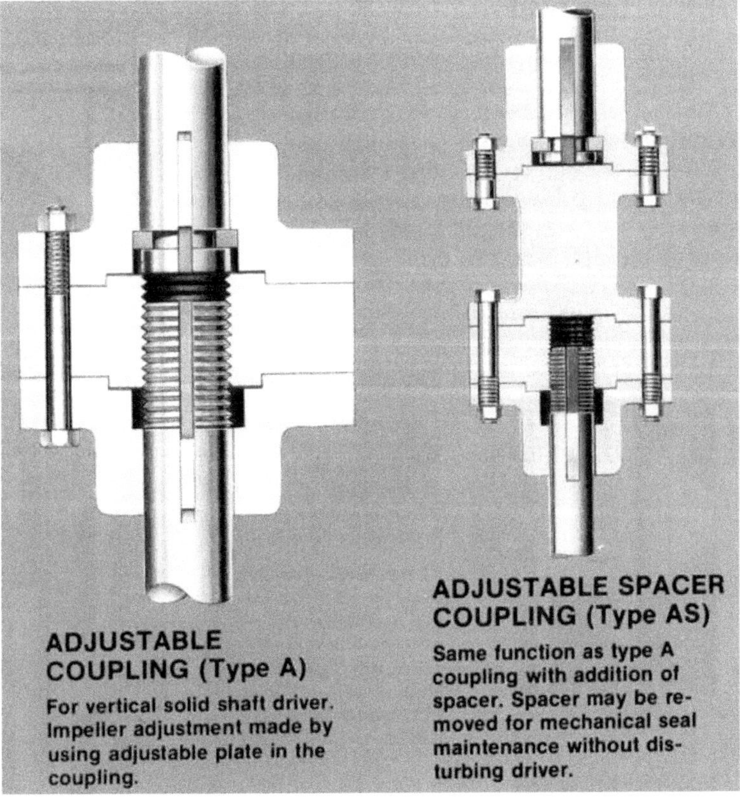

ADJUSTABLE COUPLING (Type A)
For vertical solid shaft driver. Impeller adjustment made by using adjustable plate in the coupling.
ADJUSTABLE SPACER COUPLING (Type AS)
Same function as type A coupling with addition of spacer. Spacer may be removed for mechanical seal maintenance without disturbing driver.

Figure 11-3: Adjustable plate (left) and adjustable spacer couplings for vertical pumps, right (Source: ITT/Goulds, Seneca Falls, NY 13148)

that require small front axial impeller clearances for efficient operation, which should be maintained by axial adjustment.

b. The adjustable coupling (Figure 11-3) allows axial adjustment of vertical pump shafts by rotating the nut (between the coupling halves), during final assembly. This axial adjustment can compensate for the tolerance stack-up of pump and motor components, wear of axial bowl assembly parts, and elongation of the pump shaft during operation.

c. When space is required for removal of the pump mechanical seal, without removal of the driver, an adjustable spacer coupling (Figure 11-3) should be used. Due to the extra-long spacer component, the individual components of this coupling must be manufactured to very tight tolerances, to ensure proper shaft alignment.

Hollow shaft motors (Figure 11-4) use either a rigid or a self-release anti-reverse rotation coupling located at the top of the motor, or right angle gear (see Figure 11-5). The pump shaft extends through the center of the motor (or gear) shaft. A guide bushing is recommended between the pump and hollow motor shaft, at the lower end of the motor, to improve shaft support and alignment. Pump thrust is transmitted from this internal pump shaft, through the rigid coupling at the top of the motor, to the driver (high-thrust motor or right angle gear) thrust bearing. The nut at the top of the coupling is used for axial adjustment of the pump shaft, which can be set in the field at startup.

When the total length of a vertical turbine pump exceeds approximately 20 feet, rigid couplings are also required to connect the individual line/bowl shaft sections. The simplest coupling type used for this application is the threaded coupling shown circled in Figure 11-1. Since the thread manufacturing tolerances cannot always guarantee proper shaft alignment, most vertical turbine pump (VTP) manufacturers offer optional keyed couplings. Keyed couplings with split rings for thrust transmission (or split/clamp types) reduce the risk of skewed joining of shaft sections, thereby reducing vibration and bearing wear incidents.

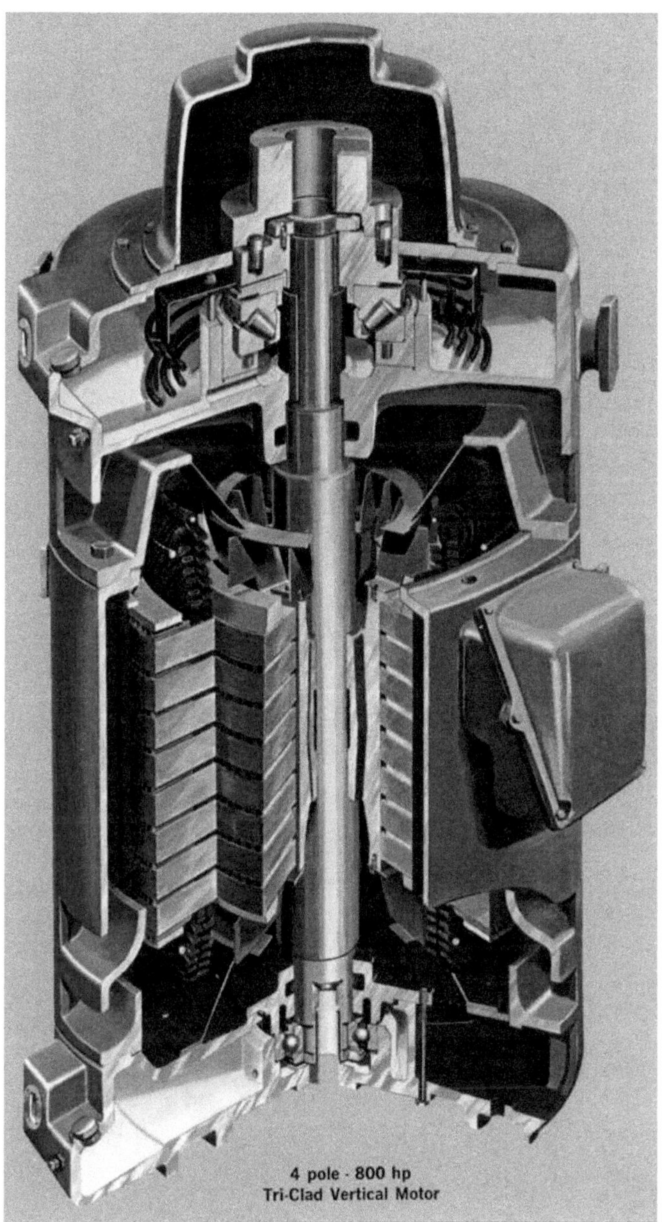

Figure 11-4: Hollow shaft motor (Source: General Electric Company, Schenectady, NY)

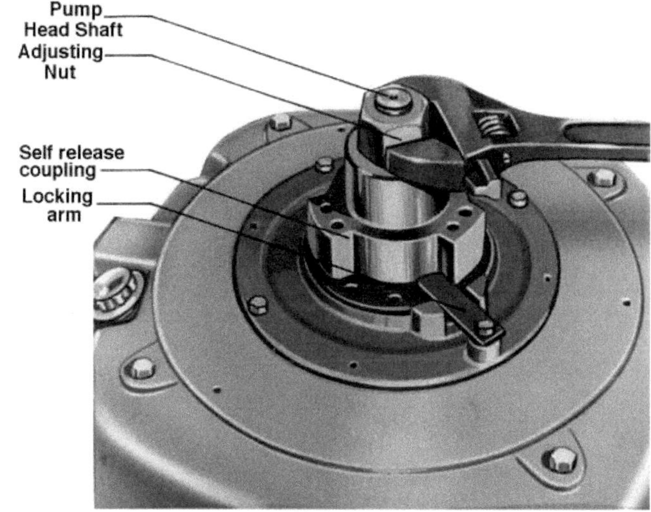

Figure 11-5: Hollow-shaft vertical motor and self-releasing anti-reverse rotation coupling

FLEXIBLE COUPLINGS

Flexible couplings may be separated into two basic types, mechanically flexible and "material-flexible." Mechanically flexible couplings compensate for misalignment between two connected shafts by means of clearances incorporated in the design of the coupling. Material-flexible couplings rely on flexing of the coupling element to compensate for shaft misalignment. A second key distinction for flexible couplings in process plants is whether or not the coupling requires periodic lubrication during service. Mechanically flexible, gear type couplings, which have been the industry standard for many years, require periodic lubrication; a maintenance cost issue which conflicts with the advances in seal and bearing-technology (regarding re-lubrication intervals). Most material- flexible couplings are non-lubricated couplings.

Mechanically Flexible Couplings

Mechanically flexible couplings compensate for misalignment between two connected shafts by means of clearances incorporated in the design of the coupling.

Gear couplings are probably the most frequently used mechanically flexible coupling configuration. Figures 11-6, 11-7 and 11-8, lower left, represent these couplings. They are "power-dense," meaning they are capable of transmitting high torque at high speeds, while remaining inherently well balanced. Axial force and moment transmission can be quite significant with gear couplings. This axial force must be absorbed by thrust bearings in either the driver or driven machine.

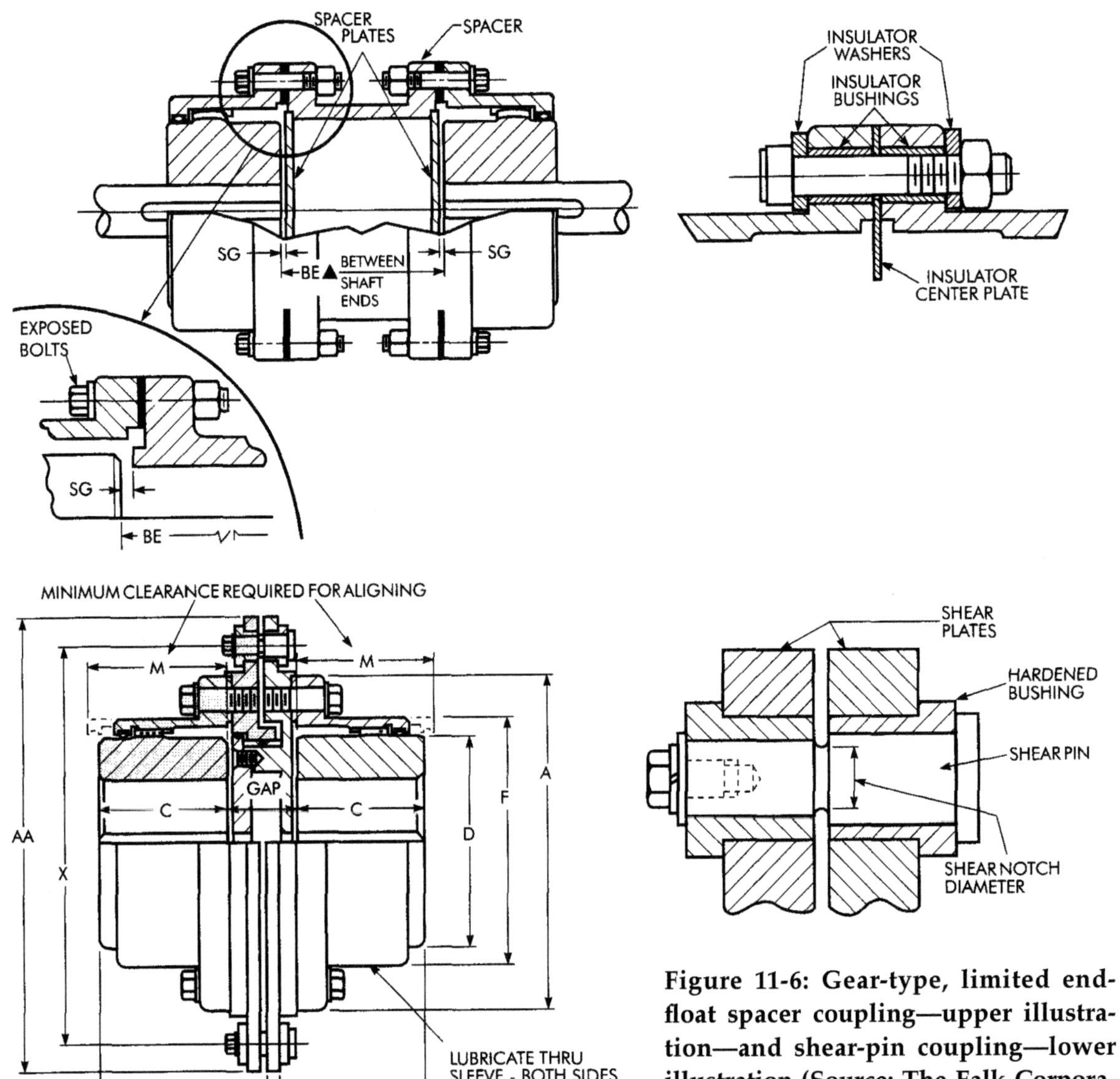

Figure 11-6: Gear-type, limited end-float spacer coupling—upper illustration—and shear-pin coupling—lower illustration (Source: The Falk Corporation, Milwaukee, Wisconsin 53208)

Special adaptations of gear couplings are shown in Figure 11-6. Spacer plates inserted, as shown in the upper left illustration, limit coupling travel so as not to axially load certain sleeve bearings in electric motor drivers. Insulator parts (upper right) are often used to prevent electric current travel from driver to driven equipment. Shear pin provisions (bottom left and right) are occasionally used as a "last chance" overload protection (Ref. 11-2).

A spacer-type gear coupling is depicted in the upper portion of Figure 11-7. Its axial travel is not limited. The lower portion of Figure 11-7 shows a spacer plate in place so as to limit motor rotor end-float.

Gear couplings transmit torque by the mating of two hubs with external gear teeth that are (most frequently) joined by flanged sleeves with internal gear teeth. Backlash is intentionally built into the gear teeth, and it is this backlash which compensates for shaft misalignment. Sliding motion occurs in a coupling of this type, and so a supply of clean lubricant (grease or viscous oil) is absolutely necessary to prevent wear of the rubbing surfaces.

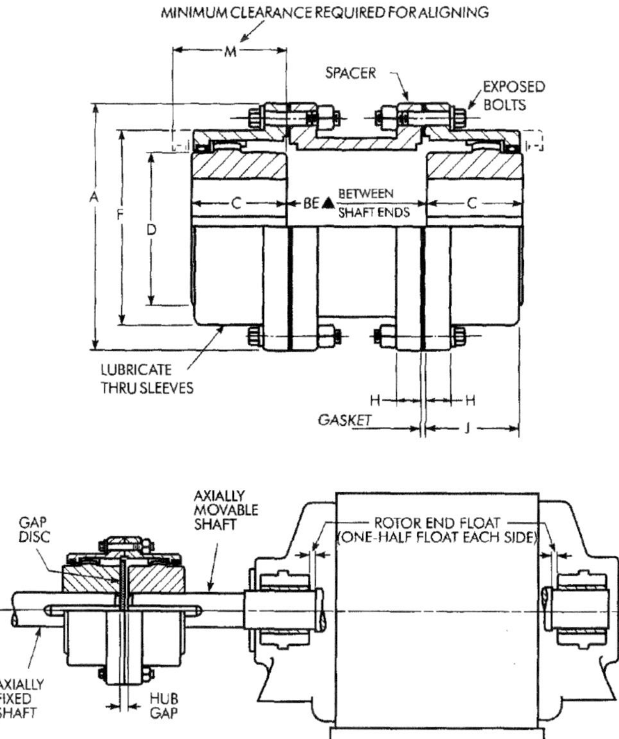

Figure 11-7: Flexible flanged gear couplings without (top) limited end float and with (bottom) limited end float (Source: the Falk Corporation, Milwaukee, Wisconsin 53208)

A mathematical expression for axial force is given in Equation 11-1:

$$F = (2)(\text{coefficient of friction})(T)/(D) \qquad \text{(Eq. 11-1)}$$

where

T = (63,025)(shaft HP)/(shaft rpm), lb-in
D = coupling pitch diameter, inches

Since the dry coefficient of friction can reach 0.3, good and effective lubrication is essential to the prevention of excessive axial forces.

General-purpose greases would have to be replaced or thoroughly replenished every three to six months. It is thus best to use a grease that is designed for couplings, especially in high-speed applications, where it might then be possible to extend lubrication intervals to once a year.

As discussed earlier in this text, grease is a blend of oil and a soap-thickener. These general-purpose compounds are not very stable at high peripheral speeds, and will eventually bleed or separate under the high centrifugal forces generated in many coupling applications. Once a grease begins to separate, the thickener accumulates in the areas where lubrication is required, and rapid wear of the contacting surfaces occurs. The oil is now free to leak out of the coupling past the seals,

Figure 11-8: Clockwise from upper left: Elastomer-in-compression, grid-spring, disc pack, and standard gear-type couplings (Source: The Falk Corporation, Milwaukee, Wisconsin 53208)

causing premature failure of the coupling. Most greases which are used as coupling lubricants were initially developed as bearing lubricants.

Bearing greases have a low viscosity and high bleed rate, which is desirable to avoid the heat caused by rolling friction. However, rolling friction is not present in couplings where the only movement is a sliding action caused by misalignment of shafts or thermal growth. Bearing lubricants are adequate only if conservative, i.e. frequent service intervals are followed. If operation cannot be interrupted to lubricate the coupling, continuous oil-lubricated couplings are available for large pump sets with pressurized, circulating oil systems.

Spring-grid couplings (Figure 11-8, upper right) combine the characteristics of mechanically flexible and material-flexible couplings. This design has two hubs, one mounted on each machine shaft. Each hub has a raised portion on which tooth-like slots are cut. A spring steel grid member is woven between the slots on the two hubs. The grid element can slide in the slots to accommodate shaft misalignment and flexes like a leaf spring to transmit torque from one machine to another. Spring-grid couplings require periodic lubrication (normally once a year) to prevent excessive wear of the grid member from the sliding action; they are generally found in low to moderate-speed, moderate power applications.

Roller-chain flexible couplings are rarely used by reliability-focused plants in process pumps and are generally limited to low-speed machinery. They are, therefore, not given further consideration in this text. Suffice it to say that a roller chain coupling employs two sprocket-like members mounted one on each of the two machine shafts and connected by an annulus of roller chain. The clearance between sprocket and roller provides mechanical flexibility for misalignment.

Material-flexible Couplings

These couplings rely on flexing of one or more coupling element to compensate for shaft misalignment. The flexing element may be of any suitable material (metal, elastomer, or plastic), which has sufficient resistance to fatigue failure to provide acceptable life. Material-flexible couplings do not require periodic lubrication. However, the flexing action can generate excessive heat (a problem with elastomer or plastic couplings), and/or cause fatigue (a problem with metals, such as steel) if loads or misalignment exceed defined limits. This will shorten the life of these couplings. Advantages of material-flexible, non-lubricated couplings are:

1. No downtime for lubrication
2. With a few exceptions generally transmit low, known thrust forces
3. They can be designed for infinite life
4. Except for certain toroidal (rubber tire-type) couplings, maintaining balance is not as difficult as maintaining balance in gear couplings.

Metal-disc, or disc-pack couplings. Disc-pack couplings are the predominant type used in process plants throughout the industrialized world. There are numerous configuration variations of this material-flexible, non-lubricated coupling. It is shown in the lower right portion of Figure 11-8 and again in the two-part illustration of Figure 11-9. In installations with close spacing between shaft ends, this coupling uses a single set of discs. For applications using a spacer, the coupling employs two sets of thin sheet-metal discs bolted to the driving and driven hub members. The coupling illustrated here was specifically designed to maximize its misalignment capability, as well as ensuring that it could be fitted in the space envelope of an existing gear coupling. It should be noted that the membranes are positioned behind the hub, thus increasing the distance

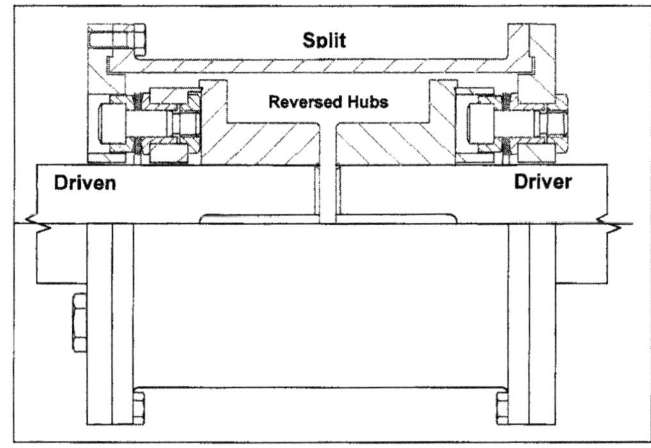

Figure 11-9: Modern non-lubricated disc pack (membrane) coupling (Source: John Crane/Slough, UK)

between membrane banks, as well as maximizing the possible shaft diameter.

Each set of discs is made up of a number of thin, contoured high-strength alloy steel laminations, which are individually flexible and compensate for shaft offset or angular misalignment by means of this flexibility. A number of high-strength drive bolts solidly connect the discs alternating the attachment between the driving and driven hub flanges. The discs may be stacked together as required to obtain the desired torque transmission capability (Ref. 11-3).

Although generally preferred for pumps by reliability-focused users, flexible disc pack couplings are not immune to mistreatment and misapplication that might cause metal fatigue. As is the case with all couplings, useful life is influenced by operation at certain misalignment angles. Manufacturers often list angles that are dimensionally achievable without interference, but expect the user to operate the equipment at much lower misalignment. Operation at maximum listed offset values may result in premature fatigue failure.

Disc-pack couplings with spacers (also called the center member) should incorporate a captured center member feature. These couplings (Figure 11-9) are safer than competing models lacking this feature since, in the event of failure of the stack of laminated discs, the center member cannot be flung out of the assembly.

Another example of an all-metal material-flexible (non-lubricated) coupling is the contoured diaphragm coupling (Ref. 11-4). A spacer-type cutaway is shown in Figure 11-10. This coupling is similar in function to the metal-disc coupling in that the single or twin disc assemblies flex to accommodate misalignment. Diaphragm couplings have the best-understood dynamic behavior, but at a premium cost. They are primarily used in turbo-compressors and turbo-generators but have found application in large pumps as well. Contoured diaphragm couplings are usually designed for full design capacity at maximum allowable deflection.

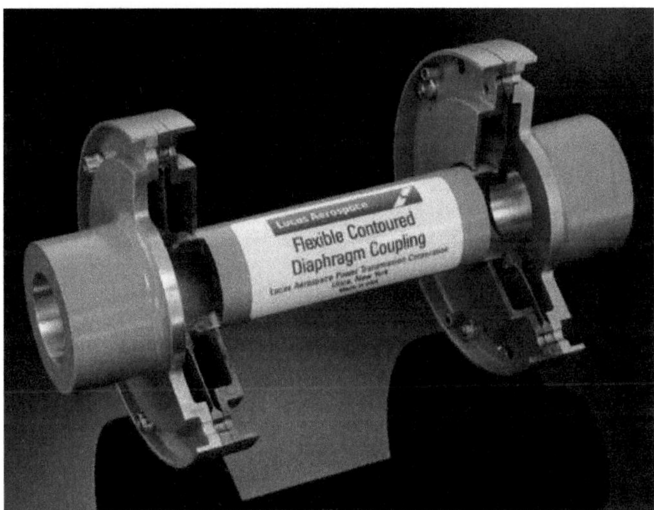

Figure 11-10: Flexible contoured diaphragm coupling (Source: Lucas Aerospace, Utica, NY 13503)

The principal performance characteristics of alloy steel disc, contoured diaphragm and gear couplings are compared in Table 11-1.

Table 11-1: Performance comparison of disc, diaphragm and gear couplings

	Disc	*Diaphragm*	*Gear*
Speed capacity	High	High	High
Power-to-weight ratio	Moderate	Moderate	High
Lubrication required	No	No	Yes
Misalignment capacity at high speed	Moderate	High	Moderate
Inherent balance	Good	Very Good	Good
Overall diameter	Low	High	Low
Normal failure mode	Abrupt	Abrupt	Progressive
Overhung moment	Moderate	Moderate	Very low
Generated moment, misaligned, with torque	Moderate	Low	Moderate
Axial movement capacity	Low	Moderate	High
Resistance to axial movement			
—sudden	High	Moderate	High
—gradual	High	Moderate	Low

Other comparative parameters are given in Table 11-2. It should be noted, however, that all data are relative. Chain couplings—typically only used on relatively small low to medium speed pumps that operate no more than 2,000 hours per year, may have to be replaced after 3-5 years of installation. Diaphragm couplings in high-speed, high torque utility feedwater service, may have to be examined after 5 years or approximately 43,000 hours of continuous operation. Properly selected disc pack couplings may require replacement of disc packs after 8 years of installation. As is so often the case, there is no substitute for experience and all of the possible iterations and qualifiers could fill a book.

Elastomer-in-compression couplings. Material-flexible couplings employing elastomeric materials are numerous (Refs. 11-3 and 11-5). Elastomers generally do not have a well-defined fatigue limit. Also, no one type or composition of elastomer is immune to chemical attack from at least some of the thousands of fluids served by

Table 11-2: Flexible coupling comparison chart

COUPLING TYPE	AXIAL FORCES GENERATED	RELATIVE COST	LUBRICATION REQUIRED	ESTIMATED SERVICE LIFE YEARS
Mechanically Flexible				
Gear	Med-High	Medium	Yes	3-5
Chain	Low	Low	Yes	3-5
Grid Spring	Medium	Medium	Yes	3-5
Metallic Material-Flexible				
Disc	Low-Med	High	No	4-8
Diaphragm	Low	High	No	5
Elastomeric Material-Flexible				
Jaw	Medium	Low-Med	No	3-5
Bonded Tire (Urethane)	Low	Low-Med	No	2-3

process pumps. Most elastomers suffer some form of degradation from heat and require de-rating at elevated temperatures. Heat, of course, develops in the material when the coupling flexes, and claims that couplings will operate at severe misalignments are often very misleading. Anything other than moderate misalignment will severely limit coupling life. Coupling degradation can be observed from powdered elastomers collecting beneath the operating coupling.

The elastomeric cog belt coupling shown in the upper left of Figure 11-8 applies compressive forces to the cogs. This, and the similar rubber jaw coupling, are commonly used on low-power drive systems. The heart of either coupling type is either a cog belt or a central spider member with several radial segments extending from a central section. The hubs, which are mounted on driving and driven shafts, each have a set of jaws, corresponding to the number of cogs or spiders on the flexible element. A cog belt or spider fits between the two sets of jaws and provides a flexible cushion between them. This cushion transmits the torque load and accommodates a measure of misalignment.

Split-sleeve elastomeric couplings. A second group of elastomeric couplings employs a split sleeve (boot) element bonded to split shoe components that are attached to the shaft hubs by radial bolting. This allows field adjustment for different shaft gaps. Each of the two elastomeric omega-shaped coupling halves (Figure 11-12) is placed in shear. Examination under a stroboscopic light will show it to twist.

It is thought that this coupling type typically allows greater shaft misalignment and yet, lower reaction forces at the bearings. Also, the coupling is able to cushion shock loads and dampen torsional vibration. But these desirable attributes come at a price. Thermographic examination indicates that severe deflection generates more heat. Plants with dependable maintenance and component replacement statistics have noted higher replacement requirements for couplings with elastomers in shear than for comparably rated disc pack couplings.

Elastomeric toroid ("Inverted Tire," Ref. 11-6) couplings are represented by Figure 11-13. The fabric-reinforced flexing element is operating in the shear mode. Centrifugal force tends to impart an axial load on the connected shafts, pushing them apart. The opposite is true with elastomeric toroid configurations that represent a true tire cross-section. Here, centrifugal force would impart an axial pulling force on the connected shafts.

In addition to temperature limitations, elastomer performance is often affected by environmental compatibility, including—sometimes—tolerance of leaking fluids. Table 11-3 compares the environmental compati-

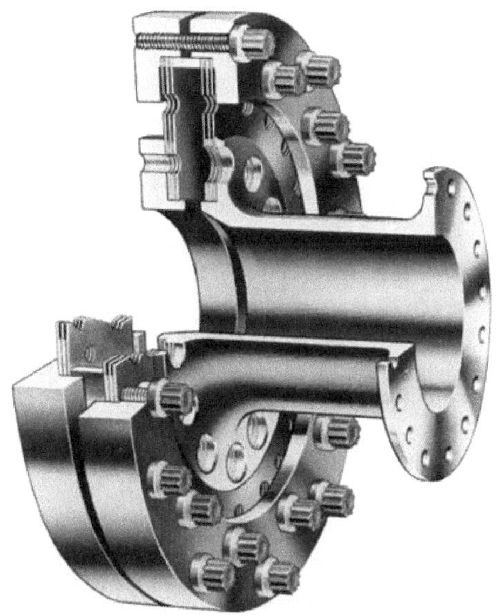

Figure 11-11: "Flexxor" non-lubricated multiple disc coupling (Source: Coupling Corporation of America, York, Pennsylvania)

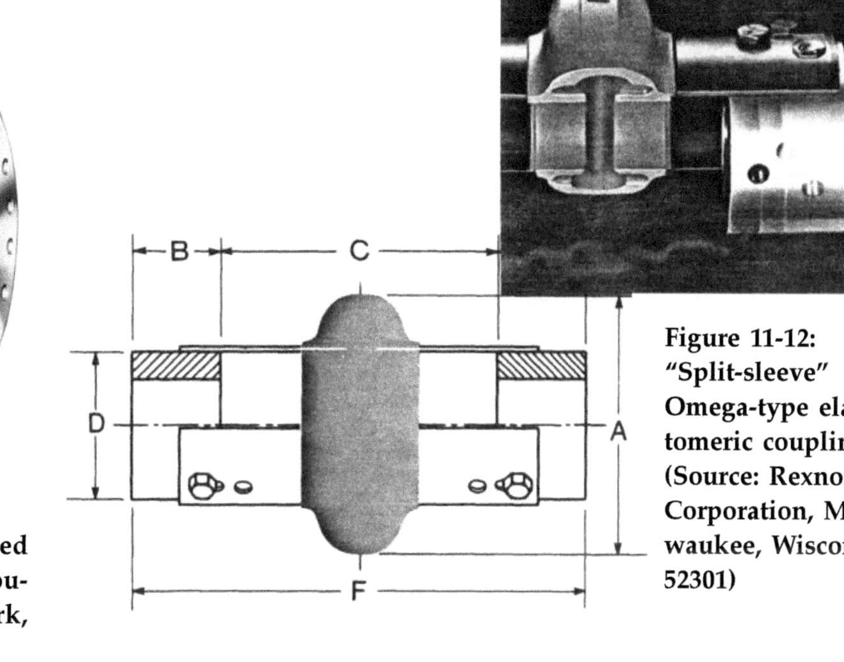

Figure 11-12: "Split-sleeve" Omega-type elastomeric coupling (Source: Rexnord Corporation, Milwaukee, Wisconsin 52301)

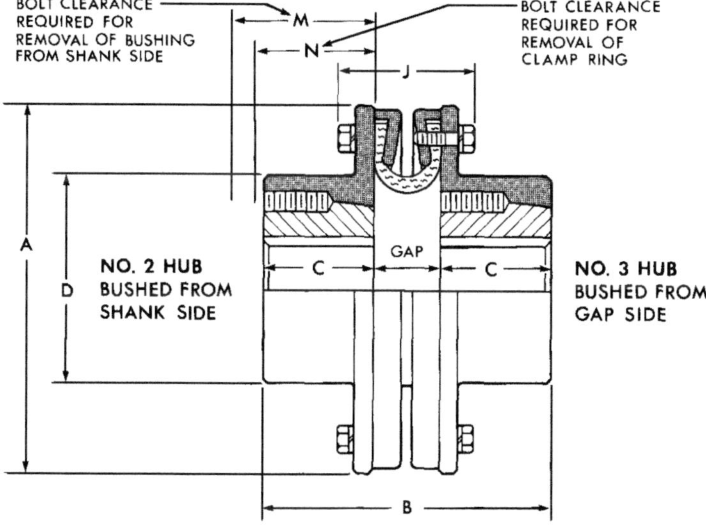

Figure 11-13: "Inverted toroid" elastomeric coupling (Source: The Falk Corporation, Milwaukee, Wisconsin, 53208)

bility of two common coupling elastomers, polyurethane and polyisoprene rubber (Ref. 11-7).

COUPLINGS FOR CENTRIFUGAL PUMPS

Properly selected and installed, disc pack couplings with captured center members are almost always the most cost-effective choice. The center member, also called a spool piece, is sized for a shaft-end to shaft-end separation of approximately 7 inches (~178 mm). In case of disc pack failure, which rarely ever happens on properly aligned pumps, the "capture feature" prevents the spool piece from being ejected. Moreover, by using a stroboscopic light, the deformation-related condition of disc pack couplings can be monitored while the equipment is operating.

Regardless of lubrication method employed, refinery experience indicates that gear couplings tend to lock up after a period of operation. The resulting thrust forces are acting on the bearings and overload failures are likely to occur. Lock-up often occurs even though the coupling teeth appear damage-free.

Elastomeric couplings sooner or later suffer from aging or other environmentally induced degradation. Obviously, elastomers in compression are a bit less risky than elastomers in tension and torsion. In addition to misalignment and environmental influences, the useful life of couplings is further influenced by the number of starts and stops.

Suffice it to say that while no one type of coupling is without its advantages and disadvantages, reliability-focused users will overwhelmingly agree that properly selected and installed disc pack couplings are the top choice for centrifugal pumps.

Regardless of coupling type ultimately used, proper assembly tolerances and sound mounting and

Table 11-3:
Typical relative performance of two elastomers

Environment	*Relative Polyurethane Rubber*	*Performance of Polyisoprene*
Abrasion	Excellent	Excellent
Acids, dilute	Fair	Good
Acids, concentrated	Poor	Good
Alcohols	Fair	Good
Aliphatic Hydrocarbons	Excellent	Poor
Gasoline, fuel	Excellent	Poor
Alkalies, dilute	Fair	Good
Alkalies, concentrated	Poor	Good
Animal and vegetable oils	Excellent	Fair
Aromatic Hydrocarbons	Excellent	Poor
Benzol, toluene	Poor	Poor
Degreaser fluids	Good	Poor
Heat aging	Good	Good
Hydraulic fluids	Poor	Poor
Low temperature embrittlement	Excellent	Good
Oil	Excellent	Poor
Oxidation	Excellent	Good
Ozone	Excellent	Poor
Radiation	Good	Good
Silicate and phosphate	Poor	Poor
Steam and hot water	Poor	Unknown
Sunlight (aging)	Excellent	Poor
Synthetic lubricants	Poor	Poor
Water swell	Excellent	Good

dismounting procedures are essential. Sledge hammers have no place in coupling-related work and the unacceptability of the hub shown in Figure 11-14 merits no further comment.

Reliability-focused users will insist on coupling hubs that incorporate puller holes to accommodate pull-off fixtures or suitable disassembly tools. Moreover, the craftsmen employed at reliability-focused user facilities will pay close attention to the fits and tolerances recommended in Chapter 15 of this text (see page 444).

Figure 11-14: Damaged coupling hub

Chapter 12

Pump Condition Monitoring Guidelines

Process pump users often seek a method of determining the optimum time for overhaul of a pump based on energy savings or other considerations. Once developed, such a tool should be widely used by maintenance engineers and managers in their role of managing assets to provide capacity for production, and energy efficiency to save operating expense or even to minimize greenhouse impact. This optimization approach can also be applied to any item in a plant where deterioration results in loss of efficiency and energy consumption can be measured or estimated.

VIBRATION BASICS

All rotating machines, including pumps, vibrate to some extent due to response from excitation forces, such as residual rotor unbalance, turbulent liquid flow, pressure pulsations, cavitation, and/or pump wear. Further, the magnitude of the vibration will be amplified if the vibration frequency approaches the resonant frequency of a major pump, foundation and/or piping component. The issue of interest is not whether or not the pump vibrates, but:

- Is the amplitude and/or frequency of the vibration sufficiently high to cause actual or perceived damage to any of the pump components, or
- Is the vibration a symptom of some other damaging phenomenon happening within the pump.

Various industry organizations, such as the Hydraulic Institute (ANSI/HI 9.6.4, Ref. 12-1), and American Petroleum Institute (API-610) have set vibration limits to help guide users to avoid excessive levels of vibration in pumps. But long before industry standards were developed, individual reliability professionals and multinational pump user companies implemented daily machinery condition logs which guided operating personnel by listing acceptable, reportable, and mandatory shut-down levels of vibration (Figure 12-1). These experience-based values culminated in the bearing life vs. vibration approximations for general-purpose machinery, Figure 12-2 (HPB, 1986).

However, machinery vibration and its measurement are complex issues and require some clarification. Typical considerations might include:

1. Bearing cap vibration be measured and/or analyzed by using displacement, velocity or acceleration to evaluate the health of the equipment? The primary measure of bearing cap vibration used in industry today is velocity.

2. Should the "total all-pass" or the "filtered" frequency be used? Most industry specifications and standards use "total all-pass" vibration values to identify problem pumps, with filtered values reserved for analysis of vibration problems.

3. Should RMS (root mean square) or peak-to-peak values be measured or specified? The Hydraulic Institute has chosen RMS acceptance limit values, since most vibration instruments actually measure vibration in RMS terms, and calculate peak-to-peak values if required. API, on the other hand, uses peak-to-peak. RMS values are roughly 0.707 times zero-to-peak values.

   However, this relationship applies only for vibration consisting of a single sine wave. For more complex waveforms, this conversion does not yield correct results.

4. What is the acceptable vibration amplitude (new, or repair levels) for a particular application? Should acceptance limits change along with overall pump power and flow-rate regions? The Hydraulic Institute bases acceptable vibration limits on pump type, power level, operating range, and whether measured at the factory or in the field. API gives

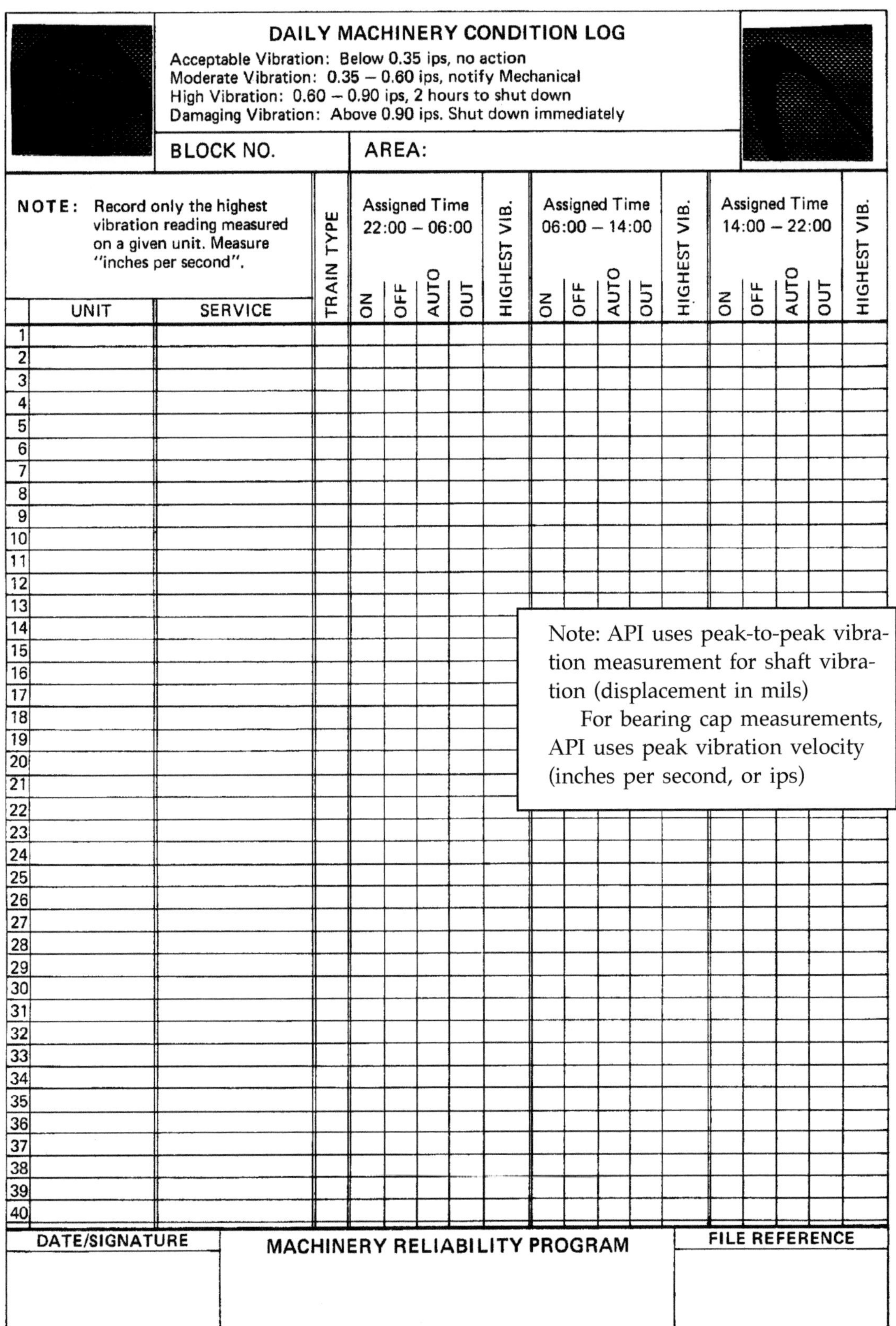

DAILY MACHINERY CONDITION LOG

Acceptable Vibration: Below 0.35 ips, no action
Moderate Vibration: 0.35 – 0.60 ips, notify Mechanical
High Vibration: 0.60 – 0.90 ips, 2 hours to shut down
Damaging Vibration: Above 0.90 ips. Shut down immediately

BLOCK NO. AREA:

NOTE: Record only the highest vibration reading measured on a given unit. Measure "inches per second".

	UNIT	SERVICE	TRAIN TYPE	Assigned Time 22:00 – 06:00 ON	OFF	AUTO	OUT	HIGHEST VIB.	Assigned Time 06:00 – 14:00 ON	OFF	AUTO	OUT	HIGHEST VIB.	Assigned Time 14:00 – 22:00 ON	OFF	AUTO	OUT	HIGHEST VIB.
1																		
2																		
3																		
4																		
5																		
6																		
7																		
8																		
9																		
10																		
11																		
12																		
13																		
14																		
15																		
16																		
17																		
18																		
19																		
20																		
21																		
22																		
23																		
24																		
25																		
26																		
27																		
28																		
29																		
30																		
31																		
32																		
33																		
34																		
35																		
36																		
37																		
38																		
39																		
40																		

DATE/SIGNATURE	MACHINERY RELIABILITY PROGRAM	FILE REFERENCE

Note: API uses peak-to-peak vibration measurement for shaft vibration (displacement in mils)

For bearing cap measurements, API uses peak vibration velocity (inches per second, or ips)

Figure 12-1: Manual vibration data log format used in the 1970's

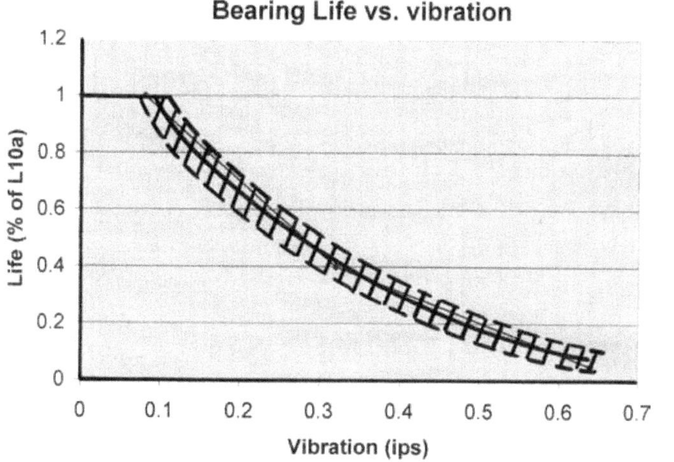

Figure 12-2: Approximate bearing life-vibration relationship for general purpose machinery

different limits for the "preferred" and "allowable" operating regions. A vibration velocity of 0.15 ips represents a "rule-of-thumb," as-new acceptance value. Moreover, this applies to operation in the preferred operating range and with clear liquids. Power levels are limited to 268 hp (200 kW) and low cavitation or suction turbulence.

5. How do factory test stand vibration measurements compare with field site values? What is the effect of foundation stiffness/grouting? Generally, the stiffness of a field pump foundation is much higher than found on a factory test stand, especially if the pump base plate is grouted (see Chapter 3). That is why the Hydraulic Institute vibration standard allows higher test stand values (up to 30% higher for field values). For vertical turbine pump installations, it is especially important to know the actual foundation stiffness to avoid high vibration from operation at a structural resonant frequency.

6. How much has the vibration amplitude and/or frequency changed during the life of the machine? It is helpful to have an as-new vibration signature taken and kept on file for future use.

7. Location of vibration measurements: The vibration transducers should be mounted in such a way as to not adversely affect measurement accuracy. In general, vibration probes should be located near the center of the bearing housing of between-bearing pumps, and near the outer bearing of end suction pumps. Horizontal and vertical dry pit pump vibration measurements are normally taken on the bearing housings (near the outer or uppermost bearings). All measurements should be taken in the horizontal, vertical and axial planes, with the maximum value used as the acceptance criterion. Vertical turbine pump vibration measurements are taken near the top of the motor support. Probes should not be located on flexible panels or cylinder walls, such as motor end covers.

Not to be overlooked are special vibration monitoring approaches that are particularly well-suited for process pumps. One such approach involves the SKF enveloping process (Figure 12-3). The relevant hand-

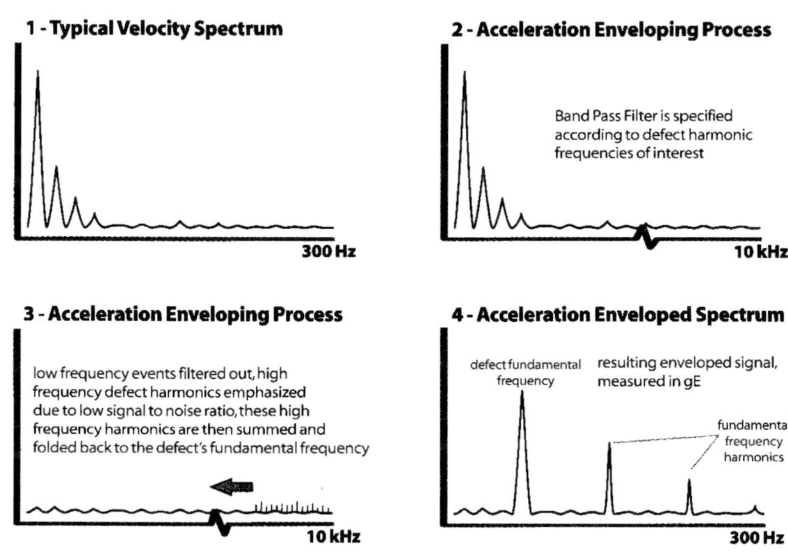

Figure 12-3: Acceleration enveloping process of vibration monitoring (Source: SKF Condition Monitoring, San Diego, California)

1. A typical Velocity Spectrum process does not always reveal possible bearing defects.
2. The acceleration enveloping process isolates bearing fault signals by filtering out the machine's higher amplitude, lower frequency vibration noise (typically everything under 10-15 times shaft speed), thereby greatly improving the signal-to-noise ratio of the bearing defect. Then, in the isolated bearing frequency range, the enveloping process looks for higher harmonics of repetitive, impulse signals created by bearing defects.
3. These defect signals are detectable in the defect's early stages and typically occur in very high frequency ranges, anywhere from five to 40 times the fundamental defect frequency.
4. Once these harmonics are detected, the enveloper sums the signals together, then folds the enhanced signal back to the defect's fundamental frequency range. The result is a frequency spectrum with a much more obvious, enhanced event displayed at the defect's fundamental fault frequency and at evenly spaced harmonics.

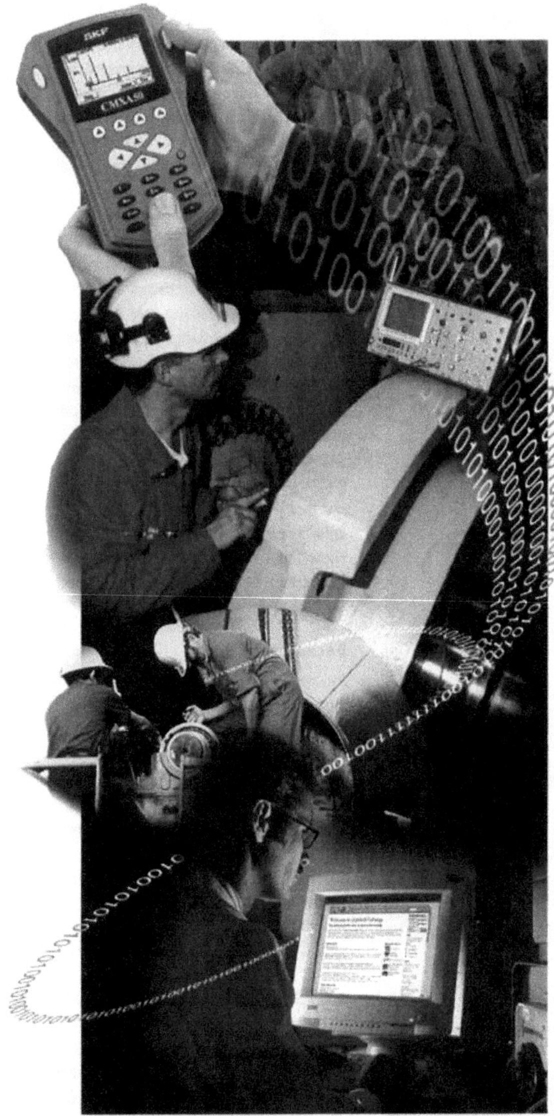

Figure 12-4: Versatile monitoring devices are part of an integrated industrial decision support system (Source: SKF Condition Monitoring, San Diego, California)

held data collection and analysis unit is shown at the top of Figure 12-4. This data collector, and others like it, represent versatile bearing distress monitoring tools.

Figure 12-5 shows shock pulse monitoring (SPM) and manual data collection in progress. In simple terms, the SPM method detects the development of a mechanical shock wave caused by the impact of two masses. At the exact instant of impact, molecular contact occurs and a compression (shock) wave develops in each mass. The SPM method is based on the events occurring in the mass during the extremely short time period after the first particles of the colliding bodies come in contact. This time period is so short that no detectable deformation of the material has yet occurred. The molecular contact produces vastly increased particle acceleration at the impact point. The severity of these impacts can be plotted, trended and displayed (see Figure 12-6).

SKF, a bearing manufacturer with manufacturing facilities world-wide, is using several of their data acquisition and analysis instruments (see top of Figure 12-4) as tools in their "@ptitude" (*sic*) industrial decision support system. Making extensive use of data banks and software, vibration data comprise just one of many inputs being integrated into one easy to use application. In essence, the system uses a structured approach to capturing and applying knowledge—both real-time and on-demand.

Another major contender is the Vibcode and Vibscanner family of German manufacturer Prueftechnik and its Miami-based U.S. collaborator Ludeca, Figure 12-7. These combination data collectors and analyzers come with ready-made software packages that start with vibration and temperature monitoring, but also incorporate speed measuring and field balancing capabilities that are of great value in a conscientiously applied pump life extension program.

ITT/Goulds has developed a variable frequency intelligent drive with a monitoring system (called "PumpSmart®") that is capable of adjusting the pump speed to react to fault occurrences, such as a closed suction valve, cavitation and pump wear.

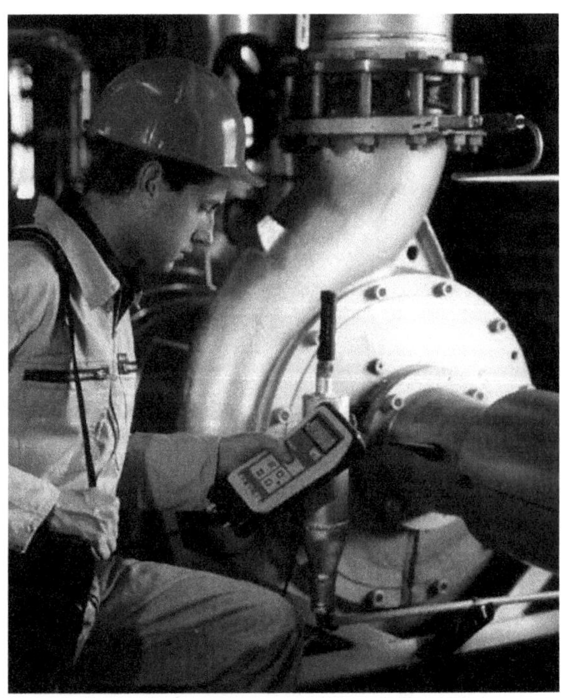

Figure 12-5: SPM data collection on a process pump (Source: SPM Instruments, Inc., Marlborough, Connecticut)

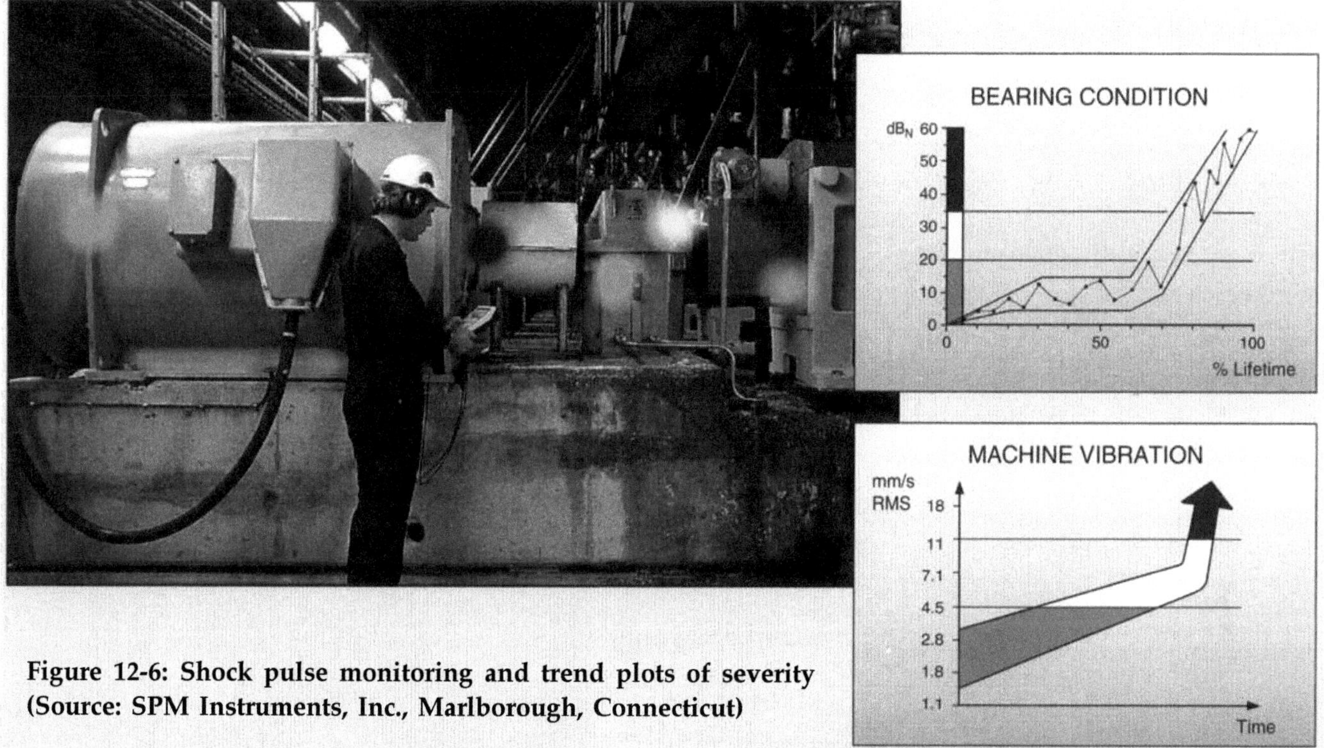

Figure 12-6: Shock pulse monitoring and trend plots of severity (Source: SPM Instruments, Inc., Marlborough, Connecticut)

PREDICTIVE CONDITION MONITORING AND FAULT TOLERANT INTELLIGENT DRIVES/CONTROLS

New intelligent pump solutions have continued to evolve thanks to the rapid advancement of electronic data processing and communications. What not long ago was thought to be too expensive, not reliable, and technically not possible, is rapidly becoming a reality. Technological advances in devices such as digital signal processors and wireless communication systems are enabling the development of effective intelligent pump solutions. These devices can be decentralized systems, local to the pump and outside of the main distributive control system, and can contain algorithms that have been developed to provide intelligent solutions, specifically designed for a unique class of pumps or application.

Today there are both custom and pre-engineered systems available which have this improved functionality that can positively impact key life cycle cost elements. In addition to saving energy by varying pump speed, these new drives/controls can also monitor a variety of critical pump conditions and either automatically make the required adjustments and/or wirelessly (or by hard wire) transmit this information for further potential action. In addition, they can provide efficient/reliable control for multi-pump operation, can reduce constant on-off cycling in wet well applications, and can even automatically sequence pump impeller rotational direction to unclog debris.

Condition Monitoring

These de-centralized data acquisition devices, which are local to the pump sub-system, gather system data either through the plant network, from directly connected sensors and/or from data available on the drive (Ref. 12-10). These data acquisition devices have high speed microprocessors capable of sampling data, so that flow, pressure, speed, torque, vibration spectral analysis and/or motor current signature analysis can run locally and in conjunction with diagnostic algorithms designed specifically for the pump system. The diagnostic results from these de-centralized systems can be fed back into the plant maintenance system, or to any remote location, so that planned maintenance activities can be deployed effectively. In addition, the capability exists on some of the newer types of data acquisition devices to implement local automated control to protect the pump system against premature failure and/ or to optimize pump system efficiency.

ROTOR BALANCING

All impellers, irrespective of their operational speed, should be dynamically ("spin") balanced before

Figure 12-7: Pruftechnik's "Vibcode" and "Vibscanner" devices collect and analyze a wide range of data and are supported by superior software (Source: Ludeca, Miami, Florida 33172)

installation, either single or two plane. Two-plane balance is required for wide impeller, when the impeller width is greater than 1/6th of the impeller diameter. ISO balancing criteria are illustrated in Figure 12-8.

There is no doubt that dynamic balancing of the three major rotating pump components, shaft, impeller and coupling, will increase mechanical seal and bearing life. All couplings of any weight or size should be balanced, if they are part of a conscientious and reliability-focused pump failure reduction program. Couplings that cannot be balanced have no place in industrial process pumps.

The preferred procedure of balancing a rotating unit is to balance the impeller and coupling independently, and then the impeller and coupling on the shaft as a single unit. Another method is to balance the rotor one time only as an assembled unit.

Multistage pump rotors, both horizontal and vertical, are best not assembled and balanced and then disassembled for re-assembly. Often, more problems are caused by the disassembly than are caused by the component balance. The static (single plane) force in the balance is always the more important of the two forces, static and dynamic (couple force). If balancing of individual rotor components is chosen, it is best to use a tighter tolerance for the static (single plane) force. In theory, if all the static force is removed from each part, there should be very little dynamic (couple) force remaining in the rotor itself.

For impellers operating at 1,800 rpm or less, the ISO 1940 G6.3 tolerance is acceptable. For 3,600 to 1,800 rpm, the ISO G2.5 is a better tolerance. Both are displayed on the balance tolerance nomograms for small (Figure 12-9) and large machinery rotors (Figure 12-10). Generally tighter balance tolerances (G1.0) are not warranted unless the balancing facility has modern, automated balancing equipment that will achieve these results without adding much time and effort. Using older balancing equipment may make the G1.0 quality difficult and unnecessarily costly to obtain and duplicate. Also, factory vibration tests have, at best, shown insignificant reductions in pump vibration with this tighter balance grade.

Balancing machine sensitivity must be adequate for the part to be balanced. This means that the machine is to be capable of measuring unbalance levels to one-tenth of the maximum residual unbalance allowed by the balance quality grade selected for the component being balanced.

Rotating assembly balance is recommended when practical and if the tighter quality grades, G2.5 or G1.0, are desired. Special care must be taken to ensure that keys and keyways in balancing arbors are dimensionally identical to those in the assembled rotor. Impellers must have an interference fit with the shaft, when G1.0 balance is desired. Although looseness between impeller hub and shaft (or balance machine arbor) is allowed for the lesser balance grades, it should not exceed the values given in Table 12-1 for grades G2.5 or G6.3:

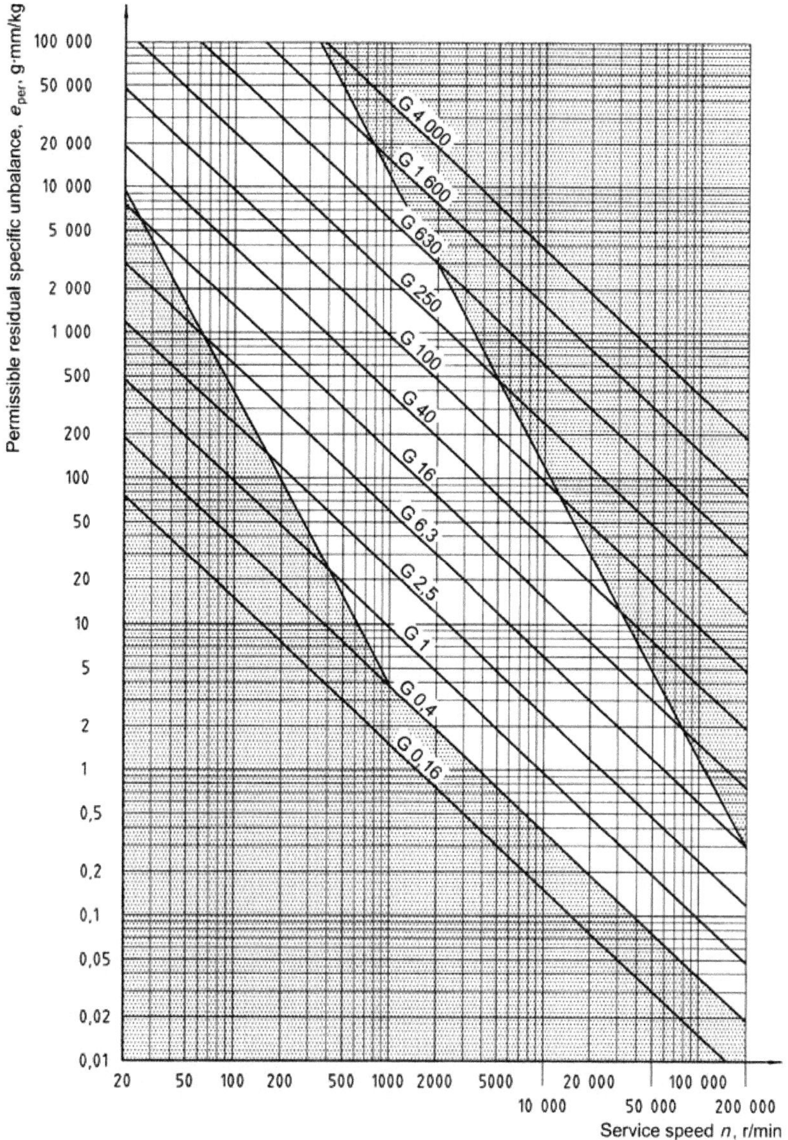

NOTE: The light area is the generally used area, based on common experience.

Figure 12-8: Permissible residual specific unbalance based on balance quality grade G and service speed *n*. (Source: ISO Standard 1940)

VIBRATION ACCEPTANCE LIMITS

Hydraulic Institute Standard ANSI/HI 9.6.4 (www.Pumps.org) presents the generally accepted allowable pump "field" vibration values for various pump types (see Table 12-2). The standard is based on RMS total, or all-pass vibration values (Ref. 12-1). The standard states that factory or laboratory values can be as much as 30% above these field limits, depending on the rigidity of the test stand. The ANSI/ASME B73 standard accepts 27% and 17% above the HI 9.6.4 values for factory tests performed on chemical end suction pumps, depending on the power level. HI also includes the API-610 values for end suction refinery pumps (in RMS terms), although the API document requires that these acceptance values be demonstrated on the factory test stand.

The HI standard also states that these values only apply to pumps operating under good field conditions, which are defined as:

1. Adequate NPSH margin (see Chapter 5).
2. Operation within the pump's preferred operating

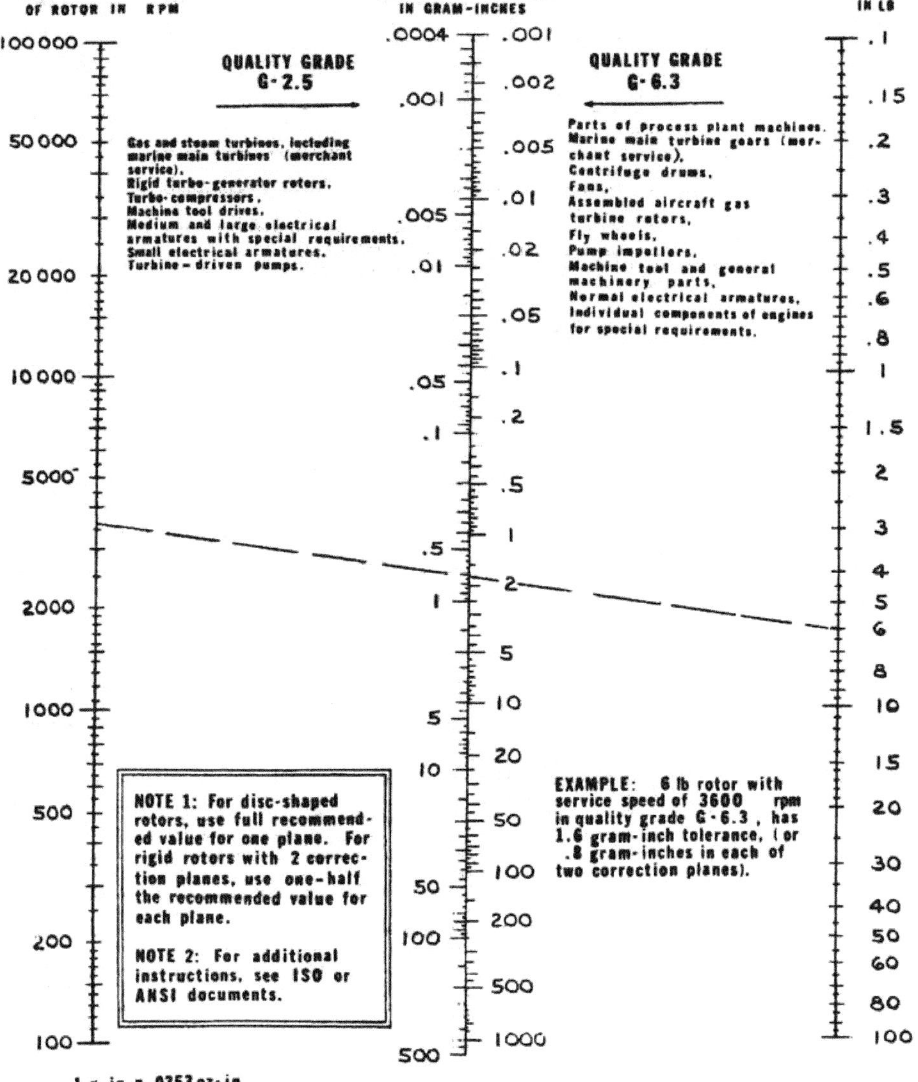

Figure 12-9: Balance tolerance nomogram for G2.5 and G6.3, small rotors (Source: Bloch, Heinz P., and F.K. Geitner, "Machinery Component Maintenance and Repair," 2nd Edition, 1990, Gulf Publishing Company, Houston, Texas, ISBN 0-87201-781-8)

region (between 70% and 120% of BEP)—see also Chapter 5.

3. Proper pump/motor coupling alignment (Ch. 3).
4. Pump intake must conform to ANSI/HI 9.8 (pump intake design).

The HI standard further bases the allowable vibration levels on pump power, Table 12-2.

It should also be noted that the acceptable vibration values for slurry and solids handling pumps are about double the values given for horizontal clean liquid pumps.

Once a pump is accepted and commissioned, somewhat higher total (all-pass) vibration values are usually accepted before further follow-up and analysis are deemed appropriate. As a general rule, follow-up is recommended if vibration levels increase to twice the "field" acceptance limits (or initial actual readings).

CAUSES OF EXCESSIVE VIBRATION

Once a pump has been determined to have a high "total/all-pass" vibration level, the next step is to identify the cause. This is the time to obtain a filtered vibration analysis (see Figure 12-11A).

You cannot just assume that the rotor is out of balance (which could be the case). There are many other potential culprits. Machinery vibration problems often result from the interaction between an exciting force (hydraulic or mechanical) and the associated structural and/or hydraulic resonance frequency response. The

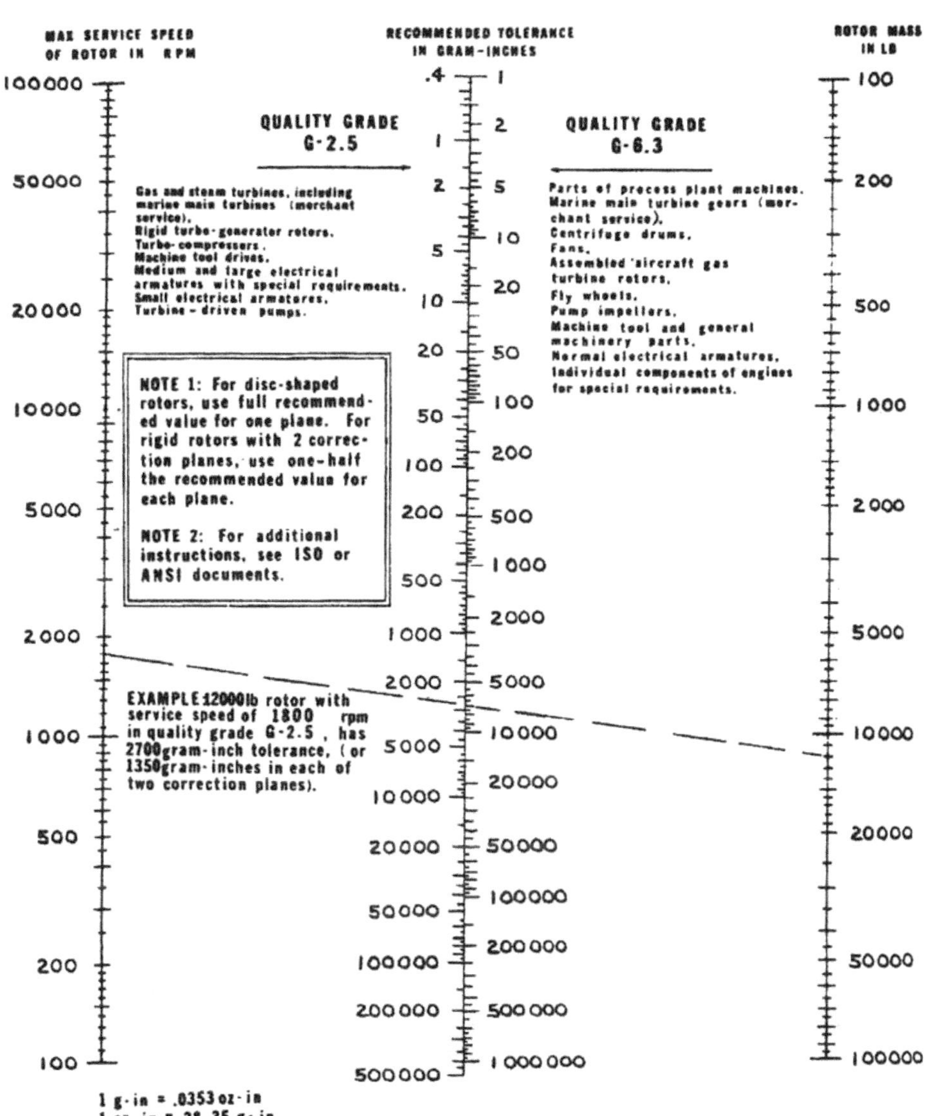

Figure 12-10: Balance tolerance nomogram for G2.5 and G6.3, large rotors (Source: Bloch, Heinz P., and F.K. Geitner, "Machinery Component Maintenance and Repair," 2nd Edition, 1990, Gulf Publishing Company, Houston, Texas, ISBN 0-87201-781-8)

Table 12-1: Maximum Diametral Hub Looseness

Impeller Hub Bore	*Maximum Diametral Looseness*	
	< 1,800 rpm	*1,800-3,600 rpm*
0 — 1.499 in.	. 0015 in.	. 0015 in.
1.5 — 1.999 in.	. 0020 in.	. 0015 in.
2.0 in. and larger	. 0025 in.	. 0015 in.

stronger the exciting force, and/or the closer these exciting force(s) are to a component natural frequency, the greater the vibration amplitude. The object for any pump vibration analysis should be to find the root cause of the excessive vibration, and then determine how to correct it. Basic vibration types to evaluate include: Rotor lateral vibration (x,y or z axis), and torsional vibration (commonly occurring with reciprocating engine drives). Structural lateral vibration is quite common with long-shafted vertical pumps, although not limited to these structures.

Table 12-2: Allowable pump field-installed vibration values

PUMP TYPE	Less than this HP	Vibration RMS value, in/sec	This HP and above	Vibration RMS value, in/sec
End Suction ANSI B73 (within POR)	268	.15	268	.19
End Suction ANSI B73 (outside POR but within AOR)	268	.19	268	.25
Vertical Inline, Sep. Coupled ANSI B73.2 (within POR)	268	.15	268	.19
Vertical Inline, Sep. Coupled B73.2 (outside POR but within AOR)	268	.19	268	.25
End Suct., API 610, Preferred Operation Region (POR)	**All**	**.12**	**All**	**.12**
End Suct., API 610, Allowable Operation Region (AOR)	**All**	**.16**	**All**	**.16**
End Suct. Paper Stock (Tested with clear liquids, within POR)	268	.15	268	.19
End Suct. Paper Stock (Tested with clear liquids, outside POR but within AOR)	268	.19	268	.25
Vertical Turbine Pump, and Mixed Flow Propeller, Short Set (within POR)	268	.13	268	.17
Vertical Turbine Pump, and Mixed Flow Propeller, Short Set (Outside POR, but within AOR)	268	.17	268	.22
Between Bearings, Single & Multistage (within POR)	268	.15	268	.19
Between Bearings, Single & Multistage (Outside POR, but within AOR)	268	.19	268	.25
Horizontal Solids-Handling (within POR)	33	.25	100	.31
Horizontal Solids-Handling (Outside POR, but within AOR)	33	.33	100	.40
Vertical Solids-Handling (within POR)	33	.30	100	.34
VerticalSolids-Handling (Outside POR, but within AOR)	33	.39	100	.44
Slurry (within POR)	All	.25	All	.25
Slurry (Outside POR, but within AOR)	All	.32	All	.32

Table 12-3: Sources of Specific Vibration Excitations

FREQUENCY	*SOURCE*
0.1 × Running Speed	Diffuser Stall
0.8 × Running Speed	Impeller Stall (Recirculation)
1 × Running Speed	Unbalance or Bent Shaft
1 × or 2 × Running Speed	Misalignment
Number of Vanes × Running Speed	Vane/Volute Gap & Cavitation

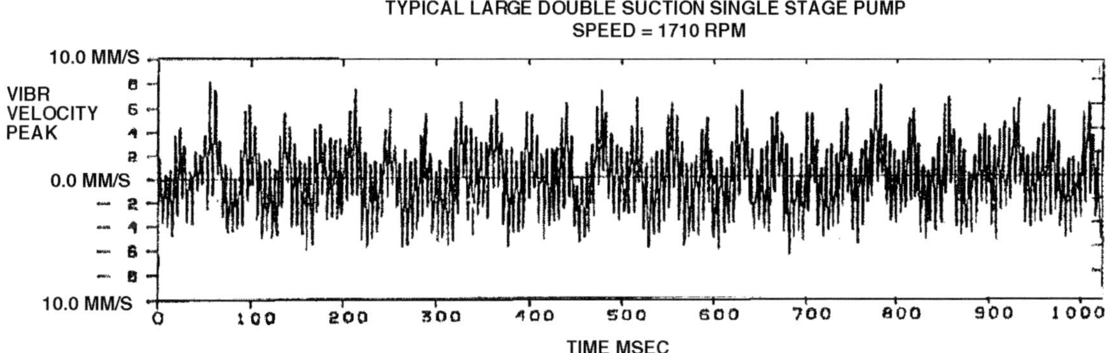

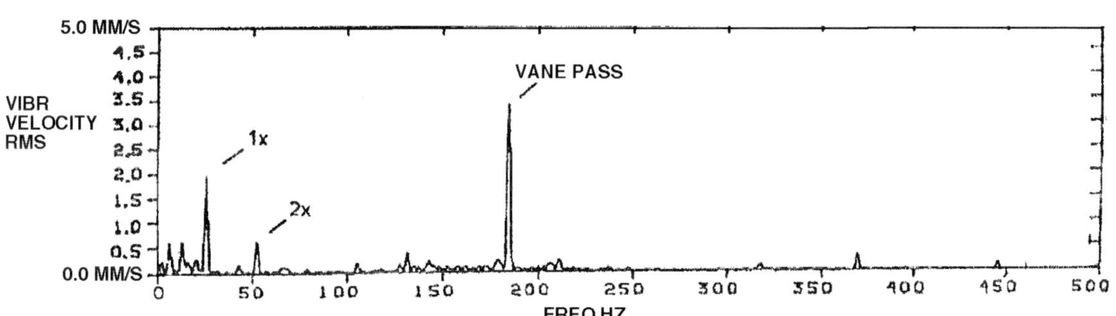

Figure 12-11A: Frequency spectrum showing (top) time-velocity plot and, (bottom) frequency vs. vibration amplitude plot

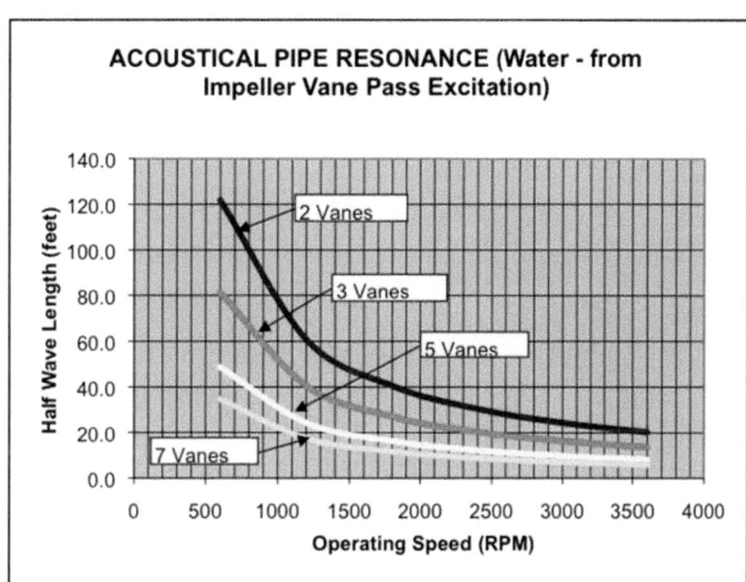

Figure 12-11B: Acoustical Pipe Resonance Half Wave Length, based on Pump Speed and Number of Impeller Vanes

CAUSES OF EXCESSIVE VIBRATION EXCITATION:

There are many potential vibration excitation sources, which can often be identified through a filtered vibration analysis. Figure 12-11A shows some of these filtered vibration peaks, as multiples of the pump running speed and number of impeller vanes. A further breakdown of potential excitation sources, with their specific frequency signatures (as multiples of the pump running speed), is shown in Table 12-3. It should be noted that the "diffuser stall," "recirculation" and "cavitation" sources, listed in Table 12-3, are hydraulic in nature, with the remaining being mechanical. The most common causes of excessive pump vibration excitation are:

1. Rotor unbalance, caused by new residual impeller/rotor unbalance, or unbalance caused by impeller metal removal (wear).
2. Shaft (or coupling) misalignment.
3. Liquid turbulence due to operation too far away from the pump best efficiency flow rate, and/or operation in the low flow pump suction recirculation region (see Figure 12-12).
4. Cavitation due to insufficient NPSH margin, especially with "High Suction Energy" pumps (see chapter 5).
5. Pressure pulsations from impeller vane/casing tongue (cutwater) interaction, in high discharge energy pumps.
6. Poor pump suction or discharge piping (having turns or fittings too close to the pump), can also cause increased vibration, normally by increased local velocities, which in turn, increases cavitation, and/or causes turbulence within the pump.
7. Bearing wear: Rolling element bearings have distinct vibration signatures based on the number of bearing balls or rollers.
8. Opening up of impeller wearing ring clearances, which can reduce the NPSH margin by shifting the pump operating flow point to higher levels.
9. Broken rotor bars on electric motors, which will generate specific frequencies.

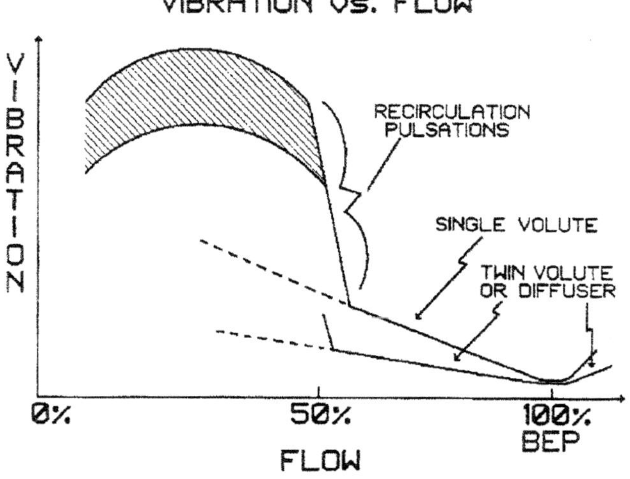

Figure 12-12: Hydraulic Institute vibration values apply only within preferred operating region of pump, i.e. vibration due to recirculation may greatly exceed anticipated vibration amplitudes

Hydraulic Pump Vibration Excitation—Amplitude

The amplitude of hydraulic pump excitations, especially those at blade pass frequencies, is typically a function of one of two pump energy factors: "High *Suction* Energy" and/or "High (*discharge*) Energy." According to the Hydraulic Institute, pumps with heads greater than 650 ft/stage that require more than 300 hp/stage, have high (discharge) energy, while high (and very high) suction energy (suction specific speed x inlet tip speed x specific gravity) has been defined in Chapter 5. It should be noted that the excitation levels related to high suction energy are also a function of the pump NPSH margin (NPSHA/NPSHR), see Chapter 5. The pressure pulsation excitations generated from either high (discharge) energy, or high (or very high) *suction* energy pumps, both occur at impeller vane-pass frequency, and each can also cause destructive acoustic pipe resonant vibration under the right conditions. Generally speaking, the higher the suction and/or discharge energy level the greater the vibration amplitude.

The most common pump *suction* hydraulic exiting forces come from turbulence or cavitation within the pump suction, which can become a problem when the suction energy at the pump impeller inlet is high enough. One of the ways to reduce pump suction energy is to reduce the pump speed with a VFD, to achieve low suction energy. However, a VFD can also make it much more likely that a pipe, foundation or pump resonance will be excited, so it can be a double-edged sword.

Discharge pressure pulsations are typically caused by the interaction between the different pressures (and velocities) on the leading and trailing surfaces of the impeller vanes, at the impeller O.D. and the casing cutwater (volute tongue). The higher the pump head and the smaller the gap between the impeller O.D. and the volute tongue, the greater the discharge pressure pulsations. These discharge pressure pulsations also occur at vane pass frequency.

RESONANCE RESPONSE:

Amplified resonant vibration response root causes are generally more complex to analyze. They typically result from pump operation at speeds (running or vane pass) close to a mechanical or hydraulic resonant frequency of a major pump, foundation or pipe component. This is of special concern with variable speed, large multistage horizontal and/or vertical pumps. A margin of safety should be provided between the pump/vane-pass speed/frequencies, and any major structural (and/or hydraulic) natural frequencies. Typical acceptable margins are in the range of 15-25%. The magnitude of the vibration response can be amplified 2.5 times or higher at or near a component natural (critical), or resonant frequency.

In addition to *structural* resonant frequencies found in pump suction and/or discharge piping, pump and motor foundations, and in vertical turbine pump installations, structural resonances are also quite common in large multistage horizontal pumps.

Acoustic Resonance Response:

Acoustical pipe resonances (Ref. 12-11), which can be excited by pressure pulsation from the pump (see Figure 12-11A), can also cause excessive pump and/or pipe vibration. This occurs when a high (suction or discharge) energy frequency excitation (typically vane-pass), corresponds with the frequency that will generate a standing acoustic wave inside a fluid filled channel.

Although acoustic pulsations are not a frequent or common source of excessive vibration in pumping systems, their occurrence is increasing with the growing popularity of variable speed pump drives. When resonant pulsations do develop in a liquid pumping system, they tend to be high in amplitude and can cause severe vibration problems. Pre-installation analyses and field testing may need to be performed to avoid an acoustic resonance problem.

Acoustic resonance can occur in the suction and/or discharge piping, in the long crossover of multistage pumps, and in other fluid filled channels, in response to a pump hydraulic excitation frequency. The phenomenon takes place when the return of a reflected pressure wave, generated by a periodic excitation, coincides with the generation of the next pressure pulse. In that case, a standing wave forms inside the fluid filled channel.

The channel length, for acoustic resonance, is determined by a reflective condition at the end of the channel opposite to the induction of the pressure wave. The channel length may be equal to quarter, half, or full wavelengths and multiples thereof. Normally only the first and second harmonics need to be considered. Typical reflection points include orifice plates, reducers or increasers (with a 50% or more velocity change), tees and partially closed gate or check valves. Long radius elbows do not reflect waves, while short radius elbows and vaned elbows may reflect waves.

SOME CORRECTIVE MEASURES:

Effective corrective measures that address resonance caused high vibration problems, are ones that detune the resonance conditions, by providing a minimum 15 to 25 percent separation margin between the pump excitation and piping acoustic and/or structural resonance frequencies. The following detuning actions should be considered:

- Change the vane count of the impeller (if the pump manufacturer offers this option) feeding the offending channel (pipe length). This changes the excitation frequency and therefore the wavelength (see Figure 12-11B).
- When a VFD is controlling the pump speed, it may be possible to "lock out" the offending speed ranges from continuous operation.
- If feasible, change the physical length of the offending (resonance) pipe.
- Install an acoustic filter (when the issue is an acoustic resonance), but filters typically must be matched to the system and resonant frequency.

If detuning is not feasible (or is the issue), the applicable excitation levels can also be reduced by implementing one or more of the measures listed below:

- If the excessive vibration is at running speed, balance the rotor, as discussed in the prior section of this chapter.
- If the excess vibration is at vane-pass frequency and the pump has high suction energy, reduce the pump suction energy level by reducing the pump speed, and/or increasing the impeller vane overlap, if it is less than 15 degrees. Pressure pulsation levels are proportional to the square of the speed.
- Also if the vibration is at vane-pass, and the pump has high suction energy, increase the NPSH Available (or reduce the pump NPSHR) to reduce cavitation (see chapter 5 for desired minimum NPSH Margin Ratios).

- Cut back on the volute cutwater tips (preferably with an angle cut) to increase the B-gap (see page 411), which should reduce the discharge excitation levels, if the pump has high discharge energy and the excessive vibration is at vane pass.
- Minimizing the impeller vane outlet thicknesses by means of overfilling, again for high discharge energy pumps with vane-pass vibration.

USING PREDICTIVE MAINTENANCE AS YOUR DECISION MAKING TOOL*

Pump Condition Monitoring Examined

The extent and effects of internal wear in centrifugal pumps vary with the nature of the liquid pumped, the pump type and its operating duty. Some pumps last for years, others for only months.

Overhauling of pumps on a fixed time or schedule basis is rarely the most cost-effective policy. Use of condition monitoring ensures that pump overhauls or performance restoration efforts are done when really necessary. However, despite the many excellent pump textbooks, there is little information available on how to apply condition-based maintenance to pumps.

Monitoring methods should be chosen that would detect each of the various failure modes that can be foreseen to occur:

- Vibration monitoring and analysis (probably the most widely applied method of condition monitoring for rotating machines in general, and suited to detect such faults as unbalance, misalignment, looseness)
- On large pumps, sampling and analysis of lubricants for deterioration and wear debris (relevant for bearings/lubrication system faults)
- Electrical plant tests (relevant for motor condition)
- Visual inspection and Non-Destructive Testing (particularly relevant for casing wear)
- Performance monitoring and analysis (relevant for pump internal condition).

For critical machines, more than one method of condition monitoring may be justified. This segment will demonstrate use of vibration and performance analysis as inter-related and practical examples of condition monitoring.

The Head-flow Method of Pump Condition Monitoring

The most useful condition monitoring method is by head-flow measurement, because in addition to pump deterioration, it detects any changes in system resistance. The head-flow measurement method can be used for all pumps where flow, or a repeatable indicator of it, can be quantified.

Throttling the pump to obtain points over the full flow range is not necessary. Some points near the normal operating duty point are sufficient to reveal the effects of wear, usually shown by the head-flow curve moving towards the zero flow axis by an amount roughly equal to the internal leakage flow. This is demonstrated by "test points-worn pump" on Figure 12-13.

A series of test readings at steady conditions at about 15 second intervals is sufficient; the average values are being plotted. Speed must be measured for variable speed pumps, and the head-flow data corrected to a standard speed (Ref. 12-2).

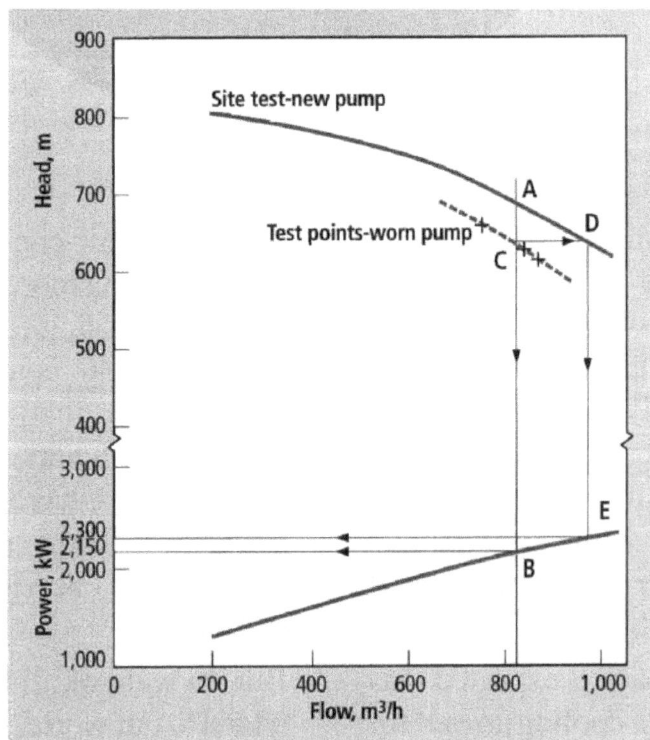

Figure 12-13: Head-flow of new pump and head-flow points from worn pump

*This major segment was contributed by Ray Beebe, Senior Lecturer at Monash University, Gippsland School of Engineering, Churchill, Victoria 3842, Australia. It is based on testing and analytical work performed under his guidance.

Field tests sometimes give results slightly different from the manufacturer's shop tests because actual site conditions for flow and pressure measurement are rarely as required by the various standards for pump testing. However, for monitoring, we are seeking relative changes rather than absolute accuracy. Non-intrusive ultrasonic flowmeters are applicable in most cases. A permanent flowmeter installed as part of a pump's minimum flow protection or process measurement can be used, provided its long-term condition is considered to be constant, or it can be inspected regularly.

Such performance information shows the extent to which a pump has deteriorated, and pumps can be prioritized for overhaul on the basis of their relative wear.

Shut-off Head Method

Measuring head at zero flow is a simple test. It is only possible where it can be tolerated; it thus excludes high-energy or high-specific-speed pumps, where power at shut-off is greater than that at duty point.

With the discharge valve closed fully for no longer than 30 seconds, suction and discharge pressures are read when steady. The liquid temperature is also needed to find the density, which is used to convert the pressure readings into head values.

Wear of vane outer diameters will show readily, as the head-flow curve of a worn pump moves towards the zero flow axis. To show sealing wear ring degradation, the pump head/flow curve needs to be relatively steep. (Note that if the pump has a rising curve, internal leakage will initially give an increase in shut-off head).

Figure 12-14 shows the trend in degradation of a 230 kW pump over three years. Wear amplitude is expressed at duty point flow, as the percentage reduction in total head compared with the new datum condition. This is usually derived from head-flow tests near duty point, but can also be obtained using the shut-off head test where this is allowable.

Please note that such testing requires that two frequently ignored conditions be met:

1. Pressure gauge connections must be provided on the pump suction and discharge sides

2. All pressure gauges must be tested and calibrated before data taking. Using a brand new gauge is no guarantee that readings will be correct.

Thermodynamic Method

Another pump monitoring method is to measure the temperature rise of the liquid through the pump, which reflects pump inefficiency. Since the differential temperature is very small, great care is required to measure it. Any recirculation effects at pump inlet and outlet must be eliminated, and tests are not possible at very low or zero flows. Efficiency can be calculated from the inlet temperature, differential temperature and head. If efficiency changes with time, comparisons can be made on efficiency vs. head plots. For high head pumps, an allowance must be made for the isentropic temperature rise that occurs as a result of pressure increase (Ref. 12-3).

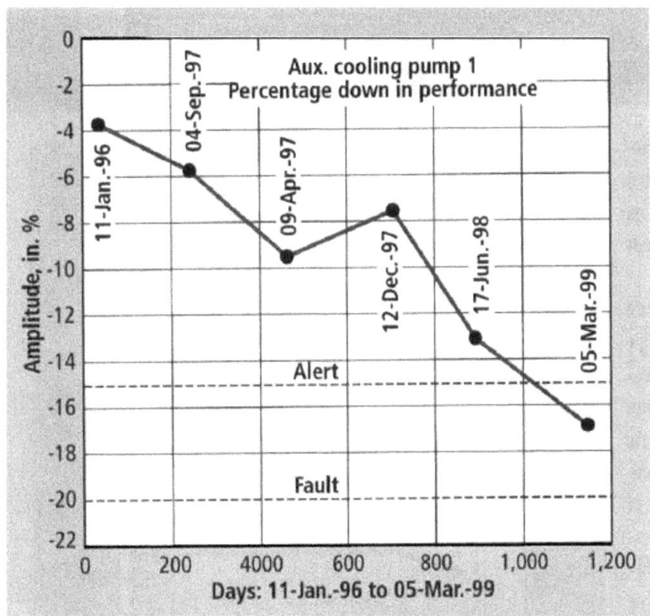

Figure 12-14: Measured degradation of a 230 kW pump by head-flow testing

A commercially available device is widely used in the UK water industry (Ref. 12-4). Pressure taps at suction and discharge are required to be two diameters away from pump flanges. These taps allow for the installation of pressure/temperature probes. Clamp-on type detectors are placed to measure motor power. Pump efficiency is then found from precise head and temperature rise measurements through the pump. From assessment of motor losses, power absorbed by the pump is computed. From all these data, the pump flow can be found.

For condition monitoring, tests at around normal operating point are usually sufficient. The thermodynamic method would be more attractive economically if no special tapping points were required. Research at Monash University on high head pumps using special semi-conductor temperature probes on the piping outside surface, covered with insulation, gave usable results, provided the pump is being operated at steady conditions for 30 minutes to allow the piping temperature to stabilize (Ref. 12-5).

Percent efficiency for pumps in water service up to 54°C is given by the empirical formula given below, which includes a correction for the isentropic temperature rise (total head is in kPa, temperatures in °C). See Ref. 12-6 for discussion.

$$\text{Pump efficiency} = \frac{100}{[1-0.003(\text{inlet T}-2) + 4{,}160\,(\text{Temp. Rise/Total Head})]}$$

Balance Flow Measurement

Multistage pumps with the impellers facing in the same direction usually have a balance disc or drum arranged such that final stage discharge pressure counteracts the axial thrust on the shaft line. This makes it possible to monitor pump condition by measuring the leak-off from the balance device (Ref. 12-7). The basis is that if increased wear in the annular space to the balance device is evident from increased leak-off flow, then the interstage clearances are also worn.

As the leak-off line is quite small compared to the pump main flow piping, a permanent flowmeter is relatively inexpensive. For some years, overhauls have been scheduled on this basis on a number of boiler feed pumps. Flows are read manually, and trends plotted using a database program (Figure 12-15). Note that here the balance flow of 15 l/s corresponds to about 10% of the duty flow, and about 250 kW of extra power. When added to the likely internal recirculation, this would mean that an even larger proportion of the power absorbed is being wasted. These pumps are variable speed, and tests show that the measured flows must be corrected in direct proportion to the speed.

On a set of pumps of another design elsewhere, both head-flow and balance flow were measured for some years, but no correlation was found.

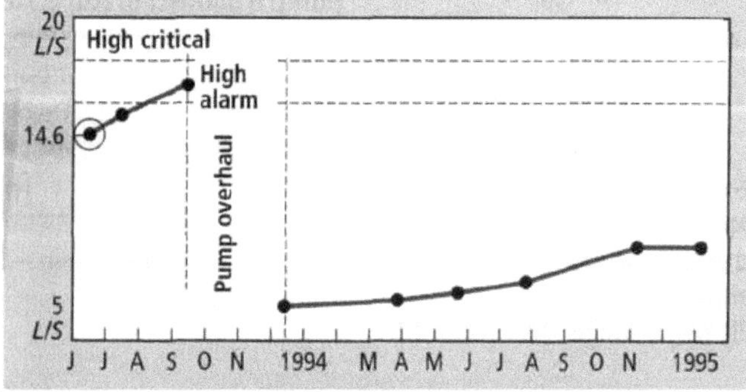

Figure 12-15: Condition monitoring of a high energy multi-stage pump by measurement of balance device leak-off flow (flows are corrected to a standard pump speed)

On yet another 11-stage pump, the head-flow performance was tested as being well below the datum curve. As the pump was dismantled, measurements showed that interstage clearances were not worn. A condition monitoring credibility crisis was averted when the balance seat area was reached and found to be severely eroded. Balance flow had obviously been very high. For the best monitoring, both head flow and balance flow should be measured, particularly if the balance area can be separately dismantled in the field.

Optimum Time for Overhaul

The most economic time to restore lost performance by overhaul will vary with the circumstances. If the *deterioration is constant over time,* then a cash flow analysis can be done to ensure that the investment in overhaul will give the required rate of return. This is the same process as used in deciding on any investment in plant improvement.

If the *deterioration rate is increasing with time,* then the optimum time for overhaul will be when the cumulative cost of increased electricity consumption equals the overhaul cost.

The method is now described for some plausible scenarios and actual situations.

Pump Deterioration Results in Reduced Plant Production

Where the cost of overhaul is insignificant in proportion to the cost of lost production cost, prompt overhaul is usually simply justified at a convenient "window."

Pump Which Runs Intermittently to Meet a Demand

In a pumping installation such as topping up a water supply tank or pumping out, deterioration will cause the pump to take more time to do its duty. The extra service time required therefore results in increased power consumption that can be related to the overhaul cost.

Pump Deterioration Does Not Affect Plant Production, at Least Initially: Constant Speed, Throttle Valve Controlled Pump

The internal wear does not cause any loss in production from the plant, as the control valve opens more fully to ensure that pump output is maintained. Eventually, as wear progresses, pump output may be insufficient to avoid loss of production, or the power taken will exceed the motor rating.

Earlier, Figure 12-13 showed the head-power-flow site test characteristics of such a pump. Its output is controlled using a throttle control valve. The duty flow is 825 m^3/h, and the duty point in the new condition is "A." The power absorbed by the pump is read off the power-flow curve as 2,150 kW, "B." The power-flow curve should ideally be found on site, but factory test data may have to suffice.

After some service, the plotted "test points-worn pump" indicate internal wear. When worn to this extent, the operating point moves to "C," as the system resistance curve lowers when the throttle valve is opened further.

The increased power required in the worn condition can be estimated by extending from the head-flow curve at constant head from the operating point to "D," and then intersecting the power-flow curve for new condition at constant flow, "E." Follow the arrowed line in Figure 12-13. This assumes that the original curve still represents the flow through the impellers (of which less is leaving the pump to the system). Of course, the power could be measured on test at extra expense if the pump was motor-driven.

In our example, power required for this duty in the worn condition is shown in Figure 12-13 by the projection from the duty flow of 825 m^3/h to the test curve to find 640 m head, then across to the "site test-new pump" curve, then down to the power curve, to find 2,300 kW.

The extra electric power consumption is therefore 2,300 − 2,150 = 150 kW which, divided by the motor efficiency (here it is 90%), will yield 167 kW.

If the wear ring clearances are known by previous experience of correlation with measured performance, or if the pump is opened up already, the extra power consumed likely to be saved by overhaul can be estimated (Ref. 12-8).

Using this method, a number of pumps of varying wear conditions could be prioritized for maintenance, based on their increased power consumption and their relative costs of overhaul, i.e. the cost/benefit ratio.

Finding the Optimum Time for Overhaul from Head-flow Data

For this example, the test points were obtained following 24 months of service since the pump was known to be in new condition; an overhaul would cost $50,000; electricity costs 10¢/kWh, and the pump is in service approximately 27% of the time.

Tests showed that the cost of deterioration had reached 167 kW × 0.10 × 720 × 0.27 = $3,240/month (taking an average month as 720 hours).

Since the time now is 24 months, $3,240 ÷ 24 gives the *average deterioration cost* rate as $135 per month.

The optimum time for overhaul can be calculated (Ref. 12-9), from

$$T = [(2)(O)/C]^{1/2}$$

where:

O = cost of overhaul
C = cost rate of deterioration

The result here is T = 27.2 months.

But it is better to calculate and plot the average total cost/month values for a range of times. Seen clearly will be the cost impact of doing the repairs at some other time, such as at a scheduled plant shutdown.

Calculating Average Cost Per Month:

In this example, take the time at 22 months.

Average monthly cost to cover overhaul:
$50,000/22 = $2,273/month

Average extra energy cost:
$135 × 1/2 × 22 = $1,485/month

Total average cost/month is the sum of these two figures =
$3,758/month

Repeat this calculation for several months, perhaps using a spreadsheet, and look for the minimum total cost, which is at 27.2 months. If plotted as cost/month against time, the resulting curves will show the overhaul cost per month dropping with time, with the lost energy cost increasing with time. (The time value of money could also be taken into account, if required). Usually the total cost curve is fairly flat for +/- 20% or so.

If the overhaul were delayed until, say, 30 months, then the accumulated cost of lost energy would have reached $135 × 1/2 × (30)^2 = $60,750. At 27.2 months, the cost is $135 × 1/2 × (27.2)^2 = $49,939. The cost of delaying overhaul is thus the difference, $10,811.

Note that this calculation is only correct *if the wear progresses at a uniformly increasing rate with time*, but, as Figure 12-14 shows, this is not always so. Information may not be available to make any other assumption, but decision makers have to start somewhere! There are other formulas for nonlinear rates of change (Ref. 12-9).

Note that some relatively small pumps may never justify overhaul on savings in energy use alone, but may be justified on reduced plant production rate.

When Pump Deterioration Does Not Affect Production

For a pump where the speed is varied to meet its desired duty, the effect of wear on power required is much more dramatic than for the case of a constant-speed, throttle-controlled pump. This is because the power increases in proportion to the speed ratio cubed.

Unless pump output is limited by the pump reaching its maximum speed, or by its driver reaching its highest allowable power output, no production will be lost. However, power consumed will increase more dramatically for a given wear state than for a constant-speed pump.

To estimate power required in the worn state, the head-flow curve must be drawn for the *current higher speed in the new condition.* Select a head-flow point on the original new condition curve, and correct it to the higher speed. Multiply the flow by the speed ratio and multiply the head by the square of the speed ratio. Repeat this for some other points at flows above duty flow to draw the new condition head-flow curve.

Follow the same method and calculations as before to find the time for overhaul at minimum total cost. The operating point is projected from the worn curve to the new curve at the same speed as the worn curve. Figure 12-16 shows the performance of a variable speed pump. When new, operation at 1,490 rpm meets the desired duty flow at operating point "A," requiring 325 kW power, point "B." After some time in service, internal leakage and "roughness" has increased such that the pump must run at 1,660 rpm to meet the required duty—still point "A."

To estimate the power required now, the head-flow curve must be drawn for the higher speed in the new condition. Several head-flow points are selected and corrected for the higher speed: multiply each flow by the speed ratio, and multiply each matching head by the speed ratio squared. This will result in the head-flow curve at 1,660 rpm in the new condition.

The head at the duty flow -point "A"- is projected across to meet the head-flow curve at 1,660 r/min (new condition, line "C" in Figure 12-16). Projection downward at constant flow leads to the increased power required at 425 kW. The extra power demand is 31%!

The same calculations as before are followed to find the time for overhaul for minimum total cost.

Optimization Using Shut-off Head Test Results

The shut-off head test information can also be used to estimate power used in the worn state, and do the optimization calculations explained previously.

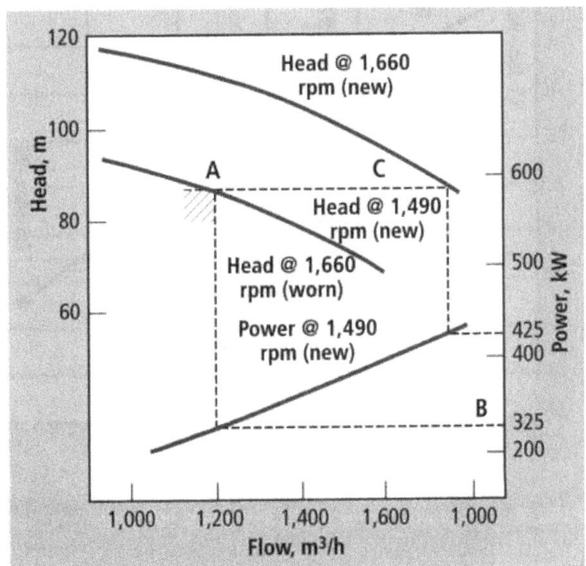

Figure 12-16: Head-flow-power characteristics of new variable speed pump, and head-flow points from worn pump

Head-power-flow characteristics in the "new" state are needed as before, and the operating point must be known. Note power required at operating point as before. Make an overlay trace of the head-flow curve in the new condition. Place it over the "new" curve and move to the left horizontally until the curve cuts the head axis at the shut-off head value obtained on the test. The trace is now in the position of the "worn" head-flow curve that is being experienced. Exactly the same process can be followed as explained previously.

Worn Vertical Multistage Pump

Both vibration and performance analysis were used to help solve a case where a pump experienced excessive vibration. This involved a 9-stage mixed flow pump, rated at 70 l/s with a head at 155m, operating at 1,480 rpm. Only the motor and the pump discharge piping were visible above the plant floor, with the base flange bolted to the floor.

For condition monitoring by performance analysis, a permanently installed orifice plate flowmeter was used. An ultrasonic flowmeter showed close agreement. Vibration measurements were taken in the horizontal and axial directions on the motor flange.

The pump had not yet been added to the routine vibration program. High vibration was reported and vibration measurements showed 14 mm/s (0.55 ips) horizontal vibration, all at 7.5 Hz. This level is not acceptable for reliable service. (See page 344.)

Lower down the casing, the vibration was half

this level, but mostly at 50 Hz, with some at 7.5 Hz. No vibration was evident at the running speed frequency of 25 Hz. More frequent monitoring commenced, and a week later, the vibration had increased to 22 mm/s (0.87 ips), all at 6.5 Hz. The performance at the usual operating point had also decreased by 18%.

Since an alternative flow path was available without reducing production, albeit reducing plant efficiency, the pump was removed for overhaul. All impellers and wear rings were found damaged beyond repair, and the shaft was bent.

After repairs, the highest vibration level was below 1 mm/s (0.04 ips) at 25 Hz and 50 Hz frequencies, and tested performance had also returned to the datum level. Routine measurements were commenced at three-month intervals.

It was suspected that the base mounting bolts had worked loose sufficiently to allow the pump to vibrate at its natural frequency of about 7.5 Hz. This was thought to have resulted in internal damage.

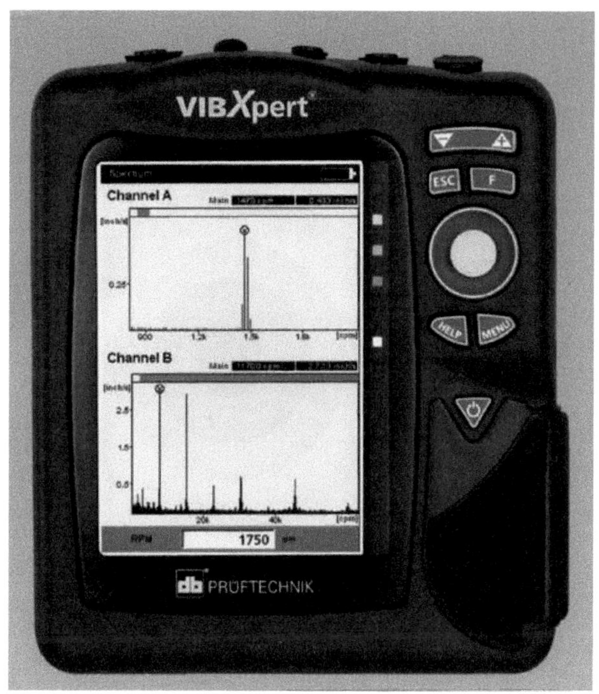

Figure. 12-17: Modern light-weight data collector and vibration analyzer (Source: Prueftechnik, Germany, and Ludeca, Inc., www.ludeca.com)

Understand Your Measurement Options

The authors believe that Ray Beebe's work at Monash University shows what can be accomplished by careful measurements on a variety of pump types. The rapid expansion of plant-wide distributive control systems (DCS) has opened another opportunity.

It is true that few process pumps (if any) are equipped with calibrated flowmeters, temperature sensors and pressure gauges. However, a process unit can be modeled to an extent where inlet and discharge pressures as well as power consumption of each pump for a given set of operating conditions can be predicted. Only a few selected parameters need to be continuously monitored in order to detect unusual events that might indicate the need for further troubleshooting. Data from pressure transmitters, thermocouples and electric motor current transmitters can now be added to plant control computer networks. Calculation of the cost of throughput reduction, or power waste, can readily be integrated in the DCS software. There is no need to wait for the detection of a fire or excessive vibration levels to initiate troubleshooting.

New data collectors and vibration analyzers enter the market every year and the user/owner of process pumps should keep informed on available models, configurations and options. Figure 12-17 represents just one of these, a lightweight, two-channel FFT data collector/vibration analyzer for monitoring and analyzing machine condition. As a data collector, this instrument records all forms of machine vibrations, bearing conditions, process data and visual inspection information. Collected machinery data are stored on mass storage cards and then transferred to suitable maintenance software for further evaluation, report generation and archiving.

Chapter 13

Pump Types and Materials (Special Reliability Considerations)

There are literally thousands of pump models and configurations in use today and it is certainly beyond the scope of this text to list them all, or to explain the myriad of variations available in the marketplace. However, experience shows that there are functional and application issues in certain pump categories. To the extent that these issues have reliability or cost justification impact, they deserve our examination and consideration.

PETROLEUM REFINERY SERVICE PUMPS (API-610)

Many refinery applications deal with flammable fluids. Safety and reliability are of foremost importance in these services, which led to the development of the API Standard 610. Although not a legally binding document, this well-known industry standard covers the basic, experience-based requirements that impart reliability to process pumps. It is of greatest importance in severe pump applications.

One of the most severe pump applications is "hot oil" (300°F/149°C to 850°F/399°C) service, which creates the following critical pump conditions:

Figure 13-1: Multistage pump in pipeline service. Note API-compliant centerline support (Source: Ruhrpumpen, Germany)

- High thermal expansion of pump components. The pump supports must maintain coupling alignment at these elevated temperatures. Casing centerline support, without a separate support of bearing housing, is normally used to address this problem (see Figure 13-1).

- High heat conductivity to pump bearings. In applications involving sleeve bearings, the oil may have to be kept cool through the use of cooling coils (Figure 13-2). Note, however, that
  (a) rolling element bearings do ***not*** require cooling if a lubricant of sufficiently high viscosity is chosen (Refs. 13-1 and 13-2).
  (b) synthetic lubricants are ideally suited for high-temperature pump bearings
  (c) since the early 1970's, thousands of high temperature pumps have been in successful high MTBF services after all cooling water was removed from their oil sumps and/or from their respective cooling water jackets.
  (d) cooling water jackets that surround or are in close proximity to bearing outer rings cool primarily the bearing outer ring, while the bearing inner ring remains at a higher temperature. This causes bearing-internal clearances to vanish and the bearing will experience excessive preload. Bearings surrounded by cooling water jackets are almost certain to fail prematurely (Ref. 13-2).

- Stuffing boxes may be provided with cooling jackets to maintain reasonable seal environment

Figure 13-2: Typical oil sump cooling provision

temperatures. However, these jackets are often too small to do much good. External cooling of the seal flush, or use of high-temperature bellows seals is inevitably more effective and is often the wisest choice (Ref. 13-2).

- Hot oil pumps deal with liquids near their boiling points. This may require pump impellers with low NPSH requirements, however, pumps with lower NPSHr accommodate only a narrow allowable operating flow range and are notorious for internal recirculation.
- Vapor buildup often occurs in the pump suction. Hot oil pumps must allow self-venting of vapors.
- These pumps typically handle flammable and/or toxic liquids. Since escaping liquid will often autoignite, mechanical seal selection must be driven by safety and reliability considerations.
- Many hot oil pumps have relatively high head requirements. Unless multi-staging is employed, hot oil pumps must operate at two pole speeds (3,600 or 3,000 rpm) to generate the necessary head with a single stage pump. This further aggravates the NPSH situation.
- Low lubricity liquids (specific gravities from 0.3 to 1.3). Mechanical seals must be chosen specifically for the oil, temperature, pressure and speed (see Chapter 8).

Cooling Water Issues Explained

Decades of solid experience with literally thousands of pumps have shown that cooling water is not needed for the majority of centrifugal pumps in process plants worldwide. Regardless of convention and tradition, there is no net advantage to the use of cooling water in pump bearing housings equipped with rolling element bearings.

In any case, for rolling element bearings that are properly installed and loaded per manufacturer's allowable guidelines, cooling the oil is neither needed nor advantageous. It may only be necessary to choose a higher-viscosity oil, if indeed warranted.

Cooling Water is For Sleeve Bearings. The correct oil viscosity is needed for reliable oil application and long bearing life. In other words, an oil ring will not function the same in lubricants with substantially different viscosities. Also, the oil film thickness developed in the bearing will differ for different oil viscosities.

It is intuitively evident that some heat is being conducted to pump bearings. Moreover, oil shear in bearings produces heat. Therefore, in may applications involving *sleeve* bearings, the oil may indeed have to be kept cool through the use of cooling water jackets that surround the bearings, or by cooling water coils immersed in the oil. Cooling thus tends to ensure that the correct lube oil viscosity is being maintained. Nevertheless, there are many sleeve bearings that will simply not require cooling water. As explained in Ref. 13-2, they can be identified by temporarily shutting off cooling water in a controlled test during which the steady-state oil temperature of a premium ISO Grade 32 synthetic is observed to remain below approximately 170°F (77°C).

To restate, while cooling may still be needed for *sleeve bearings* lubricated by oil rings, using cooling water jackets on *rolling element bearings* can prove disastrous to bearing life. Cooling water jackets that only *partially* surround bearing outer rings have often restricted the uniform thermal expansion of operating bearings and have been known to force bearings into an oval shape. There have been many instances where, as a result, bearing operating temperatures were *higher with* "cooling," and *lower after* the cooling water supply was disconnected and the jackets left open to the surrounding atmosphere.

Similarly, cooling water flowing through a coil immersed in the lube oil has also been known to cause problems. Adherence to this "traditional" cooling method very often invites vapor condensation in bearing housings. Unless the bearing housings are hermetically sealed (Ref. 13-2), moist air fills much of the bearing housing volume. Upon being cooled, the air sheds much of its water vapor in the form of liquid droplets. Since this condensate causes the lube oil to degrade, cooling the bearing environment can be indirectly responsible for reduced bearing life in pumps.

Regrettably, some pump manufacturers and installation contractors have been painfully slow in endorsing the deletion of cooling water. Others have been equally slow advocating superior synthetic lubricants and certain highly advantageous application methods. Where cost-justified, advantageous application could refer to pure oil mist ("dry sump") for both pump and driver. Pure oil mist certainly represents effective lubrication for both operating pumps and, especially, the protection of non-running pumps against harmful environments (Ref. 13-3).

In any case, for rolling element bearings that are properly installed and loaded per manufacturer's allowable guidelines, cooling the oil is neither necessary nor helpful. Instead of following misguided tradition, and for bearing housings from which cooling liquid has been removed, reliability-focused users choose the right synthetic base stock (generally diester or diester/PAO blend formulations, Refs. 13-2 and 13-3). Also, reliability-focused users select the correct viscosity and, as was discussed earlier, avoid the use of oil rings because of their serious limitations at today's higher operating speeds.

No longer using cooling water on hundreds of pumps with rolling element bearings in a petrochemical plant was estimated to save $120,000 in water consumption, piping maintenance and water treatment cost per year. In addition, deleting cooling water was thought to prevent at least three pump failures per year (at $ 6,700 each). The yearly savings were estimated to exceed $140,000 at this facility. This is again one instance where the implementation of a cost saving measure has positive reliability impact and instantaneous payback. All it takes is for reliability professionals to become familiar with decades of well-documented prior experience and the underlying scientific principles. Both are very easy to explain to plant personnel. If necessary, a simple well-instrumented test will convince even the most traditional doubters of the laws of physics.

Fortunately, the latest issues of API-610 recognize the above. This American Petroleum Institute Standard contains the collective judgment of pump users and pump manufacturers. While this standard has established minimum specifications for the design features required for centrifugal pumps used for general refinery service, it rarely discusses trade-offs and upgrade features of interest to leading edge pump reliability professionals. This is why Corporate Standards and supplemental specifications are often found necessary.

Figure 13-3 depicts an API-610 end suction centrifugal pump with an inducer in front of the impeller.

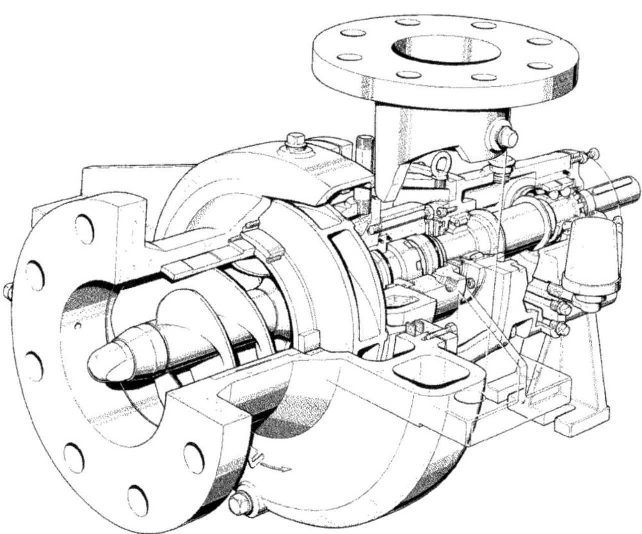

Figure 13-3: End suction API-610 pump with inducer (Source: Pompe Vergani, S.p.A., Merate/Como, Italy)

A typical API-610 vertical in-line version is shown in Figure 13-4. Both oil mist and oil sump lubrication are illustrated. Grease lubrication is discouraged on these pumps, as is bolting of the pump base to the support foundation. The entire pump is supposed to "float" with the piping, and the manufacturer's installation guidelines must be adhered to.

Other types of centrifugal pumps, in addition to "back pull-out" end suction, are also used for hot oil service and can be designed or upgraded to comply with API-610. However, it is important to note that radially split and barrel-type pumps are generally capable of handling both higher pressures and temperatures than their axially split comparison models.

A few of the many refinery pumps are shown in Figures 13-3 through 13-9. Vertical turbine pumps (Figure 13-5, Ruhrpumpen & KSB) are primarily used in low NPSH and multistage services. Between bearing axially split case double-suction pumps (Figure 13-6, Sulzer/Winterthur) are used for moderate-head, relatively high-flow services, whereas the radially split double-suction configuration of Figure 13-7 (Ruhrpumpen) is able to contain somewhat higher pressures.

Multi-stage horizontally split models (Figure 13-8, Sulzer/Winterthur, Switzerland) are used in medium pressure services. Many different flow schemes are possible for these and other multi-stage pumps. Note that the second and third impellers in this three-stage pump are mounted in mirror-image fashion so as to provide lower axial thrust.

Radially split multi-stage barrel pumps are needed for high-head applications, including, of course, many

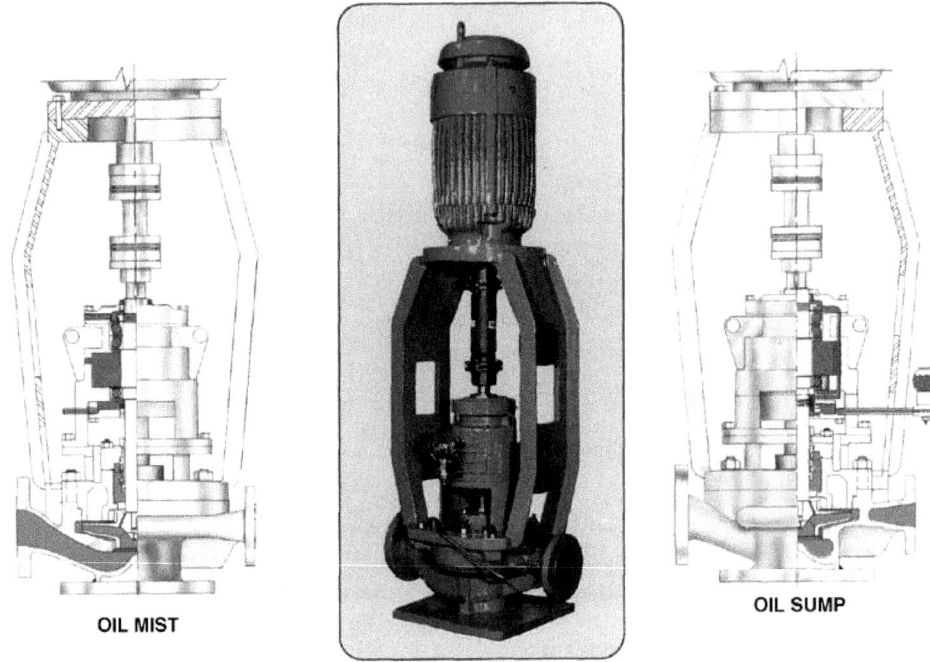

Figure 13-4: API-610 vertical inline pumps with (left) oil mist and (right) oil sump lubrication (Source: Afton Pumps, Inc., Houston, Texas 77011)

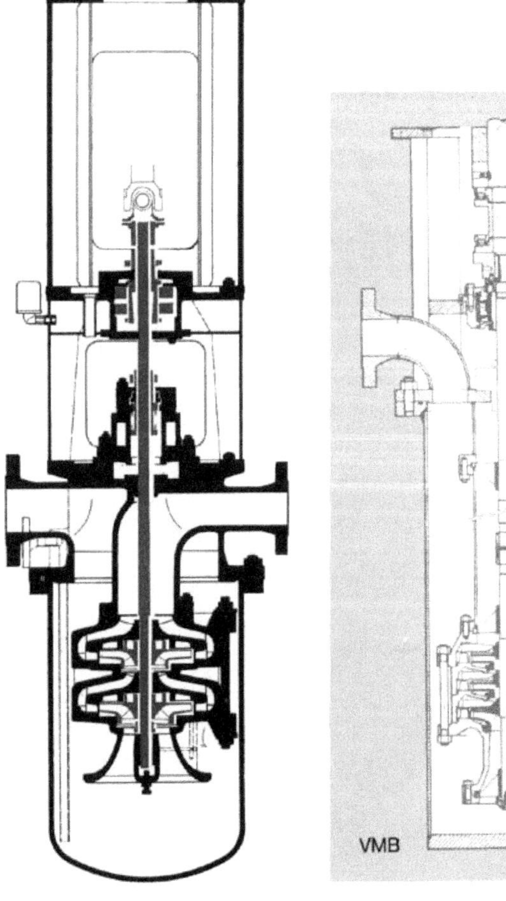

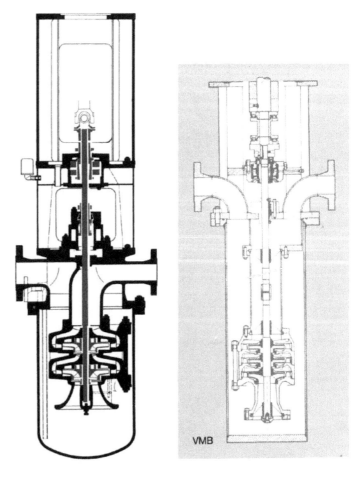

Figure 13-5: Vertical turbine pumps (VTP's) for low NPSH and cost-effective multistage applications (Sources: Ruhrpumpen, Germany, and KSB AG, Frankenthal, Germany)

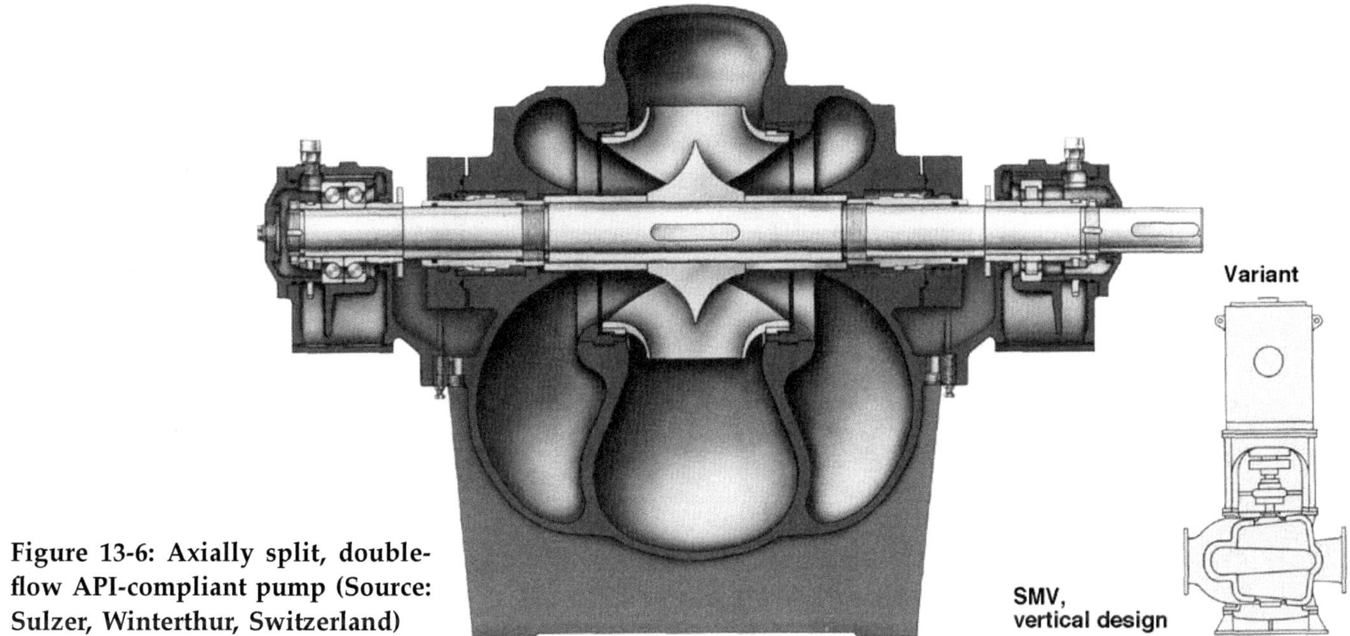

Figure 13-6: Axially split, double-flow API-compliant pump (Source: Sulzer, Winterthur, Switzerland)

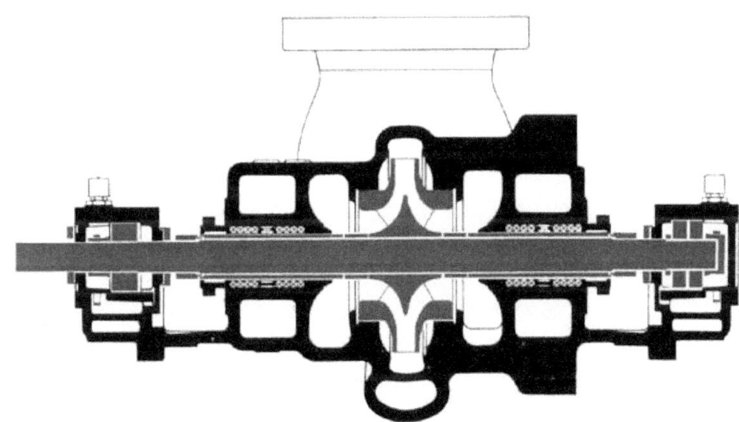

Figure 13-7: Radially split, double-flow API-compliant pump (Source: Ruhrpumpen, Germany)

low specific gravity products. Figure 13-9 depicts a three-stage pump with each impeller oriented in the same direction. Here, the manufacturer opted to use a balance piston to achieve a measure of thrust reduction. A balance line connects the space behind the balance piston to the pump inlet.

Pumps used for petroleum (hydrocarbon) services can usually survive with relatively small NPSH margin ratios, even high suction energy pumps (see Chapter 5), for several reasons:

1. Most hydrocarbon liquids have relatively low vapor volume to liquid volume ratios. This means that, if the liquid should vaporize at or near the pump suction (impeller inlet), the volume of the resulting vapor does not choke the impeller inlet passages as severely as does water vapor during cavitation. This results in a smaller drop in developed head for the same NPSH margin.

2. Less energy is released when hydrocarbon vapor bubbles collapse. The velocity generated at the instant of implosion is less, and this means less damage occurs as a result of cavitation. Avoiding cavitation is, therefore, not as important in hydrocarbon services as it is in other liquid services.

3. Hydrocarbon liquids are often a mixture of several specific hydrocarbon fractions, each with a different vapor pressure. This means that only a portion of the liquid will flash into vapor (or vapor will implode into the liquid state), at a given local inlet pressure, thus causing less pump damage.

When NPSH ratio margins are low, it is generally possible to use high, and even very high suction energy API-610 pumps in hydrocarbon liquid services. Satisfactory performance is occasionally feasible even

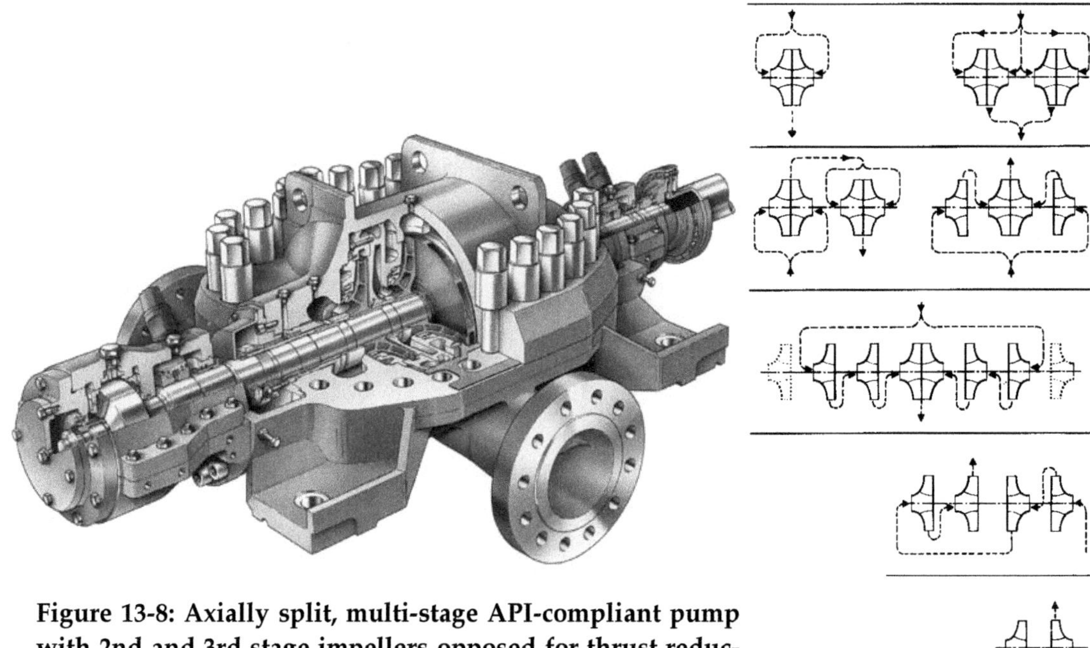

Figure 13-8: Axially split, multi-stage API-compliant pump with 2nd and 3rd stage impellers opposed for thrust reduction (Source: Sulzer, Winterthur, Switzerland)

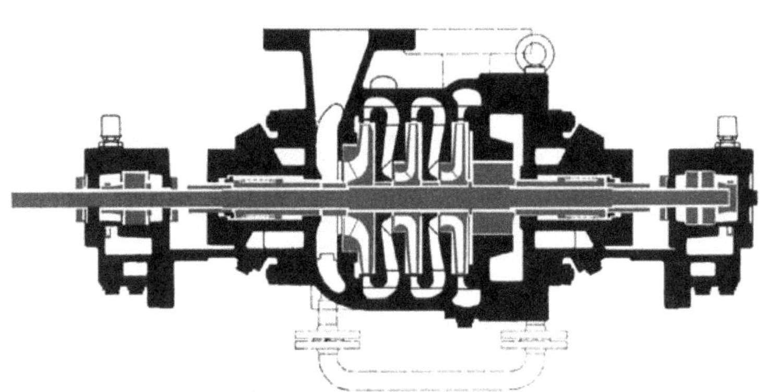

Figure 13-9: Radially split, multi-stage API-compliant barrel pump with balance piston for thrust reduction (Source: Ruhrpumpen, Germany)

though some suction recirculation exists. While factory performance tests of these pumps on water will be noisy, operation on hydrocarbon liquids may not necessarily be accompanied by unusual noise and vibration excursions.

API-610 MATERIALS OF CONSTRUCTION AND GRADE LEVELS

Many materials are utilized to satisfy refinery requirements, with cast steel (ASTM-A193 GR B7, as called for in API-610) the most common for pressure-containing components.

However, API-610 end suction refinery pumps are available in three general grade levels, which can affect the cost and reliability of the installation. The issues are how critical is the application and how much attention is given to initial cost vs. life cycle cost.

- Heavy-duty, full API-610 pumps. These pumps typically exceed the latest edition of API-610. They generally incorporate heavier shafts and bearing frames. While often highest in initial cost, heavy-duty, full API pumps should provide the longest mean time between repairs (MTBR).

- Full API-610 pumps. In general, these pumps meet, but do not exceed, API-610 requirements. They are lower in cost than heavy-duty versions, but will have greater shaft deflection when operated away from BEP.

- "Nominal" API-610 pumps. This is an unfortunate and perhaps even inappropriate designation for pumps that meet many, but not all, API-610 requirements. Configurations labeled "nominal API" generally have a back-bearing frame support. They are offered at somewhat lower cost than fully compliant API-610 pumps.

For lower temperature hydrocarbon applications (below 300°F), heavy duty ANSI B73 pumps might be acceptable and will, occasionally, be the lowest cost option (see Chapter 2).

MAGNETIC DRIVE AND POSITIVE DISPLACEMENT PUMPS

Magnetic drive end suction (Figure 13-10), canned motor (Figure 13-11), reciprocating (Figure 13-12), and various types of rotary positive displacement pumps have also been successfully used to process some of the many hydrocarbon liquids, both hot and cold. Each have certain advantages and disadvantages that merit careful investigation. It is certainly prudent to carefully review the often overly optimistic advertising claims. Moreover, just as is true for all other pumps purchased by reliability-focused users, these fluid movers should only be purchased from experienced vendors. The "seal-less" magnetic drive and canned motor pumps will be revisited a little later in the chapter.

CHEMICAL PUMPS

Chemical pumps, which handle corrosive and/or toxic liquids and slurries, are available in a variety of configurations and materials, including the seal-less pumps depicted in Figures 13-10 and 13-11. Erosion-corrosion, vapors, and solids in suspension are often encountered in chemical industry pump services. Pumps used in this industry are different from those used in other industries primarily in the materials of construction. The American Society of Mechanical Engineers has developed three ANSI standards, specifically for the chemical industry (ANSI/ASME B73.1-End Suction, B73.2 —Vertical-In-Line and B73.5 — Seal-less).

Figure 13-13 shows the most common configuration, a typical B73.1 pump. ANSI vertical in-line pumps are shown in Figure 13-14. Here, it should be noted that close-coupled units (L) are difficult to service and rigid-coupled models (M) have long, unsupported shafts which are subject to run-out, deflection and imbalance. In the interest of long bearing and seal life, integral bearing, flexibly coupled configurations (R) are usually preferred. The somewhat tall overall design "R" must ensure that resonant frequencies are avoided.

KEY SELECTION CRITERIA FOR CHEMICAL APPLICATIONS

There are many areas of concern that should be addressed at the specification stage and a good data sheet forces the specifying agency to stay focused in this regard. However, many users and their reliability staff are confronted by tasks that range from justifying repair-in-kind, to component upgrading, and even to full replacement of existing pumping equipment. Here, it might be helpful to include the following criteria in arriving at a decision:

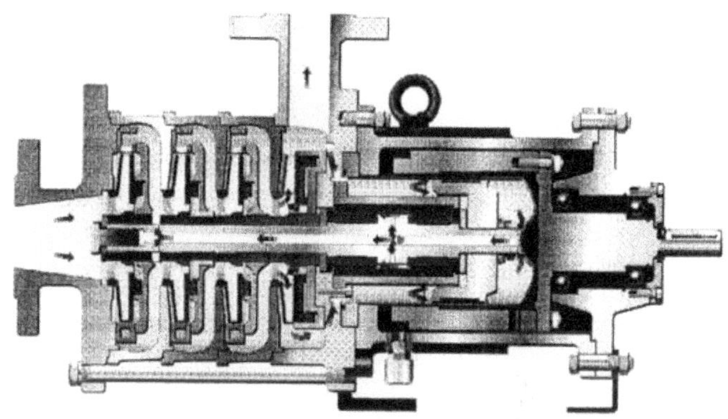

Figure 13-10: "Seal-less" multi-stage magnetic drive pump (Source: Hermetic Pumps, Houston, TX, and Gundelfingen, Germany)

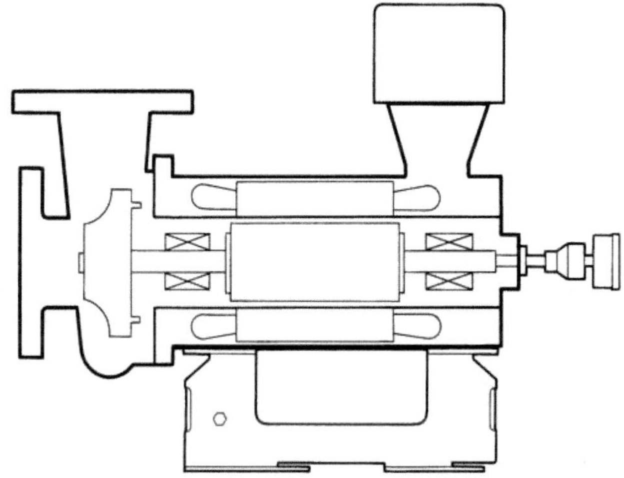

Figure 13-11: Simplified sketch of a typical canned motor "seal-less" pump (Source: Hermetic Pumps, Inc., Houston, TX, and Gundelfingen, Germany)

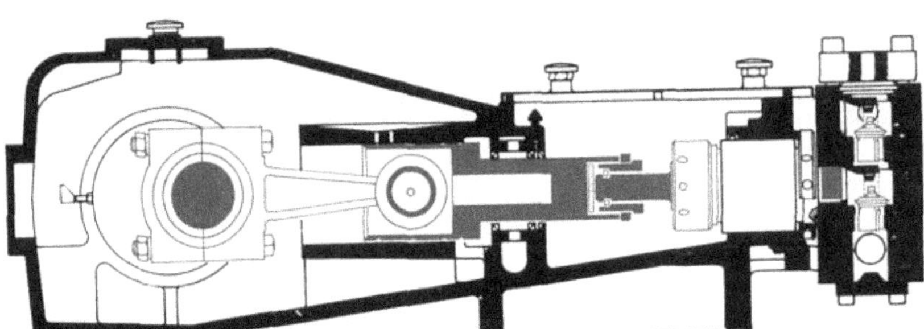

Figure 13-12: Reciprocating positive displacement pump (Source: Ruhrpumpen, Germany)

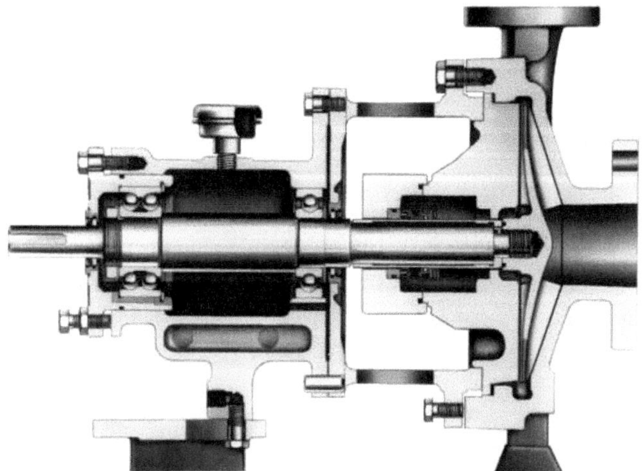

Figure 13-13: ANSI/ASME B73.1 End Suction Chemical Pump. (Source: ITT/Goulds)

- The materials of construction for the parts in contact with the liquid must be selected to offer maximum resistance to corrosion and possibly abrasion at the pumping temperature. Although ductile iron will handle some chemical solutions, most chemical pumps are made of 300 series stainless steels, duplex stainless steels, the nickel-base alloys, or the more exotic metals, such as titanium and zirconium. Pumps are also available in a whole family of plastics, including the phenolics, epoxies and fluorocarbons (see Figure 13-15).

- Shaft sealing is often the biggest concern in chemical applications, and has been given much attention for many years. Cheap seals fail often. Engineered seals and appropriate seal support sys-

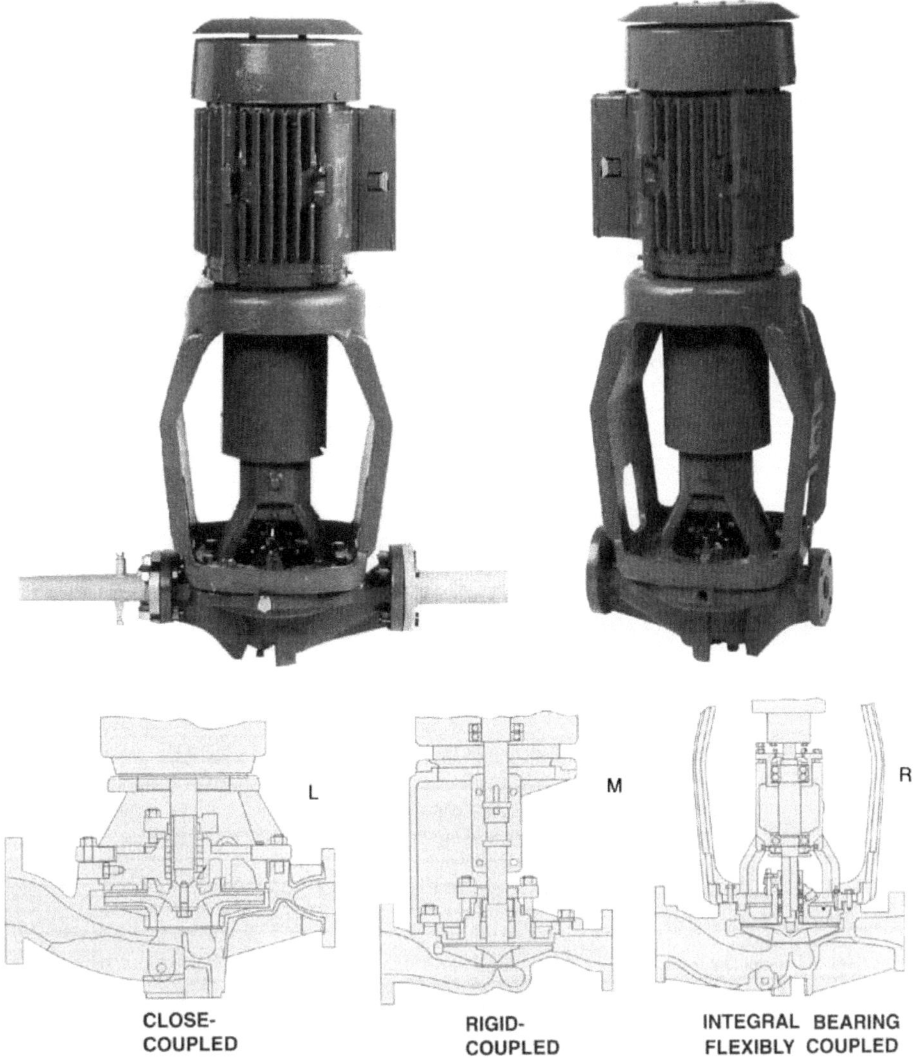

Figure 13-14: ANSI/ASME B73.2 vertical in-line pumps (Source: ITT/Goulds, Seneca Falls, NY 13148)

tems are sometimes needed and may yield rapid payback.

- Since the seal environment ("stuffing box") of many older chemical pumps was originally designed for braided packing, the available space may be insufficient to accommodate the mechanical seals needed to optimize pump MTBF.

- Older seal chambers may have to be redesigned to allow better seal cooling, removal of abrasive particles, removal of vapors, and more room for better mechanical seals (see Chapters 6 and 8).

- Mechanical seal systems have improved with better seal face materials, special multiple seal

Figure 13-15: ITT/Goulds pump with polymer composition casing mounted on a polymer base

configurations, bi-directional pumping scrolls—for better cooling of barrier fluids, and gas seals (see Chapter 8).

- Seal-less pumps (magnetic drive and canned motor) do not require a rotating shaft seal. While this eliminates one of the weakest links in a low-cost chemical pump, the internal, pumpage-wetted, sleeve bearings can be of concern to reliability-focused users. The newer seal-less pumps have become more reliable at lower cost. Some incorporate PTFE-lined ductile iron casings and covers. Increased dry running capability is claimed, using special silicon carbide internal bearings and monitoring devices.

- Water jackets, steam jackets, smothering-type glands and a wide variety of other features and geometries are available to handle liquids which may solidify at low temperatures; they can all be added to the basic layout in Figure 13-16. Multi-staging, motor-over-pump, canned motor seal-less, submersible and others can be gleaned from Figure 13-17. Also, when making a choice between canned motor and magnetic drive pumps it is always prudent to consider the advice of a vendor/manufacturer that produces and has experience with both.

CANNED MOTOR PUMPS MAKE IMPORTANT INROADS IN REFINERY SERVCES*

Magnetic drive and canned motor pumps have had relatively low acceptance rates in refineries, especially in the U.S., even though thousands of these units are successfully operating in hydrocarbon services around the world. The importance of these pumps is, however, well recognized in the efforts of the American Petroleum Institute (API). A working committee was formed in 1990 to compile data for a document covering sealless pumps in refinery services. The resulting standard (API-685) was formally adopted in 2000 and a revision to this standard was scheduled for release in 2009.

Initially Slow Acceptance Overcome

Typical canned motor pumps are constructed as shown in Figure 13-16; note that the motor rotor shares its shaft with the pump impeller (or impellers). The stator portion of the motor is separated from this rotor by a thin containment can and the stator is not contacted by the pumpage. In a canned motor pump, the bearings usually receive lubrication from the fluid being pumped and clean fluids are obviously preferred. However, contaminated fluids can be easily accommodated by first filtering and then re-injecting a fluid side-stream. Relevant provisions are incorporated in the canned motor pump illustrated in Figure 13-16.

Until the late 1990's, one of the major drawbacks to applying "canned motor" style sealless pumps in refinery services has been the unavailability of suitable motors in the larger than 300 hp category. Even though larger canned motors had been advertised, they were not considered commercially viable due their assumed cost. In addition, there has been the occasional perception that sealless pumps were not reliable, or that they were difficult, or expensive, or impossible to repair. Another sweeping (but inaccurate) judgment simply considered them a bad investment. Thanks to a large number of successful and trouble-free installations, this perception has certainly been proven incorrect.

Product-lubricated Bearings

One of the strongest impediments to the more widespread use and application of sealless pumps in oil refineries is that they will incorporate product-lubricated

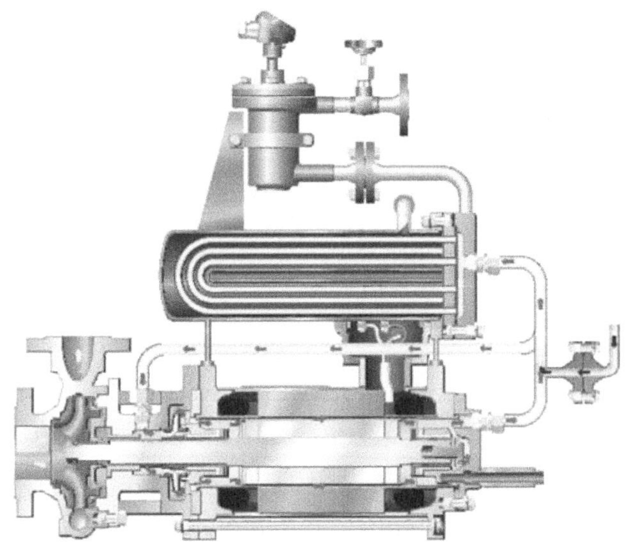

Figure 13-16: Typical high pressure canned motor pump with cooling loop for bearings. (Source: Hermetic Pumps Inc., Houston, Texas, and Gundelfingen, Germany)

*Courtesy of S. Dennis Fegan, with updates by Peter Koegl, Hermetic Pumps Inc., Houston, Texas

bearings and that abrasives in certain product streams can cause accelerated bearing wear. This is actually a question of application, since there are designs and materials available to ensure successful operation and relatively long life, even under the most severe operating conditions. The issue parallels that of mechanical seals, where certain services require creation of a suitable operating environment. The same consideration logically exists for a sealless pump where this environment must accommodate the needs of the bearings. As just one example, sealless pump components may add heat to the fluid contacting the bearings. This heat addition may now affect the vapor pressure and viscosity of the fluid; both must still be suitable for the application.

Because difficult applications drive the selection process, cooperation between pump user and manufacturer is very important. The pump manufacturer must be given as much information about the application and physical properties of the pumped fluid as possible. Also important is the transfer of information about the planned operation of canned motor pumps. It includes data on the number of starts and stops anticipated for a given period of time, startup conditions (cold start), as well as environmental issues such as cooling water quality, which could be a big factor in the overall success of a given installation and affect the type of pump offered.

A case history involving a pump handling 750°F diesel fuel at 1000 psi suction pressure and saturated with hydrogen explains the importance of cooperation and disclosure of details. When invited to bid on this service, a key manufacturer of canned motor pumps asked additional questions. It was then realized that the fluid stream contained some catalyst fines, prompting a closer review of associated requirements for handling the application. Pre-filling the motor section with clean diesel fuel solved the problem. This 100 hp unit was commissioned in 2002 and has been operating successfully since then. Two other units were subsequently commissioned in other refineries, with similar operating conditions. In other instances, the vendor opted to use a metering pump to inject a clean compatible fluid into the motor space. This then acts as a highly effective barrier fluid to prevent damage to the bearings.

Larger Units in Successful Operation

After these 100-hp units were commissioned in 2002, there arose a need for larger units requiring between 200 and 300 hp and flows between 2000 and 4000 gpm. The total developed heads ranged between 300 ft and 4000 ft, with working pressures up to 3000 psi. As of 2009, some of these units are installed in ultra-low sulfur refining applications and others in light hydrocarbon services.

Favorable internal pressure profiles are the primary reason for canned motor sealless pumps being utilized in these and other strenuous services. At no point in the combined pump and motor geometry does the fluid state have to be reduced to atmospheric pressure. In fact, most canned motor designs strive to increase the internal pressure so as to ascertain that two-phase flow does not occur. Analytical programs allow designs with pressure/temperature profiles which confirm that a pure liquid phase exists under all operating conditions.

All refinery installations require that the motors be furnished with an explosion proof label. While, as of early 2008, these labeling requirements limited the available motor nameplate rating to 350 horsepower, steps had been initiated to increase this to 700 hp by late 2008 or early 2009. Canned motors with oil-filled, pressurized stators are available; they have extended the allowable temperature range and operate with high overall efficiency. It should be noted that, by the time an all-encompassing comparison accounts for the true power consumed due to seal and coupling losses in conventional pumps, and when installation and maintenance costs are added to the equation, modern canned motor pumps often represent a surprisingly attractive choice. This attractiveness is often expressed as energy efficiency on a par with that of traditional motor-driven pump sets and cost of ownership even lower than that of conventional pumps

As a final point, axial thrust control used to represent a major challenge to producing large horsepower sealless pumps. This challenge was compounded whenever a wide range of fluids with varying physical properties had to be accommodated, as would be the case in units with low NPSHr (net positive suction head required) and steep vapor pressure curves. However, internal thrust balancing is no longer an issue for competent manufacturers. There have been dramatic improvements in the ability of sealless pumps to control axial thrust and in the availability of instrumentation to monitor this thrust. Dependable and reasonably priced "real-time" bearing monitoring has had a major beneficial impact on the increased reliability of these machines and the reduced cost of ownership.

Most certainly, canned motor pumps are coming of age. Their range of applicability has been greatly extended and they deserve to be considered by forward-looking user companies in the hydrocarbon processing and many other industries.

Figure 13-17, then, gives an overview of some of the pumps that can be produced in very many different arrangements. Among these virtually limitless configurations we find vertical and horizontal orientations, pumps with integrated cooling and filtration, positive displacement rotary pumps (Figure 13-17 [a]), the same type of pump driven magnetically (Figure 13-17 [b]), a liquid-ring magnetic drive pump (Figure 13-17 [c]), and an extended shaft vertical magnetic drive pump (Figure 13-17 [d]), shown here horizontally.

As mentioned earlier in this segment, vendor experience merits the closest possible scrutiny. The properties of the fluid being processed are of utmost importance and may qualify or disqualify seal-less pumps for a particular service. Of course, the same is true for positive displacement pumps where vendor experience is of equally great importance.

Magnetic drive pumps have very little tolerance for pipe strain. If, in a particular installation, conventional mechanical seals gave out after six months because of pipe strain, the same amount of pipe strain will probably ruin magnetic drive pumps after a few days' operation. Unless the outer rotating magnet carrier maintains concentricity with the pump shaft, the outer magnets might scrape the containment shell. This could easily cause massive, often disastrous failure.

Reliability-focused users will purchase and install every supervisory and shutdown instrumentation option available for seal-less pumps. These include, but are not limited to, thermocouples and temperature switches for over or under-temperatures, dry operation protection, high vibration or bearing wear, and axial shaft displacement.

Chemical pumps must often deal with air and vapors due to the presence of air in solution, tank unloading, and the handling of liquids near their vapor pressures. This requires special attention to pump type (self-priming, Figure 13-18, seal-less, etc.), and mechanical seal face materials, wetted sleeve bearing materials, and impeller/inducer type.

Open (Figure 13-19) or semi-open impellers can handle higher levels of entrained air or vapors (up to 10%, depending on front clearance) than closed impellers, and inducers can handle even higher gas levels (to 15%).

There are many high head, low flow rate applications in the chemical industry, which prompted the development of special low flow pumps. Some of these are geared, high-rpm three-dimensional impeller designs. Sundyne pumps are among these mature and well-understood high-speed products. Many of these are aimed at replacing traditional pump applications where long-term operation at less than BEP flow must be avoided.

More exotic specialty models include the Roto-Jet® (Figure 13-20). Liquid enters the end-bell of this pump and passes into the rotating case where centrifugal force causes the liquid to enter the rotor under pressure. The velocity energy of the liquid in the rotor is converted into additional pressure as it jets into the pickup tube. While there are some perfectly legitimate applications for these pumps, they will obviously not be too forgiving in pumpage containing solids, or in similar misapplications.

Here is the "bottom line:" Beware of claims that "one pump fits all." Beware also of the proponents of blanket campaigns to "standardize" on just one vendor, one material of construction, or one particular pump model. While these approaches might serve one company in a thousand, they would almost certainly deprive the remaining companies of reaching their true profitability potential. Remember that it should not be your goal to have a plant that just barely runs. Instead, it should be your goal to achieve best-of-class life cycle costs.

Lowest pump life-cycle costs typically go to the folks whose pump MTBF matches that of the most profitable process plants. They never buy on price alone. The most profitable plants always apply a thorough evaluation process and then buy best value.

PAPER STOCK AND OTHER LARGE PROCESS PUMPS

Pulp and paper mills use many pumps to transfer or circulate cellulose fibers suspended in water ("stock"), chemicals or solids in solution ("liquors"), and a number of residues. The latter are often waste matter in the form of slurries, although some benign residue might be clean enough to serve as water for general services. The two biggest issues in handling paper stock are abrasion from the paper stock fibers, and a high level of air entrainment. In general, the amount of air entrainment in paper stock is roughly equal to the stock consistency. Air handling, especially with medium consistency stock, thus often represents the biggest challenge.

Traditional paper stock pumps (see Figure 13-21) will handle stock up to approximately 6%, oven dry consistency. The absolute maximum limit is a function of many factors including pump design, stock fiber length, pulping process, degree of refining, and avail-

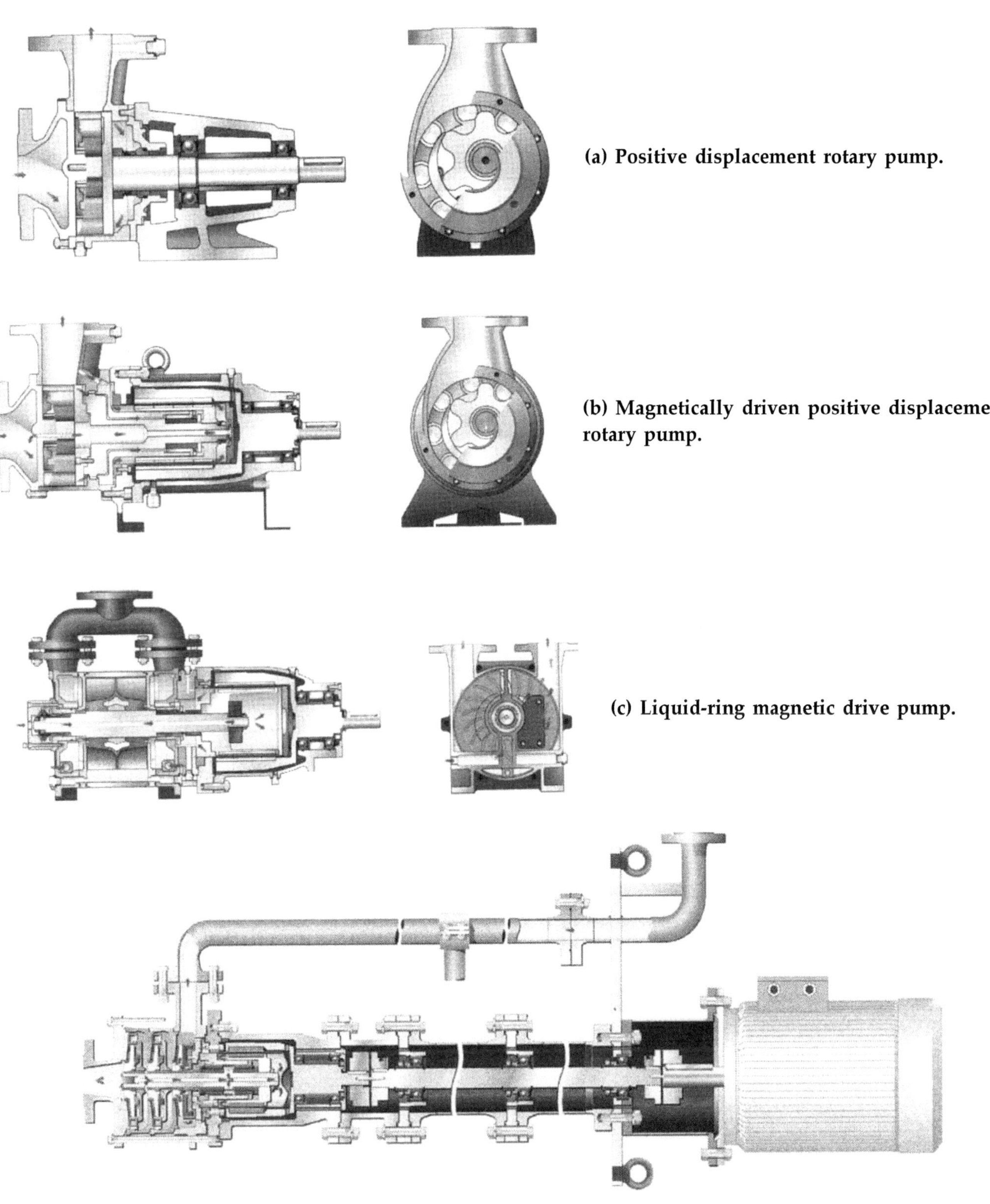

(a) Positive displacement rotary pump.

(b) Magnetically driven positive displacement rotary pump.

(c) Liquid-ring magnetic drive pump.

(d) Extended shaft vertical magnetic drive pump.

Figure 13-17: Overview of different pump configurations (Source: Hermetic Pumps, Inc., Houston, TX, and Gundelfingen, Germany)

Figure 13-18: Self-priming ANSI pump (Source: ITT/Goulds, Seneca Falls, NY 13148)

able suction head. Recent testing on various types of stock has indicated that pump performance is the same as on water for stock consistencies up to 6% oven-dried condition (Ref. 13-3).

Traditional paper stock pumps are also used for certain high flow rate chemical (and general industrial) pump applications. These flow rates would typically be in excess of those covered by ANSI/ASME B73.1. Also, the smaller, standard ANSI pumps are often used for many of the low capacity paper mill applications.

Recall, however, that paper stock is a unique

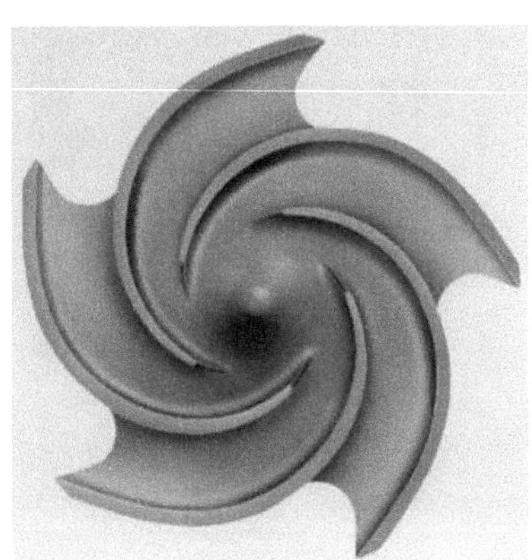

Figure 13-19: Open impeller (Source: ITT/Goulds, Seneca Falls, NY 13148)

Figure 13-21: Low consistency paper stock/large process pump (Source: ITT/Goulds, Seneca Falls, NY 13148)

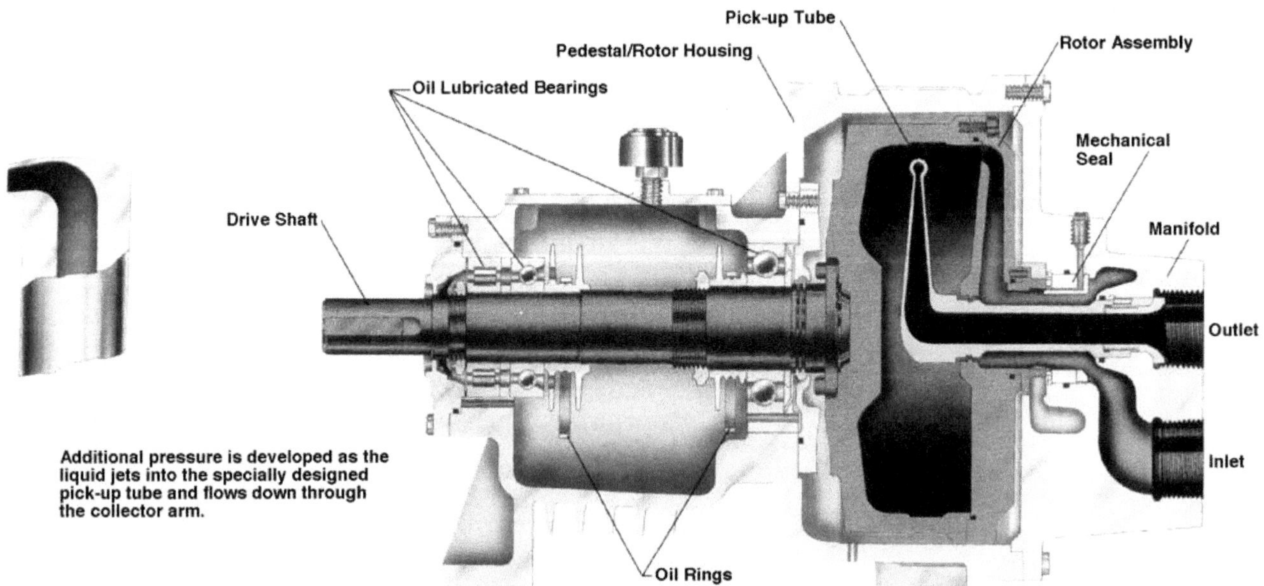

Figure 13-20: Roto-Jet® low-flow, high-head chemical pump

medium. The entrained air and paper fibers provide a cushioning effect not found in other liquids. This cushioning tends to protect against cavitation damage, even with low NPSH margin ratios. When high suction energy stock pump are applied in certain other services, say, cooling water, NPSH margins may be low and there could be suction recirculation. Since there is now no cushioning effect, the pump may experience noise, vibration and/or damage from cavitation.

Medium consistency paper stock is a term generally used to describe stock between 7% and 15% oven-dry consistency. Pumping of medium consistency paper stock with a centrifugal pump is possible, but requires a special design due to the fiber network strength and the inherently high air content (see Figure 13-22). Inducers (augers) which protrude from the suction of the pump are used to fluidize the pulp and induce flow into the pump suction.

Excessive gas entrainment is detrimental to good operation of any centrifugal pump. Accordingly, the high air content characteristic of medium consistency paper stock must be removed from the impeller inlet. This is accomplished by first centrifuging the incoming liquid toward the inlet periphery, hence, moving the air to the center of the impeller inlet. Next, the air is pulled through the balance holes in the impeller disc by a vacuum pump drawing from the backside of the impeller. Back impeller vanes, or a dynamic seal, are used to maintain a uniform air-liquid interface on the back side of the impeller. This prevents air from mixing with the liquid being discharged into the casing volute by the impeller.

The dynamic seal principle is illustrated in Figure 13-23, showing a "pneumatically activated" stand-still seal which seals against the outside environment.

These special design features and higher stock consistency do, however, create an efficiency penalty for the pump, but efficiency has for decades been a secondary consideration at some facilities. By increasing the stock consistency, the amount of water that must be transferred with the stock fibers and removed is reduced. This increases the overall process efficiency. Alternatively, suitable mechanical seals are available from some major mechanical seal manufacturers. The mechanical seal option is often preferred by reliability-focused pulp and paper mills.

Most of the pump duties in a paper mill can be accomplished by single-stage pumps operating at four-pole or lower motor speeds (i.e. 1,800/1,500 and lower rpm). For liquids other than water, two-pole motor speeds should be avoided, if possible. Lower speeds are sometimes justified because of reduced pump maintenance and greater overall reliability.

Due to the abrasive nature of paper stock, most stock pumps are made of a 300 series stainless steel,

Figure 13-22: Medium consistency paper stock pump with pulp fluidization auger (Source: ITT/Goulds, Seneca Falls, NY 13148)

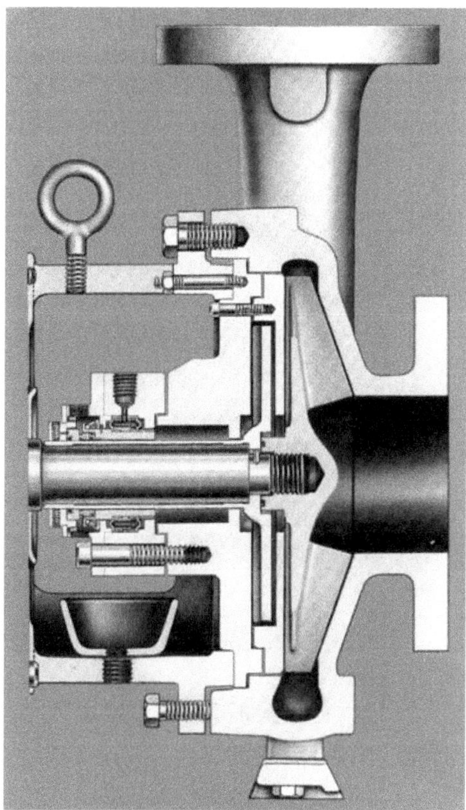

Figure 13-23: Dynamic seal (repeller seal) with stand-still provision (Source: ITT/Goulds, Seneca Falls, NY 13148)

or duplex stainless steel, with ductile iron used for less severe applications.

DOUBLE SUCTION/AXIALLY SPLIT CASE PUMPS

Double suction, between bearing, split case pumps (see Figure 13-6) are the second most popular centrifugal pump configuration, second only to the more common end suction, radial split design (see Figure 13-13).

The principal reasons for the popularity of this between-bearing configuration include:

1. The rotor can be removed by just taking off the upper casing half, without disturbing the suction or discharge piping, or moving the motor
2. Less shaft deflection due to between-bearing design
3. Lower NPSH requirement due to the fact that each impeller eye only handles one half of the total pump flow rate
4. Virtually no axial hydraulic thrust, due to the back-to-back/double suction impeller design
5. Higher efficiency because power is not lost to balance the hydraulic thrust
6. Relatively high allowable nozzle loads due to rigidity of the lower portion of the casing. There have been anecdotal claims that once nozzle loads become excessive, this would be noticed because the pump feet would "slide" on the base plate. Note, however, that such extreme nozzle loads are almost certain to result in unacceptable mechanical seal and bearing lives.

Although split case, between-bearing models are relatively well represented in some industries, these pumps are not without drawbacks:

1. Due to the large flanges required for the split casing sealing joint, these pumps are normally heavier and cost more than comparable end suction pumps, especially in higher alloy and higher pressure applications.

2. While double-suction impeller pumps have lower NPSHr values than comparable end suction pumps, the through-shaft reduces the impeller eye area. This requires an increase in the suction eye diameter (eye tip speed) and results in tighter turns for the liquid entering the impeller eye. Suction energy and required NPSH margin tend to increase as a consequence. This could be of critical importance in high and very high suction energy pumps, when operated at low NPSH margins and/or in suction recirculation (see Chapter 5).

   This phenomenon has frequently shown up in high suction energy, split case pump cooling tower services. These pumps often have low NPSH margins, operate at a most damaging temperature (100°F-120°F, or 38°C-49°C), and handle a high suction energy liquid, water. The result has been accelerated cavitation damage, even with the more cavitation resistant materials, such as stainless steel. By contrast, one of the authors recently became aware of a reliability-focused user who provided the high suction energy split case, cooling tower pumps in their facility with NPSH margin ratios in the range of 4. These pumps have been operating for 30 and 40 years without experiencing damage from cavitation

   It should be noted that some of the newer split case pumps do have improved casing inlet designs (less inlet swirl), which allows them to approach the high energy suction performance of end suction pumps. This represents a retrofit or upgrade opportunity for plants with older style pumps.

3. Double-suction pumps are more sensitive to the orientation and geometry (radius) of elbows in front of the pump inlet (see Chapter 3). Elbows installed within five pipe diameters of the pump suction flange should be perpendicular to the plan view of the shaft. This reduces the tendency of unequal flow quantities reaching the two impeller eyes (see Figure 3-5). Such an uneven flow pattern can upset the axial thrust balance, causing high bearing loads and shorter life. Uneven flows also tend to increase the NPSHr and could, simultaneously, cause high flow cavitation on one, and low flow suction recirculation on the other side of the impeller.

   It should be of interest that uneven flow distribution could also occur when double-suction pumps are mounted vertically (Figure 13-24) and an eccentric reducer is installed on the pump suction with the flat side at the top. *In this special case an eccentric reducer should not be used!*

4. Between-bearing pumps need two shaft seals

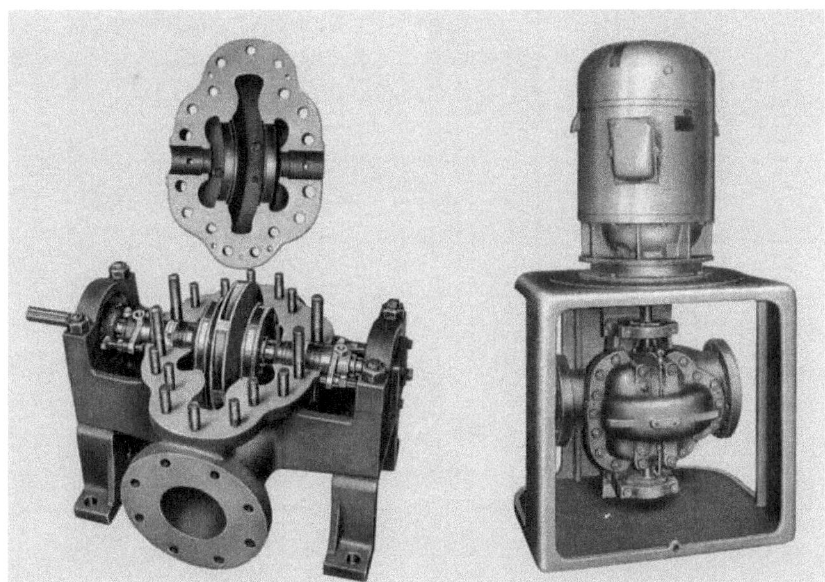

Figure 13-24: Simple double-flow pump shown with cover removed and in vertical mounting arrangement (Source: ITT/Goulds, Seneca Falls, NY 13148)

whereas only one seal assembly is required for an end suction pump. This could increase both initial and maintenance costs in services that require expensive mechanical seals.

5. The axial split complicates radial gasket sealing (such as used on mechanical seal glands), and often results in mismatch of the casing halves at the joint. This mismatch can increase the hydraulic volute friction losses, thus reducing pump efficiency.

6. Finally, since most split case pumps have stuffing boxes integral with the casing halves, large bore boxes are not normally offered, and not all mechanical seal configurations will fit into some split case pumps.

SEWAGE AND WASTE WATER SUBMERSIBLE PUMPS (SOLIDS AND NON-CLOG)

Sewage is defined as the spent water of a community, and although it is more than 99.9% pure water (in the raw sewage stage), it contains wastes of almost every form and description. It may contain bits of floating paper, garbage, rags, sticks, and numerous other items. The primary requirement of a sewage pump is to assure maximum freedom from clogging when handling large solids (frequently soft), and long stringy materials. Normally, the largest solid size (sphere diameter) that the pump can handle is specified for non-clog pumps. Inspection openings are often provided in the casing for access to the impeller.

Sewage pump applications are associated with both the collection (lift/pumping stations handling raw sewage) and treatment of sewage (handling settled sewage, service water and activated sludge). Conventional centrifugal sewage pumps cannot handle sludge with more than 2% solids. Corrosion-resistant shaft sleeves and wearing rings are desirable for maximum life. If the pumpage is corrosive and/or abrasive, the materials of construction for parts in contact with the liquid should be selected accordingly.

Conventional sewage pumps are offered in two basic configurations, dry pit (Figure 13-25) and submersible (Figure 13-26). Submersible pumps are currently more popular due to the simplified and more compact installation requirements. No pump room is needed and concern over possible flooding of a non-submersible motor is eliminated. The reliability of submersible motors and mechanical seals supplied by competent manufacturers meets all reasonable expectations.

However, care must be taken when installing a submersible motor in a dry pit application to ensure that there is sufficient cooling of the motor. The equipment manufacturer may have assumed that submersible mo-

Figure 13-25: Horizontal wastewater pump (Source: Allis Chalmers Pumps, Milwaukee, Wisconsin)

Figure 13-26: Submersible wastewater pump (Source: ITT/Goulds, Seneca Falls, NY 13148)

tors are being operated in the submerged mode and are thus being cooled by the surrounding liquid.

In order to pass large solids, non-clog pumps typically have impellers with as few as one, two, three, or four vanes, swept-back leading edges to prevent the buildup of stringy solids, and very little—if any—vane overlap. This typically results in less than optimum efficiency for the specific speed and flow rate, and lower thresholds for high and very high suction energy. Typically, these thresholds are about two-thirds of the threshold values of a normal end suction pump due to recirculation from the impeller discharge to suction (see Chapter 5).

For best reliability the maximum speed of sewage pumps should be limited to the lesser of:

- 1,750 rpm, for small pumps
- The speed that limits the suction energy for one or two vane impellers to 100×10^6 (the value of De × n × S, see Chapter 5)
- The speed that allows for a minimum "Hydraulic Institute" (Ref. 5-12) NPSH margin ratio (Low S.E. = 1.1, High S.E. = 1.3, Very High S.E. = 2.0). Note: for best reliability, these HI NPSH margin ratio values should be doubled

Many municipal lift station pumps use variable speed drives (see Chapter 4) to reduce power costs. Such lower speed operation will also improve the MTBR (mean time between repairs).

Non-submersible pumps are generally offered in horizontal and vertical configurations, with vertical designs offered with motor stands mounted on the pump (Figure 13-27), or the motors mounted many floors above the pump and connected to the pump by universal joint drive shafts to prevent possible flooding of the motor. Note that the vertical pump of Figure 13-27 and arranged for dry-pit service, employs a bottom-entry impeller. A bottom-entry layout is also used in the belt-driven pump in Figure 13-28. Direct motor drive and clean water lubricated column bearings are shown on the top-entry vertical sump pump of Figure 13-29.

Large vertical sewage pumps are prone to vibration problems, especially with variable speed operation. Structural analyses should be conducted on the motor, motor pedestal, foundation, and universal joint drive shaft (if used) to avoid operation near a resonant frequency (see Chapter 12). Proper alignment and support of universal joint drive shafts are critically important.

Non-submersible sewage pumps are often equipped with dynamic shaft seals (see Chapter 6—Dynamic Shaft Sealing) to eliminate costly replacement of mechanical seals, or packing and shaft sleeves. It also eliminates the

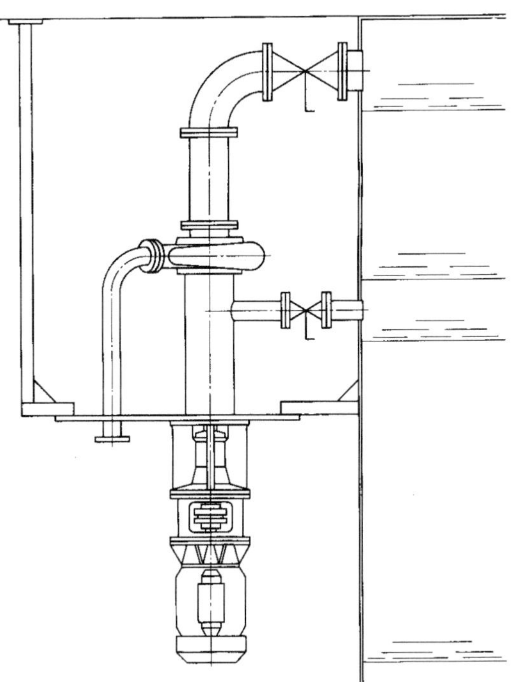

Figure 13-27: Sulzer "TC" cantilever pump in dry-pit arrangement; shaft not in contact with the medium

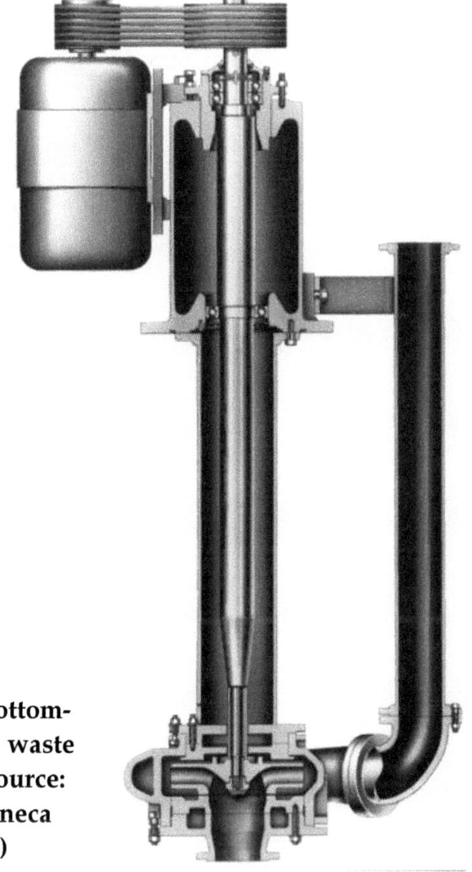

Figure 13-28: Bottom-suction vertical waste water pump (Source: ITT/Goulds, Seneca Falls, NY 13148)

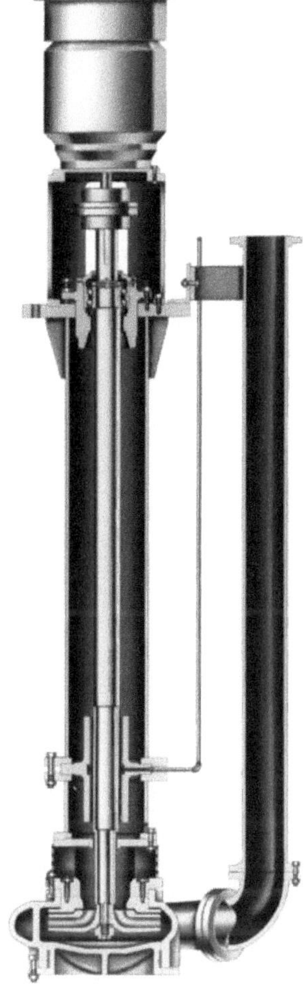

Figure 13-29: Top-suction vertical waste water pump with direct motor drive (Source: ITT/Goulds, Seneca Falls, NY 13148)

need for a source of clean water (which must be isolated from any potable water system) to flush the shaft seal.

A centrifugal submersible pump (Figure 13-26) is defined as a close-coupled impeller pump/motor unit designed to operate submerged in liquid. Submersible pumps may be operated in wet-pit or—exercising proper caution—dry-pit environments. Because the pump and motor are integral, submersible pump power and efficiency are calculated and measured differently than for conventional (non-submersible) pumps. The input power to the submersible motor and overall wire to water efficiency are normally used.

Losses attributable to electrical cabling, variable speed drives, non-sinusoidal power supplies, and externally powered motor-cooling devices are excluded when measuring motor input power (see ANSI/HI 11.6). Although these losses are excluded from the overall efficiency calculation, they may significantly increase the overall power consumption.

Additional information on other sewage pump standards can be obtained from the Hydraulic Institute (HI), American Water Works Association (AWWA), and Submersible Wastewater Pump Association (SWPA). References 13-4 through 13-6 are dealing with these issues.

SLURRY PUMPS

Slurry pumps are suitable for pumping mixtures of solids in a liquid carrier (often used as a means to transport solids), or as an incidental part of the process. The primary requirement of a slurry pump is to assure acceptable life by being resistant to abrasive wear. Slurry pumps are generally more robust than those used in clean liquid services, and often have replaceable wear parts. They normally have wetted parts constructed of hard metals (see Figure 13-30), or are rubber-lined to resist abrasive wear. Tapered roller bearings are often found on heavy-duty slurry pumps, here shown with braided packing and closed impeller with front wear surface, Figure 13-31. The rotor can be moved forward to reduce internal leakage due to wear of this sealing surface.

Upgrade opportunities exist in the areas of sealing and lubrication on some slurry pump models. Certain services (e.g. limestone slurry in desulphurization units) have benefited from mechanical slurry seal retrofits and from conversion to oil lubrication.

Figure 13-30: Hard metal slurry pump (Source: ITT/Goulds, Seneca Falls, NY 13148)

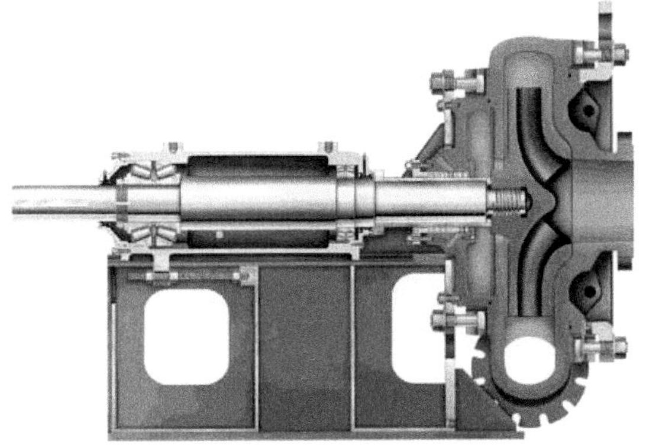

Figure 13-31: Cross-section view of slurry pump shown in Figure 13-30 (Source: ITT/Goulds, Seneca Falls, NY 13148)

Hard/white irons are used for the most erosive slurries with mildly corrosive carrier liquid. Hard duplex stainless steels, such as CdrMCu are used for moderate slurries with chemically aggressive carrier liquids. Other softer metals can be used for less severe services.

For increased wear life lined slurry pumps usually have thicker liners than other elastomer-lined pumps. The most commonly used elastomer is natural rubber, favored for its very high resilience. High resilience imparts good abrasion resistance, if particles are not too large or sharp. Various other elastomers can be used, mainly for improved chemical and/or heat resistance. Some elastomeric rubber wear parts are reinforced with ceramic beads or rods for increased wear. The liners must be bonded to a base material to prevent collapsing under low suction pressure conditions. These conditions could occur during cavitation or flow surges.

Pump wear depends on the pump design, the abrasive characteristics of the slurry, the way in which the pump is applied or selected for the duty, and the actual conditions of service. The amount of wear inside the pump varies significantly and is often location-specific. In addition to pump design and duty, wear rates depend on the velocity, concentration, size, hardness, sharpness and impact angle of the particles.

The following observations and recommendations are of interest:

- Wear is normally most severe in the impeller seal face, vane outlet and vane inlet, in that order

- The amount and location of casing wear depend on the casing design

- Wear rate increases the farther the pump is operated away from BEP

- The life of a pump handling abrasives is inversely proportional to the pump speed to the 2.5 power. A 20% drop in pump rotational speed will roughly double the life. Therefore, lower operating speeds can significantly improve life in all applications

- ANSI/HI 12.1-12.7 recommends a maximum head of 130 to 400 ft (~40 to 122 m) per stage. Its recommended maximum impeller outer diameters are based on peripheral speeds ranging from 75 to 142 ft/sec (~23-43 m/s). The standard recognizes that these guideline values are affected by the severity of the application and the materials of construction. Operation outside of these limits probably jeopardizes profitability.

- The materials of construction for the parts in contact with the liquid must be selected to offer maximum resistance to the abrasiveness of the slurry. The predominant factor causing wear must be understood and recognized before the most suitable construction materials can be selected. The rate of wear is related to materials of construction and the mixture being pumped. Abrasive wear in pumps is generalized by recognizing three types:
  a. Gouging abrasion from coarse particles impinging with high impact. Hard metal pumps are required here
  b. Grinding abrasion from particle crushing action between two surfaces. Either hard metals or rubber can be used.

c. Erosion-abrasion from free moving particles that impinge on the pump surfaces, frequently at shallow angles. Rubber-lined pumps are normally used for fine particles with low impingement angles. Hard materials resist abrasion but can be more brittle (and costly) than ductile materials. The attendant trade-offs must be weighed by the user-purchaser.

- Shaft sealing is a major concern in slurry applications; the intent is to avoid rapid wear of the seal and gland area. Precautions must be taken to prevent, or minimize, the ingress of abrasive particles. These particles could be carried into the stuffing box area by the mixture being pumped. Serious abrasion may occur in stuffing box regions away from the seal. Back (pump-out) vanes are normally fitted on the backside of the impeller shroud to reduce this risk.

- Shaft packing is the most common method of sealing slurry pumps, due to the high shaft deflections and shock loading that may damage typical hard face seal materials. Special care must be taken in the selection of the packing type and seal flush arrangement. Packing should run against a shaft sleeve that is either hardened or coated to resist wear, with a minimum hardness of 50 RC. That said, it should be noted that forward-looking users are often able to report excellent success with appropriately designed mechanical seals. However, the stuffing box region of the typical "old" slurry pump assembly may habe to be modified and/or redesigned to accommodate certain high-value mechanical seals.

- A dynamic seal (expeller/repeller) may be used to prevent abrasives from reaching the stuffing box during operation, or an expeller may be used with other means of providing back-up sealing—when the pump is shut down—such as lip seals and other proprietary devices discussed earlier in Chapters 6 and 8.

- Mechanical shaft seals are utilized, primarily for non-settling slurries, with the seals designed to handle the anticipated high shaft defections. Elastomeric or metal bellows, non-pusher type, seals are typically used for this service. However, there are notable exceptions to this general rule. A number of stationary-face mechanical seals available from AESSEAL, Eagle-Burgmann, certain mechanical seals from John Crane, also Flowserve Corporation and—undoubtedly—others incorporate a rotating hard face in very close proximity to the impeller. These successful slurry seals differ greatly from the many traditional seals that place the hard face in the seal gland and thus near the atmosphere.

- In typical non-slurry applications and services where flushing is not used, mechanical shaft seals should be installed in large bore or tapered bore seal chambers. In the case of some models, a relatively steep 30°-tapered seal bore is chosen.

- Pump piping deserves special attention in slurry systems. Well-designed piping layouts avoid excess energy consumption, settling/blockage, and wear. Factors to consider include pumpage velocity, piping arrangement, flexibility and piping materials. Piping should be arranged to avoid sudden changes in direction, low spots and areas where solids can accumulate. Inattention to these details could, obviously, result in rapid wear and blockage. Valves should be kept to a minimum. Special issues that should be evaluated when designing a slurry system are further described in ANSI/HI 9.6.6 and include:

  a. Apparent viscosity
  b. Settling vs. non-settling slurry type
  c. Carrying velocity
  d. Settling velocity
  e. Specific gravity of solids and slurry
  f. Slurry velocity region (laminar vs. turbulent)
  g. Abrasive particle size and carrier liquid corrosive characteristics

VERTICAL COLUMN TURBINE ("VTP") PUMPS

Vertical turbine "can" pump construction was shown earlier in Figure 13-5. Both suction and discharge nozzles are located above ground and both are oriented transverse to the pump shaft. VTP geometry also differs from other vertical pumps, such as vertical-in-line or vertical wet pit, by having the fluid that enters the suction nozzle travel down in the annular space between an outer and inner containment pipe, or "can." Pressurized flow leaving the last stage impeller collects concentrically around the pump shaft and moves upward in the inner column pipe.

Vertical turbine pumps were originally developed for deep-water wells, but the design was found to have certain benefits in industrial applications as well (Ref. 13-3). Among the advantages of VTP's, we find:

1. The vertical construction takes up little floor space
2. Priming problems can be avoided due to the impellers being submerged in the liquid.
3. The first stage impeller can be lowered (by increasing pit depth, if necessary) to provide the desired NPSH margin.
4. The multistage construction offers higher efficiencies on high head, lower flow applications.
5. The modular construction allows the pumps to be customized for many applications, especially when welded components are used.

However, care must be taken when applying a deep well vertical turbine pump in industrial applications. Here's why:

1. Deep well pumps rely heavily on a high axial thrust. High axial thrust often compensates for alignment and resonant frequency issues with standard well pump column shafting. This means that reductions in pump thrust, achieved by balancing the hydraulic axial thrust on the impellers, and/or lowering the rotor dead weight (short settings), can lead to vibration and/or bearing overload problems. The situation can be further complicated when VTP's operate at variable speeds.
2. Some or all of the internal sleeve bearings are lubricated / wetted by the liquid pumped, so they must be able to handle any corrosives and/or abrasives in the pumpage.
3. Extended-length components, especially the column shafting and discharge head/motor pedestal, are subject to resonant frequency/vibration problems. Since, by virtue of orientation and layout (see Figure 13-32 and 13-37), discharge heads are generally stiffer in one of the two horizontal plane directions, their respective resonant frequencies will differ.

   A vertical pump can experience excessive structural vibration of the discharge head and associated driver in the field, even though the driver has been shop tested with low vibration, the rotating components were properly balanced, and a 0.002" (0.05 mm) run-out has been achieved on a shaft system that included a precision coupling. This could be the result of a "reed" (natural) frequency of the motor, head and foundation assembly. This is occasionally caused by the stiffness of the field foundation being different from anticipated. A "reed resonance" effect will result if the natural frequency of the assembly is at or near the running frequency of the pump.
4. Operation at running frequencies (or their lower harmonics) within 15% of either of the two natural frequencies is likely to result in amplified vibration and must be avoided. When running and resonant frequencies coincide, vibration amplitudes may be so severe as to cause very rapid failure of bearings and other components.
5. Special high-thrust vertical motors, or discharge head thrust pots, are required to handle the normal high axial thrust values. European electric motors are not usually offered with high-thrust bearings capable of handling the axial thrust from standard vertical turbine pumps. So-called separate pump thrust pots may be required.

A VTP can be divided into the three basic subassemblies depicted in Figure 13-32:

- The bowl assembly
- The column assembly
- The discharge head

BOWL ASSEMBLY

Standard bowl pumping assemblies are normally multi-staged. Pump column adapters (discharge cases) and bowl assemblies in the smaller and/or lower cost VTP's are threaded, whereas they are bolted together in the larger and/or more expensive sizes. There are many options that are available on vertical turbine bowl assemblies, including wear rings, different metallurgical formulations, and so forth. Areas of concern often involve the sleeves and bearing materials. Here are some guidelines:

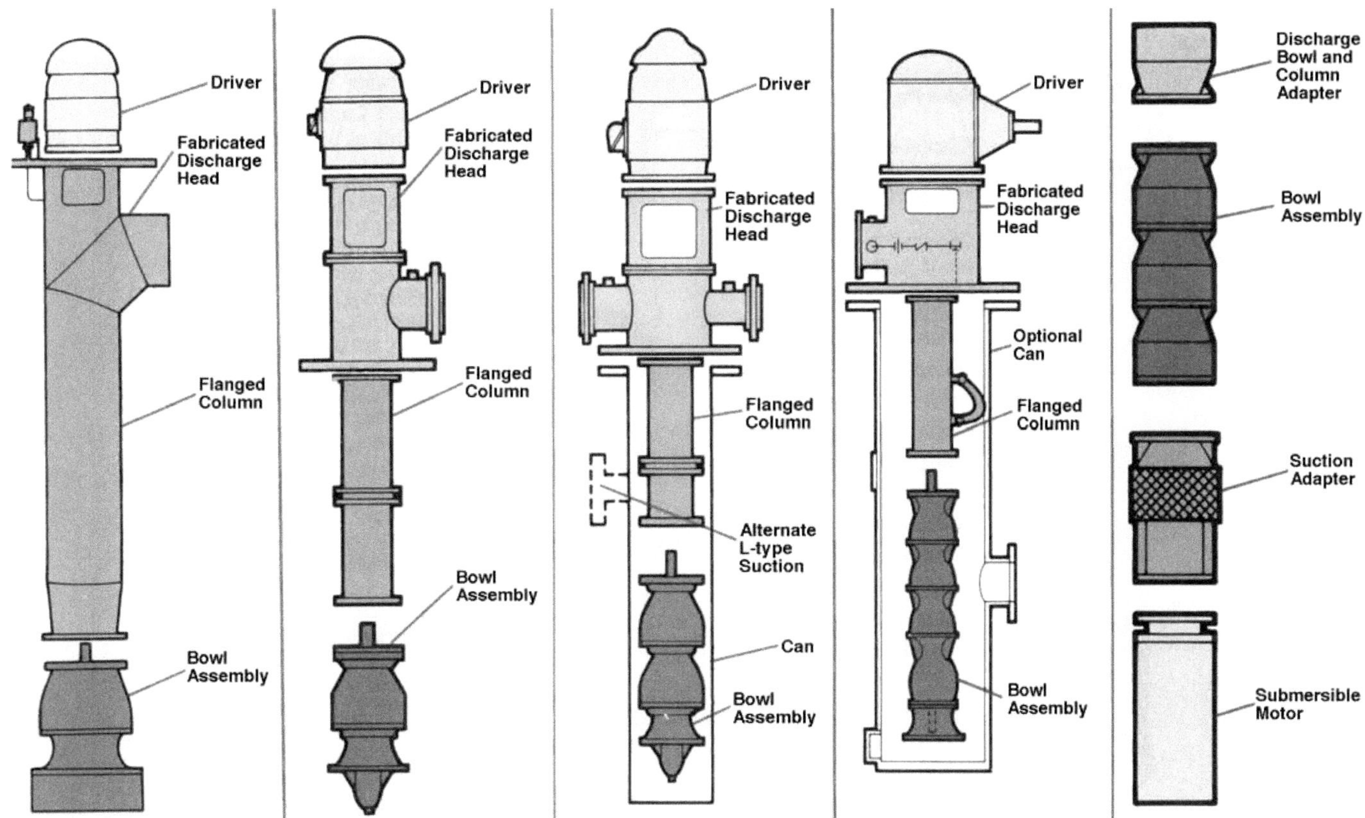

Figure 13-32: Industrial vertical turbine pumps ("canned pumps")—typical assemblies and configurations (Source: ITT/Goulds, Seneca Falls, NY 13148)

- Suction bell/inlet case — Must have minimum submergence to avoid ingesting air (see Ref. 13-4 and 13-7).

- Impeller(s) — Standard design is mixed flow (mid-range specific speed), closed or open impeller, with no axial hydraulic balance, i.e. no back wear rings or back vanes. This construction tends to improve efficiency. Figure 13-33 (closed impeller) and Figure 13-34 (open impeller) show the pressure distribution around the impellers and how the discharge pressure acting on the top shroud causes a net down-thrust.

  Figure 13-35 shows a balanced axial thrust design, which uses a back wear ring and balance holes in the impeller, to apply near-suction pressure to the top side of the impeller. Balanced impellers should be used with care to avoid column shaft resonant frequency and alignment problems.

  On many lower cost VTPs often found in agricultural (irrigation) applications, impellers are attached to the shaft by tapered collets (see lower left, Figure 13-36). However, critical industrial applications or pumps intended for long maintenance-free operation should have impellers keyed and secured to the shaft with sturdy snap rings or split rings (center left, Figure 13-36) to handle the axial thrust.

- Bowl (s) — Mixed flow/axial diffuser construction is used, with the outer diameter designed to minimize the required well bore.

- Discharge case — Connects bowl assembly to column pipe and column shaft (optional for industrial applications)

- Bowl shaft — Normally constructed from centerless ground bar stock. Shaft sleeves are not generally used under the bowl bearings, due to cost and non-standard bore requirements in the bowls and cases. From the manufacturer's point of view, sleeves also preclude larger shaft options and, as is claimed, increase maintenance costs. However, sleeves are frequently specified by reliability-focused users who may wish to upgrade to better sleeve materials in the future.

  There are occasions where it might be less expensive during overhaul for the customer to

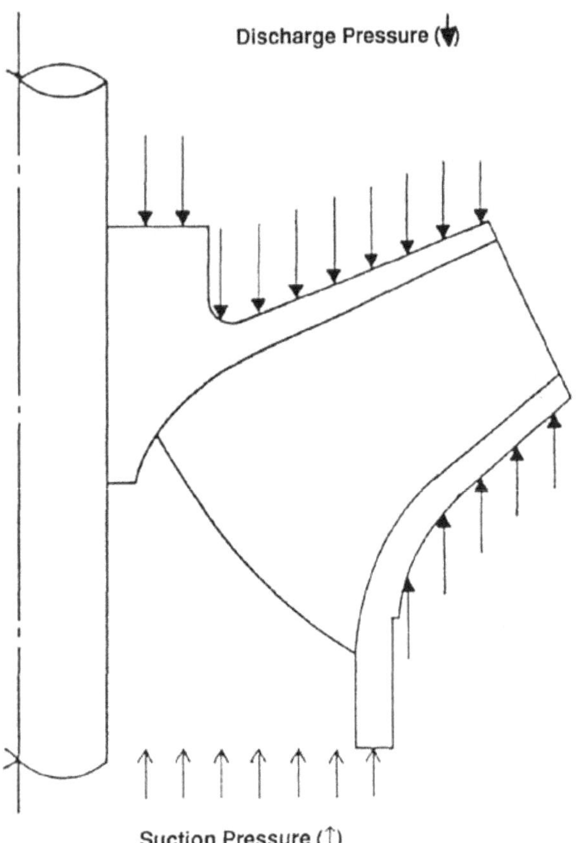

Downward axial force resulting from discharge pressure (▼) applied over surface area, less suction pressure (7) applied over the impeller eye area.

Figure 13-33: Closed impeller pressure distribution

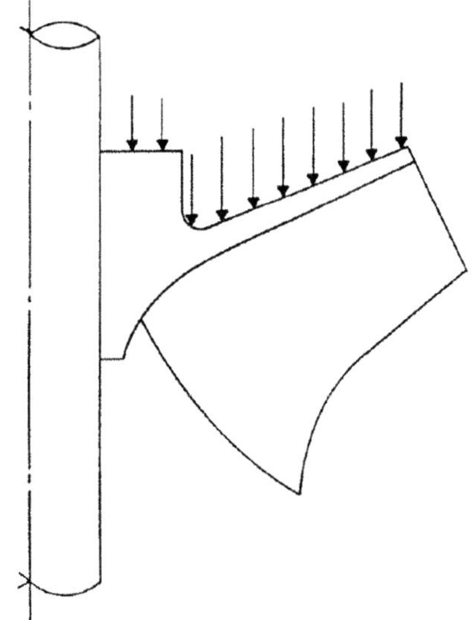

Figure 13-34: Open impeller pressure distribution

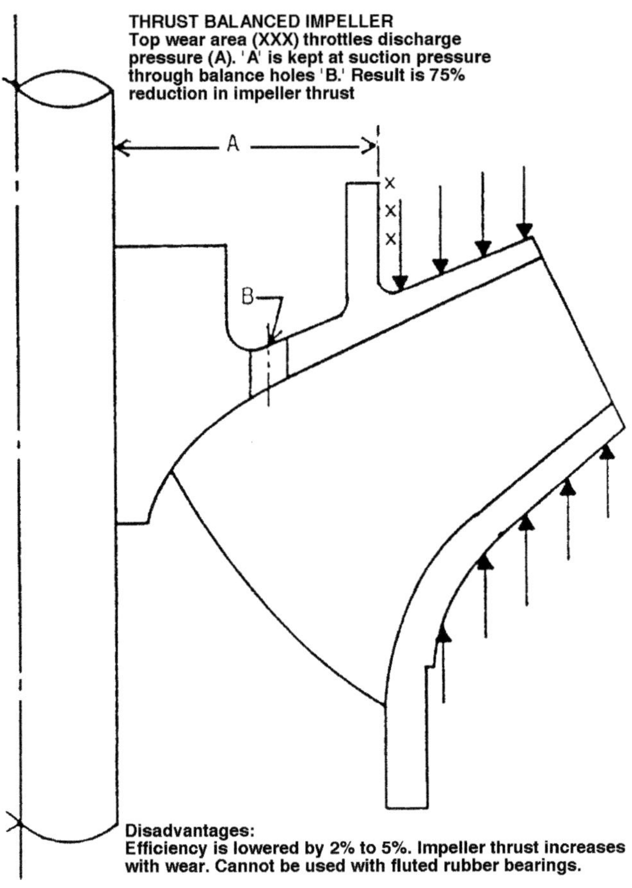

THRUST BALANCED IMPELLER
Top wear area (XXX) throttles discharge pressure (A). 'A' is kept at suction pressure through balance holes 'B.' Result is 75% reduction in impeller thrust
A
B
Disadvantages:
Efficiency is lowered by 2% to 5%. Impeller thrust increases with wear. Cannot be used with fluted rubber bearings.

Figure 13-35: Thrust balanced impeller

purchase a new shaft and standard bearings, rather than purchase a new set of sleeves and larger special bearings.

- Bowl bearings — Normally one in each bowl, suction bell and discharge case, and lubricated by the liquid pumped. The bearings in the discharge case and suction bell can easily be lubricated with a separate clean liquid. External lubrication of the bowl bearings requires expensive rifle-drilling of the bowl shaft, and is seldom done (see lower right, Figure 13-36).

VERTICAL ENCLOSED SHAFT-DRIVEN RECESSED IMPELLER SUMP PUMPS

Vertical enclosed shaft-driven sump pumps are selected in many processes where safety and reliability concerns prompt the reliability-focused to question certain less-well-designed pumps. We now find vertical enclosed shaft driven pumps applied in chemical,

CHOICE OF SEMI-OPEN OR ENCLOSED IMPELLERS

Available in alloy construction for a wide range of corrosive/abrasive services.

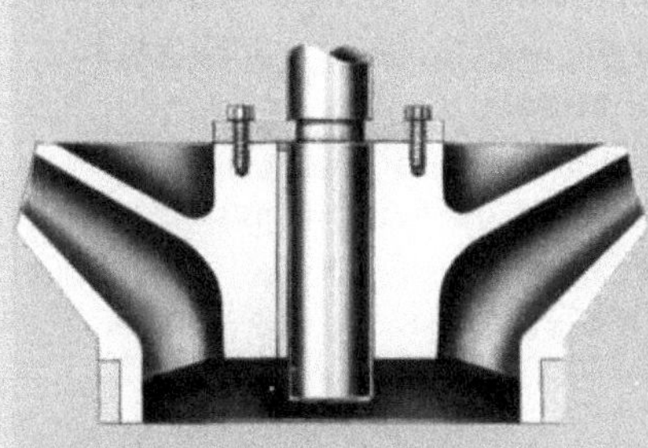

KEYED IMPELLERS

Keyed impellers are standard on 18" and larger sizes; furnished all pumps for temperatures above 180°F (82°C) and on cryogenic services. Regardless of size, keyed impellers provide ease of maintenance and positive locking under fluctuating load and temperature conditions.

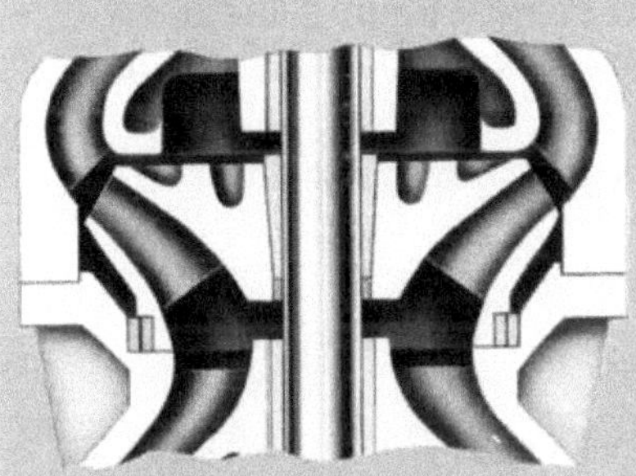

DUAL WEAR RINGS

Available for enclosed impellers and bowls; permits re-establishing initial running clearances and efficiency at lower cost. Hard facing of wear surfaces available for longer life. Wear rings can be flushed when solids are present in pumpage.

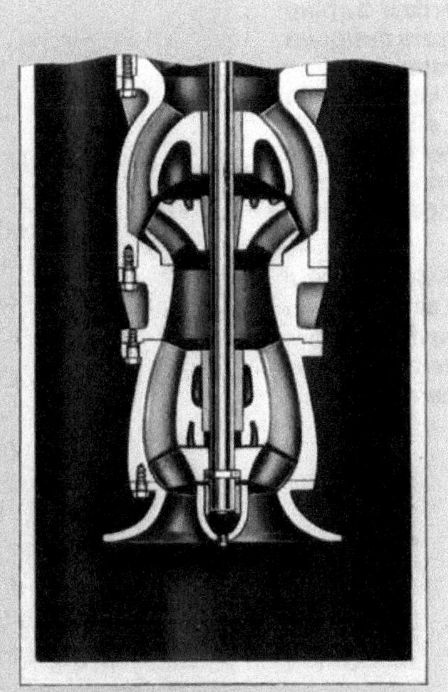

LOW NPSH FIRST STAGE IMPELLERS

For low $NPSH_A$ applications. Both large eye and mixed flow first stages available; minimizes pump length.

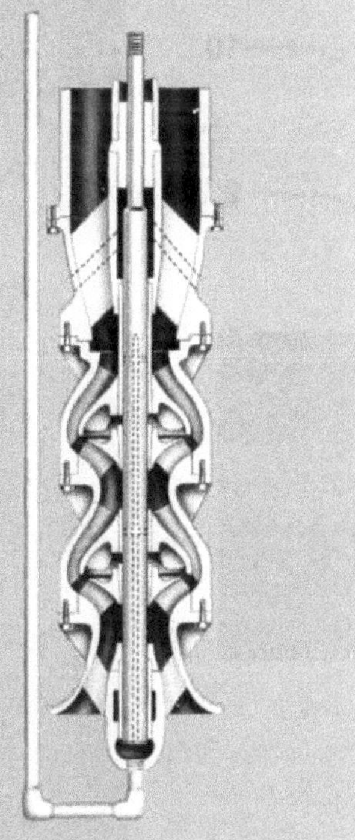

RIFLE DRILLING/ DISCHARGE BOWL

Rifle drilling of bowl shafts available for bearing protection on abrasive services.

Discharge bowl included with enclosed lineshaft construction.

Figure 13-36: Optional construction features for vertical turbine pumps (Source: ITT/Goulds, Seneca Falls, NY 13148)

petrochemical, gas production, fuel storage and other industries. They are especially important in explosion proof areas where only the best-engineered pumps merit closer consideration (Ref. 13-8).

One such thoughtfully engineered pump design (Figure 13-37) is comprised of a single stage, end suction "back pull out" type casing. Its hydraulic end is located below the liquid level and connected to the motor by means of an extended shaft. This shaft is housed and supported in a rigid tubular intermediate pipe. The design described here is certified for explosion proof areas that are classified as "Zone Zero." Its drive motor conforms to regulations found in the European Electrical Standards and described in EN 50014.

Examining Modern Vertical Pump Designs

Several other design features deserve to be highlighted after first examining a fully recessed impeller cross-section view (Figure 13-38). Understand that recessed impellers promote vortex flow. The impeller is of the semi-open type; the vanes are integral with the disk and there is no impeller shroud. On the side away from the vanes the disc incorporates integrally-cast back pump-out vanes (balancing vanes). Recessed impeller designs are available in horizontal pumps as well; the impellers are particularly suitable for handling free-flowing slurries and sludge. A modern impeller design can handle solids with dimensions essentially equal to the diameter of the discharge port.

However, explosion proofing measures sometimes require a spark arrestor in the pump discharge (Figure 13-37) and, when fitted, this arrestor restricts the permitted solids size. Selecting an appropriate suction strainer will limit the solid size to allowable diameters.

Most recessed impeller pumps rotate the liquid and solids inside the casing until the solids reach a velocity at which they exit the casing. Recirculation of solids will occur below this exit velocity; it creates wear in the casing and also increases damage to soft solids. One experienced manufacturer has overcome this problem by designing and casting the pump casing with an "axial spiral." Visualize an automobile tire to represent the basic design of a recessed impeller casing. Cutting the tire at the top and then twisting it yields a spiral. In like manner, the spiral contour helps guide solids out of the casing; it prevents pump-internal recirculation of solids.

It has been demonstrated on many occasions that this design substantially improves the true overall hydraulic efficiency of the pump. Additionally, the axial-spiral twist has greatly reduced component wear and damage to solids being pumped. As a further point of interest, the minimum flow capability of a recessed impeller pump is much lower than that of conventional radial-spiral casing design pumps. On the minus side, top centerline discharge implies a measure of vulnerability when pumping large hard solids. Solids such as rocks might, on rare occasions, smash through the casing neck. In some rock feed applications, tangential discharge (as with one noted manufacturer's first generation of "Turo"

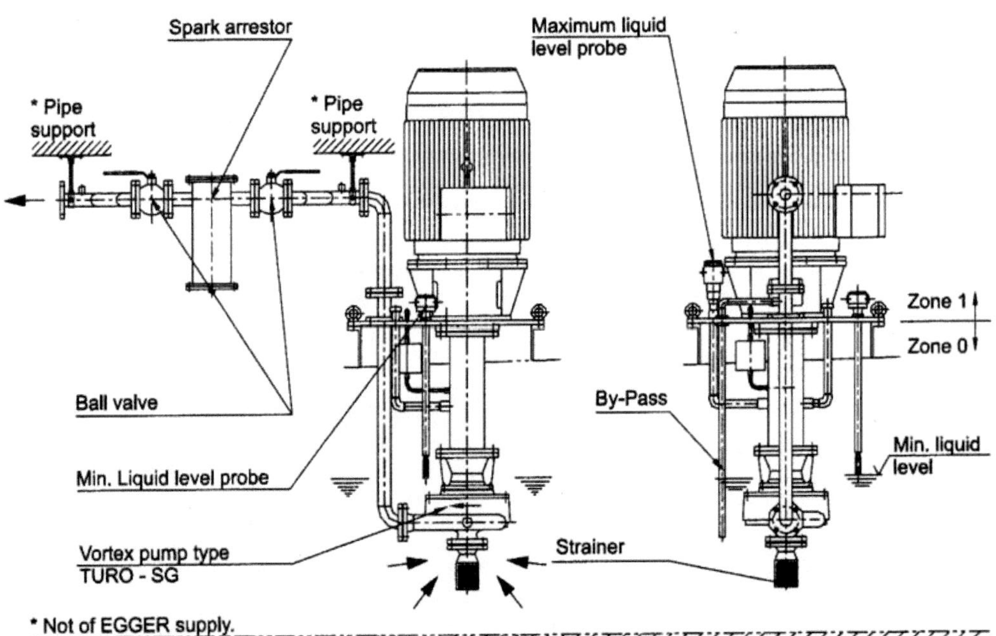

Figure 13-37: A modern vertical enclosed shaft-driven sump pump with a recessed impeller (Source: Emile Egger & Cie, Cressier/Neuchatel, Switzerland)

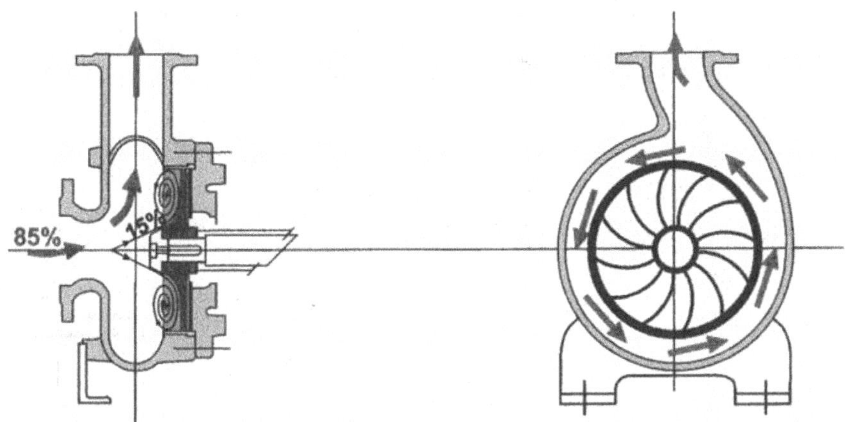

Figure 13-38: Recessed impeller principle (Source: Emile Egger & Cie, Cressier, Neuchatel, Switzerland)

pumps) might be viewed as an advantage. Where solid size is reduced (such as with the Zone Zero pumps) this vulnerability no longer exists.

So, reliability professionals are encouraged to look for unique casing design features in an advanced vertical pump. Only the most modern vertical pump casings incorporate an internal semi-axial spiral; this contour contributes to improved hydraulic efficiency while maintaining open and free flow throughout the internal spaces of these advanced recessed impeller pumps.

Look for Completeness of Installation and Ease of Maintenance

For ease of maintenance, all machined mating faces must incorporate locating registers or rabbeted fits. A tapered adapter piece forms the transition from pump to drive motor; it too has locating tabs to ensure correct and fully aligned centerlines of both motor and pump. Access ports to a non-sparking flexible coupling are provided on opposite sides of the adapter piece. These ports are designed with perforated metal safety covers or personnel guards; the operator's hands stay out of rotating parts' range.

Locating tabs and rabbeted fits go a long way towards simplifying and speeding up maintenance. Thus, when a modern vertical pump is re-assembled, its components are self-aligned by design. Impellers are typically fastened to the shaft by a parallel key and locked into position with an impeller screw and sealed washer, thereby completely sealing the shaft end from the pumped liquid.

Note the large sole plate in Figure 13-37; it also serves as a pit cover, which facilitates mounting the unit on top of a tank. The discharge pipe is secured to the sole plate by a weld neck flange; the pipe passes through the sole plate and terminates in a loose flange above the sole plate. This provision both ensures and simplifies matching the pump discharge pipe to the customer's piping.

Compliance with many existing industry regulations mandates fitting the pump with a minimum flow bypass. The bypass pipe branches off from the discharge pipe and is led back into the tank through the sole plate. The design highlighted in Figure 13-37 also includes two explosion proof-rated minimum liquid level probes mounted on the sole plate. One probe monitors the liquid level in the tank, the other monitors the liquid level in the column pipe.

The spark arrestor was mentioned earlier; note that ball valves are fitted on either side of the arrestor. Modern vertical pump designs are typically provided with a single mechanical shaft seal behind the impeller and seals are often mounted on a replaceable shaft sleeve. A bottom journal bearing is used and it, too, is mounted on a separate, replaceable shaft sleeve. When necessary to accommodate greater pump lengths, intermediate bearings are employed and located between the flange joints of the intermediate pipes. As was done with the bottom journal bearing, these intermediate bearings are also mounted on separate, replaceable shaft sleeves.

The intermediate column pipe is filled with oil; this liquid column provides lubrication to the journal bearings and also encases the drive shaft. Two angular contact ball bearings are fitted at the drive shaft end. Whenever more than one shaft section is required to accommodate the pump length, rigid intermediate flange couplings are used. To prevent fluid traveling up the shaft, a liquid thrower (flinger disc) is fastened to the shaft.

Why Recessed Impellers Are Used

There are many processes that require handling free-flowing slurries, sludge, and fibrous materials. If used in these services, a standard centrifugal pump may clog, become vapor-bound, or wear excessively. In contrast, fully recessed impellers (Figure 13-38) exhibit "gentle pumping action." Only an estimated 15% of the total throughput makes contact with the fully recessed impeller. These pumps are typically available in flow capacities approaching 100 l/sec (1,580 gpm) and heads ranging to 130 m (430 ft).

Recessed impeller pumps have been around since the 1930's. Unfortunately for the user, some manufacturers offer recessed impeller pump configurations that have not advanced from their respective configurational or hydraulic performance constraints for a generation or two. It is also fair to point out that some models require maintenance involvement to an extent that was considered acceptable decades ago, but is no longer tolerated in today's best-of-class facilities.

In essence, a number of important characteristics and advancements separate one make or design of recessed impeller from another. The very best manufacturers of vertical enclosed shaft-driven sump pumps with recessed impellers have set themselves apart from the rest by important design advances. Good commercial models first became available in the mid 1950's. Since then, even the good original designs have experienced a number of seemingly small, yet important, upgrades. Only the overall vortex-type operating principle has remained the same for Best-of-Class manufacturers. Successive iterations have consistently advanced relevant efficiency and the ability to handle solids with minimum damage to either the pump or the material being pumped.

How Recessed Impeller Pumps Often Differ

Most recessed impeller pumps rotate the liquid and solids inside the casing until the solids reach a speed at which they exit the casing. This recirculation of solids creates wear in the casing and also increases damage to soft solids. Egger has overcome this problem by designing the casing with an "axial spiral" in the casing. Visualize an automobile tire to represent the basic design of a recessed impeller casing. Cutting the tire at the top and then twisting it yields a spiral. In like manner, the spiral contour helps guide solids out of the casing; it prevents solids recirculation. The manufacturer has demonstrated on many occasions that this design substantially improves the true overall hydraulic efficiency of the pump. Additionally, the axial-spiral twist has greatly reduced component wear and damage to solids being pumped. As a further point of interest, the minimum flow capability of a recessed impeller pump is much lower than that of conventional radial-spiral casing design pumps. On the minus side, top centerline discharge implies a measure of vulnerability when pumping large hard solids. Solids such as rocks might, on rare occasions, smash through the casing neck. In some rock feed applications, tangential discharge might be viewed as an advantage.

It is thus worth understanding and considering how some recessed impeller pumps command a slight premium in initial cost. These are the ones that quickly return the incremental outlay by lower operating, maintenance, and life cycle costs.

Beware of Misunderstood Efficiency Quotes

In many cases users and engineering design contractors elect to place emphasis on pump efficiency. When asked to define efficiency, they inevitably refer to power draw. That, unfortunately, is seriously wrong. Some pumps achieve seemingly high hydraulic efficiency by simply letting the impeller edge protrude into the casing. Protruding impellers, of course, limit unimpeded passage of solids through the pump. Reliability professionals are urged to re-think what is of true importance here: the efficiency with which both liquids *and* solids are being transported. Many "old style" recessed impeller designs have simply not progressed much since their initial introduction to the marketplace. Their best operating points (BEPs) are typically in the range of 30 to 40%. On the other hand, advanced designs incorporating axial-spiral design casing internals and fully recessed impellers will have true and effective BEPs around 50 to 60%. Less energy goes into the liquid and less power is consumed to forward-feed the solids. In addition, there is less wear sensitivity with fully recessed impeller configurations.

BEARING MATERIALS

Since vertical turbine pump bearings are normally wetted by the liquid medium, major vertical turbine pump companies offer a wide variety of bearing materials to handle the various applications (see Table 13-1 for typical descriptions).

Major factors to consider when selecting bearing materials are as follows:

- Suspended solids: Determine the size and the hardness of the particles in the liquid. It is good to keep the particle size below 50 microns (.002"). Larger sizes (100-200 microns) can get into the bearings and will not pass through without caus-

Table 13-1: Typical VTP bearing materials and applications (Source: ITT/Goulds, Seneca Falls, NY 13148)

VERTICAL TURBINE BEARING MATERIAL DATA

1. Bronze-SAE 660 (Standard) #1104 ASTM-B-584-932	-50 to 250°F Min. S.G. of 0.6	General purpose material for non-abrasive, neutral pH service. 7% Tin/7% Lead/3% Zinc/83% Cu.
2. Bronze-SAE 64 (Zincless) #1107 ASTM-B-584-937	-50 to 180°F Min. S.G. of 0.6	Similar to std. bronze. Used for salt water services. 10% Tin/ 10% Lead/80% Cu.
3. Carbon Graphite Impregnated with Babbitt	-450 to 300°F All Gravities	Corrosion resistant material not suitable for abrasive services. Special materials available for severe acid services and for temp. as high as 650°. Good for low specific gravity fluids because the carbon is self-lubricating.
4. Teflon 25% Graphite with 75% Teflon	-50 to 250°F All Gravities	Corrosion resistant except for highly oxidizing solutions. Not suitable for abrasive services. Glass filled Teflon also available.
5. Cast Iron ASTM-A-48 CL30 Flash Chrome Coated	32 to 180°F Min. S.G. of 0.6	Used on non-abrasive caustic services and some oil products. Avoid water services as bearings can rust to shaft when idle. Test with bronze Bearings.
6. Lead Babbitt	32 to 300°F	Excellent corrosion resistance to a pH of 2. Good in mildly abrasive services. 80% Lead/3% Tin/17% Antimony.
7. Rubber w/Phenolic backing (Nitrile Butadiene or Neoprene)	32 to 150°F	Use in abrasive water services. Bearings must be wet prior to start-up for TPL 50'. Do not use: For oily services, for stuffing box bushing, or with hydraulically balanced impellers. For services that are corrosive, backing material other than Phenolic must be specified.
8. Hardened Metals: Sprayed on stainless steel shell (Tungsten Carbide)	All Temperatures All Gravities	Expensive alternate for abrasive services. Hardfaced surfaces typically in the range of Rc72. Other coatings are chromium oxide, tungsten carbide, colmonoy, etc. Consult factory for pricing and specific recommendation.

ing damage. Also, solids such as fly ash or silica are going to cause more damage than softer substances. General particle concentration guidelines and considerations for selecting bearings materials for handling suspended solids are:

- 0-10 ppm: Carbon-graphite, Teflon®, Dupont Vespel®
- 10-50 ppm: Bronze, nickel-impregnated carbon, PEEK, Dupont Vespel®
- 50-200 ppm: Rubber and hardened metals
- over 200 ppm: Water flush or hard face bearings

- Solids in solution: What is the chance of the solids coming out of solution? Is the pumping liquid close to its boiling point? The slight amount of heat added to the pumpage as it passes through the bearing may cause the fluid to flash, leaving solids residue. Are the solids that may come out of solution abrasive?
- Corrosion: The bearing material must be compatible with the liquid pumped
- Temperature: The rate of thermal expansion of the bearing, shaft and housings and the effect on running clearance and bearing fit must be evaluated, in addition to the temperature capability of the bearing material
- Cold flow: Some plastic bearing materials, such as Teflon®, will not maintain a press fit, and therefore, must be mechanically held in the housing. Others, such as Vespel® (Ref. 13-9), can be press-fitted as long as the manufacturer's guidelines are observed
- Specific gravity: Bearing materials with relatively high coefficients of friction (say, bronze) should not be used with liquids of a specific gravity of 0.6 and lower
- Hydraulic impeller balance: When impellers are hydraulically balanced (see Figure 13-35), avoid cutless rubber type bearings in the bowls because the flutes in the bearing will allow excessive flow into the area behind the impeller.* To compensate for this flow, the balance holes would have to be enlarged, further decreasing the efficiency

*Cutless bearings refer to rubber-lined bearings operating with water lubrication on an elasto-hydrodynamic principle.

COLUMN ASSEMBLY

The column assembly performs three functions. First, it connects the bowl assembly to the discharge head; second, it conveys the discharge flow from the bowl assembly to the discharge head, and third, it transmits the driver torque to the impeller shaft. This last function includes one or more support bearings (column bearings) for the column shaft. Column assemblies can be very short (or non-existent) when the discharge case is connected directly to the discharge head). In some instances, however, these assemblies can be hundreds of feet long. The column pipe sections (normally no longer than 10 feet, or about 3 meters) can be assembled by means of either threads or flanges.

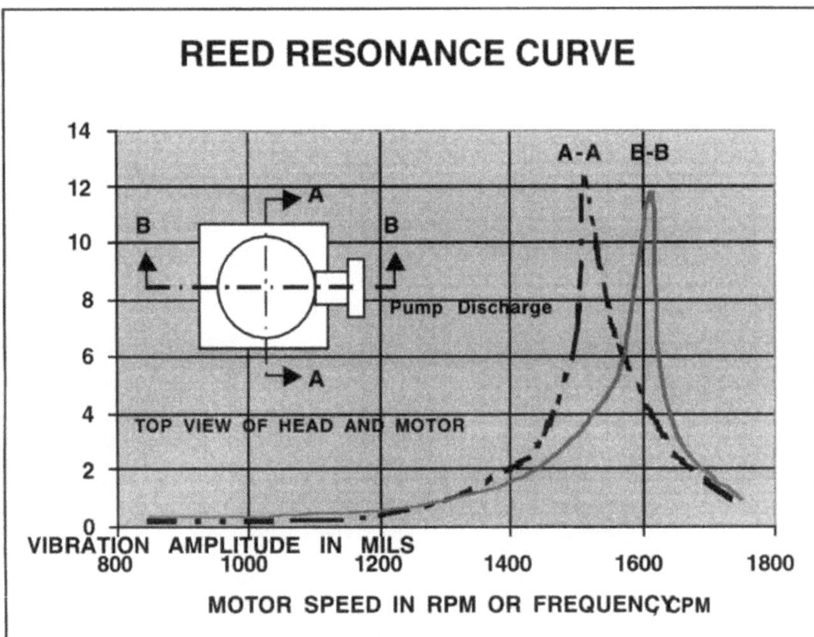

Figure 13-39: Discharge head resonance causing high amplitudes of vibration in A-A and B-B planes for a particular VTP (Source: ITT/Goulds, Seneca Falls, NY 13148)

The column shafting can be open, i.e. bearings lubricated by the fluid pumped, or enclosed, i.e. with a separate pipe around the shaft for external lubrication. Bearing brackets are located at the column pipe joints. Column shaft lengths are normally equal to the column pipe lengths, with solid shaft couplings at each connection. (Such a coupling was shown in the circled area of Figure 11-1.)

As was mentioned earlier in this segment, it is important to understand that standard threaded shaft couplings require high axial thrust to properly align the shaft. Short set applications inherently have little dead weight. If they incorporate impellers that are hydraulically balanced in the axial direction, these VTP's require upgraded industrial rigid shaft couplings (such as clamp type) to avoid excessive bearing loads from misaligned shafting.

CRITICAL SPEED

Operating VTP's at or near a shaft critical speed can generate life-shortening vibration. Resonant problems occur most often with low axial thrust and variable speed operation. For the particular pump described in Figure 13-39, operation at 1,500 or 1,600 rpm would probably prove disastrous.

Although deep well applications normally use 10 ft (~3 m) bearing spacing for open line-shaft construction, bearing spacing for industrial applications should not exceed the more conservative values shown in Figure 13-40. Here a maximum spacing of 60 inches (5 ft, or 1.52 m) would be allowed at 1,800 rpm, with shaft diameters of 1.7 inches (~43 mm) or larger.

Shaft critical speed increases with shaft diameter and the tension, i.e. axial thrust, on the shaft. Shaft critical speed decreases with increasing span length between the column bearings. This is illustrated, in Figure 13-41, for a 1.93 inch (~50 mm) diameter shaft system. A 15-25% speed separation margin away from critical shaft speeds should be observed. (See also Chapter 12—Causes of Excess Vibration).

It should be noted that there are actually more critical speeds than shown in Figure 13-41. The natural frequencies shown assume free, or knife-edge supports at the bearing locations. In reality, either end of each shaft section may or may not have a supporting moment. This depends on the direction of the displacement of the next shaft section. In essence, it is not safe to operate in most of the region between the first (one loop between supports) and second (two loops between supports) "free-free" critical speeds, unless sufficiently removed from any of the resonant frequencies. Therefore, shaft sizing and spacing should be selected to operate at least 15% below the first "free-free" critical speed. The equation used to calculate the "free-free" critical speeds (NC), as plotted in Figure 13-40, is given in Equation 13-1:

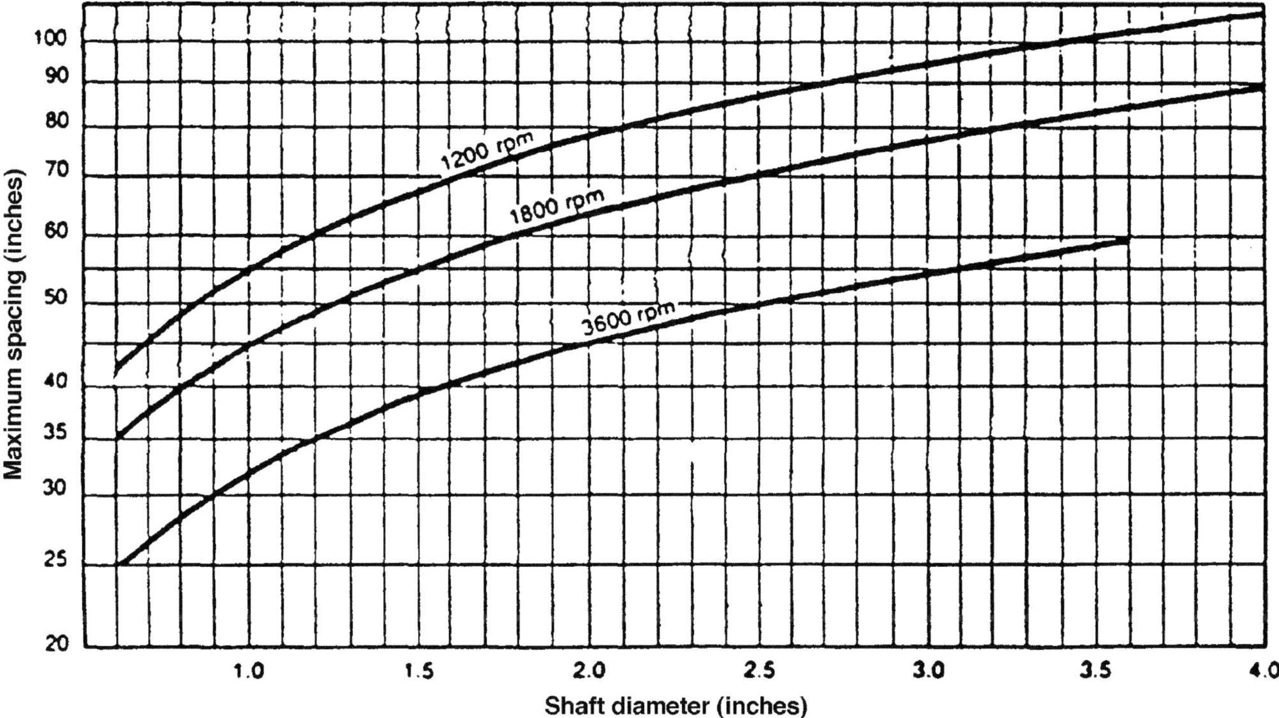

Figure 13-40. Recommended maximum bearing spacing as a function of VTP speed and shaft diameter (Source: ITT/Goulds, Seneca Falls, NY 13148)

$$NC = (30n/L)(g/w)^{0.5}[(9.87)(E)(I)(n/L + T)]^{0.5} \quad \text{(Eq. 13-1)}$$

where:

- NC = rev/sec, or cps
- n = number of deflection loops or bows between supports, i.e. (n = 1 is first critical, n = 2 is second critical)
- L = length in inches between supports (bearing span)
- g = acceleration due to gravity, 386 inches/sec^2
- E = modulus of Elasticity of Material (steel = 30,000,000 lbs/in^2)
- I = moment of inertia of cross-section in inches^4
- T = total tension (thrust) on the shaft in pounds.
- w = specific weight, lbs/in^3

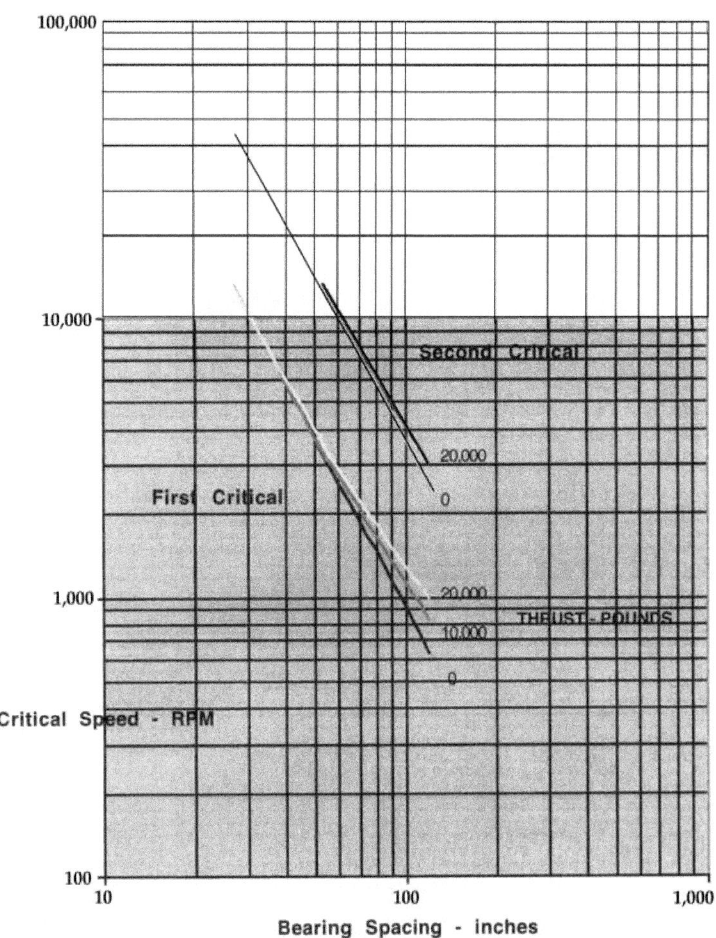

Figure 13-41: Effect of bearing spacing and axial thrust on criical speed of a 1.93 inch (~50 mm) VTP shaft. (Source: ITT/Goulds, Seneca Falls, NY 13148)

DISCHARGE HEAD ASSEMBLY

The discharge head performs such functions as

- supporting the pump driver
- accommodating an appropriate shaft seal
- providing support to the column pipe and column shaft
- directing and ducting the pumped liquid out of the pump.

Additional and optional functions include

- directing and ducting the suction liquid into canned VTPs (Figure 13-5)
- supporting and providing tension to the shaft enclosing tube (for VTPs with enclosed line shafting, Figure 13-42). Pressurized clean water (or oil in the case of a well pump) for lubricating the stuffing box and column bearings, enters through port "A," flows down the shaft through slots or clearance in the bearing, and exits out the bowl discharge case to the sump. As the pump bearings wear, flow may increase and the lube pressure may drop. This should be monitored with a pressure gauge and flow meter.
- absorbing the axial thrust of the rotor. Separate pump thrust bearings are required when the pump driver thrust bearings are not adequate. These "thrust pots" are designed in various ways and, depending on speed and load criteria discussed in earlier chapters, could be either oil (see Figure 7-42) or grease lubricated (see Figure 13-43). Pump thrust bearings must be capable of handling high down-thrusts. Nevertheless, net up-thrust may occur in short-set pumps designed for high flow rates or high discharge pressures. Up-thrust, unless accounted for in the design, can cause shaft whip or related problems.

In summary, pump thrust is often a composite of

- Impeller hydraulic unbalance thrust, a function of prevailing pressures(down-thrust)
- Dead weight of rotor (down-thrust)
- Pressure differential across stuffing box shaft sleeve (up-thrust)
- Change in liquid direction (momentum) at impeller inlet (up-thrust).

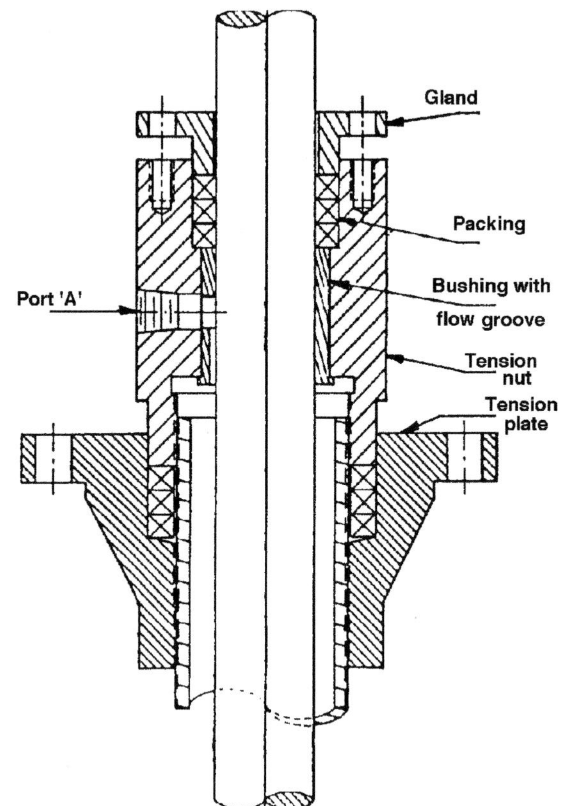

Figure 13-42: VTP with sleeve enclosing the shaft (Source: ITT/Goulds, Seneca Falls, NY 13148)

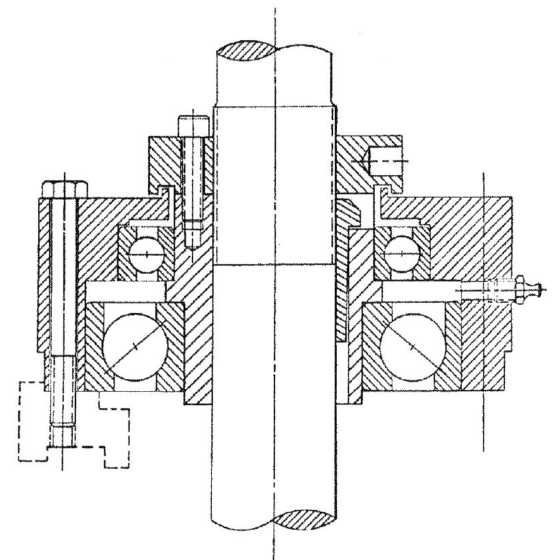

Figure 13-43: Grease-lubricated thrust pot for vertical turbine pump (Source: ITT/Goulds, Seneca Falls, NY 13148)

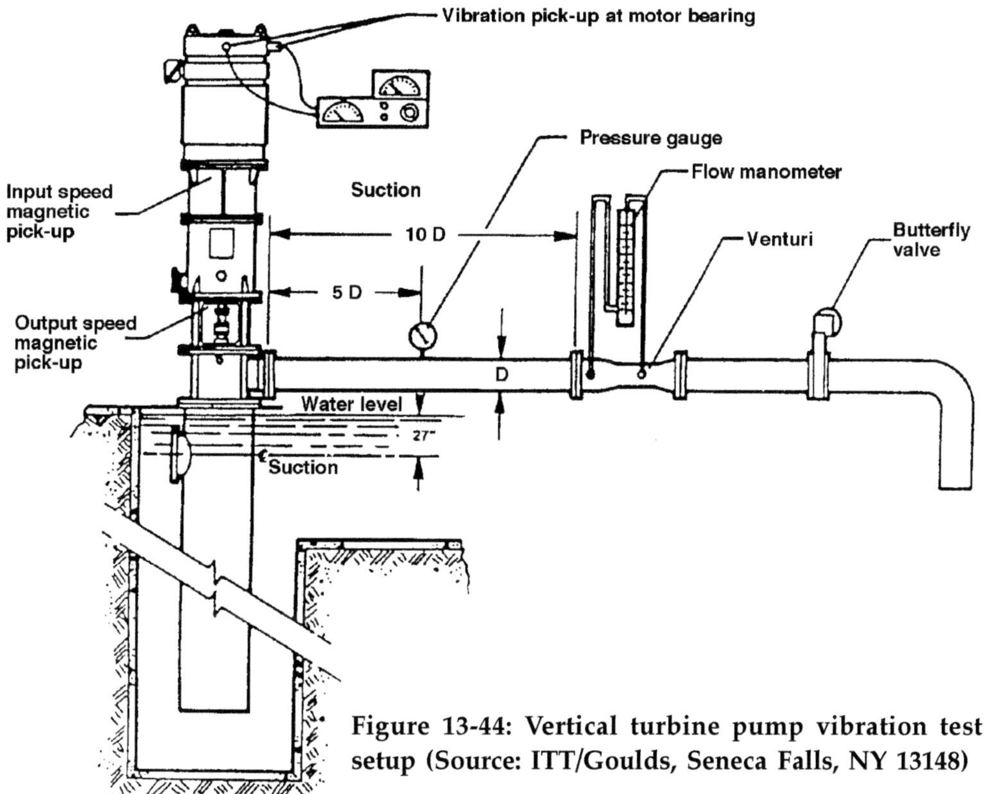

Figure 13-44: Vertical turbine pump vibration test setup (Source: ITT/Goulds, Seneca Falls, NY 13148)

The set-up for vibration testing of a vertical turbine pump is shown in Figure 13-44. Overall vibration amplitudes are typically measured in two perpendicular planes at nine different operating speeds, and eight flow points. The vibration probes are attached to the motor casing next to the upper guide bearings. Allowable vertical turbine pump field vibration values shown earlier on Table 12-2 are, however, based on vibration measurements taken at the top of the discharge head, i.e. at the motor pedestal.

SUMP DESIGN

The performance of a vertical turbine can be greatly affected by the design of the intake structure, or sump, with the degree of impact depending on the pump specific speed, size and flow rate. This was briefly described in Chapter 5.

In general, large pumps and axial flow pumps (high specific speed) are more sensitive to adverse flow phenomena than small pumps or low specific speed pumps (radial flow). Factors that should be considered in the design of a sump include (Ref. 13-7):

- Submerged vortices
- Free-surface vortices
- Excessive pre-swirl of flow entering the pump
- Non-uniform spatial distribution of velocity at the impeller eye
- Excessive variations in velocity and swirl with time
- Entrained air or gas bubbles.

Intake structures (sumps) should be designed to allow pumps to achieve their optimum hydraulic performance at all intended operating conditions. The rather detailed ANSI/HI 9.8 "Pump Intake Design" standard (Ref. 13-4) provides design guidance to this end, including minimum submergence values and recommended sump designs and parameters. It should be consulted during the installation design stage and whenever vertical pump performance problems are experienced.

SPLIT FLOW CENTRIFUGAL PUMPS

Split flow pumps should not be confused with 1950s-style overhung double-stage centrifugal pumps. While, at first glance, the two appear to be somewhat similar, split flow pumps incorporate separate discharge nozzles for the two different streams leaving the pump casing. The bending forces acting on the common shaft are within acceptable limits and attractive pump life has been demonstrated in existing applications.

The Split Flow™ pump design incorporates an auxiliary booster impeller with a separate discharge into an API Standard overhung centrifugal pump (Figure 7-12). This allows the pump to split the discharge into two separate streams. The advantages of this design are:

- Use of smaller motors—energy (kW) savings and thus carbon dioxide reduction
- Use of fewer pumps—capital savings
- Better pump-to-system hydraulic fit—extended mean-time-between-repairs (MTBR) for seals and bearings

A common design issue encountered in the refining process involves pumping a liquid to two destinations where a smaller secondary stream requires a considerably higher discharge pressure than the primary stream. An example is the overhead liquid on a naphtha splitter where liquid from the overhead receiver is used as reflux and is also pumped to a stabilizer that operates at a higher pressure than the naphtha splitter.

Conventional practice has been to install two pumps, e.g. a high-flow, low-head pump for the reflux service and a separate low-flow, high-head pump for the product/export service or to install a single pump. This single pump would typically be oversized to pump all of the liquid at the higher discharge pressure and throttle the discharge flow (by control valve) to produce the required lower head for the primary reflux stream. The conventional two-pump system requires that one pump be sized for low-flow, high-head. This frequently poses selection and reliability issues.

A Split Flow™ pump (with primary impeller) is sized for the total flow of both the primary and secondary streams at the head required for the primary stream. The auxiliary impeller, which takes suction from the primary impeller discharge, is sized for the secondary flow at the additional head required by the process. The Split Flow™ feature provides for the auxiliary impeller maximum diameter to equal the primary impeller maximum diameter. The final impeller diameters are selected based on individual stream requirements.

As of 2013, the prototype application of Split Flow™, manufactured by Flowserve, was being used extensively. As an example, in fractionator reflux service since mid-1996, one application incorporates a pair of 60 horsepower 4x6x10 pumps. The alternative common-practice oversized design would have used two 4x6x13 pumps with 125 hp motors and throttling the excess head at the primary reflux flow outlet. The Split Flow™ design facilitates conformance with impeller diameter criteria for overhung type pumps, e.g. Shell specifies a 13-inch maximum diameter for overhung pumps at 3,600 rpm.

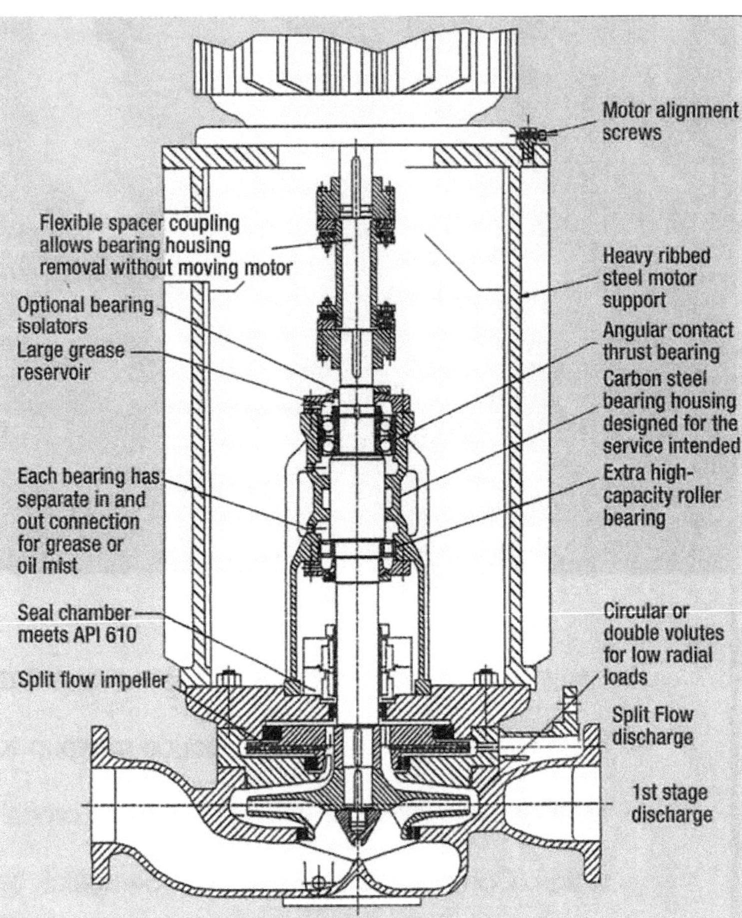

Figure 13-45: API Standard (Style OH3) pump with auxiliary impeller (Split Flow™)

API Standard 610, the premier hydrocarbon processing industry guide to centrifugal pumps, encourages innovative energy-conserving approaches to be aggressively pursued by both user and manufacturer. With this alternative approach resulting in improved energy utilization, it should be considered whenever applicable.

It has been demonstrated that the Split Flow™ pumps with smaller motors reduce energy consumption at a California refinery by 50 kW, or 440,000 kWh per year. This results in a savings of approximately

$25,000 per year. Reductions in electricity consumption also curtail CO_2 emissions. Saving 50 kW/440,000 kWh also avoid discharging 200 to 400 tons of CO_2 per year, depending on the type of fuel used for power generation.

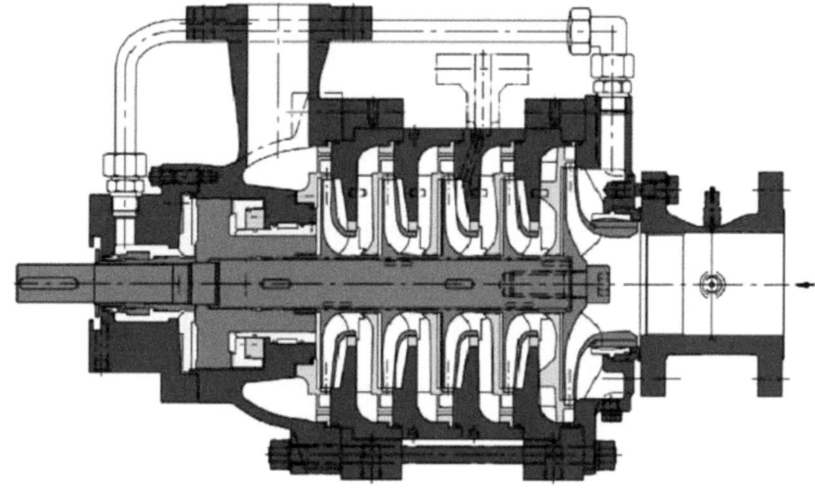

Figure 13-46: A five-stage high-performance centrifugal pump with hydraulic thrust balance. Note thrust bearing on drive side (Source: Sulzer-Bingham, Portland, Oregon)

MODERN PUMPS HAVE STIFFER SHAFTS

Some (often 1960s-vintage) multistage centrifugal pumps are designed with relatively slender shafts and operate at speeds above the so-called rotor critical. When that is the case and as the pump comes up to operating speed, there is a brief time period when the "ramping-up" speed will coincide with the rotor critical and undesirable shaft deflection will have taken place. Designs with reduced bearing spans (and thus increased ratios of shaft-diameter-to-bearing-span) have been used in modern canned motor pumps as an approach that avoids the amount or risk of shaft deflection. Furthermore, canned motor pumps are product-lubricated and require no mechanical seals. That makes them inherently less prone to experience shaft deflection.

However, while canned motor pumps are among the viable options and have been the subject of books and articles, traditional centrifugal pump designs have made progress as well. In fact, operation above rotor critical speed has also been made possible with intelligent redesigns of the more traditional centrifugal pumps. Although some of these intelligently redesigned stiff-shaft process pumps (Figures 13-46 and 13-47) are still using mechanical seals, they merit a closer look.

Figure 13-47: A six-stage high-performance centrifugal pump with hydraulic thrust balance. Note thrust bearing on non-drive side (Source: Conhagen Inc., Houston, Texas)

Note how they, like canned motor pumps (Figure 13-16), are product-lubricated. The extent to which either the pumps shown in Figures 13-46 and 13-47, or the stiff-shaft product-lubricated canned motor pumps of Figure 13-16 are best suited to serve in a given application must be determined on a case-by-case basis. That said reliability professionals must stay informed on the pros and cons of all available types of pumps.

There are also re-configured 10 stage pumps. In its re-configured version, an "old" multistage pump is equipped with hydrostatic bearings at the center stage and throttle bushing locations. Each stage is stepped and the impeller retained with split rings instead of the typical "stacked" rotor.

Pump upgrade manufacturers often coat the center stage bushing with proprietary hard coatings and have outstanding experience with well-chosen materials unless there is excessive solids ingestion. Of course, while no multistage pump will survive a serious catalyst ingestion event, there are certain important opportunities pump users should pursue in their quest for extended run lengths. Consider viewing the Sulzer, Conhagen, Hydro and similar websites as an important first step. Select from among competent OEM (original equipment manufacturers) and non-OEM providers.

Chicago-based Hydro is the largest non-OEM pump rebuilder; the company has branches in different parts of the world. They specialize in combining needed repairs with optimized upgrading. Hydro is familiar with the used and surplus equipment markets and can often convert or re-engineer pre-existing process pumps to suit tight user schedules.

When the coker jet pump in Figure 13-48 was rebuilt, the combined repair and upgrade work brought pump performance in line with the actual operational head-versus-flow performance of the customer. Rotordynamic (vibration) issues were remedied. Both hydraulic and reliability performance of this multi-stage process pump were greatly improved.

TYPICAL PUMP MATERIAL SELECTION

Suppose you were dealing with a pump that required an abnormally high number of repairs and replacement of corrosion-affected component parts. Chances are the pump was either purchased from the lowest bidder, or the corrosivity of your pumpage differs from that of the medium originally specified.

In that case the individual responsible for improving pump reliability, or the failure analyst will need guidance as to pump material selection. Although perhaps not responsible for answering questions relating to corrosion, those unfamiliar with the subject of metallurgy would benefit from the ITT/Goulds general material selection guide (Ref. 13-3) for pumps given on the next few pages. These pages—essentially Table 13-2—might thus allow pre-screening of applicable alloys and initiation of corrective pursuits if it appears that the wrong materials are involved.

Table 13-3 deals with the unique thermal expansion properties of DuPont's Vespel® CR6100, a high-performance carbon fiber composite material that has been installed in hundreds of pumps at refineries, steel mills, paper mills, chemical plants and food processing facilities in the U.S. (Ref. 13-8). It has replaced metal and other composite materials used for pump wear rings, throat bushings and line shaft bearings. (See also Chapters 6 and 15).

Figure 13-48: A completely rebuilt multi-stage coker jet pump. (Source: Hydro, Inc., Chicago, Illinois)

Table 13-2: Pump materials of construction (Source: ITT/Goulds, Seneca Falls, NY 13148)

Table 13-2A: Typical pump material selections (pg. 1 of 4)

This chart is intended as a guide in the selection of economical materials. It must be kept in mind that corrosion rates may vary widely with temperature, concentration, and the presence of trace elements or abrasive solids. Blank spaces in the chart indicate a lack of accurate corrosion data for those specific conditions. In general, the chart is limited to metals and non-metals regularly furnished by Goulds.

Code:
A = Recommended
B = Useful Resistance
X = Unsuitable

Note: Maximum temperature limits are shown where data is available. Contact a Goulds representative for temperature limits of all materials before final material selection.

A-20 — Carpenter Stainless
CD4MCu — Stainless Steel
Alloy 2205 — Stainless Steel
H-B — Hastelloy Alloy - B
H-C — Hastelloy Alloy - C
Ti — Titanium Unalloyed
316SS — Stainless Steel
'Tefzel — Ethyltetrafluoroethylene

C.I. — Cast Iron
D.I. — Ductile Iron
Steel — Carbon Steel
Brz. — Bronze
Zi — Zirconium
FRP — Fibre Reinforced Vinylester
PFA — Virgin Teflon

Corrosive	Steel C.I. D.I.	Brz	316	A-20	CD4MCu	ALLOY 2205	HAST B	HAST C	TI	ZI	TEFZEL (ETFE)	PFA	FRP
Acetaldehyde, 70°F.	B	A	A	A	A	A		A	A	A	A	A	X
Acetic Acid, 70° F.	X	A	A	A	A	A	A	A	A	A	A	A	X
Acetic Acid, <50%, To Boiling	X	B	A	A	B	A	X	A	A	A			
Acetic Acid, >50%, To Boiling	X	X	B	A	X	A	X	A	A	A	104°C	149°C	X
Acetone, To Boiling	A	A	A	A	A	A	A	A	A	A	104°C	149°C	X
Aluminum Chloride, <10%, 70° F.	X	B	X	B	X	B	A		B	A	A	A	A
Aluminum Chloride, >10%, 70° F.	X	X	X	B	X	B	A		B	A	A	A	A
Aluminum Chloride, <10%, To Boiling	X	X	X	X	X	X	A		X	A	104°C	149°C	X
Aluminum Chloride, >10%, To Boiling	X	X	X	X	X	X	A	X	X	A	104°C	149°C	X
Aluminum Sulphate, 70° F.	X	B	A	A	A	A	B	B	A	A	A	A	A
Aluminum Sulphate, <10%, To Boiling	X	B	B	A	B	A	A	A	A	A	104°C	149°C	
Aluminum Sulphate, >10%, To Boiling	X	X	X	B	X	B	B	B	X	B	104°C	149°C	
Ammonium Chloride, 70° F.	X	X	B	B	B	B		A	A	A	A	A	A
Ammonium Chloride, <10%, To Boiling	X	X	B	B	X	B	B	A	A	A	104°C	149°C	
Ammonium Chloride, >10%, To Boiling	X	X	X	X	X	X		X	X	X	104°C	149°C	
Ammonium Fluosilicate, 70° F.	X	X	X	B	X	B		X	X	X			A
Ammonium Sulphate, <40%, To Boiling	X	X	B	B	X	B	X	B	A	A	104°C	149°C	
Arsenic Acid, to 225° F.	X	X	X	B	X	B					A	A	
Barium Chloride, 70° F. <30%	X	B	X	B	X	B	B	B	B	B	A	A	A
Barium Chloride, <5%, To Boiling	X	B	X	B	X	B	B	B	A	A	104°C	149°C	
Barium Chloride >5%, To Boiling	X	X	X	X	X	X	X	X	X	X	104°C	149°C	
Barium Hydroxide, 70° F.	B	X	A	A	A	A	B	B	A	A	A	A	A
Barium Nitrate, To Boiling	X	X	B	B	B	B	B		B	B	104°C	149°	
Barium Sulphide, 70° F.	X	X	B	B	B	B			A	A	A	A	A
Benzoic Acid	X	X	B	B	B	B	A	A	A	A	A	A	
Boric Acid, To Boiling	X	X	B	B	B	B	A	A	B	B	104°C	149°C	
Boron Trichloride, 70° F. Dry	B	B	B	B	B	B	B	B					
Boron Trifluoride, 70° F. 10%, Dry	B	B	B	A	B	A		A					
Brine (acid), 70° F.	X	X	X	X	X	X		B	B		A	A	A
Bromine (dry), 70° F.	X	X	X	X	X	X	B	B	X	X	A	A	X
Bromine (wet), 70° F.	X	X	X	X	X	X		B	X	X	A	A	X
Calcium Bisulphite, 70°F.	X	X	B	B	B	B		B	A	A	A	A	
Calcium Bisulphite,	X	X	X	B	X	B		X	A	A	A	A	
Calcium Chloride, 70° F.	B	X	B	B	B	B	A	A	A	A	A	A	A
Calcium Chloride <5%, To Boiling	X	X	B	B	B	B	A	A	A	A	104°C	149°C	
Calcium Chloride >5%, To Boiling	X	X	X	B	X	B	A	A	B	B	104°C	149°C	
Calcium Hydroxide, 70° F.	B	B	B	B	B	B		A	A		A	A	A
Calcium Hydroxide, <30%, To Boiling	X	B	B	B	B	B		A	A		104°C	149°C	
Calcium Hydroxide, >30%, To Boiling	X	X	X	X	X	X		B	A		104°C	149°C	

Table 13-2B: Typical pump material selections (pg. 2 of 4)

Corrosive	Steel C.I. D.I.	Brz	316	A-20	CD4MCu	ALLOY 2205	HAST B	HAST C	TI	ZI	TEFZEL (ETFE)	PFA	FRP
Calcium Hypochlorite, <2%, 70° F.	X	X	X	X	X	X		A	A	A	A	A	X
Calcium Hypochlorite, >2%, 70° F.	X	X	X	X	X	X		B	A	B	A	A	X
Carbolic Acid, 70° F. (phenol)	X	B	A	A	A	A	A	A	A	A	A	A	
Carbon Bisulphide, 70° F.	B	B	A	A	A	A			A		A	A	
Carbonic Acid, 70° F.	B	X	A	A	A	A	A	A	A	A	A	A	
Carbon Tetrachloride, Dry to Boiling	B	B	A	A	A	A	B	B	A	A	104°C	149°C	
Chloric Acid, 70° F.	X	X	X	B	X	B	X	X			A	A	
Chlorinated Water, 70° F.	X	X	B	B	B	B		A	A	A	A	A	
Chloroacetic Acid, 70° F.	X		X	X		X			A	B	A	A	A
Chlorosulphonic Acid, 70° F.	X	X	X	X	X	X	A	A	B	X	A	A	A
Chromic Acid, <30%	X	X	X	B	X	B		B	A	A	65°C	A	X
Citric Acid	X	X	A	A	A	A	A	A	A	A	A	A	A
Copper Nitrate, to 175° F.	X	X	B	B	B	B	X	X	B		A	A	A
Copper Sulphate, To Boiling	X	X	X	B	X	B		A	A	A	104°C	149°C	
Cresylic Acid	X	X	B	B	B	B	B	B			A	A	
Cupric Chloride	X	X	X	X	X	X		X	B	X	A	A	
Cyanohydrin, 70° F.	X		B	B	B	B					A	A	A
Dichloroethane	X	B	B	B	B	B	B	B	A	B	65°C	149°C	
Diethylene Glycol, 70° F.	A	B	A	A	A	A	B	B	A	A	A	A	
Dinitrochlorobenzene, 70° F. (dry)	X	B	A	A	A	A	A	A	A	A	A	A	
Ethanolamine, 70° F.	B	X	B	B	B	B			A	A	A	A	
Ethers, 70° F.	B	B	B	A	A	A	B	B	A	A	A	A	
Ethyl Alcohol, To Boiling	A	A	A	A	A	A	A	A	A	A	104°C	149°C	
Ethyl Cellulose, 70° F.	A	B	B	B	B	B	B	B	A	A	A	A	
Ethyl Chloride, 70° F.	X	B	B	A	B	A	B	B	A	A	A	A	X
Ethyl Mercaptan, 70° F.	X	X	B	A	B	A	B	B			A	A	X
Ethyl Sulphate, 70° F.	X	B	B	A	B	A					A	A	X
Ethylene Chlorohydrin, 70° F.	X	B	B	B	B	B	B	B	A	A	A	A	X
Ethylene Dichloride, 70° F.	X	B	B	B	B	B	B	X	A	A	A	A	X
Ethylene Glycol, 70° F.	B	B	B	B	B	B	A	A	A	A	A	A	A
Ethylene Oxide, 70° F.	X	X	B	B	B	B	A	A	A	A	A	A	
Ferric Chloride, <5%, 70° F.	X	X	X	X	X	X	X	A	A	B	A	A	A
Ferric Chloride, >5%, 70° F.	X	X	X	X	X	X	X	B	B	X	A	A	X
Ferric Nitrate, 70° F.	X	X	B	A	B	A		B			A	A	A
Ferric Sulphate, 70° F.	X	X	X	B	X	B		B	B	B	A	A	A
Ferrous Sulphate, 70° F.	X	X	X	B	X	B	B	B	A	A	A	A	A
Formaldehyde, To Boiling	B	B	A	A	A	A	B	B	A	A	104°C	149°C	
Formic Acid, to 212° F.	X	X	X	A	B	A	A	A	X	A	A	A	
Freon, 70° F.	A	A	A	A	A	A	A	A	A	A	A	A	
Hydrochloride Acid, <1%, 70° F.	X	X	X	B	X	B	B	A	B	A	A	A	A
Hydrochloric Acid, 1-20%, 70° F.	X	X	X	X	X	X	B	X	X	A	A	A	A
Hydrochloric Acid, >20%, 70° F.	X	X	X	X		X	B	X	X	B	A	A	X
Hydrochloric Acid, <½ %, 175° F.	X	X	X	X	X	X	A	X	X	A	A	A	X
Hydrochloric Acid, ½-2%, 175° F.	X	X	X		X		B	X	X	A	A	A	X
Hydrocyanic Acid, 70° F.	X	X	X	B	X	B	X	X			A	A	A
Hydrogen Peroxide, <30% <150° F.	X	X	B	B	B	B	B	B	A	A	A	A	
Hydrofluoric Acid, <20%, 70° F.	X	B	X	B	X	B	X	B	X	X	A	A	
Hydrofluoric Acid, >20%, 50° F.	X	X	X	X	X	X	X	B	X	X	A	A	
Hydrofluoric Acid, To Boiling	X	X	X	X	X	X		X	X	X			
Hydrofluorsilicic Acid, 70° F.	X		X	B	X	B		B			A	A	
Lactic Acid, <50%, 70° F.	X	B	A	A	A	A	B	B	A	A	A	A	
Lactic Acid, >50%, 70° F.	X	B	B	B	B	B	B	B	A	A	A	A	
Lactic Acid, <5%, To Boiling	X	X	X	B	X	B	B	B	A	A	104°C	149°C	
Lime Slurries, 70° F.	B	B	B	B	A	B	B	B	B	B			

Table 13-2C: Typical pump material selections (pg. 3 of 4)

Corrosive	Steel C.I. D.I.	Brz	316	A-20	CD4MCu	ALLOY 2205	HAST B	HAST C	Ti	Zi	TEFZEL (ETFE)	PFA	FRP
Magnesium Chloride, 70° F.	X	X	B	A	B	A	A	A	A	A	A	A	A
Magnesium Chloride, <5%, To Boiling	X	X	X	B	X	B	A	A	A	A	104°C	149°C	
Magnesium Chloride, >5%, To Boiling	X	X	X	X	X	X	B	B	B	B	104°C	149°C	
Magnesium Hydroxide, 70° F.	B	A	B	B	A	B	B	B	A		A	A	A
Magnesium Sulphate	X	X	B	A	B	A	X	X	B	B	A	A	
Maleic Acid	X	X	B	B	B	B	B	B	A		A	A	
Mercaptans	A	X	A	A	A	A					A	A	
Mercuric Chloride, <2%, 70° F.	X	X	X	X	X	X		B	A	A	A	A	
Mercurous Nitrate, 70° F.	X	X	B	B	B	B		C			A	A	
Methyl Alcohol, 70° F.	A	A	A	A	A	A	A	A	A	A	A	A	
Naphthalene Sulphonic Acid, 70° F.	X	X	B	B	B	B	B	B			A	A	
Napthalenic Acid	X	X	B	B	B	B	B	B			A	A	
Nickel Chloride, 70° F.	X	X	X	B	X	B	A		B	B	A	A	A
Nickel Sulphate	X	X	B	B	B	B		B		A	A	A	
Nitric Acid	X	X	B	B	B	B			B	B			
Nitrobenzene, 70° F.	A	X	A	A	A	A	B	B	A		A	A	X
Nitroethane, 70° F.	A	A	A	A	A	A	A	A	A	A	A	A	X
Nitropropane, 70° F.	A	A	A	A	A	A	A	A	A	A	A	A	X
Nitrous Acid, 70° F.	X	X	X	X	X	X					A	A	A
Nitrous Oxide, 70° F.	X	X	X	X	X	X		X			A	A	
Oleic Acid	X	X	B	B	B	B	X	X	X	X	A	A	X
Oleum, 70° F.	B	X	B	B	B	B	B	B	B		A	A	X
Oxalic Acid	X	X	X	B	X	B	B	B	X	A	A	A	X
Palmitic Acid	B	B	B	A	B	A					A	A	
Phenol (see carbolic acid)											A	A	
Phosgene, 70° F.	X	X	B	B	B	B	B	B			A	A	
Phosphoric Acid, <10%, 70° F.	X	X	A	A	A	A	A	A	A	A	A	A	A
Phosphoric Acid, >10-70%, 70° F.	X	X	A	A	A	A	B	X	B	B	A	A	X
Phosphoric Acid, <20%, 175° F.	X	X	B	B	B	B	A	A	X	B	A	A	X
Phosphoric Acid, >20%, 175° F. <85%	X	X	X	B	X	B	B	X	X	X	A	A	X
Phosphoric Acid, >10%, Boil, <85%	X	X	X	X	X	X	X	X	X	X			
Phthalic Acid, 70° F.	X	B	B	A	B	A	B	B	A	A	A	A	
Phthalic Anhydride, 70° F.	B	X	A	A	A	A	A	A			A	A	
Picric Acid, 70° F.	X	X	X	B	X	B		B			A	A	
Potassium Carbonate	B	B	A	A	A	A	B	B	A	A	A	A	A
Potassium Chlorate	B	X	A	A	A	A		B	A	A	A	A	A
Potassium Chloride, 70° F.	X	X	B	A	B	A	B	B	A	A	A	A	A
Potassium Cyanide, 70° F.	B	X	B	B	B	B	B	B			A	A	A
Potassium Dichromate	B	B	A	A	A	A		B	A	A	A	A	
Potassium Ferricyanide	X	B	B	B	B	B	B	B	A	A	A	A	
Potassium Ferrocyanide, 70° F.	X	B	B	B	B	B	B	B		B	A	A	A
Potassium Hydroxide, 70° F.	X	X	B	A	B	A	B	X	B	A	A	A	A
Potassium Hypochlorite	X	X	X	B	X	B		B	A		A	A	
Potassium Iodide, 70° F.	X	B	B	B	B	B	B	B	A	A	A	A	
Potassium Permanganate	B	B	B	B	B	B		B			A	A	A
Potassium Phosphate	X	X	B	B	B	B			B	B	A	A	A
Sea Water, 70° F.	X	B	B	A	B	A	A	A	A	A	A	A	A
Sodium Bisulphate, 70° F.	X	X	X	B	X	B	B	B	B	A	A	A	A
Sodium Bromide, 70° F.	B	X	B	B	B	B	B	B			A	A	A
Sodium Carbonate	B	B	B	A	B	A	B	B	A	A	A	A	A
Sodium Chloride, 70° F.	X	B	B	B	B	B	B	B	A	A	A	A	A
Sodium Cyanide	B	X	B	B	B	B			B		A	A	
Sodium Dichromate	B	X	B	B	B	B			B		100°C	A	
Sodium Ethylate	B	A	A	A	A	A					A	A	

Table 13-2D: Typical pump material selections (pg. 4 of 4)

Corrosive	Steel C.I. D.I.	Brz	316	A-20	CD4MCu	ALLOY 2205	HAST B	HAST C	Ti	Zi	TEFZEL (ETFE)	PFA	FRP
Sodium Fluoride	X	X	B	B	B	B	X	X	B	B	A	A	
Sodium Hydroxide, 70° F.	B	B	B	A	B	A	A	A	A	A	A	A	A
Sodium Hypochlorite	X	X	X	X	X	X		B	A	B	A	A	X
Sodium Lactate, 70° F.	B	X	X	X	X	X	X	X			A	A	A
Stannic Chloride, <5%, 70° F.	X	X	X	X	X	X	B	B	A	A	A	A	A
Stannic Chloride, >5%, 70° F.	X	X	X	X	X	X	B	X	B	B	A	A	
Sulphite Liquors, To 175° F.	X	X	B	B	B	B		B	A				
Sulphur (molten)	B	X	A	A	A	A	X	A	A				
Sulphur Dioxide (spray), 70° F.	X	X	B	B	B	B		B	X		A	A	
Sulphuric Acid, <2%, 70° F.	X	X	B	A	B	A	A	A	B	A	A	A	A
Sulphuric Acid, 2-40%, 70° F.	X	X	X	B	X	B	A	A	X	A	A	A	A
Sulphuric Acid, 40%, <90%, 70° F.	X	X	X	B	X	B	A	A	X	X	A	A	X
Sulphuric Acid, 93-98%, 70° F.	B	X	B	B	B	B	B	B	X	X	A	A	X
Sulphuric Acid, <10%, 175° F	X	X	X	B	X	B	A	X	X	B	A	A	A
Sulphuric Acid, 10-60% & >80%, 175° F.	X	X	X	B	X	B	B	X	X	X	A	A	X
Sulphuric Acid, 60-80%, 175° F.	X	X	X	X	X	X	B	X	X	X	A	A	X
Sulphuric Acid, < ¾%, Boiling	X	X	X	B	X	B	B	B	X	B			
Sulphuric Acid, ¾-40%, Boiling	X	X	X	X	X	X	B	X	X	B			
Sulphuric Acid, 40-65% & >85%, Boil	X	X	X	X	X	X	X	X	X	X			
Sulphuric Acid, 65-85%, Boiling	X	X	X	X	X	X	X	X	X	X			
Sulphurous Acid, 70° F.	X	X	X	B	X	B	B	B	A	B	A	A	A
Titanium Tetrachloride, 70° F.	X		X	B	X	B		X			A	A	
Tirchlorethylene, To Boiling	B	X	B	B	B	B	B	B	A	A			
Urea, 70° F.	X	X	B	B	B	B	X	X	B	B	A	A	
Vinyl Acetate	B	B	B	B	B	B		B			A	A	
Vinyl Chloride	B	X	B	B	B	B	X	B	A		A	A	
Water, To Boiling	B	A	A	A	A	A	A	A	A	A			
Zinc Chloride	X	X	B	A	B	A	B		A	A	A	A	A
Zinc Cyanide, 70° F.	X	B	B	B	B	B	B	B	B	B	A	A	
Zinc Sulphate	X	X	A	A	A	A	X	X	A		A	A	A

Table 13-3: Z-direction thermal expansion of DuPont Vespel® CR6100 (Source: DuPont Engineering Polymers, Newark, Delaware)

Process Temperature °F	*Axial Growth at Temperature, per Inch (Based on 68°F Ambient Temperature)*
-40	-0.019
-20	-0.016
0	-0.012
20	-0.009
40	-0.005
60	-0.001
80	0.002
100	0.006
120	0.009
140	0.013
160	0.017
180	0.020
200	0.024
220	0.027
240	0.031
260	0.035
280	0.038
300	0.042
320	0.047
340	0.052
360	0.057
380	0.062
400	0.067
420	0.077
440	0.087
460	0.097
480	0.108
500	0.118

Chapter 14

Pump Failure Analysis and Troubleshooting

ORGANIZING SYSTEMATIC PUMP FAILURE REDUCTION PROGRAMS

Re-examining The Cost of Pump Failures

It has been estimated that centrifugal pump failures in the petrochemical industry cost $5,000 per average repair event. However, these are direct costs for labor and materials only. This cost does not include employee benefits, plant administration, mechanical department and reliability technician overhead, and the cost of materials procurement. Generally accepted accounting rules call for the inclusion of these costs. The all-inclusive cost of pump failures at a major U.S. refinery is given in Table 14-1. In 1983/1984 it averaged $10,287 per pump repair event! Thirty years later and after implementing a systematic program of pump failure reduction, the same refinery experienced considerably fewer than 100 pump repair events per month; their pump MTBF has increased three to four-fold. However, on average, each repair still cost in excess of $10,000.

Available statistics confirm that centrifugal pump repairs account for the bulk of maintenance expenditures in many petrochemical companies. In 1996, the average large refineries with 3,000 installed pumps

Table 14-1: Actual repair cost calculation and failure experience at a U.S. refinery, 1983/84

Total Number Pumps	=	2,754; 50% in Service = 1,377 (1984) [1,427 (1983)]
Avg. Pump Repairs/Month	=	100 (1984) inc. 40 Shop & 60 Field Repairs [90 (1983) incl. 42 Shop & 53 Field Repairs]
Avg. Pump Maint Costs/month	=	$538k (1984) Based on work order tracking [$651.7k (1983)]
Avg. Pump Repair Costs	=	$5,380 (1984) [$6,860 (1983)

In the past this refinery has included incremental burden savings in total maintenance credits for reducing pump failures. Credit work-up therefore includes:

		$ Labor	$ Mat'l	$ Total
+	Average Direct Charges Per Repair (split 50/50 labor/materials)	2,690	2,690	5,380
+	Employee Benefits (50% of labor)	1,345	——	1,345
+	Refin. Admin. & Services (9.95% of labor)	268	——	268
+	Mechanical O/H (115% of labor)	3,094	——	3,094
+	Materials Procurement (7.4% of materials)	——	200	200
	Total	$7,397	$2,890	$10,287

∴ total maintenance credits $10,287 per avoided repair.

experienced approximately 800-1,000 failure events per year. A somewhat conservative estimate assumed average direct repair costs of $5,000 per event (most of these are API pumps, whereas the $4,000-figure related to a mix favoring ANSI pumps). Again, as of early 2003, many petrochemical plants reported a more believable average cost of $10,000 after burden, shop space, field labor, engineering analysis and miscellaneous charges had been added.

Using a mean-time-between-failure (MTBF) of 18 operating months (3-year MTBF) for typical pumps, a major refinery or chemical plant many spend many millions of dollars each year for pump repairs alone. The obvious incentive to reducing pump failures is further amplified when product loss and safety considerations are given the emphasis they deserve.

Moreover, personnel not tied-up with pump *repair* tasks could be reassigned to pump monitoring, failure analysis, or upgrading tasks that add great value. It should be obvious that being pro-active, preventing failures from occurring in the first place, is worth more than doing repair work. It is, therefore, important to gain an overview of centrifugal pump failure reduction programs that can rapidly lead to improved pump service factors.

Implementing a Pump Failure Reduction Program

Using work force resources generally available at most plants, program implementation should start with the determination of problem pumps. Problem pumps might be those that failed more than three times in a running 12-month period, or those that brought down the plant, or those whose yearly repair cost exceeded "X" amount of money. Our experience shows that roughly 7% of the pump population fit in this over-three-per-year failure category. With 7% of a facility's pumps typically consuming 60% of the maintenance money going into *all* pumps, it makes much economic sense to apply the described implementation strategies first to the problem pumps, and then to expand the strategies to a wider percentage of your pumps.

The incentive to reduce pump failures is embedded in Ref. 14-1. More important, since it is both central theme and ultimate purpose of this book, it hardly merits further discussion. Although an effective program of reducing pump failures is always desirable, it should never lose sight of the ultimate goal of achieving plant safety and profitability. Here, then, are three of the key aims of such a program:

- To improve centrifugal pump reliability through more accurate determination and subsequent elimination of failure causes.
- To reduce process debits resulting from pump outages.
- To reduce centrifugal pump maintenance costs by effective analysis of component condition, replacing only those parts that actually require change-out.

However, the success of a pump failure reduction program is influenced by a number of factors. These range from administrative-managerial to technical-clerical and could include such items as management commitment, willingness of technical staff to follow-up on tasks previously handled by maintenance work forces, and clerical support for better administration and record keeping.

Definition of Approach and Goals

There are two basic implementation methods for a program of this type:

The "Headquarters" Team Approach

This method requires the formation of a team of corporate troubleshooting specialists who, over a short period of time, will make a concentrated effort to define and eliminate sources of frequent pump failures.

The Local Engineering Support Team (or Value Based Six Sigma) Approach

Here, the effort is centered around the maintenance organization, supplemented by support and guidance from a staff engineer, essentially an individual trained in pump selection and failure analysis and familiar with the six sigma process—a "Green Belt" or "Black Belt," as will be further explained below.

Although still occasionally pursued by some process plants, the corporate team approach was found to have a number of significant shortcomings:

- It lacks continuity. Team members are often unavailable if follow-up needs should arise.
- The team is sometimes considered an "outsider effort." Hence, operating personnel and field maintenance forces fail to develop the necessary rapport with the team.
- Lack of commitment and reluctance to commu-

nicate have sometimes been observed. The field forces may tend to stand back and let the "brains" struggle with the problem. Field personnel see the team as a vote of non-confidence in their own capabilities or past efforts.

- Extra manpower is generally required with this approach.

In view of these drawbacks, all facilities using process pumps would do well to actively consider pursuing the "Engineering Support/Value Based Six Sigma Team" approach for a pump failure reduction program. Let's examine how this works.

Value-based Six Sigma Improvement Process

This section outlines a centrifugal pump failure reduction program based on the popular "Value Based Six Sigma" (VBSS) improvement process. Using work force resources generally available at most plants, program implementation starts with defining the right opportunities. Specific pump reliability related actions steps are described and examples are given of typical check lists which are required to achieve the failure reduction goals.

Although relatively new, this proven improvement approach can be used as an effective pump failure reduction program. "Value Based Six Sigma" is not a quality program, it is a business initiative. It pursues quality only if it adds customer and company value. Projects are chosen based on customer feedback and impact to the bottom line. Teams are empowered and managed by the local team members. "Black Belt" (full time) or "Green Belt" (part time) six sigma specialists (trained from the local work force) are normally assigned to major teams. The idea is to design quality into the product (the pump) or process (the failure identification and elimination process), not after-the-fact inspection. The obvious goal is continuous improvement in the intertwined aspects of quality (failure elimination and up-time), and productivity (accuracy of failure definition). Six Sigma must be a top-down organization-wide initiative, which is owned by the leadership. The focus is on the process, not the people, with the primary objective being bottom line improvement. Supplier capability is critical to success.

The Six Sigma problem solving method is abbreviated as "DMAIC," indicating a Development/Definition Phase D, a Measurement Phase (M), an Analysis Phase (A), an Improvement Phase (I), and a Control Phase (C). Six Sigma problem solving is used to resolve current problems and prevent them from recurring in the new or upgraded products or processes. Aiming at reduced bottom line costs, savings can be in fixed or variable costs, but should be verifiable. Here's a closer look at the individual phases:

1. Development/Definition Phase (D)

This can be the most difficult and critical phase of DMAIC. Defining the "right opportunities" means to:

- Identify improvement opportunities (*our* MTBF vs. "theirs")
- Define customers and their requirements (operating departments want long-running pumps)
- Define project boundaries via scope, goals, or expectations. (Example: Program initially limited to pumps that have failed more than three times in a 12-month period)
- Identify support people and resources by name.

Based on analysis of frequency, cost and severity of pump failures, the plant mechanical-technical support group would designate a certain number of pumps as "problem pumps." These could all be pumps which had failed more that twice in a 12-month period, or pumps which have a history of one failure per year but cost more than $10,000 to repair. Of course, these guidelines are for illustrative purposes only and could be modified to suit any given need or situation.

A plant or operating unit should set itself an improvement goal. One typical goal would be to strive for reducing the number of "problem pumps" by 50% in one year's time, or to reduce pump maintenance expenditures by a certain amount in a designated time period. In the spirit of participative management, the initiative for goal setting and program execution should originate with field personnel rather than office or headquarters staff personnel. While being guided by staff personnel, field forces would make the program work and be credited with its successful implementation.

"Define-Action" steps are required to get the pump failure reduction program started. Management must designate responsibilities for the execution of these steps. Responsible personnel would have to:

- Designate "problem pumps," also called "bad actors," after reviewing past failure history, cost of repairs, cost of product losses, etc.

- Identify these pumps for record purposes and by actually tagging them in the field. Tagging will alert shop and field foremen to the need to arrange for a designated technician or engineer to be present whenever work is performed.

- Observe the Pareto principle. From a resource management standpoint it is important to follow this principle, better know as the 80/20 rule, which states that 80% of the gain comes from 20% of the opportunities. The team must determine which problems are the most significant, so as to allow focusing improvement efforts on the areas where the largest gains can be achieved. It is important to distinguish the "vital few" from the "trivial many."

2. Measurement Phase (M)

Recall that VBSS is a fact-based quantitative process; it requires hard data (called metrics) for analysis and to determine the degree of success. It follows that:

- The current state of the process should be measured in quantifiable terms, such as MTBF (see Chapter 16), maintenance cost, etc. *It is of indispensable importance to identify the metrics to be used for the failure reduction project.*

- A useful way to identify non-value added activities in a process is to create a process map. This process map highlights the events that do not add real value to the end product.

- A data collection plan must be developed.

- Data must be collected and the current state must be assessed.

<u>Measurement Action Steps — 1:</u>

- Update pump records to include failure history for the past 18 months.

- Develop and include step-by-step instructions for assembly, disassembly, tolerance checks, etc., as necessary. A typical model-specific shop instruction checklist is shown in Figure 14-1. Figure 14-2 illustrates a sample pump repair procedure that combines field and shop instructions.

- Follow the disassembly and reassembly of all problem pumps, and record measurements.

- Update records to include failure history for past 18 months.

It is hoped that pump maintenance data folders already exist at the plant which is interested in implementing a pump failure reduction program. If data folders are not available, this would be a good time to put them together (Ref. 14-2). They could be in any suitable format, preferably electronic, and would initially cover all problem pumps. The folders would later be expanded to cover all pumps.

Even simple forms will be of value in tracking pump failures (see Chapter 15). Computerized failure records may require more detailed forms.

Ideally, the data folders or computerized data and record bases should contain the following documents:

- API data sheets
- Performance curves
- Supplemental specifications invoked at time of purchase
- Dimensionally accurate cross section drawing
- Assembly instruction and drawings
- Seal drawings with installation dimensions and specific instructions
- Manufacturing drawings of shaft and seal gland
- Small bore piping isometric sketch
- Spare parts list and storehouse retrieval data
- Maintenance instructions, including critical dimensions and tolerances,
- Running clearances, shop test procedures for seals and bearings and
- Impeller balancing specifications
- Repair history, including parts replaced, analysis of parts condition, and
- Cost of repair
- Design change documentation
- Initial startup and check-out data
- Lubrication-related guidelines and data
- Special installation procedures
- Alignment data
- Vibration history.

Needless to say, the files should be as complete as possible, be kept current and be readily available to every person involved in the implementation of the pump failure reduction program. To reemphasize, all files should be in an electronic format, if that is at all possible at your particular site or work environment.

<u>Measurement Action Steps — 2:</u>

Once the records and data have been collected they should be reviewed for past history and the problem pumps identified, based on:

INDUSTRIAL PRODUCTS GROUP
Seneca Falls, New York 13148

Goulds Model 3196 Maintenance Checklist

OPERATION CHECKS

✓ **LUBRICATION**
Recommended Lubricants:
Oil - ISO VG100 High Quality Turbine Oil.
Grease - NGLI No.2 with Sodium or Lithium Base.

Relubrication Interval Oil:

Power End Type	Mineral Oil	Synthetic Oil
Standard	6 mo.	6 mo.
Sealed	6 mo.	24 mo.

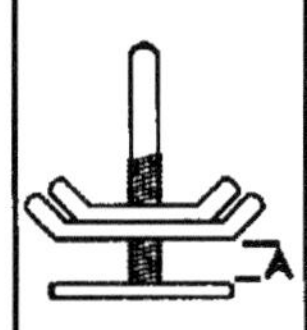

CONSTANT LEVEL OILER ADJUSTMENT			
Group	Wing Nut Level (A) in. (mm)	Bottle Size oz. (ml)	Sump Capacity oz. (ml)
ST	19/32 (15)	No.3 4 (118)	5 (148)
MT, LTC	19/32 (15)	No.3 4 (118)	13 (384)
XLT	9/16 (14)	No.5 8 (236)	71 (2099)

Relubrication Interval Grease:
Regreaseable Bearings should be greased every 3 months under normal operating conditions.

✓ **IMPELLER ADJUSTMENT**

Temperature °F (°C)	Front Clearance in. (mm)
Up to 200 (0 to 93)	0.015 (0.381)
200 - 250 (93 to 121)	0.017 (0.431)
250 - 300 (121 to 149)	0.019 (0.483)
300 - 350 (149 to 177)	0.021 (0.533)
350 - 400 (177 to 204)	0.023 (0.584)
Over 400 (Over 204)	0.025 (0.635)

✓ **ALIGNMENT**
Coupling to be aligned to within 0.002 in. T.I.R. for both parallel and angular readings.

✓ **VIBRATION**
Maximum. Vibration Level — 0.25 in./sec on the inboard and outboard bearing.

✓ **TEMPERATURE**
Normal Frame Range — 130 to 180 °F (55 to 80°C)

12/14/90 Rev.1

REBUILD CHECKS

✓ **BEARING FITS & TOLERANCES**

MODEL	ST in. (mm)	MT in. (mm)	LTX in. (mm)	XLT in. (mm)
INBOARD SHAFT O.D.	1.3781 (35.002) 1.3785 (35.013)	1.7722 (45.013) 1.7718 (45.002)	2.1660 (55.015) 2.1655 (55.002)	2.5597 (65.015) 2.5592 (65.002)
	0.0010 (0.025) TIGHT 0.0001 (0.002) TIGHT	0.0010 (0.025) TIGHT 0.0001 (0.002) TIGHT	0.0012 (0.030) TIGHT 0.0001 (0.002) TIGHT	0.0012 (0.030) TIGHT 0.0001 (0.002) TIGHT
INBOARD BEARING I.D.	1.3780 (35.000) 1.3775 (34.988)	1.7717 (45.000) 1.7712 (44.988)	2.1654 (55.000) 2.1648 (54.985)	2.5591 (65.000) 2.5585 (64.985)
INBOARD HOUSING I.D.	2.8346 (72.000) 2.8353 (72.019)	3.9370 (100.000) 3.9379 (100.022)	4.7244 (120.000) 4.7253 (120.022)	5.5118 (140.000) 5.5128 (140.025)
	0.0012 (0.032) LOOSE 0.0000 (0.000) LOOSE	0.0015 (0.037) LOOSE 0.0000 (0.000) LOOSE	0.0015 (0.037) LOOSE 0.0000 (0.000) LOOSE	0.0017 (0.043) LOOSE 0.0000 (0.000) LOOSE
INBOARD BEARING O.D.	2.8346 (72.000) 2.8341 (71.987)	3.9370 (100.000) 3.9364 (99.985)	4.7244 (120.000) 4.7238 (119.985)	5.5118 (140.000) 5.5111 (139.982)
OUTBOARD SHAFT O.D.	1.1812 (30.002) 1.1815 (30.011)	1.7722 (45.013) 1.7718 (45.002)	1.9690 (50.013) 1.9686 (50.002)	2.5597 (65.015) 2.5592 (65.002)
	0.0008 (0.021) TIGHT 0.0001 (0.002) TIGHT	0.0010 (0.025) TIGHT 0.0001 (0.002) TIGHT	0.0010 (0.025) TIGHT 0.0001 (0.002) TIGHT	0.0012 (0.030) TIGHT 0.0001 (0.002) TIGHT
OUTBOARD BEARING I.D.	1.1811 (30.000) 1.1807 (29.990)	1.7717 (45.000) 1.7712 (44.988)	1.9685 (50.000) 1.9680 (49.988)	2.5591 (65.000) 2.5585 (64.985)
OUTBOARD HOUSING I.D.	2.8346 (72.000) 2.8353 (72.019)	3.9370 (100.000) 3.9379 (100.022)	4.3307 (110.000) 4.3316 (110.022)	5.5118 (140.000) 5.5128 (140.025)
	0.0012 (0.032) LOOSE 0.0000 (0.000) LOOSE	0.0015 (0.037) LOOSE 0.0000 (0.000) LOOSE	0.0015 (0.037)LOOSE 0.0000 (0.000)LOOSE	0.0017 (0.043) LOOSE 0.0000 (0.000) LOOSE
OUTBOARD BEARING O.D.	2.8346 (72.000) 2.8341 (71.987)	3.9370 (100.000) 3.9369 (99.985)	4.3307 (110.000) 4.3301 (109.985)	5.5118 (140.000) 5.5111 (139.982)

✓ **IMPELLER BALANCE CRITERIA**
Single Plane Spin Balance - 0.01 oz.-in./lb.

✓ **INDICATOR CHECKS**
- Impeller Vane Runout 0.005 in. T.I.R. Maximum
- Shaft Straightness 0.0005 in. T.I.R. Maximum
- Shaft Runout 0.002 in. T.I.R. Maximum
- Stuffing Box Runout 0.005 in. T.I.R. Maximum

✓ **TORQUE VALUES**
Casing Bolts
30 ft-lbs(40.8 N-m) Lubricated
45 ft-lbs(61.2 N-m) Unlubricated
Adapter Bolts
30 ft-lbs(40.8 N-m) Lubricated
45 ft-lbs(61.2 N-m) Unlubricated

3196 BEARING TYPE *			
FRAME SIZE	INBOARD	OUTBOARD	
STX	6207	5306A-C3	7306 BECBM
MTX	6307	5309A-C3	7309 BECBM
LTX	6311	7310 BECBM	
XLT-X, X17	6313	5313A-C3	7313 BECBM

* SKF/MRC Bearing Designation

Figure 14-1: Model-specific shop checklist (Source: ITT/Goulds, Seneca Falls, NY, 13148)

— Cost of repairs & product losses
— Vibration history & alignment data
— Performance history (pressure, flow, cavitation)
— Lubrication data, etc.

Problem pumps should be tagged for follow-up repairs. On these pumps especially it would be mandatory to:

— Record measurements
— Review and observe the disassembly & re-assembly process
— Identify and collect historical data on pumps identical or similar to the particular problem pump.

3. Analysis Phase (A):

Once the data (metrics) have been collect, they are reviewed to determine (analyze) the current state using the following tools and processes:

- Identify root causes of variation/defects. Do not jump to the solution or solve the symptoms. Cause-and-effect (fish bone) diagrams are a good tool, with potential causes being broken down into four major groups:
  - — Methods
  - — Machines
  - — Material
  - — Personnel

A. Field disassembly

1. Remove coupling spacer.
2. Drain pump completely.
3. Disconnect all external piping (flush, steam tracing, etc.) and interfering insulation.
4. Remove casing and foot support bolts.
5. Separate casing from distance piece by means of pusher screws. Leave bearing housing connected to distance piece unless bearing replacement is contemplated.

B. Shop disassembly

1. Unscrew impeller nut and remove impeller.
2. Remove stuffing box housing by carefully applying pressure from two crowbars contacting seal plate.
3. Remove seal plate from stuffing box housing and also remove stationary and rotating seal elements.
4. Remove shaft sleeve without undue force. Apply heat, if necessary. Observe all safety rules!
5. Examine seal parts for warpage, fractures, and solids build-up on faces.

C. Shop reassembly

1. Verify flatness of seal faces to be within two light bands.
2. Measure inside diameter of throat bushing and outside diameter of shaft sleeve.
3. Machine inside diameter of throat bushing to achieve 0.020-0.030-in. clearance with shaft sleeve O.D.
4. Assemble stationary seal ring and seal plate, taking care to insert required gaskets.
5. Carefully tighten plate hold-down nuts. Use depth micrometer to verity that stationary seal ring is installed parallel to product side of stuffing box cover within .0005 in.
6. Place stuffing box cover on distance piece.
7. Place rotating seal assembly on shaft sleeve and slide over shaft. Make sure all dimensions are per CHEMEX Dwg. B-541599.
8. Insert keys.
9. Assemble impeller, taking care to place gaskets between hub and shaft sleeve, and between shaft and impeller nut.

D. Measurements required for record

Before and after replacing worn parts with new parts, please measure and record the following dimensions:

	Manufacturer's recommended value, mils	With old parts, mils	With now parts, mils
Shaft runout (impeller region)	1 mil or less		
Shaft runout (seal region)	Not to exceed 2 mils		
Seal face compression	Per CHEMEX DWG. B-541599		
Depth micrometer Check to ensure stationary seal is parallel to seal gland.	0.0005 in. or less		
Clearance between impeller disc vanes and stuffing box housing	0.032-0.047 in.		
Clearance between shaft sleeve and throat bushing	0.020 in. diametral. (max 0.032")		

E. Field reassembly

1. Back out pusher screws and bolt distance piece to casing. Make sure proper gasket is placed in between.
2. Replace foot support bolts.
3. Reconnect all auxiliary piping.
4. Refill bearing housing with Synesstic-32 lubricant.
5. Hot-align coupling to zero-zero setting (max. allowable deviation: place motor 0.002 in. higher than pump shaft) while observing dial indicators on casing.
6. Replace coupling spacer. Ascertain coupling disc stretch or compression does not exceed 0.004 in.

Figure 14-2: Sample pump repair procedure combining field and shop instructions (Source: Bloch, Heinz P. and Fred Geitner, "Machinery Failure Analysis and Troubleshooting," 3rd Edition, 1997, Butterworth-Heinemann Publishing Co., Stoneham, Massachusetts)

- Identify gaps between current performance and goal performance
- Identify the key process input variables that affect the key outputs
- Prioritize opportunities for improvement ("bad actor" pumps, 80/20 rule), and establish a time line
- Use simple tools (avoid paralysis by analysis).

Keep in mind that we are basically trying to establish the sources of variation.

Analysis action steps include identifying where within your plant or elsewhere pumps identical or similar to the "problem pumps" are operating. Compare their operating histories and examine significant deviations for clues to the source of your problem.

- Identify identical pumps in same or similar service within affiliated plants. Analyze maintenance history for these pumps.
- If no identical pumps exist, or if identical pumps are applied differently, compare construction and design details. Do not neglect installation details such as piping, etc.
- Verify that rotating elements are dynamically balanced per applicable checklist or procedure.
- Review applicable field and shop checklists for accuracy and supervisor's signatures at conclusion of work.

The results of the failure analysis should be documented on a short form similar to the one shown in Figure 14-3. Far more detailed forms have been devised and would be appropriate for plants with computerized failure records. It is important to note that this format represents the absolute "bare minimum" requirement for even small user facilities. Don't hesitate to use the more detailed forms typically found in an advanced computerized maintenance management system (CMMS).

The analysis should include:

- Review of assembly and repair processes. In turn, these may be subdivided into
  - — Field work
  - — Shop work
  - — Technical services
  - — Lubrication instructions
  - — Special commissioning instructions
  - — Installation precautions
  - — Routine surveillance instructions.
- Review of hydraulic parameters. (Note that some impellers have U-shaped NPSH curves!) The reviewer would
  - — Examine radial & axial hydraulic unbalance potential
  - — Verify that low flow rate is not a problem
  - — Determine pump specific speed
  - — Determine NPSH margin over full operating range (see Chapter 5).
  - — Examine if parallel operation is a problem
  - — Determine if there are instruments whose operating characteristics are unsuitable
  - — Evaluate susceptibility to internal recirculation-induced damage (see Chapter 5)
  - — Determine suction energy (see Chapter 5) and compare suction energy with that of other pumps. Suction energy values above (approximately) 160×10^6 for end suction pumps, 120×10^6 for between bearing/split case pumps, and 200×10^6 for vertical turbine pumps, are mechanically vulnerable unless operating with adequate NPSH margin and within the allowable operating region
  - — Verify that impeller axial thrust is within allowable limits at all pressure conditions. Examine balance hole size, wear ring diameter, stuffing box pressure. Reverse flush may not be possible if balance holes are used. Adequate delta-P is needed for reverse flush.
  - —Verify that radial clearances comply with hydraulic shock avoidance criteria (see Figure 14-4), for high (discharge) energy pumps. According to the Hydraulic Institute, pumps with specific speed values below 1,300 that generate more than 900 feet of head per stage are considered to be high (discharge) energy pumps. The differential head value for high energy drops from 900 feet at a specific speed of 1,300 to 200 feet at a specific speed of 2,700. Verify that vane tip-cutwater clearance exceeds 6%. (See page 411.)
- Conducting repair part and metallurgical reviews (see also Chapter 13)
  - — Examine component condition and initiate

EQUIPMENT RECORDS SYSTEM - MACHINERY SUBSYSTEM

REPAIR DATA INPUT FORM - VERTICAL PUMPS

Field	Columns	Entry
WORK REQUEST NO. (CARD I)	16 - 23	
REASON FOR SERVICE REQUEST	24 - 25	
DATE OUT OF SERVICE (MO.-DA.-YR.)	26 - 31	
DATE RETURN TO PROCESS (MO.-DA.)	32 - 35	
FIRST PART FAILED (PC/TP/COND)	36 - 40	
SECOND PART FAILED (PC/TP/COND)	41 - 45	
*CAUSE OF REPAIR (PRIM./SEC'ND)	46 - 49	
CORRECTIVE ACTION TAKEN	50 - 53	
WORKED BY (F, S, C)/PRIORITY	54 - 55	
M.H. REQUIRED	56 - 58	
MATERIAL COST	59 - 64	
CONTRACTOR'S NAME OR CODE/COST	65 - 71	
TRANSACTION I.D.	73 - 80	R P O I A D D ℓ
		QT / SYM / KEY
01 COUPLING (CARD 2)	16 - 20	
13 SEAL-STAT. FACE	21 - 25	
12 SEAL-ROTAT. ASSY. COMP.	26 - 30	
22 SEAL-ROTAT. FACE ONLY	31 - 35	
24 BEARING/BUSH., TOP OR MD.	36 - 40	
26 BEARING/BUSH., INTERMEDIATE	41 - 45	
28 BEARING/BUSH., BOTTOM	46 - 50	
05 SLEEVE	51 - 55	
30 IMPL. (SUCT. & SIMILAR)	56 - 60	
32 BOWL (" ")	61 - 65	
34 BUSH (" ")	66 - 70	
TRANSACTION I.D. (CARD 3)	73 - 80	R P I O A D D ℓ
42 WEAR RING, IMP. (SUCT. & SIMILAR)	16 - 20	
44 WEAR RING, BOWL (" ")	21 - 25	
46 IMPL. (OTHER)	26 - 30	
48 BOWLS (")	31 - 35	
49 BUSHINGS (OTHER)	36 - 40	
52 WEAR RINGS, IMPL (OTHER)	41 - 45	
54 WEAR RINGS, BOWL (OTHER)	46 - 50	
10 SHAFT, PUMP	51 - 55	
56 SHAFT, INTERMEDIATE	56 - 60	
60 SHAFT, TOP	61 - 65	
62 COUPLING, SHAFT	66 - 70	
TRANSACTION I.D.	73 - 80	R P I I A D D ℓ
40 CASE, BARREL (CARD 4)	16 - 20	
16 HEAD	21 - 25	
50 SECONDARY SEALING ELEMENT	26 - 30	
TRANSACTION I.D.		R P 1 2 A D D ℓ

REMARK 16 — 72 — 80

15 | XRMKADD ℓ

V.P 5 0 0 0 0 0 0 | XRMKADD ℓ

CRITICAL REMARK ONLY

*IF OTHER THAN "FIRST PART FAILED"

REASON FOR SERVICE REQUEST

01 INADEQUATE PERFORMANCE
02 BINDING OR STUCK
03 GENERAL OVERHAUL
04 LEAKING SEAL OR PACKING
07 VIBRATION OR NOISE
08 RECONDITION FROM IDLE
09 DESIGN CHANGE OR SPECIAL MOD.
24 EXTERNAL MISALIGNMENT
26 PREVIOUS REPAIR FAULTY
27 DAMAGED BEING HAULED OR INSTALL.
88 BEARING FAILURE
97 FIRE OR FREEZE DAMAGE
99 OTHERS (REMARK)

CAUSE OF REPAIR

03 GENERAL OVERHAUL
09 DESIGN CHANGE OR SPECIAL MOD.
24 EXTERNAL MISALIGNMENT
25 INTERNAL MISALIGNMENT
26 PREVIOUS REPAIR FAULTY
27 DAMAGED BEING HAULED OR INSTALL.
28 DRIVER PROBLEM
33 WRONG PART INSTALLED
41 WORN PACKING OR CARBON BOX
50 LEAKING GASKET
51 IMPROPER MATERIAL
63 TOO MUCH OR TOO LITTLE END PLAY
65 LOSS OF SEAL FLUSH
69 WRONG SIZE IMPELLER
73 NOT SUITABLE FOR PROCESS REQ.
75 IMPROPER OPERATION
80 TOO LITTLE LUBRICANT
82 IMPURITIES IN LUBRICANT
83 FAULTY CONTROLS
98 NO PROBLEM FOUND
99 OTHERS (REMARK)

CONDITION OF PARTS (KEY)

03 GENERAL OVERHAUL
10 OVERHEATED OR BURNT
12 BENT
13 BROKEN/CRACKED
16 ROUGH OR ETCHED
18 UNBALANCED
19 SCRATCHED/SCORED/GALLED
20 SEIZED
22 SPALLED
23 BAD FIT WITH MATING PART
24 EXTERNAL MISALIGNMENT
25 INTERNAL MISALIGNMENT
26 PREVIOUS REPAIR FAULTY
27 DAMAGED BEING HAULED OR INSTALL.
28 DRIVER PROBLEM
33 WRONG PART INSTALLED
35 CLEARANCES OFF
36 NORMAL WEAR
37 SEVERE WEAR
41 WORN PACKING OR CARBON BOX
50 LEAKING GASKET
51 IMPROPER MATERIAL
54 FOULED OR DIRTY
63 TOO MUCH OR TOO LITTLE END PLAY
64 CORRODED OR ERODED
65 LOSS OF SEAL FLUSH
69 WRONG SIZE IMPELLER
73 NOT SUITABLE FOR PROCESS REQ.
75 IMPROPER OPERATION
80 TOO LITTLE LUBRICANT
82 IMPURITIES IN LUBRICANT
83 FAULTY CONTROLS
98 NO PROBLEM FOUND
99 OTHERS (REMARK)

TYPE PARTS USED (SYM)

N NEW PART
R REPAIRED PART
L LOCALLY MADE PART
S CHANGED PART

Figure 14-3: "Bare minimum" repair data input form for vertical pumps

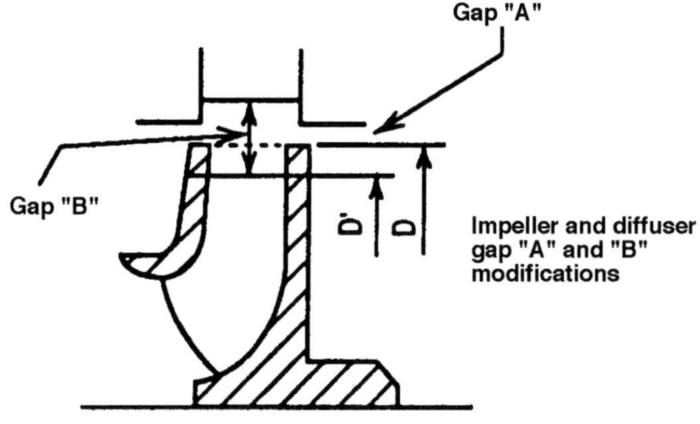

		Gap "B" Percentage of impeller radius		
Type	*Gap "A"*	*Minimum*	*Preferred*	*Maximum*
Diffuser	50 mils (0.050 in.)	4%	6%	12%
Volute	50 mils (0.050 in.)	6%	10%	12%

Gap A = Radial distance between impeller shroud O.D. and diffuser/volute wall
Gap B = Radial distance between impeller vane O.D. and diffuser/volute tongue (I.D.) $= 100\ (R_3 - R_2)/R_2$
R_2 = Radius of impeller
R_3 = Radius of diffuser volute inlet
D = Impeller shroud diameter
D′ = Impeller vane diameter after trimming

Note: If the number of impeller vanes and the number of diffuser/volute vanes are both even, the radial gap must be considerably larger (10 percent minimum.)

Figure 14-4: Recommended radial gaps for high energy pumps

comparison studies, as required.

— Verify that wear ring materials and clearances reflect best applicable experience. High temperature pump clearances may have to be as large as [(API clearance) + (0.02 × °F)] mils, diametral.
— Determine source and quality of repair parts
— Verify that part materials and tolerances/clearances are per OEM specification.
— Determine cause of part wear. Is it from:

(a) Cavitation (normally found in impeller inlet, on the visible side of the vane, a little way back from the vane leading edge).
(b) Recirculation-caused cavitation damage (found on the hidden side of the inlet portion of the impeller vane).
(c) Corrosion (normally relatively uniform)
(d) Abrasion (highest wear is found around wearing rings and in high velocity areas of the pump.)
(d) Rubbing (look for similar wear on adjacent part).
(e) High loads (accelerates bearing, wearing ring wear, and shaft fatigue failures).
(f) High temperature (reduces part strength, can cause coupling misalignment, and increases nozzle loads).

- Review of installation
  — Evidence of pipe stresses due to routing, non-sliding supports, incorrect support and hanger locations, etc. Also see Chapter 3.

  - Coupling misalignment and lockup problems. (See Chapter 11.)
  - Baseplate resonance, out-of-parallelism, grout defects. (See Chapter 3.)
  - Determine if suction and discharge piping design is adequate (straight lengths, fittings locations, size) Are suction elbows too close to double-suction pump? (See Chapter 3)

- Review of bearings and lubrication
  - Ascertain that bearing fits comply with applicable checklist
  - Verify filling notch bearings have been replaced by applicable replacement bearings. If used, spacers between duplex rows must be carefully measured
  - Duplex bearings must be matched sets, but not necessarily 40° angular contact. (See Chapter 7)
  - Verify that thermal expansion of the bearing outer race is not restricted by surrounding coolant
  - Calculate bearing L_{10} life under normal operating conditions
  - Verify that oil mist lubrication reaches every row of bearing balls. Directed mist fittings may be required instead of plain mist application fittings Consult Ref. 14-3 or Chapter 10 for specific guidance
  - Verify that motor windage does not impair effective venting of oil mist
  - Ensure that only approved grade oils are used in oil mist lubrication systems. Recall that paraffinic mineral oils and oils with high pour point can cause plugging of applicator fittings. Consider approved synthetic lube at temperatures below 25°F (–4°C).
  - ISO grade 68 or 100 mineral oil acceptable for most rolling element bearings with oil mist
  - ISO grade 32 or 68 synthetic oil also acceptable for most rolling element bearings with oil mist
  - ISO grade 32 mineral or synthetic oil acceptable for most sleeve bearings
  - Use dry sump oil mist, if available
  - Narrow, face-type bearing housing seals are useful in severe ambients or whenever lube oil contamination is to be avoided. Rotating labyrinth seals will not prevent air from entering and leaving the housing
  - Consider directed oil mist if inner race velocity exceeds 2,000 fpm (~10 m/s)

- Review of seal & quench gland design (Refs. 14-4 and 14-5)
  - Verify that API flush plans conform to applicable experience. Reverse flush requires takeoff to be at top of stuffing box. Reverse flush may not be possible if balance holes are used (see earlier comment).
  - Ascertain flush fluid has adequate lubricity, vapor pressure at seal face temperature, supply quantity and pressure, and is free from solid particles. How about during abnormal operation? At startup?
  - Tandem seal lubrication and installation details should conform to latest experience. Refer also to Chapter 8.
  - Seal installation tolerances should be verified.
  - Verify that only balanced seals are used at pressures above 75 psi (~515 kPa).
  - Ascertain that a quench steam or water is applied only if required to keep solid particles from forming external to seal components. Tackle your mechanical seal troubleshooting tasks by referring to the more detailed listing later in this chapter. Be aware that replacing 20 percent of a plant's seal population with engineered seals would probably lower yearly seal- related repairs by one-half.

- Review of safety considerations
  - If minimum flow bypasses are used, verify each pump has its own bypass arrangement
  - Do not allow excessively oversized pumps.
  - Verify isolation valves are accessible in case of fire. Are weight-closing valves possible?
  - Piping must be braced and gusseted. Flexible instrument connections may be advisable, but only if thoroughly well-engineered
  - Is on-stream condition monitoring instrumentation feasible? Should high or low frequency vibration monitoring be used? Consult Ref. 14-6 for details
  - Verify that substantial coupling guards are used.

Armed with this composite checklist and having read the reference material cited, plant engineers and mechanical technical service personnel should feel confident of their ability to diagnose recurring pump problems.

Pump Troubleshooting:
Possible Causes of Insufficient Flow

Say the symptom or deviation is "insufficient flow," page 426. You might also go directly to a symptom listing (Ref. 14-9) and note the possible causes:

1. Wear ring or between stage clearance—too large due to corrosion or repair error
2. Axial clearances too large in open or semi-open impellers
3. Corroded diffusers or cutwater
4. Impeller too small
5. Spare pump check valve stuck open
6. Malfunctioning minimum flow bypass valve
7. Cavitation—NPSHr exceeds NPSHa
8. Level in suction vessel too low
9. Entrained gas in suction
10. Air pocket in suction
11. Pump not vented properly
12. In vacuum service, air leaking through packing
13. Plugged suction screens
14. Blocked siphon breaker in suction vessel
15. Plugged suction lines
16. Closed gate valve in suction
17. Pump running backwards
18. Specific gravity lower than anticipated
19. Malfunctioning instrumentation—Flow recorder, etc.
20. Turbine steam condition inadequate—turbine runs at reduced speed
21. Worn turbine internals—turbine runs at reduced speed
22. Governor and/or linkage problem—turbine runs at reduced speed
23. Pinched or dropped wedge in a valve—downstream;
24. Process changes requiring high pump discharge pressure
25. Check valve or control valve downstream—stuck open
26. Fouled heat exchanger downstream.

Next maximize impeller performance as needed:

1. Bring the impeller middle shroud plate out to the impeller O.D. to reinforce the impeller structure.
2. Stagger the right and left side of the vane to reduce hydraulic shocks and alter the vane-passing frequency.
3. Reduce clearances to optimum between shrouds and casing. (Gap "A" on page 411)
4. Avoid even number of impeller vanes for double volute; or if the diffuser vanes are even numbered, increase the impeller side-wall thickness.
5. Impellers manufactured with blunt vane tips can also cause trouble in high discharge energy pumps by generating hydraulic "hammer" even when the impeller O.D. is the correct distance from the cutwater. The blunt tips cause disturbance in the volute. This effect may be partly or entirely eliminated by tapering the vanes by "overfiling," or "underfiling" the trailing edge, as shown earlier in Figure 4-13.

- Vertical Pump Considerations (see also Chapter 13)
  — Verify that critical speed is not a problem.
  — Check for proper lubrication of column bearings.
  — Some vertical pumps may need vortex baffles in suction bell.
  — Verify straightness of assembly and internals

4. Improvement Phase (I)

After the analysis phase of the project is completed, and the root causes and key input variables are identified, it is time to develop and implement the improvement process. This process consists of:

- Developing solutions
- Executing the solutions
- Make necessary changes
- Redesigning the process, and revising check lists

Improvement action steps continue with the development of unit-specific checklists and procedures. These, or similar guidelines must be used if the pump failure reduction objectives outlined earlier are to be reached in an expeditious and well-defined fashion. The development of pump-related field, shop, and technical staff review procedures must be entrusted to the reliability improvement team. However, the team may need substantial input from plant operating groups.

These checklists or procedures should cover topics which fall into roughly three classification categories:

- Field-work related reviews
  In the improvement phase, field-work related checklists must be developed and applied to accomplish a number of necessary steps which will lead to a reduction of problem incidents in the future. These checklists must address:

  — lubrication instructions

- routine surveillance instructions
- data taking requirements preparatory to shutdown for subsequent shop repair
- installation precautions and data requirements prior to restart after shop repair
- special commissioning instruction such as cooldown of cold service pumps, air-freeing of seal cavities, etc. A combined pump commissioning checklist/procedure example intended for field-posting is shown in Figure 14-5 (Ref. 14-6).

• Shop-work related reviews
Shop repair checklists could be either generic or model-specific. A generic checklist for bearing fits would consist of material discussed and presented earlier in Chapter 7. For instance, the shaft fit classification for pumps (k5, j5 or h5, shown in Tables 7-6 and 7-8), would be framed or highlighted in Table 14-2 (Ref. 14-8). Copies would be laminated in plastic, discussed with the shop technician group and these copies then issued for future use.

In essence, shop work instructions and procedures must be developed and conscientiously used to ascertain both quality workmanship and uniformity of product improvement efforts. Typical instructions and checklists should make maximum use of technical information which is routinely available from pump, bearing, and mechanical seal manufactures. Included in this second category are such guidelines as:

- Wear ring and throat bushing clearances
- Antifriction bearing housing fit tolerance to be used for pumps, motors and turbines, Table 7-6 (2)
- Antifriction bearing shaft fit tolerance to be used for pumps, motors and turbines (Ref. 14-8)
- Dimensional checking of mechanical seal components
- Maximum allowable rotor unbalance for pumps and small steam turbines
- Dimensions for steel bushings converting standard bearing housings to accept angular contact bearings
- Others, as required for special pumps

• Technical service reviews (failure prevention, procurement guidelines, and pre-purchase reliability assurance topics).

On the chance that existing pumps were not purchased with a view towards long-term reliability, the reliability improvement team should carefully look at our earlier chapters. There may have been significant omissions from the guidelines and recommendations made earlier in this text. In any event, technical service review checklists should provide critical mechanical and hydraulic parameters (e.g. bearing fits, piping strains, excessive suction energy), operation away from BEP (best efficiency point), inadequate NPSH margin, clearance between impeller periphery and cutwater, etc.

These checklists should also provide specific guidance on critical process parameters (solids or polymers in pumpage, non-optimized seal flush arrangements, equipment start-up and shut-down, preheating, cool-down, venting, etc.

From all of the above, it can be seen that checklist development must be a joint effort involving plant operating, process-technical, mechanical-technical support, and even pump, bearing, and mechanical seal vendor personnel. The final checklist product must also give guidance on the commissioning of vulnerable pumps. A typical example would be cold-service pumps (Figure 14-5, Ref. 14-7) that must be totally dry before cryogenic liquids can be admitted. Depending on the properties of the fluid, cold-service pumps may require a nitrogen purge to flow through the seal cavity prior to start-up. A similar approach would be needed for hot service pumps, except that, here, proper preheating of all casing internals is to be given very special attention. Unless the heating medium flows *through* the mechanical seal regions, seal failures are likely to occur at start-up.

Sample Checklists

Similarly, Figures 14-6 and 14-7 illustrate front and back of a model-specific document that falls into the shop checklist category. Attempt to find it in the manufacturer's maintenance manual or ask the manufacturers if they can provide this information for use by your shop. Develop your own, if necessary, and incorporate all pertinent data, as shown earlier in Figure 14-2.

But, since the process technicians will also benefit from checklists, the pump reliability improvement program "owner" should prevail upon other plant groups to assist in compiling checklists and guidance on pumps with critical process parameters (solids or polymers in pumpage, non-optimized seal flush arrangements, equipment start-up and shut-down, preheating, cool-down, venting, etc.). Once developed, these should be laminated in plastic and field-posted at the particular pump.

Barrier fluid system dry out procedure (mechanical technicians)*

1. Open flanges on barrier fluid return line.
2. Verify reservoir isolation valves in barrier fluid system piping are open.
3. Pump methanol into barrier fluid reservoir drain until clean methanol flows from open flanges.
4. Block reservoir drain valve and remove barrel pump hose.
5. Follow pump casing/process piping dry out procedure below.
6. Prime barrel pump with fresh barrier fluid until clean fluid flows from open flanges.
7. Reconnect hose to reservoir drain valve and unblock valve.
8. Pump barrier fluid into system until clean fluid flows from open flanges.
9. Tighten open flanges using new gasket.
10. Continue pumping barrier fluid into system until reservoir is filled.
11. Block reservoir drain valve and remove pump hose.
12. Repeat procedure for remaining barrier fluid system.

Pump casing/process piping dry out procedure (process technicians)*

1. Block all valves into pump casing, suction, and discharge process piping.
2. Open atmospheric vent valves A1 on casing drain, A2 on seal flush piping, A3 on suction piping, and A5 on discharge check valve bypass. Verify vent valve A4 on check valve bypass closed.
3. Open casing drain valve D1, verity drain valve D2 closed.
4. Open seal flush piping vent valve V1, verify vent valve V2 to reflux drum firmly closed.
5. Open methanol injection valve M1 into pump casing, and begin filling with methanol.
   CAUTION: DO NOT PRESSURIZE PUMP CASING WITH METHANOL!
6. Block individual vent valves as methanol appears. Continue filling.
7. Block methanol injection valve M1 when methanol appears at vent valve A5 on discharge check valve bypass.
8. Block vent valve A5 on discharge check valve bypass.
9. Slow roll turbine at 50-60 rpm for about 5 minutes.
10. Open atmospheric vent A2 on seal flush piping briefly to clear collected air.

Pressure out procedure (process technician)

1. Open seal flush piping vent V2 to reflux drum to pressurize pump casing.
2. Open pump casing drain D2 to cold blowdown.
3. Allow pump casing to blowdown until casing drain piping begins to frost.
4. Verify methanol completely purged by cracking vent valve Al on casing drain.
5. Block pump casing drain D1, D2 out of cold blowdown when light frost begins to form on pump casing.

Cold soak procedure (process technician)

1. Crack suction block valve B1 2-3 turns.
2. Open suction block valve B1 completely when casing is well frosted.
3. Open discharge block valve B2 completely.
4. Allow pump to cold soak for 2 hours.
5. Verify pump shaft turns freely by hand.

Startup/shutdown (process technicians-follow normal operating procedures)

---

*For vulnerable, cold-service pump. Condition: pump is completely blocked in, cleared, and depressurized.

Figure 14-5: Example of a combined pump commissioning checklist/procedure intended for field-posting near the pump (Source: Ref. 14-9)

✔ WEAR RING CLEARANCES (ENCLOSED IMPELLER ONLY)

Radial Ring Clearances - Enclosed Impellers						
Size		Impeller Ring OD		Casing Ring ID		Clearance
		Min	Max	Min	Max	
3x6-12	In	6.4771	6.4811	6.5111	6.5151	0.030-0.038
	Mm	164.52	164.62	165.38	165.48	0.76-0.96
4x6-12	In	7.3038	7.3078	7.3378	7.3418	0.030-0.038
	Mm	185.52	185.62	186.38	186.48	0.76-0.96
3x6-14	In	6.4771	6.4811	6.5111	6.5151	0.030-0.038
	Mm	164.52	164.62	165.38	165.48	0.76-0.96
4x6-14	In	7.3038	7.3078	7.3378	7.3418	0.030-0.038
	Mm	185.52	185.62	186.38	186.48	0.76-0.96
4x6-16	In	7.6975	7.7015	7.7315	7.7355	0.030-0.038
	Mm	195.52	195.62	196.38	196.48	0.76-0.96
6x8-14	In	8.2147	8.2187	8.2487	8.2527	0.030-0.038
	Mm	208.65	208.75	209.52	209.62	0.76-0.96
8x8-14	In	9.232	9.236	9.266	9.27	0.030-0.038
	Mm	234.49	234.59	235.36	235.46	0.76-0.96
10x10-14	In	10.5022	10.5062	10.5362	10.5402	0.030-0.038
	Mm	266.76	266.86	267.62	267.72	0.76-0.96
6x8-16	In	8.2147	8.2187	8.2487	8.2527	0.030-0.038
	Mm	208.65	208.75	209.52	209.62	0.76-0.96
4x6-19	In	8.2147	8.2187	8.2487	8.2527	0.030-0.038
	Mm	208.65	208.75	209.52	209.62	0.76-0.96
6x10-16	In	9.3511	9.3551	9.3851	9.3891	0.030-0.038
	Mm	237.52	237.62	238.38	238.48	0.76-0.96
8x10-16	In	10.7268	10.7308	10.7653	10.7693	0.0345-0.0425
	Mm	272.46	272.56	273.44	272.54	0.88-1.08
10x12-16	In	12.3803	12.3843	12.4189	12.4229	0.0345-0.0425
	Mm	314.46	314.56	315.44	315.54	0.88-1.08
14x14-16	In	13.5614	13.5654	13.6	13.604	0.0345-0.0425
	Mm	344.46	344.56	345.44	345.54	0.88-1.08
4x8-19	In	8.4456	8.4496	8.4796	8.4836	0.030-0.038
	Mm	214.52	214.62	215.38	215.48	0.76-0.96
6x10-19	In	9.3511	9.3551	9.3851	9.3891	0.030-0.038
	Mm	237.52	237.62	238.38	238.48	0.76-0.96
8x10-19	In	10.7268	10.7308	10.7653	10.7693	0.0345-0.0425
	Mm	272.46	272.56	273.44	272.54	0.88-1.08
10x12-19	In	12.774	12.778	12.8125	12.8165	0.0345-0.0425
	Mm	324.46	324.56	325.44	325.53	0.88-1.08
6x10-22	In	9.9416	9.9456	9.9756	9.9796	0.030-0.038
	Mm	252.52	252.62	253.38	2.5348	0.76-0.96
8x10-22	In	11.3961	11.4001	11.4346	11.4386	0.88-1.08
	Mm	289.46	289.56	290.44	290.54	0.0345-0.0425
12x14-19	In	13.9551	13.9591	13.9936	13.9976	0.88-1.08
	Mm	354.46	354.56	355.44	355.54	0.0345-0.0425
16x16-19	In	15.2539	15.2579	15.2924	15.2964	0.88-1.08
	Mm	387.45	387.55	388.43	388.53	0.0345-0.0425
10x12-22	In	12.774	12.778	12.8125	12.8165	0.0345-0.0425
	Mm	324.46	324.56	325.44	325.53	0.88-1.08
12x14-22	In	14.626	14.63	14.6645	14.6685	0.0345-0.0425
	Mm	371.5	371.6	372.48	372.58	0.88-1.08
14x16-22	In	16.5961	16.5575	16.5921	16.5961	0.0345-0.0425
	Mm	420.46	420.56	421.44	421.54	0.88-1.08
6x10-25	In	11.0812	11.0852	11.1197	11.1237	0.0345-0.0425
	Mm	281.46	281.56	282.44	282.54	0.88-1.08
8x12-25	In	12.774	12.778	12.8125	12.8165	0.0345-0.0425
	Mm	324.46	324.56	325.44	325.54	0.88-1.08
10x14-25	In	13.9551	13.9591	13.9936	13.9976	0.0345-0.0425
	Mm	354.46	354.56	355.44	355.54	0.88-1.08

Goulds 3180/85/81/86 Maintenance Checks

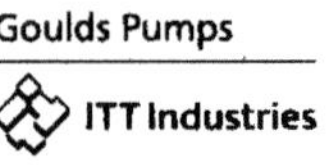

OPERATION CHECKS

✔ LUBRICATION

Recommended Lubricants:

Oil: ISO VG68 High Grade Turbine Oil [up to 180°F (80°C)]
ISO VG100 [T>180°F (80°C)]

Oil Sump Capacity		
Frame	Ounces	Liters
S	35	1
M	70	2
L	70	2
XL	105	3

Grease: NLGI No. 2 Sodium or Lithium Base [up to 230°F (110°C)]
NGLI No. 3 with oxidation stabilizers [T>230°F (110°C)]

Grease Amounts								
Frame	Initial Grease				Re-Grease [1]			
	Thrust (Angular Contact)		Radial (Cylindrical Roller)		Thrust (Angular Contact)		Radial (Cylindrical Roller)	
	Oz.	Grams	Oz.	Grams	Oz.	Grams	Oz.	Grams
S	7.0	185	6.0	165	2.5	70	2.5	70
M	10.0	290	7.0	180	4.0	115	2.5	70
L	17.0	475	10.0	280	7.0	200	4.0	115
XL	28.0	800	16.0	450	12.0	345	6.5	190

[1] Amount is based on purging half of the old grease from the housing reservoir.

Re-lubrication Interval:

Oil: Change after first 200 hours of operation for new bearings, every 3 months thereafter.

Grease: Re-grease every 3 months.

✔ IMPELLER ADJUSTMENT

Temperature, °F (°C)	Front Clearance, in (mm)
Up to 120° (50°)	0.015 (0.40)
120° (50°) - 210° (100°)	0.018 (0.45)
210° (100°) - 300° (150°)	0.020 (0.50)
300° (150°) - 390° (200°)	0.022 (0.55)
390° (200°) - 450° (230°)	0.026 (0.65)

Figure 14-6: Model-specific shop checklist—front (Source: ITT/Goulds, Seneca Falls, NY, 13148)

Systems and Specification Checklists
(see Chapter 3 and 5)

As was mentioned before, material in the technical checklist category is intended for failure analysis use by plant engineering, or technical support groups. It is a "memory jogger" and should incorporate helpful hints from the numerous publications dealing with pump reliability improvements. Reliability professionals are encouraged to review virtually any element that makes pumps and systems more reliable. Many of these were mentioned in some of the preceding chapters:

- Disallow elbow too close to suction of high suction energy and/or double flow pumps with low NPSH margins.
- Consider flow disturbances introduced by block valves mounted too close to suction of double flow pumps
- Insist on vortex breakers in suction vessels
- Verify minimum flow protection required/not required

✔ ALIGNMENT

Coupling to be aligned to within 0.002 in. TIR for both parallel and angular readings.

Cold Setting of Parallel Vertical Alignment (3180 and 3185 only)	
Temperature, °F (°C)	**Set Driver Shaft, in (mm)**
Up to 50° (10°)	0.002 (0.05) low
50° (10°) - 150° (65°)	0.001 (0.03) high
150° (65°) - 250° (120°)	0.005 (0.12) high
250° (120°) - 350° (175°)	0.009 (0.23) high
350° (175°) - 450° (220°)	0.013 (0.33) high

✔ TEMPERATURE

Temperature Limits	
Pump Configuration	**Suitability**
Grease Lube	Up to 350°F (180°C)
Oil Lube w/o Cooler	Up to 350°F (180°C)
Oil Lube with Cooler	Up to 450°F (230°C)

Note: Normal bearing operating temperatures run between 120° and 180°F (50° - 80°C). Bearing temperatures are generally 45°F (25°C) higher than the bearing housing/frame surface temperature.

REBUILD CHECKS

✔ SHAFT END PLAY

Shaft End Play, in (mm)				
Range	**Frame**			
	S	**M**	**L**	**XL**
Min	0.0	0.0	0.0	0.0
Max	0.001 (0.025)	0.001 (0.025)	0.001 (0.025)	0.001 (0.025)

✔ BEARING TYPE, FITS, AND TOLERANCES

Bearing Types		
Frame	**Inboard**	**Outboard**
S	NUP-311ECP	7311 BECBY
M	NUP-312ECP	7312 BECBY
L	NUP-314ECP	7315 BECBY
XL	NUP-317ECP	7318 BECBY

Bearing Fits and Tolerance, In (mm)				
Fit	**Frame**			
	S	**M**	**L**	**XL**
Shaft OD Inboard	2.1666 (55.032) 2.1659 (55.014)	2.3634 (60.030) 2.3626 (60.010)	2.7571 (70.030) 2.7563 (70.010)	3.3478 (85.034) 3.3470 (85.014)
TIGHT TIGHT	0.0018 (0.047) 0.0005 (0.013)	0.0018 (0.046) 0.0004 (0.010)	0.0018 (0.046) 0.0004 (0.010)	0.0021 (0.053) 0.0005 (0.013)
Bearing ID Inboard	2.1654 (55.0) 2.1648 (54.986)	2.3622 (60.0) 2.3616 (59.985)	2.7559 (70.0) 2.7553 (69.985)	3.3465 (85.0) 3.3457 (84.981)
Frame ID Inboard	4.7253 (120.023) 4.7244 (120.0)	5.1191 (130.025) 5.1181 (130.0)	5.9065 (150.025) 5.9055 (150.0)	7.0876 (180.025) 7.0866 (180.0)
LOOSE LOOSE	0.0015 (0.038) 0 (0)	0.0017 (0.043) 0 (0)	0.0017 (0.043) 0 (0)	0.0020 (0.051) 0 (0)
Bearing OD Inboard	4.7244 (120.0) 4.7238 (119.985)	5.1181 (130.0) 5.1174 (129.982)	5.9055 (150.0) 5.9048 (149.982)	7.0866 (180.0) 7.0856 (179.974)
Shaft OD Outboard	2.1660 (55.016) 2.1655 (55.004)	2.3628 (60.015) 2.3623 (60.002)	2.9533 (75.014) 2.9528 (75.001)	3.5444 (90.028) 3.5438 (90.013)
TIGHT TIGHT	0.0012 (0.030) 0.0001 (0.002)	0.0012 (0.030) 0.0001 (0.002)	0.0011 (0.028) 0 (0)	0.0019 (0.048) 0.0005 (0.013)
Bearing ID Outboard	2.1654 (55.0) 2.1648 (54.986)	2.3622 (60.0) 2.3616 (59.985)	2.9528 (75.0) 2.9522 (74.986)	3.5433 (90.0) 3.5425 (89.980)
Housing ID Outboard	4.7253 (120.023) 4.7247 (120.007)	5.1191 (130.025) 5.1185 (130.010)	6.3002 (160.025) 6.2994 (160.005)	7.4815 (190.030) 7.4806 (190.007)
LOOSE LOOSE	0.0015 (0.038) 0.0003 (0.008)	0.0017 (0.043) 0.0004 (0.010)	0.0020 (0.051) 0.0002 (0.005)	0.0024 (0.061) 0.0003 (0.008)
Bearing OD Outboard	4.7244 (120.0) 4.7238 (119.985)	5.1181 (130.0) 5.1174 (129.982)	6.2992 (160.0) 6.2982 (159.974)	7.4803 (190.0) 7.4791 (189.970)

✔ INDICATOR CHECKS

- Impeller Vane Runout: ≤ 12 006 in TIR Max
12 ≤ x ≤ 16 .007 in. TIR Max
16 ≤ x ≤ 25 .008 in. TIR Max
- Shaft Straightness - 0.0005 in. TIR Max
- Shaft Runout - 0.002 in. TIR Max
- Stuffing Box/Seal Chamber Runout - 0.005 in. TIR Max

✔ TORQUE VALUES

Torque Values, ft-lbs (N-m)		
Location		**Recommended Torque**
Impeller Nut	S & M Groups	240 (325)
	L & XL Groups	600 (800)
Frame to Seal Chamber Bolts	S & M Groups	30 (40)
	L & XL Groups	50 (65)
Brg Retainer to Housing Screw	S & M Groups	15 (20)
	L & XL Groups	20 (25)
Casing Lug Bolt	12" – 19"	125 (170)
	22" – 25"	200 (270)
Suction Sideplate Nut	12" – 16"	10 (15)
	19" – 25"	20 (25)

Figure 14-7: Model-specific shop checklist—reverse side (Source: ITT/Goulds, Seneca Falls, NY, 13148)

- Consider automatic shutdown for loss of suction in vulnerable or remotely located services. Absolutely essential with magnetic drive pumps!
- Pressurized flush and initial fill supply needed in vacuum services: verify adequacy, if installed
- Provisions for warm up or cool-down in hot or cryogenic services: verify adequacy, if installed
- Hotwell and boiler feedwater pumps: Review NPSHa, minimum flow protection, vent back to suction
- Review vapor pressure specified. Example: Summer vs. Winter operation for propane service. Vapor pressure at 25°F is 65 psia vs. 200 psia at 100°F.

Pre-assembly Inspection Checklist for Centrifugal Pumps

- Vane dimensions at impeller outlet to be identical within 0.08 (2mm). If not identical, expect hydraulic unbalance and vibration.
- Impeller liquid channels evenly distributed around shaft centerline. If casting core has shifted, could

Table 14-2: Typical shop checklist giving shaft size and corresponding bearing diameters (Source: SKF Bearing Installation and Maintenance Guide, Publication 140-710)

Bearing Bore Diameter			g6			h6			h5			j5			j6			k5		
	Inches		Shaft Dia.		Fit in 0.0001"	Shaft Dia.		Fit in 0.0001"	Shaft Dia.		Fit in 0.0001"	Shaft Dia.		Fit in 0.0001"	Shaft Dia.		Fit in 0.0001"	Shaft Dia.		Fit in 0.0001"
mm	Max.	Min.	Max.	Min.		Max.	Min.		Max.	Min.		Max.	Min.		Max.	Min.		Max.	Min.	
4	0.1575	0.1572	0.1573	0.1570		0.1575	0.1572		0.1575	0.1573		0.1576	0.1574		0.1577	0.1574		0.1577	0.1575	
5	0.1969	0.1966	0.1967	0.1964	5L 1T	0.1969	0.1966	3L 3T	0.1969	0.1967	2L 3T	0.1970	0.1968	1L 4T	0.1971	0.1968	1L 5T	0.1971	0.1969	0T 5T
6	0.2362	0.2359	0.2360	0.2357		0.2362	0.2359		0.2362	0.2360		0.2363	0.2361		0.2364	0.2361		0.2364	0.2362	
7	0.2756	0.2753	0.2754	0.2750		0.2756	0.2752		0.2756	0.2754		0.2758	0.2755		0.2759	0.2755		0.2759	0.2756	
8	0.3150	0.3147	0.3148	0.3144	6L 1T	0.3150	0.3146	4L 3T	0.3150	0.3148	2L 3T	0.3152	0.3149	1L 5T	0.3153	0.3149	1L 6T	0.3153	0.3150	0T 65T
9	0.3543	0.3540	0.3541	0.3537		0.3543	0.3539		0.3543	0.3541		0.3545	0.3542		0.3546	0.3542		0.3546	0.3543	
10	0.3937	0.3934	0.3935	0.3931		0.3937	0.3933		0.3937	0.3935		0.3939	0.3936		0.3940	0.3936		0.3940	0.3937	
12	0.4724	0.4721	0.4722	0.4717		0.4724	0.4720		0.4724	0.4721		0.4726	0.4723		0.4727	0.4723		0.4728	0.4724	
15	0.5906	0.5903	0.5904	0.5899	7L 1T	0.5906	0.5902	4L 3T	0.5906	0.5903	3L 3T	0.5908	0.5905	1L 5T	0.5909	0.5905	1L 6T	0.5910	0.5906	0T 7T
17	0.6693	0.6690	0.6691	0.6686		0.6693	0.6689		0.6693	0.6690		0.6695	0.6692		0.6696	0.6692		0.6697	0.6693	
20	0.7874	0.7870	0.7871	0.7866		0.7874	0.7869		0.7874	0.7870		0.7876	0.7872		0.7878	0.7872		0.7878	0.7875	
25	0.9843	0.9839	0.9840	0.9835	8L 1T	0.9843	0.9838	5L 4T	0.9843	0.9839	4L 4T	0.9845	0.9841	2L 6T	0.9847	0.9841	2L 8T	0.9847	0.9844	1T 8T
30	1.1811	1.1807	1.1808	1.1803		1.1811	1.1806		1.1811	1.1807		1.1813	1.1809		1.1815	1.1809		1.1815	1.1812	
35	1.3780	1.3775	1.3776	1.3770		1.3780	1.3774		1.3780	1.3776		1.3782	1.3778		1.3784	1.3778		1.3785	1.3781	
40	1.5748	1.5743	1.5744	1.5738	10L 1T	1.5748	1.5742	6L 5T	1.5748	1.5744	4L 5T	1.5750	1.5746	2L 7T	1.5752	1.5746	2L 9T	1.5753	1.5749	1T 10T
45	1.7717	1.7712	1.7713	1.7707		1.7717	1.7711		1.7717	1.7713		1.7719	1.7716		1.7721	1.7715		1.7722	1.7718	
50	1.9685	1.9680	1.9681	1.9675		1.9685	1.9679		1.9685	1.9681		1.9687	1.9683		1.9689	1.9683		1.9690	1.9686	
55	2.1654	2.1648	2.1650	2.1643		2.1654	2.1647		2.1654	2.1649		2.1656	2.1651		2.1658	2.1651		2.1660	2.1655	
60	2.3622	2.3616	2.3618	2.3611		2.3622	2.3615		2.3622	2.3617		2.3624	2.3619		2.3626	2.3619		2.3628	2.3623	
65	2.5591	2.5585	2.5587	2.5580	11L 2T	2.5591	2.5584	7L 6T	2.5591	2.5586	5L 6T	2.5593	2.5588	3L 8T	2.5595	2.5588	3L 11T	2.5597	2.5592	1T 12T
70	2.7559	2.7553	2.7555	2.7548		2.7559	2.7552		2.7559	2.7554		2.7561	2.7556		2.7563	2.7556		2.7565	2.7560	
75	2.9528	2.9522	2.9524	2.9517		2.9528	2.9521		2.9528	2.9523		2.9530	2.9525		2.9532	2.9525		2.9534	2.9529	
80	3.1496	3.1490	3.1492	3.1485		3.1496	3.1489		3.1496	3.1491		3.1498	3.1493		3.1500	3.1493		3.1502	3.1497	
85	3.3465	3.3457	3.3460	3.3452		3.3465	3.3456		3.3465	3.3459		3.3467	3.3461		3.3470	3.3461		3.3472	3.3466	
90	3.5433	3.5425	3.5428	3.5420		3.5433	3.5424		3.5433	3.5427		3.5435	3.5429		3.5438	3.5429		3.5440	3.5434	
95	3.7402	3.7394	3.7397	3.7389		3.7402	3.7393		3.7402	3.7396		3.7404	3.7398		3.7407	3.7398		3.7409	3.7403	
100	3.9370	3.9362	3.9365	3.9357	13L 3T	3.9370	3.9361	9L 8T	3.9370	3.9364	6L 8T	3.9372	3.9366	4L 10T	3.9375	3.9366	4L 13T	3.9377	3.9371	1T 15T
105	4.1339	4.1331	4.1334	4.1326		4.1339	4.1330		4.1339	4.1333		4.1341	4.1335		4.1344	4.1335		4.1346	4.1340	
110	4.3307	4.3299	4.3302	4.3294		4.3307	4.3298		4.3307	4.3301		4.3309	4.3303		4.3312	4.3303		4.3314	4.3308	
115	4.5276	4.5268	4.5271	4.5263		4.5276	4.5267		4.5276	4.5270		4.5278	4.5272		4.5281	4.5272		4.5283	4.5277	
120	4.7244	4.7236	4.7239	4.7231		4.7244	4.7235		4.7244	4.7238		4.7246	4.7240		4.7249	4.7240		4.7251	4.7245	
125	4.9213	4.9203	4.9207	4.9198		4.9213	4.9203		4.9213	4.9206		4.9216	4.9209		4.9219	4.9209		4.9221	4.9214	
130	5.1181	5.1171	5.1175	5.1166		5.1181	5.1171		5.1181	5.1174		5.1184	5.1177		5.1187	5.1177		5.1189	5.1182	
140	5.5118	5.5108	5.5112	5.5103		5.5118	5.5108		5.5118	5.5111		5.5121	5.5114		5.5124	5.5114		5.5126	5.5119	
150	5.9055	5.9045	5.9049	5.9040	15L 4T	5.9055	5.9045	10L 10T	5.9055	5.9048	7L 10T	5.9058	5.9051	4L 13T	5.9061	5.9051	4L 16T	5.9063	5.9056	1T 18T
160	6.2992	6.2982	6.2986	6.2977		6.2992	6.2982		6.2992	6.2985		6.2995	6.2988		6.2998	6.2988		6.3000	6.2993	
170	6.6929	6.6919	6.6923	6.6914		6.6929	6.6919		6.6929	6.6922		6.6932	6.6925		6.6935	6.6925		6.6937	6.6930	
180	7.0866	7.0856	7.0860	7.0851		7.0866	7.0856		7.0866	7.0859		7.0869	7.0862		7.0872	7.0862		7.0874	7.0867	
190	7.4803	7.4791	7.4797	7.4786		7.4803	7.4792		7.4803	7.4795		7.4806	7.4798		7.4809	7.4798		7.4812	7.4805	
200	7.8740	7.8728	7.8734	7.9823		7.8740	7.3729		7.8740	7.8732		7.8743	7.3735		7.8746	7.8735		7.8749	7.8742	
220	8.6614	8.6602	8.6608	8.6597	17L 6T	8.6614	8.6603	11L 12T	8.6614	8.6606	8L 12T	8.6617	8.6609	5L 15T	8.6620	8.6609	5L 18T	8.6623	8.6616	2T 21T
240	9.4488	9.4476	9.4482	9.4471		9.4492	9.4477		9.4488	9.4480		9.4491	9.4483		9.4494	9.4483		9.4497	9.4490	
250	9.8425	9.8413	9.8419	9.8408		9.8425	9.8414		9.8425	9.8417		9.8428	9.8420		9.8431	9.8420		9.8434	9.8427	
260	10.2362	10.2348	10.2355	10.2343		10.2362	10.2349		10.2362	10.2353		10.2365	10.2356		10.2368	10.2356		10.2373	10.2364	
280	11.0236	11.0222	11.0229	11.0217	19L 7T	11.0236	11.0223	13L 14T	11.0236	11.0225	9L 14T	11.0239	11.0230	6L 17T	11.0241	11.0230	6L 20T	11.0247	11.0238	2T 25T
300	11.8110	11.8096	11.8103	11.8091		11.8110	11.8097		11.8110	11.8101		11.8113	11.8104		11.8116	11.8104		11.8121	11.8112	
310	12.2047	12.2033	12.2040	12.2028		12.2047	12.2034		12.2047	12.2038		12.2050	12.2041		12.2053	12.2041		12.2058	12.2049	
320	12.5984	12.5968	12.5977	12.5963		12.5984	12.5970		12.5984	12.5974		12.5987	12.5977		12.5991	12.5977		12.5995	12.5986	
340	13.3858	13.3842	13.3851	13.3837		13.3858	13.3844		13.3858	13.3848		13.3861	13.3851		13.3865	13.3851		13.3869	13.3860	
350	13.7795	13.7779	13.7788	13.7774	21L 9T	13.7795	13.7781	14L 16T	13.7794	13.7785	10L 16T	13.7798	13.7788	7L 19T	13.7802	13.7788	7L 23T	13.7806	13.7797	2T 27T
360	14.1732	14.1716	14.1725	14.1711		14.1732	14.1718		14.1732	14.1722		14.1735	14.1725		14.1739	14.1725		14.1743	14.1734	
380	14.9606	14.9590	14.9599	14.9585		14.9616	14.9592		14.9616	14.9596		14.9609	14.9599		14.9613	14.9599		14.9617	14.9608	
400	15.7480	15.7464	15.7473	15.7459		15.7480	15.7466		15.7480	15.7464		15.7483	15.7473		15.7487	15.7473		15.7491	15.7482	
420	16.5354	16.5336	16.5346	16.5330		16.5354	16.5338		16.5354	16.5343		16.5357	16.5346		16.5362	16.5346		16.5367	16.5356	
440	17.3228	17.3210	17.3220	17.3186		17.3228	17.3212		17.3228	17.3217		17.3231	17.3220		17.3236	17.3220		17.3241	17.3230	
460	18.1102	18.1084	18.1094	18.1060	24L 10T	18.1102	18.1086	16L 18T	18.1102	18.1091	11L 18T	18.1105	18.1094	8L 21T	18.1110	18.1094	8L 26T	18.1115	18.1104	2T 31T
480	18.8976	18.8958	18.8968	18.8952		18.8976	18.8960		18.8976	18.8965		18.8979	18.8968		18.8984	18.8968		18.8989	18.8978	
500	19.6850	19.6832	19.6842	19.6826		19.6850	19.6834		19.6850	19.6839		19.6853	19.6842		19.6858	19.6842		19.6863	19.6852	
530	20.8661	20.8641	20.8652	20.8635		20.8661	20.8644								20.8670	20.8652		20.8673	20.8661	
560	22.0472	22.0452	22.0463	22.0446	26L 11T	22.0472	22.0455	17L 20T							22.0481	22.0461	9L 29T	22.0484	22.0472	0T 32T
600	23.6220	23.6200	23.6211	23.6194		23.6220	23.6203								23.6229	23.6211		23.6232	23.6220	
630	24.8031	24.8011	24.8022	24.8005		24.8031	24.8614								24.8040	24.8012		24.8063	24.8031	
660	25.9843	25.9813	25.9834	25.9814		25.9833	25.9823								25.9853	25.9833		25.9857	25.9843	
670	26.3780	26.3750	26.3771	27.3751		26.3780	26.3760								26.3790	26.3770		26.3794	26.3780	
710	27.9528	27.9498	27.9519	27.9499	29L 21T	27.9528	27.9508	20L 30T							27.9538	27.9518	10L 40T	27.9542	27.9528	0T 44T
750	29.5276	29.5246	29.5267	29.5247		29.5276	29.5256								29.5316	29.5266		29.5290	29.5276	
780	30.7087	30.7057	30.7078	30.7058		30.7087	20.7067								30.7127	30.7077		30.7101	30.7087	
800	31.4961	31.4931	31.4952	31.4932		31.4962	31.4941								31.5001	31.4951		31.4975	31.4961	
850	33.4646	33.4607	33.4636	33.4614		33.4645	33.4624								33.4657	33.4635		33.4662	33.4646	
900	35.4331	35.4292	35.4321	35.4299	32L 29T	35.4331	35.4309	22L 39T							35.4342	35.4320	11L 50T	35.4347	35.4331	0T 55T
950	37.4016	37.3977	37.4006	37.3984		37.4016	37.3994								37.4066	35.4005		37.4032	37.4016	
1000	39.3701	39.3662	39.3691	39.3669		39.3701	39.3679								39.3751	39.3690		39.3717	39.3701	
1060	41.7323	41.7274	41.7312	41.7286		41.7323	41.7297								41.7385	41.7310		41.7341	41.7323	
1120	44.0945	44.0896	44.0934	44.0908	37L 38T	44.0945	44.0919	26L 49T							44.1007	44.0932	13L 62T	44.0963	44.0945	0T 67T
1180	46.4567	46.4518	46.4556	46.4530		46.4227	46.4541								46.4629	46.4554		46.4585	46.4567	
1250	49.2126	49.2077	49.2115	49.2089		49.2126	49.2100								49.2188	49.2113		49.2144	49.2126	

expect hydraulic unbalance and vibration.

- If impeller has back wear ring, verify that each liquid channel has a balance hole. Make sure no holes are clogged.

- Radial gap between impeller vane tip and stationary vane (as a percentage of impeller radius) never less than 4% on diffuser pumps, and never less than 6% on volute pumps. This is a guide for high energy pumps, defined as pumps that exceed 900 ft. with specific speed values below 1300. *NOTE: Clearance refers to vane tip clearance, NOT impeller shroud clearance, per Figure 14-4.* (See also p. 411.)

- No loss of efficiency up to a 16% gap

- Cutwaters 180° apart on double volute pumps

- Cutwater tip width to be above 3/8" (11 mm) maximum

- Casing liquid channels to be at least 15% wider than impeller liquid channels

- Wear ring clearances per API-610 or vendor's instruction manual

- Use dissimilar non-galling materials, 150 BNH difference

- Provide grooves in one of the wear rings, preferably the softer of the two.* Groove cut to be 3/32" (2.2 mm) wide and 1/32" (0.7 mm) deep. Right-hand spiral, 3 per inch (8 mm pitch), permissible

- On vertical pumps, verify seal housing incorporates degassing hole or other means to ensure seal will always operate in liquid environment.

Issues often overlooked during specification reviews (See also Chapter 2)

- Applicability range for ANSI pumps excludes certain services (See Chapter 2)

- Corrosion properties (pump casing, mechanical seal materials

- Highlights of welding requirements

- Pump priming requirements, if applicable

- Cover specific heat-up or cool-down requirements

- Understand influence of pump-out vanes. Small clearances could lower stuffing box pressures and may have to be disallowed with some flush plans

- Totally specify mechanical seal details

- Apply 10% financial penalty if vendor drawing requirements are not complied with

Armed with these checklists, and perhaps also some of the many checklists found in Ref. 14-9, plant engineers and mechanical technical service personnel should feel confident in their ability to avoid and/or diagnose recurring pump problems.

In summary the *Improvement Phase* should:

- Upgrade deficient operating and maintenance processes & procedures

- Provide missing checklists & tabulations with critical mechanical and hydraulic data

- Upgrade deficient pump components

- Upgrade installation deficiencies

- Install continuous monitoring, if dictated

- Install automatic shutdown for loss of flow

- Change to OEM repair parts, if indicated

5. Control Phase (C)

Unlike so many other improvement processes, VBSS includes a control phase to ascertain that the gains anticipated from the implemented corrective actions are not gradually lost over time. Proper implementation of controls requires:

- Putting into place mechanisms or processes to prevent future defects

- Developing and implementing an ongoing control plan, with documentation

- Arranging for actionable responses, if the performance is unacceptable.

Control action steps require for the team to:

- Issue monthly activity summaries

---

*Right-hand spiral acceptable regardless of pump rotation, CW or CCW. Grooving of wear rings is generally recommended.

- Tabulate interim results after six months
- Report accomplishments after one year
- Develop plans for next phase of program
- Monitor progress of program and communicate results to management.

To fulfill this responsibility, the designated team representative should:

— Issue monthly activity summaries
— Tabulate interim results after six (6) months
— Report accomplishments after one (1) year
— Develop plans for next phase of program.

Program Results

The results of centrifugal pump failure reduction programs are generally very significant. One large plant reported a 29 percent reduction in failures after the first year. Although the failure reduction rates and percentages leveled off at 37 percent after three years, the ensuing savings exceeded this percentage because the reductions were concentrated on many maintenance-intensive problem pumps.

Failure rate comparisons are given in Figure 14-8 for a medium size chemical plant. They are again quite representative of typical first-year comparisons.

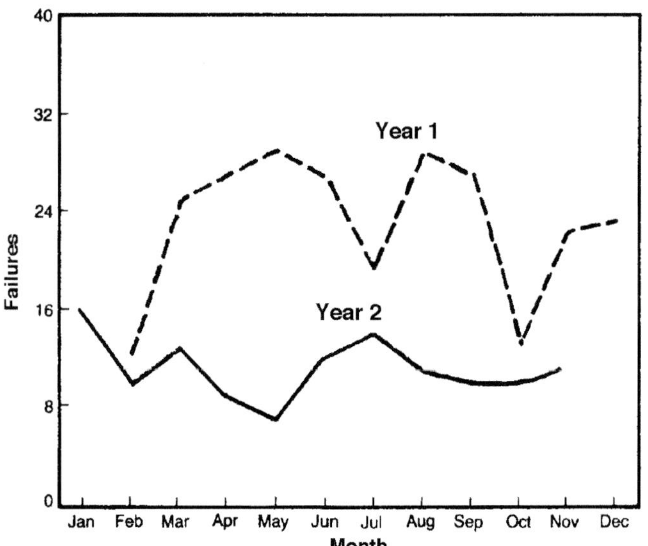

Figure 14-8: Pump failure rates experienced by a medium size chemical plant after implementing a systematic failure reduction program (Ref. 14-9)

Failure Reductions Achieved

Experience shows that significant reductions in centrifugal pump failure events are possible without, in most cases, adding to the work force of typical petrochemical plants. The key to the successful implementation of a failure analysis and reduction program can be found in management commitment, thorough involvement of plant technical personnel, and good documentation.

Management commitment manifests itself in many ways. Briefing sessions and reporting requirements can be arranged so as to lend high visibility to the program. The thoroughness of plant engineering and technical personnel is exemplified by checklist and procedure development and by their ability to convince mechanical or field work forces of the need to conscientiously follow these guidelines.

Finally, the importance of good documentation cannot be overemphasized. Computerized failure records are ideally suited to keep track of a pump failure reduction program, but the computer output can only be as good as the input. Manual data keeping is still acceptable for small plants. Close cooperation between technical and clerical staff have gone a long way towards ensuring the overall success which, happily, can be reported for centrifugal pump failure reduction programs using the approach outlined above.

Using the Kepner-Tregoe Approach to Analyze Pump Problems

The Kepner-Tregoe problem analysis concept assumes that every problem is a deviation from some standard of expected performance and that a change of some kind is always the cause of a problem (Ref. 14-9 and 14-10). Built around this premise, Kepner and Tregoe, in the late 1940's, began to teach a disciplined, well-structured approach to problem identification.

A total of seven basic concepts are stipulated and examined:

1. The problem analyst has an expected standard of performance, a "should" against which to compare actual performance.

2. A problem is a deviation from a standard of performance.

3. A deviation from standard must be precisely identified, located, and described.

4. There is always something distinguishing that has

been affected by the cause compared with that which has not.

5. The cause of a problem is always a change that has taken place through some distinctive feature, mechanism, or condition to produce a new, unwanted effect.

6. The possible causes of deviation are deduced from the relevant changes found in analyzing the problem.

7. The most likely cause of a deviation is one that exactly explains all the facts in the specification of the problem.

Using the Seven-root-cause-failure Analysis Approach

This approach has been used extensively for finding and eradicating pump failures. It has been thoroughly described in Ref. 14-9.

The seven root cause failure analysis approach accepts as its basic premise that all machinery failures, without exception, belong to one or more of the seven failure cause categories

- Faulty design
- Material defects
- Fabrication or processing error
- Assembly or installation defects
- Off-design or unintended service conditions
- Maintenance deficiencies (neglect, procedures)
- Improper operation.

The powerful and proven seven-root-cause-approach furthermore recognizes that the agents of machinery component and part failure mechanisms are always force, time, temperature, or a reactive environment.

This failure analysis and troubleshooting approach attempts to first find the root causes of failures in the categories with the highest probability ranking. Thus, if there existed many identical pumps of the same model designation and it was established that they operated reliably in identical services, the analyst would not pursue the possibility of a design defect.

However, suppose the failed pump had recently been assembled with a re-machined shaft and the primary defect found located in the bearing, one would probably focus on fabrication or dimensional processing error. Only after ascertaining that the defect was not induced by fabrication and processing errors would the analyst proceed to the next category, say, assembly-related problems.

Experienced practitioners of the seven-root-cause-analysis approach emphasize the need to:

- Start at the beginning by:
  - Reviewing the pump cross-section drawing
  - "Thinking through" how the individual parts function or malfunction
  - Understanding the process loop and process operations
- Take a systems approach. Never lose sight of the facts that:
  - The pump is only part of the overall loop
  - The part that failed is very often not the root cause of the problem and unless we find the root cause, repeat failures are likely to occur
- Collect all the pieces. The missing part may be a physical component of an observation made by someone, or a piece of information that has not surfaced earlier. Yet, any one of these missing items may contain clues that must be examined and which may have had an influence on failure cause and failure progression
- Use a calculation approach while not, of course, neglecting intuition based on your prior experience
- Fix in your mind the acronym "FRETT" and recall that machine components fail *only* because of a load-related event ("F" = force), or because of exposure to a reactive environment ("RE"), or due to an excessive number of operating cycles ("T" = time), or due to exposure to temperatures ("T") outside the range for which it had been designed.

The Cause-and-Effect Approach

Many pump problems have been solved by applying "FRETT" in conjunction with the simple, but effective, cause-and-effect methodology. This focused approach recognizes that problems have causes, or that problems are the result (effect) of something that causes them.

The cause-and-effect-approach works by moving back from the problem "P" to its preceding cause "O." This cause is now considered the effect of another cause "N" that preceded it, and so forth. An example might illustrate the approach.

The problem ("P"): Man injured → The cause ("O"): Man fell
The effect ("O"): Man fell → The cause ("N"): Man slipped
The effect ("N"): Man slipped → The cause ("M"): Leak at discharge nozzle
The cause ("M"): We are now on the component level. There are only two components at the point of leakage, a flange and a gasket.
Check for flange surface defect?
Carefully reviewed, none found!
Gasket defect? Must be due to "FRETT."

Start review!

Not "RE" (reactive environment) since same gasket material not attacked elsewhere in identical service
Not "T" (temperature) no leakage with since same or higher temperatures elsewhere
Not "T" (time) since no leakage after much longer runs elsewhere
Must therefore investigate "F"—force-related deficiencies:
Wrong stud bolt material? Missing stud bolts?
Stretched stud bolts? Insufficient torque?
Incorrect bolting sequence? etc., etc.

A thorough and amplified version of this "cause mapping" approach has been devised by "ThinkReliability" (*mark.galley@ThinkReliability.com*). ThinkReliability specializes in teaching a systems approach to analyzing failures. Cause mapping logically leads to conveying how to analyze, document, communicate and solve problems effectively.

However, regardless of which of the various troubleshooting approaches is being used, it will always be helpful to refer to troubleshooting listings. These may be "narrow" in the sense that they only list one particular symptom or deviation and its possible causes. Other approaches are very comprehensive and catalog several different symptoms and their possible causes. Both are given next.

Mechanical Seal Troubleshooting Checklist (see also Chapter 8)

This mechanical seal troubleshooting checklist is typical of the multi-symptom listing approach. Be sure not to lose sight of the goal of eliminating the root causes of mechanical seal distress. Consider upgrading whenever the economics favor doing so.

As you embark on a rigorous effort to analyze and troubleshoot mechanical seal distress, remember that certain mechanical seals may not represent an optimum selection. Replacing 20% of a plant's seal population with engineered seals would lower yearly seal-related repairs by one-half. Are you, perhaps, using a standard product where an engineered product would be far more reliable and long-term cost-effective?

Either way, start by collecting and examining all the parts of the seal, then:

— Examine the entire seal.
— Examine the wear track. Was there misalignment?
— Examine the faces for evidence of loss of fluid.
— Inspect the seal drive. Did it wear or hang up?
— Check spring condition. Corrosion? Plugging?
— Check the elastomer. Too much heat? Vaporization?
— Check for rubbing. Why did it rub?

Look for the Wear Track

- Widened track indicates component misalignment. Ask why did it happen: Bad bearings, bent shaft, shaft whip, deflection, cavitation-induced vibration, bad shaft alignment, stationary seat installation defect, severe pipe strain.
- Normal size wear track indicates good alignment, failure could have to do with clogged bellows, leakage at secondary seal, or product-related parameters.

- Narrow wear track could mean thin face was either not flat to begin with or was distorted in operation. Distortion comes from over-pressure or a basic hardware design defect. The remedy is to reduce pressure or to buy an engineered seal design.

- No wear track: Dimensional interference could have caused the entire seal to stay open, or the faces rotated in unison.

- Intermittent spots, but no complete track: Face not flat. Could be caused by tight fit, or uneven clamping. A possible solution would be to use flexibly supported face components. Note that in cryogenic applications, this symptom is the result of face freeze-up due to water being present at startup.

Look at the Seal Drive Components
- Torsional vibration and slip-stick-slip cycles can be caused by coupling misalignment.

- Lack of face lubrication, or poor lubricity.

- Excessive pressure, or incorrect springs.

- Dimensional interference.

Check Springs or Bellows
- Stress corrosion cracking or stainless steel in presence of halogens or sodium chloride; or caustic embrittlement. Material changes may be needed.

- Erosive flush may be at fault.

- Observe signs of rubbing, metal-to-metal contact, gasket-to-metal contact.

- Torsional fatigue of bellows due to slip-stick action.

- Freeze-up at seal faces.

- Clogging action brought on by spring or bellows environment. An intelligent design will avoid exposing the spring to the clogging fluid.

Look at the Faces
- Chipped edges result from slamming action or from vapor entrainment in the binder material matrix. These actions are brought on by cavitation or material defects, respectively. One possible cure is to remove heat more rapidly and effectively, another to provide a higher stuffing box pressure, or more suitable face materials.

- Heat checking: On overlay materials often the result of different thermal expansion rates between overlay and base. Could also be due to inadequate heat removal, in which case the above fixes should be considered.

- Broken surfaces are often traceable to heat shock or cold shock. This can result from hot oil flooding the seal cavity, of water hitting a hot stuffing box, or inadequate material selection, mechanical impact (shock), and failure to allow radial expansion of stationary faces.

- Erosive wear is often the result of erosive or high-pressure, locally concentrated flush streams. More uniform, deep wear is probably due to embedded abrasives—a prominent occurrence in outside-mounted seals handling abrasive-containing fluids. Bellows and stationary seals avoid many of these problems, as do certain quenching arrangements.

- Coking of bellows will result if hot oil leaks across the faces and comes into contact with air. This can be prevented by the use of stationary seals, steam quench, or cushioned seal faces (Grafoil® Cushion).

Examine the Secondary (Elastomer) Seal Components
- Swelling, flaking, and gummy feel: all point to chemical incompatibility.

- Hardening, cracking, and permanent deformation usually result from excessive heat.

- Verify absence of rubbing contact at shaft, sleeve, gland, throat bushing, and stuffing box interior cavity. Look for telltale signs, such as an oxide layer, grooving, fretting, galling.

Make Dimensional Check of Gland
- Look for lantern ring connection line protrusions
- Check for gasket interference
- Observe registration fits (rabbetting)
- Stuffing box-to-shaft concentricity
- Set screw protrusions

A final comment relating to mechanical seals: Use the selection strategy outlined earlier in Chapter 8. Insist on making contact with reputable manufacturers and disclosure of where, under virtually identical service conditions, a certain seal type has performed well. Consider using that seal. Install it properly, implementing the proper flush plan and flush flow. This text contains several other references to factors that influence seal life. Be sure not to overlook them.

Pump Troubleshooting Guides

A comprehensive, "multi-symptom" troubleshooting guide for end suction pumps, Table 14-3, will be of help in many situations. Originally provided by Igor Karassik and later (2003) updated by Steven Hrivnak, the troubleshooting guide lists the various major symptoms of pump distress and then describes both probable causes and recommended remedies. Like the preceding tabulations and checklists, this guide should speed-up the troubleshooting process. (For a pump assembly dimension checklist, refer to Chapter 15).

Table 14-3: Comprehensive troubleshooting guide for end suction centrifugal pumps (Source: Igor Karassik, updated by Steven Hrivnak, Tennessee Eastman Corporation, Kingsport, Tennessee)

Comprehensive Troubleshooting Guide-Horizontal End Suction Centrifugal Process Pumps

Symptom
No Liquid Delivery

Possible Causes	*Possible Remedies*
Suction and/or discharge valve(s) closed or partially closed	❒ Open valves
Supply tank empty	❒ Refill supply tank ❒ Install double mechanical seal and barrier system ❒ Install shutdown instrumentation
Insufficient immersion of suction pipe or bell, vortexing	❒ Lower suction pipe or raise sump level ❒ Reduce flow rate
Wrong direction of rotation	❒ Check rotation with arrow on casing—reverse polarity on motor ❒ Note: If impeller unscrews, check for damage
Speed too low	❒ Correct speed ❒ Check records for proper speed
Strainer or flame arrestor partially clogged	❒ Inspect and clean ❒ Check orientation ❒ Properly sized? ❒ Remove if start up strainer is no longer needed
Check valve plugged or installed backwards	❒ Unplug or repair check valve ❒ Reinstall in proper orientation
Obstructions in lines or pump housing	❒ Inspect and clear ❒ Improper piping ❒ Check for loose valve seat

Comprehensive Troubleshooting Guide-Horizontal End Suction Centrifugal Process Pumps

Symptom
No Liquid Delivery

Possible Causes	*Possible Remedies*
	❐ Thaw frozen lines
Pump impeller clogged	❐ Check for damage and clean
Pump not primed	❐ Fill pump & suction piping completely with liquid ❐ Remove all (air/gas) from pump, piping and valves ❐ Eliminate high points in suction piping ❐ Check for faulty foot valve or check valve, air vent
Pump is cavitating (symptom for liquid vaporizing in suction system), suction recirculation, discharge recirculation	❐ If pump is above liquid level, raise liquid level closer to pump, or lower the pump ❐ If liquid is above pump, increase liquid level elevation or increase suction pipe size ❐ Change pump size or speed ❐ Check for pipe restrictions ❐ Check air leakage through packing ❐ Install full port valve ❐ Check boiling point margin (flash point) Reduce piping losses by modifying improper piping ❐ Compare flow to BEP ❐ Check NPSHa/NPSHr margin
Air/gas entrainment in liquid	❐ Check for gas/air in suction system/piping ❐ Install gas separation chamber in suction tank/line ❐ Check well pipe: too short, or missing ❐ Check for air leaks through gaskets, packing or seals ❐ Check for air leaks in suction pipe ❐ Open air vent valve
Mismatched pumps in parallel operation	❐ Check design parameters ❐ If pumps are properly matched, check for matching piping
Pump too small (total system head higher than design head of pump)	❐ Decrease system resistance to obtain design flow ❐ Check design parameters such as impeller size, etc. ❐ Increase pump speed ❐ Install proper size pump

Comprehensive Troubleshooting Guide-Horizontal End Suction Centrifugal Process Pumps

Symptom
Low Flow

Possible Causes	*Possible Remedies*
Suction &/or discharge valve(s) closed or partially closed	❒ Open valves
Wrong direction of rotation	❒ Check rotation with arrow on casing—reverse polarity on motor ❒ Note: If impeller unscrews, check for damage
Speed too low	❒ Correct speed ❒ Check records for proper speed
Obstructions in lines or pump housing	❒ Inspect and clear ❒ Improper piping ❒ Check for loose valve seat
Strainer or flame arrestor partially clogged	❒ Inspect and clean ❒ Check orientation ❒ Properly sized? ❒ Remove if start up strainer is no longer needed
Pump impeller clogged	❒ Check for damage and clean
Impeller installed backwards (double suction pumps only)	❒ Inspect
Wrong impeller size	❒ Verify proper impeller size
Check valve plugged or installed backwards	❒ Unplug or repair check valve ❒ Reinstall in proper orientation
Internal wear (reduces throughout capability)	❒ Check impeller clearance ❒ Check for pipe strain ❒ Check for cavitation ❒ Check for corrosion wear, casing wear ❒ Pump metallurgy too soft for abrasives
Air/gas entrainment in liquid	❒ Check for gas/air in suction system/piping ❒ Install gas separation chamber in suction tank/line ❒ Checking for well pipe, too short, or missing ❒ Check for air leaks through gaskets, packing, or seals ❒ Check for air leaks in suction pipe ❒ Open air vent valve

Comprehensive Troubleshooting Guide-Horizontal End Suction Centrifugal Process Pumps

Symptom
Low Flow

Possible Causes	*Possible Remedies*
Pump is cavitating (symptom for liquid vaporizing in suction discharge recirculation)	❒ If pump is above liquid level, raise liquid level closer to pump or lower the pump ❒ If liquid is above pump, increase liquid level elevation or increase suction pipe size ❒ Change pump size or speed ❒ Check for pipe restrictions ❒ Check air leakage through packing ❒ Install full port valve ❒ Check boiling point margin (flash point) ❒ Reduce piping losses by modifying improper piping ❒ Compare flow to BEP ❒ Check NPSHa/NPSHr margin
Insufficient immersion of suction pipe or bell, vortexing	❒ Lower suction pipe or raise sump level ❒ Reduce flow rate
Viscosity too high, 500 cps most pumps, 1000 cps maximum under special designs	❒ Heat up liquid to reduce viscosity ❒ Increase size of discharge piping to reduce pressure loss ❒ Use larger driver or change type of pump ❒ Slow pump down
Mismatched pumps in parallel operation	❒ Check design parameters ❒ If pumps are properly matched, check for matching piping
Pump too small (total system head higher than design head of pump)	❒ Decrease system resistance to obtain design flow ❒ Check design parameters such impeller size, etc. ❒ Increase pump speed ❒ Install proper size pump

Comprehensive Troubleshooting Guide-Horizontal End Suction Centrifugal Process Pumps

Symptom
Intermittent Flow

Possible Causes	*Possible Remedies*
Supply tank empty	❒ Refill supply tank ❒ Install automatic refill system

Comprehensive Troubleshooting Guide-Horizontal End Suction Centrifugal Process Pumps

Symptom
Intermittent Flow

Possible Causes	*Possible Remedies*
	❒ Install electrical shutdown
Obstructions in lines or pump housing	❒ Inspect and clear ❒ Improper piping ❒ Check for loose valve seat
Mismatched pumps in parallel operation	❒ Check design parameters ❒ If pumps are properly matched, check for matching piping
Air/gas entrainment in liquid	❒ Check for gas/air in suction system/piping ❒ Install gas separation chamber in suction tank/line ❒ Checking for well pipe, too short, or missing ❒ Check for air leaks through gaskets, packing, or seals ❒ Check for air leaks in suction pipe ❒ Open air vent valve
Insufficient immersion of suction pipe or bell, vortexing	❒ Lower suction pipe or raise sump level ❒ Reduce flow rate
Pump is cavitating (symptom for liquid vaporizing in suction discharge re-circulation)	❒ If pump is above liquid level, raise liquid level closer to pump or lower the pump ❒ If liquid is above pump, increase liquid level elevation or increase suction pipe size ❒ Change pump size or speed ❒ Check for pipe restrictions ❒ Check air leakage through packing ❒ Install full port valve ❒ Check boiling point margin (flash point) ❒ Reduce piping losses by modifying improper piping ❒ Compare flow to BEP ❒ Check NPSHa/NPSHr margin
Process changes	❒ Verify if system changes exceed pump and piping design—resize pump and piping

Comprehensive Troubleshooting Guide-Horizontal End Suction Centrifugal Process Pumps

Symptom
Insufficient Discharge Pressure

Possible Causes	*Possible Remedies*
Speed too low	❒ Correct speed ❒ Check records for proper speed
Wrong direction of rotation	❒ Check rotation with arrow on casing. May have to reverse wiring on motor ❒ Note: If impeller unscrews, check for damage
Air/gas entrainment in liquid	❒ Check for gas/air in suction system/piping ❒ Install gas separation chamber in suction tank/line ❒ Check well pipe: too short, or missing ❒ Check for air leaks through gaskets, packing, or seals ❒ Check for air leaks in suction pipe ❒ Open air vent valve
Wrong impeller size	❒ Verify proper impeller size
Internal wear (reduces pump performance capability)	❒ Check impeller clearance ❒ Check for pipe strain ❒ Check for cavitation ❒ Check for corrosion wear and casing wear ❒ Pump metallurgy too soft for abrasives
Impeller installed backward (double suction pumps only)	❒ Inspect
Obstructions in lines or pump housing	❒ Inspect and clear ❒ Improper piping ❒ Check for loose valve seat
Pump impeller clogged	❒ Check for damage and clean
Pump too small (total system head higher than design head of pump)	❒ Decrease system resistance to obtain design flow ❒ Check design parameters such as impeller size, etc. ❒ Increase pump speed ❒ Install proper size pump

Comprehensive Troubleshooting Guide-Horizontal End Suction Centrifugal Process Pumps

Symptom
Short Bearing Life

Possible Causes	*Possible Remedies*
Bearing failures	❒ Inspect parts for defects-repair or replace Have bearing manufacturer analyze failed bearings and make recommendation ❒ Check lubrication procedures ❒ Check for contaminated lubricant (e.g., water) ❒ Check for over-lubrication ❒ Check for under-lubrication ❒ Verify (mineral) oil temperature less than 180°F (83°C)
Unbalance-Driver	❒ Run driver disconnected from pump unit — perform vibration analysis
Pump is cavitating (symptom for liquid vaporizing in suction system), suction re-circulation, discharge re-circulation	❒ If pump is above liquid level, raise liquid level closer to pump or lower the pump ❒ If liquid is above pump, increase liquid level elevation or increase suction pipe size ❒ Change pump size or speed ❒ Check for pipe restrictions ❒ Check air leakage through packing ❒ Install full port valve ❒ Check boiling point margin (flash point) ❒ Reduce piping losses by modifying improper piping ❒ Compare flow to BEP ❒ Check NPSHa/NPSHr margin
Unbalance-Pump	❒ Balance impeller
Misalignment	❒ Check angular and parallel alignment between pump & driver ❒ Check and eliminate any pipe strain ❒ Eliminate stilt-mounted baseplate ❒ Check for loose mounting ❒ Eliminate rigid conduit connection ❒ Check for thermal growth
Bent shaft	❒ Check TIR at impeller end (should not exceed 0.002"). Replace shaft & bearings if necessary
Casing distorted from pipe strain	❒ Orientation of bearing adaptor ok? ❒ Check for misalignment of pipe

Comprehensive Troubleshooting Guide-Horizontal End Suction Centrifugal Process Pumps

Symptom
Short Bearing Life

Possible Causes	*Possible Remedies*
	❒ Check pump for wear between casing and rotating elements ❒ Analyze piping loads ❒ Check for pipe supports ❒ Check for proper spring hanger setting ❒ Is suction piping supported within 1 to 3 feet of the pump? Is vertical piping supported from above using pipe hangers or spring hangers? Verify proper support with Engineering group
Inadequate grouting of base or stilt-mounted	❒ Check grouting. Is it cracked, crumbling, air voids, etc. Was it grouted to current industry practices? Consult Process Industry Practice RF-IE-686 ❒ If stilt-mounted, grout baseplate
Pump too large (total system head lower than design head of pump)	❒ Increase system resistance to obtain design flow ❒ Check design parameters such as impeller size, etc. ❒ Decrease pump speed ❒ Install proper size pump
Pump too small (total system head higher than design head of pump)	❒ Decrease system resistance to obtain design flow ❒ Check design parameters such as impeller size, etc. ❒ Increase pump speed ❒ Install proper size pump

Comprehensive Troubleshooting Guide-Horizontal End Suction Centrifugal Process Pumps

Symptom
Short Mechanical Seal Life

Possible Causes	*Possible Remedies*
Evaporation or solidification in stuffing box and on seal faces	❒ Install double seal & barrier system ❒ Keep stuffing box at proper temperature
Improper operation procedures	❒ Verify that Operations does not start up & shut down pump improperly ❒ Discontinue dead-heading pump

Comprehensive Troubleshooting Guide-Horizontal End Suction Centrifugal Process Pumps

Symptom **Short Mechanical Seal Life**	
	❒ Avoid running pump dry ❒ Gravity drain through pump ❒ Have purge system valved out ❒ Work with Operations to change bad habits or work with engineering to design around
Misalignment	❒ Check angular and parallel alignment between pump & driver ❒ Eliminate stilt-mounted baseplate ❒ Check for loose mounting ❒ Eliminate conduit and piping strain ❒ Check for thermal growth
Inadequate grouting of base or stilt-mounted	❒ Check grouting. Is it cracked, crumbling, air voids, etc. Was it grouted to current industry practices? Consult Process Industry-Practice RF-IE686 ❒ If stilt-mounted, grout baseplate
Casing distorted from pipe strain	❒ Check orientation of bearing adaptor ❒ Check for misalignment of pipe ❒ Check pump for wear between casing and rotating elements ❒ Analyze piping loads ❒ Check for pipe supports ❒ Check for proper spring hanger setting ❒ Is suction piping supported within 1 to 3 feet of the pump? Is vertical piping supported from above using pipe hangers or spring hangers? Verify proper support with engineering.
Bent shaft	❒ Check TIR at impeller end (should not exceed 0.002"). Replace shaft & bearings if necessary
Unbalance —Pump	❒ Balance impeller
Wrong impeller size	❒ Verify proper impeller size
Pump too small (total system head higher than design head of pump)	❒ Decrease system resistance to obtain design flow ❒ Check design parameters such as impeller size, etc. ❒ Increase pump speed ❒ Install proper size pump
	❒ If pump is above liquid level, raise liquid level closer to pump or lower the pump

Comprehensive Troubleshooting Guide-Horizontal End Suction Centrifugal Process Pumps

Symptom **Short Mechanical Seal Life**	
Pump is cavitating (symptom for liquid vaporizing in suction system), suction recirculation, discharge recirculation	❒ If liquid is above pump, increase liquid level elevation or increase suction pipe size ❒ Change pump size or speed ❒ Check for pipe restrictions ❒ Check air leakage through packing ❒ Install full port valve ❒ Check boiling point margin (flash point) ❒ Reduce piping losses by modifying improper piping ❒ Compare flow to BEP ❒ Check NPSHa/NPSHr margin
Viscosity too high, 500 cps most pumps, 1000 cps maximum under special designs (product not lubricating seal faces)	❒ Heat up liquid to reduce viscosity ❒ Install seal flush ❒ Install double seal & barrier system
Supply tank empty	❒ Refill supply tank ❒ Install double seal & barrier system ❒ Install electrical shutdown
Improper mechanical seal	❒ Check mechanical seal selection strategy
Mismatched pumps in parallel operation	❒ Check design parameters ❒ If pumps are properly matched, check for matching piping
Pump too large (total system head lower than design head of pump; too much or too little flow causes shaft vibration and short seal life)	❒ Increase system resistance (add orifice or restrict discharge valve ❒ Check design parameters such as impeller size, etc. ❒ Decrease pump speed ❒ Install proper size pump
Air/gas entrainment in liquid	❒ Check for gas/air in suction system/piping ❒ Install gas separation chamber in suction tank/line ❒ Check well pipe: too short, or missing ❒ Check for air leaks through gaskets, packing, or seals ❒ Check for air leaks in sealing system ❒ Open air vent valve
Unbalance —Driver	❒ Run driver disconnected from pump unit—perform vibration analysis

Comprehensive Troubleshooting Guide-Horizontal End Suction Centrifugal Process Pumps

Symptom **Vibration and Noise**	
Possible Causes	*Possible Remedies*
Pump is cavitating (symptom for liquid vaporizing in suction system), suction re-circulation, discharge re-circulation	❐ If pump is above liquid level, raise liquid level closer to pump or lower the pump ❐ if liquid is above pump, increase liquid level elevation or increase suction pipe size ❐ Change pump size or speed ❐ Check for pipe restrictions ❐ Check air leakage through packing ❐ Install full port valve ❐ Check boiling point margin (flash point) Reduce piping losses by modifying improper piping ❐ Compare flow to BEP Check NPSHa/NPSHr margin Check pump suction energy
Suction &/or discharge valve(s) closed or partially closed	❐ Open valves
Misalignment	❐ Check angular and parallel alignment between pump & driver ❐ Check and eliminate any pipe strain ❐ Eliminate stilt-mounted baseplate ❐ Check for loose mounting ❐ Eliminate rigid conduit connection ❐ Check for thermal growth
Inadequate grouting of base or stilt mounted	❐ Check grouting. Is it cracked, crumbling, air voids, etc. Was it grouted to current industry practices? Consult Process Industry Practice RE-IE-686 ❐ If stilt-mounted, grout baseplate
Coupling problems	❐ Check for proper grease ❐ Check for proper sizing ❐ Check for contoured key ❐ Use Class 1 alignment
Bearing failures	❐ Inspect parts for defects-repair or replace. Have bearing mfr. analyze failed bearings and make recommendation ❐ Check lubrication procedures ❐ Check for contaminated lubricant (e.g., water) ❐ Check for over-lubrication

Comprehensive Troubleshooting Guide-Horizontal End Suction Centrifugal Process Pumps

Symptom
Vibration and Noise

Possible Causes	*Possible Remedies*
	❒ Check for under-lubrication ❒ Verify (mineral) oil temperature less than 180°F (83°C)
Pump impeller clogged	❒ Check for damage and clean
Bent shaft	❒ Check TIR at impeller end (should not exceed 0.002"). Replace shaft & bearings if necessary
Check valve plugged or installed backwards	❒ Unplug or repair check valve ❒ Reinstall in proper orientation
Obstructions in lines or pump housing	❒ Inspect and clear ❒ Improper piping ❒ Check for loose valve seat
Strainer or flame arrestor partially clogged	❒ Inspect and clean ❒ Check orientation ❒ Properly sized? ❒ Remove if startup strainer is no longer needed
Insufficient immersion of suction pipe or bell, vortexing	❒ Lower suction pipe or raise sump level ❒ Reduce flow rate
Air/gas entrainment in liquid	❒ Check for gas/air in suction system/piping ❒ Install gas separation chamber in suction tank/line ❒ Check well pipe: too short, or missing ❒ Check for air leaks through gaskets, packing, or seals ❒ Check for air leaks in sealing system ❒ Open air vent valve
Pump too large (total system head lower than design head of pump)	❒ Increase system resistance to obtain design flow Check design parameters such as impeller size, etc. ❒ Decrease pump speed ❒ Install proper size pump
Pump too small (total system head higher than design head of pump)	❒ Decrease system resistance to obtain design flow ❒ Check design parameters such as impeller size, etc. ❒ Increase pump speed ❒ Install proper size pump

Wrong impeller size	❐ Verify proper impeller size
Mismatched pumps in parallel operation	❐ Check design parameters If pumps are properly matched, check for matching piping
Unbalance —Driver	❐ Run driver disconnected from pump —perform vibration analysis
Unbalance —Pump	❐ Balance impeller

Comprehensive Troubleshooting Guide-Horizontal End Suction Centrifugal Process Pumps

Symptom
Power Demand Excessive

Possible Causes	*Possible Remedies*
Motor tripping off	❐ Check starter ❐ Check heater elements or relay settings ❐ Decrease impeller size ❐ Increase motor size if too small for impeller ❐ If operations has increased flow or changed the liquid being pumped, resize pump
Speed too high	❐ Correct speed ❐ Check records for proper speed
Wrong impeller size	❐ Verify proper impeller size
Pump not designed for liquid density being pumped	❐ Check design specific gravity ❐ Check motor size ❐ Check coupling size
Pump too large (total system head lower than design head of pump)	❐ Increase system resistance to obtain design flow ❐ Check design parameters such as impeller size, etc. ❐ Decrease pump speed ❐ Install proper size pump
Bearing failures	❐ Inspect parts for defects-repair or replace. Use Bearing Failure Analysis Guide ❐ Check lubrication procedures ❐ Check for contaminated lubricant (e.g., water) ❐ Check for over-lubrication ❐ Check for under-lubrication ❐ Verify (mineral) oil temp. less than 180°F (83°C)

Comprehensive Troubleshooting Guide-Horizontal End Suction Centrifugal Process Pumps

Symptom
Power Demand Excessive

Possible Causes	*Possible Remedies*
Rotor impeller rubbing on casing or seal cover	❒ Loose impeller fit ❒ Wrong rotation with threaded impeller—impeller unscrewing ❒ Bent shaft ❒ High nozzle loads ❒ Internal running clearances too small ❒ Low flow operation below the "minimum allowable operating" region
Liquid viscosity too high	❒ Heat up liquid to reduce viscosity ❒ Use larger driver or change type of pump ❒ Slow pump down

By way of "pictorial summary," we direct the reader's attention to Figure 14-9, showing a "Centrifugal Pump Repair Procedures" billboard. In the 1960's and early 1970's, this classroom-size checklist was located in a large refinery pump repair shop. While a few of its then-current guidelines and reminders are a bit outdated by today's standards, it shows the extent to which management and technical people were committed to excellence. Needless to say, this is one of the pump users that achieved quantum improvements in pump reliability by providing the right teaching tools to its work force.

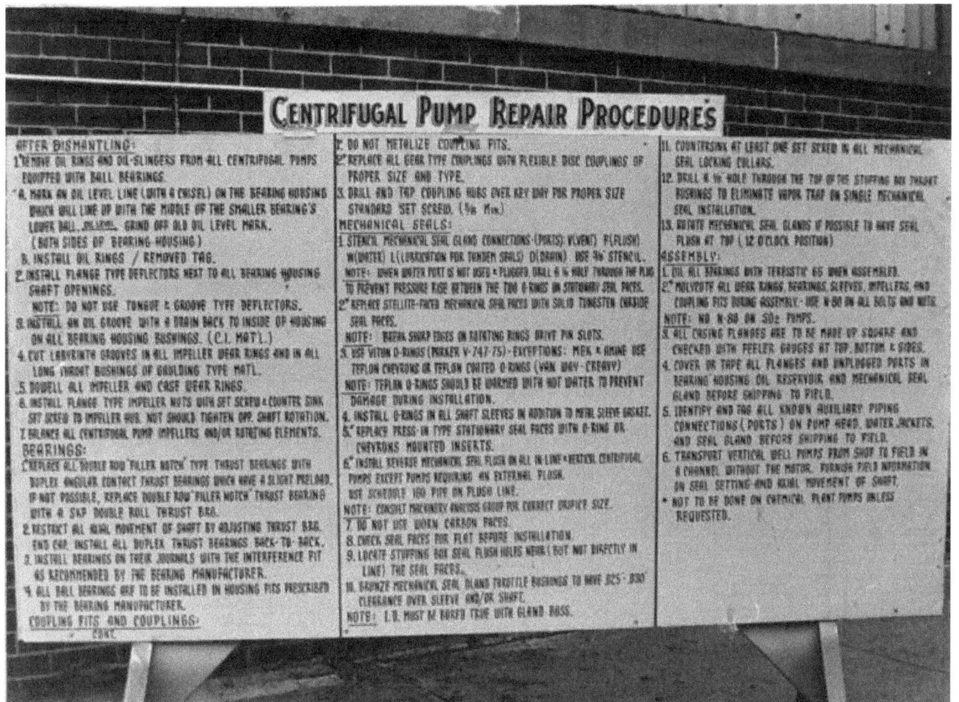

Figure 14-9: Shop instruction at a U.S. refinery workshop, 1960's vintage

Chapter 15

Shop Repair, Spare Parts Availability & Procurement

Shop repair and spare parts issues are intertwined. There are geographic locations where pump users prefer to make their own replacement parts. Some do so effectively and save money in the process, other locations do so ineffectively and fare poorly in terms of pump MTBF and plant reliability. Since parts need to be in hand before shop repair can be effected, we will deal first with spare parts availability and procurement issues.

OEM vs. Non-OEM Spare Parts

Decades ago, it was inconceivable to consider purchasing pump repair and replacement parts from parties other that original equipment manufacturers, or OEM's. How did we get to the point of buying these parts from non-OEM spare parts producers? To a large extent, original equipment manufactures are to blame for the popularity of non-OEM parts (Ref. 15-1). OEM's have not always been responsive to critical user needs. Perhaps the user's systems were down and mechanical work forces were waiting for delivery of repair parts. The OEM was often unable to respond quickly due to internal production, stocking and scheduling problems. Also, higher overhead costs and an industry practice of selling new pumps at near (or below) cost forced OEM's to rely on relatively high repair part margins to reach corporate profitability goals. On occasion, this has led to very high costs for OEM spares. Then there were instances where the OEM purchases pump shafts and other key components from non-OEM or "aftermarket" suppliers. Once the more astute pump users realized this, they were, understandably, reluctant to pay the OEM's *seemingly* inflated prices.

Be this as it may, the situation has promoted the growth of small non-OEM parts replicators that have lower overhead and are often more responsive to user needs and emergencies, at least for the more popular parts. For a while, OEM non-responsiveness forced many users to carry high inventories of repair parts, and/or to manufacture certain parts in their machine shops, such as shafts, impeller and casing wear rings, and certain shaft sleeves.

Users themselves have sometimes undermined the OEM's by cutting back on inventories. Moreover, users have often waited for a failure to occur before contacting OEM's and non-OEM's to check on the availability of the required parts. In turn, this has frequently prompted users and non-OEM's to copy the failed parts. Some users and parts replicators have even created and shared data banks and drawing files for future use.

In recent years, some major equipment OEM's have been fighting back in a variety of ways. They have been trying to improve order entry and production efficiencies, placing consigned repair part stocks at key user locations, offering to manage user pump maintenance, and becoming more competitive on pricing. Very often, then, the pump user has a choice of buying from either party.

Factors to Consider

Buying on the basis of cost and delivery alone is not the best approach for reliability-focused pump users. There are critical *functional* factors to consider, other than price and delivery, when making the decision on whether to purchase OEM or non-OEM parts:

- The use of non-OEM parts will affect the manufacturer's warranty.
- Non-OEM parts may not meet OEM performance specifications (see Figures 15-1 through 15-14). Proper impeller vane contours are difficult to copy, and subtle changes can have a large effect on pump performance.
- Since non-OEM parts are often copied from worn OEM parts, critical fits and tolerances on copied parts may not meet OEM specifications, which can lead to premature failures. For example, if a shaft fillet radius is too small, it will increase the shaft stress concentration and could lead to a fatigue failure. If a shaft fillet radius is too large, it may interfere with proper assembly stack up. Machin-

ing marks under a shaft lip seal may cause it to leak in operation. Oversized ball bearing fits on the shaft can cause excessive pre-loads which drastically shorten bearing life.

- Non-OEM's do not have R&D facilities, so they cannot keep up with the latest part changes (mechanical and metallurgical improvements).

- Most pump OEM's meet ISO 9001 quality control, while non-OEM's cannot meet these requirements without OEM drawings.

- Metallurgies and material mechanical properties my not meet OEM specifications, possibly increasing corrosion rates or reducing pressure ratings.

There are also non-functional advantages to using OEM's for repair parts:

- Many non-OEM's only offer the most popular repair parts, so that users will still have to deal with OEM's for low volume parts. Sometimes, the user misses out on OEM quantity discounts.

- OEM's offer application expertise, upgrade recommendations and recommended parts lists.

Consult Checklist before Deciding Where to Purchase

The following checklist items should be consulted before making critical repair part sourcing decisions.

1. Will use of the parts affect the manufacturer's pump warranty and increase your product liability?

2. Does the supplier accept responsibility for the exacting tolerances that maintain the precise fits required to ensure a pump's original high efficiency and long life?

3. Are the parts made from original engineering drawings or are they imprecise copies made from OEM parts?

4. Do the parts incorporate the latest design changes or refinements made to improve pump operating efficiency, long life and safe operation?

5. Does the supplier have a stringent quality control program that assures uniform quality of each and every part?

6. Is the metallurgy the same as OEM parts?

7. Does the supplier offer a choice of metallurgies for different applications?

8. Does the supplier have an ongoing research and development program, which provides continuing improvements in design, materials, and construction?

9. Does the supplier offer total service by providing low volume as well as high volume replacement parts?

10. Can the supplier stock parts that can be shipped the same day your order is received?

11. Does the supplier offer a system of recommended parts that allow you to reduce inventory and lower your costs?

12. Is the salesperson interested only in selling parts or does he/she offer the application expertise that can help solve pump problems?

How to See the Difference

Figures 15-1 through 15-14 show key dimensions where non-OEM parts may not comply with the OEM specification requirements. While the illustrations are depicting typical ANSI pump impellers (Figures 15-1 through 15-12), hook sleeves (Figure 15-13) and shafts (Figure 15-14), the underlying principles are valid for most other pump types as well. It should be noted that impellers are difficult to copy with reasonable accuracy. Subtle dimensional differences can result in large changes in pump performance. The impeller eye area, inlet throat area (between vanes), vane thickness and inlet vane angle determine the NPSH requirement of the pump.

The notes on Figures 15-1 through 15-12 explain what could happen when these areas deviate from the design norm. Impeller vane length, discharge throat area, discharge vane tip thickness and discharge vane angle will affect both developed head and pump efficiency. Even the way an impeller is de-burred and/or balanced can affect pump head and efficiency. The height and angle of the back pump-out vanes can affect axial thrust, hence, bearing life, and pump efficiency. Even the stuffing box pressure may be influenced by incorrect height—hence, clearance—of back pump-out vanes. Finally, impeller bore tolerance and balance will affect pump vibration and mechanical seal life.

Insufficient or excessive eye diameter or depth to the nose from the leading edge of the vane tips can reduce hydraulic performance and especially affect NPSH and efficiency (horsepower) characteristics.

Figure 15-1: OEM vs. non-OEM pump impeller eye diameter comparison (Source: ITT/ Goulds, Seneca Falls, NY 13148)

Improper angle can disturb inlet area causing a detrimental effect on hydraulic performance, including $NPSH_R$.

Figure 15-2: Checking the impeller inlet vane angle (Source: ITT/ Goulds, Seneca Falls, NY 13148)

Shaft sleeves that are out-of-specification can increase vibration and reduce mechanical seal and packing life. Improper inside chamfers, undercuts, and I.D. tolerances can result in sleeve run-out, vibration and significant seal life reduction. Lack of O.D. chamfers, which remove sharp edges, can damage secondary sealing members on shaft seals. Undercut areas prevent the sleeve from seizing on the shaft, but may not be appropriate where the resulting stress-riser effect jeopardizes pump life. Burnishing of the sleeve O.D. smoothes and hardens the surface for maximum packing and sleeve life.

Shafts that are not machined to intended dimensions and tolerances can fail prematurely, reduce ball bearing life, reduce lip seal life, and increase vibration. Long radius fillets minimize shaft stress, which reduces the risk of fatigue failures and ensures maximum shaft life. Radius undercuts at bearing fit shoulders avoid interference with the bearing I.D. but, again, must be avoided in instances where the resulting stress riser effect would mandate de-rating the allowable torque input. Machining marks, inadequate surface smoothness and wear tracks can cause seal leakage. Here, too, experience-based decisions are needed.

Too thin a vane will reduce its mechanical strength and enlarge the effective flow passageway causing increased flow and horsepower consumption and too thick a vane will result in reduced opening between vanes creating a smaller liquid passageway and an effectively smaller pump. Also check for bumps or irregularities along the vane resulting from improper segmented pattern removal. These will cause internal vortices which will block the flow and reduce capacity.

Figure 15-3: Checking the impeller vane depth (Source: ITT/Goulds, Seneca Falls, NY 13148)

A narrow opening will result in a smaller liquid passageway creating a smaller pump (lower capacity) and too large an opening will increase capacity and consequently consumed power.

Figure 15-4: Making vane width checks (Source: ITT/Goulds, Seneca Falls, NY 13148)

SHOP REPAIR PROCEDURES

Pump manufacturers usually supply pump maintenance manuals with detailed assembly and disassembly instructions that are either generic, or specific to a particular pump style and model. Nevertheless, a number of important checks should be performed by users whose goal it is to systematically eradicate some of the often overlooked quality checks.

Checking for Concentricity and Perpendicularity

Experience shows that after years of repairs, many pumps are due for a series of comprehensive dimensional and assembly-related checks. As a minimum, every pump that is labeled a "bad actor" and considered part of your initial pump failure reduction program should be given the checks described in Figure 15-15. The dimensional "before vs. after" findings listed in Figures 15-16 a/b/c should then be recorded in either the (preferred) electronic, or, as a minimum, paper format. Users that do not take time to record these pump repair data will find it very difficult to reach the desired failure reduction objectives. (Note that certain seal-related dimensions may not apply to cartridge seals)

Incorrect vane length will have an effect on performance in that the inlet and discharge angles could possibly be affected as well as vane curvature and the flow channel area progressive throughout the vane passage. Incorrect "wrap" of vanes and discharge angles will affect produced head.

Figure 15-5: Checking distance between vanes throughout length (Source: ITT/Goulds, Seneca Falls, NY 13148)

A shallow vane will result in a smaller liquid passageway resulting in an effectivelly smaller pump (reduced capacity), and a larger vane depth results in too large a passageway to match the original pump design characteristics (i.e., capacity and horsepower will be greater).

Figure 15-6: Observing total vane length (Source: ITT/Goulds, Seneca Falls, NY 13148)

Shaft Straightening Procedures

Due to the long shaft lengths used with vertical turbine pumps, shafts must often be straightened prior to the initial assembly, or during repair. Two methods are commonly used, heat straightening and mechanical straightening. The shaft is set up across a set of rollers in both cases (Figure 15-17). With the heat method, heat is applied to the high points in a circular path until the area turns red, and then immediately quenched with water. The process is repeated, until the shaft run-out is found to be within specification, after it has cooled.

The mechanical method can achieve a total indicator reading (TIR) straightness of .0005″ per foot, about 0.04mm/m. A press is used to gradually apply load to the shaft at the high point, until there is plastic deformation (Figure 15-18). Caution should be used not to bend the shaft too much beyond what is necessary to achieve a run-out of .0005″ per foot. This process should be repeated until the entire shaft is within specification. With the mechanical method, the shaft must be rechecked after a 72 hour aging period, to ensure that the residual stress does not distort the shaft.

Check diameter, width and depth of groove to assure proper O-ring sealing, This is critical as a poor seal will result in possible shaft corrosive damage, and difficult and costly disassembly procedures.

Figure 15-7: Checking the O-ring groove (Source: ITT/ Goulds, Seneca Falls, NY 13148)

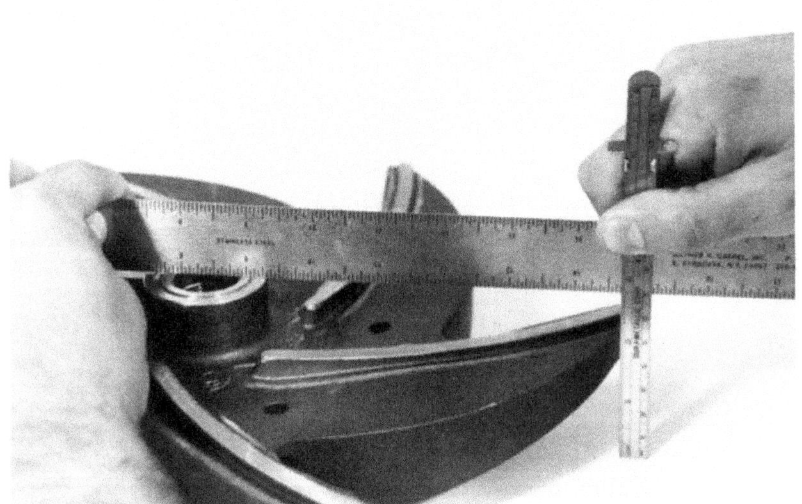

This will determine the axial clearances (or lack of) in the casing to assure proper hydraulic operation. Excessive front clearance will cause gross reduction in capacity, head and efficiency. An excessive rear clearance may render the pump-out vanes ineffective increasing axial thrust, stuffing box pressure and solids in the packing. Too little clearance may cause rubbing and serious mechanical failure.

Figure 15-8: Determining axial impeller width (Source: ITT/Goulds, Seneca Falls, NY 13148)

Assembly Dimension Checklist

Note: All of the following refer to typical refinery-grade process pumps. For nomenclature, please refer to earlier illustrations depicting back pull-out pumps.

Unless a process pump manufacturer gives specific and different values or measurements for a particular make, size, or model, experience shows the following guidelines to be useful and valid. Perhaps your pump shop would benefit from making it a habit to use and apply this assembly dimension checklist. Some of the listed diametral clearance and/or interference tolerances will be more strict than what certain pump manufacturers allow for internal cost reasons, but then again, we see the need to improve on some products.

In any event, we encourage you to make copies, laminate them in plastic, and to either hand them to each of your shop technicians, or post them near the mechanics' or repair technicians' work stations.

01. Radial ball bearing I.D. to shaft fits: 0.0002"-0.0007" interference

02. Radial ball bearing O.D. to housing fits: 0.0002"-0.0015" clearance

Insufficient depth will cause impeller to bottom-out on the shaft before O-ring seal face contacts the shaft sleeve resulting in leakage. Same damage as in 15-1. Too deep a bore will result in mechanical weakening of the impeller and reduce corrosive resistance life at the impeller eye.

Figure 15-9: Verifying impeller bore and threads (Source: ITT/Goulds, Seneca Falls, NY 13148)

Insufficient vane height will reduce effectiveness of the pump-out vanes, resulting in higher axial thrust and stuffing box pressure and ineffective ejection of solids from the mechanical seal/packing chamber.

Figure 15-10: Examining pump-out vane height (Source: ITT/Goulds, Seneca Falls, NY 13148)

03. Back-to-back mounted thrust bearing I.D. to shaft fits: 0.0001″-0.0003″ interference

04. Back-to-back mounted thrust bearing O.D. to housing fits: 0.0001″-0.0015″ clearance

05. Shaft shoulders at bearing locations must be square with shaft centerlines within 0.0005″

06. Shaft shoulder height must be 65-75% of the height of the adjacent bearing inner ring

07. Sleeve to shaft fits are to be kept within 0.0010″-0.0015″ clearance

08. Impeller to shaft fits, on single-stage, overhung pumps, are preferably 0.0000″-0.0005″ clearance fits. However, some pump manufacturers allow diametral clearances from 0.0015″ to 0.0025,″ depending on impeller diameter and pump speed (see Table 12-1).

09. Impeller to shaft fits, on multistage pumps, sometimes require interference fits and must be checked

The leading edge of the vane profile or curve must match the casing or sideplate mating profile to assure proper running clearance. Without this proper "fit," maximum performance can never be attained either in head, capacity, and/or efficiency. <u>This is critical</u>.

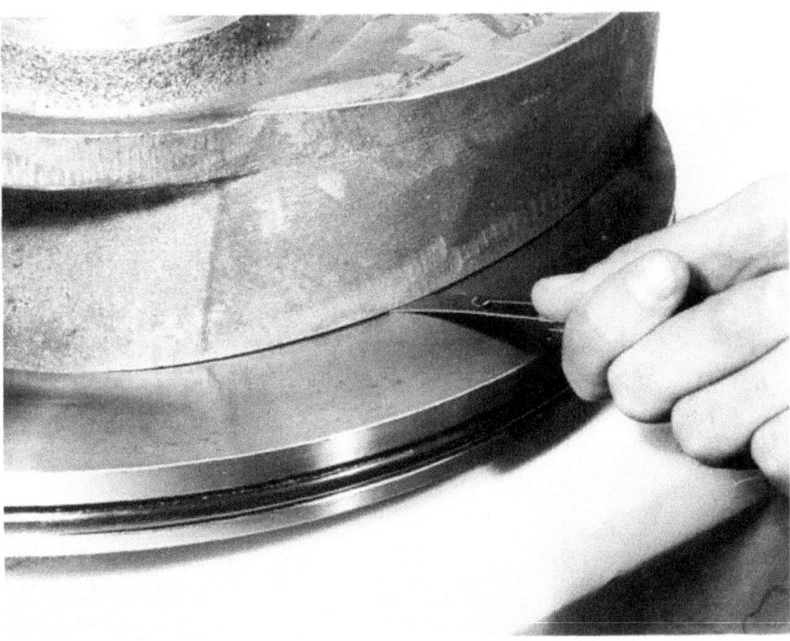

Figure 15-11: Verifying vane profile, or leading edge (Source: ITT/Goulds, Seneca Falls, NY 13148)

Excessive vane T.I.R. will make proper adjustment of impeller front clearances impossible, resulting in poor hydraulic performance.

Figure 15-12: Examining pump-out vane run-out (Source: ITT/ Goulds, Seneca Falls, NY 13148)

against the manufacturer's or reliability group's specifications

10. Keys should be hand-fitted(!) in keyways with 0.0000"-0.0002" interference. There are sound, experience-based reasons why we advocate "tightening" the manufacturer's allowable 0.001" clearance

11. In view of (10), above, keys should be *hand-fitted* to a "snug fit"

12. (a) Throat bushing to case fit is to be 0.002"-0.003" interference, depending on diameter
    (b) Throat bushing to shaft fit is to be 0.015"-0.020" clearance
    (c) Throat bushing to shaft fits of inline pumps (depending on shaft size) where the throat bushings may act as intermediate bearings, will have clearances ranging from 0.003" to 0.012"

13. Weld overlays can be substituted for metallic impeller wear rings. When separate wear rings are used, the impeller fit should be 0.002"-0.003" interference (Note: Applies to metallic parts only).

1. Inside chamfer and under-cutting mandatory to insure proper seating of sleeve on shaft. Improper seating can result in sleeve runout and drastic seal life reduction.
2. OD chamfers remove sharp edges which can damage secondary sealing members on mechanical seals.
3. Undercut area prevents seizing on shaft or pinching of sleeve to shaft by mechanical seal set screws.
4. Surface finish is ground and polished to extend packing and mechanical seal life. Burnishing both smoothes and hardens surface of sleeve.
5. ID tolerances at these locations provide snug sleeve-to-shaft fit, negate vibration, optimize mechanical seal life.

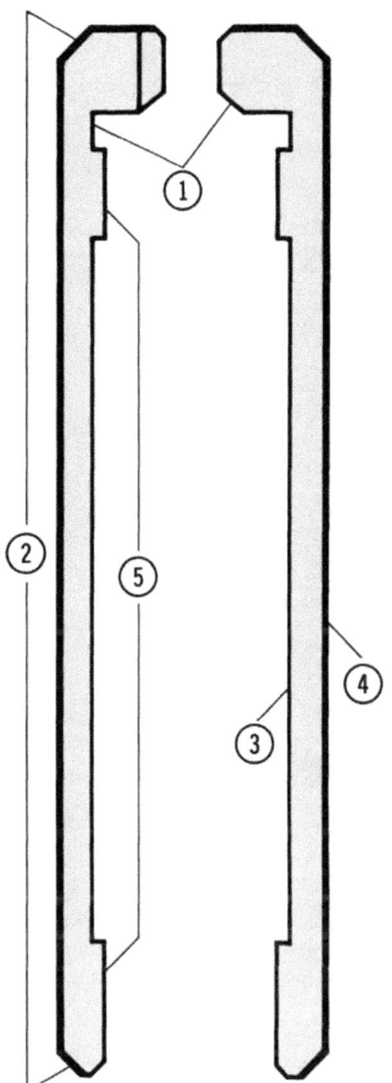

Figure 15-13: Examining shaft sleeve condition (Source: ITT/Goulds, Seneca Falls, NY 13148)

14. Impeller wear rings should be secured by either dowelling, set screws that are threaded in the axial direction partly into the impeller and partly into the wear ring, or tack-welded in at least two places

15. Clearance between impeller wear ring and case wear ring should be 0.010″ to 0.012″ plus 0.001″ per inch up to a ring diameter of 12 inches. Add 0.0005″ per inch of ring diameter over 12 inches. Use manufacturer's guidelines when applying high-performance polymers such as PEEK or Dupont Vespel®, good up to 500°F—note that tighter clearances are allowed! But tighter clearances cause changes in thrust load (axial load). Verify that bearing loads are still okay.

16. For pumping temperatures of 500°F and over, add 0.010″ to the wear ring clearance. Also, whenever galling-prone wear ring materials such as stainless steel are used, 0.005″ are added to the clearance.

17. Impeller wear rings should be replaced when the new clearance reaches twice the original value

18. Case wear rings are not to be bored out larger than 3% of the original diameter

19. Metal case ring-to-case interference should be 0.002″-0.003,″ depending on diameter

20. Case rings should either be doweled or spot welded in 2 or 3 places

21. Metal oil deflectors should—if possible—be fitted with an O-ring, and should be mounted with a shaft clearance of 0.002″-0.003″

More sophisticated measuring tools may be required to do a complete check on the shaft. Some critical dimensions and observations are:

1. Impeller thread depth and diameter to assure a snug fit of the impeller and assure sealing against the shaft sleeve.

2. Proper diameter and depth of sleeve drive pin ensures positive locking and drive of sleeve.

3. Long radius fillets minimize stress, combats fatigue failure and ensures maximum shaft life.

4. Large chamfers and perpendicular shoulder insures proper seating of the sleeve. Reduces runout and optimizes seal life. Check proper location of all shaft "steps" by comparing our shaft to theirs.

5. Ground surfaces at oil seal fits to assure optimum seal life.

6. Exact bearing fit diameter to assure proper assembly fit to bearing. Again shoulder must be square for correct bearing alignment to the shaft.

7. Radius undercuts at bearing fit shoulders to avoid interference with the bearing ID.

8. Snug keyway fit at coupling to assure proper coupling engagement.

9. Obviously with micrometers critical shaft OD's could be checked. Areas such as bearing fits, impeller fits, coupling fits, and sleeve fits.

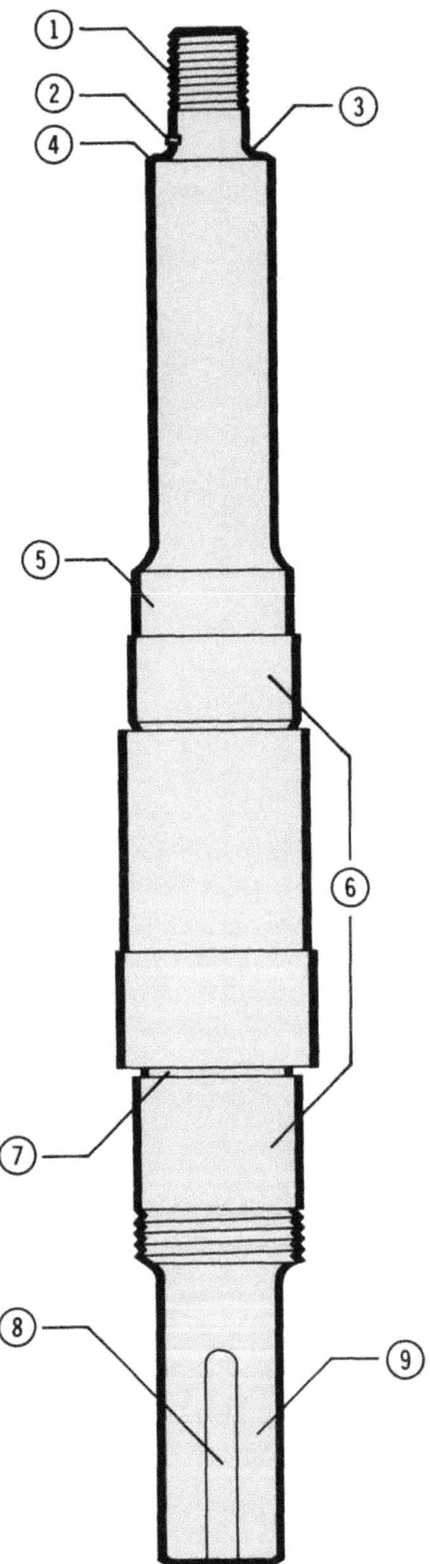

Figure 15-14: Verifying pump shaft dimensional accuracy (Source: ITT/Goulds, Seneca Falls, NY 13148)

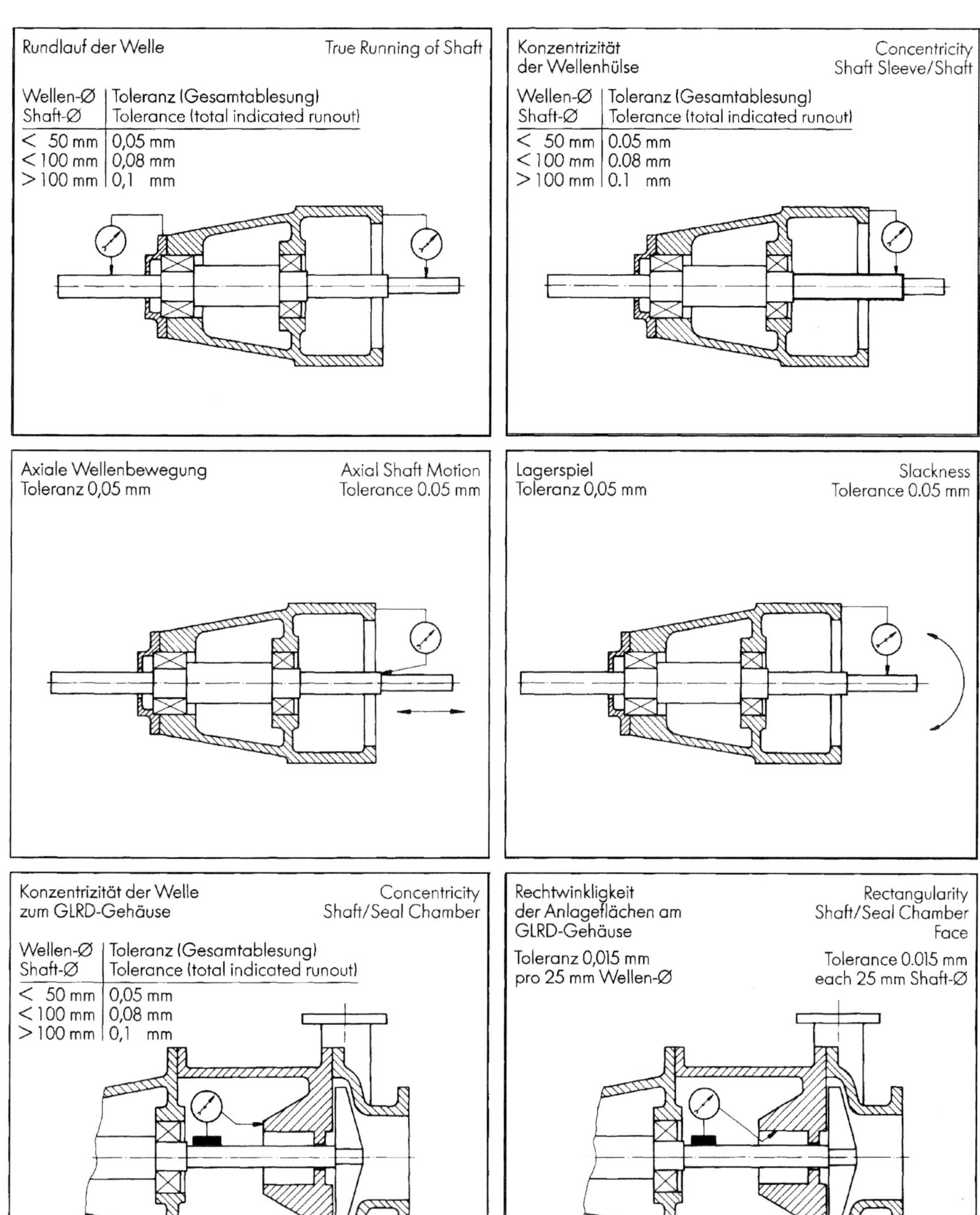

Figure 15-15: Assembly checks (Source: Pacific-Wietz, Dortmund, Germany)

PLANNER ____________
MTS CONTACT ____________
PUMP YARD NO. ____________
UNIT ____________
ITEM NO. ____________

PUMP REPAIR DATA FORM

DATE ____________
SHOP NAME ____________
FOREMAN'S NAME ____________
CRAFTSMAN'S NAME ____________
____

For additional details, use back page.

SHAFT

*I.B. refers to coupling end of pump shaft.

DIMENSION LOCATION	BEFORE	AFTER	ABNORMAL CONDITION/CORRECTIVE ACTION
O.B. Bearing Fit			
I.B.* Bearing Fit			
Impeller Fit			
Coupling Fit			
Coupling Bore			
Sleeve Fit O.B.			
Sleeve Fit I.B.			
Keyways and Threads			
Shaft Run Out			
Shoulder Run Out			
Sleeve Bore I.B.			
Sleeve Bore O.B.			
Sleeve Diameter I.B.			
Sleeve Diameter O.B.			

IMPELLER

DIMENSION LOCATION	BEFORE	AFTER	ABNORMAL CONDITION/CORRECTIVE ACTION
Impeller O.D.			
Impeller I.D.			
Hub Run Out			
Wear Ring O.D. – I.B.			
Wear Ring O.D. – O.B.			
Balance			

Figure 15-16a: Pump repair data form, part 1

BEARING HOUSING

DIMENSION LOCATION	BEFORE	AFTER	ABNORMAL CONDITION/CORRECTIVE ACTION
O.B. Bearing Housing Fit			
I.B. Bearing Housing Fit			
Depth of Thrust Bearing Fit			
Depth of Radial Bearing Fit			
Length Thrust Bearing Cap Boss			
Length Radial Bearing Cap Boss			
Bore of I.B. Bearing Cap			
Bore of O.B. Bearing Cap			
Housing to Case Alignment Fit			
Sleeve Bearing Bore I.B.			
Sleeve Bearing Bore O.B.			
O.B. Oil Ring I.D. x O.D. x W.			
I.B. Oil Ring I.D. x O.D. x W.			
Manufacturer and Bearing Number I.B. end			
Manufacturer and Bearing Number O.B. end			

SEAL AND GLAND

DIMENSION LOCATION	BEFORE	AFTER	ABNORMAL CONDITION/CORRECTIVE ACTION
Gland Alignment Boss O.D.			
Throttle Bushing Bore			
Gland Bore			
Locking Collar Bore			
Seal Ring Faces	X	X	
Stationary Ring Faces	X	X	
Seal Setting Dimension			

Figure 15-16b: Pump repair data form, part 2

CASE - HEADS - STUFFING BOX

DIMENSION LOCATION	BEFORE	AFTER	ABNORMAL CONDITION/CORRECTIVE ACTION
Case Alignment Fits			
Inner Head Alignment Fits			
Outer Head Alignment Fits			
Stuffing Box Alignment Fit			
Stuffing Box I.D.			
Case Ring Bore I.B.			
Case Ring Bore O.B.			
Case Ring Diameter I.B.			
Case Ring Diameter O.B.			
Throat Bushing Bore I.B.			
Throat Bushing Bore O.B.			
Throat Bushing Diameter I.B.			
Throat Bushing Diameter O.B.			

CAUSE OF FAILURE IF IT CAN BE DETERMINED FROM INSPECTION OF PARTS:

MISCELLANEOUS PARTS

PART/DIMENSION LOCATION	BEFORE	AFTER	ABNORMAL CONDITION/CORRECTIVE ACTION
Gland Bolts	X	X	
Oil Flinger (Deflector)	X	X	

Figure 15-16c: Pump repair data form, part 3

Figure 15-17: Checking shaft straightness
(Source: Byron Jackson Division of Flowserve Pumps, Kalamazoo, Michigan

Figure 15-18: Mechanical shaft straightening method
(Source: Byron Jackson Division of Flowserve Pumps, Kalamazoo, Michigan)

22. On packed pumps, the packing gland typically has a shaft clearance of 1/32" (0.8 mm) and a stuffing box bore clearance of 1/64" (0.4 mm)

23. On packed pumps, the lantern ring clearance to the shaft is typically 0.015"-0.020" (~0.4-0.5 mm)

On packed pumps, lantern ring clearance to stuffing box bore is typically 0.005"-0.010" (~0.1-0.2 mm)

24. Coupling-to-shaft fits are different for pumps below 400 hp (~300 kW) driver rating and pumps with drivers of 400 hp and higher: The below 400

hp pumps are often supplied with clearance fits of about 0.002." To upgrade your pumps, employ fits metal-to-metal to about 0.0005" clearance. For 400 hp and larger models, interference fits ranging from 0.0005"-0.002" (0.012-0.05 mm) are recommended and routinely implemented by reliability-focused repair facilities

Taper bore coupling and hydraulic dilation fits should be as defined by either the manufacturer or your in-plant reliability group.

25. On pumps equipped with mechanical seals, the seal gland alignment boss to stuffing box dimension should have 0.002"-0.004" (0.05-0.1 mm) clearance

26. (a) Seal gland throttle bushing to shaft clearance should typically be 0.018"-0.020," unless the pump is in "hot" service.

    (b) For "hot" service applications, a somewhat larger clearance may be specified by the manufacturer or your plant reliability group.

    (c) Seal locking collar to shaft dimensions are typically 0.002"-0.004" clearance

27. On cartridge seals, the manufacturer has set the correct seal spring compression. On non-cartridge seals—which are generally avoided by reliability-focused users—the recommended, and typical, spring compressions are:

(a) for 7/8" (22 mm) long springs—3/16" (~5 mm);
(b) for 3/4" (19 mm) medium-length springs—5/32" (4 mm); and
(c) for 1/2" (short) springs—1/16" (1.5 mm)

28. On all mechanical seals, face flatness should be 3 light bands minimum

29. Heads, suction covers, adapter pieces (if used), and bearing housing to case alignments should have 0.001"-0.004" (0.02-0.1 mm) clearance

SHOP TOOLS AND EQUIPMENT

The use of proper shop tools and shop equipment is of critical importance to reliability-focused pump users. While it is surely beyond the scope of this text to explain all shop tools and their proper use, two examples will highlight the issue.

Take, for instance, the collet driver in Figure 15-19. Unless this special tool is used for vertical turbine pumps (VTP's) equipped with tapered collets, the impeller will not be properly secured to the pump shaft.

Likewise, unless a shop uses the proper heating technique for bearings and coupling hubs, it will be near-impossible to achieve quality workmanship. A modern eddy current (induction) heater (Figure 15-20) is high on the list of necessary shop equipment.

MOVING FROM REPAIR-FOCUS TO RELIABIITY-FOCUS

This chapter would be incomplete if we neglected to recognize that in the real world environment the reliability professional is rarely in a position to implement best practices by himself. There is always a management component involved and managers often pursue short-term interests only. Short-term interests are inevitably repair-focused.

However, consistently good performance and high profitability require that industrial enterprises totally abandon their repair focus and unequivocally embrace the reliability-focused approach. Modern, reliability-focused plants must adhere to a well-formulated or even formalized management philosophy. This is an indispensable requirement if tangible and lasting equipment reliability improvement results are expected.

Adapting the thinking of W. Edwards Deming, the noted American statistician whose teachings on quality and profitability were often neglected at home, but venerated in post WW II Japan, we give the following advice to the interested manager:

- Create constancy of purpose for improvement of product, equipment and service. Implement whatever organizational setup is needed to move from being a repair-focused facility to a reliability-focused facility. Do this by teaching your reliability workforce to view every maintenance event as an opportunity to upgrade.

- Never allow costly experimentation, "re-inventing the wheel," when there is proof that a good technical text or an experienced mentor could point the way to a proven solution.

- Unless your problem pump is indeed the only one in the world delivering a particular liquid from

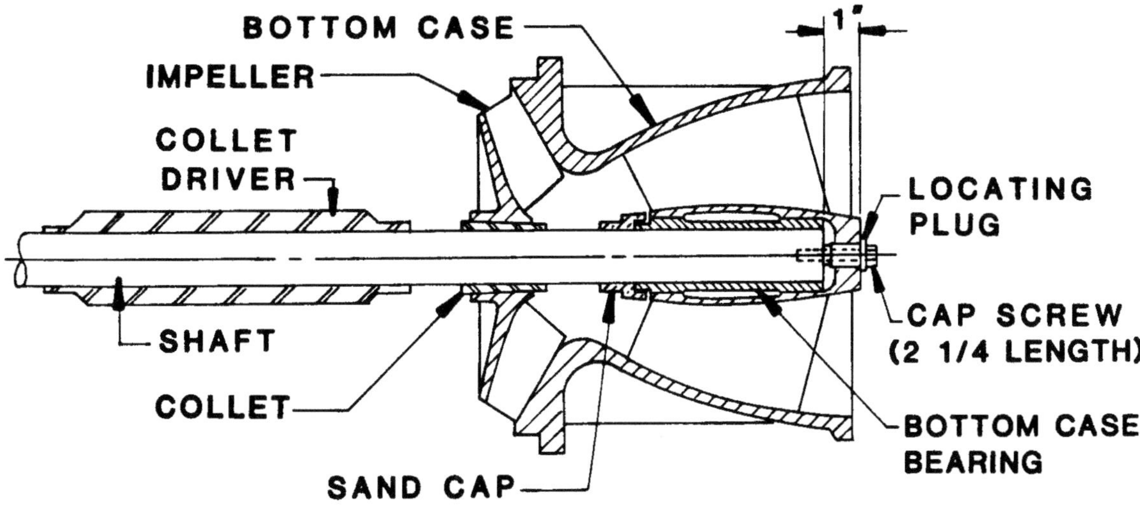

Figure 15-19: A collet driver is used to secure tapered collets to a VTP shaft (Source: Byron Jackson Division of Flowserve Pumps, Kalamazoo, Michigan)

"A" to "B," insist on determining the operating and failure experience of satisfactory pumps and mechanical seals elsewhere.

- Upgrading must result in downtime avoidance and/or maintenance cost reductions. Insist on being apprised of both feasibility and cost justification of suitable pump upgrade measures.

- Adopt a new philosophy that makes mistakes and negativism unacceptable. Ask some serious questions when a critical process pump repair isn't done right three times in a row.

- Ask the responsible worker to certify that his or her work meets the quality and accuracy requirements stipulated in your work procedures and checklists.

- End the practice of awarding business to outside shops and service providers on price alone. Ask your reliability staff to use, acquire or develop, technical specifications for critical or high reliability components. These specifications *must* be used by your purchasing department. Accept cheaper substitutes only if it can be proven that their life-cycle costs are lower than those of the high reliability components specified.

Figure 15-20: Eddy current (induction) heater (Source: Prueftechnik A.G., D-85737 Ismaning, Germany)

- Constantly and forever improve the system of maintenance, quality and responsiveness of outsourcing service providers. You must groom in-house reliability specialists competent to gage the adequacy of maintenance quality and outsourcing services.

- Insist on daily interaction of process/operating, mechanical/maintenance, and reliability/technical workforces. Institutionalize root cause failure analysis and make joint RCFA sessions mandatory for these three job functions. Do not accept this interaction to exist only in the form of e-mail!

- Institute a vigorous program of training and education. For decades, the mechanic/machinist has been allowed to find and replace the defective pump component. He or she has thus become a parts-changer. Many machinists, mechanics and technicians have been allowed to become entirely repair-focused. Train your engineers, technicians and maintenance work force members to become reliability-focused! Repair-focused plants will perish!

- Institute leadership. Give guidance and direction. Impart resourcefulness to your reliability professionals. Become that leader or appoint that leader. The leader must be in a position to outline and delineate the approach to be followed by the reliability professional in, say, achieving extended pump run length—the subject of this book!

- Drive out fear. Initiate guidance and action steps that show personal ethics and evenhandedness that will be valued and respected by your workforce.

- Break down barriers between staff areas. Never tolerate the ill-advised competition among staff groups that causes them to withhold pertinent information from each other.

- Eliminate numerical quotas. No reasonable person will be able to solve 20 elusive pump problems in a 40-hour week. If a problem is worth solving, it's worth spending time to solve the problem. Until you have groomed a competent and well-trained failure analysis team, consider engaging an outside expert on an incentive-pay basis.

- Remove barriers to pride of workmanship. Don't convey the message that jobs must be done quickly. Instead, instill the drive to do it right the first time and every time. To that end, make available the physical tools, written procedures, work process definitions and checklists used by Best-of-Class companies.

- Institute both fairness and accountability at all levels. As a shop foreman, superintendent or manager, take the lead. Eliminate roadblocks and impediments to progress. Realize what you are trying to do—increasing plant-wide pump MTBF—has long since been accomplished elsewhere. With good leadership, your organization can also achieve this goal.

COMBINING PUMP MAINTENANCE WITH SYSTEMATIC PUMP UPGRADING*

Like virtually all other industrial machines, pumps require periodic maintenance for dependable long-term operation. In fact, the very term "maintenance" is defined as keeping machines in the as-designed or as-purchased and manufactured condition. At issue is whether the pump owner's profitability objectives are best served by "maintaining only," or by judiciously combining maintenance and upgrading tasks.

There are compelling reasons to combine maintenance and upgrading. Two reasons among many are the need to optimize energy efficiency and the need to eliminate repeat failures. Repeat failures cost money, siphon off human resources and are almost always the precursor to far more serious events. The question is, however, whether an intelligent and well thought-out combination of maintenance and upgrading should be entrusted only to the original equipment manufacturer (OEM), or if qualified and highly competent non-OEMs should be considered as well.

Irrespective of OEM or non-OEM, vendor design and shop competency must be determined through diligent and repeatable audit and assessment efforts. These efforts need to be framed in written guidelines and, as will be seen, must utilize many of the elements brought out in the earlier chapters of this book. While, for ease of reading, we often only mention the non-OEM, we firmly believe that periodic assessments and competency audits apply equally to the OEM.

OEM and Non-OEM considerations

Experience shows that a highly qualified independent rebuild shop with demonstrated capabilities and experienced personnel can offer high-quality upgrades that improve both uptime and efficiency. Such a shop can do so consistent with current system performance requirements. With the considerable consolidations in the pump industry, the distinct possibility exists that the OEM is not able to offer the same engineering competence he previously had and that independent shops should be considered. This segment of our text deals with a case study and details where such upgrading was being planned, implemented, and verified to have had the desired results. It further explains the role played by competent pump rebuild shops (we chose to call them "CPRS") in these important endeavors.

Our work supports the premise that rebuilding a vintage process pump to original OEM specifications rarely makes sense given current pump rebuilding technology and changes to the system performance that occur over time. We find compelling reasons to systematically upgrade the efficiency and potential run length of small and large centrifugal pumps. For smaller pumps, this upgrading is primarily aimed at the drive

*Segment based on a co-authored presentation (ASME POWER2008-60065) by Heinz Bloch, Robert Bluse (Pump Services Consulting, Golden, CO 80401) and James Steiger (HydroAire, Inc., Chicago, IL 60607). Hydro Aire, Inc., is referenced and gratefully acknowledged as a premier CPRS with world-class standing.

end; it can be done routinely and on an attrition basis. By "attrition basis" we mean that every time a small pump enters the shop, certain routine upgrades must be performed. Consider them as failure risk reduction steps that must become routine. These routine steps certainly harmonize with the desire to view every intervention or repair event as an opportunity to upgrade. In contrast, the upgrading of larger pumps involves work on both the hydraulic-end and drive-end (power end) portions of the machine. On larger centrifugal pumps this upgrading must be pre-planned and then carried out during a future maintenance outage. Both hydraulic and mechanical (power end) upgrading make much sense and will be highlighted here.

Pump Manufacturers' Consolidations May Affect Users

Starting with the late 1980's considerable consolidation and re-shuffling has affected the pump industry. What used to be big names among the manufacturers have vanished from the scene. The situation is particularly evident in size categories around 100 kW and larger—perhaps up to 50,000 kW in high-pressure utility water services. It is in this wide range of sizes that the loss of legacy brand experience makes itself felt. Some OEM's are no longer staffed with predominantly experienced personnel and the loss has repercussions. At times the now often less knowledgeable OEM employees are ill-equipped to work with the owner-operators of these pumps. Yet, regardless of surviving OEM or newer CPRS involvement, owner/user cooperation and interaction are critically important. Essential OEM and non-OEM capabilities must be discussed and guidance mapped out. In other words, defining pre-repair and post-repair mechanical reliability and hydraulic efficiency achievements are of the essence.

Competent Pump Repair Shops Utilize Upgrade Options that Extend Uptime and Save Operating Cost

Consider a large power plant with two 10,000 kW boiler feed pumps that operate in parallel. These two pumps are scheduled for repair during the next scheduled shutdown. There may be a tendency to interpret this as traditional maintenance work that could be accomplished within the scheduled time and would cost $200,000. But what if a CPRS could be found and it could be ascertained that this CPRS could perform a combined maintenance and upgrade job? Suppose the upgrade would result in an efficiency gain of 2% and power is worth $0.07/kWh. The savings would amount to $245,280 per year. Suppose further that the CPRS would charge $300,000 to do the work; that's an incremental cost of $100,000 which would return handsomely, regardless of the precise calculation method employed.

A well-informed pump user will have captured much pertinent information on his pumps. The user will have implemented appropriate operating routines and will understand their impact on pump reliability (Ref. 15-2). Most important from repair and upgrade points of view, the reliability-focused user will have failure frequency data relating to his pumps. These data and an understanding of what caused a given pump deficiency will enable both user and competent pump rebuild shop to point out and explain, specify or recommend a number of appropriate options. Once cost-effective options are selected, the competent pump rebuild shop should be asked to implement such measures as upgrading of sensitive components, avoidance of vulnerable lubricant application methods, and so forth.

Based on understanding what failed and why it failed, a reliability-focused user will take steps to authorize and implement *routine* shop upgrades. We define as routine upgrades those done on every important pump that enters a competent repair facility. Appendix 2 lists many of these measures and action items (Refs. 15-5 and 15-6).

Routine shop upgrading measures are rarely pursued by an OEM, whereas the CPRS will be eager to explain and advocate them. There are many items or measures that could be implemented as a matter of routine; they could be obtained from this textbook. Among them are the ones restated in the remaining segment of this chapter; they are further explained in conjunction with the cited illustrations:

1. Because the *unbalanced* constant level lubricator will not maintain proper oil level when housing-internal pressure differs from that of the surrounding atmosphere, the unbalanced model is discarded and a balanced model incorporating a sight glass is installed. The balance line is routed to the top of the bearing housing at the former location of the housing vent; the vented "breather" is now discarded (Figure 9-19).

2. The new *balanced* constant level lubricator is always mounted on the "up-arrow" side shown in the vendor literature (Figure 9-18). With this (proper) mounting direction, oil levels will be more consistently maintained.

3. Whenever possible, oil rings are being replaced with suitable flinger discs (Figure 9-26). The CPRS knows that, unless a shaft system is truly horizontal, oil rings often "run downhill" and abrade upon making contact with adjacent components (Figure 9-27). Also, unless they are truly concentric and immersed to just the right depth in a lubricant with narrowly maintained viscosity, oil rings may not perform as expected. Flinger discs have a metal hub and are set-screwed or suitably fastened to the shaft. Sometimes, the actual disc is made of a suitable elastomer (Figure 9-25) or a flexible metal; in either case its lowermost 3/8-inch portion is immersed in the lube oil. To be considered suitable, the manufacturer-endorsed peripheral speed limitation must be observed.

4. On larger bearings and in installations where circulating lube oil is preferred but "unaffordable," the CPRS may advocate conversion to direct oil spray or oil jet lubrication (Figure 9-15) via an external pump (Figure 9-29). This can be done with a setup that draws oil from the bearing housing oil sump and then forces this oil into spray nozzles (Figure 9-30).

5. Pumps with dry sump oil mist previously applied at the center of the bearing housing are being modified so as to apply oil mist per API-610 8th and later Editions, i.e. between the bearing protector seal and the bearing (Figure 9-11).

6. Unless shaft surface speeds exceed 10 m/s (~2,000 ft/min), certain pumps and small steam turbines are being retrofitted with dual-face magnetic bearing housing seals (Figure 6-11). Once converted, the bearing housing is now quasi hermetically sealed—nothing goes in or out. The bearing housing end cap is painted with white spray paint so that any (highly unlikely) oil leakage will show up easily.

7. Modern rotating labyrinth bearing protector seals are installed, unless dual-face magnetic seals are both available and cost-justified. Carefully note that advanced bearing housing protector seals will not allow dynamic O-rings to make shearing contact with the edges of O-ring grooves. To enhance stability, the rotors of advanced bearing housing protector seals are clamped to the shaft with dual O-rings, Figure 9-36.

8. Unless oil rings ("slinger rings") are used, in which case a thinner oil may be needed, the CPRS may explain where ISO Grade 68 diester or PAO (polyalpha-olefin) synthetic lubricants should be considered (Table 7-2 and Appendix 2). An aluminum or stainless steel label stating oil type is affixed to the top of pumps so converted.

9. Cooling water is removed from all centrifugal pumps with rolling element bearings. If desired by the client, a CPRS can provide written explanations why such pumps achieve longer bearing life (Appendix 2).

10. The shaft interference fit for back-to-back angular contact bearings is carefully measured and verified not to exceed 0.0003" on shafts up to and including 80 mm diameter. The resulting reduction in bearing operating temperature is then explained. Liberal use is made of the checklists found in Chapter 15 and the CPRS utilizes supplementary checklists as well. All applicable checklists are freely explained to the client.

Of course, pump repair and rebuilding efforts often go beyond just the routines that we have described above. Repair scopes differ from pump to pump and must be defined if the highly desirable goals of uptime extension and failure risk reduction are to be achieved.

Competent Pump Repair Shop Facility Assessment

Let us assume the reader or a responsible reliability professional represents a plant seeking a capable alliance partner willing to maintain, repair and upgrade the facility's pumps. The first order of business would be to engage in a well-focused assessment.

There are many formats that will allow pump user-owners to gauge or assess the competence of any repair facility. Indeed, true reliability engineering includes making an assessment of potential bidders for both new and old equipment (Refs. 15-3 & 15-4). One such format starts with a general listing of items, names, and similar logistical and general information. It progresses to specialization reviews and 60 or more additional questions that will be of real interest. It investigates

- In what types of equipment repairs does this shop specialize?
- With which OEM pumps and models do they have experience?
- What is this facility's annual revenue stream?

- What is the annual employee turnover rate?
- What sort of technical training is available to the co-workers?
- Does this shop have training records and where are they located?
- What are plans to continue to keep and attract qualified co-workers?
- How many shifts do they operate?
- How easy is it to switch to 24-hour emergency coverage, if required?
- What is the education and discipline of the engineers? (Mechanical, hydraulic, or other discipline)

These data entries might represent the obvious, but there's much more to it. For instance, to make a relevant assessment, the experience background of the CPRS employees needs to be reviewed. Some of must be explored in personal interviews that take a bit of time. It will be worth the effort!

Human Assets and Their Experience Levels

- Description of Position--Years of Experience--Comments
- Manager(s)
- Shop Superintendent(s)
- Repair Coordinator(s)
- Foremen
- Buyer(s)
- Quality Assurance and Control Inspectors
- Machinists, Mechanics, Welders
- Field Service People, Engineers, Sales Persons

A typical checklist for reviewing CPRS personnel and their experience qualifications

Know the CPRS's Customer Base and Their Satisfaction

Other elements deserving of review deal with the CPRS facility's customer base and their satisfaction. During an evaluation, the user-owner may wish to explore—

- Who are this shop's top ten customers?
- What markets do they represent? (e.g. refining, pipeline, power, other)
- How long have these top ten customers been among the top 10? (ask for explanation of variances)
- How does the CPRS measure customer satisfaction? Is it transparent or called "highly confidential"?
- Does the CPRS measure productivity, safety, and quality, and are the charts visible?
- What is their rework/scrap rate in terms of percentage of total sales?
- Does the CPRS document NCR's (non-conformance reports)? What does the CPRS do with these NCR's?
- Does this CPRS have a process for continuous improvement to reduce rework?

Compare the Safety, Environment, Cleanliness, Order and Capability

Virtually any comparison can again be done in a suitable tabular format (Table 15-1) and here is just one of many possibilities:

- What is the square footage (floor space) of the shop?
- What is the square footage of the office?
- Is the air quality good? Is air circulation sufficient?
- Is the production area adequately lit?
- What safety programs are in place?
- Does the CPRS have a safety manager on site? (Explain). Who is responsible for safety?
- What is the OSHA Recordable Injury Incident rate achieved by this CPRS?
- What is the general state of cleanliness in the shop?
- Do tools, equipment, etc., have their own place and is everything stored in its proper place?
- Are there preventive maintenance (PM) records of shop machinery and who is performing these PMs?
- What is the condition of the CPRS's major machine tools? How old are they?

Understand the Scheduling and Visual Management Systems

It is intuitively evident that planning and scheduling efforts need to have focus. To flush out vulnerabilities, we might ask:

- What electronic production scheduling tools are in place? Are they being used?
- Who has schedule responsibility? How often are production meetings taking place and who attends these?
- What is the CPRS's on-time delivery performance? How is this measured?
- What happens when a delivery is in jeopardy?

Table 15-1: Typical comparison chart layout

Machine Tool(s)	No. at this Site	Size / Capability	Condition
Manual Lathes			
Horizontal Boring Mill(s)			
Vertical Boring Mill(s)			
Bridgeport-Type Milling Machines			
CNC Equipment			
Welding Equipment			
Balance Machine(s)			
Hydraulic Torque Winch(es)			
Hardness Tester(s)			
Cleaning Capability			
Hydrostatic Test Gear			
Heat Treat Capability			
Micrometers			
Contour Measuring Machines			
NDE Capability			
Overhead Crane(s)			
Forklifts			

What is the process used to notify the customer and to improve schedule?

- What is the repair process flow? Is it visible? Are shop co-workers trained on their respective roles and responsibilities?
- Are there any "bottlenecks" or excess work in process at any one machine or work station?
- Are there computer terminals on the shop floor that feed the scheduling system?
- Is the plant laid out in a continuous flow or does work in process travel back and forth from workstation to workstation?
- What is the level of communication on the shop floor?
- Are shop travelers/routers used and are they signed off at required checkpoints?

Quality Assurance and Quality Control Systems

Here are some worthy questions that are part of the Q/A and Q/C assessment:

- What quality certifications does this shop have? (ISO, Mil Spec)
- Is there a vision and mission statement? Is it visible and displayed in the facility?
- Is there a designated quality manager and to whom does that person report?
- What systems/processes are in place to ensure the requirements are clearly defined and adhered to?
- Are non-conformance reports written and what is the process to ensure no further non-conformances are likely to occur?
- How is quality measured and are there charts/graphs that are visible in the facility?
- What is the process to communicate special requirements to ensure they are incorporated into the repair process?
- Does this facility have an effective "Root Cause Failure Analysis" process and how does it work?

Documentation Management

- Does this facility have standardized receiving inspections? As-found reports? As-built reports? Balance reports? Repair process and flow charts?
- Does this facility have a digital camera and software to include pictures on the repair reports?
- How long does it take to complete an inspection and as-found report? Repair quotation? How long does it take to complete the final as-built report after the repair process is completed?
- What is the preferred method to communicate/transmit documentation packages? (electronic, paper, other)
- Are recommended upgrades (hydraulic, mechanical, etc.) well defined and is an ROI (return-on-investment) calculation used to determine payback?
- What is the flow process for engineering reviews and work scope requirements?
- At what point are these communicated to the customer?

Outside/Procured Services Must be Defined

- How does the repair shop manufacture parts? What is the process used to ensure dimensional and metallurgical conformance?
- Where does the shop procure its castings? What process is being followed to procure these castings?
- What other services does the repair shop contract out? (Investigate heat treat, metal spray, chrome plating, heat treating, non-destructive examination or NDE)
- Have these suppliers been surveyed by the repair shop to ensure they will provide conformance to the specification?
- Who has the responsibility for final QA/QC of materials and services procured from third parties or outside vendors?
- How long has the relationship existed between the outside supplier and the repair shop for each service?
- What has been the historical quality and delivery performance of the outside shops? How is this measured and what records are kept?

CPRS Assessment Scoring Matrix

Once an assessment is made, it is important that each surveyed category is measured. Follow-up is needed to ensure that conformance criteria are met. Any categories found to be unacceptable need to be revisited and changes made to bring them up to either full conformance or a defined level of acceptability. A scoring matrix (Table 15-2) will help.

Needless to say, the scoring matrix can be expanded to include other items of interest. The "comments" segment lends itself to cataloging highly detailed information.

How the CPRS Assists in Pump Repair Scope Definition

The competent pump rebuild shop ("CPRS") has both the tools and the experience needed to define a work scope beyond the routine upgrading we spelled out above. The CPRS takes a lead role in defining the repair scope and all parties realize that reasonably accurate definitions will be possible only after first making a thorough "Incoming Inspection." On a written form or document, on both paper and in the computer memory, the owner-customer, manufacturer, pump type, model designation, plant location, service, direction of rotation and other data of interest are logged in, together with operating and performance data. The main effort goes into describing the general condition of a pump and this effort might be followed by a more detailed description of the work. Either way, it constitutes the condition review.

Condition reviews include photos of the as-received equipment and close-up photos of parts and components of special interest. End floats, lifts and other detailed measurements are taken and recorded on a dimensional log sheet both before and after total dismantling. Components are marked or labeled, and hardware is counted and cataloged. Bearings, bushings and impellers are removed. Bead blasting, steam or other cleaning methods are listed and a completion date for these preliminary steps is agreed upon. It should be noted that only now would a competent shop consider it time to arrive at the next phase in more closely defining the scope of its repair and upgrading efforts. An example of combining repairs and upgrading might deal with wear materials.

Table 15-2: CPRS scoring matrix

Category	Does not meet requirements	Meets requirements	Exceeds requirements	Comments
Human assets and experience				
Customer base and satisfaction				
Safety, environment, capability				
Scheduling system, visual flow				
Quality and quality systems				
Documentation management				
Outside / procured services				

An Upgrade Example: Wear Materials for Improved Energy Efficiency

As discussed in Chapter 6, fluid processing industries have embraced the use of current generation composite materials (Figure 6-28) in centrifugal pumps to increase efficiency, improve MTBR (mean-time-between-repairs), and reduce repair costs. One such material that has been used successfully by major refineries is a proprietary reinforced carbon fiber fluoropolymer resin, Vespel-6100® (Refs. 15-7 to 15-10). It is a composite material with uniquely low coefficient of expansion and superior temperature stability. As was also discussed in Chapter 6 and Refs. 6-8 and 6-13, the Vespel polymer composite has replaced traditional metal and previous generation composite materials in pump wear rings, throat bushings, line shaft bearings, inter-stage bushings, and pressure reducing bushings. Although not acceptable in abrasive-containing services, the properties of this polymeric material eliminate pump seizures and allow internal rotating-to-stationary part clearances to be reduced by 50% or more. Composite wear materials are included in the 9th (2003) and later editions of the American Petroleum Institute's Centrifugal Pump Standard, widely known as API-610.

Knowledge of this low-expanding, high temperature capability material is a must for CPRS facilities. Correctly applied, these proprietary materials have proven to eliminate pump seizures, provide dry-running capability, and greatly reduce the severity of damage from wear ring contact. Using Vespel® the risk of damaging expensive parts in relatively clean pumping services up to about 450 °F is certainly reduced and a CPRS must be able to assist in calculating the merits of this upgrading approach. The value of efficiency gains in pumps is of great importance and is a criterion for the label CPRS. A CPRS will verify that axial (thrust) loads on pump bearings are still within limits after reducing internal clearances.

Recommended Radial Gap Guidelines for Pumps

It has been observed that not every OEM delivers pumps with internal clearances designed and manufactured for optimized hydraulic and mechanical performance. Optimization of both may result in giving up a small gain in efficiency.

From an overall reliability point of view, adhering to the radial clearance criteria referred to as Gap "A" and Gap "B" (Figure 14-4) is very important. Again, so as to assess if the proposed shop qualifies as a CPRS, ask about impeller design and manufacturing practices. It should be noted that, if the number of impeller vanes and the number of diffuser/volute vanes are both even, the radial gap must be considerably larger, say 10 percent minimum.

The repair facility must know about steps to maximize impeller performance:

1. Bring the impeller middle shroud plate out to the impeller O.D. to reinforce the impeller structure.
2. Stagger the right and left side of the vane to reduce hydraulic shocks and alter the vane-passing frequency.
3. Reduce clearances to optimum between shrouds and casing (Gap " A")
4. Avoid even number of impeller vanes for double volute; or, if the diffuser vanes are even numbered, increase the impeller sidewall thickness.
5. Impellers manufactured with blunt vane tips can also cause trouble by generating hydraulic "hammer" even when the impeller O.D. is the correct distance from the cutwater. The blunt tips cause disturbance in the volute. This effect may be partly or entirely eliminated by tapering the vanes by "overfiling," or "underfiling" the trailing edge, as described in Figure 4-29.

Case History: An Ingersoll-Rand Model 6CHT 9-stage Boiler Feedwater Pump

The importance of a CPRS's knowing these guidelines is best explained by an example. We chose a case which involved a 9-stage boiler feedwater pump that had been previously repaired and was subsequently shipped and reinstalled at the owner's site. When the field installation crew was ready to hand the pump over to operations, the rotor would not turn and there was zero axial float. The owners lost little time shipping the critically important pump to the CPRS.

When the pump reached the CPRS's workshop, a work scope had to be developed on an expedited basis. The CPRS first disassembled and inspected the components. It was immediately realized that the channel rings had to be investigated and mapped. Typical mapping data were captured in a number of computer-generated sketches; Figures 15-21 and 15-22 serve as typical examples.

Both figures are part of the action sequence that followed this mapping.

1. Set up each channel ring and perform a T.I.R. (total indicator reading) inspection
2. Remove the old alignment ring from stages 7, 8 and the last stage. Manufacture three new align-

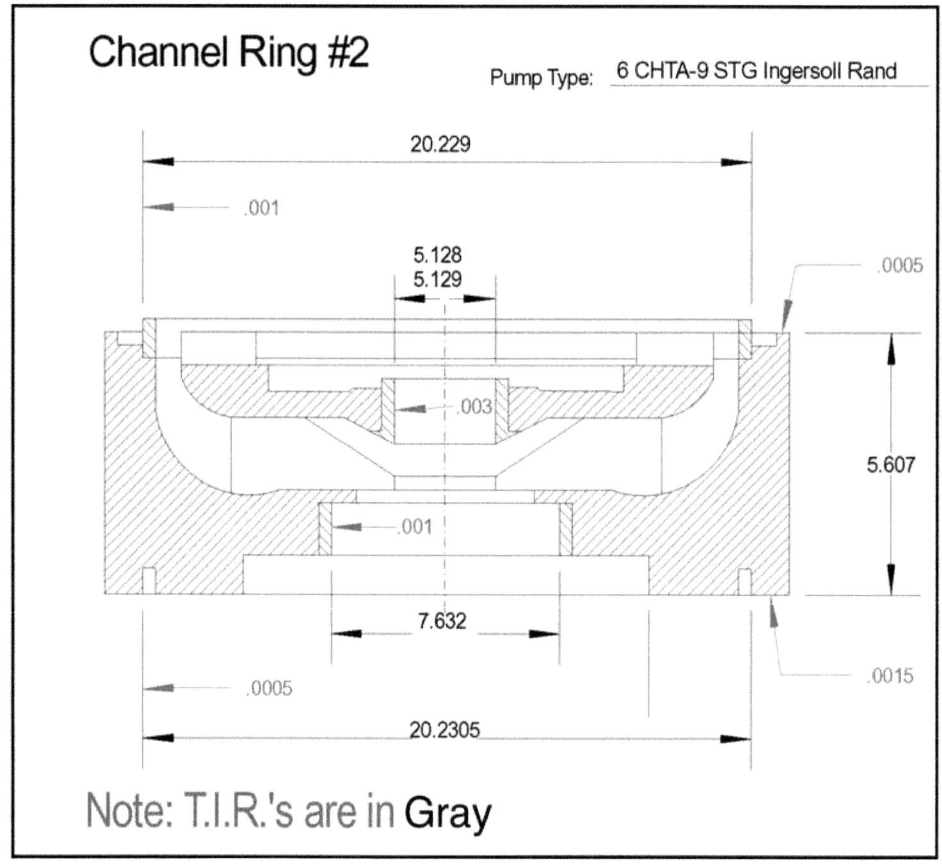

Figure 15-21: Dimensional mapping of a channel ring as carried out by a CPRS. This valuable recording of data is typically done as part of repair scope definition

ment rings.

3. Install the new alignment rings.
4. Set up and finish machine the inter-stage fit diameters to obtain the proper fit up.
5. De-burr the channel rings.
6. Made note that, on the last stage channel ring, the 8th stage vane tips were protruding into the impeller diameter area.

The CPRS knew that protruding vane tips (Figure 15-23) are unacceptable and must have caused interference contact. Indeed, that correlated well with the owner-operators complaint after the pump had been repaired by "others" and it was realized that the rotor could not be moved after the initial, defective, repair.

Note that the CPRS recorded the "as found" and "as repaired" conditions. The vane contour was cut back and ground the to the clearance dimensions mentioned in the Gap "A" and Gap "B" table given above. The element was assembled, using all the existing components except for the gaskets. On the discharge head, the CPRS set up and performed a T.I.R. (total indicator reading) inspection of the critical diameters and faces. New gaskets were supplied for the element assembly and preparations made for quick shipment to the owner-operator.

This CPRS quoted a delivery estimate of 2-3 days and asked for payments to be effected net 15 days, F.O.B. its Chicago facility. Upon completion of the project, the CPRS offered to submit a final report showing documentation in accordance with its quality assurance (Q/A) program. This would include, as applicable, test reports for:

- Material and heat treating certification
- Concentricity reports of sub-assemblies
- In-process dimensions and N.D.E. (non-destructive examination) reports
- Rotor balance reports
- Reports of final internal clearances and rotor movements

The client agreed with the proposal and the CPRS proceeded with the repairs. Moreover, the CPRS re-

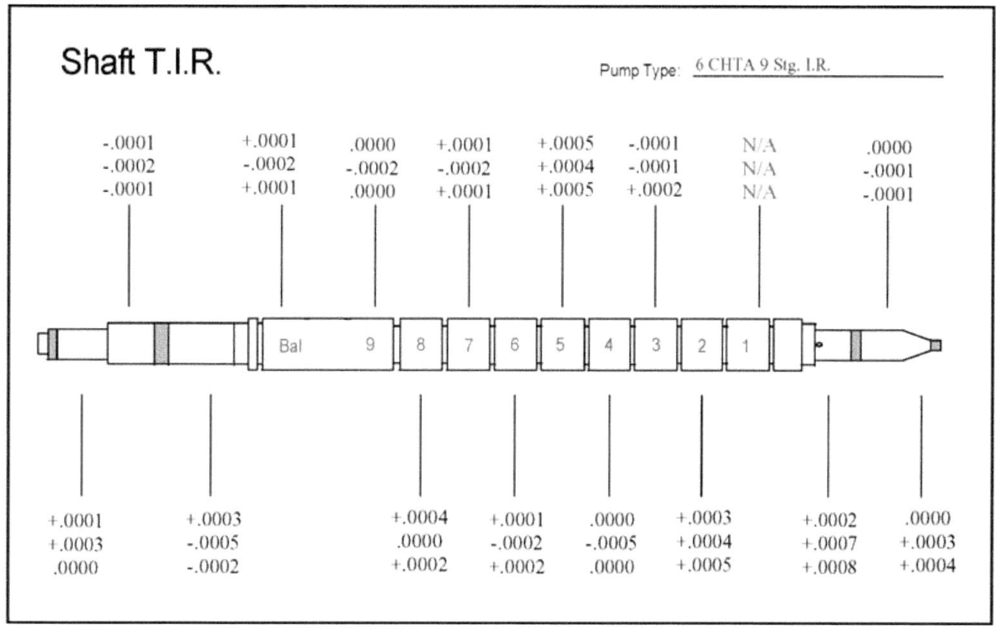

Figure 15-22: Shaft dimensional mapping performed as part of receiving inspection and pump repair scope definition

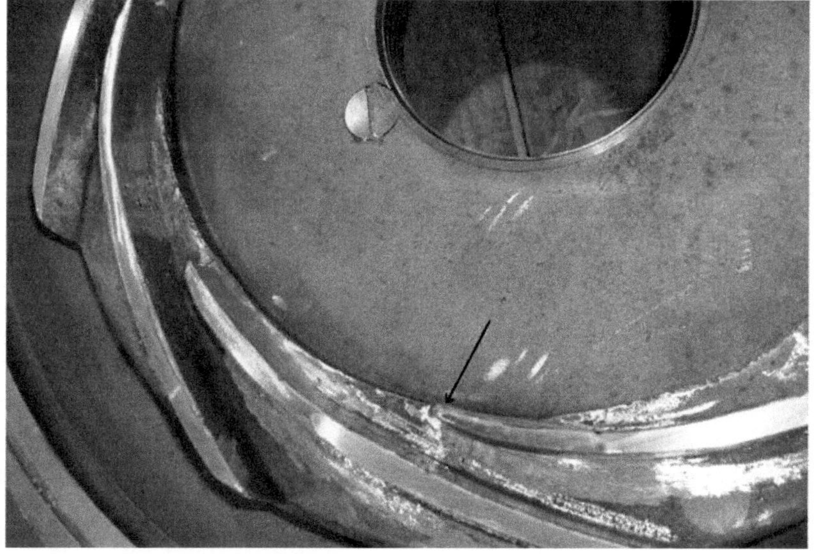

Figure 15-23: Protruding diffuser vane tips discovered during pump dismantling and repair scope definition

affirmed its standard policy to warrantee refurbished equipment and parts of its own manufacture against defects in material and workmanship under normal use. A typical warrantee period for parts and services is one (1) year from the date of initial startup, but not exceeding five (5) years from date of shipment, provided that final alignment, lifts and floats are witnessed by a CPRS Service Technician. It is only fitting that any deviation from the stated policy must be authorized in writing by the CPRS.

Shop Repair Procedures and Restoration Guidelines are Needed—Even by the Best

Pump manufacturers usually supply pump maintenance manuals with detailed assembly and disassembly instructions that are either generic, or specific to a particular pump style and model. A number of important checks should be performed by the CPRS for users whose serious goal it is to systematically eradicate failure risk. Both CPRS and user have responsibilities in ascertaining that all quality checks are performed with due diligence.

Experience shows that after years of repairs, many pumps are due for a series of comprehensive dimensional and assembly-related checks. As a minimum, every critical pump deserves much scrutiny. After the well-known dial indicator checks are complete, the dimensional "before vs. after" findings should be recorded in either the (preferred) electronic, or, as a minimum, paper format. Users that do not take time to record these pump repair data will find it very difficult to reach their desired failure reduction objectives.

Understand Spare Parts Availability and Procurement Decisions

At the beginning of this chapter, we had listed factors to consider when making the decision to buy OEM or non-OEM spare parts. When working with a CPRS, this decision can be left to the CPRS. A precondition is warranty coverage. The CPRS must warranty all workmanship and parts in the same manner that an OEM would.

The objection that non-OEM parts might not meet OEM performance specifications deserves to be questioned. Note that the CPRS may often be in a position to improve this original performance and will give proof of such claims. As to the risks incurred when non-OEM parts are copied from worn OEM parts, remember the meaning or definition of CPRS. Such an entity is aware of these risks and will use its pump design and manufacturing experience to avoid these issues.

While it is true that non-OEM's generally do not have R&D facilities, realistically, the CPRS does not tread into R&D territory and makes no claims to provide warranties on unproven solutions. Still, the design experience and response time of a CPRS can be superior to that of the OEM. Also, while a non-OEM's metallurgies and material mechanical properties may, in the past, not have met OEM specifications, this is not true of an experienced CPRS. Moreover, while many parts-oriented non-OEM's sometimes offered only the most popular repair parts, this is no longer an issue with a good CPRS; good CPRS providers are not "parts-oriented."

Consult Repair and Restoration Guidelines before Deciding Where to Purchase

Using the CPRS assessment scoring matrix discussed earlier in this chapter, a CPRS will have been pre-selected before a repair incident arises. Still, when the repair-and-upgrade opportunity presents itself, it is incumbent upon both parties to agree on work scope and critical repair part sourcing decisions. In other words, at that time more definitive repair and restoration guidelines should be consulted in a meeting of user-owner and CPRS representatives. Some of the data provided earlier will have been rolled into a good CPRS's operating mode, or reflect in the manner the CPRS is conducting his business.

Jointly with the CPRS, the user makes critically important repair part sourcing decisions. The right choice will lower overall maintenance costs by improving equipment MTBF. The ultimate effect will be reduced life cycle costs for the pump.

Since the CPRS performs all-encompassing upgrade work, its personnel will, among other tasks, analyze a number of hydraulic issues and do the following:

- Calculate suction specific speed
- Calculate suction energy
- Calculate NPSH margin
- Compare susceptibility to internal recirculation-induced damage
- Verify that radial clearances comply with hydraulic shock avoidance criteria for high (discharge) energy pumps.
- Accept that the Hydraulic Institute considers as high energy pumps ones with specific speed values below 1,300 that generate more than 900 feet of head per stage.
- Realize that the differential head value for high energy drops from 900 feet at a specific speed of 1,300, to only 200 feet at a specific speed of 2,700.
- Consider both impeller and diffuser gap "A" and "B" modifications mentioned earlier in the CPRS reporting documents, also explained in Ref. 15-5. These gap values were based on Dr. Elemer Mackay's widely reported groundbreaking work with EPRI, the Electric Power Research Institute. At least one CPRS makes strong use of this body of work.

Quite obviously the CPRS must have troubleshooting know-how and must willingly use it to solve the owner-operator's problems. There are many good texts that explain pump troubleshooting; dozens are listed in the various references to this chapter. One powerful reason is traced to system-related deficiencies that need to be corrected and these are intermixed with the other items. It makes much sense to work with a CPRS that will assist in delineating and explaining all needed corrections. These might well include corrections that must be addressed *in spite of falling outside of the CPRS's work scope.* Recall that our text gave checklists on, say, dimensional deviations, or procedural oversights that might cause a pump to produce insufficient flow. A qualified CPRS should be able to point to those or similarly detailed checklists that will be used to assist the client in troubleshooting a repair. In essence, the CPRS becomes the pump user's resource. In turn, the pump user understands that the CPRS performs a premium repair that is well worth premium pricing. Explaining how to extend pump life and avoid future repairs is worth money and to even remotely assume that this cost-saving information will be provided by the lowest bidder is an unrealistic expectation.

In conclusion, the CPRS might tutor the client in looking for tell-tale signs of problems on each component and offer guidance on the applicability or availability of superior repair and upgrading techniques. Rebuilding a vintage process pump to original OEM specifications often makes no sense given current pump rebuilding technology and changes to the system performance that occur over time. A highly qualified independent rebuild shop with guidelines, checklists, procedures and a willingness to cooperate with the owner-user is called a CPRS. Its competent and experienced personnel can verifiably offer high-quality upgrades that improve both uptime and efficiency consistent with current system performance requirements.

With the considerable consolidations in the pump industry, the distinct possibility exists that the OEM is not able to offer the same engineering competence he previously had. If one simply makes these statements, the issues can be debated for a long time. However, assessing vendor competence and making comparisons using the various points made in this segment of our text tend to become more objective. To the pump user, such assessments may be worth a small fortune in maintenance cost avoidance, more efficient operation, and pump run length extension.

Chapter 16

*Failure Statistics and Component Uptime Improvement Summary**

The majority of our readers no doubt examined this text because they realize that pumps are critical and essential fluid movers. In the process industries, pump failures can have all kinds of consequences in terms of cost and safety. While pump uptime is often linked to systems parameters and hydraulic performance, pump reliability is also dependent on the performance of such important secondary components as seals, couplings and bearings. This is why our final chapter looks at the influence of secondary components on overall pump reliability, sets targets for life times, and summarizes ways of improving the performance of those components. Fittingly, the chapter concludes with case histories that show real improvements in practice. It was contributed and updated (in 2005) by John Crane, Slough, UK.

PUMP MTBF AS A FUNCTION OF COMPONENT MTBF

In addressing the issue of mechanical reliability in mechanical systems, it is important to understand some of the terminology and recognize some of the pitfalls in the application of reliability measures.

Mechanical reliability is the probability that a component, device or system will perform its prescribed duty without failure for a given time when operated correctly in a specified environment.

Component MTBF, or mean time between component failures, is defined as:

M = [(Total number of components)/(Total number of failures)] x elapsed time

For example, if 10 in 1,000 components fail within 5,000 hours:

M = (1,000/10) x 5,000 = 500,000 hours

When applying MTBF to pumps, seals and other components, much care is required. Classically, MTBF is applicable to constant failure rate situations, whereas, in reality, failures occur more in line with a different set of rules. This is approximated in the so-called "Bathtub Curve" (Figure 16-1).

The significance of the curve can easily be understood by referring to an automobile tire. If the tire fails quickly due to a manufacturing defect or damage while being mounted on a wheel, that is an "infant mortality." If it fails due to a chance encounter with a nail in the road, that is rated a "chance failure." If the tire survives long enough to lose all of its useful tread, then that is a "wear-out" failure and the lifetime of the tire will have been maximized. It is a clear objective that virtually all pump failures should be of the "wear-out" type.

In the real world, failures in a given situation tend to be a mixture of the three types so care is required in

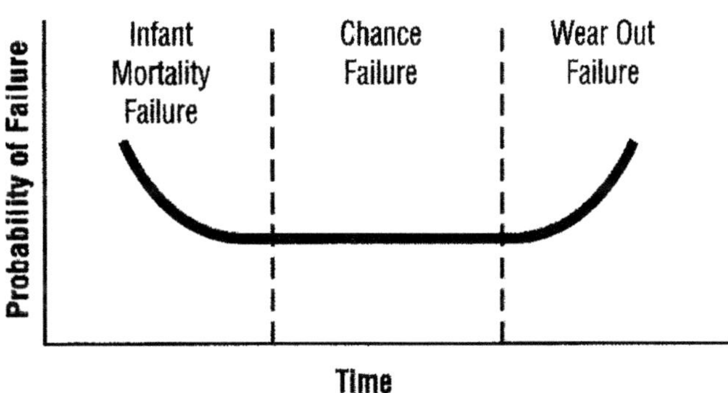

Figure 16-1: The "bathtub" curve

*Condensed and adapted, by permission, from Neil M. Wallace, John G. Evans & Peter E. Bowden's presentation "Improving Pump Reliability from its Secondary Components." Reproduced with permission of the Turbomachinery Laboratory (*http://turbolab.tamu.edu*). From Proceedings of the Seventeenth International Pump Users Symposium, Turbomachinery Laboratory, Texas A&M University, College Station, Texas, pp. 171-186, Copyright 2000

the interpretation of MTBF data. Much can be gained by analyzing and understanding which types of failure are actually taking place by reference to the "hazard rate function."

The hazard rate function z(t) is a measure of the probability that a component will fail in the next time interval given that it has survived to the beginning of that time interval. In the infant mortality area, z(t) is decreasing; in the "chance failure" area, it is constant; and in the "wear out" area, it is increasing.

In looking at measuring MTBF of pumps and seals at a given plant, there are complications in that some pumps are spared (typically 60% in a refinery) and also because some pumps have different numbers of seals. It is, therefore, error-prone to make too many assumptions.

EXAMINING PUMP REPAIR RECORDS AND MTBF

In the preface to this text, we had already alluded to pump failure statistics. These failure statistics are often translated into MTBF. For what it's worth and so as not to get enmeshed in arguments, many of the best practices plants in the time period of the early 2000's simply took all their installed pumps, divided this number by the number of repair incidents, and multiplied it by the time period being observed. For a well-managed and reasonably reliability-focused U.S. refinery with 1,200 installed pumps and 156 repair incidents in one year, the MTBF would be (1,200/156) =7.7 years. The refinery would count as a repair incident the replacement of any parts—including coupling parts—regardless of cost. In this case, a drain plug worth $1.70 or an impeller costing $5,000 would show up the same way on the MTBF statistics. Only oil replacement or oil changes would not be counted as repairs.

The total repair cost for pumps at a best-practices plant would include all direct labor, materials, indirect labor and overhead, administration cost, the cost of labor to procure parts, and even the prorated cost of pump-related fire incidents. This was discussed in the lead pages of Chapter 14; it merits being mentioned again in conjunction with the stated average cost of pump repairs: $10,287 in 1984 and $11,000 in 2005. We believe this indicates, in relative terms, a repair cost reduction, because a 2005-dollar bought considerably less than the 1984-dollar. It can be reasoned that predictive maintenance and similar monitoring having led to a trend towards reduced failure severity.

Using the same bare-bones measurement strategy and from published data (Refs. 16-1 through 16-7), as well as observations made in the course of performing maintenance effectiveness studies and reliability audits in the late 1990's and early 2000's, the mean-times-between-failures of Table 16-1 have been estimated.

GORDON BUCK'S STUDY OF PUMP STATISTICS

In early 2005, Gordon Buck, the John Crane Company's Chief Engineer for Field Operations in Baton Rouge, Louisiana examined the repair records for a number of refinery and chemical plants to obtain meaningful reliability data for centrifugal pumps. A total of 15 operating plants having nearly 15,000 pumps were included in the survey. The smallest of these plants had about 100 pumps; several plants had over 2,000 pumps. All plants were located in the United States.* Also, all

Table 16-1: Pump Mean-Times-Between-Failures

• ANSI pumps, average, USA:	2.5 years
• ANSI/ISO pumps average, Scandinavian P&P plants:	3.5 years
• API pumps, average, USA:	5.5 years
• API pumps, average, Western Europe:	6.1 years
• API pumps, repair-focused refinery, developing country:	1.6 years
• API pumps, Caribbean region,	3.9 years
• API pumps, best-of-class, U.S. Refinery, California:	9.2 years
• All pumps, best-of-class petrochemical plant, USA (Texas):	10.1 years
• All pumps, major petrochemical company, USA (Texas):	7.5 years

plants had some sort of pump reliability program underway. Some of these programs could be considered as "new"; others as "renewed" and still others as "established". Many of these plants, but not all, have an alliance contract with John Crane. In some cases, the alliance contract included having a John Crane technician or engineer on-site to coordinate various aspects of the program.

The pumps were mostly a mixture of API and ANSI designs. All editions of API 610 are represented but most of the pumps were probably 5th or 6th edition designs. The vast majority of the pumps were single stage, overhung process pumps but virtually every pump type was probably represented in the database. The exact distribution of pump types, designs and specifications is not known.

Single, dual non-pressurized and dual pressurized seal arrangements are included in the data. The vast majority are single seals. Although it is safe to assume that the newer pumps use cartridge seals, many pumps, old and new, use component seals. Pusher and non-pusher (bellows) seal types are included in the data.

MTBR was calculated based on the total pump population as well as on a plant basis. The exact definition of a "failure" or "repair" varied somewhat; however, refineries and chemical plants were selected for the database on the basis of having reasonably well defined programs. In general, the definition for MTBR is based on Process Industry Practices, IPI REEE002, March 1998.

Several different databases were used to compile these statistics. The distribution of pump and seal design as well as application details can be estimated as follows: About 60 to 70% of the pumps are based on API 610 specifications. Approximately 60 to 70% of the seals are single seals; 10 to 20% are dual non-pressurized seals and 20 to 30% are dual pressurized seals. Whether in an API or ANSI pump, about 55 to 65% are cartridge arrangements. With respect to operating conditions, 85 to 95% of the sealing pressures are less than 300 psig and less than 400 F. With respect to shaft speed, 55 to 65% are at 3600 rpm and 35 to 45% at 1,800 rpm.

Most of the refinery pumps in this survey are at least 30 years old and many are older. In fact, since many of these pumps are based on API 610, 5th and 6th editions, their current owners would not purchase them in 2005.

*Footnote: Material to next subheading was contributed by Gordon Buck and is reproduced by permission

The ANSI pumps in the survey were somewhat newer than the API pumps; however, ANSI designs have also improved in recent years. In particular, many of the ANSI pumps in the survey are believed to have small cross-section seal chambers whereas API 682 requires enlarged seal chambers.

If the aggregate of all pumps and failures is considered as one large mixed-process plant, the gross average MTBR is 45.2 months. The aggregate of all refinery pumps and failures is 49.0 months MTBR. The aggregate of all chemical plant pumps and failures is 41.3 months MTBR.

Another way of looking at the data is to take the average of the MTBRs for each plant; that is, to average the averages. In that case, the average of the refinery MTBRs is 51.8 months with a range of 40.2 to 61.1 months; the standard deviation is 8.2 months. The average of the chemical plant MTBRs is 45.4 months ranging from 19.4 to 69.0 months; the standard deviation is 14.6 months.

Although many refineries and chemical plants would be pleased to have these MTBR averages, these values are actually lower than expected and certainly lower than is possible. As noted previously, many of these plants have relatively new alliance contracts with John Crane with the goal of improving MTBR. A review of these contracts shows that MTBR is indeed improving; however, the reported averages are valid for this point in time.

It can be argued that a reasonable pump MTBR goal for an existing refinery following established "best practices" is about five years. A new refinery should be able to attain six, perhaps even eight, years MTBR. A chemical plant, especially an older one, will probably have a lower pump MTBR than a refinery. Even higher MTBR should certainly be possible by paying close attention to well-known best practices. At the same time, lower MTBR is likely if sloppy operation and maintenance are accepted. The keys to high pump MTBR include:

1. purchase, operate and maintain pumps to API 610, 10th Edition
2. purchase, operate and maintain seals to API 682, 3rd Edition
3. establish a plant-wide pump reliability program, including a rotating equipment engineer
4. establish pump and seal training programs for operators and mechanics.

SEAL MTBF EXPLORED

On the seals issue, some pumps have two seals (between bearings pumps) while others (overhung pumps) have only one. '"Between bearings pumps" constitute around 15% of all pumps in a typical refinery.

Some people measure seal MTBF as:

MTBF(a) = [(Total number of pumps)/(Total number of failures)] x review period

Others do it more correctly as:

MTBF(b) = [(Total number of seals)/(Total number of failures)] x review period

Typically, MTBF(b) ~ MTBF(a) x 1.15

To be totally correct, it should be measured as:

MTBF(c) = [(Total number of running seals)/(Total number of failures)] x review period.

Typically, MTBF(c) ~ MTBF(b) x 1/(1 + 0.6) ~ MTBF(b) x 0.625

In the case of bearings, there are normally two radial bearings per pump whether it be a between bearings or overhung design. In that case:

MTBF(b) = MTBF(a) x 2, and

MTBF(c) = MTBF(b) x 0.625

Finally, in the case of couplings, there is normally one coupling per pump. Therefore:

MTBF(b) = MTBF(a), and

MTBF(c) = MTBF(b) x 0.625

This may seem a little "picky" but nevertheless important if meaningful comparisons are to be made. It is always important to state the basis on which MTBF is calculated and there is perhaps a clear need for a standard definition to avoid misinterpretation. Then again, why get "hung up" on MTBF statistics after having seen, from the preceding chapters of this text, that there is still some" ripe, low-hanging fruit" to be harvested in your plant? A reliability-focused plant will fix the obvious before aiming for high-tech solutions. Even the smartest high-tech solution will be wasted where the basics are not observed or not understood.

That said, it is also important to understand the dependence of pump MTBF on the MTBFs of its components in what is most commonly a "series system." (In some situations e.g. a tandem mechanical seal, failure of the primary seal might not mean failure of the pump. This is because the secondary seal may be capable of allowing the pump to continue operating for a substantial period of time. That is a parallel system and is generally outside the scope of intent of this discussion.)

Series Systems

In the "pump world," a series system can be schematically represented as shown in Figure 16-2. Needless to say, this system can be expanded to include piping, baseplate, alignment and a veritable host of other issues that can influence pump reliability. Remember that these were the subject of hundreds of the preceding pages.

So, if any component fails, then the pump becomes unserviceable and must be repaired. The pump, thus, has one effective MTBF, based on the MTBFs of the individual components. This can be illustrated using public domain data on ANSI pump reliability, Table 16-2. These data should be viewed as "minimum expectation."

If X and Y are two independent events, and P_X is the probability that X will occur, and P_Y is the probability that Y will occur, then the probability that both events will occur is given by:

$$P_{(XY)} = P_X \cdot P_Y \qquad \text{Eq. 1}$$

The reliability of a series system or probability of survival of the system is the probability of all the components surviving, since a failure of only one component means overall system failure. For constant values of λ (chance failures), it can be shown that:

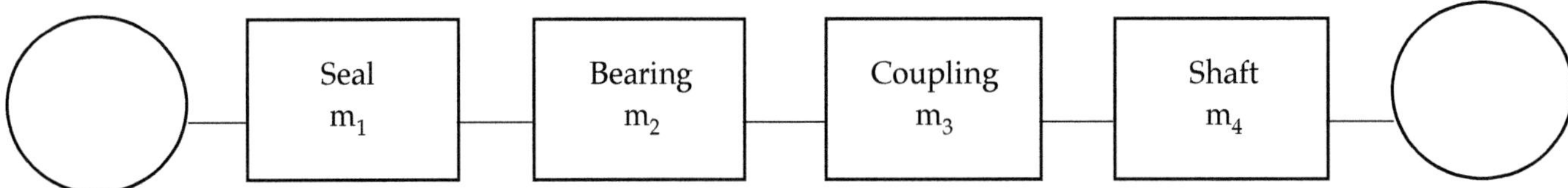

Figure 16-2: Schematic showing a pump as a series system

$$\text{MTBF, } m_s = \frac{1}{l_s} = \frac{1}{(l_1 + l_2 + l_3 + \text{etc})} = \frac{1}{\left(\frac{1}{m_1} + \frac{1}{m_2} + \frac{1}{m_3} + \text{etc}\right)} \quad \text{Eq. 2, RULE 1}$$

Table 16-2: ANSI pump reliability data

Component	MTBF (years)
Mech seal	1.2 (m_1)
Ball bearing	3.0 (m_2)
Coupling	4.0 (m_3)
Shaft	15.4 (m_4)

However, the data normally produced for a pump is *not* at constant failure rate and the following equation is commonly, alternativly used.

$$\frac{1}{m_s^2} = \frac{1}{m_1^2} + \frac{1}{m_2^2} = \frac{1}{m_3^2} = \frac{1}{m_4^2}$$

$$\frac{1}{m_s^2} = \frac{1}{1.2^2} + \frac{1}{3^2} = \frac{1}{4^2} = \frac{1}{15.4^2} \quad \text{Eq. , RULE 2}$$

Therefore, m_s = 1.07 yrs. (Note m_s = 0.68 yrs, using Rule 1)

Using Eq. 3, the effect of varying seal life can be seen in Table 16-3. Taking the maximum reasonable values for bearing and coupling life of 10 and 20 years, respectively, and proposing seal lives of three and five years, the effect on pump life, using Rule 2, can be seen in Table 16-4.

Table 16-3: Effect of varying seal life (Ref. Eq. 3)

Mechanical Seal Life	*Pump Life Rule 1*	*Pump Life Rule 2*
0.8	0.527	0.758
1.2	0.675	1.070
1.6	0.785	1.326
2.0	0.871	1.523
2.4	0.939	1.687
2.8	0.995	1.809
3.2	1.040	1.905

MTBF SIMULATIONS AND SEAL MTBF LEVELS CURRENTLY BEING ACHIEVED

A very convenient and interesting way to investigate and understand the dependency of system MTBF on its components is through the use of some 'in-house' simulation software. The illustrations show how average-performing components (Figure 16-3a) will result in a pump lifetime of only 15 months, whereas, to achieve a pump lifetime of 51 months (Figure 16-3b), top performance would be required from all pump components.

Table 16-4: Pump life as a function of seal life

	Life (years)	
Seal	3	5
Bearing	10	10
Coupling	20	20
Shaft life	15.4	15.4
Pump	**2.797**	**4.2**

In 1976, a survey of seal life was conducted by the British Hydraulic Research Group (BHRG); it covered over 5,000 pumps in a range of process industries. Average seal lives were only around 12 months, and the users were very dissatisfied and demanded improvements. The sealing industry responded.

Many companies, particularly in the oil and gas industries, started working very closely with the seal vendors in setting up seal MTBF improvement programs. The authors' company, the United States branch of John Crane, took a very proactive part and cooperated in over 30 major refinery surveys. The year-by-year results from 21 of these are depicted in Figure 16-4. These data have been combined with equivalent data from the user, which made it possible to generate the overall seal MTBF chart. (Adjustments had to be made to combine the data since the vendor MTBF was based on the number of pumps, whereas the user data were based on the number of seals. The final data are based on the number of seals and required multiplying the vendor MTBF's by 1.15.)

The chart in Figure 16-5 shows combined MTBF values based on over 12,000 seals in 36 refineries. It includes upper and lower quartile values that are useful for suggesting target performance values (Table 16-5). The average MTBF improvement rate in Figure 16-5 is 30 months in six years, or five months per year. That should be regarded as the base rate that is the average of a large number of plants, not all of whom were on improvement programs. Furthermore, one should expect large percentage improvements initially as the worst "bad actors" are dealt with.

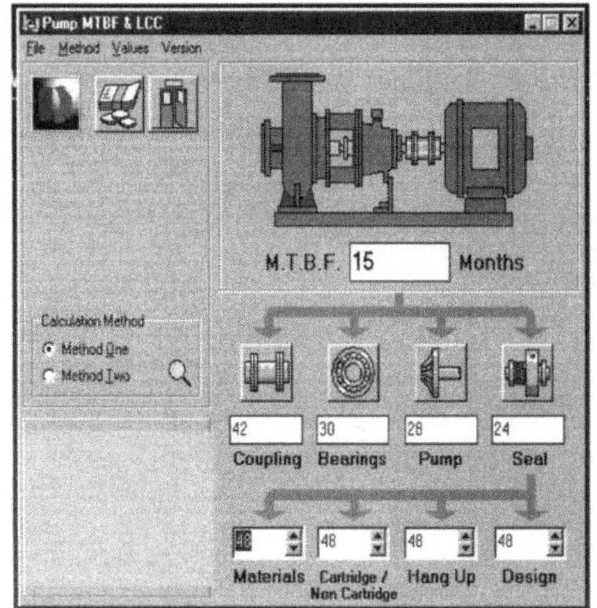

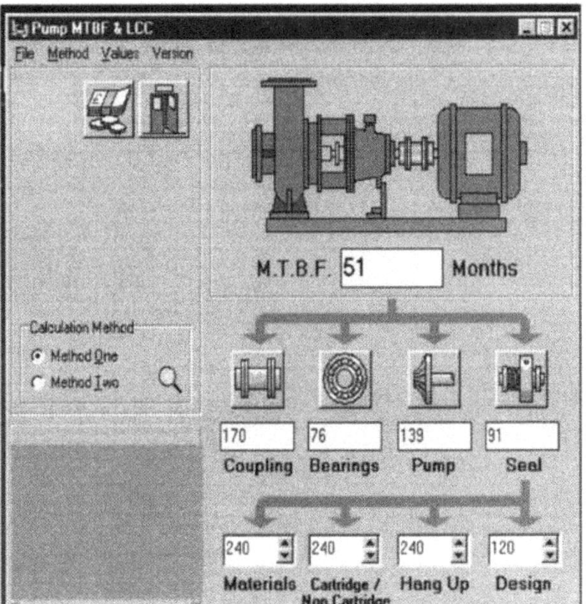

Figures 16-3a and 16-3b: Pump MTBF Simulation

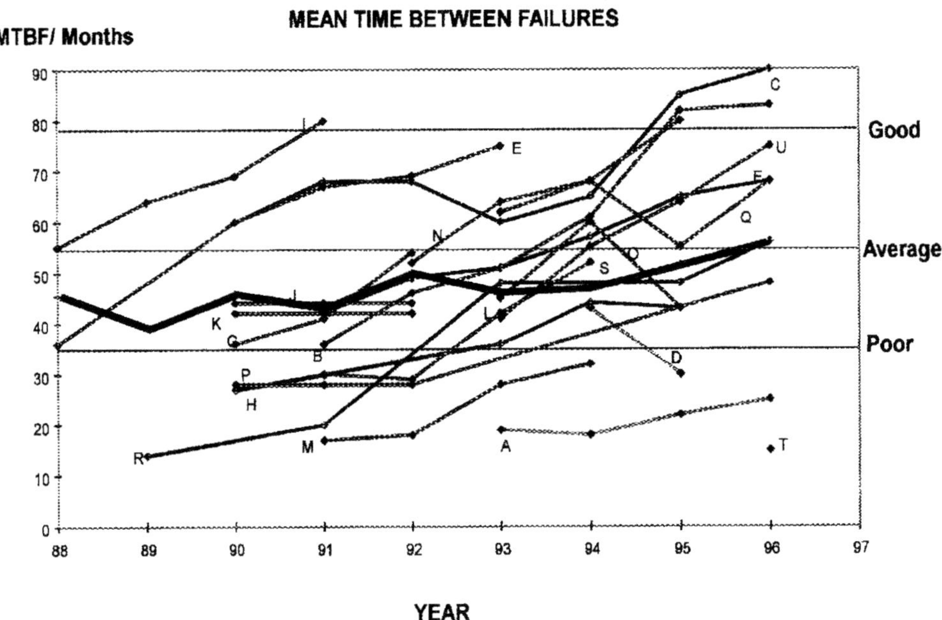

Figures 16-4: Seal MTBF Results of 21 Major Refinery Surveys

Table 16-5: Suggested refinery seal target MTBFs

Target for seal MTBF in oil refineries	
Excellent	>90 months
Very good	70/90 months
Average	70 months
Fair	62/70 months
Poor	<62 months

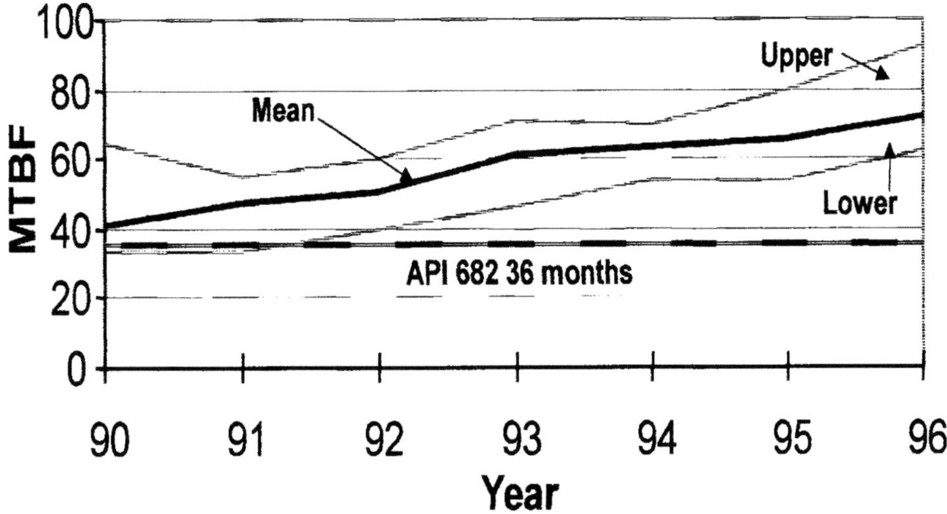

Figure 16-5: Summary seal MTBF data for 36 refineries

Not all plants are refineries, however, and different results can be expected elsewhere. In chemical plants, pumps have traditionally been "throw away" items as chemical attack limited life. Things have improved in recent years, but the limited space available in DIN and ASME stuffing boxes does limit the type of seal that can be fitted to more compact and simple versions. Lifetimes in chemical installations are generally believed to be around 50 to 60 percent of the refinery values.

TARGET PUMP AND COMPONENT LIFETIMES

Based on the lifetime levels being achieved in practice in 2000 and combined with the known "best practice" as outlined elsewhere in this text, the target component lifetimes of Table 16-6 are recommended and should be considered readily achievable:

Table 16-6: Realistic target pump and component lifetimes
(Note that "target" is less than "best actually achieved")

Seals		
	Refineries	Chemical and other plants
Excellent	90 months	55 months
Average	70 months	45 months
Couplings		
All plants	Membrane type	120 months
	Gear type	> 60 months
Bearings		
All plants	continuous operation	60 months
	spared operation	120 months
Pumps		
Based on series system calculation		48 months

For instance, using the "series system" calculation, the generalized target lifetime for a pump in the typical refinery is 48 months.

It should again be emphasized that many plants are achieving these levels. Nevertheless, to reach these pump lives the pump components themselves must be operating at the highest levels. An unsuitable seal with a lifetime of one month or less will have a catastrophic effect on pump MTBF as will a badly performing coupling or bearing.

HOW COMPETENT MANUFACTURERS IMPROVE COMPONENT RELIABILITY

Analysis Techniques

The importance of having an accurate theoretical model for the simulation of product behavior cannot be overstated. Finite Element Analysis (FEA) has played a major part in this field in recent years and is widely used, often on a case-by-case basis. Of even more importance is a model that can look at all aspects of product behavior under a wide variety of both static and transient conditions. One such model for mechanical seals has been created and developed over many years to the point where it can accurately predict behavior and performance for seals operating in liquids and gases and with special interface lubrication features.

Typical inputs and outputs for such analyses are:

Inputs	*Outputs*
Operating conditions	Deformations
Sealed fluid characteristics	Heat generated/temperature rise
Materials of construction	Film stability/pressure distributions
Seal face groove geometry	Leakage

Conventional FEA analysis has been widely used for looking at the influence of mechanical and thermal distortions in mechanical seals. Moreover, FEA analysis has been successfully extended to transient dynamic conditions.

Combining steady state and dynamic analyses has proved invaluable in understanding seal behavior at start-up, shut-down and under varying operating conditions. This is where problems often occur and were not previously predictable. Indeed, these analysis tools are very important in three areas:

- New product development (minimizing testing)
- Predicting operation in difficult conditions and optimizing the design for those conditions.
- Problem solving on existing troublesome applications

COUPLINGS

It would not be fair to say that gear couplings fail every time they encounter lubricant deficiencies. Very often, however, the need to lubricate does require the pump to be stopped outside a normal shut-down opportunity.

A refinery in the Middle East underwent a major conversion from soft packing to mechanical seals and from gear couplings to non-lubricated disc (membrane) couplings. It had been the practice to clean, inspect, and regrease gear couplings every four to six months. After conversion to the membrane coupling shown earlier in Figure 11-9, membrane couplings were replaced very infrequently and, using failure rate data from Ref. 16-1, could be said to have an MTBF in the vicinity of 30 years.

The cost of coupling ownership must be calculated properly by including both maintenance and replacement costs in any calculation to show real-life cost benefit. As far as this aspect of reliability is concerned, flexible element couplings have two very significant benefits:

- They require no lubrication.
- They have a very long theoretical life.

Additional benefits can include low imposed loads on the shafts, which extends bearing life, and no wearing parts, which means retention of dynamic balance and low vibration for extended mechanical seal and bearing life. Note that the BHRG survey mentioned earlier and conducted in 1976 identified vibration as the number one cause of reduced seal life and increased leakage levels.

The main enemy of membrane couplings is excessive misalignment, which can result in fatigue failures. However, some flexing element couplings—couplings with metal membranes—are able to take typically 0.5 degrees per membrane bank (one millimeter of lateral misalignment per 100 mm of shaft separation), which is many times greater than the level couplings can be easily aligned to. The modified Goodman diagram, shown in Figure 16-6, illustrates the margins normally applied to metal membrane couplings, which means they can be expected to run maintenance-free for many years.

This issue was confirmed in Ref. 16-1, which compared gear couplings with non- lubricated couplings. At that time it was concluded that "the importance of proper alignment between pump and driver is best demonstrated by the failure statistics of several refinery units. The replacement rate for metallic disc-pack couplings of the type shown is generally fewer than 3 per 100 pumps per year." This implies a coupling MTBF of over 30 years for properly selected and well-aligned non-lubricated membrane couplings!

In short, metal membrane couplings can reasonably be expected to last the life of the plant, unless they are overloaded or badly aligned on installation or due to some change such as foundation movement. Clearly, however, a bearing failure can result in severe coupling damage.

BEARINGS

We must also return to bearings within the context of this chapter dealing with failure statistics and component uptime improvement. As was alluded to in Chapter 7, a rolling element bearing has a finite life and eventually will fail due to fatigue, even if operated under ideal conditions. The operational life of a rolling element bearing, limited by fatigue, is a statistically calculable parameter. The required service life of a machine can therefore be matched to the bearings. Provided that a large enough bearing can be fitted and that satisfactory operational conditions are maintained, service lives of many years are the norm. Satisfactory application of the correct lubricant is one of the chief prerequisites here and the flaws highlighted in Chapter 9 merit close attention by reliability-focused users.

In any event, the fatigue life is governed by operating speed and radial and axial loads applied to the bearings. The generally accepted method of defining it is the L10 life. The "L10 life" is the expected number of cycles, or hours ("L10h"), without evidence of fatigue that 90 percent of a group of apparently identical bearings will achieve when operating under the same conditions of load and speed. The basic calculation for this is described in ISO 281 (1990). Basic rating life in milions of revolutions:

$$L_{10} = \left(\frac{C^P}{P}\right) \text{ or } L_{10h} = \left(\frac{1{,}000{,}000}{60n}\right) L_{10} \text{ hours}$$

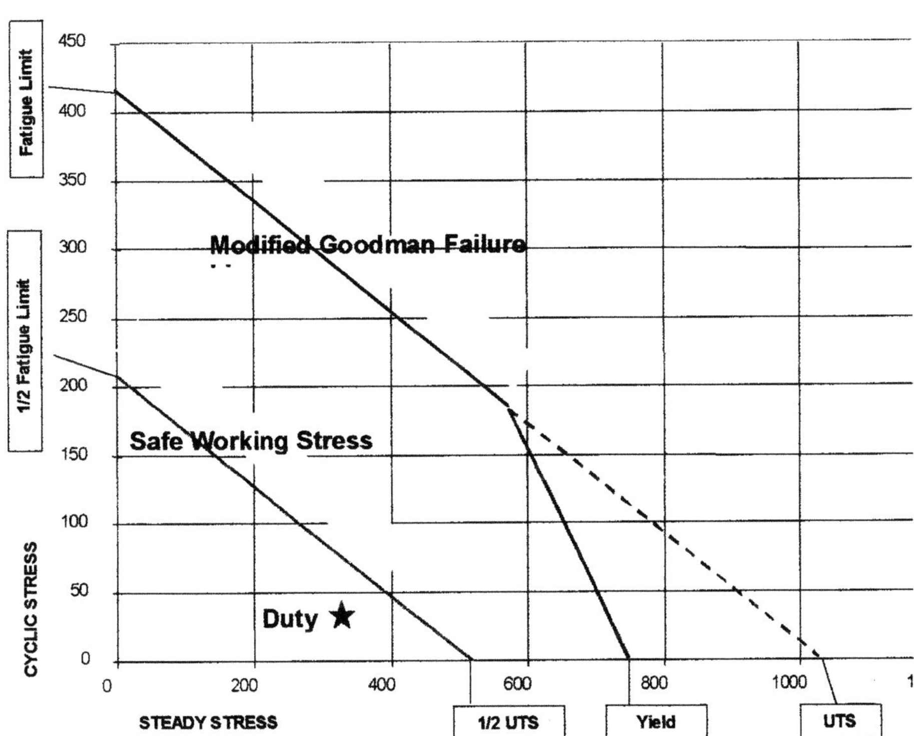

Figure 16-6: Modified Goodman Diagram for AISI 301 Half Hard Membranes

for ideal conditions, where:

C = the basic dynamic load rating depending on the bearing design
P = the equivalent combined dynamic bearing load
p = the exponent = 3 for ball bearings and = 10/3 for roller bearings
n = the speed in rpm

Unfortunately, most industrial installations are not exposed to ideal conditions, and bearings often fail well before reaching their theoretical design life. To ensure reasonable practical operating periods, rotating machinery standards based on operating experience require L10h bearing life to be typically as follows:

- 25,000 hours for spared machines in general operation
- 40,000 hours for non-spared machines with two years or more continuous operation
- 100,000 hours for high reliability, remote, and unattended continuous operation

API 610 (1995) specifies a three-year design life for bearings:

- 25,000 hours continuous operation
- 16,000 hours at maximum axial and radial load

Nevertheless, published statistics indicate that:

- Two-thirds of all rolling element bearings that had to be replaced in service failed prematurely.
- One-third of bearings fail due to fatigue spalling. These tend to be the longest- running bearings where operational conditions have been consistently satisfactory.
- One-sixth of bearings fail due to incorrect fitting, incorrect selection, damage due to external causes such as vibration or electric currents, or overloading.
- One-third of bearings fail early due to lubrication problems, usually because of an insufficient quantity of grease, or loss of oil from the bearing.
- One-sixth fail due to contamination entering the bearing. Solid contaminants cause surface damage by indenting, wear, and early fatigue by bridging the oil film. Moisture can cause corrosion and can reduce the effective viscosity of the lubricant, again causing wear and reducing the fatigue life.

The effect of water contamination on basic bearing life was dealt with in Table 2-1 and Chapter 6. It can also be seen in Figure 16-7 which shows that if the water content of an uninhibited mineral oil increases through contamination from the nominal 100 ppm to 400 ppm the bearing life will be cut in half. Similarly, solid contaminants will reduce the life by varying amounts depending on the amount, distribution in the oil film, and particle size relative to the oil film thickness. Again, a previous segment of this text, Chapter 9, addressed these issues in Tables 9-4 through 9-6.

Some bearing manufacturers have refined the basic bearing life calculation to include factors that account in some measure for the degree of contamination. However, these factors show what improvement in rating life can be achieved with known, controlled levels of contamination. They still depend upon the bearing being protected from the environment by effective sealing. Chapters 6 and 9 contain updates on the subject of sealing bearing housings.

As was brought out earlier, traditional lip seals will fail to prevent water ingress into the oil during heavy rain or pump wash down. Damage to the seal usually ruins the bearing. Modern rotating labyrinth seals address the problem and help extend bearing life dramatically as a consequence. However, face-type bearing housing seals (Chapter 9) eliminate the air gap of rotating labyrinth seals and represent the latest technology. It should be noted that API 610 9th Edition section 5.10.2.7 stipulates use of labyrinth or magnetic type seals, but excludes lip seals.

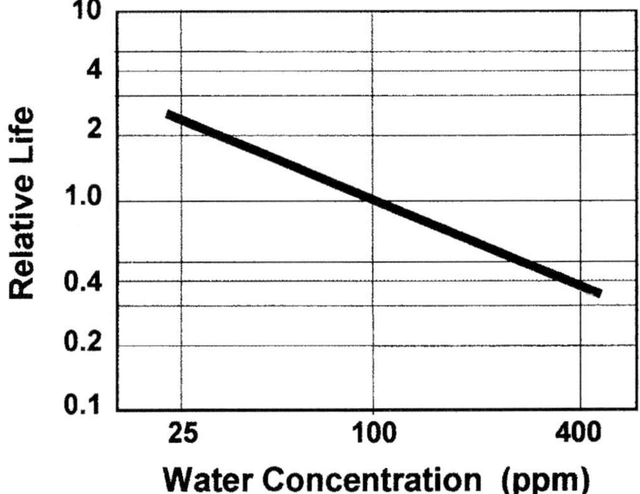

Figure 16-7: Effect of Water Contamination on Basic Bearing Life

Overloading of pump bearings is prevented by selecting a suitable type with sufficient load capability, but bearing size is often limited by the operating speed and lube application, as well as by space considerations. It thus becomes necessary to control and limit the loads and moments that may be imposed upon pump bearings. In addition to the basic rotating weight distribution of the machines, significant forces can be applied to bearings by misalignments between coupled machines. Careful consideration of the type, weight, and especially flexibility of the coupling selected can minimize these imposed loads and moments.

SUPPLIER/CUSTOMER RELATIONSHIPS

At the conclusion of Chapter 8, this text addressed the importance of partnership to the success of mechanical seal failure reduction and pump operating and maintenance cost optimization endeavors. In the quest to achieve higher reliability of rotating equipment and the attendant benefits in cost reductions and increased availability, two things have become abundantly clear to the authors and these sentiments and experiences can be summarized as given here:

- First, the equipment operator must be committed to a reliability improvement program in the long term. To use the old adage, the output is directly dependent on the input – too many times operators have embarked on ambitious and exciting programs only for their attention to wander after the first few months, with little achieved. Support from the top is, of course, a fundamental necessity.

- Second, an effective, open and non-confrontational partnership (often informal) between the supplier and operator is just as important. The benefits can be manifold. Too much rotating equipment in operation today was specified many years ago with consumable parts having been replaced by more of the same.

But times change and progress is made. In 2013, most suppliers can offer more modern and reliable alternatives and new preferred materials. They can bring their wide experience to bear and, given both opportunity and incentive, can offer techniques and technologies that can provide solutions for those difficult "bad actors." It is surprising how much mutual benefit can be derived from these collaborative situations when approached the right way.

But let's be realistic. Not every supplier is equipped or prepared to take part in such a relationship, which is why it is equally important to make the right choices in the first instance rather than demanding nothing but lowest first cost on new plant projects. The old wisdom "you get what you pay for—if not immediately, then certainly eventually," will always be true. There will be no exceptions to this statement.

A Simple Methodology for Achieving Improvements in Pump Components.

A systematic and sustained approach is an essential prerequisite of achieving performance improvements in an existing plant. There are many alternative methods, but they all depend on the basic elements of establishing current performance, setting goals, establishing and implementing an improvement plan, monitoring progress and being prepared to continue and repeat the process on a *continuing* basis.

As shown earlier in this chapter, improved reliability of seals, couplings, and bearings leads to improved reliability of pumps, which, in turn, leads to reductions in pump operating costs, the objective of the exercise. However, although extremely important, it must be remembered that it is not the only way to achieve the goal. Other measures should be taken to reduce running cost, reduce inventory, etc., and these issues need to be dealt with as well. On the question of improving reliability, however, the authors recommend following a logical sequence of events:

- Understand what is being (and what can be) achieved in relevant or equivalent situations
- Set appropriate scope or targets and time-scales for your own plant
- Determine what is being achieved at your plant
- Enlist the help of your suppliers, as appropriate
- Conduct a survey
- Analyze the data
- Set up a performance improvement plan
- Identify the bad actors
- Adjust the plan with revised targets and time-scales, as necessary
- Set up a sensible monitoring and reporting system
- Collect and analyze data regularly
- Compare to target
- Revise the plan and repeat the process

Data Collection and Analysis

A well-designed and user-friendly data collection and analysis system is very important in underpinning any serious reliability improvement program. Again, the very simplest and basic ones were described in the earlier chapters of this handbook. That is where we made the point that without data it is not possible to perform meaningful root cause analyses. Without data, the user is stuck in the repair-focus mode. He will forever struggle with repeat failures and will fall further and further behind the reliability-focused user-competitor.

That is not to imply that all is lost. Many plant operators have their own systems and many of them work extremely well to the extent that they are used on a continuous basis and provide all the information required. Some, however, are too complicated and are not used to any serious extent. Some plants, particularly older ones, don't have any suitable systems at all and that is where alternative systems available from some suppliers and other third parties provide a solution.

One such system can be described briefly as follows. At the lowest level, it acts as a repository for pump or rotating equipment data, facilitating population management and standardization. The kind of data stored includes equipment details, seal chamber details, duty conditions, pumped product details, details of the mechanical seal, power transmission coupling, bearings and any other pertinent data (Figures 16-8 and 16-9).

At the next level, it can log and provide failure history with every pump failure logged and reported on and the history maintained of all failure reports. Data stored includes failure logs, observations, detailed failure reports and recommendations for rectification.

The real power of such a system is in using it as part of the reliability management process where it provides root cause analysis capability, which drives the needed reliability improvement actions and provides focus and benchmarking.

The results are reflected in Figures 16-10 and 16-11, showing impressive seal MTBF and reliability. While the operator/seal vendor partnership was informal but effective, the greatest credit must go to the operators who were determined to reach their own "pacesetter" targets. These were, themselves, derived from Figures 16-4 and 16-5 and reflect an earlier presentation at the Texas A&M Pump Symposium in 1998 (Ref. 16-7).

The cooperation initiative alluded to in these illustrations has continued over the years and performance can be seen to be still world-class despite a small reduction in MTBF over time. Here, up-to-date figures illustrate MTBF levels still running at over 10 years. The cause of the small MTBF reduction can be attributed to operations-related issues, "non-failure" replacements and worn-out seals rather than design issues. The increase in operations-related issues may well have its origin in decisions that must be addressed by plants that are maturing with the years.

This remarkable performance at the Star refinery, now known as the Alliance Refinery Company PU—North (ARC PUN), is a confirmation of many of the issues raised in this chapter, but also in this text. Clearly then, these issues deal with having the desire to achieve high reliability, taking the trouble to set targets, identifying and fixing bad performers, putting a system in place for data logging and monitoring trends and keeping the whole process going for such a long period of time. A

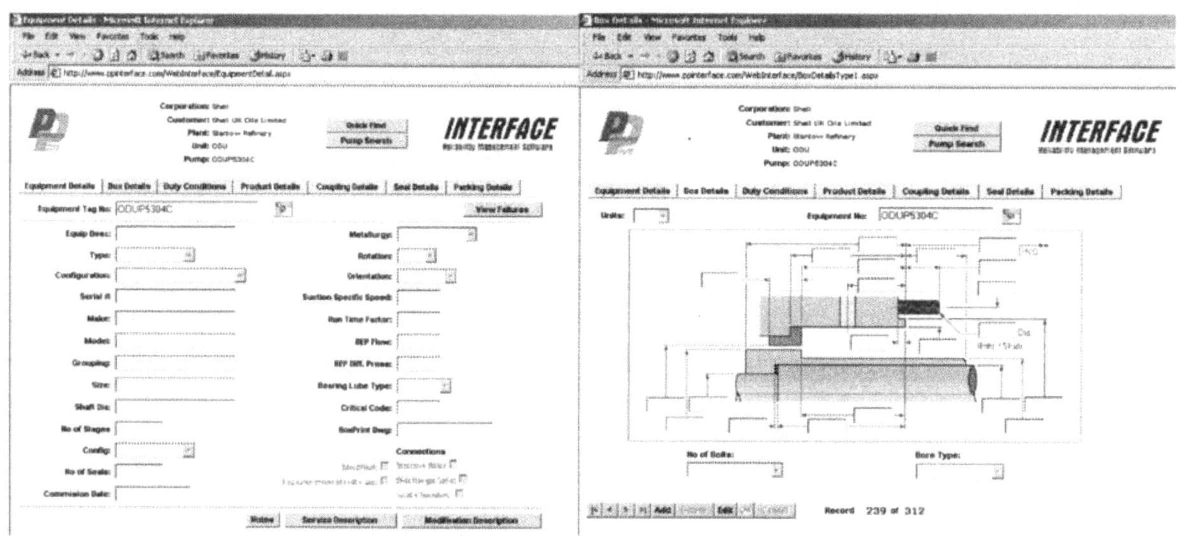

Figure 16-8: Typical base data

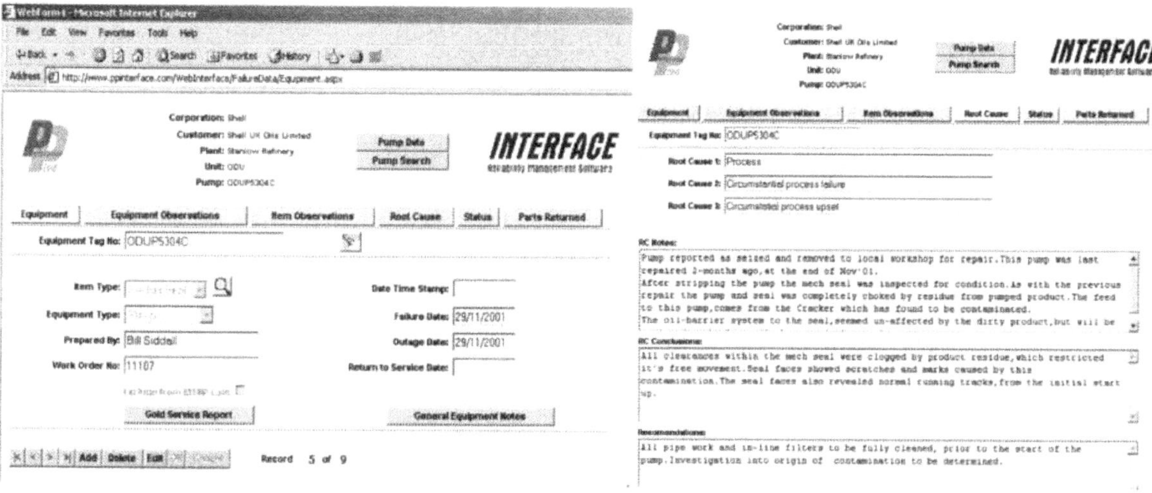

Figure 16-9: Failure data

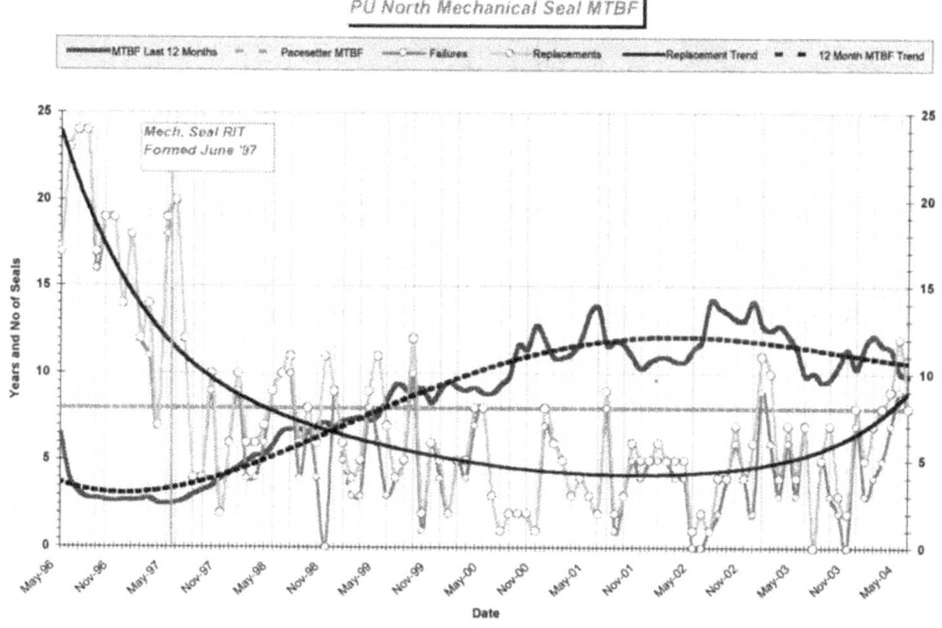

Figure 16-10: MTBF vs. time

lesson to many of us!

And so, in conclusion, component performance has improved dramatically in the time period since, say, 1985. Better designs, new technologies, new materials and a better understanding of what the needs are and how to satisfy them—all these and a few others have contributed to where best-of-class companies are today, in 2012 or 2013.

To take advantage of known performance enhancements in new plant as well as in upgrade situations on existing facilities, care must be taken to look at real total life cycle costs and not just first costs. Many process plants are designed to operate for twenty or thirty years. Good decisions made at the outset will pay handsome dividends in the long term.

But most existing plants also profess to desire freedom from avoidable maintenance and unscheduled outages. They, too, must avail themselves of what we have been discussing here. Indeed, there are equally important options now available for upgrading old equipment with benefits that show surprisingly low payback periods. To understand what those options are and how to implement them can result from diligent reading and follow-up. Don't neglect any aspect of our preceding discussions and admonitions. Pay attention to the reliability-focused specification and selection criteria mentioned in the introductory chapters. Understand the importance of conscientiously following appropriate installation procedures, work processes and carefully formed and managed supplier/operator partnerships, both formal and informal. As a result, you can make real and tangible improvements in plant safety, reliability, uptime and profitability.

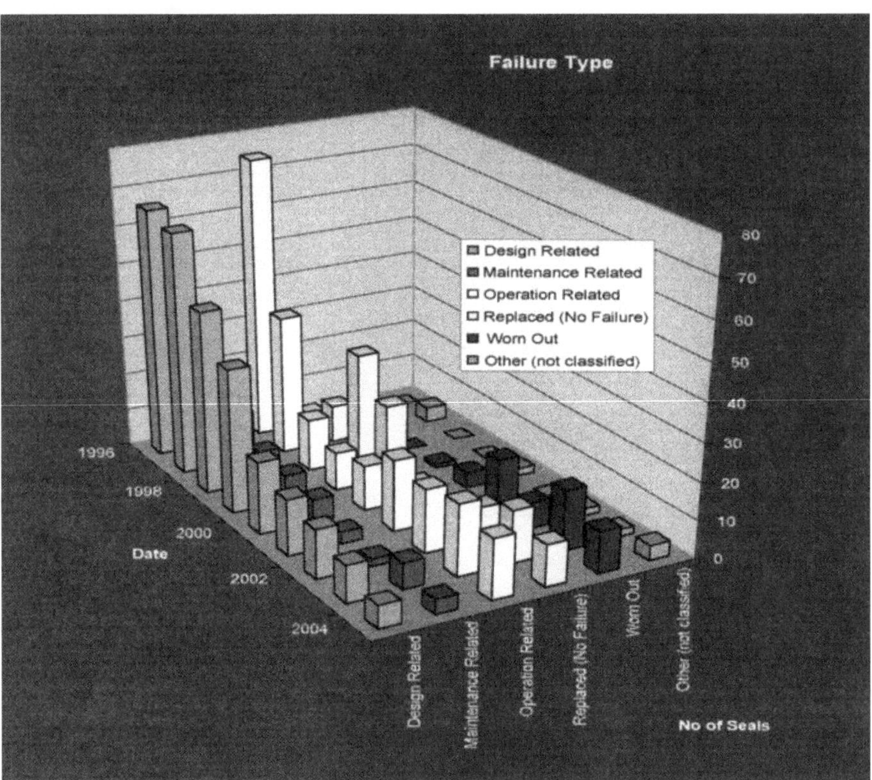

Figure 16-11: Seal failure causes

Chapter 17

Breaking the Cycle of Pump Repairs

There is, perhaps, no better summary than looking back on what this text will have conveyed to the attentive reader:

- How the cycle of pump repairs can be broken.
- What it will be worth to no longer have to contend with frequent failures, or
- Why unexpected or randomly occurring failures should be unacceptable to modern industry.

Reliability professionals in process plants are never pleased with unexplained failures. They have long realized that most centrifugal pump styles, models and configurations are relatively simple machines. It has been said centrifugal pumps represent basically the least complex machine (next to electric motors) routinely found in modern industry.

Whenever possible, pump owner-operators link failure occurrences and failure numbers. As they examine this linkage they note how detailed observations might complement rigorous statistics. Of the numerous process centrifugal pumps undergoing repair right this very minute, an estimated 90% have failed randomly before. Chances are that some have run just fine for two or three years after initial commissioning. Then there may have been a repair event and afterwards the pumps were never quite the same; they might have failed twice a year since the first repair.

Chances are other pumps failed frequently or randomly, perhaps once a year, and did so from the time they were originally started up years ago. That brings up such questions as: Could it be we don't really know why many process pumps are failing? Could it be we just don't give pumps the attention they deserve? Is it because everybody's priorities are elsewhere? Or are there perhaps elusive failure reasons, i.e., factors overlooked by all parties?

Fortunately, improvement is both possible and cost-justified. Allowing repeat failures on process pumps rarely makes economic sense. Simple benefit-to-cost or life cycle analyses will easily demonstrate that the pursuit of remedial action greatly benefits users. This text dealt with elusive pump failure causes and highlighted these in the many pages of this book. Elusive failure causes are recapitulated or re-stated in this summary chapter.

EXAMINING THE COST OF FAILURES

One way of exploring the value of extending pump mean-time-between-failures (MTBF) is to examine the likely savings if we could improve the MTBF from presently 4.5 years to a projected 5.5 years. Say, a facility has 1,000 pumps; that's 1,000/4.5 = 222 repairs before and 1000/5.5 = 182 repairs after understanding and solving the problem. Avoiding 40 repairs at $ 6,000 each is actually a very low estimate, but would be worth $240,000. Avoiding repairs frees up manpower for other tasks: At 20 man-hours times 40 incidents times $100 per hour, reassigning these professionals to other repair avoidance tasks would be worth at least $80,000.

There is also one ~$3,000,000 fire per 1,000 pump failures. It was confirmed in 2009 and 2012 by two different corporations. Avoiding 40 repairs would be worth 40/1,000 x $ 3,000,000 = $120,000 and, together, the three items ($240k, $80k and $120k) add up to $440,000.

Although $ 6,000 was used for repair cost avoidance calculations earlier, an average API pump repair at a Texas refinery costs slightly over $10,200; a refinery in Mississippi reported $11,000. These estimated numbers have not changed in almost 30 years; the widespread use of predictive maintenance methods (PdM) has resulted in catching incipient failures before they become severe failures. If the incremental cost of upgrading during the next repair adds $2,000 to the repair bill and avoids even a single failure every 3 or 4 years over the 30-year total life of a pump, the pay-back will have been quite substantial. It would be reasonable to assume 8 avoided repairs at $6,000 give payback of 48k/2k = 24:1.

The quoted repair cost numbers of close to $11,000 reflect what needs to be considered in a pump repair cost calculation: Direct labor, direct materials, employee benefits at roughly 50% of direct labor, refinery administration and services costs at close to 10% of direct labor, mechanical-technical service personnel overhead costs amounting to ~115% of direct labor, and materials procurement costs from 7% to 8% of materials outlay were mentioned in our text. Disregarding the true cost of failures or repairs is likely to deprive some users of seeing the true benefit-to-cost ratio associated with pump upgrades.

We could examine other ways to calculate as well. It would be reasonable to assume that implementing a component upgrade (generally the elimination of a weak link) extends pump uptime by 10%. Implementing 5 upgrade items yields 1.1^5 = 1.61— a 61% mean-time-between-repair (MTBR) increase. Or, say, we gave up 10% each by not implementing 6 reasonable improvement items. In that instance, 0.9^6= 0.53, meaning that the MTBR is only 53% of what it might otherwise be. That might explain industry's widely diverging MTBRs. The MTBR-gap is quite conservatively assumed to range from 3.6 years to 9.0 years in U.S. oil refineries and, as of 2013, no well-informed pump professional has disagreed with this range of MTBR numbers.

PUMPS HAVE A DEFINED OPERATING RANGE

The onset of pump problems is not the same for different pumps, or for different services. Attempts to identify best practices are to be commended. Both Paul Barringer and Ed Nelson contributed to Figure 17-1, the typical HQ curve. They plotted eight traditional non-BEP problem areas on that curve. The plot supports the notion that pump reliability can approach zero as one operates farther away from the best efficiency point, or BEP.

> **HIGHLIGHT 1: Stay well inside the defined operating range. Safe operating margins are the key to failure avoidance**

The implications of Figure 17-1 are captured in *Highlight 1*. Just because pumps are able to run at lower than BEP flows does not mean that it's good to operate there. Compare it to a vehicle able to go 12 mph in 6th gear, or 47 mph in 1st gear. It can be done, but will likely prove costly if done for very long. Pioneering efforts to define minimum allowable flows can be traced back decades and attention is drawn to the sketch, Figure 17-2, originally published by Irving Taylor in 1977 (Ref. 17-2). His work is worth mentioning because he approximated in a single illustration what others have tried to convey in complex words and mathematical formulas. Taylor deserves much credit because he kept the average user in mind.

Irving Taylor's trend curves of probable NPSHr for minimum recirculation and zero cavitation-erosion in water, Figure 17-2, are sufficiently accurate to warrant the attention of reliability professionals who wish to work within safe margins. Hundreds of references exist on the subjects of cavitation and internal recirculation; stable pump operation is always the central aim. However, the actual NPSHr needed for zero damage to impellers and other pump components may be many times the number published in the manufacturer's literature. The manufacturers' NPSHr plots (lowermost curve in Figure 17-2) are commonly based on observing a 3% drop in discharge head or pressure. Taylor's plot

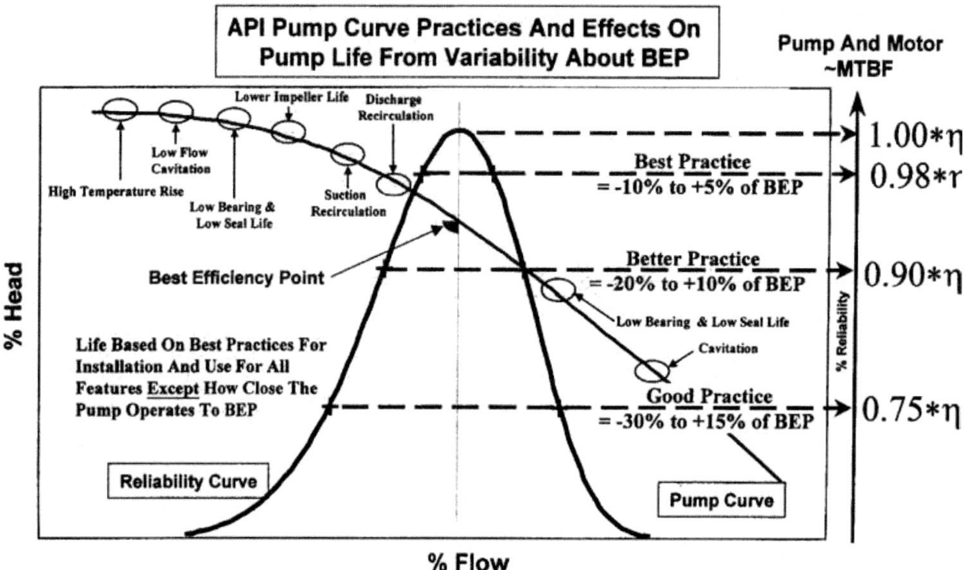

Figure 17-1: Barringer-Nelson curves show reliability impact of operation away from BEP (Ref. 17-1)

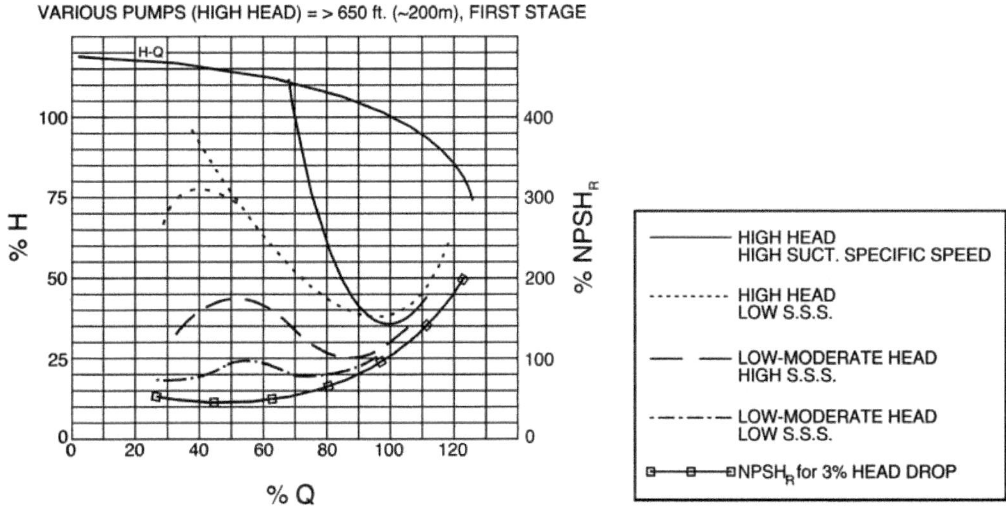

Figure 17-2: Pump manufacturers usually plot only the NPSHr trend associated with the lowermost curve. At that time a head drop or pressure fluctuation of 3% exists at BEP flow (Taylor, 1977; Ref. 17-2). BEP stands for "Best Efficiency Point"

places the Q = 100% intersect at an NPSHr = 100% of the manufacturer's stated value. Unfortunately, whenever this 3% fluctuation occurs, a measure of damage may already be in progress. It is prudent to assume a more realistic NPSHr and to provide an NPSHa in excess of this likely NPSHr. Doing so builds a certain margin of safety into the pump and reduces the risk of catastrophic failure events.

There are hydrocarbon services where an NPSHa surplus of just 1 ft over NPSHr will be sufficient to avoid cavitation. However, there are also services, such as Carbamate, where a 25 ft surplus is not nearly enough. Therefore, Taylor's trend curves are considered general approximations for prudent users. He indicated that users should enlist the help of competent pump manufacturers and experienced design contractors to agree on NPSH multipliers or bracket the right NPSH margins for a particular liquid or pumping service.

Taylor gave a demarcation line between low and high suction specific speeds (Nss) at somewhere between 8,000 and 12,000. His data were corroborated in surveys taken after 1977 by Jerry Hallam, then a young engineer at the Amoco oil refinery in Texas City; there were also a few other observers. Typical readings pointed to 8,500 or 9,000 as numbers that deserve attention. If pumps with Nss numbers higher than approximately 9,000 are operated at flow rates much higher or lower than BEP (Best Efficiency Point), the life expectancy or repair-free operating time of these pumps will be reduced. Whether these reductions will amount to 10% off normal or 60% off normal is the subject of much debate and requires reviews on a pump-specific basis.

While no rigorous Nss value exists, cautious reliability professionals observe safe margins. Many users choose Nss = 9,000 as the limit for flows away from BEP. There are, however, some pumps (including certain high-speed Sundyne designs) that will operate quite well with Nss-values higher than 9,000. But these are special cases and a close pump-user-to-pump-manufacturer relationship is needed to shed light on long term experience.

This text, however, alerted the reader to more precise—and certainly more current—predictors of the potential for cavitation erosion damage in centrifugal pumps. In particular, the segments "Suction Energy" and"NPSH Margin" in Chapter 5 conveyed solid information on what we called the pump "Suction Energy Ratio" (Figure 5-12). This parameter takes into account pump specific speed (Ns), plus the inlet tip speed and liquid specific gravity, coupled with an "NPSH Margin Ratio" which is based on pump type. The text explained that "Low Suction Energy" pumps are not prone to experience cavitation damage, even in the relatively low flow regions. In contrast, "High Suction Energy" pumps will suffer a measure of cavitation damage, especially when operating in the low flow suction recirculation region, where insufficient NPSH margins can further exacerbate matters. These important margins are discussed in our text; the detailed calculations in Chapter 5 support the trends of percent NPSHr versus percent of flow Q at BEP, shown in Figure 17-2. Observing these trends is of special importance for "High" and "Very High Energy" pumps.

GETTING INTO MECHANICAL ISSUES

The vulnerability of operating process pumps in parallel is not always appreciated by pump purchasers, although API-610 clearly advises against parallel operation for pumps with relatively flat performance curves. Reasonable, yet general, specifications require a 10% minimum head rise from BEP to shut-off.

There are problems with short elbows near the suction nozzle of certain pumps and flow stratification and friction losses are sometimes overlooked. Some sources advocate a minimum of 5; others advocate a 10-diameters equivalent of straight pipe run at the pump suction. (See also Tables 3-1 and 3-2, Chapter 3.) Together, pump parallel operation and piping issues make up our *Highlight* 2.

> **HIGHLIGHT 2: Avoid parallel operation unless head rise from BEP to shutoff is 10%. Beware of close elbows and wrong elbow orientation in double-flow pumps. Install eccentric pipe reducers with the correct orientation**

The tie-in between the lack of conservatism in piping and issues of less-than-adequate pump reliability is tenuous; still, the multipoint trouble illustration in Figure 17-1 is of interest here. Suffice it to say that tight-radius elbows and incorrect pipe reducer orientation can quickly wreck certain pump configurations (Ref. 17-3). Neglecting piping issues can be a costly mistake, but ours is primarily a pump text. Pulling piping into place at a pump nozzle can cause edge-loading of the pump's bearings, which will lead to premature bearing failures. Did soil settlement under pipe supports play a role in misalignment? Concrete driveway sections misalign a few scant years after construction, so why would pipe supports in a user facility still be vertical decades after they were first installed? Soil settlement and the stability of pipe supports should be included in periodic plant audits. By listing *Highlight* 2 the authors want to make users aware of hydraulic and flow separation issues. The flow velocity at the small-radius wall of an elbow will differ from the velocity at the large-radius wall and flow straighteners may be required in a particular service.

DEVIATIONS FROM BEST AVAILABLE TECHNOLOGY

User plants will usually get away with one or two small deviations from best available technology. But when three or more deviations occur, failure risks usually increase exponentially. That said there are a number of reasons why a few well-versed reliability engineers are reluctant to accept pumps that incorporate the drive end shown in Figure 17-3 (Ref. 17-4). The short overview of reasons is that truly focused professional take seriously their obligation to consider the actual, *lifetime-related* and not just *short-term*, cost of ownership. For instance, specialists realize that the bearing housing in Figure 17-3 may work initially and then fail prematurely. Vendor-manufacturers may not be aware of features they had overlooked. Note that this bearing housing is shown exactly as originally provided by the pump manufacturer; it includes several risk-increasing features. Allowing these features to exist will sooner or later hurt the profitability of users and vendors alike. All are related to lube application and process pump users should pay very close attention to these and other lube application matters. These matters lead us to *Highlight 3*, followed by a description of vulnerabilities.

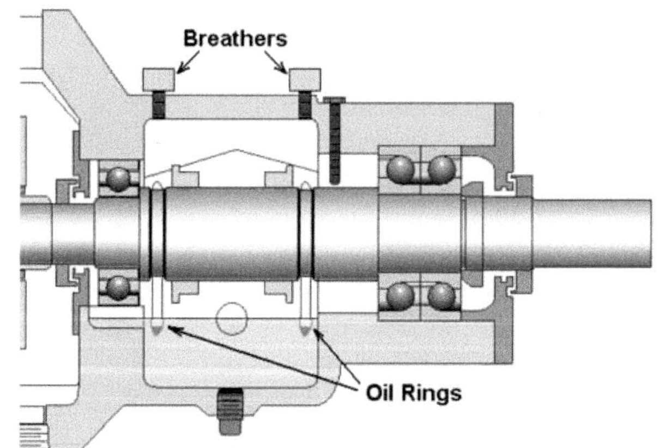

Figure 17-3: A bearing housing with several potentially costly vulnerabilities

> **HIGHLIGHT 3: Oversights can affect the adequacy of lube application**

Upon examining Figure 17-3 a careful reader can be certain of five facts:

- In Figure 17-3, oil rings are used to lift oil from the sump into the bearings. These oil rings tend to skip and jump at progressively higher shaft surface speeds, or if not perfectly concentric, or if not operating in perfectly horizontal shaft systems.
- As the pump is transported from shop to field, an oil ring can become dislodged and get caught between the shaft periphery and the tip of the long limiter screw.
- The back-to-back oriented thrust bearings of Figure 17-3 are not located in a cartridge. This limits flinger disc dimensions (if they were to be retrofitted) to no more than the housing bore diameter.
- Bearing housing protector seals are missing from the picture in Figure 17-3. Once added, bearing protector seals in this housing could change the flow of venting air.
- Although the bottom of the housing bore (at the radial bearing) shows the needed oil return passage, the same type of oil return or pressure equalizing passage seems to have been left out near the 6- o'clock position of the thrust bearing. A small pool of oil can accumulate behind the thrust bearing and this oil will probably overheat. Detrimental carbon debris could form.
- No particular constant level lubricator is shown in Figure 17-3 and there is uncertainty as to the type or style of constant level lubricator that will be provided. Unless specified, OEMs may opt to provide the least expensive constant level lubricator.

It should be noted that the angular contact thrust bearings in Figure 17-3 will usually incorporate cages (ball separators) that are angularly inclined, which means they are arranged at a slant. These cages often act as small impellers (Ref. 17-5), and impellers promote flow from the smaller towards the larger of the two diameters. This is more readily evident from Figure 17-4, and particular attention should be given to windage created by the impeller-like air flow action of an inclined bearing cage. In many cases, the pump is designed with an oil ring to the left of this bearing. While the design intent is for oil to flow from left to right, windage from an inclined cage will act in the opposite direction.

HIGHLIGHT 4: Windage in angular contact ("AC") bearings can oppose oil flow

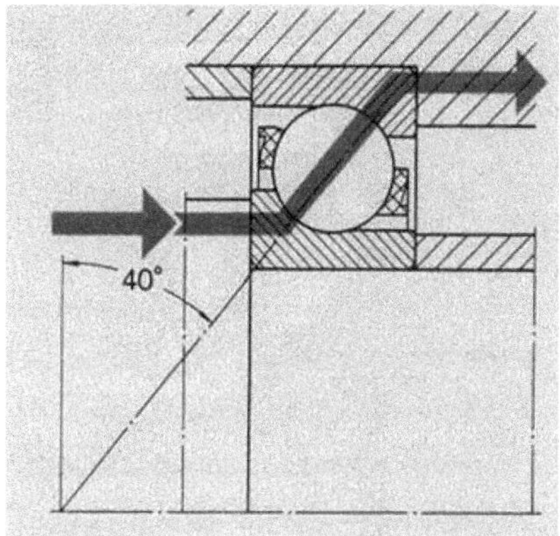

Figure 17-4: Attempts to apply lubricant in the direction of the arrow (oil flow from left to right) meet with windage (air flow right-to-left) from an inclined cage. The two directions often oppose each other (Ref. 17-5)

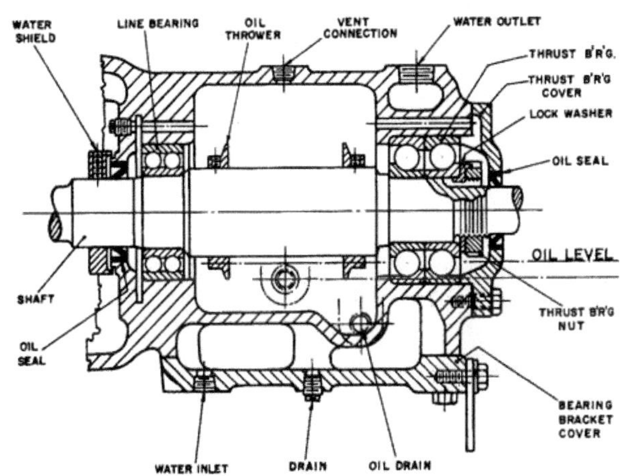

Figure 17-5: The oil level in this 1960s-vintage housing was set for low-to-moderate speed pumps. Oil throwers create a spray that overcomes windage; the two throwers also prevent oil stratification. Pressure equalization passages are drilled near the top of all bearings (Ref. 17-6)

Windage is listed as *Highlight 4*. Alleviating or eliminating windage is an important step in reducing the risk of premature bearing failure. This risk is greater in some bearing housing configurations, or with particular means of lube application. Accordingly, even purely precautionary steps to abate unequal pressures inside a bearing housing make much sense. Unless there is pressure equalization, oil flow may be disturbed.

Note how carefully the now defunct Worthington Pump Company ascertained that pressures on each side of a bearing were equalized. Worthington went through the trouble of drilling balance holes right above the bearings, Figure 17-5. There needs to be pressure equalization on each side of a bearing—it's all about risk reduction (Refs. 17-4 and 17-6).

By way of overview, we note that one of the oldest and simplest methods of oil lubrication consists of an oil bath through which the rolling elements will pass during a portion of each shaft revolution (Figure 17-5).

However, this "plowing through the oil" may cause the lubricant to heat up significantly and should be avoided on susceptible process pumps. There's excessive heat generation risk whenever dn, the mean distance from diametrically opposite rolling element centers as expressed in mm x rpm, exceeds a particular number. That 6-digit number ranges from 150,000 to perhaps 300,000. It is predetermined by bearing manufacturers who estimate at what point churning and heat buildup will exceed desired limits. The manufacturers then advocate lowering the oil level so that it no longer contacts the rolling elements. In essence, as a certain "dn" threshold is exceeded, some other means of lifting oil into the bearing must be chosen (Ref. 17-1).

Aiming to stay within the inch-system preferred by pump users in the United States, a number of users and bearing manufacturers found that the ratios of bearing outside diameters (OD) to bearing inside diameters (ID) are similar in the bearing sizes typically used in process pumps. This allows users to focus on a simplified approximation, DN, the product of shaft diameter (D, inches) times revolutions-per-minute (N, rpm). Whenever DN exceeds 6,000 and so as to avoid risking excessive heat buildup, oil levels reaching the ball center or the lower third of the lowermost rolling element are considered a churning risk. In that case, some other means of lifting oil into the bearing are chosen.

Note also the cooling water jacket in Figure 17-5. Bearing housing cooling is not needed on process pumps which incorporate rolling element bearings. Cooling is harmful if it promotes moisture condensation (water cooling coils) or restricts thermal expansion of the bearing outer ring (water cooling jacket). In 1967, these concerns were seen to influence pump reliability. The jacketed cooling water passages in Figure 5 were from then on left open to the ambient air environment. The decision to delete cooling water from pumps with rolling element bearings was first implemented in 1967 at an oil refinery in Sicily. The owner's engineers had recorded bearing lube oil in four identical pumps reaching an average of 176 degrees F with cooling water in the jacketed passages. Without cooling water, the lube oil averaged 158 degrees F, which is 18 degrees F cooler. The bearings now lasted much longer. These findings and experiences were shared with all those that were willing to read, or willing to listen (Ref. 17-7). At least 20 reference texts could be cited.

As of 2013, not everybody has acted on the message. That is why cooling water issues are listed here as *Highlight 5*. Water is becoming an increasingly precious resource and cooling water is very often responsible for actually reducing (!) the life of rolling element bearings in process pumps. There are two possible reasons for that, and more information can be found in Refs. 17-1 and 17-4:

(a) If a cooling water jacket surrounds the bearing, heat input and thermal growth may take place in the bearing inner ring, while the outer ring (due to its then lower temperature) may not expand as much. Consequently, the rolling element clearances may be reduced to the point of excessive preload and early bearing failure.

(b) If a cooling water coil is provided in the oil sump, the "air space" region in the upper two-thirds of the bearing housing may become cold enough for water vapors to condense. Water may then contaminate the lubricant and shorten bearing life.

HIGHLIGHT 5: Cooling water can cause bearings to run hot

MORE ON LUBRICATION AND BEARING DISTRESS

Only 9% of all bearings actually reach their as-designed life, and lubrication-related issues are often at fault in the estimated 50%-60% of pump failures that involve bearing distress. Obviously, having the correct oil level should be a consideration in bearing housings with rolling element bearings. Oil level settings are part of our progressive investigation of elusive failure causes in process pumps. Understanding where to set oil levels is *Highlight 6*.

HIGHLIGHT 6: Understand where to set oil levels and how pressure balance is needed to maintain an oil level setting

The traditional oil sump was depicted (in Figure 17-5) with the lubricant reaching to about the center of the lowermost bearing elements. This arrangement works well at low shaft surface velocities. To gain reliability advantages, synthetic lubricants, oil mist application (called "oil fog" in some languages) and *liquid-oil jets* (also known as "oil spray") are often used. Oil jet lubrication existed before the development of plant-wide oil mist systems (Ref. 17-8).

Circulating systems also merit consideration in certain high-load or very large pumping services. Generally speaking, circulating systems are selected for large pumps utilizing sleeve bearings. In these systems, the oil can be passed through a heat exchanger before being returned to the bearing. However, regardless of lube application method on rolling element bearings, cooling will not be needed as long as high-grade mineral or synthetic lubricants are utilized (Refs. 17-4 and 17-7). High-performance mineral oils developed after 2010 are deliberately mentioned here. Since 2011, some of these have become more cost-effective in situations where synthetics may be "overkill"—among them oil mist systems in the U.S. Gulf Coast region.

Irrespective of base stock and oil formulation, the required lubricant viscosity is a function of bearing diameter and shaft speed. Technical reasons are described in numerous books and articles, among them Refs. 17-4, 17-9, and 17-10. Most process pump bearings will reach long operating lives if the oil viscosity (at a particular operating temperature) is maintained in a range from 13 to 20 cSt (Ref. 17-11). Whenever oil rings are used to "lift" the oil from sump to bearings, the need to maintain a narrow range of viscosities takes on added importance (Ref. 17-12). In the special case of the same bearing housing containing both rolling element and sliding bearings, it will be prudent to address the implications of (some) oil rings not being able to function optimally in the higher viscosity (ISO Grade 68) lubricant that's often needed for rolling element bearings. The oil ring may have been designed to cater to sleeve bearings, which normally work best with a lower viscosity lubricant, but VG 32 *mineral* oils are rarely a best choice for rolling element bearings in pumps. High performance *synthetic* VG 32 is usually the best choice where different bearing styles share the same housing (Ref. 17-4). In these situations, synthetics are not "overkill."

In all cases overheating the oil must be avoided, but synthetics can tolerate much higher temperatures than mineral oils. Whenever large bearings are used at operating speeds from 3,000 to 3,600 rpm, allowing oil levels to reach the center of the lowermost bearing ball or roller is to be avoided. Because the "plowing effect" of rolling elements in a flooded sump produces frictional power loss and heat, an oil level below that indicated in Figure 17-5 is then often chosen and provisions are made to "lift" the oil. A widely accepted empirical rule calls for lower oil levels and "lifting" whenever DN > 6,000 (in this expression D = shaft diameter, inches, and N = shaft rpm). Another, separately derived empirical rule, allows shaft peripheral velocities no higher than 2,000 fpm in bearing housings where the oil sump level is set to reach the center of the lowermost rolling element.

In oil mist lubrication systems it is generally understood that with shaft surface velocities in excess of 2,000 fpm, windage effects are opposing the flow of oil mist. As this is being observed, uninformed or baffled oil mist users have, in some cases, reverted back to conventional oil lubrication. In sharp contrast, reliability-focused users have, for many decades, installed directed oil mist reclassifiers to overcome windage at >2,000 fpm. The mist dispensing opening in these reclassifiers is located ~0.2-0.4 inches from the rolling elements. Thousands of these have been supplied and used with total success. This information is available from dozens of texts and articles (Refs. 17-10, 17-13, and 17-14).

Again, once the shaft peripheral velocity exceeds 2,000 fpm, the oil level should be no higher than a horizontal line tangent to the lowermost bearing periphery. This means there should be no contacting of the oil level with any part of a rolling element and oil "lifting" is needed.

Assume that Figures 17-3 and 17-6 represent situations where DN > 6,000. Therefore, and because initial cost was to be minimized, either oil rings (Figure 17-3) or shaft-mounted flinger discs (Figure 17-6) were chosen. Both arrangements are available to lift the oil, or to somehow get the oil into the bearing by creating a random spray. Shaft-mounted flinger discs (Figure 17-6) are well represented in many European-made pumps. If properly designed, their operating shaft peripheral speed range exceeds that of oil rings.

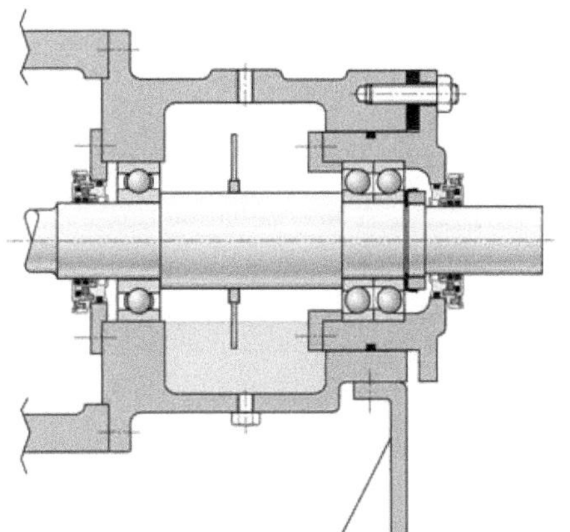

Figure 17-6: A bearing housing with a cartridge containing the thrust bearing set. The bearing housing bore is slightly larger than the diameter of the steel flinger disc, making assembly possible. The drawing does not show the needed oil return passage at the 6-o'clock bearing positions (Refs. 17-4 and 17-11).

TWO DIFFERENT DN-RULES EXPLAINED

When determining oil level settings, either of two empirical rules could be applied. To illustrate *Rule (1)*: A 2-inch bore bearing at 3,600 rpm, with its DN value of 7,200, would operate in the risky or ring instability-prone zone > 6,000. Equipment with a 3-inch bore bearing operating at 1,800 rpm (DN = 5,400) might use oil rings without undue risk of ring instability. In another example, using *Rule (2)*: A 3-inch (76 mm) diameter bearing bore at 3,600 rpm would operate with a shaft peripheral velocity of (3.14D/12)(3,600) = 2,827 fpm (~14.4 m/s), which would disqualify oil rings from being considered for highly reliable pumps. The fact that a pump manufacturer can point to satisfactory test stand experience at higher peripheral velocities is readily acknowledged, but field situations represent the "real world" where shaft horizontality and oil viscosity, depth of oil ring immersion, bore finish and out-of-roundness are rarely perfect. We can thus opt for using either the *DN < 6,000* or the *surface velocity < 2,000 fpm*, or the lesser of these two "real-world" rules-of-thumb.

Either way, the vendor's test stand experience is of academic interest at best. Pump manufacturers test under near-ideal conditions of shaft horizontality, oil ring concentricity and immersion, oil level and lubricant viscosity. Users might ask how often they have observed non-round oil rings, or rings that have shaft radius wear marks (from shaft fillet radii) on one side of the ring. If the answer is "never," perhaps another look will be warranted. For the reliability-focused, the wide-ranging field experience that led to these two rules-of-thumb will govern over all else.

The cartridge approach shown in Figure 17-6 has been in use for an estimated 50 or 60 years on thousands of open-impeller ANSI pumps because it facilitates impeller position adjustment in the axial direction. The same cartridge approach may be needed to dimensionally accommodate flinger discs (Figure 17-6) instead of vulnerable oil rings (Figure 17-7). Of course, cartridge-mounted bearings are a cost-adder and you may hear claims that the benefit-to-cost-ratio will not justify upgrading to cartridges. However, with the average API pump repair costing slightly over $10,200 at a Texas oil refinery and $11,000 at an oil refinery in Mississippi, we might be surprised at the payback multiplier. Even a single avoided failure over the 30-year total life of a pump will probably pay for it many times over.

THE TROUBLE WITH OIL RINGS AND CONSTANT LEVEL LUBRICATORS

Issues with oil rings are found in many scholarly works (Refs. 17-15 through 17-18). On a website post in September 2012, the Malaysian equivalent of OSHA alerted us to catastrophic failures brought on by oil rings (www.dosh.gov.my, Ref. 17-19). All of these sources observed problems with oil rings, although an industry source opined (in 2011) that "ring lubrication is an accepted practice and it would take user consensus to damn it." But, history shows us that innovations are rarely driven by consensus. If they were, the Wright Brothers would have worked on repeat pump and bicycle repairs instead of developing a powered flying machine.

Meanwhile, keep in mind that this tutorial is for the reliability-focused. Nothing will convince those who accept without questioning dozens of repeat failures of centrifugal pumps at their plants. Many illustrations of failed oil rings are available. Studies, observations and measurements have shown their field reliability in process pumps out of harmony with the quest for higher reliability and availability. Work described in Refs. 17-12 and 17-15 recommends oil ring concentricity within 0.002 inches. However, in 2009, shop measurements were performed by the author at a pump user's site in Texas. The oil rings measured in 2009 exceeded the 0.002-inch allowable out-of-roundness tolerances by a factor of 30 (Ref. 17-20).

Experience shows that oil rings are rarely the most dependable or least-risk means of lubricant application. They tend to skip around and even abrade (Figure 17-7) unless (a) the shaft system is truly horizontal, (b) ring immersion in the lubricant is correct, (c) ring eccentricity is within limits, (d) surface finish is within range, and (e) oil viscosity is within tolerance. Needless to say, for each of these 5 parameters to be within close limits in actual operating plants is as rare as ostrich hatchlings walking away from a duck's eggs. While deviations in one parameter might have only a small negative effect on bearing life, having three or four out-of range parameters will have more serious consequences.

HIGHLIGHT 7: Flinger discs can outperform oil rings. Oil rings must be concentric within 0.002 inches and maintaining that degree of concentricity mandates a stress-relieving step before finish-machining

Highlight 7 is frequently heeded by the reliability-focused; they often specify and select pumps with flinger discs. Flinger disc-equipped pumps might be important for shipboard services. Although sometimes used in slow speed equipment to merely prevent temperature stratification of the oil (see Figure 17-5), the larger diameter flinger discs (Figure 17-6) serve as efficient (non-pressurized) oil distributors at moderate speeds. Of course, the proper flinger disc diameter must be chosen and solid steel flinger discs should be preferred over plastic materials. Insufficient lubricant application results if the diameter is too small to dip into the lubricant; conversely, high operating temperatures can be caused if the disc diameter is much too large or if no thought was given to its overall geometry.

Many European-made pumps incorporate flinger discs ("oil throwers"), and so does at least one U.S. manufacturer. Flexible flinger discs have been used to enable insertion in some "reduced cost" designs, i.e., configurations where the bearing housing bore diameter is smaller than the flinger disc diameter. To accommodate the preferred solid steel flinger discs, bearings must be cartridge-mounted (Figure 17-6). Using a cartridge design, the effective bearing housing bore (i.e., the cartridge diameter) is made large enough for passage of a steel flinger disc of appropriate diameter. Over the years and in the pursuit of reducing manufacturing cost there have been attempts to get around the use of oil rings. Roll pins inserted transversely in pump shafts and flexible (plastic) flinger discs have brought mixed results and marginal improvement at best and negative bearing reliability at worst. Cheap discs pushed on the shaft became a source of failure and were disallowed by API-610 in about 1995. Unsuitable plastics and disc configurations chosen without the benefit of sound engineering practices have also not been sufficiently reliable.

We estimate the incremental cost (comprising material, labor, CNC production machining processes) of an average-size (30 hp) process pump with cartridge-mounted bearings at $300. The value of even a single avoided failure was earlier estimated at over $10,000 and the benefit-to-cost ratio would thus exceed 33-to-1.

HIGHLIGHT 8: Oil rings can become unstable; skip, scrape, abrade

The shortcomings of oil rings were known in the 1970s, *Highlight 8*. A then well-known pump manufacturer claimed superior-to-the-competition products. This manufacturer's literature pointed to an "anti-friction oil thrower [i.e., a flinger disc], ensuring positive lubrication to eliminate the problems associated with oil rings." An illustration was shown earlier (Figure 9-23).

Figure 17-7: Oil rings in as-new ("wide and chamfered") condition on left, an abraded, meaning badly worn and now without chamfer ring is on the right side. Record both before versus after widths (*Highlight* 9, also Ref. 17-4)

About two decades later, in 1999, at least one major pump manufacturer saw fit to examine the situation more closely. In a comprehensive paper the manufacturer described remedial actions which included Grade 46 oil viscosity and oil rings made of high performance polymers (Ref. 17-21). However, the problem did not go away. Users in Canada reported that black oil persisted, and so did repeat failures, even after adopting non-metallic oil rings.

Black oil can easily be traced to one of two origins. A simple analysis will point to either overheated oil (i.e., carbon) or will detect slivers of elastomeric "dynamic" O-ring material from components that operate too close

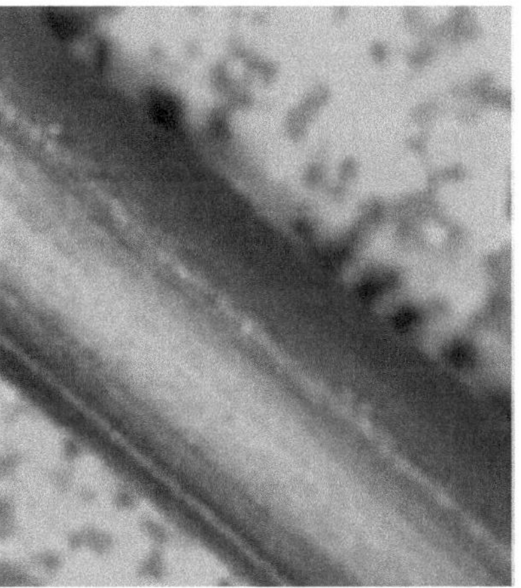

Figure 17-8: O-ring damaged by close contact with sharp edged rotating part

to sharp-edged O-ring grooves (Figure 17-8).

HIGHLIGHT 9: Measure oil rings new and after use. Any width difference must be attributed to ring abrasion which, of course, limits bearing life

Unidirectionality of constant level lubricators must be observed; it is described in at least one manufacturer's literature (Ref. 17-22). Observant reliability professionals have noted that caulking (where transparent bottles meet die-cast metal bases) will, over time, develop stress cracks (fissures). Rain water can then reach the lubricant via capillary action. Accordingly, bottle-type constant level lubricators are a preventive maintenance item; they should be replaced after 4 or 5 years of service (Ref. 17-4).

Note also how, in Figure 17-9, the oil level in the bearing housing is no longer reaching the rolling elements. This constant level lubricator lacks pressure balance. Any pressure increase in the space above the liquid oil will drive the oil level down. For a while, the top layer of oil will overheat; carbon will form and black oil will appear in the glass bulb. Increasing temperature in the closed space causes a further pressure increase and the oil level decreases even more. Oil then no longer reaches the rolling elements and another bearing failure is likely.

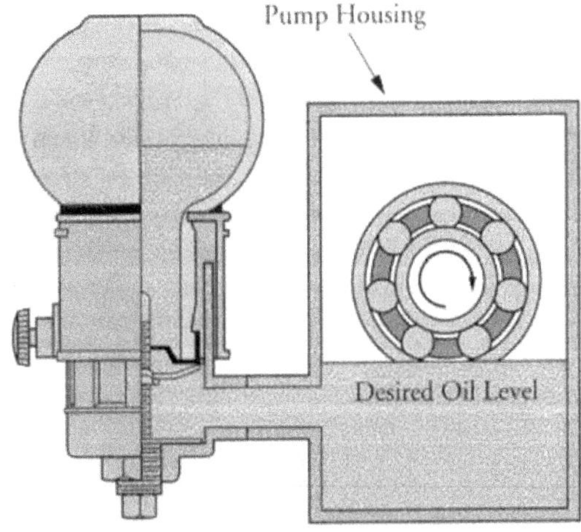

Figure 17-9: Pressure non-balanced constant level lubricator

The seemingly similar lubricator in Figure 17-10 is configured for a balance line which ensures that the oil levels in the die-cast lubricator support (or at the edge of the slanted tube shown in this illustration) and in the pump bearing housing are always exposed to the same pressure (Ref. 17-22). Undersized balance lines can exist; either a generous diameter hard pipe or a suitably sized stainless steel hydraulic balance line is favored. If constant level lubricators cannot be avoided, a pressure-equalized model or arrangement (Figure 17-10) is recommended.

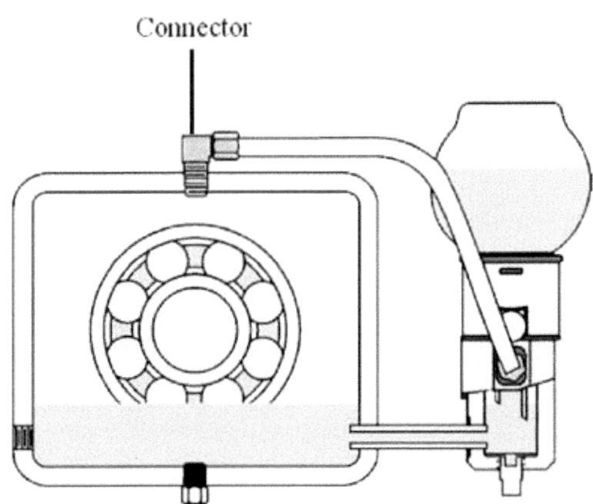

Figure 17-10: Pressure-balanced constant level lubricator. A relatively large diameter balance line is needed to prevent drops of oil from obstructing the line.

Again, bearing distress is inevitable if a constant level lubricator fails to maintain the desired oil level. An incorrect level setting can be caused by a number of factors. It will be clear from Figure 17-9 that even small increases in the bearing housing-internal pressure can hasten the failure risk. Suppose there is heat generation and because of the addition of bearing protector seals the air no longer escapes and there's a lack of housing-internal pressure balance. Perhaps the reasons why Worthington had included housing-internal balance holes in Figure 17-5 have been forgotten, but these balance holes served a purpose. The result of heat generation might well be that the housing-internal pressure goes up. As the housing-internal pressure rises ever so slightly, it will exceed the ambient pressure to which the oil level at the wing nut or slanted tube in the bulb holder portion of the constant level lubricator in Figure 17-9 is exposed. According to the most basic laws of physics, a pressure increase in the bearing housing causes the oil level near the bottom of the bearing inner ring shoulder (Figure 17-9) to be pushed down. Lubricant will no longer reach the bearing's rolling elements; oil turns black, and the bearing will fail quickly and seemingly randomly. A summary is captured in *Highlight 10.*

To re-state: At DN > 6,000 and to satisfy minimum requirements in a reliability-focused plant environment, a stainless steel flinger disc fastened to the shaft will often perform well. Such a disc will be far less prone to cause unforeseen outages than many other presently fa-

HIGHLIGHT 10: Constant level lubricators must be installed on the up-arrow side. If the shaft rotation is clockwise, the up-arrow is on the left. If it rotates counterclockwise, the up-arrow is on the right.
<<>> Reliability-focused users insist on pressure-balanced models.
<<>> PM-driven replacements are needed because caulking has finite life.
<<>> Declining oil levels can rapidly cause formation of overheated "black oil."

vored methods. Remember that traditional oil rings will abrade and slow down if they contact a housing-internal surface. They are sensitive to oil viscosity and depth of immersion, concentricity and RMS surface roughness.

If you upgrade to flinger discs, you are accepting the findings of the legacy manufacturer whose advertisement was illustrated earlier in Figure 9-23. That manufacturer's findings were backed by facts. Still, it must be ascertained that flinger discs are used within their applicable peripheral velocity so as to contact the oil and fling it into the bearing housing (Ref. 17-2). The flinger disc O.D. must exceed the outside diameter of the thrust bearing and this dimensional requirement strongly favors placing the outboard (thrust) bearing(s) in a separate cartridge. Providing such a cartridge will add to the cost of a pump, as will the cost of a well-designed flinger disc. However, in most cases, the incremental cost will be considerably less than what it would cost to repair a pump just once.

BEARING HOUSING PROTECTOR SEALS

At the risk of stating the obvious: Let's be sure the lube in a pump's bearing housing is kept clean. Even the most outstanding lubricant cannot save a bearing unless the oil is kept clean. This is where bearing housing protector seals are of value (Ref. 17-4).

Lubricant contamination originates from a number of possible sources and can also be a factor in "unexplained" repeat failures. Unless process pumps are provided with suitable bearing housing seals, an interchange of internal and external air (called "breathing") takes place during alternating periods of operation and shutdown. Bearing housings "breathe" in the sense that rising temperatures during operation cause air volume expansion, and decreasing temperatures at night or after shutdown cause air volume contraction. Open or inadequately sealed bearing housings promote this back-and-forth movement of moisture-laden and dust-containing ambient air. But, simply adding bearing protector seals could change windage or housing-internal pressure patterns in unforeseen ways. This, too, we must recognize as a potential source of "unexplained" failures in housings without internal balance holes (see Figure 17-5).

Ideally, housings should not invite breathing and the resulting contamination. There should be little or no interchange between the housing interior air and the surrounding ambient air. The breather vents shown earlier in Figure 17-2 can often be removed and plugged. Don't be shocked by that statement. Many hundreds of millions of refrigerators and automotive air conditioning systems operate with neither vents nor breathers. Conceivably, old-style bearing housing seals allow O-rings to contact a groove, as depicted in Figure 17-11. Contact with sharp-edged grooves invites dynamic O-rings to scrape.

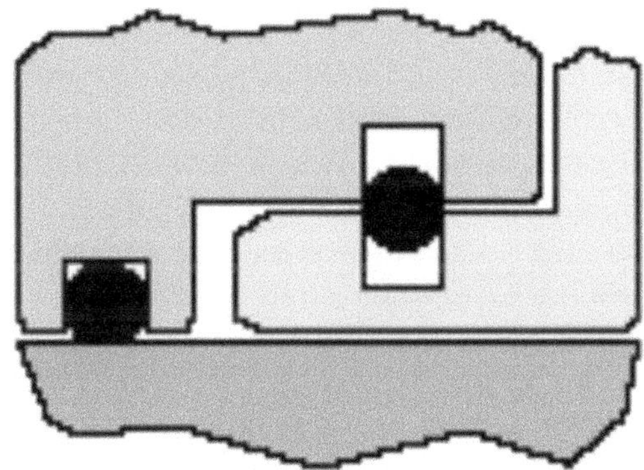

Figure 17-11: Visualizing component damage risk. Note that the sketch does not replicate an actual product. The illustration merely highlights what can happen with bearing protector seal designs which incorporate sharp-edged grooves. It reminds us that we should become familiar with how parts work and how they might fail.

Abraded elastomer shavings can contaminate the lubricant and cause oil to change color, *Highlight 11*. Also, using only a single O-ring for clamping the rotor to the shaft makes the rotor less stable than if two rings are used for clamping duty.

Visualize rotor instability by mentally removing the stationary component in Figure 17-11. The rotor pivots around the clamping O-ring and destructive vibration would occur at high speeds. We could study the rotor dynamics of such a situation or we might reach the same common-sense conclusion by giving it some thought. Two clamping O-rings will provide more stability than one single clamping O-ring.

HIGHLIGHT 11: Dynamic O-rings in contact with sharp corners will fail prematurely

In essence, bearing housing protector seals can greatly improve both life and reliability of rotating equipment by safeguarding the cleanliness of the lubricating oil. However, these protector seals add little value if oil contamination originates with oil ring wear, or if pressure-unbalanced constant level lubricators are used that allow air and moisture to intrude, or if the oil is not kept at the proper level, or if the bearing housing design disregards windage concerns, or if water enters into the oil.

We know all about see-through containers at the bottom of the pump bearing housing. However, by the time water becomes visible in such a "sludge cup container," the saturation limits of oil-in-water will have been exceeded and much damage could have been done to the bearings. We can deduce that free water in the oil is a symptom of not having the right bearing housing protection. Our reliability focus should be on treating the root cause, not the symptom. We should prevent water from reaching the bearings in the first place. These proactive and precautionary thought processes are at the core of this tutorial on pump failure prevention.

RANKING THE DIFFERENT LUBE APPLICATION PRACTICES

Although oil ring lubrication is widely used, it is relatively maintenance-intensive and ranks last from the author's experience and risk reduction perspective. Next, flinger discs have been used for many decades and allow operation at higher DN values than oil rings. Because they are firmly clamped to the shaft there is far less sensitivity to installation and maintenance-related deviations. On the other hand, non-clamped flinger discs were tried a few decades ago, and with very disappointing results. API-610 disallows push-on flingers and some other low-cost oil application components (Ref. 17-23).

HIGHLIGHT 12: Pure oil mist represents many decades of fully proven superior technology

Plant-wide oil mist lubrication systems are ranked ahead of flinger discs. Oil mist has proven superior to conventional lubricant application since the late 1960s. Pump bearing failure reductions ranging from 80 to 90% have been reported by Charles Towne of Shell Oil, and many others (Refs. 17-25 through 17-29). Charles Towne performed tests on identical process units at Shell Oil and deserves much credit for seminal work on the subject. See *Highlight 12* for a summary.

The highly beneficial in-plant, real-life results reported by Towne (Shell Oil) refer to pure oil mist, not purge mist. Pure oil mist is an oil-air mixture with a volumetric ratio of 1: 200,000. The oil is atomized to globule form and carried by the air, applied in modern plants as shown in Figure 17-12. The same illustration, Figure 17-12, could be used to depict liquid oil spray. Liquid oil spray is sometimes called "jet oil" lubrication (Refs. 17-4 and 17-8) and differs from oil mist.

HIGHLIGHT 13: Bearing manufacturers rank spraying liquid oil into a bearing's cage a bit higher than the widely practiced (and quite cost-effective) oil mist (oil fog) application

These facts and findings are summarized in *Highlight 13*. And so, with regard to the introduction of liquid (not misted) lube oil into rolling element bearings, Figure 17-13 implies a number of very important recommendations for the truly reliability-focused:

- It hints that pump bearing housings need not be symmetrically configured. (Asymmetry is visualized by looking into the housing. In Figure 17-14, the distance to the right edge of the bearing housing is not the same as the distance to the left edge of the bearing housing. The additional volume thus gained will accommodate a small oil pressurization pump; this small pump is to be arranged inside the process pump's bearing housing).
- A box-like geometry with a flat cover and ample space to incorporate a wide range of oil pumps is feasible. Box-like bearing housings for process pumps would open up a host of new and inventive solutions. These might incorporate shaft-driven or other reliable self-contained means or oil application pumps (Ref. 17-30). The oil application pump would possibly take suction from an increased-size oil sump
- The main process pump shaft need not be in the geometric center of the box.

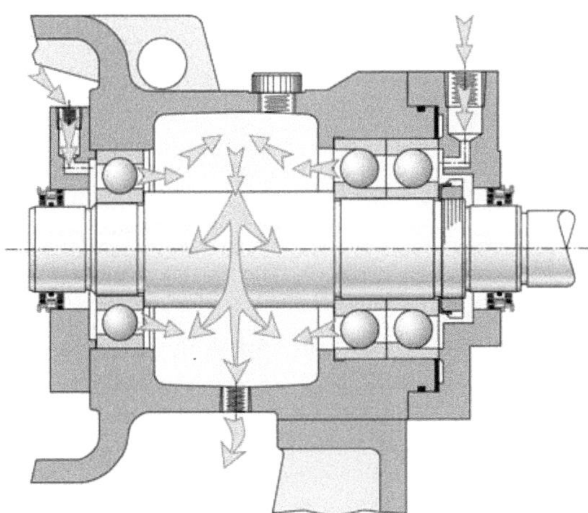

Figure 17-12: Oil *mist* lubrication applied to a pump bearing housing in accordance with API-610, 10th Edition (Ref. 17-14). With oil *spray* lubrication, liquid oil would enter at the nozzles. Note dual *mist* (or, for spray lube application, dual *liquid oil*) injection points. Observe dual-face bearing housing seals that prevent oil *mist* (or oil *spray*) from escaping to atmosphere.

Figure 17-13: Drive arrangements similar to the highly reliable mechanical governor drive in this small steam turbine are suggested for bearing housing-internal lube oil pumps in process pumps.

- Flat surfaces would invite clamp-on, screw-in or flange-on oil pumps
- Oil pressurized by the oil application pump would be routed through a filter and hydraulic tubing to spray nozzles incorporated in the end caps. Therefore, the cross-section view of a bearing housing with oil spray would be identical to the one shown for oil mist in Figure 17-12 (Ref. 17-14)
- Internal pressure equalization and windage issues would never again be a concern.
- The incremental cost of superior bearing housings would be more than matched by the value of avoided failures

In Figure 17-12 and with either oil mist or oil spray there would be no oil rings, flinger discs, or constant level lubricators. Because the mist (or spray) application nozzles shown here are relatively close to the bearings oil mist flow, or the stream of liquid oil, will overcome windage. While this jet oil or oil spray lube application method seems like a bold idea, the method is extensively documented by MRC and SKF, also in at least 7 of our many reference texts (among them Ref. 17-7). This lubrication method is very often used in high performance aircraft.

- The duty imposed on self-contained oil spray pumps (a small pump inside the bearing housing of a process pump) would be quite benign compared to other known, reliable, shaft-driven pumping technologies or services
- Oil filtration would be easy
- The elimination of oil rings and constant-level lubricators would be a very positive reliability improvement step
- Part of the energy requirement of an oil application pump would be re-gained in the form of reduced bearing frictional losses

With spray lubrication, much needed oil application innovation would benefit the drive end and thousands of repeat failures of pumps would no longer occur. However, as of today, little interest has been shown by manufacturers and users to redesign pump bearing housings.

The market drives these developments, *Highlight 14*. If buyers and pump owners tolerate repeat failures and the manufacturers benefit from the sale of spare parts, it will be business as usual. Still, and at the risk of stubbornly bucking the trend: As responsible engineers, we should advocate changes in mindsets. As realists we are under no illusions as to where some users and manufacturers will be when the dust settles: We will never convince or even reach some of them. All we wanted to do is explain matters to those whose reliability focus extends beyond "business as usual" and who are interested in pushing for lower-risk oil application alternatives.

One of the most straightforward ways to drive a housing-internal oil pump could be modeled on the right-angle worm drives typically found in small steam turbines. While the arrangement shown in Figure 17-13 is associated with a mechanical governor, it is shown here as but one of many highly reliable options that merit consideration for small oil pumps that take suction from the process pump's oil sump and pressurize it.

Among the possibilities worthy of examination is reconfiguring the portion of the equipment shaft which is located between the radial bearing and the thrust bearing. It might be possible to "re-contour" this shaft section to become the rotor of a progressive cavity pump. After routing the pressurized lube oil exiting from such a housing-internal pump through a downstream spin-on filter, the pressurized oil would be sprayed into nozzles which direct the oil into the process pump bearings.

HIGHLIGHT 14: Advocate risk reduction—self-contained pump bearing lubrication

Recall again that all bearing manufacturers consider spraying liquid oil into the rolling elements the best possible lubrication method. Since 1958 (and, possibly, even earlier) bearing manufacturers have ranked an oil spray (liquid "oil jet") ahead of oil mist lubrication and far ahead of oil rings (slinger rings). That's a compelling fact which should not be ignored.

Motivated reliability professionals and informed users can avoid these and will appreciate recommendations on failure risk reduction. For the truly reliability-focused pump users, a number of conclusions and upgrade recommendations may be of interest:

1. Discontinue using maintenance-intensive oil rings and, if possible, constant level lubricators.
2. As a matter of routine, the housing or cartridge bore should have a passage at the 6 o'clock position to allow pressure and temperature equalization and oil movement from one side of the bearing to the other. Note that such a passage was shown in Figure 17-3 for the radial bearing, but not for the thrust bearing set.
3. With proper bearing housing protector seals and the right constant level lubricators, breathers (or vents) are no longer needed on bearing housings. The breathers (or vents) should be removed and one of the openings in Figure 17-3 can often be plugged.
4. If constant level lubricators are used, a pressure-balanced version should be supplied and its balance line should be connected to the closest breather port.
5. Bearings should be mounted in suitably designed cartridges and loose slinger rings (oil rings) should either be avoided or, in some high DN cases, disallowed.
6. Suitably designed flinger discs should be secured to the shaft whenever the oil level is lowered to accommodate the need to maintain acceptable lube oil temperatures (i.e., for pumps operating with DN-values in excess of 6,000).
7. Modern and technically advantageous versions of bearing housing protector seals should be used for both the inboard and outboard bearings. Lip seals are not good enough, and neither are outdated rotating labyrinth seal designs.
8. Understand that the implementation of true reliability-thinking must strongly support moves away from traditional bearing housings. These moves should push for exploration of the alternatives alluded to in Figures 17-13 and 17-14.

Knowledgeable engineers can show that some widely accepted pump components tend to malfunction in the real world. Moreover, as industry often moves away from solid training and from taking the time needed to do things right, designing-out risk and designing-out maintenance become attractive propositions.

DEMAND BETTER PUMPS AND PAY FOR VALUE

Competent reliability professionals understand they must study and understand the risk, and explain to managers the risk of making purchasing decisions on price alone. Risk abatement may include adding more members to one's maintenance workforce and to schedule more frequent pump downtime events. Another risk abatement strategy might be to develop and issue well thought-out procurement and installation specifications. Issuance of these specifications would be followed by up-front machinery quality assessment

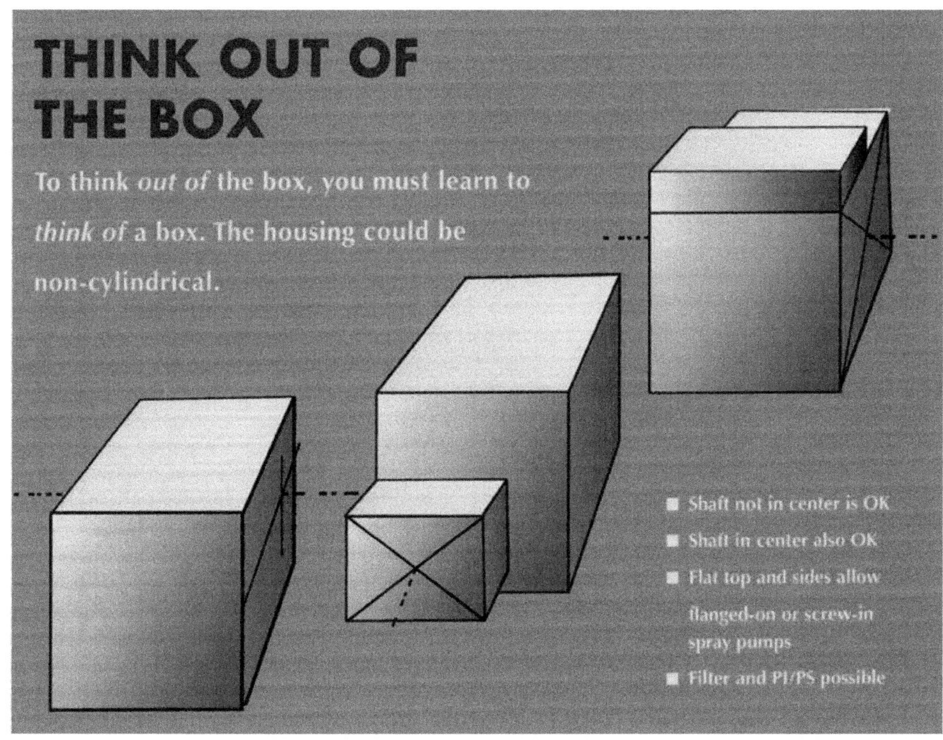

THINK OUT OF THE BOX
To think out of the box, you must learn to think of a box. The housing could be non-cylindrical.
Shaft not in center is OK
Shaft in center also OK
Flat top and sides allow flanged-on or screw-in spray pumps
Filter and PI/PS possible

Figure 17-14: Proposed new generations of bearing housings. The intent is to eliminate oil rings and constant level lubricators. The process pump bearing housing should incorporate an oil pump that will create the pressurized spray deemed most advantageous by all world-scale rolling element bearing manufacturers

(MQA), which typically adds 5% to the budgeted cost of pumps.

Another mandatory and clearly related pump failure risk reduction step would be to ascertain, before allocation or approval of budgets, that these include the pricing of *long-term reliable* pumps, and not the pricing for *least initial* cost pumps.

The accuracy of the information in this book can be confirmed by consulting the various references. Organizations intent on systematically reducing repeat pump failures may wish to pay close attention to this text; it will certainly enable them to move ahead of organizations that remain complacent or indifferent. As more and more reliability-focused managers see fit to implement guidelines and outright disincentives to specifying, purchasing or tolerating process pumps that fail repeatedly and often catastrophically, our message will have success.

In all, pump reliability professionals must not lose sight of their charter and mission. Working in harmony with basic science and achieving high pump reliability and availability are essential. If the least expensive initial cost process pump were also the most reliable, its manufacturer would have driven the competition off the field. The industrialized world has many pump manufacturers and some of these operate on business models which combine low price with high maintenance. Others pursue a higher pricing structure for products which allow reduced maintenance frequencies. Quite obviously, each has its place.

Appendix 1

Pump Start-up and Shutdown Procedures

Note: These procedures are of a general nature and may have to be modified for non-routine services. Review pertinent process data, as applicable. Re-write in concise sentences if instructions are to become part of checklists which operators are asked to have on their person while on duty.

A. Starting Centrifugal Pumps

1. Arrange for an electrician and a machinist/specialist to be present when a pump is initially commissioned. Ascertain that large motors have been checked out.

2. For low and medium specific speed pumps (below values of 4,000 to 6,000) close the discharge valve and open the suction valve. The almost closed discharge valve creates a minimum load on the driver when the pump is started. Experience has shown that the discharge valve is often very difficult to open against pressure. Assuming that the motor inrush current allows and that the motor will not kick off, the discharge valve may be just "cracked" about 1/8 open—before the pump is started. The input power required at shut-off on higher specific speed pumps (values above 4,000 to 6,000) may equal or exceed the power with the discharge valve open. For starting against a closed discharge valve, a minimum flow bypass must be open.

3. Be sure the pump is primed. Opening all valves between the product source and the pump suction should get product to the suction, but does not always ensure that the pump is primed.

Only after ascertaining that fluid emissions are not hazardous or are routed to a safe area, open the bleeder valve from the pump casing until all vapor is exhausted and a steady stream of product flows from the bleeder. It may be necessary to open the bleeder again when the pump is started, or even to shut down and again bleed off vapor if pump discharge pressure is erratic.

Note: Priming of a cold service pump may have to be preceded by "chill-down." A cold service pump is one that handles a liquid which vaporizes at ambient temperatures when under operating pressures. Chilling down of a pump is similar to priming in that a casing bleeder or vent valve is opened with the suction line open. There are three additional factors to be considered for cold service pumps:

- Chilling a pump requires time for the pump case to reach the temperature of the suction fluid.

- The chill-down vents are *always* tied into a closed system.

- On pumps with vents on the pump case and on the discharge line, open the vent on the discharge line first for chill-down and then open the pump case vent to ensure that the pump is primed.

  Should it be necessary to have a cold service pump chilled-down and ready for a quick start, e.g., refrigerant transfer pumps during unit start-up, then the chill-down line can be left cracked open to get circulation of the suction fluid.

4. If a minimum flow bypass line is provided, open the bypass. Be sure minimum flow bypass is also open on spare pump if it starts automatically.

5. Never operate a centrifugal pump without liquid in it.

6. Check lube oil and seal pot level.

7. Start the pump. Confirm that the pump is operating by observing the pressure gauge. If the discharge pressure does not build up, stop the pump immediately and determine the cause.

8. Open the discharge valve slowly, watching the pressure gauge. The discharge pressure will probably drop somewhat, level off and remain steady. If it does not drop at all, there is probably a valve closed somewhere in the discharge line. In that case, close the discharge valve. Do not continue operation for any length of time with discharge valve or line blocked.

9. If the discharge pressure drops to zero or fluctuates widely, the pump is not primed. Close the discharge valve and again open the bleeder from the casing to exhaust vapor. If the pump does not pick up at once, as shown by a steady stream of product from the bleeder and steady discharge pressure, shut down the pump and driver and check for closed valves in the suction line. A dry pump will rapidly destroy itself.

10. Carefully check the pump for abnormal noise, vibration (using vibration meter) or other unusual operating conditions. An electrician and machinery engineer should be present when pumps are initially commissioned.

11. Be careful not to allow the bearings to overheat. Recheck all lube oil levels.

12. Observe whether or not the pump seal or stuffing box is leaking.

13. Check the pump nozzle connections and piping for leaks.

14. When steady pumping has been established, close the start-up bypass and chill-down line (if provided) and check that block valves in minimum flow bypass line are open.

B. Watching Operation of Centrifugal Pumps

1. Especially during commencement of pumping, but also on periodic checks, note any abnormal noises and vibration. If excessive, shut down.

2. Note any unusual drop or rise in discharge pressure. Some discharge pressure drops may be considered normal. When a line contains heavy, cold product, and the tank being pumped contains a lighter or warmer stock, the discharge pressure will drop when the line has been displaced. Also, discharge pressure will drop slowly and steadily to a certain point as the tank level is lowered. Any other changes in discharge pressure while pumping should be investigated. If not explainable under good operating conditions, shut down and investigate thoroughly. Do not start up again until the trouble has been found and remedied.

3. Periodically check oil mist bottom drain sight glass for water and drain, if necessary.

4. On open oil mist systems, regularly check both oil level and oil mist vapor flow from vents or labyrinths.

5. Seal oil pots need to be checked regularly for correct level. Refill with fresh sealing liquid (oil, propylene glycol or methanol, as specifically required and approved for the particular application).

6. Periodically check for excessive packing leaks, mechanical seal leaks, or other abnormal losses. Also, check for overheating of packing or bearings. Do not trust your hand, use a surface pyrometer instead. Note that excessive heat will cause rapid failure of equipment and may result in costly and hazardous fires.

7. Operate all spare pumps at least once a week to prevent the bearings from seizing and to ensure that the pump will be operable when needed. Alternatively, "Best Practices Plants" have determined monthly switching of "A" and "B" pumps to be the most appropriate and cost-effective long-term operating mode.

8. When a pump has been repaired, place it in service as soon as possible to check its operation. Arrange for a machinist to be present when the pump is started up.

C. Shutting Down Centrifugal Pumps

1. For low or medium specific speed pumps (below values of 4,000 to 6,000) close the discharge valve. This takes the load off the motor and, if check valves leak, may prevent reverse flow through the pump.

2. Shut down the driver.

3. If pump is to be removed for mechanical work, close the suction valve and open vent lines to flare or drain as provided. Otherwise leave suction valve open to keep pump at correct operating temperature.

4. Shut off steam tracing, if any.

5. Shut off cooling water, sealing oil, etc., if pump is to be removed for mechanical work.

6. At times, an emergency shutdown may be necessary. If you cannot reach the regular starter station (in case of fire, for example), stop the pump from the starter box, which is located some distance away, and is usually accessible. If neither the starter station nor the starter box can be reached, call the electricians. Do not, as a part of regular operations, stop pumps from the starter box. Use the regular starter station instead.

Appendix 2

A "Hundred-plus Points" Summary of Pump Reliability Improvement Options and Considerations

Use this summary as a reminder of what is contained in the Pump User's Handbook. This summary serves as a checklist of options and considerations that will have to be practiced and implemented if long-term pump reliability is to be achieved.

A. Cooling Issues

1. Do not allow pedestal cooling of centrifugal pumps, regardless of process fluid or pumping temperature. Calculate and accommodate thermal growth by appropriate cold alignment offset. Consider using hot alignment verification measurements.

2. Do not allow *jacketed* cooling water application on bearing housings equipped with rolling element bearings. Note that water surrounding only the bearing outer ring will often cause bearing-internal clearances to vanish, leading to excessive temperatures, lube distress, and premature bearing failure.

3. Understand and accept the well-documented fact that on pump bearing housings *equipped with rolling element bearings,* it is possible to delete cooling of any kind. Simply change over to the correct type (synthesized hydrocarbon lube preferred) and generally somewhat more viscous, grade of lubricating oil. However, be mindful of the fact that oil rings ("slinger rings") will often work only with thin (typically ISO Grade 32) lubricants.

4. Recognize that cooling coils immersed in the oil sump may cool not only the oil, but also the air floating above the oil level. The resulting moisture condensation can cause serious oil degradation. This is one of several reasons to avoid cooling of pump bearing housings, if at all possible.

5. Recognize that *stuffing box cooling* is generally ineffective. If the seal must be cooled, consider introducing a flush stream that has been cooled in a pump-external heat exchanger or cooler before entering into the seal cavity. Air heat exchangers are still available as of 2013.

6. In water-cooled *sleeve bearing applications* where cooling water may still be needed, excessively cold cooling water will often cause moisture

condensation. Even *trace amounts* of water may greatly lower the capacity of lubricants to adequately protect the bearing. Close temperature control is therefore important.

B. Bearing Issues

7. Do not use filling-notch bearings in centrifugal pumps. Replace with dimensionally identical Conrad-type bearings.

8. All bearings will deform under load. On thrust bearings that allow load action in both directions, deformation of the loaded side could result in excessive looseness — hence, skidding — of the unloaded side. This skidding may result in serious heat generation and thinning-out of the oil film. Metal-to-metal contact will now destroy the bearing. Always select bearing configurations that limit or preclude skidding.

9. The API-610 recommended combination of two back-to-back mounted 40-degree angular contact bearings is not always the best choice for a particular pump application. Matched sets of 40-degree and 15-degree angular contact bearings (SKF/MRC's "PumPac" brand) are designed to prevent detrimental skidding and may thus be more appropriate for a given service application.

10. Thrust bearing axial float, i.e. the total amount of movement possible between thrust bearing outer ring and bearing housing end cap, should not exceed 0.002 inch (0.05mm). Some hand fitting or shimming may be necessary to thus limit the potentially high bearing-internal acceleration forces.

11. Consider replacing old-style double-row angular contact bearings (bearings with one inner and one outer ring) with newer double-row angular contact bearings that have two brass cages and one outer and two inner rings each. Note that axial clamping of the two inner rings would be necessary, but that the newer bearings would resist skidding.

12. Observe allowable assembly tolerances for rolling element bearings in pumps with alloy steel shafts:
    —Conrad, single angular contact and double-row bearings: bore-to-shaft: 0.0002-0.0007" (0.005-0.018 mm) interference fit.
    Note that this fit applies to conventional alloy steel shafts, but different (generally lighter) fits must be used with stainless steel shafts
    —Bearing outside diameter-to-housing fit: 0.0007-0.0015" (0.018-0.037 mm) loose fit

    It should be noted that fit dimensions different from the above will be required for stainless steel shafts due to greater thermal expansion.

13. Be selective in using bearings with plastic cages in pumps:
    —Plastic cages tend to be damaged unless highly controlled mounting temperatures are maintained
    —Plastic cage degradation will not show up in typical moderate-cost vibration data acquisition and analysis.

14. Be certain to use only precision-ground, matched sets of thrust bearings in either back-to-back thrust or tandem thrust applications. Matched bearings must be furnished by the same bearing manufacturer. Verify precision grinding by observing that appropriate alphanumerics have been etched into the wide shoulder of the outer ring.

15. Use radial bearings with C3 clearances in electric motors so as to accommodate thermal growth of the hotter-running bearing inner ring. Note that synthetic "EM" greases are preferred for electric motors.

16. Investigate column bearing materials upgrade options on vertical deep well pumps. Note that Vespel is good for wear rings and throat bushings, but not sleeve bearings.

17. Become familiar with advanced bearing housing protector seals and use these to prevent lubricant contamination.

C. Lubrication Issues

18. On sump-lubricated pumps, do not permit oil levels higher than through the center of rolling elements at the 6 o'clock position.

19. On pumps with dn-values in excess of 160,000, allowing lube oil to reach even this position may result in excessive heat generation. For these, the oil level may have to be lowered further and oil rings may have been chosen by the pump manufacturer. Oil rings ("slinger rings") are very often less-than-ideal.

20. Except for moderate load applications in relatively cool ambient environments, ISO Grade 32 and lighter lubricants are not suitable for rolling element bearings in centrifugal pumps. A thicker ISO Grade 68 lubricant will perform better in the majority of rolling element bearings in pumps.

21. Bearing housing cooling may still be needed in pumps equipped with sleeve bearings. The majority of these pumps require ISO Grade 32 lubricants. Close viscosity control may have to be maintained for satisfactory long-term lubrication.

22. Observe required ISO viscosity grades in moderate climates (Europe, the Americas, Pacific Rim, Australia): rolling element bearings—ISO 68, synthesized hydrocarbon optional; sleeve bearings—ISO 32, synthesized hydrocarbon optional; combining both bearing types in the same housing—ISO 32 PAO or dibasic ester synthesized hydrocarbon mandatory.

23. In bearing housings with both rolling element bearings (preferred lube viscosity ISO 32) and sleeve bearings (which may require ISO Grade 68 lubricants), consider satisfying both needs by using ISO Grade 32 or 46 PAO or diester-based synthesized hydrocarbon oils.

24. Next to oil-jet lubrication, dry-sump oil mist applied in though-flow fashion per API 610 8th Edition represents the most effective and technically viable lubrication and bearing protection method used by reliability-focused industry.

25. Be aware of upgrade and conversion options whereby a simple and economical inductive pump, or a small canned motor pump, can serve as the source of a continuous stream of pressurized lube oil. Used in conjunction with a spin-on filter, the resulting clean stream of lubricant can be directed at the bearing rolling elements for optimum effect.

26. Certain grease formulations cannot be mixed with other grease types. Incompatible greases often enter into a chemical reaction that renders them unserviceable in less than one year.

27. Over-greasing of electric motor bearings is responsible for more bearing failures than grease deprivation. Practicing proper regreasing procedures is essential for long bearing life.

28. Life-time lubricated (sealed) bearings will last only as long as enough grease remains in serviceable condition within the sealed cavity. Whenever "dn," the product of bearing bore (mm) times shaft rotational speed (rpm), exceeds 80,000, reliability-focused plants consider it no longer economical to use life-time lubricated bearings in continuously operating industrial machinery.

29. Grease replenishing intervals depend on bearing inner ring bore dimension and shaft rotational speed. Reliability-focused user plants consider dn = 300,000 the maximum for grease lubrication of electric motors and other machines in continuous service. It has been reasoned that beyond this dn-value (d = bearing bore, mm; n = shaft rotational speed, rpm), grease replenishing intervals become excessively frequent and oil lubrication would be more economical.

30. Realize that oil ring lubrication no longer represent state-of-art. Oil rings are alignment-sensitive and tend not to perform dependably if one or more of the following requirements are not observed:
    - Unless the shaft system is absolutely horizontal, oil rings tend to "run downhill" and make contact with stationary components. Ring movement will be erratic and ring edges will undergo abrasive wear.
    - The product of shaft diameter (inches) and shaft speed (rpm) should be kept below 8,000. Thus, a 3-inch shaft operating at 3,600 rpm (dn = 10,800) would not meet the low-risk criteria!
    - Operation in lubricants that are either too viscous or not viscous enough will not give optimized ring performance and may jeopardize bearing life.

31. Flinger spools fastened to pump shafts often

perform considerably more reliably. Consider retrofits using flingers made with metal cores to which properly designed (!) elastomeric discs are attached. Better yet, specify pumps with thrust bearings inserted into a bearing cartridge. The outside diameter of this cartridge must be larger than the flinger disc so as to allow assembly. Now the flinger disc could be made of solid stainless steel. It should be immersed in the oil roughly $3/8^{th}$ inch (~11 mm) below the oil level.

32. If the use of oil rings is unavoidable, be aware that a 30 degree angle between the contact point at the top of a shaft and points of entry into the oil represents proper depth of immersion. Too much immersion depth will cause rings to slow down, whereas insufficient depth tends to deprive bearings of lubricant.

33. Oil rings with circumferentially machined grooves will provide increased oil flow. However, so as to function without risk of skipping and abrading, oil rings will require a completely horizontal shaft system—rarely possible as long as shims are used to obtain shaft alignment.

34. Consider quasi-hermetically sealing bearing housings to preclude ingress of atmospheric contaminants and egress of lubricating oil. Install bulls-eye sight glasses on well-protected pump bearing housings.

35. On grease-lubricated couplings, verify that only approved, homogeneous, suitably formulated coupling greases are used. At elevated speeds, most greases centrifuge apart and separate into their respective oil and thickener constituents

36. Do not allow coupling greases to be used on electric motor bearings, and vice versa.

37. Fully consider vulnerabilities of unbalanced constant level lubricators. Use pressure-balanced models only.

38. Mount constant level lubricators on the correct side of the bearing housing. Observe "up-arrow" provided by manufacturer of constant level lubricator. Incorrect mounting will lead to greater disturbances around the air/oil interface in the surge chamber of constant level lubricators, i.e., they will rupture the meniscus. Correct mounting reduces the height difference between uppermost and lowermost oil levels. In other words, it ensures a more limited level variation. Recognize caulking has a finite life, which requires many constant level lubricator types to be subject to preventive maintenance intervention.

39. Verify that relubrication and grease replenishment procedures take into account that:
    - —Mixing of incompatible greases will typically cause bearing failures within one year.
    - —Attempted relubrication without removing grease drain plugs will cause the grease cavity to be pressurized. Over-greasing will cause excessive temperatures. On shielded bearings, cavity pressurization tend to push the shield into contact with rolling elements or bearing cage, causing extreme heat and wear. Yet, (and while counter-intuitive), single-shielded bearings should be mounted with the shield adjacent to the grease cavity.

D. Mechanical Seal Issues

40. Select mechanical seal types, configurations, materials, balance ratios and flush plans that have been certified to represent proven experience and to give extended seal life.

41. Except for gas seals ("dry seals"), mechanical seals must be operated so as to preclude liquid vaporization between faces. However, the seal environment cannot be effectively cooled by using cooling water in a jacketed seal chamber. External flush cooling is far more effective.

42. Mechanical seals with quench steam provisions are prone to fail rapidly if quench steam flow rate or pressure are not kept sufficiently low. The installation of small diameter fixed orifices is very often needed to prevent excessively large steam quench amounts.

43. Select the optimum seal housing geometry and dimensional envelope to improve seal life. Recognize that slurry pumps generally benefit from *steeply tapered* seal housing bores. The traditional concentrically bored stuffing box environment

does not usually represent the optimum configuration and neither does a very *shallow* taper.

44. Avoid inefficient pumping rings on dual seals. Instead, use the far more efficient bi-directional pumping device arrangement.

45. On hot service pumps, follow approved warm-up procedures. Verify that seal regions are exposed to a through-flow of warm-up fluid, i.e. are not dead-ended.

46. Understand the difference between conventional mechanical seals (seals where the flexing portion rotates) and stationary seals (where the flexing part is stationary). Maximum allowable speeds and permissible shaft runouts are lower for conventional seals.

47. Flush plans routing liquid from stuffing box back to suction are reliable only if the pressure difference exceeds 25 psi (~172 kpa).

48. Fluid temperatures in seal cavity must be low enough to prevent fluid vaporization occurring in seal faces.

49. Whenever setscrews ("grub screws") are needed, use only hardened stainless steel set screws so as to reduce corrosion risk.

E. Hydraulic Issues

50. Ascertain that pumps operating in parallel have closely matched operating points and share the load equally. Examine slopes of performance curves for each. Understand that differences in internal surface roughness may cause seemingly identical pumps to operate at different flow/pressure points.

51. Ascertain that centrifugal pumps are not operating in the low-flow range where impeller-internal flow recirculation is likely to exist.

52. On pumps with power inputs over 230 kW, verify that "gap A," the radial distance between impeller disc tip and stationary parts is in the range of only 0.050-0.060 inches (1.2-1.5 mm). Pump-internal recirculation is thus kept to a minimum. On reduced diameter impellers, this would imply that trimming is done only on vane tips!

53. On pumps with power inputs over 230 kW, ascertain that "gap B," the radial distance between impeller vane tips and cutwater, is somewhere in the range of 6% of the impeller radius. This will reduce vibratory amplitudes occurring at blade passing frequency.

54. Ensure hydraulically-induced shaft deflections in single volute pumps are not excessive. This may require restricting the allowable flow range to an area close to BEP (Best Efficiency Point).

55. Except for "Split Flow™" pumps, operate two-stage overhung pumps only at flows within 10% of BEP. Recognize severe shaft deflection and risk of shaft failure due to reverse bending fatigue when operating far away from BEP.

56. Check impeller specific speed vs. efficiency at partial flow conditions. Consider installation of more suitable impellers for energy conservation.

57. Consider installing in the existing pump casing an impeller with different width, or with different impeller vane angle, or different number of vanes, or combinations of these. Observe resulting change trends in performance curves.

58. Consider changing slope of performance curve by inserting a restriction bushing in the pump discharge nozzle.

59. Review if NPSH gain by cooling the pumpage is feasible and economically justified.

60. Consider extending the allowable flow range by using an impeller with higher (!) NPSHr. Verify that NPSHa exceeds the NPSHr of the new impeller.

61. Use a ratio NPSHa/NPSHr of 3:1 or higher for carbamate and similar difficult services.

62. Be aware of pre-rotation vortices and their NPSHr-raising effects on mixed flow pumps.

63. Consider use of inducer-style pumps but realize that these often lower the NPSHr only in a nar-

row range of flows. Outside of that range, the NPSHr may actually become greater.

64. Consider use of a vertical column pump or placing the pump below grade if NPSHa-gain is needed. Consider recessed impeller pumps for slurries.

65. Consider the effects on performance curve that could result from:
    — vane underfiling
    — vane overfiling
    — "volute chipping"

66. Calculate axial thrust values and verify adequacy of thrust disc or balance piston geometry. Modify disc or piston diameter as required.

67. Consider opening of existing impeller balance holes if thrust must be reduced to extend bearing life.

68. Review and implement straight-run requirements for suction piping near pump inlet flanges entry. Aim for a straight pipe run as indicated earlier in this text—as a more vague rule of thumb allow at least 5 pipe diameters between an elbow and the pump suction nozzle.

69. Realize that two elbows in suction piping at 90 degrees to each other tend to create swirling and pre-rotation. In this case, use ten pipe diameters of straight run piping between the pump suction nozzle and the next elbow.

70. On top suction pumps, maintain a ten pipe diameter straight pipe length between suction block valve and pump suction nozzle.

71. Verify that eccentric reducers in suction lines are installed with flat side at top so as to avoid air or vapor pockets.

72. Use pumping vanes or suitably dimensioned impeller balance holes to reduce axial load acting on thrust bearings.

73. Oblique trimming of impeller vanes exit tips will reduce the severity of vibratory amplitudes at blade passing frequencies and their harmonics.

F. Mechanical Improvement Options

74. Calculated shaft deflections should not exceed 0.002 inch (0.05 mm) over the entire operating range of the pump.

75. Implement suitable vortex breaker baffles on large vertical sump pumps. Seek guidance from applicable Hydraulic Institute standards.

76. Implement wear ring modifications (circumferential grooving) to reduce severity of rub in the event of contact due to excessive shaft deflection or run-out.

77. Examine need for occasional measures to cure plate-mode or impeller cover vibration. Consider "scalloping" of the impeller cover (removal of material between vane tips at the impeller periphery). This may be necessary to avoid impeller vibration, other than unbalance-related vibratory action.

78. Consider providing generous fillet radii at shaft shoulders in contact with overhung impellers to avert reverse bending fatigue failures.

79. Verify shaft slenderness is not excessive. On API 5th Edition pumps, the stabilizing effect of packing may have been lost when converting to mechanical seals. Therefore, throat bushings may have to be replaced by minimum clearance (0.003 in/in, or 0.003 mm/mm shaft diameter) shaft support bushings. These high performance plastic (Vespel® or Graphalloy) bushings should be wider than the customary open-clearance throat bushings originally installed.

80. Verify acceptability of equipment spacing in pump pits and also ascertain conservatism of sump design.

81. Verify absence of shaft critical speeds on vertical pumps. Insist on conservative bearing spacing.

82. Consider hollow-shaft drivers on vertical pumps and use suitable reverse rotation prevention assemblies.

83. Ascertain acceptability of equipment spacing in pump pits and also verify that the sump design meets applicable Hydraulic Institute guidelines.

84. Verify absence of shaft critical speeds on vertical pumps. Insist on conservative bearing spacing.

85. Consider hollow-shaft motor drivers on vertical pumps and use suitable reverse rotation prevention assemblies.

86. Beware of exceeding the rule-of-thumb maximum impeller diameter for overhung pumps, 15 inches (381 mm).

87. Consider in-between-bearing pump rotors whenever the product of power input and rotational speed (kW times rpm) exceeds 675,000.

88. Verify acceptability of equipment spacing in pump pits and also ascertain conservatism of sump design.

89. On hollow-shaft drivers with reverse rotation prevention assemblies, verify lubrication provisions.

G. Installation Issues

90. Eliminate shaft misalignment by allowing foot-mounted pumps to reach equilibrium temperature. Only then secure or mount the bearing housing support bracket.

91. Do not allow shaft misalignment to exceed the limits plotted by competent alignment service providers. Use 0.5 mils per inch (.0005 mm per mm) of shaft separation (DBSE, or distance between shaft ends) as the maximum allowable shaft centerline offset.

92. Use ultra-stiff, epoxy-filled, formed steel baseplates, whenever possible.
    - Proceed by first inverting and filling with epoxy grout to become a monolithic block.
    - After curing, turn over and machine all mounting pads flat and co-planar within 0.0005 inch per foot (0.04 mm/m).
    - Next, install complete baseplate on pump foundation, anchor it and level to within same accuracy.
    - Finally, place epoxy grout between top of foundation and the rim portion of the space beneath monolithic baseplate.

93. On welded baseplates, make sure that welds are continuous and free of cracks.

94. On pump sets with larger than 75 kW drivers, ascertain that baseplates are furnished with two tapped positioning block screws per each casing foot, i.e. two screws ("jacking bolts") per mounting pad. Pad heights must be such that at least 1/8 inch (3 mm) stainless steel shims can be placed under driver feet.
    - Provide these positioning block locations with 0.100" shims between the tip of each jackbolt and the machine foot to be aligned. Be sure to remove these shims after completing the alignment task so as to accommodate differential thermal expansion occurring on different parts of the pump set. The 0.100 inch shim(s) can later be used as gages to determine how much thermal movement has occurred in operation.

95. Only now mount pump and driver. Use laser-optic or similarly accurate alignment device. Do not allow piping to be pulled into place by anything stronger than a pair of human hands.

96. While connecting piping to pump nozzle, observe dial indicators placed at nozzles in x-y-z directions and at bearing housing (s) in x-y directions. Dial indicator movements in excess of 0.002 inches (0.05 mm) are not acceptable.

97. Investigate magnitude and direction of thermal expansion of pipe and verify that pipe growth does not load up either pump suction or discharge nozzles.

98. Verify that vertical in-line and certain canned motor pumps are free to respond to pipe movement. These pumps should not be bolted to the foundation.

99. Realize the high probability of foundation having settled and/or pipe supports and pipe hangers no longer having the intended spring force action. Do not allow piping to cause distortion of pump casings
    - Piping fitups must have enough "give" such that two hands pushing on pipe will allow bolting pipe flange to pump nozzle.
    - Once bolting is torqued, there should be less

than 0.002 inch movement at any external part of the pump. Install temporary dial indicators to verify (see item 96, above).

100. Install temporary suction strainer; verify approved construction features and materials.

H. Couplings and Other Issues

101. Examine couplings for adequacy of puller holes or other means of hub removal. For parallel pump shafts with keyways, use 0.0-0.0005" (0.0-0.0012 mm) total interference. Use one of several available thermal dilation methods to mount hubs on shafts.

102. On grease-lubricated couplings, use only approved coupling grease. Non-lubricated captured center member disc pack or diaphragm couplings are preferred. See item 35, above.

103. Avoid elastomeric couplings for large pumps. Note that:
   - —Toroidal ("tire-type") flexing elements will exert an axial pulling force on driving and driven bearings. Also, they are difficult to balance.
   - —Polyurethane flexing elements fare poorly in concentrated acid, benzol, toluene, steam, arctic climates and certain other environments
   - —Polyisoprene flexing elements fare poorly in gasoline, hydraulic fluids, sunlight (aging), silicate, cold temperatures and certain other environments.

104. Disallow loose-fitting keys for coupling hubs. Hand-fit keys to fit snugly in keyway. On all replacement shafts, machine radiused keyways and modify keys to match radius contour. Use only hardened stainless steel set screws.

105. Use only limited end-float couplings if drive motors are equipped with sleeve bearings.

Appendix 3

Supplementary Specification for API-class Pumps

The following material was derived from various non-proprietary user company and contractor sources and represents an "industry-composite" example of a supplementary specification for Heavy Duty Pumps. Similar formats were found of significant value at the time this revised and updated text was being published.

Caution: *The authors are expressly asking you NOT to duplicate this example specification supplement for your plant unless you have verified its applicability and understand the meaning of each and every clause.*

INTRODUCTION

Note that Table 1 records the vendor drawing and data requirements. See text for guidance on typical requirements, or compile your own list. Owner input is required as indicated.

1.0 **GENERAL**

1.1 **Scope**

1.1.1 a. (Exception) This supplementary specification covers the mandatory requirements for heavy duty centrifugal pumps. Any deviation from this specification must be approved by the Purchaser in accordance with the procedures specified in [indicate Purchaser's Engineering Practice].

b. All requirements specified in this supplementary specification are additions to those of API Standard 610, unless specifically noted as exceptions. Paragraph numbers in this supplementary specification which do not appear in API Standard 610 are new paragraphs. Paragraph numbers may not be consecutive and all API 610 paragraphs not mentioned herein shall be considered to be verbatim requirements of this specification.

c. The Purchaser's requirements for all paragraphs in API-610 which require a purchaser's decision (bulleted paragraphs) are indicated on the Data Sheets or in this supplementary specification. An asterisk (*) indicates that a decision by the Owner's Engineer is required or that additional information is furnished by the Owner's Engineer.

d. Vendor qualification. Pumps shall be supplied by Vendors qualified by experience to manufacture the units proposed. To qualify, Vendor must have at least two pumps of comparable rating and design performing satisfactorily in a similar service. Special attention to Vendor qualification should be given for pumps rated over 500 HP (~375 kW), pumps handling cold liquids at –50°F (–46°C) or lower, or pumps with a high suction energy value.

1.2 **Alternative Designs**

1.2.1 (*) a. Where the Vendor qualification of Paragraph 1.1.1d prevents the application of the latest technology, the Vendor shall submit to the Purchaser an alternative proposal incorporating such features for review and approval by the Owner's Engineer. The proposal shall specifically identify the undemonstrated features and state their advantages.

(*) b. Proposals to furnish pumps that do not meet all requirements of this standard, based on specific nonflammable, non-toxic service conditions, shall be submitted to the Purchaser for approval by the Owner's Engineer.

1.4 **Definition of Terms**

The following additional terms are used in this supplementary specification.

1.4.58 Bubble point temperature is the temperature at which the first bubbles appear in a liquid when heated at constant pressure.

1.4.59 Product temperature margin (PTM) is the difference between the bubble point temperature and the actual temperature of the liquid at a given pressure.

1.4.60 Inspector [indicate Purchaser's definition]

1.4.61 Owner [indicate Purchaser's definition]

1.4.62 Owner's Engineer [indicate engineer appointed by Purchaser]

1.4.63 Purchaser—The party placing a direct purchase order. The Purchaser is the Owner's designated representative.

1.5 **Referenced Publications**

STANDARDS AND PUBLICATIONS

1.5.4 The referenced Corporate Engineering Practices, and any other attached to the requisition shall apply to the extent referenced herein or on the Data Sheets. The Data Sheets and Inspection Checklists from API-610 shall be used in conjunction with this document.

2.0 **BASIC DESIGN**

2.1 **General**

2.1.4 Pumps with constant-speed drivers shall be capable of at least a 5% head increase or decrease at rated conditions, either by trimming impeller(s) or with different impeller(s).

2.1.5 (*) For hydrocarbons, NPSHr shall be less than NPSHa by at least 3 ft (1 m) unless approved by the Owner's Engineer.

2.1.9 (*) Pump suction energy ratio (actual suction energy divided by high suction energy gating value for pump type) shall not exceed 1.3 without written approval of the Owner's Engineer. See also paragraph 1.1.1d.

2.1.11 (*) a. The head curve for the pump shall be continuously rising from the specified capacity point to the shut-off point. The pump head at shut-off shall be between 110 and 120 percent of the head at the specified capacity point. Pumps whose shut-off head exceeds 120 percent of the

Corporate Engineering Practices (EP's)

EP 1-2-3	Documentation Format Requirements
EP 1-2-4	Grouting and Baseplates for Structural Steel and Equipment
EP 1-2-5	Compact and Extended Body Gate and Globe Valves
EP 1-2-6	Piping Fabrication
EP 1-2-7	Piping for Rotating Equipment
EP 1-2-8	General Purpose Steam Turbines
EP 1-2-9	Pressure Vessels for Non-Process Services
EP 1-3-1	TEMA Shell and Tube Heat Exchangers for Non-Process Services
EP 1-3-2	Induction Motors NEMA Frame
EP 1-3-3	Induction Motors above 200 HP
EP 1-3-4	Synchronous Motors

ASME Codes

Sec VIII Pressure Vessels, Section VIII, Division 1
B31.3 Process Piping

API Publications

Std. 610 Centrifugal Pumps for Petroleum, Heavy Duty Chemical, and Gas Industry Services

head at the specified capacity point may be submitted for approval by the Owner's Engineer. Application of a discharge orifice as a means of providing continuous head rise to shut-off shall require written approval by the Owner's Engineer.

b. Pumps for parallel operation shall have 8-10% equal head rise to shut-off.

(*) c. Low capacity, high-head pumps may be exempt from the requirements of 2.1.11.a and 2.1.11.b with the approval of the Owner's Engineer.

2.1.14 The maximum allowable sound pressure level is 85 dBA measured at 3 feet from any equipment surface. Test data on similar pumps shall be made available upon request.

2.1.17 Pumps shall be designed to operate continuously without the use of cooling water.

2.1.29 Unless otherwise specified, the pump and auxiliaries shall be suitable for unsheltered outdoor installation at [indicate Purchaser's Engineering Practice that applies to site].

2.1.30 Diffuser or double volute pumps are required for pumps with discharge nozzle sizes of 4 inch NPS or larger. Impellers with an even number of vanes are not acceptable on double volute pumps.

2.2 **Pressure Casings**

2.2.4 Areas of vertically suspended, double-case and horizontal multistage pumps which are normally subjected to suction pressure shall be designed for the same pressure as discharge areas of the pump.

2.2.14 (*) Vertical turbine pumps shall have columns and bowls of the bolted design. Screwed bowl sections may be supplied only if specifically approved by the Owner.

2.3 **Nozzles and Pressure Casing Connections**

2.3.2 **Suction and Discharge Nozzles**

2.3.2.1 (Exception) All pumps shall have suction and discharge flanges of equal rating. The ANSI permissible hydrostatic test pressure of suction flanges shall be equal to, or greater than, the pump casing hydrostatic test pressure.

2.3.3 **Pressure Casing Connections**

2.3.3.6 (Exception) All pipe nipples shall be Schedule 160.

2.3.3.7 (Exception) Plugged openings that are not required shall be seal welded except in mechanical seal gland plates.

2.3.3.10 a. All vents and drains shall be valved, either flanged or socket welded.

2.3.3.12 a. Valved vent and drain connections shall comply with the requirements of Owner's applicable Engineering Practice [Purchaser will state number].

b. Valves for vents, drains, and clean-out connections shall be as follows:

1. Either a [indicate preferred make] or [indicate alternative make or model] Class 800 extended body valve, compact body valve with a Schedule 160 nipple, or flanged valve per [indicate applicable Engineering Practice number] shall be used.

2. For alloy pump cases, the valve and connection material shall be the same as that of the pump case.

3. All piping between the valve and the pressure casing shall be socket welded or butt-welded. Butt-welds are required for design pressures above 1,000 psi (6,900 kPa).

4. The socket weld at the connection of extended body valves to fittings or the pump case shall have a complete bridge weld from fitting or casing to the body of the valve.

5. Hydrotesting of the valved connections shall be made by hydrotesting the pump casing at the specified pressure with the valve(s) fully attached and completed. The hydrotest certification should note the inclusion of the valve(s) in the test.

2.5 **Rotors**

2.5.2 (Exception) Colleted impellers are not permitted

2.5.3 (Exception) Solid hub impellers are required.

2.5.7 (Exception) The shaft stiffness for all pumps shall limit the total shaft deflection under the most severe conditions over the complete head-capacity curve to .002 inch at the face of the stuffing box or seal chamber and to less than one-half the minimum diametrical clearance at all bushings and wear rings. The shaft shall be designed to transmit momentarily at least four times the rated torque of the pump driver.

2.6 **Wear Rings and Running Clearances**

2.6.1 Unless otherwise specified, cast iron shall not be used as an impeller wear ring material. Vespel® wear rings are preferred. If metallic wear rings are supplied, circumferential grooves, 0.03" deep and 0.06" wide, shall be machined into the rotating wear ring periphery.

2.8 **Dynamics**

2.8.1.1 Full rotordynamic (lateral and torsional) analyses are required for pumps exceeding 4,000 RPM or 1,000 hp.

2.8.2 **Torsional Analysis**

2.8.2.1 (Exception) A torsional vibration analysis is required when the train includes a gear.

2.8.2.2 No torsional natural frequency shall be closer than 5 percent to twice any shaft speed.

2.8.2.6 If a torsional analysis is required per paragraph 2.8.2.1, a detailed report shall be furnished.

2.8.3 **Vibration**

2.8.3.10 The Vendor shall demonstrate that the pump can operate at any capacity from 120 percent of the rated capacity to the quoted minimum continuous stable flow without exceeding the vibration limits given in paragraph 2.8.3.7.

2.9 **Bearings and Bearing Housings**

2.9.1 **Bearings**

2.9.1.5 Unless better bearings are available, ball-type thrust bearings shall have machined bronze cages. The adequacy of all thrust bearings shall be verified with skid prevention of the unloaded bearing of prime importance. In case of potential problems, the vendor shall provide suitably preloaded and matched ball-type thrust bearing sets with dissimilar contact angles (e.g. 40°/15°).

Sets of properly preloaded, back-to-back mounted 15° or 29° angular contact bearings may be offered for double-flow impellers located between two bearing housings.

2.9.2 **Bearing Housings**

2.9.2.2 Constant level lubricators shall be TRICO Optimatic, or owner-approved equivalent fully balanced type, and installed with balance line connected to a fitting at the top of the bearing housing. Lubricators must be installed on the "up-arrow" side of the housing, as recommended by TRICO Manufacturing Company.

2.9.2.6 As a minimum requirement, provide rotating labyrinth style bearing isolators at all bearing housings. If specified, bearing housing seals shall be of the spring or magnet-activated face type so as to provide the most effective hermetic sealing possible.

2.10 **Lubrication**

2.10.3 Unless otherwise specified, pumps with rolling element bearings shall be designed for future conversion to pure oil mist lubrication using ISO Grade 68 synthetic (PAO or diester) lube oils.

2.10.4 Provide API-610 pressurized lube-oil systems on pump trains over 500 HP. Provide API-614 lube-oil systems on pump trains over 1000 HP and on all non-spared critical trains.

2.11 **Materials**

2.11.1 General

2.11.1.1 Twelve percent chrome castings shall be per ASTM A487, Grade CA6NM.

Austenitic stainless steel shall be per ASTM A351 with liquid penetrant examination to S.6.

Monel castings shall be per ASTM A494 grade M-30C with weldability test S.2 and liquid penetrant examination S.5.

2.11.2 **Castings**

2.11.2.2 Chaplets in casings other than cast iron that are not completely fused into pressure castings or other defects shall be replaced by weld metal equivalent to the casing composition. No other repair method is permitted. Casings shall be heat treated following any major repairs.

2.11.3 **Welding**

2.11.3.5.5 PWHT is required for all carbon and ferritic alloy steel pressure-containing components that are welded and/or weld repaired, when the weldment is exposed to a process containing wet H_2S. The PWHT procedure outlined in ASME Code Section VIII, paragraphs UW40, UW49, UHA32, and UCS56 shall be followed, except that the notes in Tables UHA32 and UCS56 do not apply.

All welds, regardless of type or size, that are exposed to wet H_2S shall be PWHT at a minimum temperature of 1150°F. External attachments or seal welded threaded connections on P-1 Group 1 and 2 materials do not require PWHT.

2.11.4 **Low Temperature**

(Exception) Pressure casing components, either cast, forged, or welded, shall meet the following impact requirements:

Note (1): In the notation 15/12, the first number is the minimum average energy of the three specimens while the second number is the minimum for any one specimen in the impact determination.

a. The minimum CET for any part of a casing shall apply to the entire casing.

b. No impact tests are required for a particular pressure containing component if the maximum casing working pressure generates a stress less than 25 percent of the minimum specified yield stress for the material as determined by the Manufacturer.

CET (°F)	Maximum Casing Working Pressure	Impact Requirements [See Note(1)]
< 20 °F	All	15/12 ft-lbf
> 20 °F to < 60 °F	> 1000 psi	15/12 ft-lbf
> 20 °F to < 60 °F	< 1000 psi	None
> 60 °F	All	None

(*) c. All impact test details shall be approved by the Owner's Engineer.

3.0 **ACCESSORIES**

3.1 **Drivers**

3.1.1 Unless otherwise specified, drivers shall be induction motors designed per [insert Purchaser's Engineering Practice] if 200 HP or smaller, and per [insert Purchaser's Engineering Practice] if larger than 200 HP. Synchronous Motors shall be in accordance with [insert Purchaser's Engineering Practice]

3.1.4 Motors shall not be overloaded at the end of the pump curve.

3.1.5 The motor shall be designed in accordance with the following:

a. Electrical characteristics -3 phase/60 cycle with 460 volts for motors less than 100 HP and 2400 volts for 100 HP and larger

b. Starting conditions -70% of rated voltage

c. Type of enclosure -Totally enclosed fan cooled (TEFC), unless otherwise specified

d. Sound pressure level- 85 dBA at 3 feet

e. Area classification -Class I, Group D, Division 2 unless specified otherwise

f. Type of insulation -Class F V PI with a Class B temperature rise

g. Rotor material
400 HP and larger—copper bars
300 HP and 350 HP—copper bars
if intermittent duty
350 HP and smaller—cast aluminum

h. Service factor—1.0 unless specified otherwise

i. Ambient temperature—Design 40°C, actual 20° to 105°F (–7°C to 41°C)

j. Elevation [insert installation site-specific information]

k. Temperature detectors -Two 100 ohm platinum RTD's per phase for motors 125 HP and larger, 1 RTD per bearing when sleeve bearings are furnished

l. Vibration sensors -2 eddy current proximity probes per bearing for motors 1,000 HP and larger equipped with sleeve bearings

3.1.6 Reduced voltage at starting shall be 70% of rated voltage.

3.1.9 Steam turbines shall conform to [insert Purchaser's Engineering Practice], unless otherwise specified.

3.1.13 For vertical pumps, the Vendor shall assemble, align and dowel the motor or gear to the pump in its shop to assure proper unit fit-up and shaft mating.

3.2 **Couplings and Guards**

3.2.2 When the coupling type is not specified by the Owner, a "Metastream Type T" non-lubricated disc type coupling with stainless steel disc packs shall be supplied by the pump Vendor.

3.2.4 (Exception) Couplings shall be mounted on the shaft with a keyed straight cylindrical fit having a 0.0005 to 0.001 inch interference between the hub and the shaft.

3.2.5 The coupling to shaft juncture shall be capable of withstanding a momentary torque four times the rated torque without yielding.

3.2.8.1 API-671 governs in applications over 750 HP and for unspared critical machines.

3.2.12 Coupling guards for horizontal pumps shall be fabricated from aluminum sheet or expanded steel sheet. Aluminum guards shall be provided with a hinged door for inspection of coupling flexible elements.

3.3 **Baseplates**

3.3.3 (Clarification and addition) Baseplates shall be inverted, suitably cleaned and primed before being filled with epoxy. After this, the baseplate is to be righted again and all the pads machined flat and parallel. Centers of mounting pads shall be at the correct relative elevation within 0.0005 inch per foot of separation between pads (i.e., two pads three feet apart shall be a maximum of 0.0015 inch from the correct elevation).

Additionally, each pad shall be level in all directions within 0.002 inch per foot. Space shall be provided at each pad for a precision level for two- plane leveling of the baseplate with the rotating equipment mounted. These requirements shall be met in a relaxed (non-clamped) state.

3.3.4 All shims shall straddle hold down bolts.

3.3.14 Alignment positioning jackscrews shall be provided for all equipment feet, regardless of equipment weight.

3.3.21 Mounting plates for vertical pumps shall be per the following:

Double-casing vertical pumps shall have a steel mounting plate that completely surrounds the outer casing and is attached to it by a continuous weld. The steel mounting plate shall have a 3/4 inch (~19 mm) pipe connection to vent the space between the outer barrel and the foundation. The plate shall be rectangular with radiused corners. Foundation bolts shall not be used to secure the flanged pressure-casing joint.

A rabbetted fit, with excess clearance shall be provided for the mounting flanges between the pump and direct mounted driver. Driver/pump alignments shall be achieved via dowels per paragraph 3.1.13.

3.3.21.4 All vertical pumps shall have sufficient anchor bolting to withstand nozzle reaction forces due to start-up conditions.

3.4 **Instrumentation**

3.4.3 Provide vibration probes per paragraph 3.4.3.1 and bearing RTDs per paragraph 3.4.3.2 on pump trains exceeding 1,000 HP and/or 4,000 RPM, and on all unspared critical trains.

3.5 **Piping and Appurtenances**

3.5.1 **General**

3.5.1.4 Barrier/buffer fluid reservoirs shall be designed for mounting off the pump baseplate and shall be supplied by the Purchaser along with connecting tubing and instrumentation.

3.5.1.6 h. Piping shall be designed to facilitate disassembly and reassembly of the equipment.

i. Vendor's piping shall terminate with a flanged connection of a line rating at least equal to the design pressure and design temperature rating of the equipment.

3.5.1.7 Piping design, materials, joint fabrication, and inspection shall be in accordance with ASME B31.3 and the additional requirements of [insert Purchaser's Engineering Procedure]. Threaded or slip-on flanges are not acceptable.

3.5.1.15 Threaded connections shall be made up without the use of PTFE tape.

3.5.2 **Auxiliary Process Fluid Piping**

All orifice plates shall be installed in flanged connections and have a tab, marked with the orifice size, extending from the flange where they are installed to indicate their location.

3.5.2.10.1 Flanges are required.

3.5.2.11 (*) Mechanical seal flush lines in flammable or hazardous service shall be piped unless tubing is approved by the Owner's Engineer.

3.7 **(New Section) Pressure Vessels and Heat Exchangers**

(*) Unless otherwise approved by the Owner's Engineer, pressure vessels and heat exchangers associated with rotating equipment shall be designed, fabricated, inspected and tested in accordance with [insert Purchaser's Engineering Procedure(s)]

4.0 **INSPECTION, TESTING, AND PREPARATION FOR SHIPMENT**

4.1 **General**

4.1.1 The term Inspector, as used in this document, refers to the designated Owner's Representative.

4.2 Inspection

4.2.1 General

4.2.1.3 As a minimum, inspection shall include: Verification of the equipment dimensions, compliance with baseplate machining tolerances, examination of test data and checking preparation for shipment. For services with pumping temperatures 500°F or higher, the Inspector shall verify that the internal clearances conform to 2.6.4.2.

4-2.2.1 Pressure boundary components of pumps in flammable or toxic service shall be inspected per the requirements of paragraph 4.2.2-4 (magnetic particle inspection) or paragraph 4.2.2.2 (radiography).

4.3 **Testing**

4.3.1 **General**

4.3.1.1 As a minimum, inspection and testing of all pumps shall be conducted according to the completed Inspection Checklist from API-610 and the following:

NOTES:

(1) Hydrostatic tests will be witnessed for Monel casings.
(2) Witnessed performance test shall be specified if the pump is high-capacity over 1,000 HP, over 4,000 RPM, and/or for critical service.

(*) (3) If the NPSH required for the pump differs from the specified available NPSH by 6 feet or less, an NPSH suppression and performance test is required. For pumps requiring NPSH testing, the test of one pump in each service shall be witnessed by the Inspector. For the balance of identical pumps, witnessing is not required; however, certification of test results is required.

(4) If specified by the Owner's Engineer.

(*) (5) Each sleeve bearing-equipped pump receiving a witnessed performance test shall have a witnessed bearing inspection by the Inspector following the final performance test.

4.3.2 Hydrostatic Test

4.3.2.1 a. (Exception) Suction sections of multistage horizontal and double-case pumps shall have a hydrostatic test pressure equal to the maximum casing discharge pressure. Vertical pumps shall be tested at the full hydrostatic test pressure from the suction flange to the discharge flange.

c. Mechanical seal flushing liquid coolers shall

Inspection or test	Required	Witnessed	Certified Data
Hydrostatic Test	Yes	No (1)	Yes
Performance Test	Yes	No (2)	Yes
NPSH Test	(3)	(3)	Yes
Inspection by Purchaser	(4)		
Dismantled Inspection	(5)		

be hydrostatically tested to at least the same pressure as the pump casing.

4.3.3 **Performance Test**

4.3.3.2.1 All pumps shall be operated for at least one hour at rated speed and capacity. One of the data points shall be the rated point.

(*) When specified by the Owner's Engineer, the seal chamber pressure shall be measured at each point during the performance test. The Vendor shall ensure that the measured seal chamber pressures, when corrected for rated suction pressure and specific gravity, are consistent with those assumed in designing the mechanical seal (see 2.7.1.12). Corrective steps needed to improve the seal design or seal chamber conditions shall be mutually agreed upon by the Vendor and Owner's Engineer.

4.3.3.3.3 (*) (Exception) In no case shall the minus tolerance allowance for shut-off head exceed two percent. The shut-off head with the positive tolerance allowance included shall not exceed 120 percent of the head at the rated capacity, unless previously approved by the Owner's Engineer.

4.4 **Preparation for Shipment**

4.4.3.4 (Exception) Unless otherwise specified, the rust preventive applied to unpainted exterior machined surfaces shall be of a type:
(1) to provide protection during outdoor storage for a period of 12 months exposed to a normal industrial environment, and
(2) to be removable with mineral spirits or any Stoddard solvent.

5.0 **SPECIFIC PUMP TYPES**

5.2.6 **Lubrication**

5.2.6.2 b. If cooling is permitted, (i.e. for pumps equipped with sleeve bearings only), materials of construction for coolers in salt water cooling service shall be as shown on top of page 517.

5.2.6.4 Lube oil pumps shall be IMO-type twin-screw construction. Pumps enclosed in the lube-oil reservoir shall have steel or ductile iron cases.

5.2.6.7 Non shaft driven horizontal pumps shall have positive suction heads and suction lines sloped to vent to the reservoir.

5.2.6.8 (*) Motor drives for lube oil pumps shall be specified by the Owner's Engineer.

5.3 **Vertically Suspended Pumps**

5.3.7.3.5 A separate sole plate is required. The underside shall be suitably prepared for epoxy grout.

6.0 **VENDOR'S DATA**

6.2 **Proposals**

6.2.3 **Technical Data**

p. Wear ring clearances and diameters for proposed pumps for pumping temperature 500°F (260°C) or higher.

q. Design details of all coolers.

r. Minimum case thickness and amount of corrosion allowance.

s. For pumps rated over 500 HP, pumps handling cold liquids at —50°F or lower, or pumps with suction energy ratio over 1.3, include proof of compliance with 1.1.1.d.

t. Vendor's proposals for spare parts shall include proposed method of protection from corrosion during shipment and subsequent storage.

u. Vendor's proposals shall specify the maximum dynamic and static pressure ratings of the mechanical seal.

v. Vendor's proposal shall state the minimum flow rate recommended for sustained operation (more than 1,000 hours per year) on the specified fluid.

w. Alternative Vendor proposal provided per paragraph 1.2a shall identify undemonstrated features and their advantages.

Items	**Material Description**
Tubes	ASME SB338 Gr. 2 titanium
Tube Sheets	Carbon steel with SB265 Gr. 1 titanium cladding on tube side (1/2 inch minimum thickness after machining)
Baffles, Tie-Rods & Spacers	Carbon Steel
Channel	AL-6XN with 1/16 inch corrosion allowance for barrel and partitions with carbon steel slip-on flanges, gasket surfaces weld overlaid with 1/4 inch Inconel 625
Channel Cover	Carbon steel with Gr. 1 titanium cladding (5/16 inch minimum thickness after machining)
Floating Head	Carbon steel with 1/4 inch minimum thickness Inconel 625 weld overlay on wetted surface, pass partitions of Inconel 625
Shell	Carbon steel

DOCUMENTATION REQUIREMENTS FOR CENTRIFUGAL PUMPS

Item	*Description*		*As-Built*
1	Certified dimensional outline drawing	Equipment Owner/Purchaser to indicate applicable document in which drawing and/or instruction manual format (paper, electronic, etc.) will be listed	Yes
2	Cross-sectional drawing and bill of materials		Yes
3	Shaft seal drawing and bill of materials		Yes
4	Coupling assembly drawing and bill of materials		Yes
5	Primary and auxiliary sealing schematic and bill of materials		Yes
6	Cooling or heating schematic and bill of materials		Yes
7	Lube-oil schematic and bill of materials		Yes
8	Lube-oil system arrangement drawing and list of connections		Yes
9	Lube-oil component drawings and data		Yes
10	Electrical and instrumentation schematics and bill of materials		Yes
11	Electrical and instrumentation arrangement drawing and list of connections		Yes
12	Performance curves		Yes
13	Vibration analysis data		Yes
14	Damped unbalanced response analysis		Yes
15	Lateral critical speed analysis		Yes
16	Torsional critical speed analysis		Yes
17	Certified hydrostatic test data		Yes
18	Weld procedures		Yes
19	Performance test data		Yes
20	Certified rotor balance data for multistage pumps		Yes
21	Residual unbalance check		Yes
22	Rotor mechanical and electrical runout		Yes
23	Data sheets		Yes
24	Clearances		Yes
25	Installation, operation and maintenance manuals		Yes
26	Spare parts recommendations		Yes

Note: If the document cannot be provided in the format as indicated, the Manufacturer shall consult with the Owner's Engineer for an acceptable alternative.

6.2.4 **Curves**

Quoted efficiency and power shall take into account the increased clearance required for pumps in hot services.

6.3 **Contract Data**

6.3.2 **Drawings**

Cross-sectional dimensioned drawings of the stuffing box, seal, gland, shaft sleeve, and proposed flushing arrangement for the mechanical seal shall be provided.

7.0 **GUARANTEE AND WARRANTY**

7.1 **Mechanical**

Unless exception is recorded by the Vendor in his proposal, it shall be understood that the Vendor agrees to the guarantees and warranties specified in Items 1 and 2 below:

1. All equipment and component parts shall be warranted by the Vendor against defective materials, design, and workmanship for one year after being placed in service (but not more than 18 months after date of shipment).

2. If any performance deficiencies or defects occur during the guarantee and warranty period, the Vendor shall make all necessary alterations, repairs, and replacements free of charge, free on board factory. Field labor charges, if any, shall be subject to negotiation between the Vendor and the Purchaser.

7.2 **Performance**

The equipment shall be guaranteed for satisfactory performance at all operating conditions specified on the data sheet. Field checks on performance, when made by the Purchaser, shall be made within 60 days of initial operation.

8.0 **TABLE(S)**

Refer to preceding page 517.

Appendix 4

Specification and Installation of Pre-grouted Pump Baseplates

This appendix, or standard procedure, outlines the requirements for specifying and installing pre-grouted machinery baseplates. A typical application for this standard procedure would be the installation or retrofit of an ANSI or API pump. However, this standard does not cover the installation requirements for machinery mounted on sole plates.

1.0 **Purpose**

The purpose of this standard procedure is to provide specific requirements for pre-filling any machinery baseplate, or—in particular—a pump baseplate, with epoxy grout, and machining the mounting surfaces of the baseplate after the grout has cured. Additionally, this standard procedure outlines the requirements for installing the pre-grouted baseplate in the field, utilizing a special grouting technique for the final grout pour. This special technique makes use of a low viscosity epoxy grout for the final pour. The technique greatly reduces the field costs associated with traditional installation methods.

By utilizing this standard procedure, baseplate mounted machinery can be installed with zero voids, eliminate the possibility of expensive field machining, and reduce field installation costs by 40% to 50%.

2.0 **Specification of Pre-Grouted Baseplates**

2.1 The underside of the baseplate to be pre-grouted must be sandblasted to white metal to remove all existing paint, primer, or scale.

2.2 Any tapped bolt holes that penetrate through the top of the baseplate, such as the coupling guard hold down bolts, must be filled with the appropriate sized bolts and coated with never-seize to create the necessary space for bolt installation after grouting of the baseplate.

2.3 Anchor bolt or jack bolt holes, located inside the grouted space of the baseplate, must have provisions for bolt penetration through the baseplate after grouting.

2.4 If the baseplate has grout holes and/or vent holes, these holes must be completely sealed prior to grouting.

2.5 All pre-grouted baseplates will be filled with catalyzed epoxy grout or a premium non-shrink cement grout.

2.6 Once the baseplate has been filled with epoxy grout, the grout must be completely cured before any machining is performed.

2.7 The machining of the baseplate must be set up to assure that the baseplate is under no stress or deformation.

2.8 Prior to machining, the baseplate must be adjusted and leveled to assure that no more than 0.020″ of metal is removed at the lowest point.

2.9 The baseplate will have two (2) mounting surfaces for the driver, and two (2) to four (4) mounting surfaces for the driven equipment. The flatness tolerance for all these mounting surfaces will be 0.001″ per ft. The finished surface roughness must be no more than an 85P profiled surface.

2.10 The two (2) mounting surfaces for the driver must be co-planar within 0.002.″ The two (2) to four (4) mounting surfaces for the driven equipment must also be co-planar within 0.002.″ The original dimensional relationship (elevation) between the driver mounting surfaces and the driven mounting surfaces must be maintained to within 0.020.″

2.11 Once the machining process has been completed, an "as machined" tolerance record must be taken, and provided with the pre-grouted baseplate.

3.0 General Field Grouting Requirements

3.1 The epoxy grout utilized for the final field grout pour is a low viscosity epoxy grout. This grout has a special aggregate and has the consistency of thin pancake batter. This allows for a very thin final grout pour, with the optimum vertical thickness being 3/4″ (~19 mm).

3.2 All grout material components must be stored in a dry and weatherproof area in original unopened containers. Under no circumstances should grouting components be stored outside subject to rain or under a tarpaulin with no air circulation.

3.3 For optimum handling characteristics precondition the resin and hardener to a temperature between 64°F and 90°F (18°C and 32°C).

3.4 The work area, including foundation and machinery must be protected from direct sunlight and rain. This covering (shading) should be erected 48 hours prior to alignment and grouting, and shall remain until 24 hours after placement of the grout, by which time the grout will have cured and returned to ambient temperature. The shading is also to prevent the foundation from becoming wet. It is important that the concrete remain dry prior to grouting.

3.5 Grouting shall be scheduled to take place during early morning or afternoon hours depending on the surface temperatures.

3.6 Just before starting the grouting operation, the temperature of the concrete foundation and machinery shall be taken using a surface thermometer. Ideal surface temperatures shall be between 70°F and 90°F (21°C and 32°C).

4.0 Foundation Preparation

4.1 The concrete must be chipped to expose a minimum of 50% aggregate so as to remove all laitance and provide a rough surface for bonding. Hand chipping guns only will be used: No jackhammers will be permitted. If oil or grease are present, affected areas will be chipped out until free of oil or grease.

4.2 The concrete to be chipped should not extend more

than 2″ outside the "foot print" of the pre-grouted baseplate. Low viscosity epoxy grout can only be poured up to a 2″ depth, and should not extend more than 2″ from the edge of the baseplate. By limiting the chipped area of the concrete to just outside the foot print of the baseplate, simple forming techniques can be utilized.

4.3 After chipping, the exposed surface must be blown free of dust and concrete chips using oil and water free compressed air from an approved source. Concrete surface can also be vacuumed.

4.4 After the foundation has been chipped and cleaned, adequate precautions must be taken to ensure there is no contamination of the concrete surfaces. To prevent debris, loose materials, or parts, from falling on the top of the concrete, properly cover the workspace with polyethylene sheet.

4.5 The foundation bolt threads must be protected during the grouting operations.

4.6 As regards the bolts ("Drawing #1"), which will be tensioned after grouting, care must be taken to prevent the bolt surfaces from coming in contact with the epoxy grout. All anchor bolts should have grout sleeves, which must be filled with a nonbonding material to prevent the epoxy grout from filling the grout sleeve. This can be accomplished by protecting the anchor bolt beforehand between the top of the grout sleeve and the underside of the baseplate by wrapping the bolt with foam insulation, Dux-Seal or other nonbonding material.

5.0 **Pre-grouted Baseplate Preparation**

5.1 Prior to positioning the baseplate over the foundation, the bottom side of the pre-filled baseplate must be solvent washed to remove any oil or other contaminants from the surface. After the surface has been cleaned, sand the surface to break the glaze of the epoxy grout.

5.2 Vertical edges of the baseplate that come in contact with the epoxy grout must be radiussed/cham-

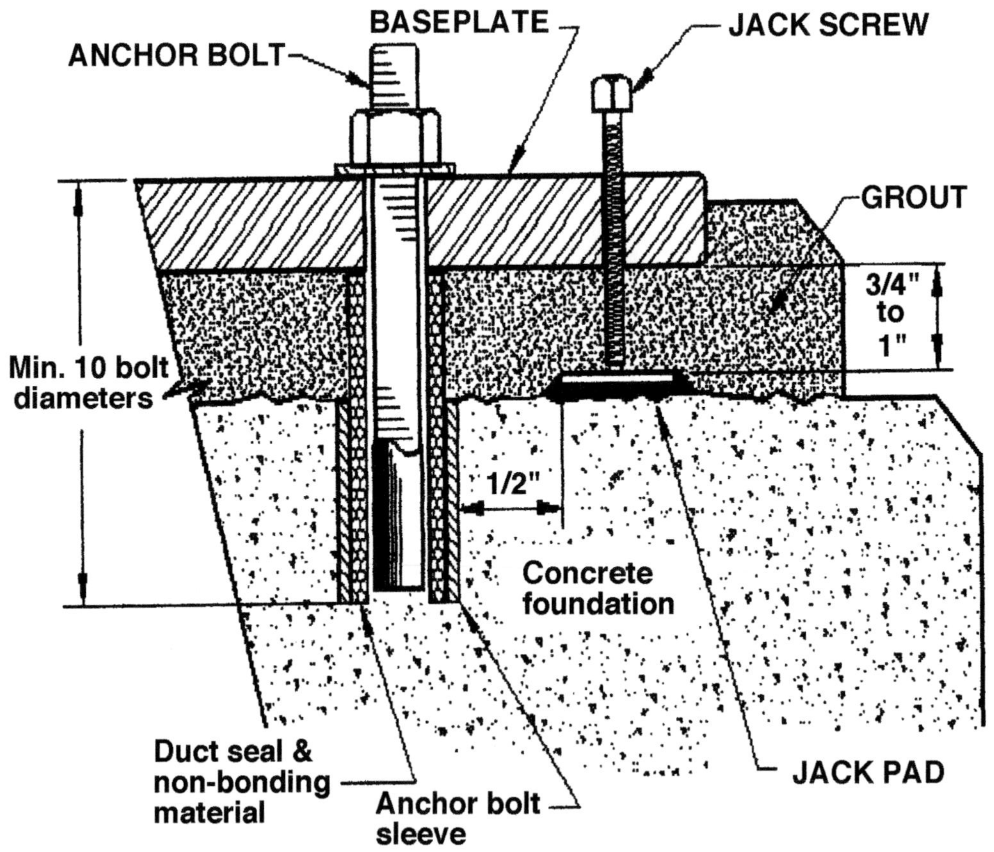

Drawing #1. Anchor bolt detail

fered to reduce stress concentration in the grout.

5.3 Vertical jackscrews should be provided at each anchor bolt. The jackscrews will be used to level the pre-grouted baseplate. These jackscrews will be removed after the low viscosity epoxy cures, generally 24 hours after placement at 78°F (26°C).

5.4 Leveling pads should be used under each jackscrew to prevent the baseplate from "walking" while leveling the baseplate. The pads will remain in the grout, and must be made from stainless steel. The pads must have radiused edges and rounded corners to reduce stress concentrations in the grout.

5.5 With the jackscrews and leveling pads in place, level the pre-grouted baseplate to 0.002"/foot for API applications and 0.005"/foot (0.12 mm/m) for ANSI applications.

5.6 After the baseplate has been leveled, the jackscrews must be greased or wrapped with Dux-Seal to facilitate their removal once the grout has cured. Wax is not a suitable releasing agent.

6.0 **Forming**

6.1 Low viscosity epoxy grout should only be poured up to a 2" depth, and should not extend more than 2" from the edge of the baseplate. The optimum pour depth is 3/4" to 1". The best wood forming material for this product is a "2 by 4."

6.2 Any wood surface coming in contact with epoxy grout, shall be coated three times with paste floor wax. (Liquid wax or oil is not acceptable as an alternative). All forms must be waxed three times *before* the forms are placed on the foundation (note "Drawing #2"). Do *not* wax the wood surface that comes in contact with the foundation. This may prevent the silicone sealant from sticking to the form board and forming a proper seal. Care should be taken to prevent any wax from falling on either concrete foundation or baseplate.

6.3 In most cases, there is very little room between the side of the baseplate and the edge of the foundation. To help position the form boards, it is best to fasten the boards together with wood screws or nails. One side of the form should leave an opening between the board and the baseplate that measures 1" to 2" (see "Drawing #3"). The other three sides should have a separation of 1/2" to 3/4". The larger side will be used to pour the low viscosity epoxy grout.

6.4 Forms shall be made liquid tight: Silicone sealant that does not cure to a hard consistency is best

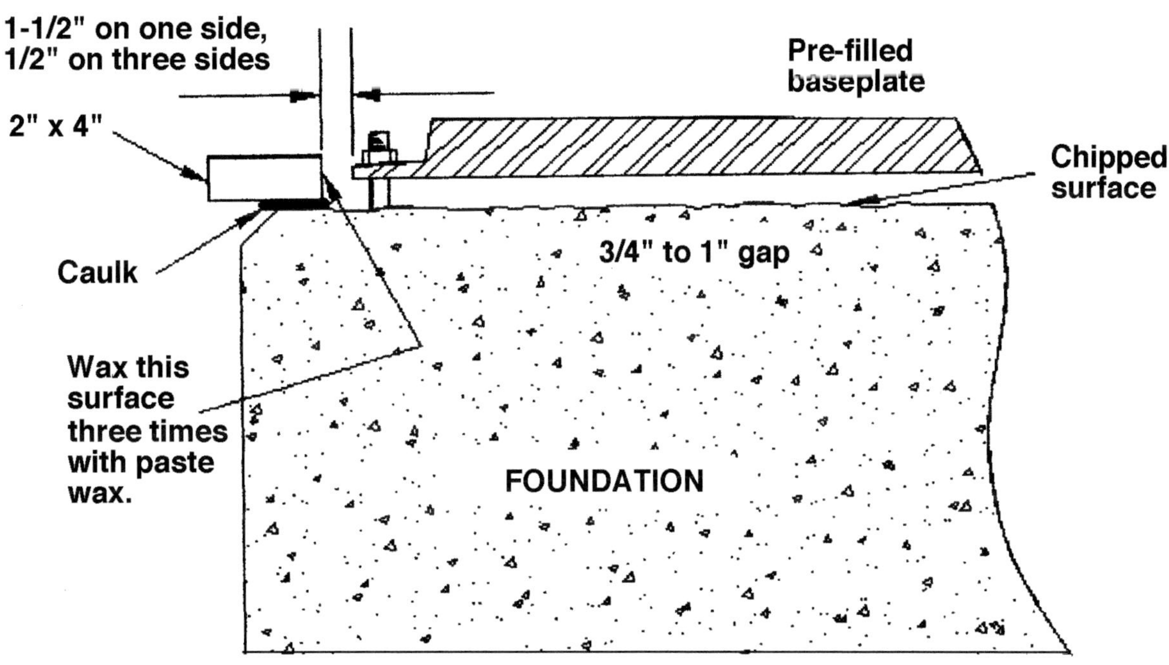

Drawing #2. Foundation detail

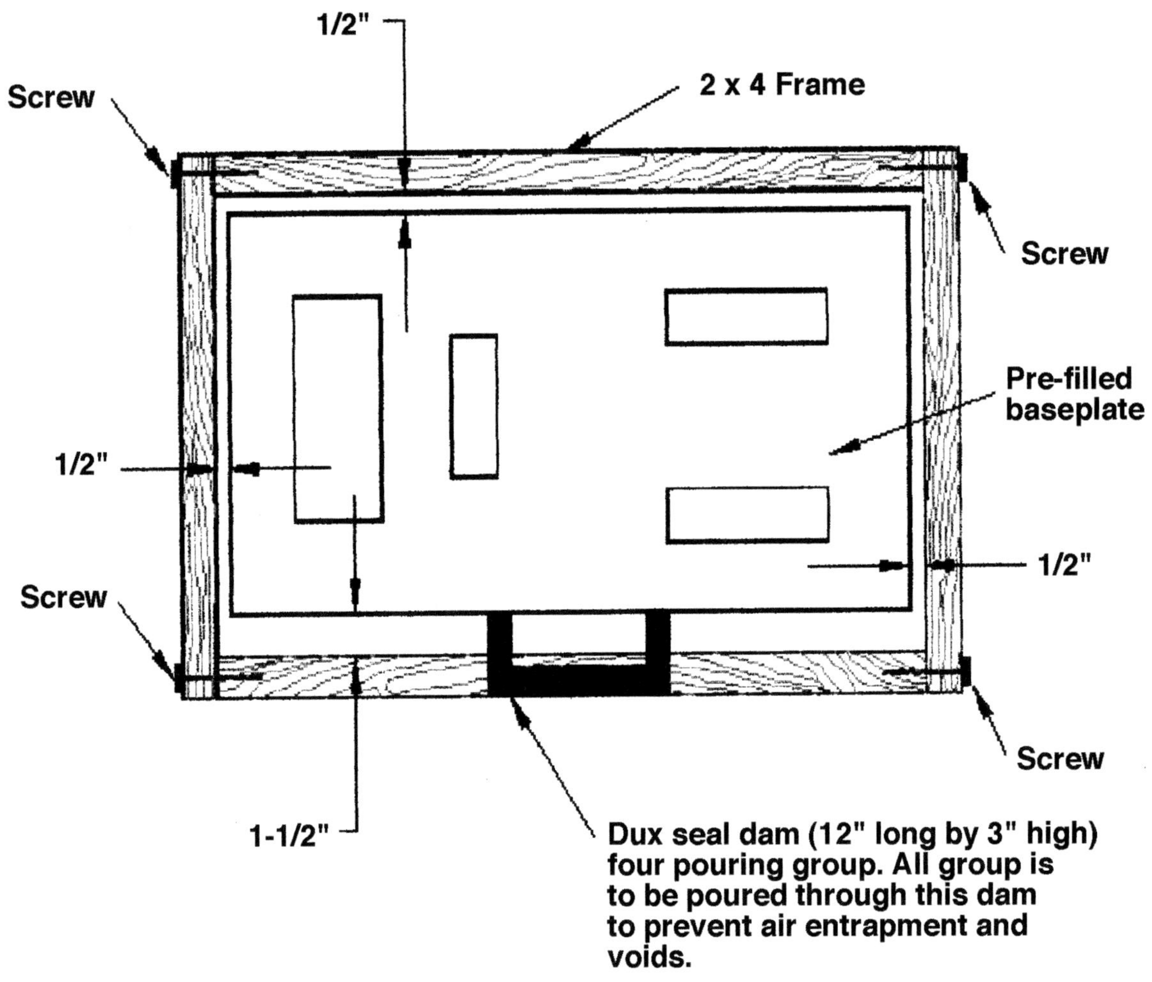

Drawing #3. Form detail

suited for this application. A sealant that remains pliable will facilitate easy removal. The best approach is to apply the sealant directly on the foundation, where the front edge of the form will fall, and then press the form down to create the seal. Check for cracks and openings between the form and the foundation, and apply additional sealant where needed. Allow at least an hour for the sealant to cure before pouring the grout.

6.5 Because of the small depth of the epoxy grout pour, it will be very difficult to use a chamfer stripe to create a bevel around the outside of the form. The best approach is to use a grinder after the grout has cured to create a bevel.

6.6 Once the form boards are in place, a small "head box" can be made using blocks of duct seal. To help create a slight head for the grout, build a duct seal dam on the side of the baseplate with the larger opening. The dam should be about 3" tall, 12" long, and form a rectangle by connecting the two short sides directly to the baseplate. The end result will be a 3" head box that will be used to pour and place all the epoxy grout. The best location for the head box is the mid-point of the baseplate.

7.0 Grouting Procedure for Low Viscosity Epoxy Grout

7.1 The required number of units of epoxy grout, including calculated surplus, should be laid out close to the grouting location. The 1/2" drill and mixer blade should be prepared for the grouting operation.

7.2 Low viscosity epoxy grout is a three-component, high-strength, 100% solids epoxy grouting compound. The resin, hardener, and aggregate are supplied in a 6 gallon (~23 l) mixing container. One unit produces 0.34 cubic feet of grout.

7.3 The hardener will be poured into the can containing the resin. Using a 1/2" hand drill, 200-250 RPM, with a "Jiffy-type" mixer blade, gradually blend the mixture. Care should be taken not to whip air into the mixture. The resin and hardener must be mixed for three minutes. Pour the resin-hardener combination into the mixing container and gradually mix the aggregate into the resin-hardener mixture. Use the mixer blade to blend the material until the aggregate is completely wet.

7.4 The pouring must be carefully managed starting with the inside forms and working toward the outside forms. Start pouring the grout from the initial location and do not move along the forms until the grout has made contact with a pre-calculated percentage of the underside of the baseplate.*

7.5 No mechanical vibrators or strapping should be used to place the grout under the baseplate. Low viscosity epoxy grout has exceptional flow characteristics and can be placed with adequate hydraulic head.

7.6 Cure time for low viscosity epoxy grout is 24 hours at 76°F (26°C). Protect recently poured grout from any sudden temperature change and direct sunlight by providing a shade over the work site.

7.6 Forms must be left in place until the grout has cured. The surface of the grout should be firm and not tacky to the touch.

7.8 Immediately after the grouting process is complete, all tools and mixing equipment must be cleaned with plant pressure water and nozzle.

7.9 Before the grout starts to harden, the excess grout should be scraped off with trowels and the baseplate washed with a wet rag or a solvent, such as WD-40.

*Remember that the inside of the baseplate is filled with epoxy. Therefore, it will not be necessary to make the present low-viscosity grout fill the entire volume, or displace all the air! The grout area needs to be less than 50 psi. Example: 16,000 lbs on 320 in^2 = 50 psi. Hence, grout areas over 320 in^2 would be large enough here.

8.0 All unused mixed epoxy material and cleanup residue should be disposed of in accordance with local laws and ordinances. The mixed epoxy material is non-hazardous, and should be compatible with general waste.

8.1 **Inspection**

Inspection Check List

Inspection of Work Site— Check for:

- Proper shading
- Preparation of concrete, baseplate, jackscrews, leveling pads
- Wood forms properly waxed and sealed
- Foundation bolts wrapped & sealed

Before Mixing—Check for:

- Mixing equipment clean
- Surface temperature of epoxy grout components (<90°F or 32°C)
- Ambient Temperature (<95°F or 35°C)

While Mixing—Check for:

- Slow drill motor rpm's to avoid entrapping air
- Resin and hardener mixed 3 minutes (use the timer)

Before Pouring—Check for:

- Temperature of concrete foundation (<95°F or 35°C))
- Temperature of the machinery baseplate (<95°F)

While Pouring—Check for:

- Continuous operation
- Adequate head to fill corners inside baseplate

After Pouring—Check for:

- Ambient temperature for the record
- Maintaining head until the grout starts to set

Curing—Check for:

- Work site kept shaded for 24 hours to avoid sharp temperature increase
- Formwork left in place until grout is no longer tacky to the touch

Appendix 5

Protection of Mechanical Seals in Non-operating Pumps

The protection of liquid or dry gas type mechanical seals in pumps that are stored or "mothballed" for prolonged periods of time deserves attention. The precautionary recommendations of A.W. Chesterton Company are given here as a general guide.

In general, protection by blanketing with oil mist is superior to all other methods. See Chapter 10 for details.

1. **Storage**

Chesterton recommends, when possible, to remove the mechanical seal from the pump. The company correctly points out that mechanical seals are assembled and tested in a clean room environment.

Left unprotected in the field, both internal as well as external contaminants, such as airborne dust, can accumulate in the critical sealing areas causing leakage or damage to the seal on start-up. (The authors believe this to be equally true for liquid and dry gas seals).

The seal should be labeled to identify the materials of construction, then packaged and stored in a controlled dry environment. If the seal has previously been in operation, or has been in contact with fluids, it should be returned to the manufacturer for inspection and/or repair.

2. **Pumps where fluid has been introduced**

- Valve-off the pump and drain fluid from the pump casing
- Remove seal environmental controls and drain fluids, including barrier fluid, from the seal
- Clean seal chamber(s) with a solvent compatible with the seal materials to remove all possible residues and drain from the pump and seal. Rotate shaft by hand during this process
- Plug all seal ports
- Mask the opening between the shaft or seal sleeve and the gland to protect the seal from environmental contamination
- Tag the equipment with the date of storage.

3. **Pumps where no fluid has been introduced**

- Remove seal environmental controls
- Plug all seal ports
- Mask the opening between the shaft or seal sleeve and the gland to protect the seal from environmental contamination
- Tag the equipment with the date of storage.

4. **Restarting equipment**

- Check with the seal manufacturer to assure the date of storage in conjunction with the materials of construction does not exceed the shelf life of the mechanical seal
- Remove masking from the opening between the shaft or seal sleeve and the gland
- Valve out the pump and remove plugs from seal ports and flush with a solvent compatible with the seal materials to remove all possible residues. Rotate shaft by hand during this process
- Drain fluid from the pump casing
- Reconnect seal environmental controls and/or plug seal ports
- Open suction and crack open discharge valves
- Vent the seal chamber to allow the seal to become surrounded by liquid
- Start the pump, using the checklist/procedure of Appendix 1 of this text.

Appendix 6

Explaining the Ingress Protection Code

The Ingress Protection Code ("IP" for short) refers to the different levels of protection an enclosure provides and gives a means of classifying the degrees of protection from dust, water and impact. It is widely used in Europe and relates to thorough third-party verification of the containment effectiveness of the tested device. Originally intended for electrical equipment and enclosures, the application parameters fully cover bearing protector seals for rotating machinery (see Figure 9-36).

The code is separated into two characteristic numerals; the first numeral indicates the protection of hazardous parts and equipment against ingress of foreign solid objects. The second numeral indicates protection of equipment against harmful ingress of liquids.

If a particular test is conducted to examine only the protection against solid objects, the letter "X" is placed in the IP column describing protection against liquids. Conversely, if the particular test is conducted only to examine the protection against liquids, the letter "X" is placed in the IP column describing protection against solid objects (as, incidentally, was done on the certificate reproduced on page 526).

FIRST NUMBER (Column I)
Protection against solid objects

IP	Tests
0	No protection
1	Protected against solid objects over 50 mm e.g. accidental touch by hands
2	Protected against solid objects over 12 mm e.g. fingers
3	Protected against solid objects over 2.5 mm (tools and wires)
4	Protected against solid objects over 1 mm (tools, wires and small wires)
5	Protected against dust – limited ingress (no harmful deposit)
6	Totally protected against dust

SECOND NUMBER (Column II)
Protection against liquids

IP	Tests
0	No protection
1	Protected against vertically falling drops of water
2	Protected against direct sprays of water up to 15° from the vertical
3	Protected against sprays up to 60° from the vertical
4	Protected against water sprayed from all directions – limited ingress permitted
5	Protected against low pressure jets of water from all directions – limited ingress permitted
6	Protected against strong jets of water e.g. for use on ship decks – limited ingress protected
7	Protected against the effects of temporary immersion between 15 cm and 1m. Duration of test 30 minutes
8	Protected against long periods of immersion under pressure

It should be noted that the ingress protection code is designed to examine if a product is suitable for a specific requirement. Reliability-conscious users call up IP testing as a basic requirement for environmental testing against external influences such as water and dust ingress and against contact with live parts.

As an example, the test certificate reproduced below indicates that, in 2010, this particular bearing protector seal earned the performance category IP 66. (Several years earlier it had been awarded performance category IP 55, indicating "protected against dust—limited ingress—no harmful deposit" and "protected against low pressure jets of water from all directions—limited ingress permitted").

Later, the first column protection designation (i.e.,

formerly IP 5) was upgraded to IP 6. Likewise, the most recent IP protection-against-liquid test result (second column protection: IP 6) indicates that the bearing protector seal at issue will effectively defend bearing housing interiors against strong jets of water. The bearing housing protector seal at issue is shown in Figure 9-36; it was first marketed in 2006. In line with the explanations given above, as of 2010 its rating became IP 66.

IP 66 indicates a superior degree of protection, a level or an extent that has been achieved by few such products. The most demanding reliability-focused industrial process plants and maritime fleet owners desire this degree of performance for modern bearing housing protector seals.

IP test certificate for an advanced bearing housing protector seal

TEST CERTIFICATE
ISSUED BY SIRA TEST & CERTIFICATION LIMITED

TEST FOR THE INGRESS PROTECTION OF A LABTECTA BEARING ISOLATOR

Supplier:	AESSEAL plc Mill Close Templeborough Rotherham S60 1BZ
Model or Type Identification:	1.750" diameter Labtecta bearing isolator
Standard:	BS EN 60529:1992 Incorporating Amendments Nos 1 and 2
Deviations from Standard:	None
ST&C Test Procedure:	LOP 220
ST&C Test Reports:	09/0443, N58D20672A, 10/0090 and N58D21827A
Samples Delivery Date:	23 July 2009 and 17 February 2010
Tests Conducted Between:	23 July 2009 and 17 to 18 February 2010

This certificate refers to the performance of the test sample when tested against the agreed programme. It does not imply that any other samples or products necessarily comply with the requirements of the test programme.

Sira Test & Certification Limited being a UKAS accredited Test House in accordance with ISO/IEC 17025 has tested the above bearing isolator, and has found it to comply with the requirements of the Ingress Protection Code: IP 66.

Dated 5 March 2010

S P Cork
Laboratory Manager

This certificate may only be reproduced in its entirety and without change

Page 1 of 1

Certificate No: Sira 58D21827ALab
Form 6240, Issue 8

a CSA International company

Sira Test & Certification Ltd **Registered in England:** **Registered Office:**

References

INTRODUCTION (REFERENCES "I")

1. Chemical Processing, January 2003
2. Bloch, H.P. and F.K. Geitner; "Machinery Failure Analysis and Troubleshooting," Butterworth-Heinemann, Stoneham, Massachusetts, 3rd Edition, 1997, ISBN 0-88415-662-1)

CHAPTER 1

1. "Pump Life Cycle Costs: A Guide to LCC Analysis for Pumping Systems," 2001, Hydraulic Institute, Parsippany, NJ
2. Bloch, H.P., "Use Equipment Failure Statistics Properly," Hydrocarbon Processing, January 1999
3. Branham, Douglas C.; News Flash (*dcbranham@lsc.com*), dated 4/22/2005
4. Karassik, Igor J.; " Centrifugal Pump Clinic" p. 334, ISBN: 0-8247-1016-9, Marcel Dekker, Inc., New York, NY
5. Sofronas, Anthony; "Analytical Troubleshooting of Process Machinery and Pressure Vessels: Including Real-World Case Studies," pp.153, ISBN: 0-471-73211-7, John Wiley & Sons, New York, NY
6. Sofronas, Anthony; "Problems with a blocked-in centrifugal pump," *Hydrocarbon Processing*, September 2009, pp. 116

CHAPTER 2

1. Bloch, Heinz P., "How to Select a Centrifugal Pump Vendor," Hydrocarbon Processing, June 1978
2. Bloch, Heinz P., "How to Buy a Better Pump," Hydrocarbon Processing, January 1982
3. Bloch, Heinz P., "Machinery Reliability Improvement," Gulf Publishing Company, Houston, Texas, 1982. Also revised 2nd & 3rd Editions, ISBN 0-88415-661-3
4. Bloch, Heinz P., "Optimized Lubrication of Antifriction Bearings for Centrifugal Pumps," (ASLE Paper 78-ASLE-1-D-2), April 1978
5. Bloch, Heinz P., "Upgrading of Centrifugal Pumps: A Status Report," Hydrocarbon Processing, February 1992
6. Bloch, Heinz P., and Don Johnson, "Downtime Prompts Upgrading of Centrifugal Pumps," Chemical Engineering, November 25, 1985
7. Dufour, John W., and W.E. Nelson, "Centrifugal Pump Sourcebook," McGraw-Hill, Inc., New York, NY 1992, ISBN 0-07-018033-4
8. Bloch, Heinz P., and F.K. Geitner, "Introduction to Machinery Reliability Assessment," Van Nostrand-Reinhold, New York, NY 1990; also revised 2nd Edition, Gulf Publishing Company, Houston, Texas, 1994, ISBN 0-88415-172-7
9. Bloch, Heinz P., "Practical Lubrication for Industrial Facilities," Fairmont Publishing Company, Lilburn, GA 2000, ISBN 0-88173-296-6
10. Noria Corporation, Tulsa, Oklahoma, Sales Literature. Also available from Royal Purple Lubricants, Porter, Texas
11. Eschmann, Hasbargen, Weigand, "Ball and Roller Bearings," John Wiley & Sons, New York, NY 1985, ISBN 0-471-26283-8
12. Bloch, Heinz P., "Defining Machinery Documentation Requirements for Process Plants," (ASME Paper No. 81-WA/Mgt-X, presented at ASME Winter Annual Meeting, Washington, D. C., November 15-20,1981)
13. Bussemaker, E.J., "Design Aspects of Base Plates for Oil and Petrochemical Industry Pumps," IMechE (UK) Paper C45/81, pp. 135-141

CHAPTER 3

1. "Centrifugal Pumps—Vertical Nozzle Loads," ANSI/HI 9.6.2, Hydraulic Institute, Parsippany, NJ, 2001

2. Budris, Allan R., "Designing Pump Piping -Protecting Pump Performance and Reliability," Chemical Processing, August 2002

3. "Piping for Rotodynamic Pumps" (Centrifugal Pumps), HI 9.6.6, Draft Document, Hydraulic Institute Pump Piping Working Group, Parsippany, NJ, 2003

4. American National Standard for "Centrifugal and Vertical Pump NPSH Margins," ANSI/HI9.6.1, 1998, Hydraulic Institute, Parsippany, NJ.

5. Barringer, P., and Todd Monroe, 1999, "How to Justify Machinery Improvements Using Reliability Engineering Principles," Proceedings of the 16th International Pump Users Symposium, Turbomachinery Laboratory, Texas A&M University, College Station, Texas 77843

6. Stay-Tru® Company, Houston, Texas. Sales Literature

7. Myers, R., 1998, "Repair Grouting to Combat Pump Vibration," Chemical Engineering

8. Machine Support, Inc., Cheasapeake, VA 23320, and Ridderskerk, 2984 Netherlands; Standard Sales Literature

9. Bloch, Heinz P., "Update Your Alignment Knowledge," Pumps & Systems, December 2003

10. Bloch, Heinz P., and F.K, Geitner, "Machinery Component Maintenance and Repair," 2nd Edition, Gulf Publishing Company, Houston, Texas 1990; ISBN 0-87201-781-8

11. "Laser-Optic Instruments Improve Machinery Alignment," Oil and Gas Journal, October 12, 1987

12. Bloch, Heinz P., "Laser-Optisches Maschinenausrichten," (Antriebstechnik, Vol. 29, Nr. 1, June 1990, Germany

13. Bloch, Heinz P., "Laser Optics Accurately Measure Running Shaft Alignment," (Oil & Gas Journal, November 5, 1990)

14. Bloch, Heinz P., "Use Laser-Optics For On-Stream Alignment Verification," (Hydrocarbon Processing, January 1991)

15. "Easy-Laser®," Damalini AB, Molndal, Sweden. Marketing Bulletin 05-0227

16. Bloch, Heinz P., and F.K. Geitner; "Maximizing Machinery Uptime," Elsevier & Butterworth-Heinemann, Stoneham, Massachusetts (ISBN 0-7506-7725-2)

17. *Marks' Standard Handbook for Mechanical Engineers*, 7th Edition (1969), McGraw-Hill Book Company, New York, NY

18. www.mcnichols.com/products/wiremesh

19. Karassik, Igor J. "Centrifugal Pump Clinic," 2nd Ed., Marcel Dekker, Inc. (1989)

CHAPTER 4

1. "Cameron Hydraulic Data," 1958, Ingersoll -Rand Company, Phillipsburg, NJ

2. "Goulds Pump Manual -GPM6," 1995, Goulds Pumps, Inc., Seneca Falls, NY 13148

3. Bloch, Heinz P. and C. Soares, "Process Plant Machinery," 2nd Edition, 1998, Butterworth-Heinemann Publishing Company, Stoneham, Massachusetts; ISBN 0-7506-7081-9

4. Budris, Allan R., "How Parallel pumps can be Pushed into Damaging Low Flow Pump Internal Suction Recirculation," *WaterWorld*, March 2011.

5. Budris, Allan, R., "How to Optimize the Overall Efficiency when Pumps are Operated as both Pumps and Power Recover Turbines in the Same System," *WaterWorld, October 2010.*

CHAPTER 5

1. Bush, Fraser, Karassik, "Coping with Pump Progress: The Sources and Solutions of Centrifugal

Pump Pulsations, Surges and Vibrations." Pump World, 1976, Volume 2, Number 1, pp. 13-19 (A publication of Worthington Group, McGraw-Edison Company).

2. McQueen, R., "Minimum Flow Requirements for Centrifugal Pumps." Pump World, 1980, Volume 6, Number 2, pp.10-15

3. Budris, Allan R., Eugene P. Sabini & Barry Erickson, "Pump Reliability -Hydraulic Selection to Minimize the Unscheduled Maintenance Portion of Life-Cycle Cost," Pumps & Systems, November 2001

4. "Allowable Operating Region," ANSI/HI 9.6.3-1997, Hydraulic Institute, Parsippany, New Jersey

5. Makay, Elmer and Olaf, Szamody, "Survey of Feed Pump Outages," FP-754 Research Project 641 for Electric Power Research Institute, Palo Alto, CA

6. Budris, Allan R., and Philip A. Mayleben, "The Effects of Entrained Air, NPSH Margin and Suction Piping on Centrifugal Pumps," 1997, Proceedings of the 15th International Pump Users Symposium, Texas A&M University, Houston, Texas

7. Budris, Allan R., Barry Erickson, Francis H. Kludt & Craig Small, "Consider Hydraulic Factors to Reduce Pump Downtime," Chemical Engineering, January 2002

8. Karassik, Igor J., "So, You Are Short On NPSH?" presented at Pacific Energy Association Pump Workshop, Ventura, CA, March 1979.

9. Hydraulic Institute, NPSH Margin Work Group, Chairman: Allan R. Budris, "Cavitation problems?" Plant Services, August 1997.

10. American National Standard for "Centrifugal and Vertical Pumps for NPSH Margin," ANSI/HI9.6.1, 1998, Hydraulic Institute, Parsippany, New Jersey.

11. "Pump Intake Design Standard," ANSI/HI 9.8, 1998, Hydraulic Institute, Parsippany, New Jersey.

12. Budris, Allan R., "How to Identify and Avoid Damage from Low Flow Pump Internal Suction Recirculation," *WaterWorld*, June 2010.

13. Budris, Allan R., "The Right Coating can Improve Pump Impeller Cavitation Damage Resistance at a Reasonable Cost," *WaterWorld*, April 2012.

Bibliography/Supplementary Reading:

Bush, Fraser, Karassik, "Coping with Pump Progress: The Sources and Solutions of Centrifugal Pump Pulsations, Surges and Vibrations." *Pump World*, 1976, Volume 2, Number 1, pp. 13-19 (A publication of Worthington Group, McGraw-Edison Company).

Fraser, W.H., "Flow Recirculation in Centrifugal Pumps," Proceedings of the 10th Turbomachinery Symposium, Texas A&M University, College Station, Texas, December,1981.

CHAPTER 6

1. Bloch, H.P., "Improving Machinery Reliability," Gulf Publishing Company, Houston, TX, 1st Edition, 1982, ISBN 0-87201-376-6

2. Karassik, Igor J., "So, You Are Short On NPSH?" presented at Pacific Energy Association Pump Workshop, Ventura, CA, March 1979.

3. Ingram, J.H., "Pump Reliability—Where Do You Start," Paper presented at ASME Petroleum Mechanical Engineering Workshop and Conference, Dallas, TX, September 13-15, 1981

4. Bloch, H.P., "Optimized Lubrication of Antifriction Bearing for Centrifugal Pumps," ASLE Paper No. 78-AM-1D-1, presented in Dearborn, Michigan, April 17, 1978.

5. McQueen, R., "Minimum Flow Requirements for Centrifugal Pumps," *Pump World*, 1980, Volume 6, Number 2, pp. 10-15.

6. Bloch, Heinz P., and Don Johnson, "Downtime Prompts Upgrading of Centrifugal Pumps," *Chemical Engineering*, November 25, 1985

7. "Goulds Pump Manual -GPM6," 1995, Goulds Pumps, Inc. , Seneca Falls, NY 13148

8. Bloch, Heinz P., "Twelve Equipment Reliability Enhancements With a 10:1 Payback," presented at NPRA Reliability and Maintenance Conference, New Orleans, LA, May 2005

9. Adams, Erickson, Needelman and Smith, (1996) Proceedings of the 13th International Pump User's Symposium, Texas A&M University, Houston, TX, pp. 71-79)

10. Urbiola Soto, Leonardo, "Experimental Investigation on Rotating Magnetic Seals," Masters' Thesis, Texas A&M University, College Station, Texas, 2001/2002

11. Bloch, Heinz P., "Slinger Rings Revisited," *Hydrocarbon Processing*, August 2002

12. Bloch, Heinz P., "Centrifugal Pump Cooling and Lubricant Application—A Technology Update," Texas A&M 22nd International Pump Users Symposium, Houston, Texas, March 2005

13. "Vespel® CR-6100 Applications Manual," DuPont Engineering Polymers, Newark, Delaware.

14. Bloch, Heinz P. and F.K. Geitner, "Major Process Equipment Maintenance and Repair," Gulf Publishing Company, 3rd Edition, 1997, p.32, ISBN 0-88415-663-X

CHAPTER 7

1. ANSI/ASME Standard B73.1 1984, Specification for Horizontal End Suction Centrifugal Pumps for Chemical Process, American National Standards Institute, American Society of Mechanical Engineers.

2. API Standard 610, American Petroleum Institute, Centrifugal Pumps for General Refinery Service, Seventh Edition, February 1989.

3. Bloch, H.P., "Large Scale Application of Pure Oil Mist Lubrication In Petrochemical Plants," ASME Paper 80-C2/Lub 25, presented at Century 21 ASME/ASLE International Lubricaton Conference, San Francisco, CA, August 18-21, 1980.

4. Bloch, H.P. and Shamim, A.; (1998). "Oil Mist Lubrication—Practical Applications," ISBN 0-88173-256-7.

5. Bloch, H.P.; (2000). "Practical Lubrication for Industrial Facilities," The Fairmont Press, Inc., Lilburn, GA, 30047, ISBN 0-88173-296-6.

CHAPTER 8

1. Safematic Oy, Muurame, Finland (Sales Literature, 1996)

2. Burgmann Seals America, Inc., Houston, Texas 77041. Also, Feodor Burgmann Dichtungswerke GmbH, D-82502 Wolfratshausen, Germany; Design Manual 14

3. Borg-Warner Seals, Temecula, California (Sales Literature, 1988)

4. Pacific-Wietz GmbH, Dortmund, Germany (Sales Literature, 1990)

5. A.W. Chesterton, Stoneham, Massachusetts 02180 (Marketing Literature, 2001)

6. Flowserve Corporation, Kalamazoo, Michigan 49001 (Marketing Bulletins)

7. Will, T.P.; "Effects of seal face width on mechanical seal performance—hydrocarbon tests," presented at ASME/ASLE Lubrication Conference, Hartford, Connecticut, 1983.

8. Bloch, H.P., "Improving Machinery Reliability," First Edition, Gulf Publishing Company, Houston, Texas, 1982 (deleted from some of the later editions).

9. Bloch, H.P., and Schuebl, W., "Reliability improvement in mechanical seals," Proceedings of the 2nd International Pump Users Symposium, Texas A&M University, Houston, Texas, 1985.

10. Schoepplein, W.; "Mechanical seals for aqueous media subject to high pressure," 8th International Fluid Sealing Conference, Durham, Great Britain 1978.

11. Surface Technologies, Ltd., www.surface-tech.com, 2003.

12. Bloch, H.P. and Geitner, F.K.; " Machinery Failure Analysis and Troubleshooting," First Edition, Gulf Publishing Company, Houston, Texas, 1983, page 615.

13. AESSEAL plc, Rotherham, UK and AESSEAL Inc., Rockford, Tennessee; www.aesseal.com, *(website data)*

14. Predictions of Accelerated Climate Change, Cox P.M., Betts R.A., Jones C.D., Spall S.A. & Totterdell I.J. (2000); "Acceleration of global warming due to carbon-cycle feedbacks in a coupled climate model;" *Nature*, Vol. 408, pp.184-187

15. Pumps - Shaft Sealing Systems for Centrifugal and Rotary Pumps, International Standard API 682 (ISO 21049), American Petroleum Institute. 3rd Edition, September 2004

16. Energy Cost Reduction in the Pulp and Paper Industry – An Energy Benchmarking Perspective, D.W. Francis, M.T. Towers and T.C. Browne Pulp and Paper Research Institute of Canada The Office of Energy Efficiency of Natural Resources Canada, 2002

17. AESSEAL plc., Rotherham, UK, and Rockford, TN; "API Flush Plans," laminated booklet

18. Water: How to Manage a Vital Resource, Angel Gurría, Secretary-General, OECD; OECD FORUM 14-15 May 2007, Paris, www.oecd.org

19. BNXS01: Carbon Emission Factors For UK Energy Use (Presently under Consideration) Defra Market Transformation Program; Version 2.2, 10 July 2007

20. Bloch, H.P., and Geitner, F.K.; Major Process Equipment Maintenance and Repair," 3rd Edition, 1997, Gulf Pubishing Company, Houston, Texas; pp. 85-87, ISBN 0-088415-663X

21. API Standard 682/3rd Edition, September 2004; also, ISO 21049: "Shaft Sealing Systems for Centrifugal and Rotary Pumps" American Petroleum Institute, Washington DC

22. API Standard 682/3rd Edition, Clause 7.3.4.2.1

23. API Standard 682/3rd Edition, Figs. 3 & 4

24. API Standard 682/3rd Edition, Clause 7.3.4.3

25. API Standard 682/3rd Edition, Clause 7.3.4.2.1 NOTE

26. API Standard 682/3rd Edition, Clause 7.3.1.1 NOTE 1

27. API Standard 682/3rd Edition, Clause 6.1.1.7

28. API 682/2nd Edition, July 2002; note regarding Clause 7.3.4.2.1

29. API Standard 682/3rd Edition, Annex G

30. Carmody, C.; Roddis, A.; Amaral Teixeira, J.; Schurch, D.; "Integral pumping devices that improve mechanical seal longevity"; presented at the 19th International Conference on Fluid Sealing, Poitiers, France, 25-26 September 2007

31. API Standard 682 3rd Edition, clause 8.6.2.3

32. API Standard 682 3rd Edition, clause 8.2.3

33. API Standard 682 3rd Edition Annex F

CHAPTER 9

1. Bloch, Heinz P., "Practical Lubrication for Industrial Facilities," Fairmont Publishing Company, Lilburn, GA, 2nd Ed., 2009, ISBN 0-88173-579-5.

2. SKF USA, General Catalog, 2000.

3. "Diagnosing lube oil contamination, " Plant Services, April, 2000, pp. 107-114.

4. Wilcock, Donald F., and E.R. Booser, 1957, "Bearing Design and Application," McGraw-Hill Book Company, New York, NY 10121.

5. Miannay, Charles R., "Improve Bearing Life with Oil Mist Lubrication," Hydrocarbon Processing, May 1974.

6. Bloch, Heinz P., "Dry-Sump Oil Mist Lubrication for Electric Motors," Hydrocarbon Processing, March 1977.

7. The Barden Bearing Company, Commercial Sales Bulletin, 1990.

8. Bloch, Heinz and Shamim, A.; (1998). "Oil Mist Lubrication—Practical Applications," ISBN 0-88173-256-73.

9. Bloch, Heinz P., "Bearing Protection Devices and Equipment Reliability: Constant Level Lubricators," Pumps & Systems, September 2001.

10. "Bearing Protection Devices and Equipment Lubricators: What is Really Justified?" Pumps & Systems, October 2001.

11. "Advanced lubricants produce savings," Maintenance Technology, April 1997.

12. Orlowski, David C. "Gibit Gambits," Volume 1, No.71, June 13, 1989.

13. Urbiola Soto, Leonardo, "Experimental Investigation on Rotating Magnetic Seals," Masters' Thesis, Texas A&M University, 2001/2002.

14. Bloch, Heinz P., and Don Johnson, "Downtime Prompts Upgrading of Centrifugal Pumps," Chemical Engineering, November 25, 1985.

15. Eschmann, Hasbargen, Weigand, "Ball and Roller Bearings," John Wiley & Sons, New York, NY 1985, ISBN 0-471-26283-8.

16. "Aluminum Soap Greases Formulated with Synthesized Hydrocarbons," Royal Purple Lubricants, Porter, Texas 77565.

17. Bloch, Heinz P., and Luis Rizo, "Lubrication Strategies for Electric Motor Bearings in the Petrochemical and Refining Industries," Paper No. MC 84-10, Proceedings of the NPRA Mainteance Conference, February 14-17, 1984.

18. Bloch, Heinz P., "Basic Lubrication Knowledge Challenged," Pumps & Systems, March 2002.

19. Bloch, Heinz P., "Automatic Lubrication as a Modern Maintenance Strategy," World Pumps, September, 1996.

20. Bloch, Heinz P., "Automatic Lubrication is Key Part of Modern Maintenance Programs," Pulp & Paper, October, 1996.

21. Bloch, Heinz P. "Automatic Lubrication in Maintenance Cost Reduction," Maintenance Technology, January, 1997.

22. Bloch, Heinz P., "Automatic Lube Greases Skids for Profit," Power Engineering, February, 1997.

23. Bloch, Heinz P., "Twelve Equipment Reliability Enhancements With a 10:1 Payback," presented at NPRA Reliability and Maintenance Conference, New Orleans, LA, May 2005.

24. Adams, Erickson, Needelman and Smith, (1996) Proceedings of the 13th International Pump User's Symposium, Texas A&M University, Houston, TX, pp. 71-79.

25. Bradshaw, Simon; "Investigations into the Contamination of Lubricating Oils in Rolling Element Pump Bearing Assemblies," (2000), Proceedings of the 17th International Pump User's Symposium, Texas A&M University, Houston, Texas.

26. Bloch, H.P.; "Pump Wisdom—Problem Solving for Operators and Specialists"; (2011), John Wiley & Sons, Hoboken, NJ

27. Noria Corporation, Tulsa, OK, as reported in Ref. 9-28 ("Cleanliness Grid")

28. ISO 4406 Cleanliness Level Standards, as cited in Royal Purple Product Guide, Porter, TX

29. Marketing Literature, Royal Purple Ltd., Porter, TX

CHAPTER 10

1. Lubrication Systems Company of Texas, Inc., Accessories Brochure.

2. Bloch, H.P. and Shamim, A.; (1998). "Oil Mist Lubrication—Practical Applications," ISBN 0-88173-256-73.

3. Bloch, H.P.; (2000). "Practical Lubrication for Industrial Facilities," The Fairmont Press, Inc., Lilburn, GA, 30047, ISBN 0-88173-296-6.

4. Bajaj, Kris K., (1988)."Oil Mist Lubrication of Shaker Screen Bearings," Lubrication Engineering, October, 1988.

5. Bloch, H.P.; (1980). "Large Scale Application of Pure Oil Mist Lubrication in Petrochemical Plants," ASME Paper 80-C2/Lub-25 (1980).

6. Ward, T.K., (February 1996), "1995 Refinery Oil Mist Usage Surveys," Lubrication Systems Company Sales Literature.

7. Stewart-Warner Corporation, (1982), "Oil Mist Lubrication Systems for the Hydrocarbon Processing Industry," p. 4.

8. Lubrication Systems Company of Texas, Inc., "Design and Installation of Lubrimist Equipment."

9. Bloch, H.P., (1980). "Sampling and Effluent Collection for Oil Mist Systems," Practicing Oil Analysis, May-June 2002, pp. 26-33.

10. Miannay, C.R., (May 1974). Improve Bearing Life with Oil Mist Lubrication," Hydrocarbon Processing, pp.113-115.

11. Bloch, H.P., (1985). "Storage Preservation of Machinery," Proceedings of the 14th Texas A&M Turbomachinery Symposium, October 1985.

12. Bohn, Ed, "GM Invests in Lube Program Upgrades," (2002) Lubrication Magazine, January-February, 2002.

13. Morrison, F.R., Zielinsky, J., James, R., "Effects of synthetic fluids on ball bearing performance," ASME Publication 80-Pet-3, February, 1980.

14. Pinkus, O., Decker, O., and Wilcock, D.F. "How to save 5% of our energy," Mechanical Engineering, September 1997.

Supplementary Bibliography for Oil Mist

- Bloch, Heinz P., "Dry Sump Oil Mist Lubrication for Electric Motors," *Hydrocarbon Processing*, March 1977
- Bloch, Heinz P., "Oil Mist Lubrication Cuts Bearing Maintenance" *Plant Services Magazine*, November 1983
- Bloch, Heinz P.; "Benefits of Oil Mist Lubrication for Electric Motor Bearings" *Plant Services Magazine*, April 1986
- Bloch, Heinz P.; "Preservation by Oil Mist Application," *Plant Services Magazine*, November 1987
- Bloch, Heinz P.; *Oil Mist Lubrication Handbook*, 1987; Gulf Publishing Company, Houston, Texas
- Bloch, Heinz P.; "Oil Mist Lubrication: Is it justified and how should it be executed in the 90s;" *Hydrocarbon Processing*, October 1990, pp.25
- Bloch, Heinz P.; "Best-of-Class" Lubrication for Pumps and Drivers," Pumps & Systems, April 1997
- Bloch, Heinz P.; "Update Your Oil Mist Lubrication Knowledge," *Hydrocarbon Processing*, February 2003
- Bloch, Heinz P., "Lubrication Strategies For Electric Motors," *Machinery Lubrication Magazine*, January 2004
- Bloch, Heinz P.; "Oil Mist Lubrication for Electric Motors," *Hydrocarbon Processing*, August 2005
- Bloch, Heinz P.; "Applying Oil Mist," *Lubrication & Fluid Power*, February 2005
- Bloch, H.P. and Geitner, F.K.; "Machinery Uptime Improvement," 2006; Elsevier-Butterworth-Heinemann, Stoneham, MA (ISBN 0-7506-7725-2)
- Bloch, H.P. and Ehlert, Don; "Get The Facts on Oil Mist Lubrication" *Hydrocarbon Processing*, August 2008
- Bloch, Heinz P.; *Pump Wisdom*, John Wiley & Sons, Hoboken, NJ; 2011
- Budris, Allan; "Bearing, Lubrication Issues that can Reduce Pump Life Cycle Costs," *Water World*; Vol. 26, Issue 8, p.14, August 2010
- Budris, Allan; "Optimizing Bearing Lubrication can Extend Bearing Life, Reduce Costs," *Water World*, October 2011
- Clapp, A.M., "Fundamentals of Lubrication Relating to Operations and Maintenance," Proceedings of Texas A&M International Turbomachinery Symposium, 1972
- Ehlert, Don C.; "Centralized Lube Systems Often Best" *Plant Services*; 6/1991
- Ehlert, Don C.; "Bearing Lubrication Trends and Tips" *Pumps & Systems*; 12/1993
- Ehlert, Don C.; "Lubrication Made Simple" *Pumps & Systems*; April 2005
- Ehlert, Don C.; "Getting the Facts on Oil Mist Lubrication," Texas A&M Middle East Turbomachinery Symposium, 2011
- Ehlert, Don C.; "Consider Closed-Loop Oil Mist Lubrication," *Hydrocarbon Processing*; June 2011

- Ehlert, Don C.; "Consider Updating Your Lubricant System" *Hydrocarbon Processing;* May 2012
- Ehlert, Don C.; "Eliminate Risks in Bearing Lubrication" *World Pumps;* July/Aug. 2012
- Ehlert, Don C.; "The Ultimate Oil Filter" *Pump Engineer;* January 2013
- Hartman, John, "Oil Mist System Provides Rapid Payback," *Hydrocarbon Processing Magazine,* August 2004
- Honeycutt, Jerry, "Reducing Motor Bearing Failures," *Machinery Lubrication Magazine,* May 2002
- McNally, Bill; McNally Institute, "Why Pump Bearings Fail," April 2012
- McNally, Bill; McNally Institute, "What is Wrong with Centrifugal Pumps," 3/2010
- Noria Corp., "Redefining Bearing Failure," *Reliable Plant Magazine,* November 2008
- Radu, Ciprian; "The Most Common Causes of Bearing Failure and the Importance of Bearing Lubrication," *RBK Technical Review,* February 2010
- Rehmann, Chris, and Bloch, H.P.; "Closed Oil Mist Lubrication is Best Available Technology," *Machinery Lubrication Magazine,* November 2010
- Shamim, A., and Kettleborough, C.F., Tribological Performance Evaluation of Oil Mist Lubrication," *ASME Journal of Energy Resources Technology,* Vol. 116, No. 3, pp. 224-231; 1994
- Shamim, A., and Kettleborough, C.F., "Aerosol Aspects of Oil Mist Lubrication Generation and Penetration in Supply Lines," presented at the Energy Resources Technology Conference and Exhibition, Houston, Tribology Symposium ASME, PD - Vol. 72, pp. 133-140; January 1995
- Smith, E.A., "Electric Motor Bearing Lubrication Faces New Challenges," *Machinery Lubrication Magazine,* June 2001
- Towne, C.A., "Practical Experience with Oil Mist Lubrication," *Lubrication Engineering,* Vol. 39 (8), pp.496-502; 1983
- Underwood, John, "Grease-lubricated Electric Motors—a New Perspective," *Machinery Lubrication Magazine,* January 2008

CHAPTER 11

1. Karassik, Igor J., et al; "Pump Handbook," 2nd Edition, 1986, Mc Graw-Hill Book Company, NY, NY. 10121, ISBN 0-07-033302-5.
2. Calistrat, Michael M., "Flexible Couplings," Caroline Publishing Co., 1994, Houston, TX 77245, ISBN 0-9643099-0-4.
3. Lovejoy Inc., "The Coupling Handbook," Marketing Literature, 2000.
4. Lucas Aerospace, Utica, NY 13503, Marketing Literature, 1999.
5. Mancuso, Jon R., "Couplings and Joints," 1986, Marcel Dekker, Inc., New York, NY 10016, ISBN 0-8247-7400-0.
6. Falk Corporation, Milwaukee, Wisconsin 53208, Marketing Literature, 2002.
7. Rexnord Corporation, Engineering Bulletin, 1984, Milwaukee, Wisconsin 53201.

CHAPTER 12

1. "Centrifugal/Vertical Pump Vibration," ANSI/HI 9.6.4, 2000, Hydraulic Institute, Parsippany, New Jersey.
2. APMA, Australian Pump Technical Handbook (1987).
3. Beebe, R.S., "Machine condition monitoring," MCM Consultants, 2001 Edition.
4. Yates, M., "Not Just the Yatesmeter," World Pumps, December 1992.
5. Beebe, R.S., "Thermometric testing of high energy pumps using pipe surface measurements," Third ACSIM (Asia-Pacific Conference on Systems Integrity and Maintenance), Cairns, Australia, 2002.
6. Whillier, A., "Site testing of high-lift pumps in the South African mining industry," IMechE Paper C155/72, Conference on Site Testing of Pumps, London, 1972, pp. 209-217.

7. Karassik, I. J., and T. McGuire (Eds.), Pump Handbook, McGraw-Hill (1998).

8. Stepanoff, A. J., "Centrifugal And Axial Flow Pumps," Wiley (1957), and Figure 1-77A of the PDF figures on www.pumps.org.

9. Haynes, C. J. and M.A. Fitzgerald, "Scheduling Power Plant Maintenance Using Performance Data," ASME Paper 86-JPGC-Pwr-63, 1986.

10. Budris, Allan R., "Reduce Pump Costs with Predictive Condition Monitoring and Fault Tolerant Intelligent Drives/Controls," *WaterWorld,* April 2010.

11. Budris, Allan R., "Accoustical Resonance can Cause Excessive Pump and Associated Pipe Vibration," *WaterWorld*, February 2013.

CHAPTER 13

1. Bloch, H.P., "Mechanical Reliability Review of Centrifugal Pumps for Petrochemical Services," presented at the 1981 ASME Failure Prevention and Reliability Conference, Hartford, Connecticut, September 20-23, 1981.

2. Bloch, H.P., "Improving Machinery Reliability," 3rd Edition, Gulf Publishing Company, Houston, Texas, 1998, ISBN 0-88415-661-3

3. "Goulds Pump Manual — GPM6," 1995, Goulds Pumps, Inc., Seneca Falls, NY 13148

4. "Pump Piping for Rotodynamic (Centrifugal) Pumps," HI 9.6.6, 2003, Draft, Hydraulic Institute, Pump Piping Working Group, Parsippany, NJ.

5. "Centrifugal/Vertical Pump Vibration," ANSI/HI 9.6.4, 2000, Hydraulic Institute, Parsippany, NJ.

6. "Centrifugal/Vertical Nozzle Loads," ANSI/HI 9.6.2 — 2001, Hydraulic Institute, Parsippany, NJ.

7. Sprinker, E.K. and Paterson, F.M., "Experimental Investigation of Critical Submergence for Vortexing in a Vertical Cylindrical Tank," ASME Paper 69-FE-49, June, 1969

8. Bloch, H.P. and Ron Franklin; "What's new in vertical enclosed shaft-driven recessed impeller pumps"; *Hydrocarbon Processing,* May 2009

9. "Vespel® CR-6100 Applications Manual," DuPont Engineering Polymers, Newark, Delaware. Also marketing publication VCR6100PUMP, December 2002.

Additional Reading on Split Flow™ Options:

1. United States Patent No. 5,599,164, February 4, 1997 "Centrifugal Process Pump with Booster Impeller"; http://www.uspto.gov/web/offices/com/sol/og/index.html

2. Kane, Les; (Editor); "Pump Innovation Reduces Motor Size" HP Innovations, Hydrocarbon Processing, January 1998; http://splitflowpumps.com/articles/hcp.htm

3. William E. Murray; "API Process Pump Innovation Results in Significant Cost Savings," Pumps and Systems, July 1998; http://splitflowpumps.com/pdf/psm_7-98.pdf

4. Sloley, Andrew W., and William E. Murray; "The Split Flow™ Option: Pump Innovations Reduce Motor Size and Capital Cost," Presented at the Chemical Engineering Exposition and Conference, June 1999; http://splitflowpumps.com/about/option.htm

5. William E. Murray and Van Wilkinson, "Seal Life for Unconventional Refinery Process Pump," User Case Study, Presented at the 17th International Pump User's Symposium, March 6-9, 2000; http://splitflowpumps.com/pdf/caseStudy1.pdf

6. Francis Lee Smith, Ph.D., "Split Flow™ Process Pumps Save Energy," Application for DOE NICE3 Program, January 2002; http://splitflowpumps.com/about/nice3.htm

7. Bloch, Heinz P.; "Split Flow™ Inline Vertical Pumps: A New Option," (HP In Reliability) Hydrocarbon Processing, February 2002; http://splitflowpumps.com/pdf/hcp-0202.pdf

8. Bruce T. Murray and James Sweeney; "Carbon Dioxide Reduction: Split Flow™ Can Save from

233 to 433 Metric Tonnes of CO_2 per Year," February 2008; http://splitflowpumps.com/about/co2.htm

9. API Standard 610 (American Petroleum Institute, Washington, D.C.); Centrifugal Pumps for Petroleum, Heavy Duty Chemical, and Gas Industry Services.

10. Split Flow™ website; http://splitflowpumps.com

CHAPTER 14

1. Bloch, H.P., "Setting Up a Pump Failure Reduction Program," (presented at 38th ASME Petroleum Mechanical Engineering Workshop, Philadelphia, Pennsylvania, September 12-14, 1982).

2. Bloch, H.P., "Machinery Documentation Requirements for Process Plants," ASME Paper 81-WA/Mgt-2, presented in Washington D.C., Nov.1981.

3. Bloch, H.P., "Large Scale Application of Pure Oil Mist Lubrication In Petrochemical Plants," ASME Paper 80-C2/Lub 25, presented at Century 21 ASME/ASLE International Lubrication Conference, San Francisco, CA, August 18-21, 1980.

4. Bloch, H.P., "Improve Safety and Reliability of Pumps and Drivers," Hydrocarbon Processing, January through May, 1977, five-part series.

5. Eeds, J.M., Ingram, J.H., Moses, S.T., "Mechanical Seal Applications-A User's Viewpoint" -Proceedings Sixth Turbomachlnery Symposium Texas A& M University, 1977, pp. 171-185.

6. James, R., "Pump Maintenance," Chemical Engineering Progress, February 1976, pp. 35-40.

7. Sangerhausen, C.R., "Mechanical Seals in Light Hydrocarbon Service: Design and Commissioning of Installations for Reliability," presented at ASME Petroleum Mechanical Engineering Conference, Denver, Colorado, September 14-16, 1980.

8 SKF Bearing Company, King of Prussia, Pennsylvania, General Engineering Data Manual, 1991.

9. Bloch, H.P. and F.K. Geitner; "Machinery Failure Analysis and Troubleshooting," Butterworth-Heinemann, Stoneham, Massachusetts, 3rd Edition, 1997, ISBN 0-88415-662-1).

10. Kepner & Tregoe, "The Rational Manager," McGraw-Hill Book Company, NY, NY, 1965.

CHAPTER 15

1. Bloch, Heinz P., "Buying Replacement Parts from Non-OEM Equipment Manufacturers," Hydrocarbon Processing, May 1997.

2. Perez, Robert X.; "Operator's Guide to Centrifugal Pumps," 2008, Xlibris Corporation, (ISBN 978-1-4363-3985-8)

3. Bloch, H.P., 1988, *Practical Machinery Management for Process Plants Vol.1: Improving Machinery Reliability, Second Edition*, Gulf Publishing Company Houston, Texas (ISBN 0-87201-455-X).

4. Bloch, H.P. and Geitner, F.K., 1985, Practical Machinery Management for Process Plants, Volume 4: *Major Process Equipment Maintenance and Repair;*: Gulf Publishing Company, Houston, Texas, (ISBN 0-88415-663-X).

5. Nelson, W. Ed, and John Dufour; 1993, *Centrifugal Pump Sourcebook*, McGraw-Hill Publishing Company, New York, NY, (ISBN 0-07-018033-4), pp. 186-188.

6. Lobanoff, V.S. and Ross, R.R., 1992, *Centrifugal Pumps: Design and Application, Second Edition*, Gulf Publishing Company Houston, Texas, (ISBN 0-87201-200-X).

7. Corbo, M.A., Leishear, R.A., and Stefanko, D.B., 2002, *"Practical Use of Rotordynamic Analysis to Correct a Vertical Long Shaft Pump's Whirl Problem,"* Proceedings of the 19th International Pump Users Symposium, Turbomachinery Laboratory, Texas A&M University, College Station, Texas, pp. 107-120.

8. Corbo, M.A. and Malanoski, S.B., 1998, *"Pump Rotordynamics Made Simple,"* Proceedings of the 15th International Pump Users Symposium, Turbomachinery Laboratory, Texas A&M University, College Station, Texas, pp. 167-204.

9. Komin, R.P., 1990, *"Improving Pump Reliability in Light Hydrocarbon and Condensate Service With Metal-Filled Graphite Wear Parts,"* Proceedings of

the Seventh International Pump Users Symposium, Turbomachinery Laboratory, Texas A&M University, College Station, Texas, pp. 49-54.

10. Pledger, J.P., 2001, *"Improving Pump Performance and Efficiency With Composite Wear Components,"* World Pumps, Number 420.

CHAPTER 16

1. Bloch, Heinz P., "Improving Machinery Reliability," 3rd Edition, Gulf Publishing Company, Houston, Texas, 1998, ISBN 0-88415-661-3
2. Bloch, Heinz P., and Abdus Shamim, (1998) *Oil Mist Lubrication: Practical Applications*, Fairmont Press, Inc., Lilburn, GA, 30047; ISBN 0-88173-256-7, Fig. 9-7, p.109
3. Shelton, Harold L., "Estimating the lower explosive limits of waste vapors," *Environmental Engineering*, May-June 1995, pp. 22-25
4. Lilly, L.R.C., (1986) *Diesel Engine Reference Book*, Butterworth & Co, London, U.K., ISBN 0-408-00443-6, p.21/3
5. Davidson J. *"The reliability of mechanical systems"*; 1994 (Mechanical Engineering Publications, London)
6. Bloch, Heinz P.; "Improve safety and reliability of pumps and drivers," *Hydrocarbon Processing*, February 1977
7. Wallace, Neil, and M.T.J. David; "Pump reliability improvements through effective seals and coupling management," presented at 15th International Pump Users Symposium, Texas A&M University, Houston, Texas, 1998

CHAPTER 17

1. Barringer and Associates, Inc; (www.barringer1.com)
2. Taylor, Irving; "The Most Persistent Pump-Application Problems for Petroleum and Power Engineers," ASME Publication 77-Pet-5 (Energy Technology Conference and Exhibit, Houston, Texas, September 18-22, 1977)
3. Karassik et al; "Pump Handbook," 2nd Edition (1985), McGraw-Hill, New York, NY, ISBN0-07-033302-5
4. Bloch, Heinz P.; *Pump Wisdom*, (2011), John Wiley & Sons, New York, NY, (ISBN 9-781118-041239)
5. SKF Americas, *General Bearing Catalog*, (1990) Kulpsville, Pennsylvania
6. Worthington Pump Company, *Pump Operation and Maintenance Manual*, 1968
7. Bloch, Heinz P.; *Improving Machinery Reliability*, 3rd Edition, (1982, 1998) Gulf Publishing Company, Houston, Texas
8. MRC Bearings General Catalog 60, *TRW Engineer's Handbook*, 2nd Edition (1982), p. 197
9. Eschmann, Hashbargen and Weigand; *Ball and Roller Bearings: Theory, Design, and Application*, (1985) John Wily & Sons, New York, NY; ISBN 0-471-26283-8
10. Bloch, Heinz P.; *Practical Lubrication for Industrial Facilities*, 2nd Ed., (2009), The Fairmont Press, Lilburn, GA, 30047 (ISBN 088173-579-5)
11. SKF USA, Inc.; *Bearings in Centrifugal Pumps*, (1995), Publication 00-955, Second Edition
12. Wilcock, Donald F., and E. Richard Booser; *Bearing Design and Application*, (1957), McGraw-Hill Publishing Company, New York, NY
13. Bloch, H.P., *Oil Mist Lubrication Handbook*, 1st Edition, Gulf Publishing Company, Houston, (1987)
14. Bloch, Heinz P. and Abdus Shamim; *Oil Mist Lubrication—Practical Application* (1998), The Fairmont Press, Lilburn, GA, 30047 (ISBN 088173-256-7)
15. Baudry, Rene A., and Leonid M. Tichvinsky; "Performance of Oil Rings," *Mechanical Engineering*, 1937, Volume 59, 89-92; *ASME, Journal* of Basic Engineering; (1960) 82D, pp. 327-334
16. Heshmat, Hooshang, and O. Pinkus; "Experimental Study of Stable High-Speed Oil Rings." (1984) American Society of Mechanical Engineers (Paper); also (1985) *Journal of Tribology*, Volume 107 (1), pp. 14-22

17. Hersey, M.D., Discussion of Performance of Oil Rings," *Mechanical Engineering*, 1937, Volume 59, 291

18. Urbiola Soto, Leonardo; "Experimental Investigation on Rotating Magnetic Seals," Masters Thesis, Texas A&M University, 2001/2002

19. Government of Malaysia, Department of Occupational Safety and Health http://www.dosh.gov.my/doshv2/index.php?option=com_content&view=arti-cle &id=424%3Afire-at-oil-refinery&catid=84%3Asafety-alerts&Itemid=118&lang=en

20. Bloch, Heinz P.; "Deferred Maintenance Causes Upsurge in BFW Pump Failures," *Hydrocarbon Processing*, May 2011 [based on consulting work by the author]

21. Bradshaw, Simon; "Investigations into the Contamination of Lubricating Oils in Rolling Element Pump Bearing Assemblies," (2000), Proceedings of the 17th International Pump User's Symposium, Texas A&M University, Houston, Texas

22. TRICO Manufacturing Corporation, Pewaukee, Wisconsin; Commercial Literature, 2008; also www.tricocorp.com)

23. American Petroleum Institute, Alexandria, VA, API-610, *Centrifugal Pumps*, 10th Edition, 2009

24. Towne, Charles A., "Practical Experience with Oil Mist Lubrication," *Lubrication Engineering*, Vol. 39 (8), pp. 496-502 (1983)

25. Miannay, Charles R. "Improve Bearing Life," *Hydrocarbon Processing*, (May 1974)

26. Shamim, Abdus, and Kettleborough, C.F., Tribological Performance Evaluation of Oil Mist Lubrication," *ASME Journal of Energy Resources Technology*, Vol. 116, No. 3, pp. 224-231 (1994)

27. Shamim, Abdus, and Kettleborough, C.F., "Aerosol Aspects of Oil Mist Lubrication Generation and Penetration in Supply Lines," presented at the Energy Resources Technology Conference and Exhibition, Houston, Tribology Symposium ASME, PD - Vol. 72, pp. 133-140 (January 1995)

28. Ehlert, Don, "Getting the Facts on Oil Mist Lubrication," Texas A&M Middle East Turbomachinery Symposium, (2011)

29. Ehlert, Don; "Consider Closed-Loop Oil Mist Lubrication;" *Hydrocarbon Processing*, June 2011

30. Bloch, Heinz P.; "Inductive Pumps Solve Difficult Lubrication Problems;" *Hydrocarbon Processing*, September 2001

Index

D

E

F

S

T

W

X

Y

Z